Regional Geology Reviews

The Geology of series seeks to systematically present the geology of each country, region and continent on Earth. Each book aims to provide the reader with the state-of-the-art understanding of a regions geology with subsequent updated editions appearing every 5 to 10 years and accompanied by an online “must read” reference list, which will be updated each year. The books should form the basis of understanding that students, researchers and professional geologists require when beginning investigations in a particular area and are encouraged to include as much information as possible such as: Maps and Cross-sections, Past and current models, Geophysical investigations, Geochemical Datasets, Economic Geology, Geotourism (Geoparks etc), Geo-environmental/ecological concerns, etc.

More information about this series at http://www.springer.com/series/8643

Dimitrios I. Papanikolaou

The Geology of Greece

Springer

Dimitrios I. Papanikolaou
Faculty of Geology and Geoenvironment
National and Kapodistrian University of Athens
Athens, Greece

ISSN 2364-6438 ISSN 2364-6446 (electronic)
Regional Geology Reviews
ISBN 978-3-030-60730-2 ISBN 978-3-030-60731-9 (eBook)
https://doi.org/10.1007/978-3-030-60731-9

This Springer imprint is published by the registered company Springer Nature Switzerland AG
The registered company address is: Gewerbestrasse 11, 6330 Cham, Switzerland

Preface

The Geology of Greece was my major teaching responsibility and research focus for about 40 years in the Department of Geology at the National and Kapodistrian University of Athens. My first book regarding the Geology of Greece published in 1986 (Eptalofos Publ. Co., Athens, 240 p., in greek) was focused in assisting the students of Geology to refine the basic aspects about the geology of Greece, incorporating the «theoretical» part together with the description of the geological formations and the overall geodynamic and paleogeographic evolution. It should be noted that the Geology of Greece is a complex and rather challenging task, incorporating a Late Paleozoic—Early Cenozoic orogenic history of a Tethyan segment and an active subduction and orogenic arc in the Eastern Mediterranean. My second book in 2015 (Patakis Publ. Co., Athens, 443 p., in greek) was revised and extended, aiming also in assisting all the geoscientists to learn about the more recent developments of geology and their implications in the geological structure of Greece. Thus, topics related to the tectono-stratigraphic terranes, seismic tomography, geodetic measurements (GPS), extensional detachments and thermo-chronometry were also included. Additionally, a long list of references was included, aiming to cite the publications that documented and demonstrated a variety of new data and interpretations regarding the geology of Greece and to assist readers in retrieving original data. Some more recent publications and review papers were also cited, in order to provide a more updated literature on several critical thematic issues. Representative geological maps for almost every tectonic unit of the Hellenides have been selected, modified and included in the description of the units. This publication is, in fact, the second edition of the 2015 book, published in english with important new topics, such as the separate new chapter on Neotectonics. Thus, the creation of the Aegean plate at the southern border of the Eurasian plate during Miocene, the distinction of the Northern and Southern Hellenides, the description of the extensional detachments, the mantle flow dynamics in the Mediterranean, the paleomagnetic rotations and the description of the submarine and active volcanoes were included as separate subchapters. Additionally, an updated list of cited references up to September 2019 was included. The evolution of ideas from the «geosyncline period» of the 1960s to the «plate tectonics period» of the 1970s and the «tectono-stratigraphic terranes period» of the 1980s and 1990s has been presented in several parts of the book, either regarding the general geological knowledge or the specialized applications in the geology of Greece. Another goal was to shed light on old publications in order to incorporate old novel papers published in greek or other foreign languages that formulated the general aspects of the geological evolution of Greece. Unfortunately, these papers are not usually accessible in the present-day Internet searching facilities and risk to be ignored by the younger scientists, who do not follow the «classical» library setting. This is more accentuated for the papers published in greek, which are often considered as grey literature.

Scientific discussions, exchange of ideas and thematic collaborations with a plethora of colleagues from Greece and abroad throughout my research activities helped me understand the state of the art on several scientific issues. Especially, during the 20 years working in IGCP projects and particularly as Project Leader of IGCP no 276 «Paleozoic Geodynamic domains and their Alpidic evolution in the Tethys» (1987–1997) I had the opportunity to travel to a large number of highly important geological sites along the Tethyan belt and to collaborate with specialists from many countries. Thus, I benefited from discussions and collaboration with: B. C. Burchfiel, L. Royden and B. Reilinger from MIT (Boston), B. C. Blake (California), J. Rodgers (Yale), H. Masson, A. Baud, A. Escher, A. Steck and G. Stampfli (Lausanne), V. Dietrich (Zurich), J. Aubouin, M. Bonneau and L. Jolivet (Paris), C. Fourquin and H. Bergougnan, (Reims), T. Druitt (Clermont–Ferrand), D. Richter and K. Reicherter (Aachen), J. Makris and Ch. Huebscher (Hamburg), V. Jacobshagen (Berlin), St. Duerr (Frankfurt), B. Stoeckhert (Bonn), G. Roberts and C. Tzedakis (London), A. Robertson (Edinburgh), F. Sassi (Padova), I. Finetti (Trieste), G. Bonardi (Napoli), W. Cavazza (Bolonia), C. Sengor, Y. Yilmaz, A. Okay, T. Taymaz and N. Ocakoglu (Istanbul), E. Demirtasli, E. Bozkurt and C. Goncuoglu (Ankara), D. Kozhoukharov, E. Kozhukharova, I. Zagorcev, I. Haidoutov and Z. Ivanov (Sofia), S. Karamata and M. Dimitrievic, (Belgrade), R. Stojanov (Skopje), S. Kovacs and E. Marton (Budapest), F. Ebner (Graz), K. Petrakakis (Wien), J. Vojar and E. Vojarova (Bratislava), S. Adamia (Tbilisi), A. Saadalach (Algers), M. Julivert (Barcelona) and N. Morner (Stockholm).

In Greece, I benefited from discussions and collaboration with my supervisor Ilias Mariolakos and my old colleagues and friends Christos Sideris, Nikos Skarpelis, Spyros Lekkas, Victor Sabot, George Kalpakis, George Migiros, Evangelos Lagios and Taxiarchis Papadopoulos. Several old students of mine have been my close collaborators and friends such as Efthymis Lekkas, Stelios Lozios, Vangelis Logos, Ioannis Fountoulis, Maria Triantaphyllou, Paraskevi Nomikou, Stephanos Kilias, Emmanuel Vassilakis, Kostis Soukis, Michalis Diakakis and Leonidas Gouliotis. Several younger students are my new collaborators who have also contributed to the translation of the greek text, the preparation and reproduction—modification of the numerous figures of my book, such as Elina Kapourani, Stavros Birbilopoulos, Dimitra Boundi, Danae Lambridou and Spyros Mavroulis.

Finally, I want to express my love and gratitude to my wife Virginia and my children and colleagues Ioannis and Maria, who have supported me during all these years of my wandering around the geology of Greece. Especially Ioannis has devoted several months in reviewing and editing the final english text and the new topics that were inserted up to the last minute.

I cannot forget my early research period in the 1970s when my father Ioannis joined me in the field, taking notes of my numerous measurements of structural data and became an amateur geologist.

Athens, Greece
April 2021

Dimitrios I. Papanikolaou

Contents

About the Author

Dimitrios I. Papanikolaou is Emeritus Professor of Geology specialized in Structural Geology and Tectonics, Geology of Greece, Marine Geodynamics and Neotectonics at the National and Kapodistrian University of Athens. He studied Natural Sciences (B.Sc. 1971) and Geology (B.Sc. 1976) and obtained his Ph.D. in Geological Sciences (1978) in the University of Athens. He was elected successively Lecturer, Assistant Professor, Associate Professor and in 1993 Full Professor of Geology at the University of Athens. He did his post-doctoral research in the University of Lausanne (1979–1981). He acted as visiting Professor in the University of Reims (1982–1983) and the University of MIT in Boston (2003) and provided lectures in several other Institutions worldwide for shorter periods. Throughout his carrier he was teaching structural geology and Tectonics as well as The Geology of Greece (for over 40 years) supported by fieldtrips. He was the Director of the post-graduate M.Sc. programs «Prevention and Management of Natural Hazards» (2008–2016) and «Oceanography» (2007–2016). He was elected President of the Geological Society of Greece (1988–1996) and of the Carpatho-Balkan Geological Association (1993–1995). He was the Project Leader of IGCP 276 of UNESCO/IUGS «Paleozoic geodynamic domains and their alpidic evolution in the Tethys» (1987–1997). For several years he served as President of the Earthquake Planning and Protection Organization of Greece (1993–1998), as General Director of the Hellenic Centre for Marine Research (1994–2000) and as Secretary General for Civil Protection in the Ministry of Interior (2000–2002). He is associate editor in several international scientific journals and has edited several special volumes, particularly regarding the Geology of the Aegean. He has published more than 300 papers in various international and Greek scientific journals.

List of Figures

1 Greece Within the Alpine Orogenic System

1.1 The Alpine Orogenic System

The surface of the Earth is divided: (1) into oceanic domains, underlain by oceanic crust, the oldest of which today can be traced back to the early Jurassic in the Western Pacific Ocean, and (2) into continents, underlain by continental crust, which involves a variety of rocks from the oldest, dating back to the Archaean, to the newest, dated as Upper Cenozoic found in the most recently formed mountain ranges. The geometry of the oceans is simple, represented as a symmetrical zone around the middle, at the mid-ocean ridge, where new oceanic crust is constantly being created, especially in cases of new oceans, such as the Atlantic, in which tectonic processes have not greatly affected the original geometry. In contrast, a large-scale complex and chaotic image characterizes the continents. These have resulted from the successive tectonic processes of the Earth, from the initial creation of the very first crust up till today. Thus, continents are usually divided into two uneven sections: (i) A main section of mild morphology, consisting of rock formations of the Paleozoic or Precambrian, which has been tectonically inactive for many decades or even hundreds of millions of years. This section corresponds to the interior of the plates, composed of continental crust—former cratons—in which ancient, usually crystalline rocks have participated in ancient orogenic procedures by creating mountain ranges, which in turn disappeared and were leveled, due to the effect of exogenous factors. (ii) A smaller section of acute morphology, creating one or more mountain ranges—which are usually present along the plate boundaries of the continents—and which is tectonically active. This section includes the highest mountain ranges on Earth, due to the fact that the intense and exclusive activity of endogenous forces along these narrow bands of the Earth's surface does not manage to balance out with the exogenous forces of weathering, erosion and sediment transport, whose intensity is uniformly homogeneous throughout the whole surface of the Earth.

This set of mountain ranges, which have been created in the recent geological past, mainly during the Cenozoic era, and where energy is released along the convergent plates («orogeny in the making»), constitute the Alpine Orogenic System (Fig. 1.1). These are the present day tectonically active convergent and collisional zones, creating the orogenic arcs along the main plate boundaries in each subduction zone. Therefore, collaboration between geologists and geophysicists is necessary along the global Alpine orogenic system that will lead to the extraction and comparison between geological—geotectonic and geophysical data, something not feasible in older orogenic mountain chains, which are now inactive, with no possibility of analysis of several geophysical parameters, such as i.e. the seismicity.

Unlike the Alpine system, the earlier orogenic systems can be observed in the interior of the plates of continental crust, such as Europe. The successive earlier orogenic systems in Europe are the Variscan or Hercynian system for those mountain ranges formed during the Late Paleozoic, the Caledonian system for those ranges formed during Early Paleozoic (mainly during the Silurian era), and other older orogenic systems dating back even earlier in the Precambrian. The aforementioned orogenic systems were named after distinctive mountain ranges of corresponding eras of tectogenesis, which can be observed in Europe. The classical image of the geotectonic subdivision of Europe by Stille (1924) is indicative of the above mentioned systems (Fig. 1.2) and separates Europe into: (i) a primary core, which became a compact crystalline continental crust during the Precambrian and was named **Ur-Europa** or **Archaeo-Europa**, (ii) a younger crustal area adjoined to the previous core to the west, in Scandinavia and the British Isles, which originated from the Caledonian orogen, and was named **Palaeo-Europa**, (iii) a crustal area tangent to the previous ones along their border to the south, which was formed during the Late Paleozoic with the Variscan orogen, thus creating **Meso-Europa** (as well as the Urals to the east), (iv) a younger crustal area adjacent to the previous ones to the south, along the Mediterranean Sea, which participated and still continuous to participate in the formation of the mountain belts of the Alpine orogen, especially during the

D. I. Papanikolaou, *The Geology of Greece*, Regional Geology Reviews,
https://doi.org/10.1007/978-3-030-60731-9_1

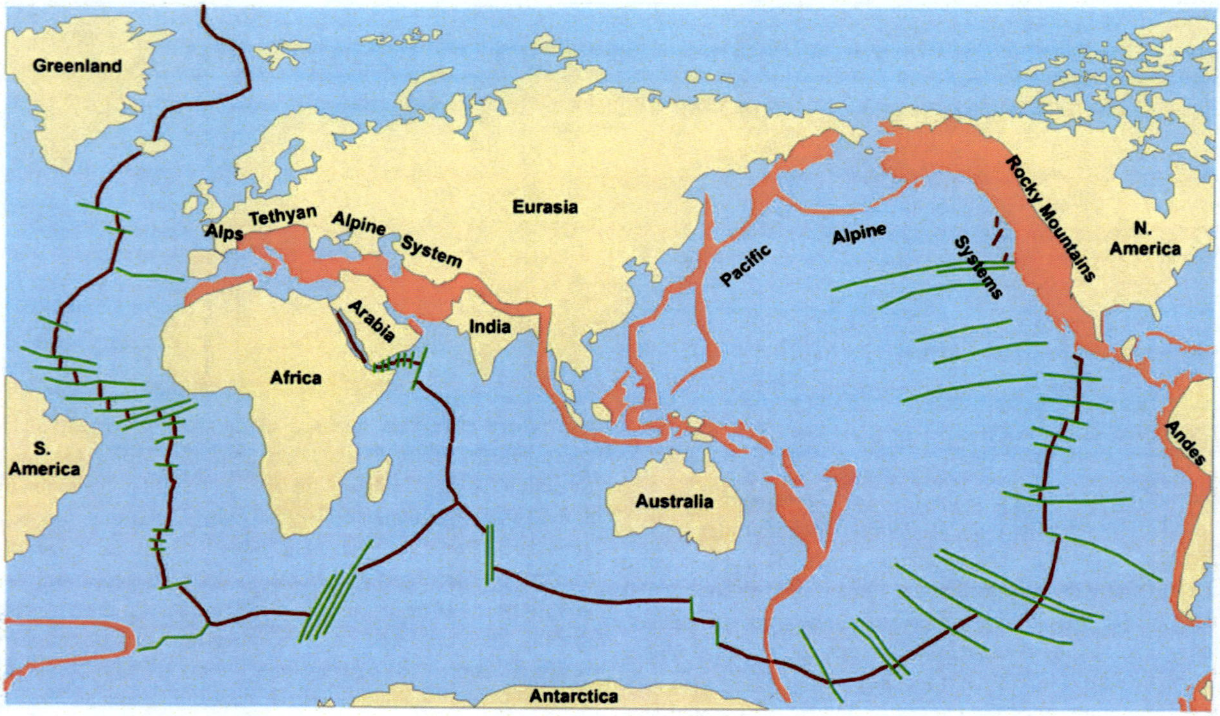

Fig. 1.1 The Alpine Orogenic System of the Earth (highlighted in red) has been formed along convergent and collisional plate boundaries. The Tethyan Alpine System is formed along the convergent and collisional zones of Eurasia to the north, and the African-Arabian-Indian-Australian plates of the former Gondwana supercontinent to the south. On the contrary the Circum Pacific Alpine orogenic systems are formed along the convergent zones of the oceanic plates of the Pacific with the surrounding continental plates of Eurasia, Australia, North America and South America

Cenozoic, representing the current active southern margin of the Eurasian continental plate, which is called **Neo-Europa**.

This example of Europe, where a small Precambrian cratonic block evolved into the contemporary Eurasian plate, shows how continental areas are developed with continuous accretions of orogenic chains with new continental crust, a process described by Stille (1924,1936) as **continental growth**. At the same time it emphasizes the relationship between the current physico-geographical image of a continental region and the age of its tectogenesis, expressed by the intense relief of the Alpine landscape of Neo-Europa, in contrast to the mild flattened terrains of the Central–Northern Europe, developed above Meso-Europa, Palaeo-Europa and Archaeo-Europa.

Greece is a typical area of the Alpine Tethyan system and of Neo-Europa, as a continuation of the Alpine orogenic chain, which, through various mountain ranges, including the Dinarides, the Hellenides, the Taurides, the Iranides, and the Afganides, reaches the mountainous complex of the Himalayas. This is one of the two main mountain belts of the Alpine system, which includes the highest peaks in the world, resulting from the collision of two continental plates, that is Eurasia to the north and parts of the former Gondwana to the south, i.e. mainly Africa, Arabia, and India. This Alps–Himalayas mountain chain was created by the folding of sediments that were deposited during the Mesozoic and part of the Cenozoic into a large ocean, separating the two supercontinents, Eurasia and Gondwana, named **Tethys** by Suess (1885–1909). Tethys was a deity of Greek mythology, daughter of Heaven (Uranus) and Earth (Gaia), and wife of Oceanus, with whom she gave birth to the Oceanides and the gods of the rivers. Thus, the particular Alpine–Himalayan mountain range, running parallel to the equator, is identified with, the term **Tethyan Orogenic System**, separating it from the other mountain ranges surrounding the Pacific Ocean, which are called **Circum Pacific Orogenic Belts**. Similarly, the Circum Pacific Orogenic Belts are divided into separate segments with a different geotectonic setting, such as the following distinct systems: (i) Andean, along the South American Pacific Coast, resulting from the convergence of the oceanic Pacific plate with the South American continental plate, (ii) the Rocky Mountains, along the North American Pacific coast, resulting from the convergence of the oceanic Pacific plate with the continental North American plate, part of which is now substituted by the strike slip zone of the San Andreas fault, (iii) the Japanese–Indonesian

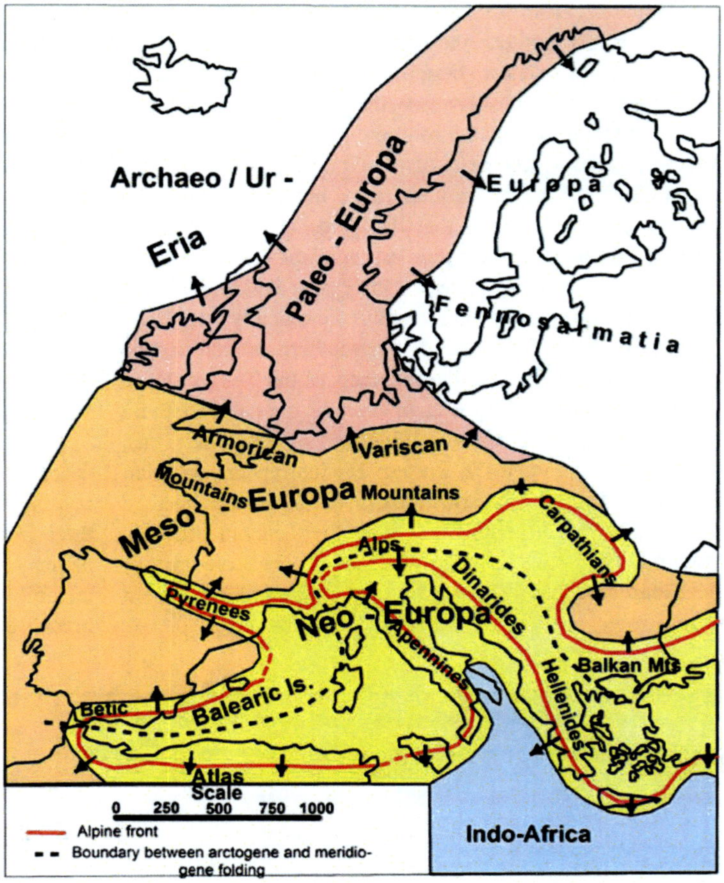

Fig. 1.2 Geotectonic subdivision of Europe in different orogenic systems, which have gradually expanded the continent from the original core of Archaeo-Europa to the current Neo-Europa, with the addition of Palaeo-Europa in the Silurian and of Meso-Europa in the Permian (according to Stille 1924, modified)

island arcs, resulting from the convergence of the oceanic crust of the Pacific Ocean with parts of the Eurasian Plate or its commonly complex margin, including some microplates with oceanic crust, such as the Philippines, and (iv) the New Zealand Arc, resulting from the convergence between the oceanic Pacific Ocean and the eastern margins of the Australian plate. The above mentioned Alpine orogenic systems should also include the Indonesian orogenic system, resulting from the convergence of the oceanic crust of the Indian Ocean with the southeastern edge of the Eurasian plate.

The difference between the current geotectonic state of the Tethyan orogenic system and the rest of the Pacific orogenic systems is enormous, since convergence is observed only between a small residual part of the Tethys Ocean and Eurasia in the Eastern Mediterranean Sea. Indeed, collision has prevailed all over the remaining system between Eurasia and the southern continental plates of Africa, Arabia and India. On the contrary, in the Circum Pacific Systems convergence between oceanic plates and continental plates dominates with the exception of the Philipinnes, where both converging plates are oceanic. The term Alpine system s.s. (sensu stricto) is also used, i.e. in the narrow sense of the term, especially for the Tethyan System, while Alpine system s.l. (sensu lato) is used, in the broad sense of the term, for all of them combined.

1.2 Greece Within the Tethyan Orogenic System

The Tethyan orogenic system comprises almost linear segments, as well as areas with curvilinear segments, which constitute specific orogenic arcs. As determined by the mountain ranges that create the individual linear or curvilinear sections of the Tethyan system, the sections of the orogenic system are either simply called with a characteristic geographical name as e.g. the Dinarides, the Iranides, etc., or as distinct orogenic arcs, such as the Western Alps Arc, the Carpathian Arc, the Hellenic Arc, etc.

In addition to the distinction of successive segments along the belt, there is also an important distinction across the orogenic system, separating it into two branches. In other words, the orogenic system can be seen as a dual mountain range, which runs along the entire Alpine zone in parallel arrangement. From the current geographical layout, the distinction of a northern and southern branch is made. There is also a distinction based on the paleogeographic-geodynamic criteria between the Eurasian margin of Tethys in the north, and the southern margin of Tethys in the south, comprising the belts of the Gondwana derived margins (African, if we are only interested in the Mediterranean region). It is noteworthy that the northern branch can be observed throughout the entire length of the Alpine zone in direct contact with the Eurasian Plate over which the rocks of the Tethys Ocean are usually tectonically emplaced (Fig. 1.3). On the other hand, the southern branch can be observed either on top of the remnants of the Tethys Ocean in the Eastern Mediterranean or directly over the southern continental plates in the remaining part of the zone. Thus, with the exception of the Eastern Mediterranean, collision between Eurasia and various parts of Gondwana can be observed along the entire length of the Tethyan orogenic system, (Fig. 1.3). This means that the only preserved part of the Gondwana passive margins occurs along the southern shores of the Eastern Mediterranean (Libya, Egypt, Palestine, Lebanon, Syria) (see No 7, in Fig. 1.3).

A more detailed examination of the various mountain ranges of the Tethyan system in the Mediterranean enables us to distinguish the two branches (Fig. 1.4). The northern branch comprises mainly the Betics in the Iberian Peninsula, the Pyrenees, the Alps (mostly their northern part of the Helvetics), the Carpathians, the Balkanides, the Pontides, the Caucasus, and other mountain ranges (e.g. Elburz) towards the east, up to the Himalayas. The southern branch mainly includes the Rif, the Atlas, the Sicily, the Apennines, the Southern Alps, the Dinarides, the Hellenides, the Taurides, the Iranides and further mountain ranges to the east, up to the Himalayas.

It is important to note that apart from the generally accepted different origin of the two branches from the two continental margins of the Eurasian and Gondwana plates, there is also a different tectonic asymmetry (vergenz), evident by the asymmetry of the folding and the direction of tectonic transport of the tectonic nappes. Thus, the northern branch is dominated by northward movement, i.e. rocks of

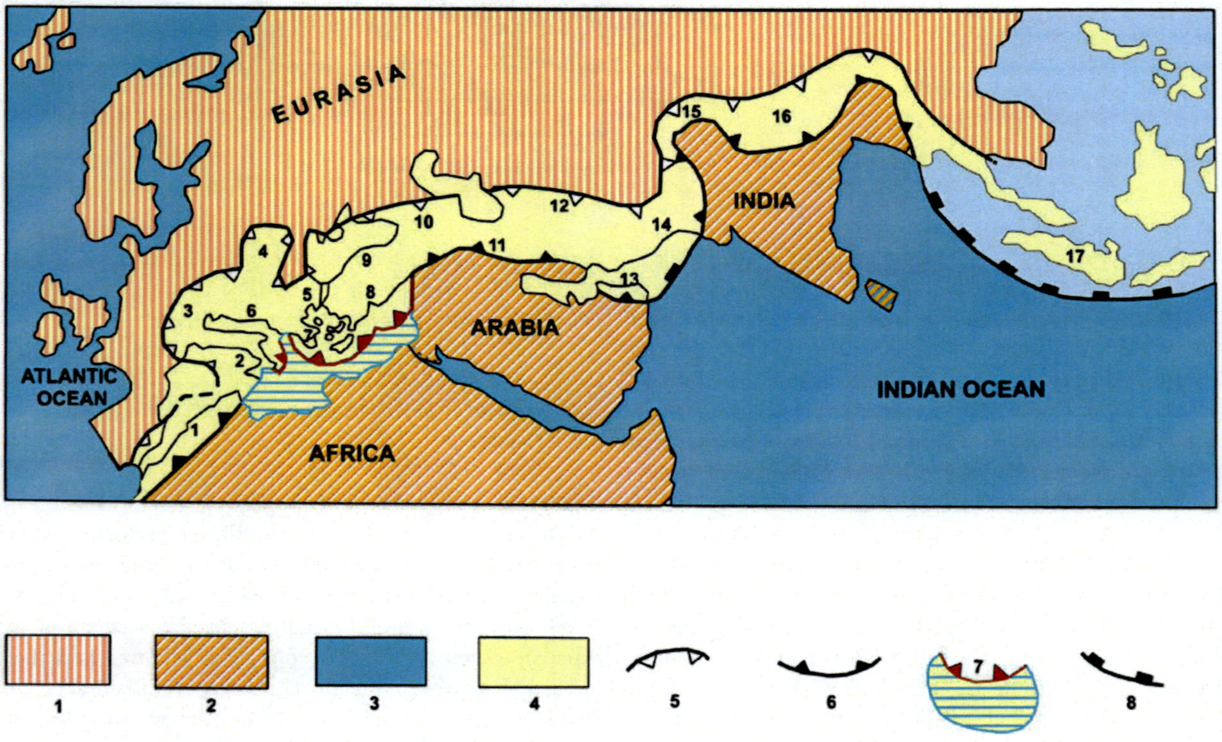

Fig. 1.3 The Hellenic arc is the only part of the Gondwana margin that is not yet crushed between the two plates. Collision of the continental plates of Gondwana with Eurasia has occurred throughout the remainder of the Tethyan orogenic system, except for the subduction of the Indian Ocean, which opened later to the south of Tethys. 1: Eurasian continental plate. 2: Continental plates of the former Gondwana. 3: Oceanic crust of Atlantic and Indian oceans. 4: Folded Alpine sediments. 5: Tectonic front of the northern branch. 6: Tectonic front of the southern branch. 7: Subduction zone of the Tethyan oceanic remnants, shown by blue stripes. 8: Subduction zone of the Indian ocean. 1: Atlas, 2: Apennines, 3: Alps, 4: Carpathians, 5: Balkanides, 6: Dinarides, 7: Hellenides, 8: Taurides, 9: Pontides, 10: Caucasus, 11: Zagros, 12: Afganides, 13: Oman, 14: Macran, 15: Karakorum, 16: Himalayas, 17: Indonesia

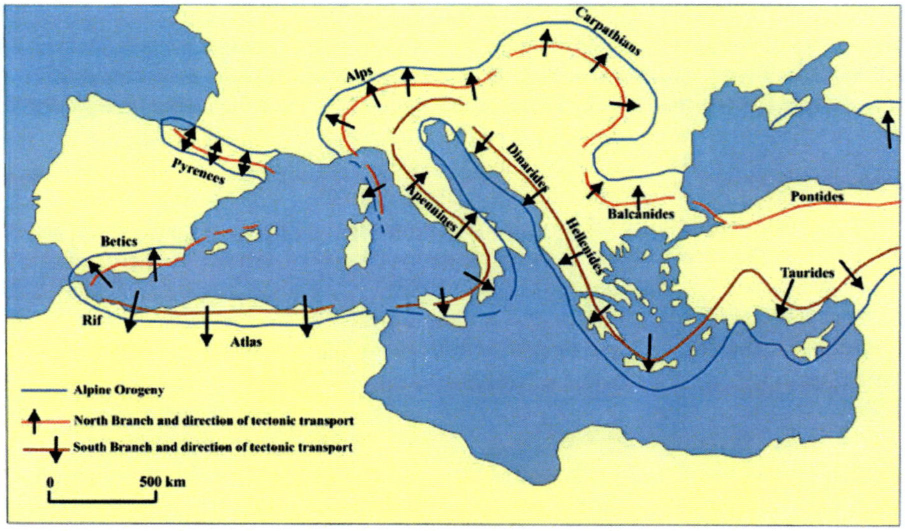

Fig. 1.4 The two branches of the Alpine Tethyan system in the Mediterranean region, with opposite directions of tectonic transport (vergenz)

the Eurasian margin and of the Tethys Ocean have moved onto the Eurasian plate, comprising the foreland (especially in the Northern Alps and the Carpathians). On the contrary, a movement to the south is observed in the southern branch, i.e. rocks of Tethys and of its southern margin have been emplaced either on other Tethyan members (the Apennines, Dinarides, Hellenides) or on the African plate (Rif and Atlas) or on the Arabian plate (eastern Taurides and Iranides), which also form the foreland of the southern branch.

This double asymmetry of the two branches of the Alpine system was modeled by Kober (1933) in the geometric image of a mushroom (Fig. 1.5), strongly influenced by Argand's tectogenetic views (1924) about crushing of the sediments of the Alpine "geosyncline" (i.e. the Tethys Ocean) from the approach of the cratonic masses of Eurasia and Gondwana (i.e. the current continental plates).

The intermediate zone between the two branches of the Alpine system consists either of ductile ophiolitic rocks and associated sediments (e.g. Axios/Vardar), which form an ophiolite suture zone, or of crystalline masses, which were considered to be **intermediate mountains** (zwischengebirge), such as Rhodope. These intermediate crystalline masses, forming the hinterland for each branch, were considered as the impact moles on which the soft mobile sediments of the geosynclines were smashed. Today, the overall geotectonic approach has changed and there are either supporters of a single main southward tectonic movement with only a minor secondary deformation of the northern branch

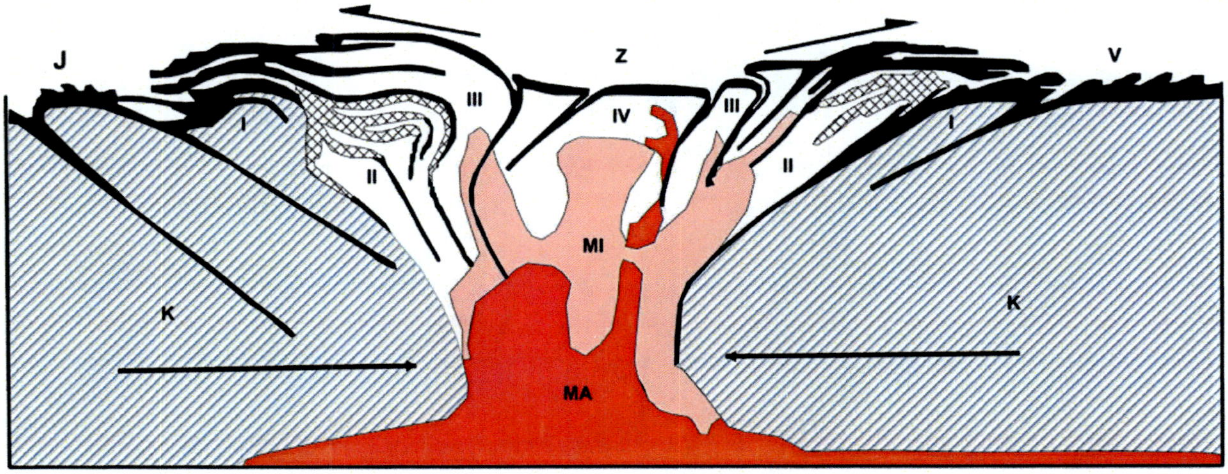

Fig. 1.5 The double asymmetry of the Tethyan Orogenic System in the shape of a mushroom (after Kober 1933, modified). I: external, II: metamorphic, III: central, IV: internal, K: cratonic masses, Z: intermediate mountains, MA: granitic magma, MI: migmatites, V: foreland, J: Jura

towards the north, mainly during the collisional phase, as supported by Suess since (1885–1909), or supporters of various bilateral tectonic movements, due to the presence of intermediate microplates, which substitute the former intermediate crystalline masses. Understanding the existence and provenance of the intermediate crystalline masses within Tethys Ocean was also a major concern in the late 1980s, following the plate tectonics theory, in order to accept its extension to the theory of the **tectono-stratigraphic terranes** in Europe (e.g. Papanikolaou and Sassi 1989).

In conclusion, Greece, i.e. the Hellenides, which also form the present day Hellenic Arc, belong to the Alpine system and more specifically to the southern branch of the Tethyan orogenic system.

The term Hellenides, established by Kober (1928), includes the part of the southern branch of the Alpine system that forms the Hellenic arc. It extends from the transverse tectonic zone of Scutari–Pec to the north, on either side of which different geotectonic units meet up, until the Antalya Gulf in Southwestern Anatolia to the southeast, where following an acute acme, the direction of the orogenic belt changes eastwards leading to the creation of the Taurus Arc (Fig. 1.6).

The Hellenic Arc is approximately 1,500 km long. It follows a NNW-SSE morphotectonic trend in Albania and Mainland Greece, then subsequently bends to E-W from Kythira to Crete, then to the NE-SW direction in the Dodecanese and Lycia of Southwest Asia Minor, reaching Isparta and Antalya in the intersecting zone between the Hellenic and Taurus Arcs (Monod, 1976). Thus, the two branches of the arc form a right angle. The most important feature of the Hellenic Arc is that it is the only segment along the whole Tethyan System where the orogenic process due to plate convergence is still active and exhibits all the geodynamic characteristics of a developing orogenic arc. This is due to the fact that the Eastern Mediterranean and especially the Ionian Sea, is the last remnant of the Tethys Ocean, which has not participated entirely yet in the Alpine orogeny and which also includes the undeformed passive margin of the African plate. The boundary between the oceanic basin of the Eastern Mediterranean and the continental margin of the African plate is traced along the current coastline of Libya and Egypt (see also Chap. 3).

It should be noted that the previously described **orogenic arc of the Hellenides** (Kober 1928) has been developed from the Early Mesozoic to the Late Miocene along the **active European margin**, whereas the present day **Hellenic arc and trench system** (LePichon and Angelier 1979) is limited to a small arcuate segment of the previous arc. To the north it is defined by the Kefallonia–Preveza transform fault and to the east by the complex structures developed sub-parallel to the western Anatolia coastal zone. This definition mainly based on the seismicity and later also on the GPS velocities, outlined the modern **Aegean microplate** (McKenzie 1970, 1972; Le Pichon et al. 1995; Kahle et al. 1995; Reilinger et al. 1997) (Fig. 1.7). This major difference has justified the previous distinction of the **Alpine tectonics** of the Hellenides from the **neotectonics** in the Hellenic arc/Aegean area (e.g. Aubouin 1974; Angelier 1979; Mercier 1979).

The above peculiarity of the Hellenic Arc allows further research to be contacted on the various geodynamic

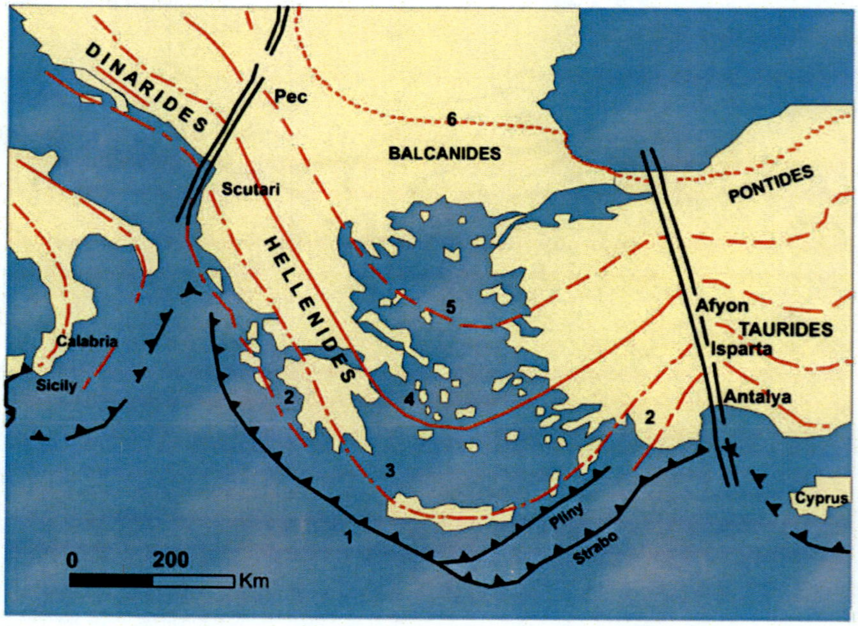

Fig. 1.6 The Hellenic Arc as defined by the transverse tectonic zone Scutari–Pec with the Dinarides to the north and by the acme of the orogenic arc at Antalya with the Taurides towards the east. 1: Hellenic Trench, 2: Geotectonic trend of the Ionian, 3: G.t. of the Pindos, 4: G.t. of the Eastern Greece, 5: G.t. of the Axios, 6: G.t. of the Balcanides–Pontides

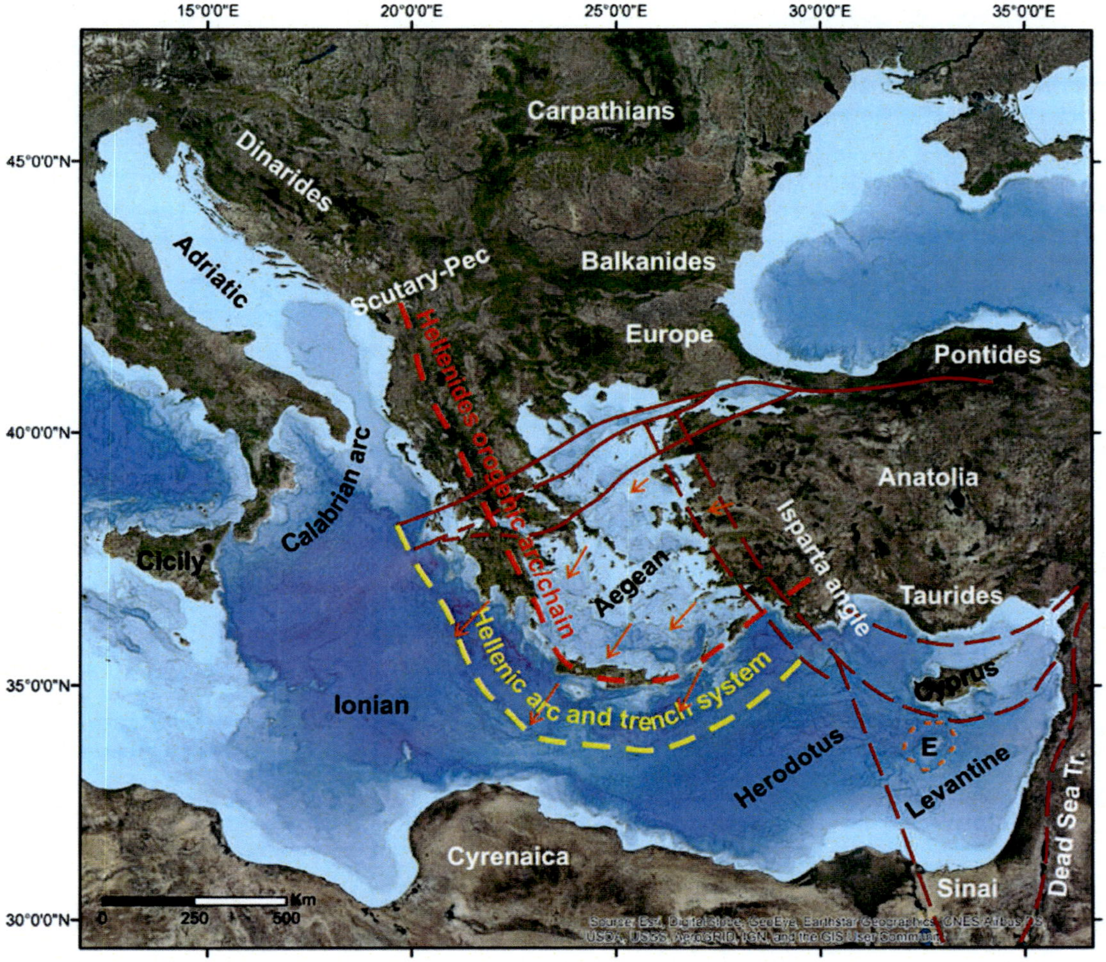

Fig. 1.7 The Hellenides orogenic arc and the modern Hellenic arc and trench system, which, since the Late Miocene, is limited to a small segment of the previous structure at the front of the newly formed Aegean microplate

phenomena related to orogeny, with the elaboration of actualistic models, which can be further compared to the corresponding results from scientific research in the Circum Pacific orogenic systems and particularly the Japanese Islands. This explains the great interest on the geology of Greece from a plethora of research groups from different countries since the nineteenth century.

References

Angelier, J. 1979. Néotectonique de I' arc Egéen. Soc. Geol. Nord, Publ., 3, 1–417.

Argand, E. 1924. La tectonique de I'Asie. Congrès Géol. Intern., Bruxelles.

Aubouin, J. 1974. Des tectoniques superposées et leur signification par rapport aux modèles géophysiques. L'exemple des Dinarides, paléotectonique, tectonique, tarditectonique, néotectonique. Bull. Soc. géol. France, XV, 426–460.

Kahle, H. G., Muller, M. V., Geiger, A., Danuser, G., Mueller, S., Veis, G., Billiris, H. & Paradisis, D. 1995. The strain field in NW Greece and the Ionian islands : results inferred from GPS measurements. Tectonophysics, 249, 41–52.

Kober, L. 1928. Der Bau der Erde. Berlin.

Kober, L. 1933. Die Orogentheorie. Berlin.

Le Pichon, X. & Angelier, J. 1979. The Hellenic Arc and Trench system: a key to the neotectonic evolution of the Eastern Mediterranean area. Tectonophysics, 60, 1–42.

Le Pichon, X. Chamot-Rooke, N., Lallemant, S., Noomen, R. & Veis, G. 1995. Geodetic determination of the kinematics of central Greece with respect to Europe: Implications for eastern Mediterranean tectonics. Journal of Geophysical Research: Solid Earth, 100(B7), 12675–12690.

McKenzie, D.P. 1970. Plate tectonics in the Mediterranean Region. Nature, 226, 239–243.

McKenzie, D.P. 1972. Active tectonics of the Mediterranean Region. Geoph. J.R. Astron.Soc., 30, 109–185.

Mercier, J.L 1979. Signification néotectonique de l'Arc Egéen. Une revue des idées. Rev. Géol. Dyn. Géogr. Phys., 21, 1, 5–15.

Monod, O. 1976. La «courbure d' Isparta» : une mosaique de blocs autochthones surmontes de nappes composites á la jonction de l' arc hellenique et de l' arc taurique. Bull. Soc. Geol. France, 18, 521–531.

Papanikolaou, D. & Sassi, F.P. 1989. Palaeozoic Geodynamic Domains and their Alpidic evolution in the Tethys: A brief outline of the IGCP No 276 proposal. IGCP 276, Newsletter 1, Sp. Publ. Geol. Soc. Greece, 1, 7–10.

Reilinger, R. et al. 1997. Global positioning system measurements of present-day crustal movements in the Arabia-Africa-Eurasia plate collision zone. J. Geophys. Res., 102, 9983–9999.

Stille, H., 1924. Grundfragen der vergleichenden Tektonik. Berlin.

Stille, H., 1936. Wege und Ergebnisse der geologisch-tektonischen Forchungen. 25Jahre Kaiser Wilhelm—Gesellschaft zur Forderung der Wiss., 11, Die Naturw.

Suess, E., 1885–1909. Das Antlitz der Erde. Vol. 1–3, Leipzig.

2 Organization and Evolution of the Tethyan Alpine System

2.1 How, When, Where and Why Tethys Ocean Was Created and then Disappeared

The existence of Tethys Ocean is justified on the basis of the outcrops of marine sediments of various stratigraphic-sedimentary facies, from the coastal neritic facies to the abyssal one, of the Mesozoic and partially the Cenozoic era. These outcrops can be observed all along the Tethyan orogenic system, with extraordinary similarities of specific stratigraphic formations that can be observed in e.g. the Alps, the Hellenides, the Iranides, but also the Himalayas. The fact that the stratigraphic sequences of the various units of Tethys usually initiate from the sediments of the Upper Paleozoic or the Triassic, creates a time frame for the beginning of the creation of this large paleogeographic unit. The fact that the Tethyan sediments can be observed today crushed between the two continental plates of Eurasia and Africa-Arabia-India (=Gondwana), geotectonically defines the paleogeographic area of the Tethyan development in an oceanic basin of unknown width between the two continents (Argand 1924). The width of Tethys can be estimated either by: (i) unfolding the sediments, restoring the tectonically emplaced nappes to their original position and calculating how many and which units have been probably disappeared, either due to the subduction and destruction of the lithosphere or due to the erosion of newly emerging mountain ranges, which is a very complex process, leading to doubtful results, or (ii) using the results of paleomagnetic research, whose output may also be questionable, due to uncertainties on the exact age and tectonic position of the measured samples. Irrespective to the forthcoming discrepancies and disagreements regarding the possible extent and opening of Tethys, it is certain that this ocean was created during the Late Paleozoic–Triassic between Gondwana, which still included all its parts, namely South America–Africa–Arabia–India–Madagascar–Australia–Antarctica, and Laurasia which included North America–Greenland and Eurasia, the latter been created during the Latest Paleozoic by the amalgamation of Europe and Asia (Siberia) along the collision zone of the Ural Mountains.

The growth of Tethys is related to the divergence of the two plates Gondwana and Laurasia during the Triassic–Jurassic, while its gradual reduction started mainly since the Late Jurassic–Early Cretaceous. The compression initiated due to the convergence of the two continents, which led to a collisional phase during the Late Cenozoic, resulting to its gradual closure and disappearance.

A wealth of data support the view that Tethys had a triangular shape, closing westwards near the Iberian Peninsula and opening eastwards and especially after the Iranides. A schematic image of the global tectonic evolution from the Pangaean period in the Late Paleozoic–Early Triassic up to a foreseeable situation after fifty million years is depicted in Fig. 2.1 (Dietz and Holden 1970; Scotese 2001). It is noteworthy that the closure of Tethys Ocean resulted from the convergence—collision of successive fragments of Gondwana, moving towards Eurasia, the latter remaining almost intact during the same period.

2.2 What Was the General Paleogeographic Organization of Tethys?

The first major question regarding the organizational structure of Tethys is whether it corresponded to a simple shape, such as the actualistic model of the Atlantic Ocean, or another more complex shape, such as the actualistic Caribbean and Indonesian models. The answer to this question is rather easy; Tethys was not defined by a simple Atlantic-type organization model. However, specifying the complexity of Tethys is extremely difficult and there is no widely accepted model so far (actualistic or not).

Nevertheless, let's first take a look as to why the simple model of the Atlantic type is rejected. The new data that came to light in the 1960s and shaped the—then—fresh theory of the lithospheric plate tectonics are sufficient to know the type, thickness, and age of the sediments of the

D. I. Papanikolaou, *The Geology of Greece*, Regional Geology Reviews,
https://doi.org/10.1007/978-3-030-60731-9_2

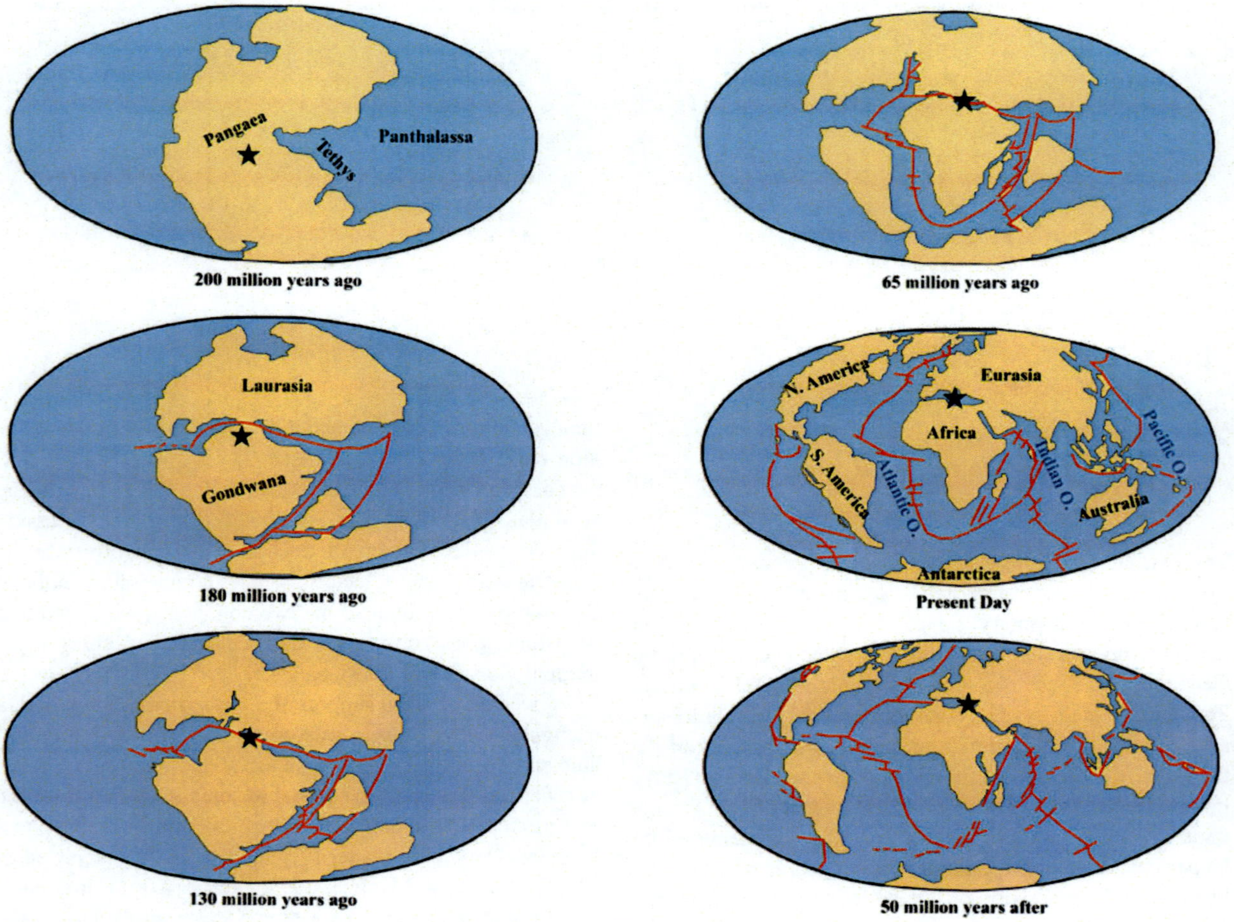

Fig. 2.1 A schematic representation of the gradual formation of the Earth's surface over the last two hundred million years and a forecast for the next fifty million years starting today (after Dietz and Holden 1970, modified). An asterisk marks the position of Greece in successive eras

Atlantic sea bottom, as well as the structure of the underlying oceanic crust. Thus, it is well documented that around the centre of the axial symmetry of the Atlantic there is the mid-ocean ridge, where divergence of the plates and creation of new oceanic crust can be observed. Moving away from the mid-ocean ridge to the two continents, Europe to the east and North America to the west, we gradually encounter older rock formations of the oceanic crust and, needless to say, older sediments above them, reaching the continental platforms of the two continents, at 70–100 Ma, i.e. the Late Cretaceous. Therefore, the overall picture of the Atlantic is a very satisfying symmetry on either side of the mid-ocean ridge, considered today as the cicatrix of the Atlantic. In this actualistic model it is possible to distinguish between identical physico-geographical regions and, consequently, environments—sedimentological facies on either side of the median ridge, making the distinction of (Fig. 2.2): (1) external neritic areas on the continental platforms of each of the two passive continental margins, (2) continental slopes starting from the continental platform to the seabottom of the oceanic basin for each continent, where slope sediments and pelagic sedimentation prevails, (3) wide abyssal–pelagic zones for each section of the ocean on either side of the mid-ocean ridge, and (4) an axial area along the divergent zone of the plates, that is along the centre of symmetry that defines the mid-ocean ridge, where basaltic submarine volcanism dominates. Assuming that we could close the Atlantic Ocean today, it would result into a collisional mountain range, showing a two-sided facies symmetry on both sides of the ophiolitic suture zone up to the two continental margins.

Unlike the Atlantic model shown above, the variety of the paleogeographic units of Tethys, as expressed by the sedimentary facies and the accompanying magmatic rocks, shows that Tethys was characterized by multiple alternations of neritic platforms with shallow-water carbonate sedimentation (e.g. Tripolis, Parnassos) and basins with pelagic and/or abyssal sedimentation (e.g. Pindos, Maliac) (Dewey

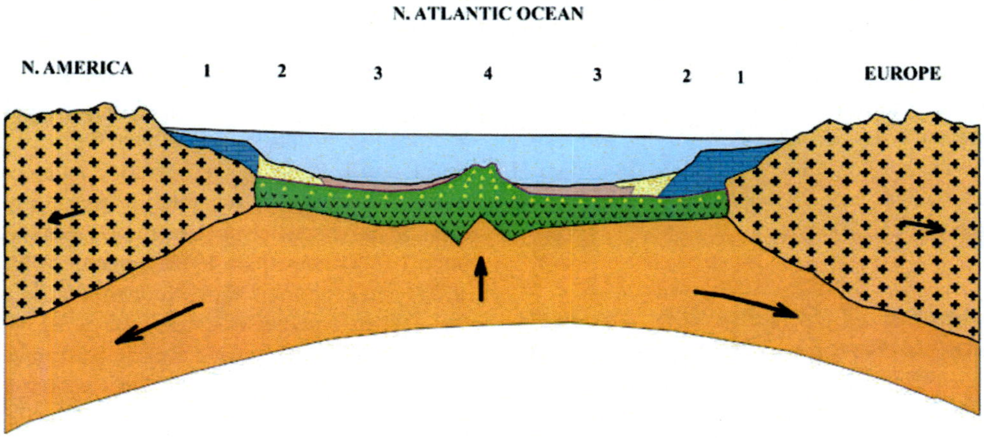

Fig. 2.2 Schematic representation of the physico-geographic environments–facies along a simplified cross-section of the Atlantic Ocean, showing a bilateral symmetry on either side of the mid-ocean ridge (from Papanikolaou 1986b, modified). 1: coastal and neritic facies of continental platforms, 2: transitional slopes facies, 3: pelagic–abyssal facies, 4: volcano-sedimentary abyssal-pelagic facies in the mid-ocean ridge area

et al. 1973). Furthermore, a complex paleogeographical organization is present, as well as remnants or large pieces of pre-Alpine continental crust and/or pre-Alpine orogenies between the various paleogeographical units of Tethys. These features show that Tethys Ocean involved highly heterogeneous geotectonic domains between the two former super continents–plates of Gondwana and Laurasia.

In general, it is accepted that the northern margin of Tethys was developed on an already tectonized margin of the Eurasian plate due to the Variscan orogeny, while the southern margin of Tethys was developed over a stable area during almost all of the Phanerozoic, where neither the Variscan nor the Caledonian orogeny had any effect, with the exception of Northwest Africa - Maghreb, where Variscan basement is present. This area of Gondwana was tectonized during the so-called Pan-African Orogeny, which was terminated with the late-orogenic granites of Cambrian age (e.g. Sinai), following the previous emplacement of Neoproterozoic ophiolitic rocks and eclogites recently discovered also in the Menderes massif (Candan et al. 2016). This stability of the Gondwanian margin in Africa and Arabia is shown by the continuous sedimentary sequences of Cambrian to Cretaceous–Eocene age, observed in the Taurides (Sengor et al. 1984a, b).

Between the above two continental margins the problem of the intermediate pre-Alpine crystalline massifs remains (e.g. Rhodope, Pelagonian), which cannot be geotectonically intergraded into one of the two continents. At the same time their presence affects the whole picture of the paleogeographical organization of Tethys in the form of independent microplates/terranes between the two large plates, which are in this way separated from two or more oceanic basins together with their mid-ocean ridges.

Thus, we have reached the issue of the existence of one or more ophiolitic sutures of Tethys, which occupied the scientific community in the 1980s (Robertson and Dixon 1984, and the relevant volume of the Eastern Mediterranean from the Edinburgh Conference, 1982/1984), which was re-addressed more recently (e.g. Bortolotti et al. 2013; Nirta et al. 2018). The initial viewpoint of one ophiolitic suture was based on the plate tectonics theory, though with a static view of the inner geometry of the Europe–Tethys–Africa/Arabia system. On the contrary, the viewpoint of more ophiolitic sutures involved a complex internal organization, with the presence of continental microplates, which divided the oceanic space of Tethys in smaller oceanic basins. These microplates soon acquired the particular kinetic features of the tectono-stratigraphic terranes, already proposed since the early 1980s in the Western United States (Coney et al. 1980; Howell 1980; Howel and Jones 1983; Zen 1983; Jones et al. 1983; Schermer et al. 1984). It is interesting to note that the new theory of the tectono-stratigraphic terranes, which completes the theory of the plate tectonics, did not find a warm response in Europe, where the views regarding the organization and development of the Tethyan Alpine system had been consolidated and conservative. Thus, -in a way- we had a repetition of the laborious process of spreading and enforcing the new theory, like almost a century ago with the theory of the tectonic emplacement of nappes. In both cases, the main focus was how to explain the presence of "exotic" domains in regional geology, which could only be justified by long-distance transport (see Schardt's «les regions exotiques du versant N des Alpes suisses», 1898 as well as the «exotic terranes» of Coney et al. 1980). Thus, from the static approach of the oceanic opening and development between the two continental plates of the 1970s, we passed in the 1980s-1990s to the kinetic approach of the creation of oceanic basins in between the lithospheric fragments of the terranes, that were drifting to the north, with a different number of oceanic basins developed in various parts along the Tethys (Papanikolaou 1989, 1997, 2013; Bonardi et al. 2001; Haydoutov 2002; Von Raumer et al. 2003; Moix et al. 2008; Himmerkus et al. 2007, 2009).

2.3 The Number of Ophiolite Suture Zones of Tethys—Terrane Tectonostratigraphy

When the concept of Tethys was firstly defined by Suess (1885–1909), it referred only to the physico-geographical element of an ocean, as an offshore area of a basin with marine sedimentation–geosyncline. After the new plate tectonics theory, the oceans were considered to be developed above oceanic crust, which comprises all the plutonic and volcanic rocks of basic composition, overlying the ultra-basic rocks of the upper mantle, defined as an ophiolitic complex or just ophiolites. Thus, the geological definition of the term ocean today generally corresponds to a region underlain by oceanic crust.

The existence of ophiolites in the area of the Tethyan Orogenic System in the form of tectonic nappes between the deformed sediments indicates that Tethys definitely had oceanic crust. Some of the ophiolitic outcrops of the Hellenides are universally regarded as typical cases of oceanic crust or as characteristic tectonic nappes (Moores 1969; Smith 1971; Zimmerman 1972; Dercourt 1972; Hynes et al. 1972) (e.g. Vourinos, Othris, Northern Pindos, as well as the ophiolites of Troodos Mountain in Cyprus).

Although the existence of oceanic crust and the activity of a mid-ocean ridge for some time period are generally established for Tethys, it is not certain whether we had only one oceanic basin or more. The problem arises from the fact that by observing the geotectonic maps of the Tethyan system two, three, or more linear ophiolite belts can be distinguished (Fig. 2.3). Thus, there is a school of researchers who considered that all occurrences–ophiolite belts originate from the same ocean, merely constituting tectonic repetitions, almost as the repetitions of layers due to folding or thrusting in a simple geological map (e.g. Ricou et al. 1984).

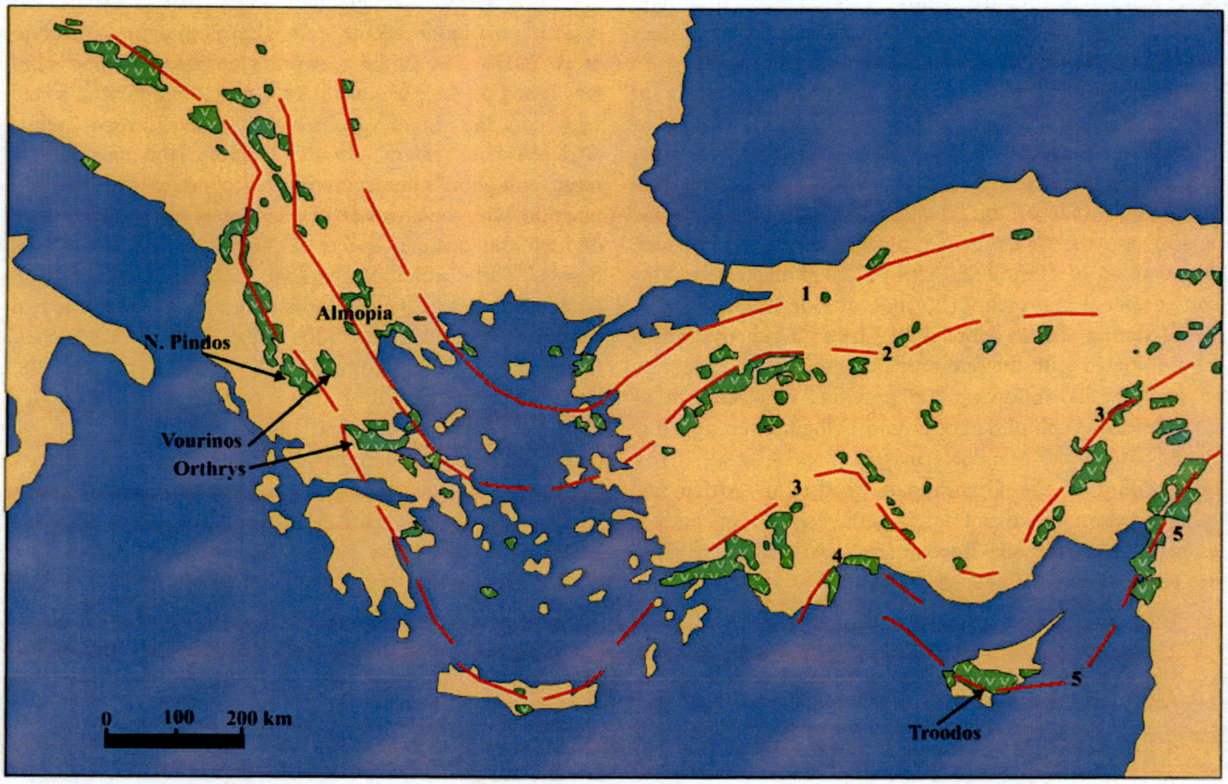

Fig. 2.3 Simplified map, showing the major ophiolite outcrops in the Eastern Mediterranean. Five ophiolitic belts are distinguished, each of them representing a possible ophiolite suture zone of Tethys. 1: Intra-Pontide ophiolitic belt, 2: Axios/Vardar–Izmir–Ankara ophiolitic belt, 3: Pindos–Othris–Lycian nappes ophiolitic belt, 4: Antalyan ophiolitic belt, 5: Troodos–Baer Bassit–Hatay ophiolitic belt

On the contrary, other researchers considered each ophiolite belt as a different ophiolite suture zone and therefore accept the existence of an independent oceanic basin for each belt (e.g. Sengor et al. 1984a, b).

In the case of Greece, two major ophiolite belts are distinguished; those of Axios/Vardar and Northern Pindos–Othris–Argolis–Crete. On the contrary, in Anatolia there is the Intra-Pontide zone in the north (1) and towards the south the Izmir–Ankara (2), Lycian (3), Antalya (4) and additionally Troodos/Cyprus (5) zones. The prevailing opinion was that the main Tethyan ophiolite suture zone was the Vardar/Axios ophiolitic belt, extended eastwards to the Izmir–Ankara ophiolitic belt. The choice of one or more major suture zones had consequences both on the concept/extent of the two continental margins and on the existence and number of the microplates in the intermediate oceanic domain. Thus, in the case of one suture the entire volume of the folded rocks north of the suture belongs to the northern margin, while the entire volume of the remaining folded rocks–ophiolitic outcrops included–south of the suture belongs to the southern margin. On the contrary, in the case of more sutures, the two margins of Eurasia to the north and Gondwana to the south are confined between the boundaries of the large plates and the corresponding first nearest ophiolitic belt, while all the other rocks are parts of the intermediate microplates/terranes and their margins. Since all intermediate crystalline pre-Alpine massifs are almond-shaped, the resulting organization of Tethys becomes extremely complex, especially when the proposed period of each oceanic basin's history is also considered (Papanikolaou 2009).

2.4 The Age of the Tethyan Ophiolites

The issue of the age of the ophiolites is one of the most discussed issues with opposing arguments. This is due to the difficulty of dating the ophiolites by the use of common geological methods, as they are found in allochthonous nappes and the associated sediments are usually detached from the ophiolite basement through tectonic decollement. A major parameter of the problem of the ophiolites, related to the unravelling of the history of each oceanic basin, is that during the subduction of the oceanic plate, the crust penetrates deep into the subduction zone and disappears into the asthenosphere/mesosphere. Only when the oceanic crust is detached and smashed between the two converging plates its re-exposure to the surface is possible in the form of a tectonic nappe, a phenomenon known as **ophiolite obduction** (Coleman 1971; Dewey and Bird 1971; Zimmerman 1972; Dewey 1976). Thus, only a small portion of the oceanic domain can be usually observed in the obducted ophiolite outcrops today and thus only a part of their history can be restored. This can be realized better if we compare the estimated paleogeographic dimensions of the oceans (several hundreds of km or even a few thousands of km) and their maximum dimensions based on their present outcrops within the mountain ranges (a few km or several decades of km).

At the same time, however, there is another major misunderstanding, made by habit, due to its general application to the other rock formations and due to the non thorough assimilation by the scientific community of the formation of the ophiolites within the divergent plates. Thus, almost as a rule, there is still today reference about the age of the ophiolites, meaning a common age (e.g. Lower Cretaceous or Lower Jurassic), as if it were a stratigraphic formation, pertaining the same age of deposition. Nevertheless, in the case of the ophiolites we know that it is a "diachronic stratum" of oceanic crust with lateral growth, beginning during the opening of the ocean along the continental margins—where the earliest ophiolitic rocks can be found—and ending with the termination of the activity of the mid-ocean ridge, where the younger ophiolitic rocks of the particular ocean have been formed. Consequently, some segments of the ophiolites will have a much older age than others, such as for example parts of the Atlantic oceanic floor ranging from Upper Jurassic to Upper Cretaceous or Eocene, or Miocene or even Pleistocene age. At the same time, in the Pacific ocean floor today there are older Lower Jurassic ophiolites and younger Pleistocene ophiolites. Especially when the overlain abyssal sediments are detached in a subsequent orogeny, it is extremely difficult to distinguish ophiolites of different ages. In this way the issue of the age of the ophiolites is taken to another level, because it is not so important to know the age of a particular ophiolitic segment/outcrop as it is to know the beginning and the end of the ophiolite genesis for each oceanic basin.

By using the most common geological dating techniques it is possible to approximate the age of the lavas of an ophiolite complex (Fig. 2.4), as long as their syngenetic stratigraphic relation with the adjacent abyssal-pelagic sediments, represented mainly by radiolarites and red pelites, is maintained.

Another case of ophiolite dating is the radiometric dating of the amphibolites, sporadically observed at the base of the ultramafic tectonites of the ophiolite complex (metamorphic sole), which may be associated with tectonic transform zones or with the tectonic emplacement of the ophiolites. In Greece, on the one hand, the syngenetic sediments are dated as Jurassic in some cases (Maliac, Western Thessaly, Peonia) and Upper Cretaceous in others (Arvi, Dodekanese) and on the other hand, the metamorphism of the amphibolites at the base of some ophiolite complexes display a Lower Jurassic age (Spray and Roddick 1980; Spray et al. 1984; Liati et al. 2004). Apart from the above data, there is also the dating of the tectonic emplacement of the ophiolites on the

Fig. 2.4 Impressive outcrop of basaltic pillow lavas of Upper Triassic age from Western Thessaly in the region of Smokovo

surrounding sediments of the Tethyan margins, which always postdates the age of the ophiolitic rocks (Papanikolaou 2009). Thus, for example, the Cenomanian transgression is common in eastern Greece, covering the ophiolites as well as their tectonic contact over the underlying tectonic units of the so-called internal units (Brunn 1956,1960). This fact dates only the tectonic emplacement of the ophiolites, that is to say, the particular ophiolites under the unconformity may be of any pre-Cenomanian age, while ophiolites of younger than Cenomanian age may be found in other areas of the same or of another oceanic basin. Especially in the case of the Hellenides, it is known that part of the oceanic crust of Tethys has participated in an orogenic arc since the Malm, due to the fact that blocks of ophiolites can be observed in the Malm (former schist-hornstein formations), as well as ophiolite detritus in the so-called Upper Jurassic–Lower Cretaceous flysch, observed in Beotia, Western Thessaly, Yugoslavia, etc. (Papanikolaou 1986a).

The latest dating of basaltic lava with geochemical MORB affinities (Middle Ocean Ridge Basalts) in Greece has been reported in the Arvi unit in Crete, where the syngenetic sediments contain Globotruncanes of Maastrichtian age (Tataris 1964; Bonneau 1976, 1984). However, in this case the age refers only to the volcanic rocks and not to the overall ophiolite complex.

Generally the beginning of the creation of the ophiolites in the Hellenides is considered as Late Triassic and its end as Late Cretaceous. The common assumption is that the largest part of the Tethyan ophiolites observed today in the various Alpine mountain ranges has been created during the Jurassic.

Relevant to the age of the Tethyan ophiolites is also the problem of the relationship between the generally Mesozoic Tethys and the ocean that was located approximately at its place during the Paleozoic, as shown by the ophiolites involved in the Variscan Orogeny (e.g. Caucasus), and for which the term **Palaeo-Tethys** has been used (Hsu and Bernoulli 1978; Sengor 1979; Sengor et al. 1984a, b). In this specific case it is certain that the ophiolites belonging to Palaeo-Tethys were tectonically emplaced during the Late Devonian–Early Carboniferous. Many researchers, however, accept that Palaeo-Tethys or even part of its oceanic crust continued to exist throughout the Late Paleozoic and even in an extended period of the Triassic. This creates a problem, because the time of existence of the Palaeo-Tethys remnants coincides with the time of creation of Tethys, which, in contrast, in this case is known as **Neo-Tethys**. In this already complex case two general views can be distinguished: (i) Palaeo-Tethys and Neo-Tethys were almost consecutive of one another in approximately the same geotectonic region, and (ii) Palaeo-Tethys was closing in one region while at the same time, in another position, Neo-Tethys was opening, and not necessarily in a parallel arrangement (Fig. 2.5). At the same time, the distinction of an intermediate geotectonic cycle between the Variscan and the Alpine was proposed, with the peak of the orogeny during the Late Triassic, which resulted in the so-called **Cimmerides** (Sengor et al. 1980, 1984a, b). In the case of the Cimmerides, mainly found in Turkey, the Lower Jurassic unconformably overlies Paleozoic–Triassic strata, often metamorphic and with granite intrusions, as well as ophiolites corresponding to the intra-Pontide ophiolitic belt. These ophiolitic formations probably correspond either to Palaeo-Tethys or to an intermediate ocean separating Palaeo-Tethys region to the north from the Neo-Tethys region to the south during the Permian–Triassic period.

In addition to the above distinction of the Mesozoic from the Paleozoic Tethys there are also other cases where a prefix has been used in front of the term Tethys, or other completely different terms for a specific stage of Tethys or for a particular part of it (Bizu-Duval et al. 1977; Hsu and Bernoulli 1978; Boccaletti 1979; Robertson et al. 1991). One of these cases is the term **Para-Tethys**, which corresponds to a northern branch of the Tethys Ocean, considered only as a marine physico-geographic unit not underlain by oceanic crust. This basin was developed during Oligocene–Pliocene in today's lowlands of Hungary (Pannonian Basin) and sections of Switzerland, Austria, f. Yugoslavia, Czechoslovakia and Romania (Steininger and Rogl 1984) (Fig. 2.6).

Fig. 2.5 Schematic representation of the paleogeography in the Late Triassic–Early Jurassic, highlighting the existence of the Cimmerian microcontinent between Eurasia and Gondwana, at the eastern part of Tethys. This microcontinent separates the remnants of the closing Palaeo-Tethys to the north from the opening Neo-Tethys to the south. In the western part of Tethys the rifting-opening of Neo-Tethys is beginning, as well as more to the south in the Pindos basin (after Sengor et al. 1984a, b, modified)

This vast marine palaeogeographic area is also known as the Pannonian Basin, rendering it a now viable term, whereas the term Para-Tethys tends to disappear. This large marine unit was separated from Tethys in the south during this period by the orogenic arcs of the Dinarides–Hellenides. Thus, it was not connected with the Tethyan oceanic basins, which gave rise to the present Mediterranean. Para-Tethys was basically an epicontinental marine basin, developed behind the orogenic systems of the Dinarides–Hellenides as a back-arc basin. It is of particular significance for the paleogeography and stratigraphy–paleontology of the Oligocene–Miocene, because special biocommunities have been created in this northern part of Tethys, different from those of the southern part, uprising several issues regarding stratigraphic correlations and fauna migrations (see for the RCMNS congresses on Mediterranean Neogene Stratigraphy).

2.5 The Pre-Alpine Formations and the Beginning of the Alpine Cycle

The distinction between two consecutive orogenic cycles occurring in an area is possible where distinctive rocks and structures appear with no doubt. In the case of the Alpine cycle this implies that one can observe an important stratigraphic unconformity between intensively deformed rocks of the Paleozoic from the transgressive non-or slightly-deformed sediments of the Mesozoic. That means that the Variscan structures of the Late Paleozoic (folds, nappes, schistosities etc.) do not appear in the overlying Alpine formations of Triassic or younger age. This clear image exists today in many areas of the so-called Meso-Europa, where Alpine sediments (usually of Permian or younger age) are almost horizontal, covering uplifted, folded and schistosed or metamorphosed Paleozoic rocks. On the other hand, as we enter the area of the Alpine Orogeny, the Neo-Europa, the Alpine rocks gradually tend to be so intensively deformed and transformed that distinguishing them from the underlying Variscan ones becomes either difficult or even impossible. Additionally, the Triassic unconformity, usually separating the Variscan from the Alpine cycle, often constitutes a surface of tectonic decollement during the Alpine orogeny. This phenomenon creates distinct tectonic units of the Alpine and of the Variscan cycle, thus creating problems in determining the primary relation between basement and sedimentary cover. In this way, we usually observe units cut off from their substratum in the Alpine system, which cannot shed light on the transition period from the Variscan to the Alpine cycle.

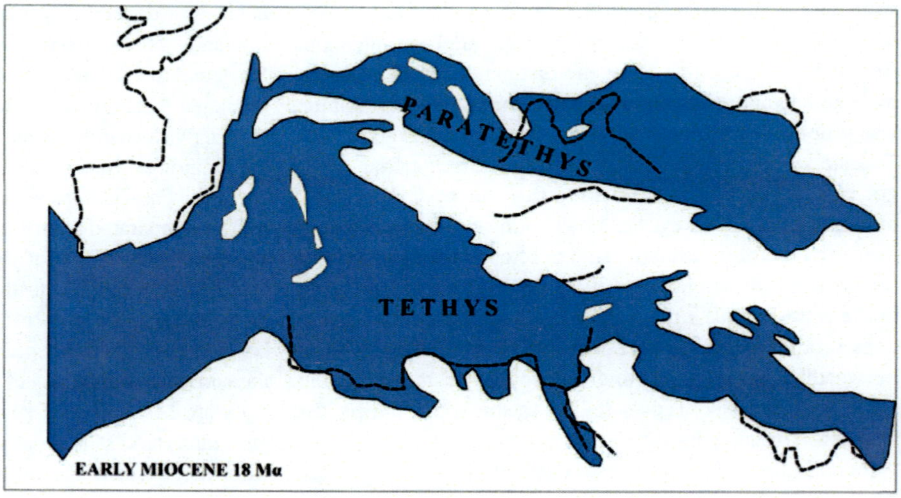

Fig. 2.6 Schematic palaeogeographical representation of Para-Tethys during Early Miocene (from Steininger and Rogl 1984, modified)

The Hellenides are a typical example of decollement of their generally Mesozoic strata from their pre-Alpine basement. It is worth noting that there is no documentation of an undisputed Triassic unconformity in Greece overlying the Paleozoic. It is indicative that a lot of Upper Paleozoic occurrences continue without a significant unconformity to the Triassic strata, while some areas considered as typical of the Triassic transgression onto the Upper Paleozoic (e.g. mount Knimis in Atalanti, Maratos 1965, 1972) were proved to be tectonic contacts of Triassic onto Triassic strata, while the rocks regarded as Triassic transgressive conglomerates were in fact volcanic breccia-conglomerates formed on the slopes of Triassic volcanic domes (Sideris 1981,1986). On the contrary, a pure angular unconformity between the Devonian and Lower Triassic is observed in the Ohrid Lake area, a few kilometers north of the Greek borders in a unit corresponding to the SubPelagonian/Almopia unit of the Internal Hellenides (Papanikolaou and Stojanov 1983).

It should be noted that there are no non-metamorphic pre-Alpine units in Greece, after the discovery that the Lower-Paleozoic fossils reported from Chios (Ktenas 1921a, 1921b, 1923; Besenecker et al. 1968) are found inside olistoliths of a wild flysch formation of Permian age (Papanikolaou and Sideris 1983,1992). The only exception is the case of Kos, where a slightly metamorphosed Paleozoic sequence crops out in mount Dikaios (Desio 1931; Papanikolaou and Nomikou 1998).

Another problem related to the distinction of the Alpine from the pre-Alpine formations is posed by the metamorphism. Previously, under the influence of Northern European experience, the metamorphic rocks of the Hellenides were collectively considered as pre-Alpine and regarded as metamorphic or crystalline massifs, that is to say, areas made of crystalline rocks that got their compact structure prior to the beginning of the Alpine cycle (Philippson 1901; Renz 1940; Trikkalinos 1955; Dercourt 1964). However, in the last decades it has been proven, both in the Alps (Pennique) and in other locations, as well as in Greece (e.g. Attica-Cyclades, Peloponnese-Crete), that many of the metamorphic rocks of the former "metamorphic massifs" are Alpine (Negris 1915; Kober 1929, 1931; Marinos and Petrascheck 1956; Papanikolaou 1979, 1986a). Thus, in many cases of metamorphic rocks where the data are not sufficient, we refer to them as possibly Alpine or possibly pre-Alpine rocks, although the increasing geochronological data obtained during the last decades, especially on metamorphism and magmatism, have helped to reduce the challenging outcrops.

Regardless of the above problems, the characterization of the Upper Paleozoic formations as either Variscan or Alpine is difficult, as there are large occurrences of molasse or molasse-like deposits of the Variscan cycle, which often form the basis of the Alpine stratigraphic sequences. This is clearly visible in the northern margin of Tethys, where red clastic sediments are found beneath the common marine sediments of Germanic facies, such as the Bundsandstein of the Lower Triassic and the overlying limestones (e.g. Muschenkalk) (Boncev 1986; Lepper et al. 2005).

Moreover, in many regions of the Alps and elsewhere (e.g. the Apennines) unique clastic formations with quartz conglomerates and contemporaneous volcanism can be observed, known as the Verrucano facies. In these cases one may wonder whether it is the end of the Variscan cycle with molasses and volcanic arcs or the beginning of the Alpine cycle, with the rifting stage. The difficulty of the problem suggests that it may be a transitional situation, especially in the area of Tethys and that it is related to the problem of the relation between Paleo-Tethys, Neo-Tethys and the Cimmerides.

Generally, however, the paleogeographical picture at the end of the Paleozoic and the beginning of the Mesozoic in the region of Tethys and especially in the area of the present eastern Mediterranean is characterized by (Argyriadis 1975): (i) continental to coastal facies with clastic formations and coal in the Eurasian margin, and (ii) marine facies with characteristic, usually dark, reef limestones with algae (e.g. Mizzia), foraminifera (e.g. Fusulina, Schwagerina, Stafella), corals, sea urchins etc., which are interstratified in clastic sequences of clay-sandstone formations, which often grade to turbiditic horizons with olistoliths on the southern margin of Tethys, on the periphery of Gondwana. The facies of the southern margin are uniformly found throughout the southern branch of the Tethyan Alpine System, from the Southern Alps–Dinarides to the Himalayas.

The dominant clastic material in the deposits of the Upper Paleozoic is generally considered to derive from the then fresh mountain ranges of Variscan Europe. Based on the magnitude of clastic sedimentation, the size of the clasts and other indicators of large or small scale transportation from the coastal transportation area to the deposition area in the Tethyan basins (e.g. the alteration of feldspars, heavy minerals, etc.) various paleogeographical subdivisions have been proposed. In the Hellenides though, there are areas where neritic sedimentation in shallow water carbonate platforms has already begun since the Late Permian, continuing with minor pauses to the Triassic (e.g. Hydra, Lesvos autochthon, Chios allochthon) (Baud et al. 1991).

In areas where no Upper Paleozoic sediments are observed, the bases of the Alpine sequences consist of Triassic volcano-sedimentary sequences, which evolve into shallow water carbonate platforms during the Middle Triassic in some areas and during the Late Triassic in some others (Papanikolaou 1989,2013) (see also Sect. 7.3).

A general conclusion about the beginning of the Alpine cycle is that, besides the two Tethyan margins/branches, the paleogeographical organization of Tethys during the Late Paleozoic–Middle Triassic is unclear and does not conform with the organization of the rest of the Alpine cycle, which begins to take its shape with the formation of distinct carbonate platforms and oceanic basins mainly during the Late Triassic (Papanikolaou 2013).

2.6 The Post-Alpine Formations and the End of the Alpine Cycle

The characterization of a formation generally as post-orogenic and specifically as post-Alpine means that the formation is subsequent to a particular structure, as shown by the unconformable overlapping of folds, faults, thrusts, nappes, schistosity, and generally structures of the underlying formations that resulted from their participation in the orogenic events. However,, it is necessary to further distinguish the actual **post-Alpine** (s.s.) formations from the molasse formations, which are also unconformably overlying on the Alpine structures, but are still considered as an integral part of the developing orogenic arc (Aubouin 1974), as this will be discussed in Chap. 4, when analyzing the mechanism of orogeny. The distinction between the above two types of unconformable sediments over the Alpine substratum is difficult at the scale of a mountain, island, or peninsula, whereas the distinction is easy at the scale of a large segment of the Alpine orogenic arc. This arises from the fact that molasse formations are deposited in basins created during the orogeny that are geometrically identical to the orogenic arc, forming the so-called arc parallel structures. On the other hand, the post-Alpine formations are deposited in basins created either outside the orogenic arc area or inside the arc but with a different geometry. This depends on the new stress field, developed due to broader geotectonic motions, such as those that led to the fragmentation of the Hellenic arc in the Pliocene–Quaternary (Papanikolaou and Royden 2007).

Another problem concerns the age of the post-Alpine formations, which is varying across the belt. This is due to the fact that the Alpine orogeny began several tens of millions of years ago in many areas of the so-called internal zones both in the Alpine system in general and in the Hellenides in particular, while in others, like the actual Hellenic arc, Alpine orogeny is still in progress and thus post Alpine formations are also in the process of being formed. Thus, the older post-Alpine sediments may be of Lower Tertiary age, while the younger ones may be of Pleistocene or Holocene age, while in many areas of the arc front their formation has not yet begun. This phenomenon is associated with the so-called **migration** of orogeny, which functions as an **orogenic wave** in geological time. This creates a gradation in each part of the Alpine system, with the earlier post-Alpine formations observed in the hinterland and the younger post-Alpine formations towards the foreland. In Greece, the post-Alpine formations generally include sediments of Neogene and Quaternary age. However, it is important to distinguish the term post-Alpine for a specific part of the orogenic arc and for a specific period of time because in the Lower Miocene, for example in the Hellenides, a syn-orogenic sedimentation of flysch was observed in some areas (e.g. in Gavrovo and Ionian), a sedimentation in molasse basins was observed in others (e.g. in the Mesohellenic trough), and post-orogenic sedimentation of pure post-Alpine formations was observed towards the interior of the arc (e.g. in Aliveri/central Evia).

2.7 The Main Geotectonic Stages of the Evolution of the Alpine Cycle and of the Tethys Ocean

The Tethyan Alpine geotectonic cycle consists of distinctive successive stages that include its growth, restriction and disappearance. A key characteristic of the Tethyan Alpine System in the Hellenides segment is that from the onset of the orogeny to the final stage of continental collision, a stable tectonic polarity is observed. In particular the European margin is constantly active with subduction processes and the creation of orogenic arcs, unlike the African margin, which remains passive throughout the whole cycle (Papanikolaou 1989,1997; Stampfli et al. 1991). The individual stress fields of the lithosphere in the Tethys area and its neighboring plates, as well as their effects on various geodynamic phenomena and the paleogeography can be distinguished into four evolutionary stages, which are the following (Papanikolaou 1986b) (Fig. 2.7):

(i) Rifting and opening of the Tethys Ocean
(ii) Partial compression and creation of the orogenic arc
(iii) Total compression and destruction of the oceanic crust
(iv) Disappearance/closure of the ocean and continental collision.

i. **Rifting and opening of the Tethys ocean stage**

This stage initiates a new geotectonic cycle, by exerting extensional stresses on a continent, such as Pangaea, with a peneplain landscape on a crust created by older orogenies. This is essentially the formation of a tectonic zone with divergent plate boundaries. Thus, the rifting stage begins like

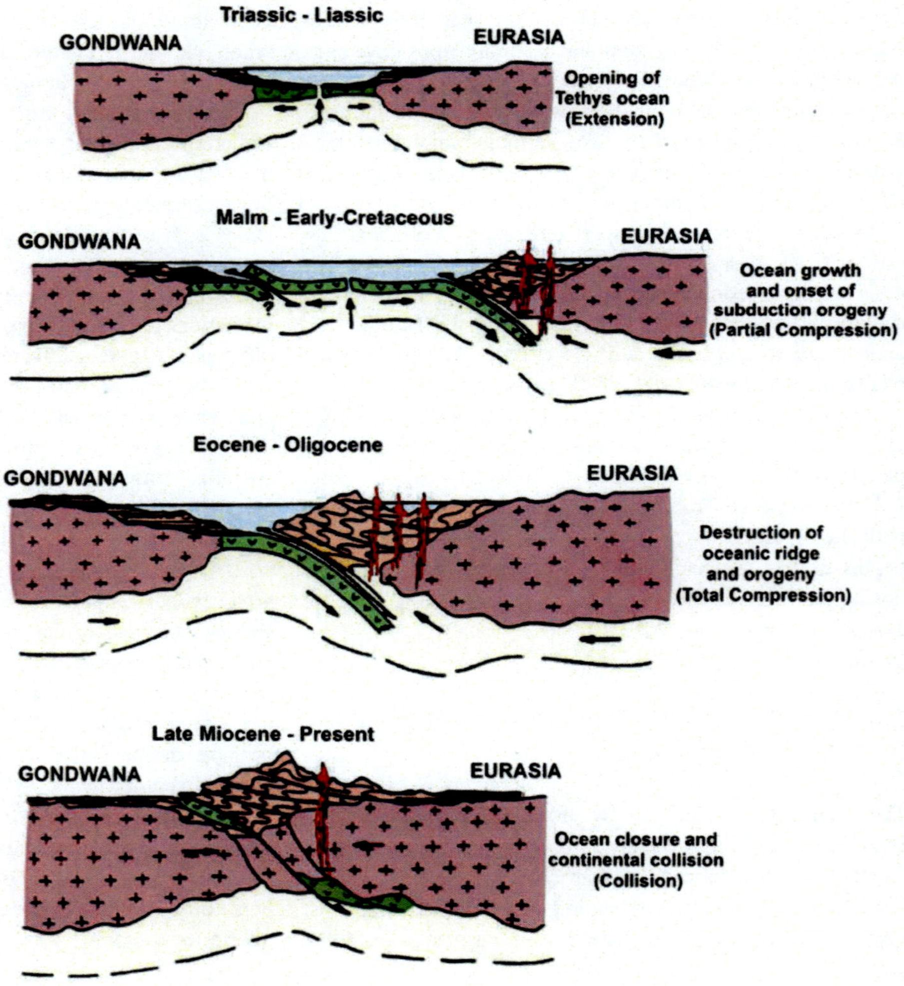

Fig. 2.7 Schematic representation of the four main stages of the geotectonic evolution of the Tethyan Alpine System (from Papanikolaou 1986b)

the present East African Rift, continuous to the embryonic stage of the Red Sea to end up in the mature stage of the Atlantic Ocean. At this stage the mid-ocean ridge is formed and becomes active with the basaltic pillow lavas and abyssopelagic sediments being deposited in the ocean basin. Coastal neritic sediments are deposited on the two continental margins of the ocean, the Eurasian in the north and the Gondwanian in the south, characterized by the absence of geodynamic phenomena and are considered as passive margins. In terms of tectonic plates, this initial stage is associated with the creation of two tectonic plates from the division of the original continent. Interestingly enough, the two new plates initially consist mainly of continental crust, but gradually they also acquire oceanic crust.

The end of this stage and the beginning of the next one is set mainly during the Malm, when the paleogeographical realm of Tethys reached its maximum size, as shown also by the paleomagnetic data, as well as by its paleogeographic organization, which comprises all of its basic pre-orogenic physico-geographical units (see also Chaps. 9 and 10). It should be noted that the main direction of the break zone in the lithosphere, which later evolved into the mid-ocean ridge of Tethys, was diagonal in relation to the pre-existing Variscan structure. As a result, the Variscan structure has been dissected into large segments of the Variscan mountain ranges on the Eurasian margin in some areas, and very small in others.

ii. **Partial compression and creation of the orogenic arc stage**

This stage sets off when the divergence of the two plates has ceased or they begin to re-approach each other, while at the same time the mid-ocean ridge is still active (Fig. 2.7). At this stage, a significant change in the geodynamic regime takes place, with an establishment of compressive stresses over one of the continental margins that causes rupture, usually along the old ocean/continent transition zone, and marks the creation of an orogenic arc. This stage signifies the onset of the, subduction of the oceanic crust of Tethys under the Eurasian margin and creation of the orogenic arc. The result is folding of the sediments, metamorphism, creation of volcanic arcs and related geodynamic phenomena, which render it an active margin, opposed to the other Africa/Arabia passive margin, where sedimentation continues as before.

This is a typical converging plate boundary stage similar to the present day Japanese type arc system, with a subduction zone under a continent. However, a secondary case of a Tonga type arc system, with an intra-oceanic subduction zone may also exist. Thus, the continuation of activity of the mid-ocean ridge, where tensile stresses are still applied, may continue to expand the oceanic space, but it can replace only part of its subducting portion in the active margin, thus driving to an overall restriction of the ocean and to a differentiation of the tectonic stresses that drive the plates (Chapple and Tullis 1977). The Indian Ocean is currently at this stage, with an active mid-ocean ridge creating new oceanic crust and at the same time, destruction of oceanic crust along the Indonesian subduction zone. Several parts of the Pacific Ocean are also at a similar stage, although this ocean is more complex.

In terms of tectonic plates, at this stage one of the two plates breaks at the boundary of oceanic/continental crust and two new plates are formed, one of which is subducting and is purely oceanic, while the other one is advancing and is purely continental. By considering also the other non-disrupted half of the ocean we now have totally three plates in the system. If, by any chance, the same procedure happens also at the other margin, then we will end up with four plates, two of which will be subducting oceanic and the other two will be advancing continental. In the case of an intra-oceanic subduction zone, then there will be four plates, of which two will be purely oceanic, one purely continental and the last one mixed.

Thus, in the region of Tethys already since Malm and during the Early Cretaceous, the Eurasian margin gradually becomes active and gives rise to the first mountain ranges of the northern Alpine branch, whereas the southern African margin remains passive. The difference between the two margins is important, because apart from the geodynamic phenomena of the orogeny the Eurasian margin experiences the destruction of oceanic crust, which following ophiolite obduction and erosion, supplies with ophiolite detritus the trenches along the convergent boundaries of the system.

The characteristic phenomenon observed throughout the whole Tethyan system as a result of this stage is the observation of orogenic unconformities with transgressive sediments of Cretaceous age (usually Upper Cretaceous) over the underlying folded sediments of Triassic-Jurassic age, often along with ophiolites and metamorphic and/or magmatic–volcanic rocks. These areas belong to the so-called internal zones of the orogenic system with varying ages, e.g. in Greece there is the Cenomanian transgression over the ophiolites and the paleofolded sediments, while in Anatolia there is the Maastrichtian transgression. Generally at this stage we have the so-called paleo-Alpine orogenic events of Stille (1936) (Kimmeric, Austrian, etc.).

iii. **Total compression and destruction of the oceanic crust stage**

This stage is characterized by an active subduction zone and orogenic arc in the active margin already formed in the previous stage but without the simultaneous existence of a mid-ocean ridge. As a result, compression is dominant throughout the system, i.e. the opposing stresses of the two continents are exerted simultaneously and deform the rocks along the subduction zone and within the orogenic arc. The ocean is now a remnant of the older ocean and is constantly reduced in size. The duration of this stage depends on the subduction rate. A modern example of this stage is the actual Hellenic arc and trench system, where the Tethyan remnant of the Eastern Mediterranean oceanic crust, belonging to the African plate, is being subducted. Based on the 4–5 cm/year current subduction rate in the Hellenic subduction zone (e.g. Mcklusky et al. 2000; Kahle et al. 1995), and the 200–250 km ocean's width between Crete and Cyrenaica, the ocean is expected to close in the next 5–6 million years.

Two tectonic plates co-exist at this stage, one continental and one mixed, due to the fact that the destruction of the mid-ocean ridge led also to the destruction of the previous interim oceanic plate. This stage has been initiated in the case of Tethys approximately during the latest Cretaceous and has been established throughout the Tethys system in the Eocene–Oligocene.

iv. **Closure/Disappearance of the ocean and continental collision stage**

At this final stage, all of the oceanic area created during the previous stages between the two continental plates disappears. Their re-approach, results in the direct transmission of the lithospheric stresses from one continent to the other, with crushing of the intermediate sediments and associated rocks of the former ocean. This stage marks the culmination of orogeny with maximum elevation of the Alpine mountain ranges, which are now **collisional ranges**. Collision of Eurasia with parts of the former Gondwana has largely occurred by the Late Miocene, when the collision of Adria with Eurasia formed the Alps, the collision of Arabia with Eurasia formed the Caucasus–Iranides and the collision of India with Eurasia formed the Himalayas. Only a few segments along the remnants of Tethys (namely the Eastern Mediterranean) are observed at the end of stage iii, where no collision has occured yet but is expected to initiate within the next 5–6 million years.

Therefore, at this final stage the pre-existing convergence zone is transformed into a collisional zone. The continental collision has led to the reformation of the stress field in the area and the creation of new kinematics of the plates in the wider region. Particularly, the effect of the termination of the oceanic subduction zone on the plate kinematics should be taken into account as the detached oceanic slab is gradually penetrating into the mantle below the asthenosphere. This essentially means the end of the Alpine geotectonic cycle, which began with the divergence of the plates and ended with their re-attachment. On the other hand it drives to a re-organization and a new mechanism of plate Kinematics, which will probably start a new cycle elsewhere. In terms of plate tectonics the termination of this stage signifies the formation of a single amalgamated continental plate.

The different timing of each stage along the various segments of Tethys is due to the diagonal geometry between its tectonic and paleogeographic organization and to the different rates of geotectonic motions, resulting from the differentiated driving forces at the lithosphere—asthenosphere boundaries. In particular, the different timing observed at the initiation of collision between the two plates is due to: (i) the existence of geometrically irregular boundaries between Tethys and the African margin, like the Great Gulf of Sirte on the Libyan coasts, and (ii) in the rifting of the Arabian Plate from the African plate along the Red Sea and its faster advance towards the north, compared to the African plate in the Late Cenozoic. Thus, nowadays the oceanic region of the Eastern Mediterranean lies outside the Alpine orogenic front, representing the main Tethyan remnant north of the remaining passive margin of Gondwana.

The subduction of Tethys, followed by the margins of parts of Gondwana under Eurasia along the southern branch of the Alpine segment of the Dinarides-Hellenides-Taurides etc. is in contrast with the allochthonous Tethyan elements of the northern Alpine branch emplaced over the Eurasian margin. This occurred during Oligocene–Pliocene in the Alps and the Carpathians, after the collision of the Adria microplate (which is a mole of Gondwana) with Eurasia (Trumpy 1980).

The previous model of the four geodynamic stages of Tethyan evolution covers the broad idea of one simple ocean, almost like the Atlantic Ocean, without the presence of microcontinents/tectono-stratigraphic terranes. In the case of Tethys, however, we do know that there were several terranes along its length, from the current Western Mediterranean to the Himalayas (see terrane maps of the final volume of IGCP No 276). Thus, the above model should be completed incorporating more stages following the terrane concept, by including the presence of continental microplates and the intermediate oceanic basins, whose number and dimensions vary according to the number and dimensions of the drifted continental fragments. In the case of the Hellenides, as it will be further analysed later (Chap. 4), there are four continental fragments, namely H1, H3, H5 and H7, which through their northward movement created respectively five oceanic basins of Tethys, namely H0, H2, H4, H6, and H8 (Papanikolaou 1989, 1997, 2009, 2013).

The final evolution model of Tethys described earlier and the incorporation of the terranes of the Hellenides is provided in six stages as follows (Fig. 2.8):

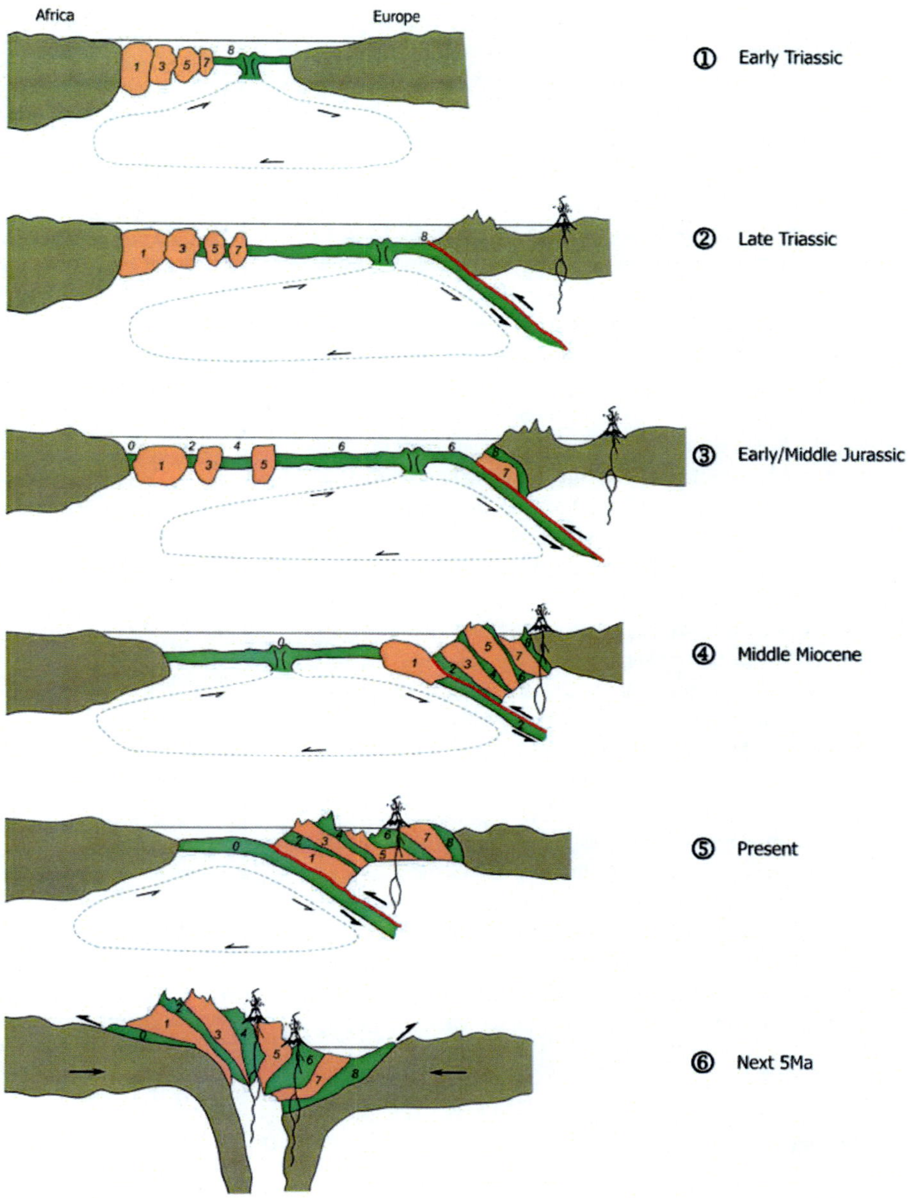

Fig. 2.8 Schematic representation of the evolutionary stages of Tethys in the Hellenides region, incorporating the tectono-stratigraphic terranes

(1) The first stage of rifting and opening of the ocean is not differentiated from the general model, except that at the same time as the opening of the main branch of Tethys in the Early–Middle Triassic (e.g. H8) there is also a widely spread extensional zone, leading to the creation of additional rift zones in the northern margin of Africa. Thus, at least four continental fragments (H1, H3, H5, H7) are distinguished. At this period, the opening of the neighboring oceanic basin to the south is also initiated (H6), while the other basins more to the south (H4, H2, H0) are in the early rifting phase (see also Chaps. 9 and 10).

(2) The second stage involves the initiation of subduction of the northernmost oceanic basin, therefore starting the partial compression, with the creation of an orogenic arc at the active margin of Europe, approximately at the Late Triassic. At the same time the gradual opening of the other basins to the south is taking place.

(3) The third stage, at the Triassic/Jurassic boundary, includes the complete subduction and closure of the northernmost basin (H8), as well as the accretion of the northernmost continental terrane (H7) in the European margin. The incorporation of the continental terrane is in fact a microcollision and is carried out with the

detachment of the upper crust from the rest of the crust that follows the base of the subducted lithosphere to great depths beneath the asthenosphere. At the same time, the subduction of the next oceanic basin (H6) along the trench has started, while the other basins further to the south are opening as the continental blocks are drifted to the north.

(4) The next stage, is a shift in time and process from the northernmost terrane H7 to the southernmost continental terrane H1, by eliminating the other intermediate stages of the intermediate terranes which would be a repetition of the previous two stages for the intermediate terranes during the Early Jurassic to Eocene. Thus, this stage refers to the last continental terrane (H1) which is subducted below the European margin, causing a microcollision with the previously accreted terranes. This stage lasts from the Late Eocene to the Late Miocene. Following the termination of this stage the last remnant of the Tethys oceanic crust is traced in the present Eastern Mediterranean basin (H0).

(5) The fifth stage, which began in the Late Miocene and continues to the present day, is the phase of the total compression of the Tethys Ocean, with the subduction of the southernmost basin, whose mid-ocean ridge is inactive. This stage is expected to terminate over the next 5–6 million years revealing the final closure of Tethys.

(6) The sixth stage, which is expected to begin in the next 5–6 million years in the Hellenides marks the final stage of collision, in which the Tethys Ocean will completely disappear, and the two continents will return to direct contact, creating collisional mountain ranges. The latter has already occurred along the largest part of the Tethyan Alpine System already since Miocene times.

References

Argand, E. 1924. La tectonique de I'Asie. Congrès Géol. Intern., Bruxelles.

Argyriadis, I. 1975. Mésogée permienne, chaine hercynienne et cassure téthysienne. Bull. Soc. géol. France, 17, 56–67.

Aubouin, J. 1974. Des tectoniques superposées et leur signification par rapport aux modèles géophysiques. L'exemple des Dinarides, paléotectonique, tectonique, tarditectonique, néotectonique. Bull. Soc. géol. France, XV, 426–460.

Baud, A., Jenny, C., Papanikolaou, D., Sideris, C., & Stampfli, G., 1991. New observations on Permian stratigraphy in Greece and geodynamic interpretation. Bull. Geol. Soc. Greece, XXV, 187–206.

Besenecker, H., Dürr, S., Herget, G., Jacobshagen, V., Kauffmann, G., Lüdtke, G., Roth, W. & Tietze, K.W. 1968. Geologie von Chios (Ägäis). Geologica et Paleontologica, 2, 121–150.

Bizu-Duval, B., Dercourt, J. & Le Pichon, X. 1977. From the Tethys Ocean to the Mediterranean Seas: A plate tectonic model of the Evolution of the western Alpine system. Intern. Symp. Struct Hist. Medit. Basins, Split 1978, Technip, 143–164.

Boccaletti, M. 1979. Mesogea and Mesoparatethys: their development at the Tethyan continental margins and their influence on the later evolution of the Mediterranean and Paratethys. Ann. Géol. Pays Hellén, hors série. I, 139–148.

Bonardi, G., Cavazza, W., Perrone, V. & Rossi, S. 2001. Calabria–Peloritani terrane and northern Ionian Sea. In: Vai and Martini (ed.) Anatomy of an orogen: the Apennines and adjacent Mediterranean basins. Kluwer Academic Publishers, Dordrecht, 287–306.

Boncev, E. 1986. Apercu general sur la tectonique des Balkans. Geol. Balc., 16.2, 3–32.

Bonneau, M. 1976. Esquisse structural de la Crète alpine. Bull Soc. Géol. France, 8, 2, 351–353.

Bonneau, M. 1984. Correlation of the Hellenide nappes in the south-east Aegean and their tectonic reconstruction. Geol. Soc. London, Sp. Publ.,17, 517–527.

Bortolotti V, Chiari, M., Marroni, M., Pandolfi, L., Principi, G. & Saccani, E 2013. Geodynamic evolution of ophiolites from Albania and Greece (Dinaric—Hellenic belt): one, two or more oceanic basins? Intern. J. Earth Sciences, 102, 783–811.

Brunn, J. 1956. Contribution à l'étude Géologique du Pinde Septentrional et d'une partie de la Macédoine Occidental. Ann. Géol. Pays Hellén., 7, 1–358.

Brunn, J.H. 1960. Les zones helléniques internes et leur extension. Réflexions sur I' orogenèse alpine. Bull. Soc. Géol. France, (7), II, 470–486.

Candan, O., Koralay, O. E., Topuz, G., Oberhansli, R., Fritz, H., Collins, A. S., & Chen, F. 2016. Late Neoproterozoic gabbro emplacement followed by early Cambrian eclogite-facies metamorphism in the Menderes Massif (W. Turkey): Implications on the final assembly of Gondwana. Gondwana Research, 34, 158–173.

Chapple, W. M. & Tullis, T.E., 1977. Evaluation of the forces that drive the plates. J.Geoph. Res., 82, 1967–1984.

Coleman, H. 1971. Plate tectonic emplacement of upper mantle peridotites along continental edges. J. Geoph. Res., 76, 1212–1222.

Coney, P. J., Jones, D. L. & Monger, J. W. H. 1980. Cordillieran suspect terranes. Nature, 288, 329-333.

Dercourt, J. 1964. Contribution à l' étude géologique d'un secteur du Péloponnèse septentrional. Ann. Géol. Pays Hellén., 15, 1–418.

Dercourt, J. 1972. The Canadian Cordilliera, the Hellenides and the sea-floor spreading Theory. Canad. Jour. Earth Sci, 9, 709–743.

Desio, A. 1931. Le isole italiane dell' Egeo. Mem. Carta geol. Ital., 24, 534 p.

Dewey, J. F., 1976. Ophiolite obduction. Tectonophysics, 31(1–2), 93–120.

Dewey, J. F. & Bird, J. M. 1971. Origin and Emplacement of the ophiolite suite: Appalachian ophiolites in Newfoundland. J. Geophys. Res., 76, 3179–3206.

Dewey, J. F., Pitmann, W. C., Ryan, W. B. F. & Bonnin, J. 1973 .Plate tectonics and the evolution of the Alpine system. Bull. Geol. Soc. Amer., 84, 3137–3180.

Dietz, R. S. & Holden, J.C. 1970. The Breakup of Pangaea. Scient. Amer., 223, 30–41.

Haydoutov, I., 2002. Peri-Gondwanan terranes in the pre-Palaeozoic basement of Bulgaria. Geol. Balc., 32, 2–4.

Himmerkus, F., Anders, B., Reischmann, T. & Kostopoulos, D. 2007. Gondwana-derived terranes in the northern Hellenides. Geol. Soc. Am. Mem., 200, 379–390.

Himmerkus, F., Reischmann, T. & Kostopoulos, D. 2009. Serbo-Macedonian revisited: a Silurian basement terrane from northern Gondwana in the Internal Hellenides, Greece. Tectonophysics, 473, 20–35.

Howell, D.G. 1980. Mesozoic accretion of exotic terranes along the New Zealand segment of Gondwanaland. Geology, 8, 487–491.

Howell, D.G. & Jones, D.L. 1983. Tectonostratigraphic terrane analysis and some terrane vernacular. Proc. Circum-Pacific Terrane Conf., Stanford Univ., 6–9.

Hsü, K. & Bernoulli, D. 1978. Genesis of the Tethys and the Mediterranean. Init. Rep.D.S.P.D..XIII, 1, 943–949.

Hynes. A. J., Nisbet, E.G., Smith, A.G., Welland. M. J. P. & Rex, D. C. 1972. Spreading and emplacement ages of some ophiolites in the Othris region (eastern central Greece). Zeit. deuts. Geol. Ges. 123, 455–468.

Jones, D.L. Howell, D.G., Coney, P.J. & Monger, J.W.H. 1983. Recognition, character and analysis of tectonostratigraphic terranes in western North America. In: Hashimoto & Uyeda ed. Accretionary tectonics in the Circum-Pacific regions: Advances in Earth & Planetary Sciences, Terra Science Publ. Co., Tokyo, 21–35.

Kahle, H. G., Muller, M. V., Geiger, A., Danuser, G., Mueller, S., Veis, G., Billiris, H. & Paradisis, D. 1995. The strain field in NW Greece and the Ionian islands: results inferred from GPS measurements. Tectonophysics, 249, 41–52.

Kober, L. 1929. Beitrage zur Geologie von Attika. Sitz. Akad. Wiss. Wien, 138, 299–327.

Kober, L. 1931. Das Alpine Europa. Verlang von Gebrüder Borntraeyer, Berlin.

Ktenas, C. 1921a. Sur la découverte du Dévonien à l' île de Chios. C.R. somm. Soc.Géol. France, p. 131.

Ktenas, K. 1921b. Sur le Carbonifere de l' ile de Chios. C. R. somm. Soc. Geol. France, p. 146.

Ktenas, K. 1923. Sur la decouverte d' un horizon a Productus cora a l' ile de Chios. C. R. somm. Soc. Geol. France, p. 206.

Lepper, J., Rambow, D & Rohling, H.G. 2005. Der Buntsandstein in der Stratigraphischen Tabelle von Deutschland 2002. Newsletters of Stratigraphy, 41 (1–3), 129–142.

Liati, A., Gebauer, D. & Fanning, C.M. 2004. The age of ophiolitic rocks of the Hellenides (Vourinos, Pindos, Crete): first U–Pb ion microprobe (SHRIMP) zircon ages. Chemical Geology, 207, 171–188.

Maratos, G. 1965. Livanates sheet. Geological Map of Greece at scale 1/50,000, IGME.

Maratos, G. 1972. The geology of Greece. Athens, 189 p (in greek).

McKlusky, S. C. et al, 2000. Global Positioning System constraints on plate kinematics and dynamics in the Eastern Mediterranean and Caucasus. J. Geoph. Res., 105, 5695–5719.

Moix, P., Beccaletto, L., Kozur, H.W., Hochard, C., Rosselet, F. & Stampfli, G.M. 2008. A new classification of the Turkish terranes and sutures and its implication for the paleotectonic history of the region. Tectonophysics, 451, 7–39.

Moores, E.M. 1969. Petrology and structure of Vourinos ophiolitic complex of northern Greece. Geol. Soc. Amer. Sp. paper, 118.

Marinos, G., & Petrascheck, W.E. 1956. Laurium. Geol. Gephys. Res., IGME, IV, 1–247 (in greek).

Negris, Ph. 1915. Roches cristallophylliennes et tectonique de la Grèce. 123 p. Athènes.

Nirta, G., Moratti, G., Piccardi, L., Montavari, D., Carras, N., Catanzariti, R., Chiari, L. & Marcucci, M. 2018. From obduction to continental collision: new data from central Greece. Geol. Mag., 155 (2), 377–421.

Papanikolaou, D. 1979. Stratigraphy and structure of the Paleozoic rocks in Greece: an introduction. In: Sassi, F.P. (Ed.), IGCP Project No. 5 "Correlation of Prevariscan andVariscan Events of the Alpine–Mediterranean Mountain Belt, 93–102.

Papanikolaou, D. 1986a. Late Cretaceous Paleogeography of the Metamorphic Hellenides. Geol. Geoph. Res., IGME, Special volume in honor of Prof. Papastamatiou, 315–328.

Papanikolaou, D. 1986b. Geology of Greece. Eptalofos Publications, 240 p. Athens (in greek).

Papanikolaou, D., 1989. Are the medial crystalline massifs of the eastern Mediterranean drifted Gondwanian fragments? In: Papanikolaou, D., Sassi, F.P. (Eds.), Sp. Publ. Geol. Soc. Greece, 1,63–90 and IGCP 276, Newsletter, No 1, Athens.

Papanikolaou, D. 1997. The tectonostratigraphic terranes of the Hellenides. Ann. Géol. Pays Hellén., 37, 495–514.

Papanikolaou, D. 2009. Timing of tectonic emplacement of the ophiolites and terrane paleogeography in the Hellenides. Lithos, 108, 262–280.

Papanikolaou, D. 2013. Tectonostratigraphic models of the Alpine terranes and subduction history of the Hellenides. Tectonophysics, 595–596, 1–24.

Papanikolaou, D. & Nomikou, P. 1998. The Paleozoic of Kos: a low grade metamorphic unit of the basement of the external Hellenides terrane. IGCP 276, Newsletter 6, Sp. Publ. Geol. Soc. Greece, 3, 155–166.

Papanikolaou, D. & Royden, L. 2007. Disruption of the Hellenic arc: Late Miocene extensional detachment faults and steep Pliocene–Quaternary normal faults—or what happened at Corinth? Tectonics, 26.

Papanikolaou, D. & Sideris, CH. 1983. Le Paléozoique de l' autochtone de Chios: Une formation à blocs de type wildflysch d' âge Permien (pro parte). C. R. Acad. Sc .Paris, 297, 603–606.

Papanikolaou, D. & Sideris, C. 1992. Introduction to the geology of Chios island—fieldguide. 6th Congress of the Geological Society of Greece on «The Geology of the Aegean» and IGCP 276, 1992 FieldMeeting, Athens, 15 p.

Papanikolaou, D. & Stojanov, R. 1983. Geological correlations between the Greek and the Yugoslave part of the Pelagonian Metamorphic Belt. In: Sassi F.P. (ed), IGCP No 5, Newsletter, 5, 146–152.

Philippson, A. 1901. Beiträge zur Kenntnis der griechischen Inselwelt. Peterm. Milt. Erganzunheft,134, 1–172, Gotha.

Renz, C. 1940. Die Tektonik der griechischen Gebirge. Pragm. Akad. Athinon, 8.

Ricou, L.E., Marcoux, J. & Whitechurch, H. 1984. The Mesozoic organisation of the Taurides: one or several ocean basins? Geol. Soc. London, Sp. Publ. 17, 349–359.

Robertson, A.H.F., Clift, P.D., Degnan, P.J. & Jones, G. 1991. Palaeogeographic and palaeotectonic evolution of the Eastern Mediterranean Neotethys. Palaeogeography, Palaeoclimatology, Palaeoecology, 87, 289–343.

Robertson, A.H.F. & Dixon, J.E. 1984. Introduction: aspects of the geological evolution of the Eastern Meditarranean. In: Dixon, J.E., Robertson, A.H.F. (Eds.), The Geological Evolution of the Eastern Mediterranean, Oxford, 551–561.

Schardt, H. 1898. Les regions exotiques du versant N des Alpes suisses. Bull. Soc. Vaud. Sci. Nat., 34, 114–219.

Schermer, E. R., Howell, D.E. & Jones, D. L. 1984. The origin of allochthonous terranes: perspectives on the growth and shaping of continents. Ann. Rev. Earth & Plan. Sci., 12, 107–131.

Scotese, C. 2001. Atlas of Earth History. Arlington, Texas.

Sengor, A.M.C. 1979. Mid-Mesozoic closure of Permo-Triassic Tethys and its implications. Nature, 279, 590–593.

Sengor, A.M.C., Yilmaz, Y. & Ketin, I. 1980. Remnants of a pre-late Jurassic ocean in northern Turkey: fragments of Permian - Triassic Palaeo-Tethys? Bull. Geol. Soc.Amer., 91, 599–609.

Sengor, A.M.C., Satir, M. & Akkok, R. 1984a. Timing of tectonic events in the Menderes Massif, western Turkey: implications for tectonic evolution and evidence for pan-African basement in Turkey. Tectonics, 3, 693–707.

Sengor, A.M.C., Yilmaz, Y., Sungurlu, O. 1984b. Tectonics of the Mediterranean Cimmerides: nature and evolution of the western termination of Palaeo-Tethys. Geol. Soc. London, Sp. Publ. 17, 77–112.

Sideris, Ch. 1986. Contribution to the knowledge of the geodynamic development during Permo-Triassic of the Eastern Greece domain. PhD Thesis, University of Athens, (in greek).

Sideris, Ch. 1981. A new perception of the Atalanti «Paleozoic». Ann. Géol. Pays Hellén., 30/2, 637–646 (in greek).

Smith, A.G. 1971. Alpine deformation and the oceanic areas of the Tethys, Mediterranean and Atlantic. Geol. Soc. Am. Bull., 82, 2039–2070.

Spray, J.G. & Roddick, J.C., 1980. Petrology and $^{40}Ar/^{39}Ar$ Geochronology of some Hellenic sub-Ophiolite metamorphic Rocks. Contrib. Mineral. Petrol., 72, 43–55.

Spray, J. C., Bebien, J., Rex, D.C. & Roddick, J.C. 1984. Age constraints on the igneous and metamorphic evolution of the Hellenic—Dinaric ophiolites. Geol. Soc. London, Sp. Publ., 17, 616-627.

Stampfli, G., Marcoux, J. & Baud, A. 1991. Tethyan margins in space and time. Palaeogeography, Palaeoclimatology, Palaeoecology, 87, 373–409.

Steininger, F. & Rögl, F., 1984. Paleogeography and palinspastic reconstruction of the Neogene of the Mediterranean and Paratethys. Geol. Soc. London, Sp. Publ. 17, 659–668.

Stille, H., 1936. Wege und Ergebnisse der geologisch-tektonischen Forchungen. 25Jahre Kaiser Wilhelm - Gesellschaft zur Forderung der Wiss., 11, Die Naturw.

Suess, E., 1885–1909. Das Antlitz der Erde. Vol. 1–3, Leipzig.

Tataris, A. 1964. About the presence of the Olonos-Pindos zone in the Symi-Viannos area (Eastern Crete) and the age of its spilites. Proc. Acad. Athens, 39 (in greek).

Trikkalinos, J. 1955. Über das Alter des metamorphen Gesteine Attikas. Ann. Géol. Pays Hellén., 6, 193–198.

Trumpy, R. 1980. An outline of the Geology of Switzerland, Wepf and Co, Basel.

Von Raumer, J., Stampfli, G. & Bussy, F. 2003. Gondwana - derived microcontinents – the constituents of the Variscan and Alpine collisional orogens. Tectonophysics, 365, 7–22.

Zen, e-an. 1983. Exotic terranes in the New England Appalachians - limits, candidates and ages; A speculative essay. Geol. Soc. Am. Mem., 158, 55–82.

Zimmerman, J. 1972. Emplacement of the Vourinos ophiolitic complex, Northern Greece. Geol. Soc. Am. Mem., 132, 225–239.

3 The Mediterranean

Today, the Mediterranean Sea is the last evolutionary stage of the Tethys Ocean, which has, for its most part, disappeared. The geology of Greece, as a Mediterranean country, is evidently related to the geology of the Mediterranean. Thus, it is essential to know, when the Mediterranean was created, whether it is a uniform physico-geographical unit, what is its relation with Tethys and which are the basic geodynamic processes today, also specifying its future. The geological and geophysical structure of the Mediterranean Basin is highly differentiated between the Western and the Eastern Mediterranean with opening of post-collisional basins in the west versus rapid convergence in the east, where the active orogenic arcs still advance above the subducting last remnants of the Tethys Ocean (Stampfli and Borel 2004). This tectonic diversity is illustrated in 8 geotraverses of the TRANSMED Atlas, crossing through the entire Alpine orogenic system, from the Betics–Rif—High Atlas in the west to the Crimea-Anatolia-Cyprus-Sinai in the east (Cavazza et al. 2004).

The physico-geographical division of the Mediterranean in regards to the Tethyan Alpine system clearly shows that it is not a uniform basin, but a complex structure with a differentiated evolutionary process, especially during the Late Cenozoic. This results from the study of the bathymetric map of the Mediterranean, which shows the main deep sea basins and their locations in relation to the two branches of the Alpine system and the two continental margins (Fig. 3.1). Thus, the two basins of the Western Mediterranean, the Balearic and the Tyrrhenian, are located within the Alpine system between the two orogenic fronts of the two branches, while the two eastern basins, the Ionian and the Herodotus/Levantine, are located south of the southern orogenic front of the Alpine system, over the African plate. Thus, the western basins have opened after the collision between Europe and Africa in the Oligocene–Miocene within the Alpine orogeny and their age is Miocene–Quaternary (Bonardi et al. 2001). On the other hand, the two eastern basins are the northernmost part of the African plate, which is in contact with the southern front of the three active alpine orogenic arcs, Calabria–Sicily, Hellenic and Cyprus. Therefore, the eastern basins have a tectonically active northern boundary, expressing the tectonic contact of the European plate with the African one, while their southern boundary forms the passive margin of the African continent, since the opening of the basin of the eastern Mediterranean during the Triassic–Jurassic. The isolated Black Sea basin in the north-east is considered as a distinctive feature of the Mediterranean, as it is an inactive isolated basin (land locked basin) since the end of the Cretaceous, a remnant of tectonic processes characterized by the presence of major strike-slip faults, observed throughout the Eastern Carpathian–Crimean–Caucasus system. A similar isolated inactive basin is observed more to the east in the South Caspian Sea, which shares a similar structure and evolution history of a back arc basin as the Black Sea (Zonenshain and Le Pichon 1986).

In the 1990s a detailed mapping of the ocean floor of the Mediterranean was accomplished by combining digital bathymetric data from various oceanographic surveys in many countries, coordinated by UNESCO. On this detailed bathymetric background, a series of other data, such as gravitational, sedimentary and seismological were presented and led to the creation of corresponding thematic maps at scale 1/2,500,000, which are used in the following chapters. The correlation and exploration of these maps allows the synthetic view of the Mediterranean both in terms of geometry and structure but also in terms of the geodynamic processes.

3.1 The Morphology of the Mediterranean

After the 1960s, it was clear that the starting point of the deep basins underlain by oceanic crust was at a depth of about 3,000 m. A distinctive trait is that the best fit of continents in the Atlantic Ocean between the Americas and Africa/Europe was made along the 3,000 m isobaths

D. I. Papanikolaou, *The Geology of Greece*, Regional Geology Reviews,
https://doi.org/10.1007/978-3-030-60731-9_3

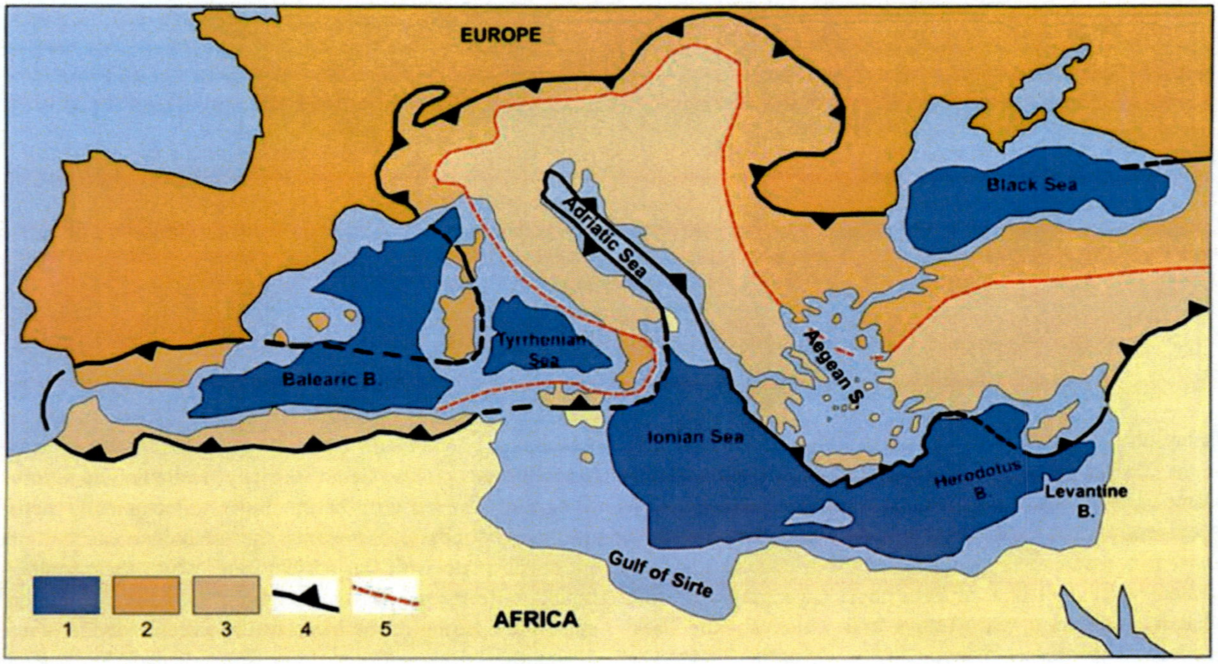

Fig. 3.1 The present large marine basins in the Mediterranean region, as defined by the 3 km bathymetric contour (from Papanikolaou 1986). The two western basins, the Balearic and the Tyrrhenian Sea, are developed inside the Alpine system, after the collision of Northwest Africa with Europe in the Neogene, unlike the two eastern basins, the Ionian and the Herodotus/Levantine, which lie to the south of the Alpine front and which opened in the Mesozoic and belong to the African plate. The Black Sea basin is of Upper Cretaceous age and it is a remnant of the strike-slip tectonism of the northern margin. 1. Mediterranean deep basins, 2: northern margin, 3: southern margin, 4: Alpine mountain chain fronts, 5: main ophiolitic suture zone of Tethys

(Bullard et al. 1965). Thus, in a bathymetric map we can delineate the possible oceanic basins at depths of more than 3,000 m and the continental slope zones up to 300–500 m of the continental shelf. Observation of the **bathymetric map** of the Mediterranean (Fig. 3.2) shows that large regions of the Mediterranean are shallow coastal areas, prolongation of the continents and epicontinental seas. The most typical cases are those of the Adriatic and the Gulf of Sirte, where the shallow sea covers the continental crust of Europe and Africa respectively. It is noteworthy that the Aegean Sea does not show such deep depths and it is expected to be underlain by continental crust. Finally, it should also be noted that the largest depths of the Mediterranean are about 5,000 m west of Pylos, in southwest Peloponnese and occur along the tectonic contact of the two plates in the Hellenic Trench. Depths around 4 km are observed all along the Hellenic trenches, with complicate structure of basins and ridges from the western Ionian islands to the southeastern Cretan and Dodekanese margins (Jongsma 1977; LeQuellec et al. 1980; Leite and Mascle 1982).

It is also noteworthy that the 3,000 m isobaths, and hence, the continental–oceanic crust boundary, is located adjacent to the coastal zone in some areas, typically on the shores of Cyrenaica and Egypt, while it is very far off the coast in others, such as the Gulf of Sirte and the Sinai region. Lastly, it is remarkable that there is a much shallower axial zone with depths of 1,500–2,000 m within the region of the Eastern Mediterranean deep basins, which divides both the Ionian and the Levantine into a north and a south trough. This creates a submarine morphological ridge along the axis of the Eastern Mediterranean, which was discovered by Heezen and Ewing (1963) and was given the name **Mediterranean Ridge** by Emery et al. (1966). Unfortunately, this purely morphological terminology has created misunderstandings, since the term “ridge” was used for mid-ocean ridges at about the same time within the plate tectonics theory. However, this is not the case with the morphological ridge of the Mediterranean since no oceanic crust is produced. Instead, we have compression and orogenic type deformation, with the formation of a submarine mountain chain, representing an accretionary prism (Finetti 1976; Finetti et al. 1991; Huguen et al. 2001, 2006).

Another unique morphological feature of the Mediterranean Sea occurs in the area of the Levantine Sea south of Cyprus, where a deep carbonate platform is observed at a depth of about 800–900 m, forming the **Eratosthenes** Seamount (Fig. 3.3.). This platform is 80 km long in the NE-SW direction and 50 km wide, forming an elongate

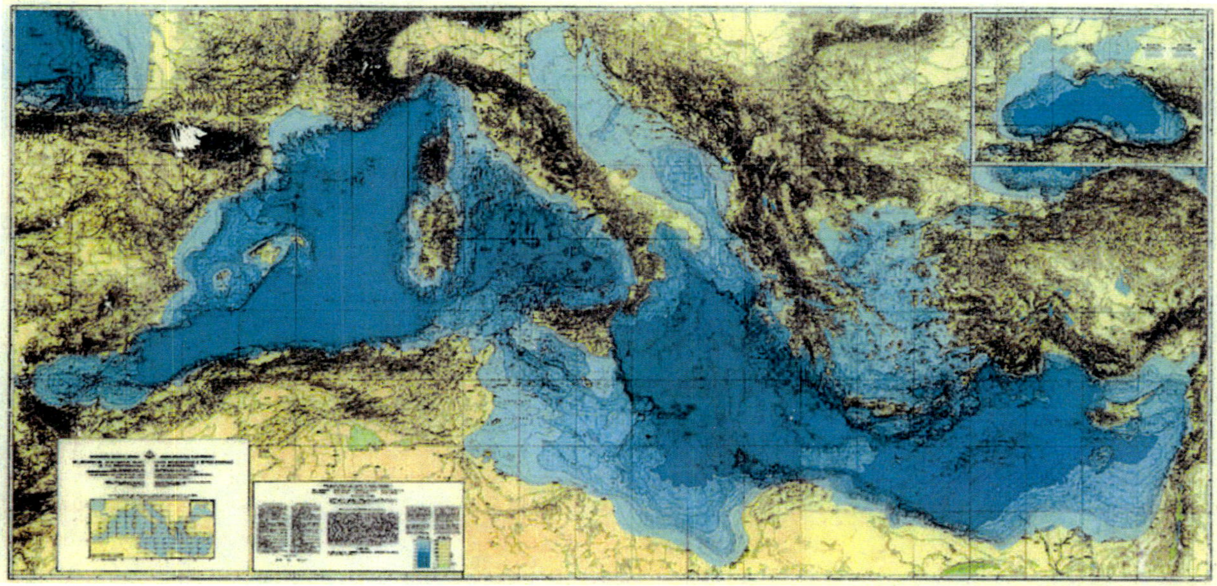

Fig. 3.2 Bathymetric map of the Mediterranean. The 3 km isobaths define approximately the continental–oceanic crust boundary in the deep basins. Large marine segments of the Mediterranean, such as the Adriatic, the Gulf of Sirte and the Aegean Sea, lie in shallow depths on continental crust (highly diminished bathymetric map of Unesco, 1997)

structure. Its surrounding depth of the Levantine basin is about 2500 m and thus, a major submarine relief of about 1500 m is suddenly forming a barrier, looking eastwards of the abyssal plain. Towards the south and southeast the huge Nile deltaic cone surrounds the Eratosthenes sea mount at about 2000–2100 m depth. Towards the north the Cyprus continental margin occurs at a distance of only 20–30 km, separated by an E-W trough of 2,500 m depth. The Eratosthenes Seamount represents an exceptional example of a separate continental fragment/terrane within the Eastern Mediterranean, which has not been accreted yet to the Alpine accretionary wedges, but it is about to beggin continental subduction beneath the Cyprus arc (Robertson 1998; Mascle et al. 2000).

3.2 The Crustal Structure of the Mediterranean

The **gravimetry** of the Mediterranean and the **Bouguer anomalies** point to a mass shortage in some areas and excess mass in others. This means that by selecting the appropriate gravity values in mgal, we can delineate areas with high gravity values featuring oceanic crust and areas with low gravity values featuring continental crust (Fig. 3.4). In between we have a transitional zone, from the characteristics of oceanic to continental crust, which is sometimes narrow and sometimes quite wide. At the same time, it is possible to locate the plate convergent zones by the differentiation of the gravimetric data in the active Mediterranean orogenic arcs (Rabinowitz and Ryan 1970).

The first conclusion is that a general contingency is noted between the gravity results and the bathymetric data previously analyzed. Thus, characteristic gravity values for oceanic crust are found in areas with a depth of more than 3,000 m, while characteristic values for continental crust can be found in the continents and in areas where prolongation of the continent into the marine area is evident from bathymetric data, e.g. in the Adriatic and Sirt. Marine seismic data have shown that the foreland of the Northern Hellenides in the Northern Ionian Sea and South Adriatic consists of continental lithosphere (Finetti and Del Ben 2005), whereas the foreland of the Southern Hellenides beneath the Ionian Sea consists of oceanic lithosphere affinity (Makris and Stobbe 1984; Finetti et al. 1991; De Woogd et al. 1992; Gesret et al. 2010).

In addition to this general contingency, there are also special cases—abnormalities where one would never expect such gravity characteristics shown on the map, like the central region of the Mediterranean Ridge and the South Levantine region. On the other hand, morphological areas with a clear narrow zone of the boundary of continental–oceanic crust in the Cyrenaica–Egypt coasts are also clear on the gravity data. At this point it should be noted that especially in the Eastern Mediterranean basins old oceanic crust can be found of Middle/LateTriassic age, about 220–230 Ma, probably the oldest in situ oceanic crust on the planet (Speranza et al. 2012). Therefore, there is a great

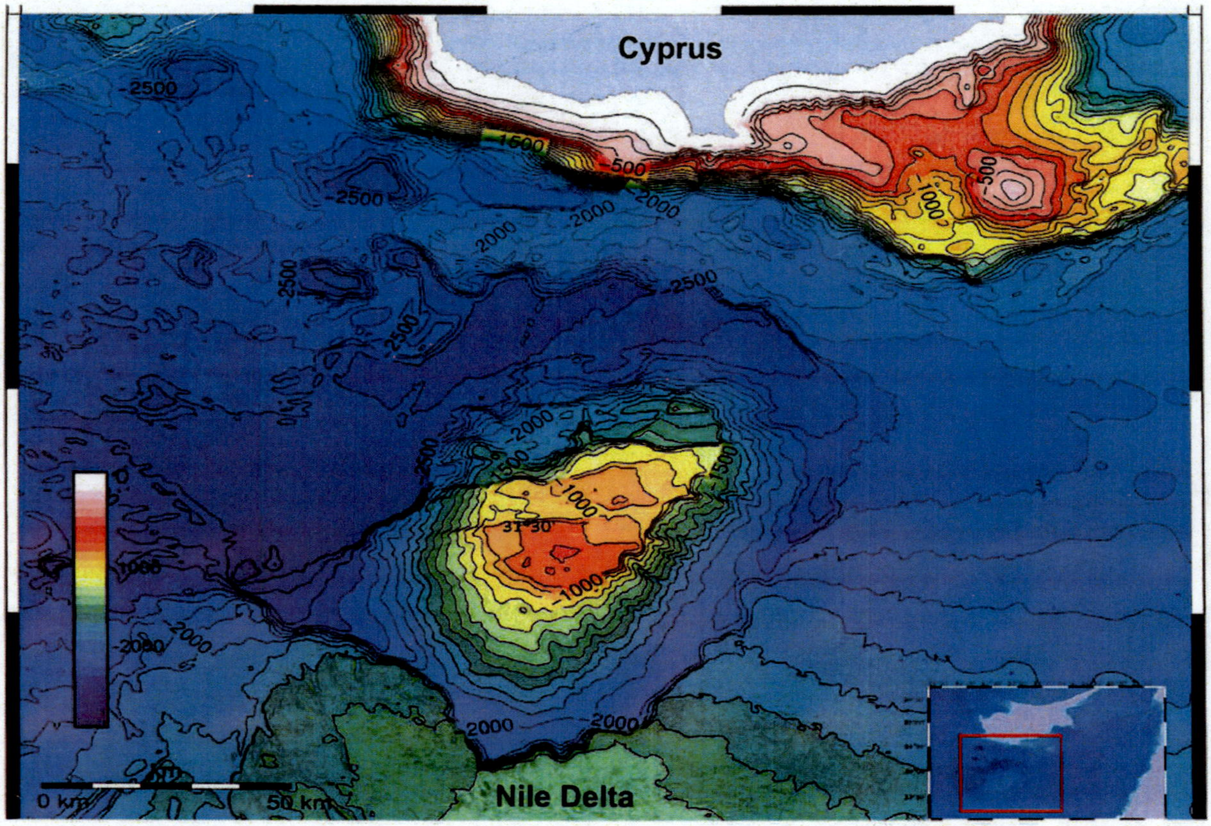

Fig. 3.3 Bathymetric map of the area south of Cyprus, where the submerged platform of the Eratosthenes Seamount is observed

sedimentary thickness below the ocean floor (Finetti 1976; Makris and Stobbe 1984; Finetti et al. 1991) which fades or even overprints the gravitational oceanic characteristics, due to the accumulation of several kilometers thick, lightweight material. It should be also noted that the existence of mafic magmatic rocks does not always correspond to oceans, but either to marginal basins of the Japanese Sea type (like the case of the Upper Jurassic Peonia mafic complex in the Hellenides), or to the mafic rocks created in rift grabens during the extensional phase of a continent. In any case, geophysical research in the Levantine basin has shown that it is underlain by thinned continental crust, which resulted from rifting without reaching the stage of sea floor spreading and oceanization, contrary to the domain north of the Eratosthenes Seamount (Netzeband et al. 2006). Additionally, recent geophysical and paleomagnetic studies have proposed the existence of Late Paleozoic oceanic crust in the Herodotus Basin, separated by a NNE-SSW trending tectonic zone from the eastwards extending Levantine Basin, which is largely underlain by thin continental crust (Granot 2016).

3.3 The Recent Sedimentation in the Mediterranean

The thickness of sediments in the Mediterranean shows great variations above the oceanic crust, which, based on seismic surveys for petroleum exploration, has been identified up to 8–10 km thick (Finetti 1976; Finetti et al. 1991). Indicatively, it is reported that the well-known Messinian salinity crisis in the Mediterranean resulted in the deposition of very thick evaporites in various basins, with thickness reaching up to 3 km (Hsü et al. 1978; Roveri et al. 2014). The existence of this seismically impermeable to the seismic waves salt layer under the Mediterranean seabed allowed the construction of a detailed map of cumulative sedimentation of the basin during the last 5.4 Ma, i.e. from the beginning of Pliocene (Fig. 3.5).

This map shows enormous differences in sediment thickness and hence in sedimentation rate throughout the Plio-Quaternary in various regions of the Mediterranean. Therefore, the so-called sediment-poor zones show a small to

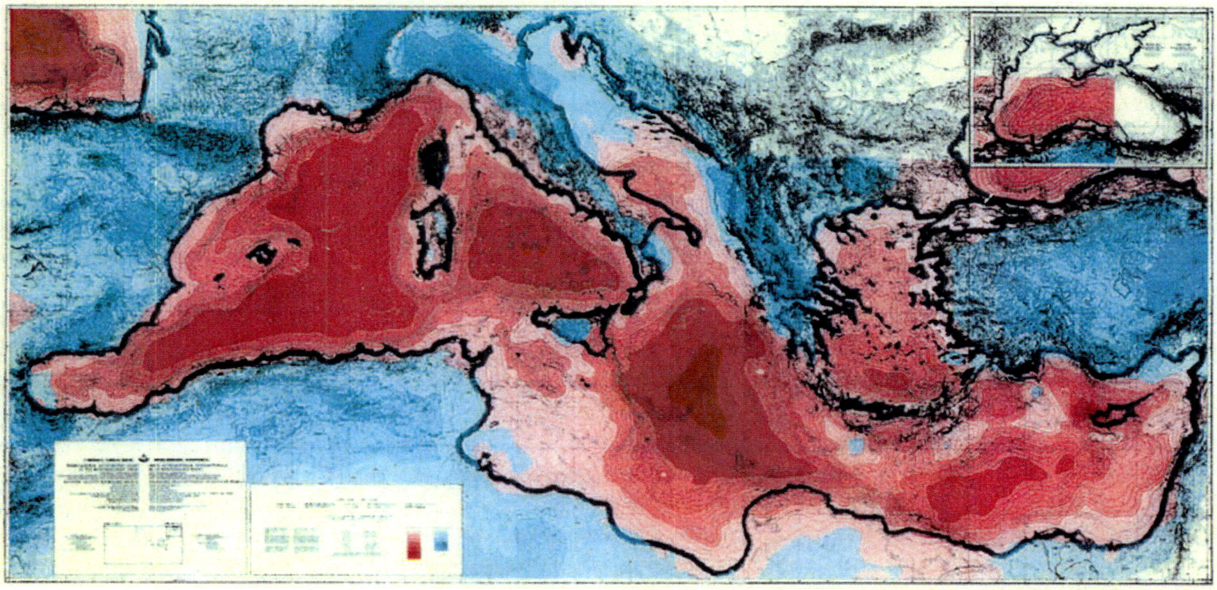

Fig. 3.4 Gravity map of the Mediterranean showing Bouguer anomalies. High values correspond to "excess mass" representing high density oceanic crust, while low values correspond to "mass shortage", i.e. to low density continental crust. Areas of high gravity values are comparable to the major deep basinal areas of Fig. 3.2 (highly diminished gravimetric map of Unesco, 1997)

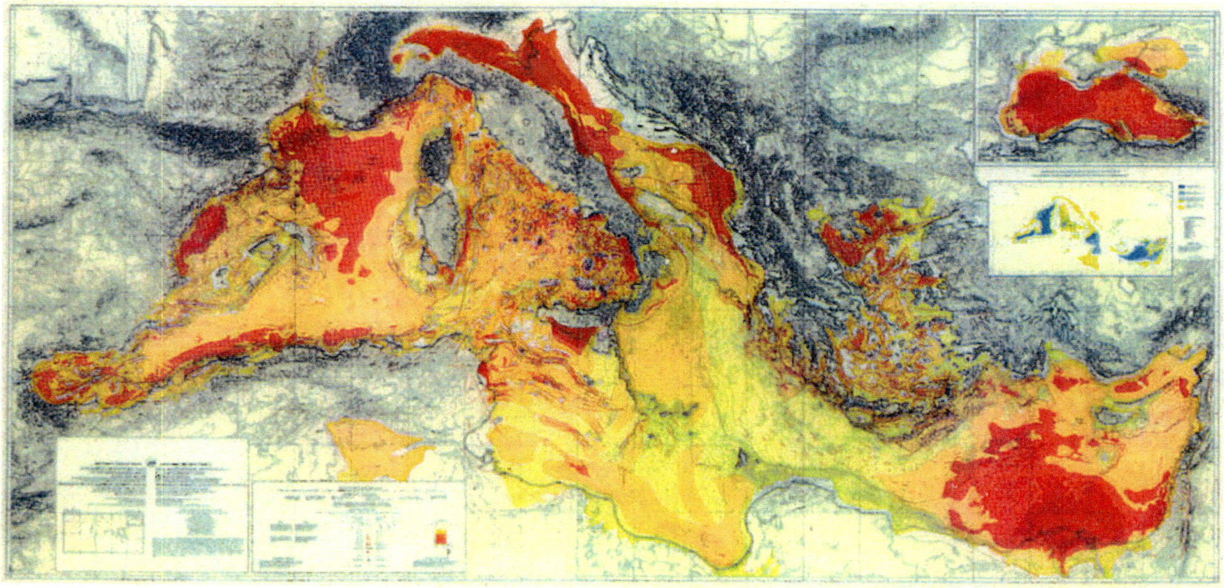

Fig. 3.5 Distribution map of the thickness of the Plio-Quaternary sediments in the Mediterranean (highly diminished sedimentary thickness map of Unesco, 1997). Great thickness is observed in areas of accumulation of clastic deltaic sediments, such as south of the Rhone, north of the Nile, in the Adriatic and the North Aegean. The deltaic prism of the Nile, 5-kilometer thick, has shaded the continental–oceanic crust boundary both from bathymetric and gravimetric points of view

minimal sediment thickness not exceeding 200 m. In contrast to these areas there are regions with extremely high sedimentation rate and cumulative thickness of a few kilometers. These large differences are due to the diversification of the supply routes of terrigenous clastic material of the sedimentary basins and are mainly found at the exits of the large deltaic prisms, which are formed at the estuaries of the large rivers of both Europe and Africa. This is characteristic in the cases of the Rhone deltaic prism in the Western Mediterranean, the Po in the Adriatic, and the Nile in Egypt/Sinai. Especially in the case of the Nile, the deltaic prism has a thickness of approximately 5 km and has, thus, differentiated both the bathymetry and the gravitational imprint. It is characteristic that the sediments of the Nile prism reach to the north up to the western coasts of Cyprus, but are absent from the top of the Eratosthenes Seamount. That means that if we wish to improve bathymetric and gravity data in this region, we would have to remove the thickness and gravitational effect on the Nile deltaic prism (bathymetric and gravity correction), which would lead to the clear picture of the ocean/continent boundary in the Southeast Mediterranean. In the Aegean, it is characteristic that large sedimentation rates are observed in the North Aegean basin/graben, with a sharp decrease to the south. This is due to the fact that all the major rivers of the South Balkan peninsula provide with their clastic material the North Aegean basin, where it is trapped and cannot expand further south, due to the great submarine cliff, formed along the southern marginal fault of the basin, which uplifts the Limnos–Sporades platform by more than 1 km (Papanikolaou et al. 2002; Papanikolaou and Papanikolaou 2007) (see also Fig. 11.18).

3.4 The Seismicity of the Mediterranean

The distribution of **seismicity** in the Mediterranean is known to vary substantially between the Western and the Eastern Mediterranean (Fig. 3.6). At the same time, lack of seismicity is observed in almost the entire African margin of the Eastern Mediterranean, while moderate seismicity is observed on the African coast of the Western Mediterranean (North Tunisia, Algeria, Rif). The bulk of the seismicity is located along the three orogenic arcs of the Eastern Mediterranean, but there is a huge difference between the Hellenic arc in the center and the adjacent arcs of Sicily–Calabria in the west and Cyprus in the east. It is also characteristic that the seismicity of the Hellenic arc is reduced north of Preveza, where it follows the contact between the European plate and the Adriatic microplate in the North Ionian and the Adriatic. That is to say, the seismicity follows the nature of the subducted oceanic crust of the Eastern Mediterranean in the Ionian Sea, which ends northwards in the area between the islands of Paxos–Corfu and Apulia in Southern Italy. It is also characteristic that the intense seismicity along the Hellenic arc penetrates eastwards into the Aegean microplate, with a clear configuration of the tectonic

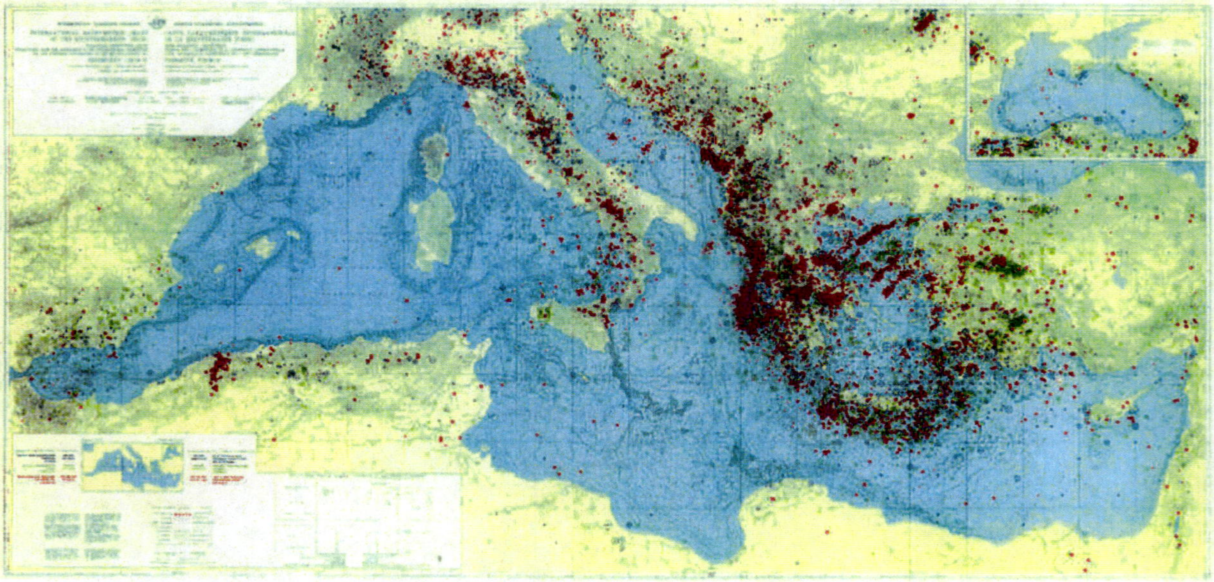

Fig. 3.6 Seismicity Map of the Mediterranean (highly diminished seismicity map of Unesco, 1997). There is a tremendous accumulation of earthquake epicenters in the Hellenic arc, with a sharp lateral attenuation both to the north (Dinarides–Southern Alps) and to the east (Cyprus arc). The seismic zone in the Apennines is of intermediate seismicity, whereas the Alps, the Pyrenees, the Betics, Rif, North Atlas, the Carpathians, and the Balkanides are regions of low seismicity. The relatively low seismicity along the North Anatolian fault compared to the North Aegean is noteworthy, where at least two rectilinear seismic zones can be distinguished (North Aegean and Skyros). The African margin of the Mediterranean from Central Tunisia, Libya, Egypt, Palestine, Libanon and Syria appears to be relatively aseismic except for the Dead Sea rift

strike slip zones of the North Aegean and Skyros basins, which continue eastwards to the North Anatolian fault.

The different geodynamic processes occurring along the northern branch from those along the southern branch of the Alpine system are illustrated by the insignificant seismicity of the northern orogenic systems, e.g. the Alps and the Carpathians, in relation to the southern ones, e.g. the Hellenides. The above picture of the seismicity in the Mediterranean is considerably different when the deep-focus earthquakes related to the active subduction zones are distinguished and analyzed. Focal mechanisms show regions affected by extension through normal faulting mainly in the back arc regions, or by compression through reverse faulting along the convergent boundaries or by vertical shear zones through strike-slip faulting mainly along the northern boundaries of the Anatolian and Aegean plates (see corresponding Chap. 10, about the Hellenic arc).

3.5 The Tectonic Setting of the Eastern Mediterranean

The tectonic structure of the two Tethyan basins in the Eastern Mediterranean shows the process of oceanic closure and help us understand the lateral transition from the collisional fronts of the Alps to the west and Caucasus to the east, towards the intermediate convergent zone along the three orogenic arcs and especially the Hellenic arc. Data from litho-seismic studies in the Mediterranean permitted the analysis of the geotectonic structure in the area between the Hellenic Arc and the African margin (Fig. 3.7), as well as the comparison of the geotectonic structure between two cross-sections in the Ionian Sea and the Libyan Sea (Finetti 1976; Finetti et al. 1991) (Fig. 3.8). The cross-section Kythera–Gulf of Sirte, of ENE-WSW direction, shows a remnant of the Eastern Mediterranean Ocean, beneath the

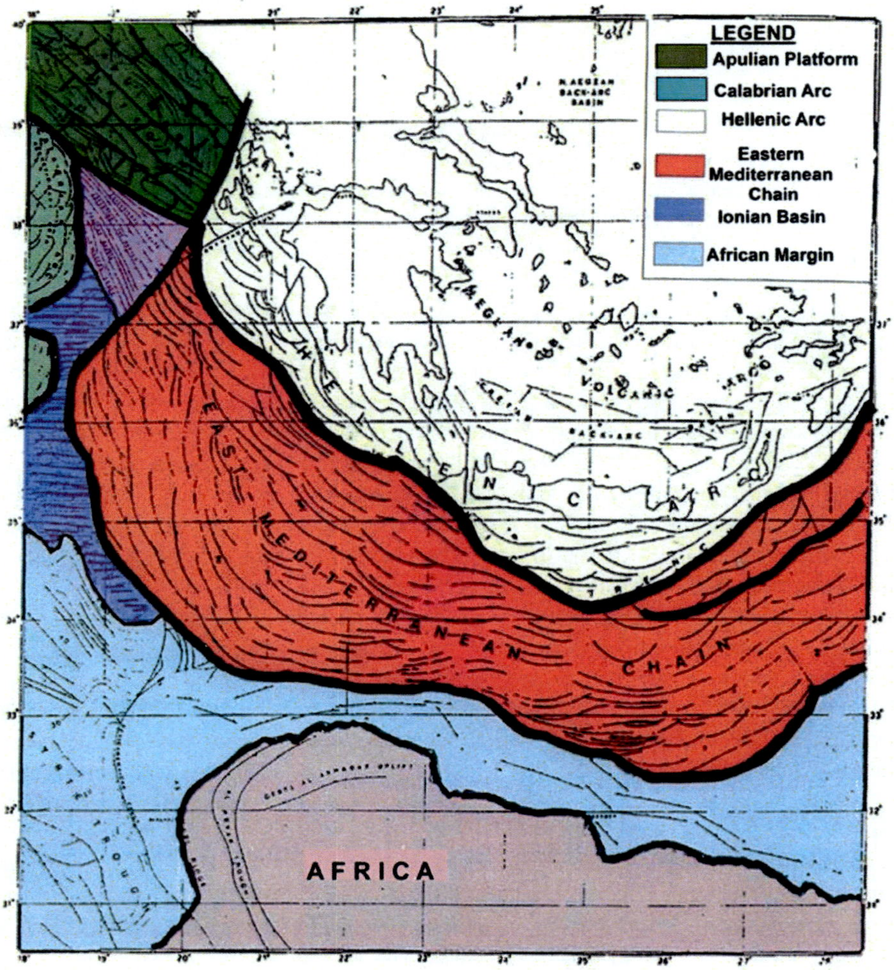

Fig. 3.7 Tectonic map of the Eastern Mediterranean based on litho-seismic cross sections (from Finetti et al. 1991, modified). In the area of the East Mediterranean Ridge intense thrusting to the south can be observed forming a submarine mountain range—the East Mediterranean Chain—within the accretionary prism in the front of the Hellenic arc

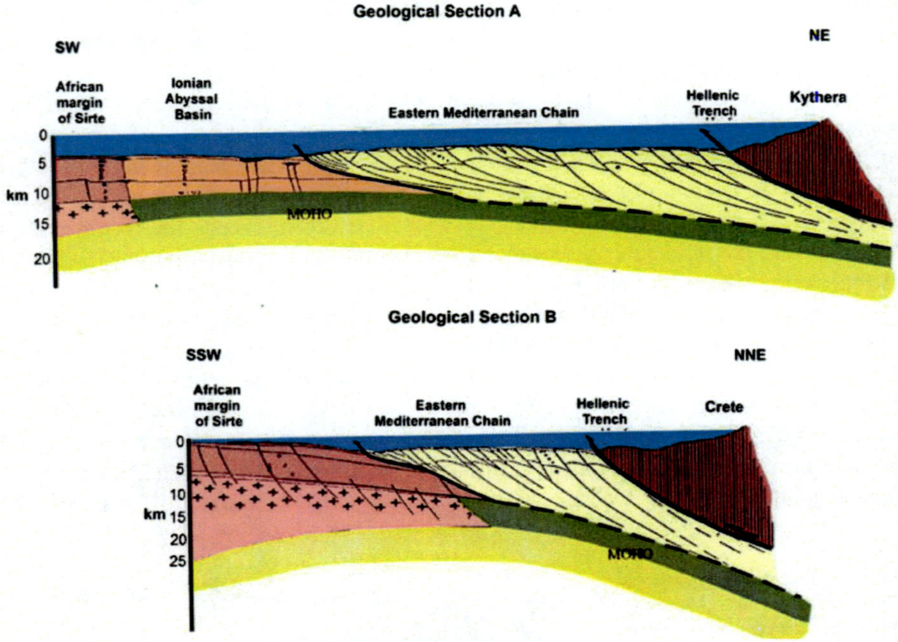

Fig. 3.8 Two simplified tectonic cross sections from the front of the Hellenic arc to the African margin (from Finetti et al. 1991, modified). The main difference is the presence of the remnant of the Ionian oceanic basin in the cross section of Kythera - Gulf of Sirte, as opposed to its disappearance—under the accretionary prism in the section of Gavdos/Crete–Cyrenaica

horizontal seabed in the abyssal plane with a depth of 3,000–3,500 m, between the Mediterranean Ridge and the African platform. The oceanic crust and the several kilometers thick sedimentary cover is underthrusted below the deformed sediments of the accretionary prism of the Eastern Mediterranean Ridge. This cross section clearly shows that we are at the end of stage iii of total compression and destruction of the oceanic crust of Tethys. It is characteristic that the abyssal plain of the Ionian oceanic basin has a mere 150 km width between the tectonic front of the two plates and the African margin in the Gulf of Sirte in Libya. At the same time, the structure of the Eastern Mediterranean Ridge corresponds to an imbricated and folded sedimentary sequence, a kind of a «fold and thrust belt», which creates a submarine mountain range in the making. The inner structure of the accretionary prism shows an asymmetry to the south and the reverse tectonic structures are clearly visible from the considerable displacements of the Messinian evaporate layer, which can be detected for the most part of the cross section. Therefore, the correct term for the Ridge of the Eastern Mediterranean is the **East Mediterranean Chain** (Finetti 1976; Finetti et al. 1991).

The second cross section of Gavdos–Cyrenaica, with an approximate N-S direction, does not show a horizontal remnant of oceanic crust between Crete and Libya. On the contrary, the structure of the submarine mountain chain is more prominent, the depth of the higher dome/anticline structure is much shallower, close to 1,500 m, and all the compressive structures of the thrusts and folds are much more intense. The front of the allochthonous sediments of the accretionary prism reaches next to the Herodotus trough, very close to the African platform, and the detection of the oceanic crust of the Ionian basin is not possible. It is highly possible that we have a tectonic detachment of the sediments of the accretionary prism from the underlying subducted plate and the oceanic crust has already penetrated under the front of the Hellenic arc along the trench. Therefore, we are no longer in the stage iii of total compression, as there is no longer any remnant of the oceanic basement. At the same time, however, the stage iv of the continental collision has not yet begun. The two continental crusts have not yet had any contact, as they are separated by a narrow zone of tectonic wedge sediments of the accretionary prism. In conclusion, the Gavdos–Cyrenaica cross section illustrates the transitional stage between total compression and continental collision.

Finally, it should be noted that immediately south of the Hellenic trenches there have been some data indicating the presence of a **backstop** within the accretionary prism (Kastens 1991; Huguen et al. 2001, 2006; Le Pichon et al. 2002; Reston et al. 2002; Mascle and Chaumillon 1998) (Fig. 3.9.). This structure is characterized by an opposite thrusting of the accretionary prism to the north along its northern zone, contrary to the general transport of the detached sediments to the south. The latter provides the possibility to distinguish an inner zone and an outer zone. The accretionary wedge comprises Pre-Messinian Miocene sediments and Messinian evaporites overlain by a very thin Plio-Quaternary cover (Kastens 1991). The pre-Messinian basement of the backstop was proposed to consist of the outermost outliers of the

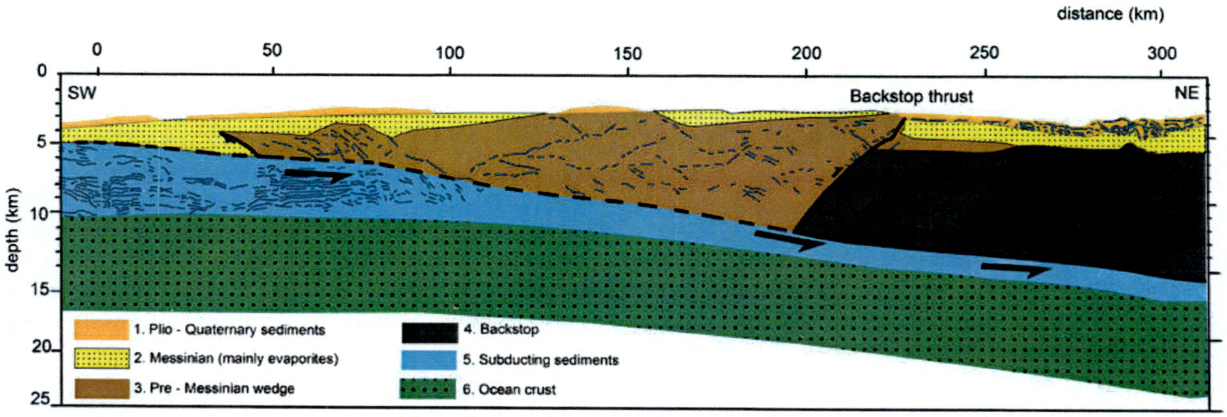

Fig. 3.9 Synthetic geological section from the IMERSE cruise multichannel profile of the East Mediterranean Ridge and backstop at the SW of the Hellenic trench in the SW Peloponnese (modified from Reston et al. 2002; Le Pichon et al. 2002)

Miocene Hellenic nappes, extending the onshore structure of southern Peloponnese and western Crete (Le Pichon et al. 2002). However, it might well represent the external wedge of the submarine lateral extension of the outer platform unit of the External Carbonate platform of the Hellenides, croping out in the Ionian islands.

The above results have shown that the front of the Hellenic orogenic wedge above the subducting African plate does not coincide with the Hellenic trenches, but is traced further southwards at the front of the accretionary wedge along the Herodotus trench, immediately north of the Cyrenaica peninsula (Fig. 3.10). The discovery of Mesozoic limestones in the Hellenic trenches (Mascle et al. 1986; Huguen et al. 2001) shows that the trenches have been developed within the outer prolongation of the Hellenic nappes and/or of the external carbonate platform of the Hellenides. Thus, they do not separate the Hellenic orogenic system from the subducting African lithosphere with its Miocene sedimentary accretionary prism. Their nature favours a major tectonic division within the frontal zone of

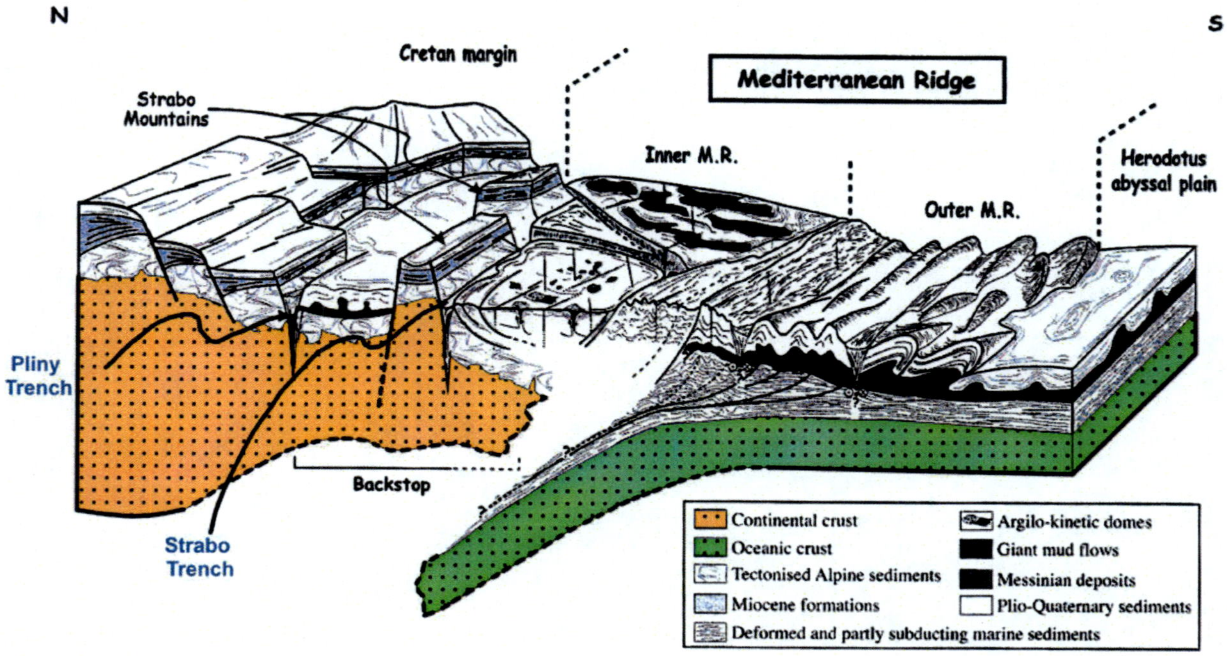

Fig. 3.10 Interpretative 3D structural sketch of the Mediterranean Ridge and Backstop from the Eastern Mediterranean, based on swath data and seismic profiles from the Prismed 2 and Prismed 1 surveys (modified from Huguen et al. 2001)

the orogenic wedge with important strike slip motion and major crustal differences on both sides of the sliding blocks (Le Pichon et al. 2019).

3.6 The Seismic Tomography of the Mediterranean

The detection of **large scale heterogeneities in the mantle** in the late 70s and 80s (Dziewonski et al. 1977; Dziewonski 1984) was followed by the development and application of the new **seismic tomography methodology** during the 1990s, offering the possibility to detect the large tectonic structures in the Earth's interior and especially the subduction zones of the lithosphere. The basic parameter of this method is the analysis of the velocity of seismic waves, which depends on the lithology. Thus, the oceanic crust has higher seismic velocity values due to its density and crystalline structure than the continental crust. By analysing the seismic waves of large global earthquakes it is possible to map the Earth's interior into zones with similar velocities and to distinguish the extremes outside the limits of the values by one or two standard deviations ($\pm 2\sigma$). The result is the imprinting of the different zones of seismic wave velocity along vertical cross sections of the Earth's interior with a deviation of ± 1–2% of the mean values. From the seismic tomographs the detection of the subducted oceanic crust is possible by using values greater than two standard deviations 2σ. Several seismic tomographs in the Mediterranean clearly show the subduction zone of the East Mediterranean Ocean in many cross sections (Spakman et al. 1988, 1993) (Fig. 3.11).

It is remarkable that subducted oceanic crust can be detected up to 1,400 km depth, where deep mantle

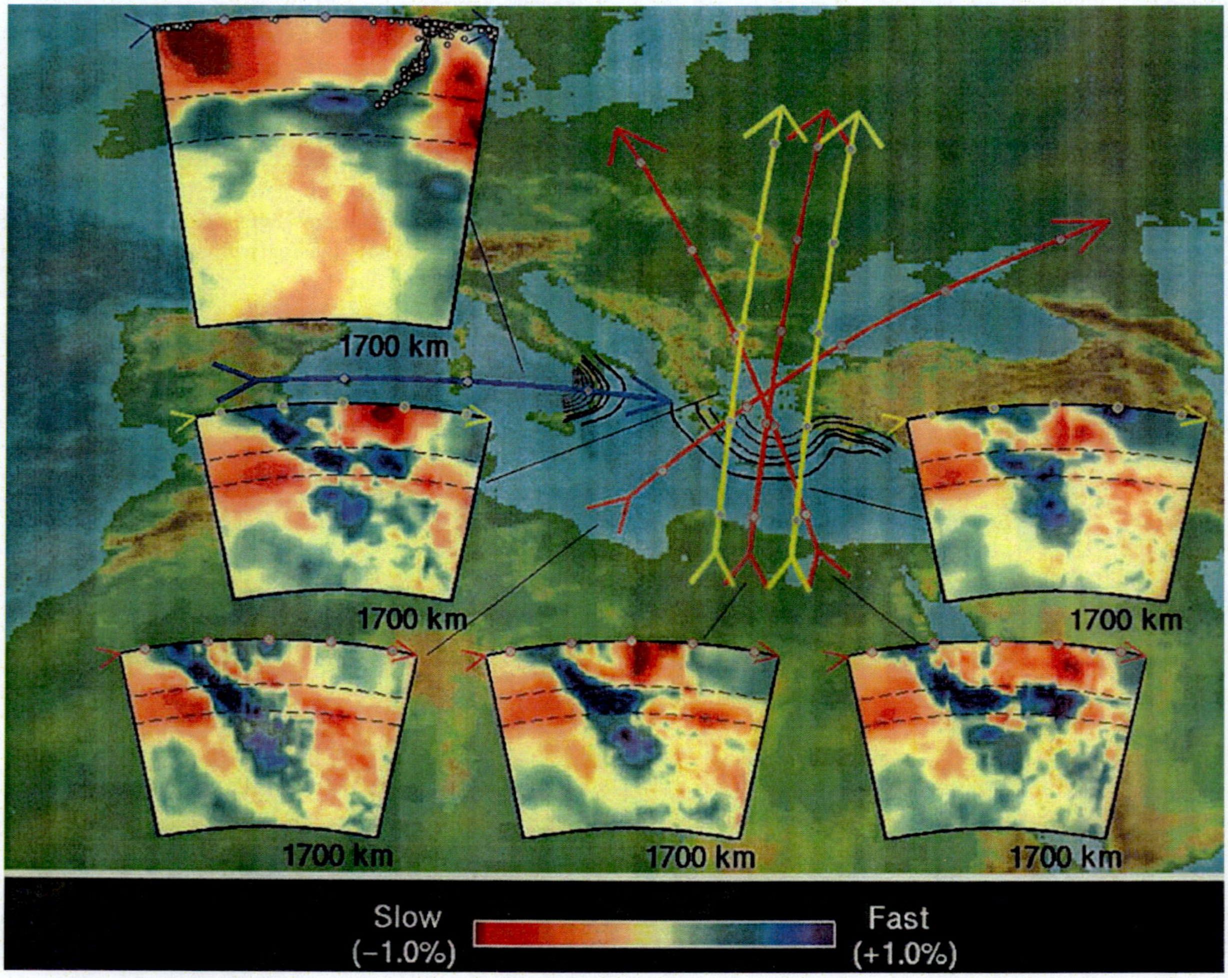

Fig. 3.11 Seismic tomographs of the Mediterranean (from Spakman et al. 1993). The detection of the subduction zones of Tethys, coloured in blue, under Europe up to a depth of 1,400 km is based on the high values of seismic wave velocities (above 1% of the mean)

circulation occurs and even deeper down to the core/mantle boundary (Van der Hilst et al. 1997). The image is indisputable up to a depth of 700 km, while there are cases where one can observe the lithosphere even beyond 1,200 km, with simultaneous bending into the mesosphere (Karason and Van der Hilst 2001). The impressive result of seismic tomography is that by using the depth of the subducted ocean, it is possible to calculate both the dimensions of the lost ocean and the minimum period of the subduction process, as well as to estimate which parts of the subducted slab along the tomography correspond to which subduction period (see also "History of the Hellenic subduction zone", Chapter 10.2). This is possible with the use of actualistic values of subduction rates, such as the current Hellenic subduction zone, which has a rate of 4–5 cm/year. This means that the crust that was subducted 10 Ma ago should be found today at a distance of about 400–500 km along the subducted slab away from the trench. Of course subduction rates depend on the type of the subducted crust with extreme values of 9–10 cm/year as a maximum, which can be currently observed in the oceanic subductions of the Western Pacific, and 0.5–1 cm/year as a minimum, which can be observed today in the continental subduction in the Adriatic (Royden and Papanikolaou 2011).

In conclusion, new data obtained from seismic tomography confirm the conclusions extracted by classical geological research from the Mesozoic up to date about the time and process of the subduction of oceanic crust and the convergence of the plates.

3.7 GPS Geodetic Measurements and Kinematics of the Eastern Mediterranean

Geodesy since the 1990s provided a new tool for studying plate kinematics and strain accumulation, offering important quantitative insights (velocities and rates) to geotectonics. The use of Global Positioning System (GPS) in geotectonics was based on the selection of fixed measurement points/stations and their re-occupation after several years in order to: (i) identify whether elastic deformation occurs at present day, (ii) reveal the kinematics, (iii) extract slip rates. If we have an increase in the distance, this is interpreted as a product of extension in the crust, and, on the other hand, if we have a decrease in the distance, this is interpreted as convergence/compression. Of course, as long as the geographical coordinates change, horizontal motion components can also be determined. Thus, in converging plates, the convergence rate can be determined in mm/year or cm/year, in divergent zones the corresponding divergence rate, and in strike-slip faults the horizontal movement, further described as clockwise/dextral or counter-clockwise/sinistral.

In the Eastern Mediterranean, the first geodetic determination of tectonic deformation in central Greece, based on re-measurements of triangulation points by Billiris et al. (1991) was followed by GPS measurements for the area of Greece by Le Pichon et al. (1995), Kahle et al (1995) and Davies et al (1997). More extended measurements for the whole Eastern Mediterranean were obtained from a combination of measurement networks in the various countries of the region, coordinated by MIT researchers in Boston (Fig. 3.12) (Reilinger et al. 1997, 2006, 2010; McKlusky et al. 2000), followed by many others (e.g. Kahle et al. 2000; Nyst and Thatcher 2004; Muller et al. 2013).

The kinematics resulting from geodetic measurements in the Eastern Mediterranean showed that the convergence rate between Europe and Africa is relatively small, at about 10 mm/year, regardless of which continent remains fixed in the analysis system. On the other hand, the convergence/collision rate between Europe and Arabia is almost double, at about 20 mm/year on the Arabia–Caucasus axis, which is further analyzed as a 10 mm/year movement of Arabia to the north and 10 mm/year of Eurasia to the south. The Caucasus Mountains and the active volcanic complexes of Mount Ararat and Lake Van are collisional mountain ranges/structures, which have been created since Miocene in the collisional zone of the Eurasia–Arabia plates. The difference of about 10 mm/year to the north of the relative movement of Arabia to Africa has been kinematically absorbed by the Dead Sea transform zone. This transform begins in the south from the Red Sea in the Gulf of Aqaba and runs along the coastal zone of the Eastern Mediterranean–Levante with a N-S direction along the Dead Sea, the Sea of Galilee, the Jordan River and further north along the grabens dividing the Lebanon mountains to the west from the Arab dessert, up to the Hatay Mountains in North Syria to the east. The overall strike-slip motion during the last 13 Ma has been estimated, based on geological parameters of about 40–60 km (Freund et al. 1970; Angelier and Le Pichon 1978). Of course, the primary result of Africa's divergence and lateral slip of Arabia from Africa is the rifting and opening of the Red Sea, the youngest ocean on the planet today. The total rifting is about 200 km wide and the movement of Arabia shows a slight left-lateral rotation. As a result the Red Sea is wider southwards towards Aden than northwards towards the Suez Canal and the Sinai Peninsula.

The kinematics of the Aegean and Anatolia microplates which are distinguished between the two large plates in the Eastern Mediterranean (McKenzie 1970, 1972) is completely different, both in relation to the kinematics of the large plates and between themselves. Thus, the Anatolia microplate moves with about 20 mm/year to the west, while the Aegean microplate moves with a more than double rate, of approximately 40–50 mm/year to the south-southwest.

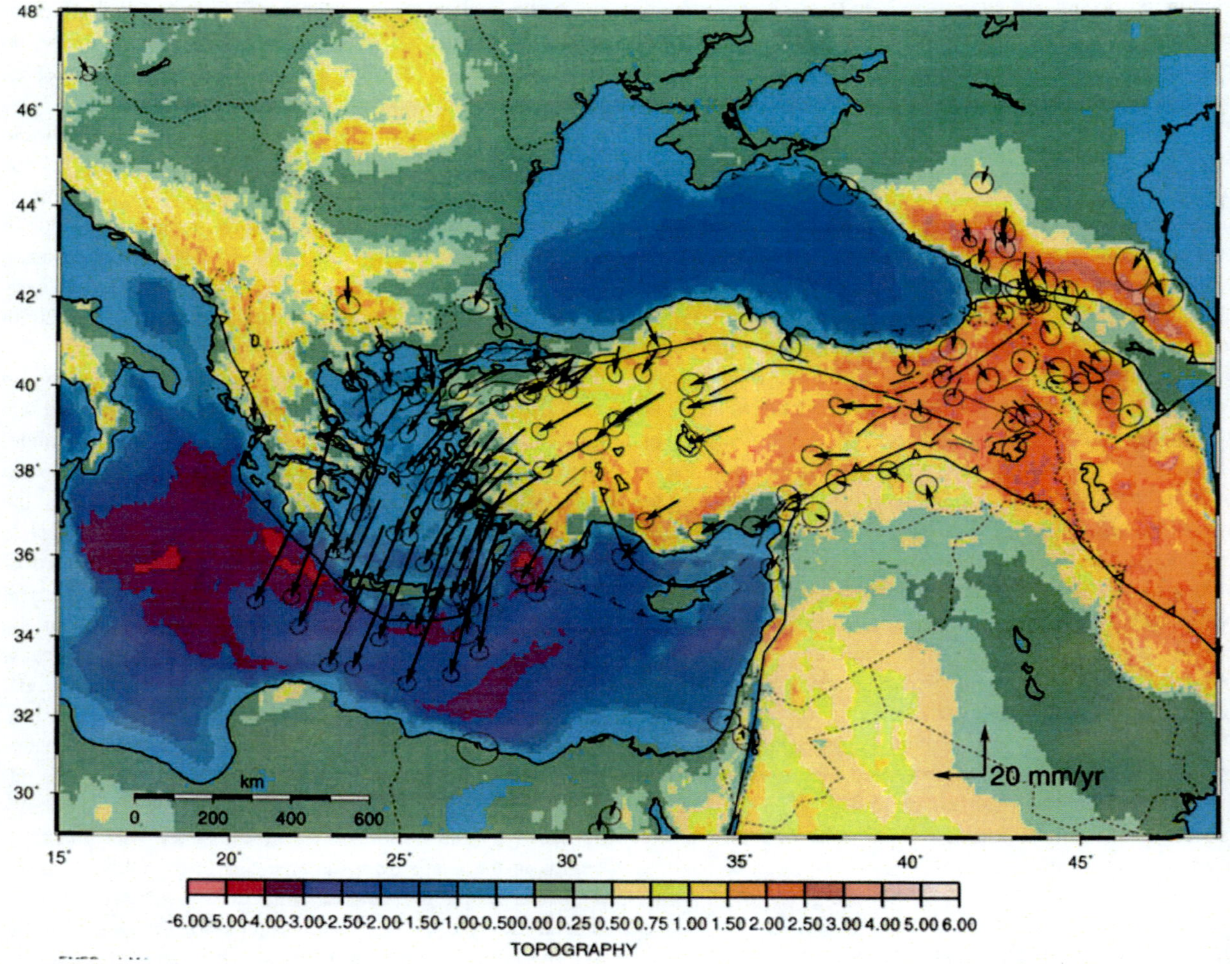

Fig. 3.12 GPS-based annual displacement velocity values in the Eastern Mediterranean (based on Reilinger et al. 1997). The North Aegean graben has an unique setting, as it lies on a microplate boundary, separating the low convergence rate of Europe to the south (10 mm/year) from the high rate of the Aegean microplate (40–50 mm/year) to the south-southwest, considering the African plate fixed

The boundaries of the two microplates are well defined in some zones and less clear in others. The northern boundary of Anatolia has been shaped since Miocene by the North Anatolian Fault, which is a large right-lateral strike slip fault (Sengor et al. 2005). This fault, which is a microplate boundary, divides Eurasia in the north, moving southwards at 10 mm/year, from Anatolia in the south, moving westwards at 20 mm/year. In Northern Asia Minor, this fault passes through the Pontides geotectonic region and its cummulative offset is 80–90 km, judging from the offset of characteristic geological formations (Armijo et al. 1999; Sengor et al. 2005). Further west, in the Sea of Marmara and the North Aegean region, the fault is divided and its main branch defines the North Aegean basin, comprising the largest active fault in the Aegean (Papanikolaou and Papanikolaou 2007). Its southern branch shapes the Skyros basin a little further south (Papanikolaou et al. 2019). Further to the west, the boundary of the Aegean microplate with Europe is represented by the **Central Hellenic Shear Zone** (CHSZ) (Papanikolaou and Royden 2007), which terminates at Central Western Greece between Preveza and the Gulf of Patras (Fig. 3.13).

The southern boundary of the two microplates is formed by the two consecutive orogenic arcs (Hellenic and Cyprus) along the tectonic boundary between the European active margin and the African plate. The western boundary of the Aegean microplate is part of the Hellenic trench in the Ionian Sea, while the eastern boundary of the Anatolian microplate is the East Anatolian Fault in Eastern Asia Minor. The boundary between the Aegean and Anatolian microplates is

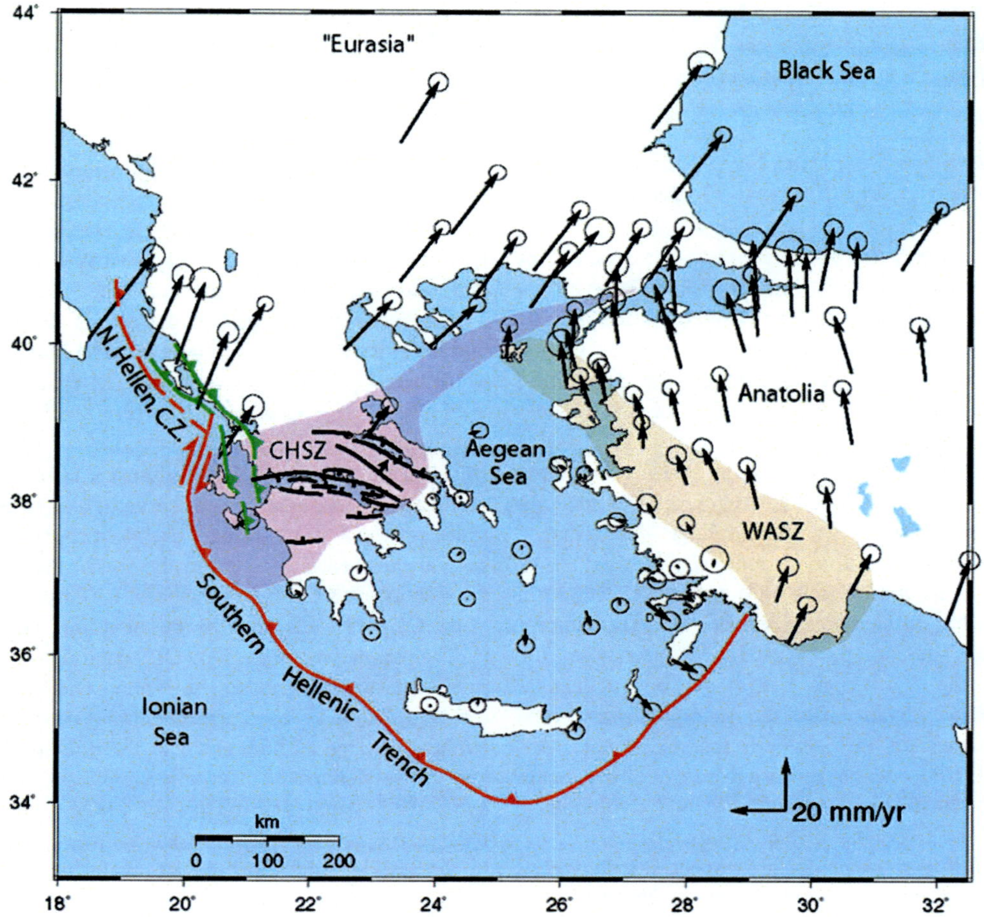

Fig. 3.13 The intermediate right-lateral Central Hellenic Shear Zone (CHSZ), transforming the relative movement between the European plate and the Aegean microplate, and the left-lateral West Anatolia Shear Zone (WASZ), transforming the relative movement between the Aegean and the Anatolia microplates (from Papanikolaou and Royden 2007). The model is based on data of McKlusky et al. 2000 and considers a fixed Aegean microplate

the least prominent and, apart from its seismic signature (McKenzie 1970, 1972), has been determined more recently mainly on the basis of geodetic data, as the **West Anatolian Shear Zone** (Papanikolaou and Royden 2007) (Fig. 3.13). This boundary, like its equivalent of Central Greece, is clearly visible only when the Aegean microplate is considered fixed, which results to their relative movements. Thus, in West Anatolia the movement is transtensional in a counter-clockwise component, while in Central Greece it is also transtensional, but this time with a clockwise component. This observation changes also the view, prevailed for several years, that the increased seismicity of the Aegean is cumulatively caused by the Hellenic arc deformation and the push of Anatolia to the west. Anatolia, on the contrary, is retarding kinematically and not only does not push the Aegean, but it pulls it back, exerting tensile stresses. The neotectonic grabens of the Western Asia Minor in Ionia and Lydia areas are caused by these same tensile stresses, as are the grabens of the Buyuk and Kuyuk Menderes Rivers in Smyrna–Ephesus (Seyítoglu and Scott 1996; Yilmaz et al. 2000). Finally, it should be noted that while Anatolia moves laterally along the right-lateral strike slip fault of Northern Anatolia, the Aegean microplate has a considerable tensile component, in the NNE-SSW direction which has created the North Aegean basins. Thus, the marginal faults of the North Aegean are oblique-normal faults, with a vertical component of several kilometers (Papanikolaou et al. 2002, 2006, 2019; Papanikolaou and Papanikolaou 2007). In conclusion, the high seismicity of the Hellenic arc should be attributed to the effect of the roll back of the retreating Hellenic subduction zone, which causes the extension in the inner part of the arc and the subsidence of the Aegean region

(Royden 1993; Jolivet and Faccenna 2000; Flerit et al. 2004; Jolivet and Brun 2010; Royden and Papanikolaou 2011; Brun et al. 2016).

3.8 The Geodynamic and Neotectonic Evolution of the Eastern Mediterranean

The relation between the Mediterranean Sea and Tethys was established during the late stages of the Tethyan evolution from the Middle Miocene onwards, when the Arabia–Eurasia collision led to the closure of the Tethys Ocean. Then, the Mediterranean was created, as the remnant of Tethys in the Eastern Mediterranean west of the collision zone. Approximately 13 million years ago, the eastern continuation of Tethys to the Persian Gulf was closed (Freund et al. 1970; Le Pichon and Angelier 1979) marking the onset of the Mediterranean. Therefore, the Mediterranean as an entity/concept is relatively young in geological terms. Since then it is an autonomous unit, although the Eastern Mediterranean has different characteristics from the Western Mediterranean. Thus, in the Western Mediterranean, the Balearic and the Tyrrhenian basins are post-collisional structures –post-Tethyan- after the incorporation of the Adria microplate into the southern margin of the Alps (Channel et al. 1979), while the eastern basins of the Ionian and the Herodotus/Levantine continue their evolutionary process since their opening in the Triassic–Jurassic as the southernmost Tethyan basins. In this respect, it is important to refer to the Permo-Triassic stages of the Pangaea configuration, when the passive African margin had been developed with rifting and probably sinistral megashear zones acting between Africa and Europe, before the opening of the South Atlantic Ocean (e.g. Garfunkel 2004).

As early as the 1970s, an attempt was made to understand the evolution of the southern European margin with the distinction of microplates, which originally was mainly based on seismic data (McKenzie 1970, 1972, 1978). At the same time, some models of lateral escape of the microplates maintained by strike slip faults were proposed, especially at the collisional boundaries of the large plates, inspired by the Himalayas and applied to the Caucasus and Anatolia (Tapponnier 1977; Tapponnier and Molnar 1976; Molnar and Tapponier 1977; Dewey and Sengor 1979). During the same period, the induction of the lateral movement of the microplates away from the front of the collision of Arabia–Eurasia was proposed as the cause of the creation of the great curvature of the Hellenic arc, with the term "Arcs induits" (Brunn 1976). The overall model of the neotectonic evolution of the Eastern Mediterranean was analysed in successive stages as a result of the escape of Arabia to the north, in relation to Africa, which, with the onset of the collision of Arabia with Eurasia in the Middle Miocene, caused the lateral escape of Anatolia to the west. This was followed by the escape of the Aegean to the south-southwest, where the oceanic remnant of the Eastern Mediterranean in the Ionian Sea was less resistant (Mercier 1979) (Fig. 3.14).

During the same time, the quantification of tectonic deformation and kinematics of the evolution of the Eastern Mediterranean begins through modeling (Fig. 3.15) (Le Pichon and Angelier 1979, 1981). The kinematic vectors result from a combination of geophysical–seismological data and geologic—neotectonic data. The kinematics began 13 Ma ago with the onset of the Arabia- Eurasia collision and the closure of Tethys to the east, when the history of the Mediterranean begins with the isolated oceanic remnant of the former Tethys Ocean in the eastern Mediterranean basin. The model shows that in the boundary of Middle to Late Miocene the Hellenic arc was almost a linear structure and gradually acquired the present curve of a right angle. At the same time, the front of the arc has progressively shifted several hundred kilometers to the south. Projection of this process towards the future suggests the onset of orogenesis at the Cyrenaica Peninsula in the next 5–6 million years.

The comparison of the new GPS data of the 1990s-2000s, presented earlier, with the Le Pichon and Angelier model (1979, 1981) is in good agreement and confirms the initial evolutionary model both as far as the directions of motions and the geodetic rates. The only significant difference concerns the internal deformation of the Aegean, where the older geological–geophysical data show a gradual increase in the rate of southward motion, with a minimum in the region of Lemnos–Northern Sporades, and a maximum in Crete, which indicates a general N-S extension of the microplate. On the contrary, the new geodetic data show the interior of the Aegean microplate practically as not deformed, with almost constant GPS rate everywhere, from Skyros–Chios–Lesvos in the north to Crete in the south (see also Fig. 3.13). This significant difference between geological and geodetic data is due to the temporal dimension, since the geological elements include the whole period of the last 13 Ma, while the geodetic data show the contemporary deformation of only one decade. There is obviously a great difference, because during this geologic period the current arc structure in the Aegean Sea was formed, with the opening of the Cretan Sea in the back arc area, which, based on the age of the sediments of the basin, was opened during the Late Miocene–Pliocene, while it seems to have slow down during the Quaternary. Therefore, while Crete has moved away from the Cyclades by a few tens of kilometers, during Late Miocene–Pliocene, there is neglisible extension at present day. A small exception concerns the eastern part of the Cretan basin northwards Karpathos and Rhodes islands, where an opening of 4–5 mm/year is actually observed (see also Fig. 3.13). Another important process deduced from the model of Le Pichon and Angelier (1979,

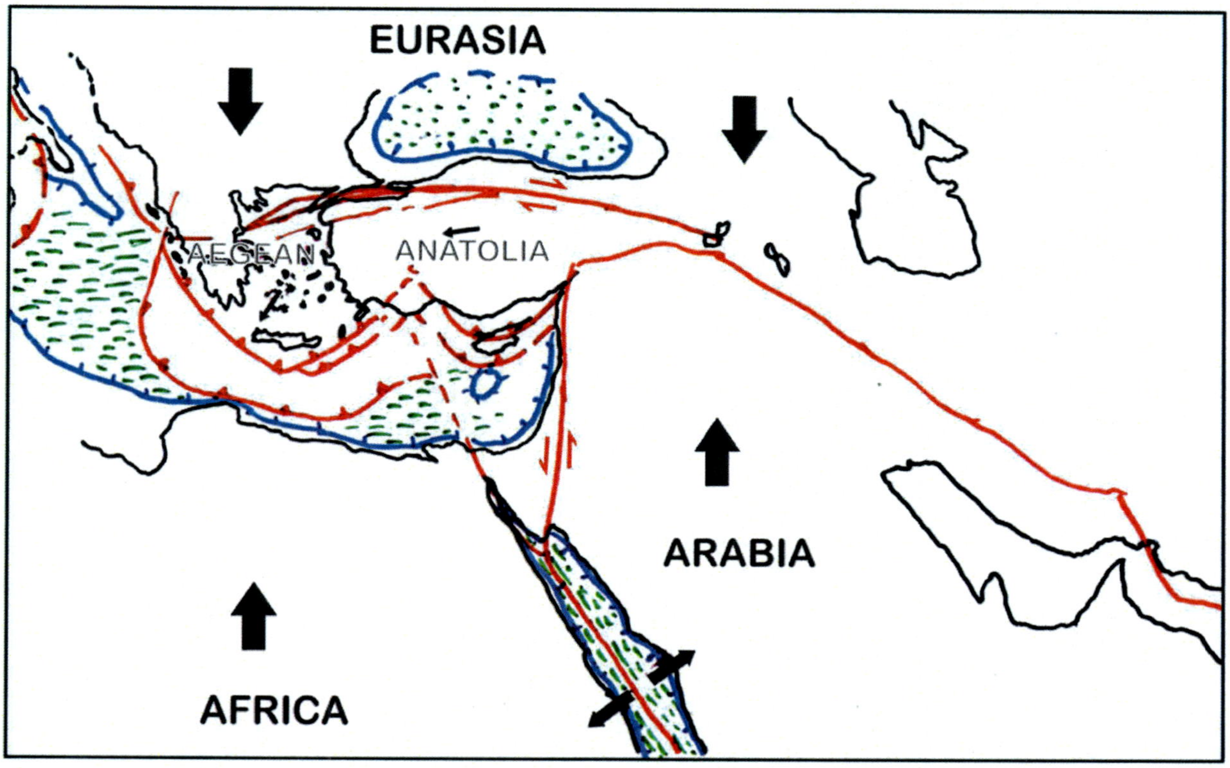

Fig. 3.14 Schematic representation of Eastern Mediterranean neotectonics based on the movements of the large plates of Europe, Africa, and Arabia and of the microplates of Anatolia and the Aegean in between (based on Mercier 1979)

1981) is the clockwise rotation of the southern Aegean area which is shown by the small vectors of the northeastern parts (Evia, Skyros, Cyclades) compared to the large vectors of the southwestern parts (southern Ionian islands, southern Peloponnese, Crete). This clockwise rotation of the Aegean microplate was later confirmed by **paleomagnetic** studies which supported a 40–50° angle since the Middle Miocene (Kissel and Laj 1988; Kissel et al. 2003; Van Hinsbergen et al. 2005, 2010) (see also Fig. 11.10). The above change of the kinematics in the Aegean area and especially the stabilization of the Aegean microplate, without active internal deformation as indicated by the GPS, has been attributed to: (i) the initiation of the modern cold oceanic subduction beneath the South Aegean which transformed the previous dynamics, together with (ii) the rotation produced between the right lateral North Anatolian strike slip zone in the north and the opposite transform motion of the Pliny and Strabo trenches at the eastern branch of the Hellenic arc and trench system in the south (Le Pichon et al. 2016) (see also Fig. 11.9). The western propagation and progressive strain localization of the North Anatolian Fault left the Aegean microplate with minor internal deformation which is regarded as the main process of this major change (Reilinger et al. 2010; Le Pichon et al. 2016). It is interesting that the European—Aegean/Anatolian tectonic boundary is proposed to be gradually developed during Plio-Quaternary, with extensional rift structures in the Marmara Sea before 4.5 Ma, similar to those of the present Corinth rift, followed by strike slip dominating kinematics at about 2.5 Ma, which have then propagated to the west up to the Kefalonia transform fault (Le Pichon et al. 2016).

The detailed marine geophysical survey of the Eastern Mediterranean showed the existence of a major 150 km long NNW-SSE tectonic zone with right lateral strike slip motion, extending from the exit of the Suez rift zone up to the west of Cyprus, delimiting the **Levantine–Sinai microplate** (Mascle et al. 2000) (Fig. 3.16). This microplate has a triangular geometry and consists of the Levantine basin, including the Eratosthenes Seamount, bordered to the east from the Dead Sea transform fault and to the north form the Cyprus arc. It is interesting that the East Mediterranean ridge stops towards the east along this tectonic zone, along which there is also a difference in the crust, with probably Paleozoic oceanic crust to the west and thinned continental crust to the east (Granot 2016). The active state of this intra-Mediterranean tectonic zone is demonstrated by the observation that the Nile deep sea fan has been deformed by this active zone/microplate boundary. The above

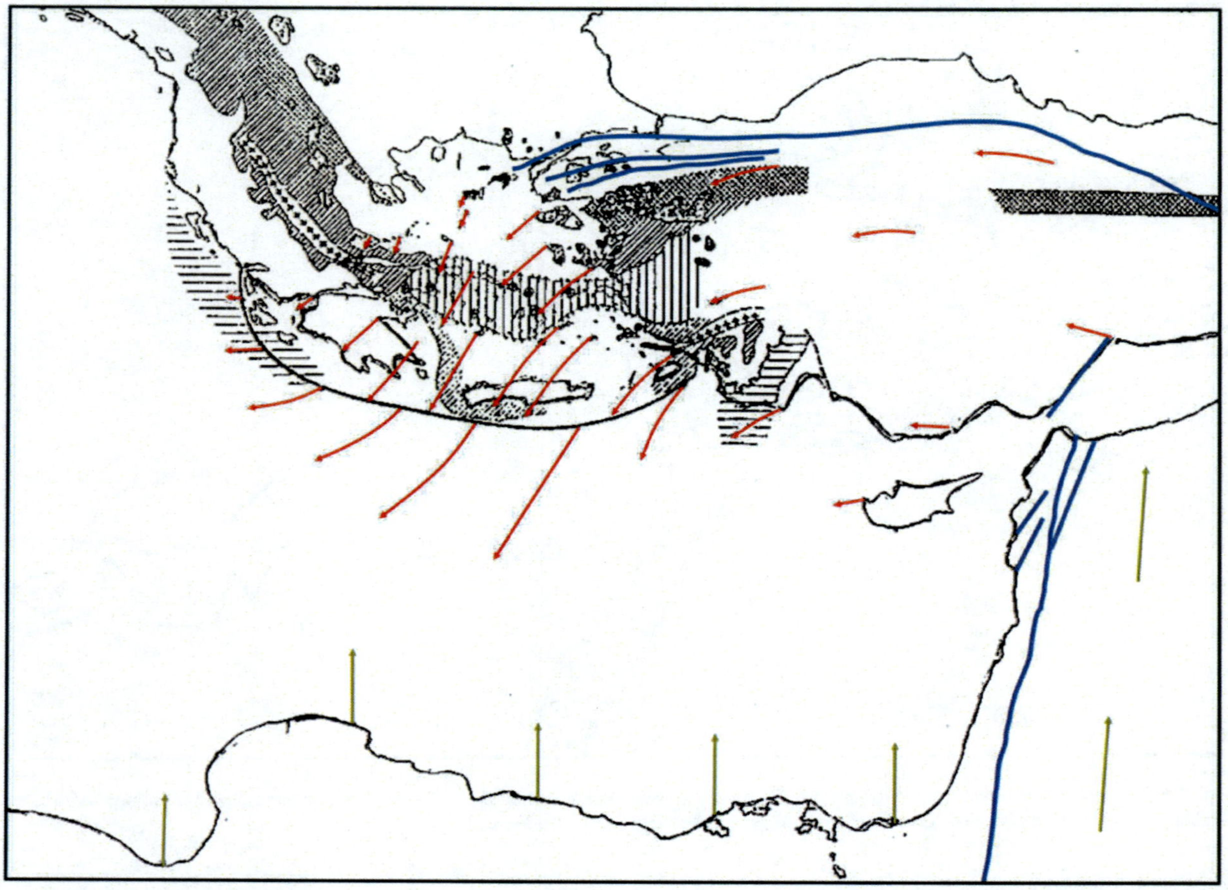

Fig. 3.15 Quantification of the tectonic deformation of the Eastern Mediterranean for the last 13 million years, since the formation of the Mediterranean after the closure of Tethys to the east (from Le Pichon and Angelier 1979, modified). The vectors in red show the total displacement and the current geography may result from the configuration of the edges of the vectors

observations suggest that inherited structures play a major role to the present day kinematics and overalll to the present day geological and tectonic setting.

In conclusion, the Eastern Mediterranean comprises a number of old inherited elements from the early stages of the Tethys Ocean and a number of new tectonic features formed since the Middle-Late Miocene. The overall tectonic structure of the Eastern Mediterranean is configurated in a synthetic tectonic map (Fig. 3.17) comprising: (i) The major continental plates of Eurasia, Africa and Arabia. (ii) The intermediate micro-plates of the Aegean, Anatolia and Levantine-Sinai. (iii) The three orogenic arcs of Sicily-Calabria, Hellenic and Cyprus. (iv) The remnant oceanic basins of Tethys in the Ionian and Herodotus, belonging to the northern part of the African plate. The Adria carbonate platform is developed north of the Ionian Basin, representing the northernmost not yet subducted part of the African plate. Its lateral continuation occurs beneath the fold and thrust belts of the Apennines and the Hellenides.

It is remarkable that the three arcs show major differences especially in the development of their volcanic arcs and back-arc basins (e.g. Transmed Atlas, Cavazza et al., editors, 2004). Thus, the Cyprus arc has the Cilicia-Adana minor back-arc basin, which is a narrow shallow (less than 1 km depth) basin underlain by continental crust but has no active volcanic arc (the previous volcanic arc was developed in Kappadokia, north of the Taurides throughout Miocene-Early Pleistocene). The Hellenic arc has the Cretan back-arc basin with depths up to 2.5 km underlain by thin continental crust of 16 km thickness and the active Aegean volcanic arc. The Sicily–Calabria arc has the Pliocene-Pleistocene oceanic back-arc basin in the Tyrrhenian Sea with depths up to 3.5 km and an active volcanic arc in the Aeolian Islands. These differences reflect the subduction process in each case with: (a) Almost inactive subduction in the Cyprus arc, because of its lateral westward escape together with Anatolia. (b) The well documented Hellenic subduction in both segments, with orthogonal convergence in the Ionian with dips from 17° to

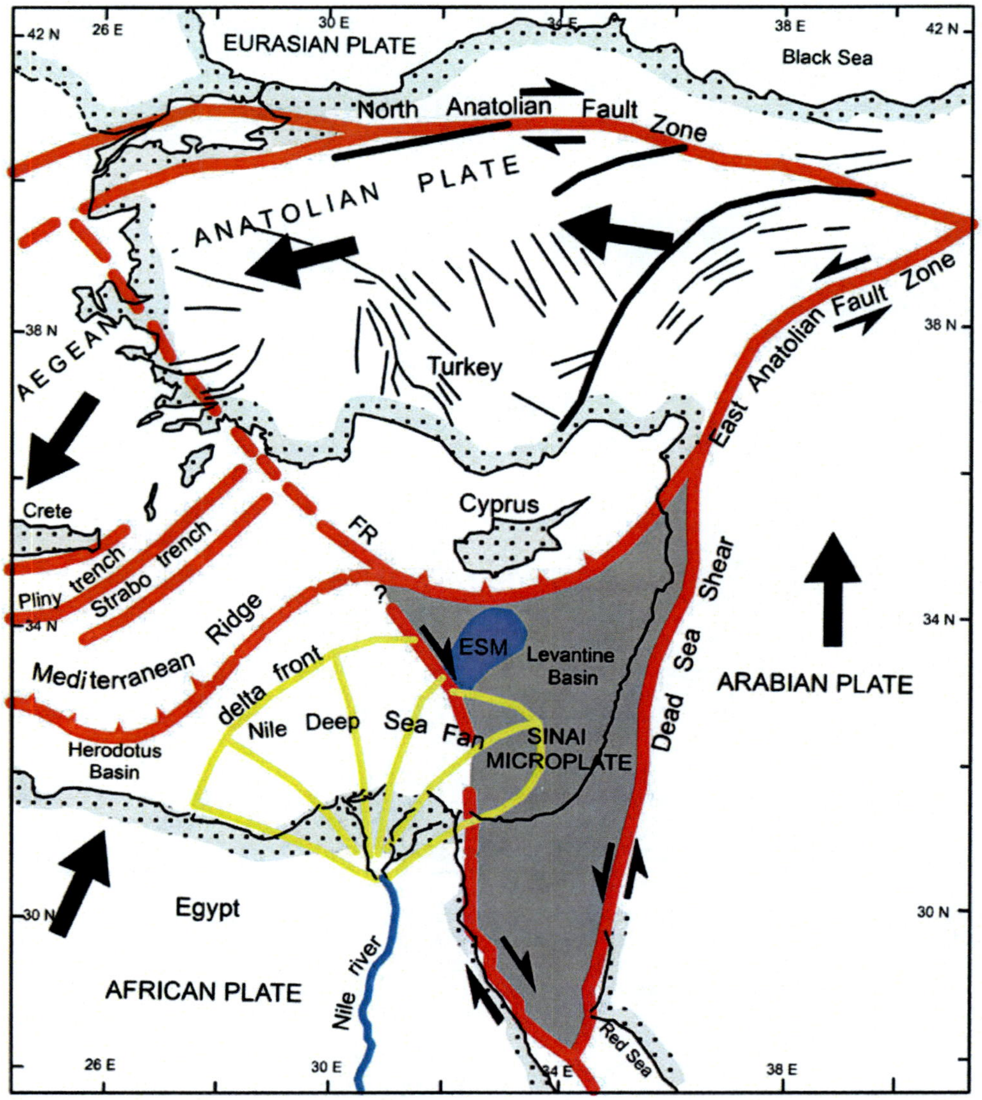

Fig. 3.16 Sketch of plate configuration in the Eastern Mediterranean showing the Levantine-Sinai microplate, south of Cyprus (from Mascle et al. 2000, modified)

40° to the northeast and oblique convergence together with left lateral strike slip motion along the Strabo and Pliny trenches with 40–45° to the north-northeast. (c) The well documented subduction in the Sicily-Calabria arc, with its highly dipping slab to the west-southwest. The general aspect of the Eastern Mediterranean tectonics indicates a compact structure with compression and important strike slip motions between the Anatolia and the Levantine-Sinai micro-plates in the east, adjacent to the collision front of Arabia/Eurasia. This regime is transformed to a more complex status with extension in the Aegean micro-plate behind the Hellenic arc, where the strike slip motions still prevail in its eastern part. In the west, the opening of the Tyrrhenian oceanic basin behind the Sicily-Calabria arc indicates the complete reversal of the collision status of the eastern segment of the Eastern Mediterranean.

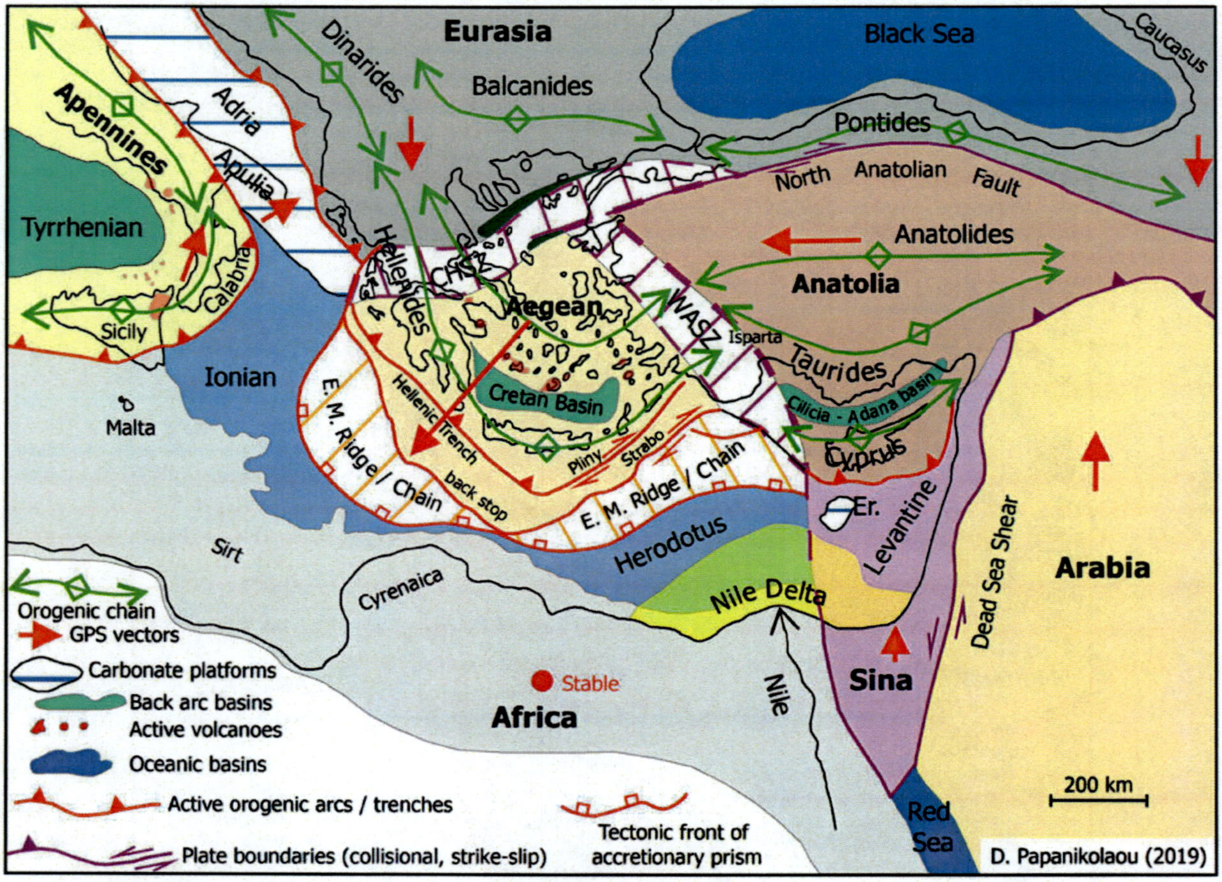

Fig. 3.17 Synthetic tectonic map of the Eastern Mediterranean. The Central Hellenic Shear Zone (CHSZ) and the West Anatolian Shear Zone (WASZ) form the dynamic boundaries between the Aegean-Anatolian-Eurasian plates. The allochthonous units in front of the Hellenic trenches extend up to the front of the East Mediterranean (E. M.) Ridge/Chain. The Levantine Basin includes the carbonate platform of the Eratosthenes sea mount (Er.). GPS vectors are shown relatively to stable Africa

References

Angelier, J. & Le Pichon, X. 1978. L'arc hellénique, clé de l'évolution cinématique de la Méditerranée orientale depuis 13 M.A. C. R. Acad. Sc. Paris, 287, 1325–1328.

Armijo, R., Meyer, B., Hubert, A. & Barka, A. 1999. Westward propagation of the North Anatolian fault into the northern Aegean: timing and kinematics. Geology, 27, 267–270.

Billiris, H., Paradissis, D., Veis, G., England, P., Featherstone, W., Parsons, B., Cross, P., Rands, P., Rayson, M., Sellers, P., Ashkenazi, V., Davison, M., Jackson, J. & Ambraseys, N. 1991. Geodetic determination of tectonic deformation in central Greece from 1900 to 1988. Nature, 350, 124–129.

Bonardi, G., Cavazza, W., Perrone, V. & Rossi, S. 2001. Calabria – Peloritani terrane and northern Ionian Sea. In: Vai and Martini (ed.) Anatomy of an orogen: the Apennines and adjacent Mediterranean basins. Kluwer Academic Publishers, Dordrecht, 287–306.

Brun, J.P., Faccenna, C., Gueydan, F., Sokoutis, D., Philippon, M., Kydonakis, K. & Gorini, C. 2016. The two - stage Aegean extension, from localised to distributed, a result of slab rollback acceleration. Canadian Journal of Earth Sciences, 53, 11, 1142–1157.

Brunn, J.H. 1976. L' arc concave zagro-taurique et les arcs convexes taurique et égéen: collision et arcs induits. Bull. Soc. Géol. France, 18, 2, 481–497.

Bullard, E., Everett, J. & Smith, A. 1965. The Fit of the Continents around the Atlantic. Philos. Trans., R. S. London, A258, 41–51.

Cavazza, W. et al, 2004. The TRANSMED Atlas: geological-geophysical fabric of the Mediterranean region. Episodes, 27, 4, 244–254.

Channel J.E.T., D' Argenio, B. & Horvath, F. 1979. Adria, the African Promontory in Mesozoic Mediterranean Paleogeography. Earth Science Reviews, 15, 213–292.

Davies, R., England, P., Parsons, B., Billiris, H., Paradissis, D. & Veis, G. 1997. Geodetic strain of Greece in the interval 1892–1992. J. Geophys. Res., 102, 24,571–24,588.

De Woogd, B., Truffert, C., Chammot-Rooke, N., Huchon, P., Lallemant, S. & Le Pichon, X. 1992. Two-ship deep seismic soundings in the basins of the Eastern Mediterranean Sea (Pasiphae cruise). Geophys. J. Int., 109(3), 536–552.

Dewey, J. F. & Şengor, C., 1979. Aegean and surrounding regions: complex multiplate and continuum tectonics in a convergent zone. Geol. Soc. Am. Bull., 90, 84–92.

Dziewonski, A. M. 1984. Mapping the lower mantle: Determination of lateral heterogeneities in P velocity up to degree and order 6. J. Geophys. Res., 89, 5929–5952.

Dziewonski, A. M., Hager, B. H. & O'Connell, R. J. 1977. Large scale heterogeneities in the lower mantle. J. Geophys. Res., 82, 239–255.

Emery, K.O., Heezen, B.C. & Allan, T.D., 1966. Bathymetry of the eastern Mediterranean Sea. Deep Sea Research and Oceanographic Abstracts, 13(2), 173–192.

Finetti, I. 1976. Mediterranean Ridge: A young submerged chain associated with the Hellenic arc. Boll. Gef. Teor. Appl., 19, 31–65.

Finetti, I. & Del Ben, A. 2005. Crustal tectono-stratigraphic setting of the Adriatic Sea from new CROP seismic data. In CROP Project: Deep seismic exploration of the central Mediterranean and Italy. Atlases 1, 519–548, Elsevier, Amsterdam.

Finetti, I., Papanikolaou, D., Del Ben, A. & Karvelis, P. 1991. Preliminary geotectonic interpretation of the East Mediterranean Chain and the Hellenic Arc. Bull. Geol. Soc. Greece, 25/1, 509–526.

Flerit, F., Armijo, R., King, G., & Meyer, B. 2004. The mechanical interaction between the propagating North Anatolian Fault and the back-arc extension in the Aegean. Earth and Planetary Science Letters, 224(3), 347–362.

Freund, R., Garfunkel, Z., Zak, I., Goldberg, M., Weissbord, T. & Derin, B. 1970. The shear along the Dead Sea Rift. Phil. Trans. Roy. Soc. London, A, 267, 107–130.

Garfunkel, Z. 2004. Origin of the East Mediterranean basin: a re-evaluation Tectonophysics, 391, 11–34.

Gesret, A., Laigle, M., Diaz, J., Sachpazi, M. & Hirn, A. 2010. The oceanic nature of the African slab subducted under Peloponnesus: Thin-layer resolution from multiscale analysis of teleseismic P- to S- converted waves. Geophys. J. Int., 183(2), 833–849.

Granot, R. 2016. Paleozoic oceanic crust preserved beneath the eastern Mediterranean. Nature Geoscience, 9, 701–705.https://doi.org/10.1038/ngeo2784.

Heezen, B.C. & Ewing, M. 1963. The Mid Oceanic Ridge. In: Hill, m. n. (Ed.), The Seas. 388-410, Interscience, New York.

Hsü, K., Montadert, L, Bernoulli. D. Cita, M.B., Erickson, A., Garisson, R.E. Kidd, R.B., Melieres, F., Müller, C. & Wright, R. 1978. History of the Mediterranean salinity crisis. Init. Rep. D.S.D. P .XIII, 1, 1053–1078.

Huguen, C. Mascle, J. Chaumillon, E. Woodside, J.M. Benkhelil, J. Kopf, A. & Volkonskaia, A. 2001. Deformational styles of the Eastern Mediterranean Ridge and surroundings from combined swath mapping and seismic reflection profiling. Tectonophysics, 343, 21–47.

Huguen, C., Chamot-Rooke, N., Loubrieu, B. & Mascle, J. 2006. Morphology of a pre-collisional, saslt-bearing accretionary complex: The Mediterranean Ridge (Eastern Mediterranean). Marine Geophysical Researches, 27, 61–75.

Jolivet, L. & Brun, J.P. 2010. Cenozoic geodynamic evolution of the Aegean. Int. J. Earth Sci., 99, 109–138.

Jolivet, L. & Faccenna, C. 2000. Mediterranean extension and the Africa - Eurasia collision. Tectonics, 19, 6, 1095–1106.

Jongsma, D. 1977. Bathymetry and shallow sructure of the Pliny and Strabo Trenches, south of the Hellenic Arc. Geol. Soc. Amer. Bulletin, 88, 797–805.

Kahle, H. G., Muller, M. V., Geiger, A., Danuser, G., Mueller, S., Veis, G., Billiris, H. & Paradisis, D. 1995. The strain field in NW Greece and the Ionian islands: results inferred from GPS measurements. Tectonophysics, 249, 41–52.

Kahle, H.G., Cocard, M., Peter, Y., Geiger, A., Reilinger, R., Barka, A. & Veis, G. 2000. GPS derived strain rate field within the boundary zones of the Eurasian, African and Arabian plates. J. Geophys. Res., 105, 23.353–23.370.

Karason, H. & Van der Hilst, R.D. 2001. Tomographic imaging of the lowermost mantle with differential times of refracted and diffracted core phases (PKP, P diff). J. Geoph. Res., 106, 6569–6587.

Kastens, K. 1991. Rate of outward growth of the Mediterranean Ridge accretionary complex. Tectonophysics, 199, 25–50.

Kissel, C. & Laj, C. 1988. The Tertiary geodynamic evolution of the Aegean arc: a paleomagnetic reconstruction. Tectonophysics, 146, 183–201.

Kissel, C., Laj, C., Poisson, A. & Gorur, N. 2003. Paleomagnetic reconstruction of the cenozoic evolution of the eastern Mediterranean. Tectonophysics, 362 (1–4), 199–217.

Speranza F., Minelli, L. Pignatelli, A. & Chiappini, M. 2012, The Ionian Sea: The oldest in situ ocean fragment of the world? J. Geophys. Res., 117. B 12101.

Le Pichon, X. & Angelier, J. 1979.The Hellenic Arc and Trench system: a key to the neotectonic evolution of the Eastern Mediterranean area. Tectonophysics, 60, 1–42.

Le Pichon, X. & Angelier, J. 1981. The Aegean Sea. Phil. Trans. R. Soc. London, A300, 357–372.

Le Pichon, X. Chamot-Rooke, N., Lallemant, S., Noomen, R. & Veis, G. 1995. Geodetic determination of the kinematics of central Greece with respect to Europe: Implications for eastern Mediterranean tectonics. Journal of Geophysical Research: Solid Earth, 100(B7), 12675–12690.

Le Pichon, X., Lallemant, S.J., Chamot-Rooke, N.,Lemeur, D., and Pascal, G. 2002. The Mediterranean Ridge backstop and the Hellenic nappes. Marine Geology, 186, 111–125.

Le Pichon, X., Şengor, A.M.C., Kende, J., İmren, C., Henry, P., Grail, C., Karabulut, H. 2016. Propagation of a strike-slip plate boundary within an extensional environment: The westward propagation of the North Anatolian fault. Canadian Journal of Earth Sciences, 53 (11), 1416–1439.

Le Pichon, X., Sengor, A. M. C. & Imren, C. 2019. A new approach to the opening of the Eastern Mediterranean Sea and the origin of the Hellenic subduction zone: Part 2: The Hellenic subduction zone. Canadian J. Earth Sciences, in print.

Leite, O. & Mascle, J. 1982. Geological structures on the South Cretan continental margin and Hellenic trench (Eastern Mediterranean). Marine Geology, 49, 199–223.

Lequellec, P., Mascle, J., Got, H. & Vittori, J. 1980. Seismic structure of southwestern Peloponnesus continental margin. Bull. Am. Assoc. Petr. Geol., 64, 242–263.

Makris, J. & Stobbe, C. 1984. Physical properties and state of the crust and upper mantle of the eastern Mediterranean Sea deduced from geophysical data. Mar. Geol., 55, 347–363.

Mascle, J. & Chaumillon, E. 1998. An overview of the Mediterranean Ridge collisional accretionary complex as deduced from multichannel seismic data. Geo-Marine Letters, 18, 81–89.

Mascle, J., Coussin, M., Fleury, J.J., Le Cleac'h, A.G. & Saint-Marc, P. 1986. Calcaires Mesozoiques dans les fosses Helleniques: implications paleogeographiques et geodynamiques. Marine Geology, 73, 323–341.

Mascle, J., Benkhelil, J., Bellaiche, G., Zitter, T., Woodside, J., Loncke, L. & Prismed II Scientific Party. 2000. Marine geologic evidence for a Levantine-Sinai plate, e new piece of the Mediterranean puzzle. Geology, 28, 779–782.

McKenzie, D.P. 1970. Plate tectonics in the Mediterranean Region. Nature, 226, 239–243.

McKenzie, D.P. 1972. Active tectonics of the Mediterranean Region. Geoph. J.R. Astron.Soc., 30, 109–185.

McKenzie, D., 1978. Active tectonics of the Alpine - Himalayan belt: the Aegean sea and surrounding regions. Geophys.J.R. Astr. Soc., 55, (1), 217–254.

McKlusky, S. C. et al, 2000. Global Positioning System constraints on plate kinematics and dynamics in the Eastern Mediterranean and Caucasus. J. Geoph. Res., 105, 5695–5719.

Mercier, J.L 1979. Signification néotectonique de l'Arc Egéen. Une revue des idées. Rev. Géol. Dyn. Géogr. Phys., 21, 1, 5–15.

Molnar, P. & Tapponier, P. 1977. Relation of the tectonics of eastern China to the India-Eurasia collision: Application of slip-line field theory to large-scale continental tectonics. Geology, 5(4), 212–216.

Muller, M. D., Geiger, A., Kahle, H. G., Veis, G., Billliris, H., Paradissis, D. & Felekis, S. 2013. Velocity and deformation fields in the North Aegean domain, Greece and implications for fault kinematics, derived from GPS data 1993–2009. Tectonophysics, 597-598, 34-49.

Netzeband, G., Goel, K., Hubscher, C., Ben-Avraham, Z., Delghani, G., Gajewski, D. & Liersch, P. 2006. The Levantine basin – crustal structure and origin. Tectonophysics, 418, 167–188.

Nyst, M. & Thatcher, W. 2004. New constraints on the active tectonic deformation of the Aegean. J. Geophys. Res., 109 (B11406), 23.

Papanikolaou, D. 1986. Geology of Greece. Eptalofos Publications, 240 p. Athens (in greek).

Papanikolaou, I. & Papanikolaou, D. 2007. Seismic hazard senarios from the longest geologically constrained active fault of the Aegean. Quaternary Intern., 171–172, 31-44.

Papanikolaou, D. & Royden, L. 2007. Disruption of the Hellenic arc: Late Miocene extensional detachment faults and steep Pliocene–Quaternary normal faults — or what happened at Corinth? Tectonics, 26.

Papanikolaou, D., Alexandri, M., Nomikou, P. & Ballas, D. 2002. Morphotectonic structure of the Western Part of the North Aegean Basin based on swath bathymetry. Marine Geology, 190, 465–492.

Papanikolaou, D., Alexandri, M. & Nomikou, P. 2006. Active Faulting in the North Aegean Basin. Geol. Soc. Amer. Sp. Paper, 409, 189–209.

Papanikolaou, D., Nomikou, P., Papanikolaou, I., Lampridou, D., Rousakis, G. & Alexandri, S. 2019. Active tectonics and sesmic hazard in Skyros Basin, North Aegean Sea, Greece. Marine Geology, 407, 94–110.

Rabinowitz, P.D. & Ryan, W.B.F. 1970. Gravity anomalies and crustal shortening in the eastern Mediterranean. Tectonophysics, 10, 585–608.

Reilinger, R. et al. 1997. Global positioning system measurements of present-day crustal movements in the Arabia-Africa-Eurasia plate collision zone. J. Geophys. Res., 102, 9983–9999.

Reilinger et al, 2006. GPS constraints on continental deformation in the Africa – Arabia – Eurasia continental collision zone and implications for the dynamics of plate interactions. J. Geoph. Res. 111, B05411.

Reilinger, R., McClusky, S., Paradissis, D., Ergintav, S., & Vernant, P. 2010. Geodetic constraints on the tectonic evolution of the Aegean region and strain accumulation along the Hellenic subduction zone. Tectonophysics, 488(1), 22–30.

Reston, T.J., Fruehn, J. & Von Huene, R. & the Imerse working group, 2002. The structure and evolution of the Western Mediterranean ridge. Marine Geology, 186, 83–110.

Robertson, A. 1998. Tectonic significance of the Eratosthenes Seamount: a continental fragment in the process of collision with a subduction zone in the Eastern Mediterranean (Ocean Drilling Program Leg 160). Tectonophysics 298, 63–82.

Roveri M., Flecker, R., Krijgsman,W., Lofi, J., Lugli, S., Manzi, V., Sierro, F., Bertini, A., Camerlenghi, A., De Lange, G., Govers. R., Hilgen, F., Hubscher, C., Meijer, P. & Stoica, M. 2014. The Messinian salinity crisis: past and future of a great challenge for marine sciences. Marine Geology, 352, 25–58.

Royden, L.H., 1993. Evolution of retreating subduction boundaries formed during continental collision. Tectonics, 12, 629–638.

Royden, L.H. & Papanikolaou, D.J., 2011. Slab segmentation and late Cenozoic disruption of the Hellenic arc. Geochemistry, Geophysics, Geosystems, 12, Q03010.

Sengor, A.M.C., Tuysuz, O., Imren, C., Sakinc, M., Eyidogan, H., Gorur, N., Le Pichon, X. & Rangin, C. 2005. The North Anatolian Fault: A new look. Ann. Rev. Earth Planet. Sci., 33, 37–112.

Seyítoğlu, G., & Scott, B. C. 1996. The cause of NS extensional tectonics in western Turkey: tectonic escape vs back-arc spreading vs orogenic collapse. Journal of Geodynamics, 22(1–2), 145-153.

Spakman, W., Wortel, M.J.R. & Vlaar, N.J. 1988. The Hellenic subduction zone: a tomographic image and its geodynamic implications. Geoph. Res. Let., 15, 60–63.

Spakman, W., Van Der Lee, S. & Van Der Hilst, R. 1993. Travel-time tomography of the European–Mediterranean mantle down to 1400 km. Physics Earth & Plan. Inter., 79, 3–74.

Stampfli, G. & Borel, G. 2004. The TRANSMED transects in space and time: constraints on the paleotectonic evolution of the Mediterranean domain. In: Cavazza, W., Roure,F., Stampfli, G., Ziegler, P. (Eds.), The TRANSMED Atlas: The Mediterranean Regionfrom Crust to Mantle. Springer, Berlin, 53–80.

Tapponnier, P., 1977. Evolution tectonique du système alpin en Méditerranée: poinçonnement et écrasement rigide - plastique. Bull. Soc. Géol. France, 19, 3, 437–460.

Tapponnier, P. & Molnar, P. 1976. Slip-line field theory and large-scale continental tectonics. Nature, 264, 319–324.

Van der Hilst, R.D., Widyantoro, S. & Engdahl, E.R. 1997. Evidence for deep mantle circulation from global tomography. Nature, 386, 578–584.

Van Hinsbergen, D.J.J., Hafkenwscheid, E., Spakman, W., Meulenkamp, J. & Wortel, R. 2005. Nappe stacking resulting from subduction of oceanic and continental lithosphere below Greece. Geology, 33, 325–328.

Van Hinsbergen, D. J., Dekkers, M. J., Bozkurt, E., & Koopman, M. 2010. Exhumation with a twist: paleomagnetic constraints on the evolution of the Menderes metamorphic core complex, western Turkey. Tectonics, 29(3).

Yilmaz, Y. et al. 2000. When did the western Anatolian grabens begin to develop? Geol. Soc. London, Spec. Publ., 173, 353–384.

Zonenshain, I. P. & Le Pichon, X. 1986. Deep basins of the Black Sea and Caspian Sea as remnants of Mesozoic back arc basins. Tectonophysics, 123, 181–211.

4 Orogenic Model

4.1 Theories of Tectogenesis and Orogeny

Until the late 1960s following the prevailing tectogenetic theories of the **geosyncline** (Dana 1873; Haug 1900; Stille 1924; 1936; Kay 1947; Peyve and Sinitzyn 1950; Aubouin 1965) orogeny was considered as a complex geodynamic phenomenon, whose explanation and logic were inconceivable, mysterious and always something indefinable was regarded as the triggering mechanism. Orogeny was considered as the final result of the geotectonic process of a geologic—geotectonic cycle, which appeared almost like a "deus ex machina" of the ancient Greek tragedies, which gave the solution to the deadlock of the drama in a supernatural way, i.e. not understood on the basis of rationalism. Thus, the explanation of orogeny was conceived following simplistic models according to the evolution of the tectogenetic theories, which were gradually evolved but, nevertheless, remained always partial conceptions. Indicatively, there have been the theories (Galanopoulos 1961; Psarianos and Manolessos 1963; Spencer 1977): (i) of the uplift (Hatton), where the triggering cause was the rise of magma from the depths, (ii) of contraction (E. de Beaumont, Dana), where orogeny was considered as a surface ripple, due to the cooling of the Earth's interior, (iii) of isostasy (Airy), where the differences in thickness and composition (especially specific weight) between different crustal blocks create corresponding equilibrium motions, based on the existence of a surface of isostatic compensation, (iv) of Wegener's continental drift, which introduced the great mobility of the various crustal blocks, with movement of the continents to various directions, sometimes in a converging course and other times divergent, (v) of collisions, introduced by Argand, through his experience in the Tethys system, which he believed that resulted from the crushing of the sediments between the two cratonic masses of Eurasia and Gondwana, (vi) of the supeficial mass movements due to gravity, with the examples of the pre-Alps of the Western Alps (Schardt, Lugeon, Gignoux, Moret), (vii) of the magmatic currents (Ampferer, Wening-Meinesz), in which upstream and downstream currents are created in the upper mantle under the oceans and continents respectively, due to different heat loss, (viii) of the oceanic trenches (Benioff), where oceanic trenches can be created along the boundaries of continents and oceans, where deep seismicity of several hundreds of kilometers is observed, as well as material melting, creating volcanoes at the surface of the Earth.

Several elements from the above theories and some other similar ones were combined, from the globally accepted nowadays **theory of lithospheric plate tectonics**, such as Wegener's theory of continental drift (Wegener 1920), Argand's collision theory (Argand 1924), Ampferer's magmatic currents in the mantle (Ampferer 1906) and Benioff's oceanic trenches (Benioff 1955). The orogenic phenomena due to surface gravitational mass movements, by Schardt, Lugeon, et al., are considered as secondary phenomena during orogeny and are created after the uplifting of mountain ranges (Schardt 1893, 1898; Lugeon 1902, 1943) due to the convergence and/or collision of the plates.

Today, all of the Earth's orogenic systems, at least those of the Mesozoic and the Cenozoic, can be explained through the motion of the present-day lithospheric plates, by calculating the segments of new oceanic crust that has been created, the segments of old oceanic crust that has been destroyed/subducted and the areas of continental crust that have been broken or amalgamated or accreted with the addition of new orogenic zones. The resulting from the plate tectonics theory orogenic mechanism, alternatively described as "mountain building", has been analyzed in various models since the early 1970s (Dewey and Bird 1970, 1971).

The present distribution of orogeny is determined by the convergent/collisional zones of the tectonic plates (see Fig. 1.1). On the contrary, in divergent zones no orogeny is observed, whereas in the intermediate cases of horizontal offset with lateral strike slip motion of the plates, the orogenic phenomena are limited, unless there is a converging component.

In the tectonic plates theory we can implicitly accept that it combines: (i) the mobility of the lithospheric plates above

D. I. Papanikolaou, *The Geology of Greece*, Regional Geology Reviews,
https://doi.org/10.1007/978-3-030-60731-9_4

the asthenosphere, related to Wegener's continental drift, (ii) the creation of orogeny in the zones where convergence and collision of tectonic plates are present, in relation with the collision of the Tethyan system from Eurasia and Gondwana of Argand, and (iii) the explanation of the mechanism of the tectonic plates movement above the asthenosphere, with the creation of new oceanic crust in the mid-ocean ridges and its destruction in the subduction zones, which occur in Benioff's oceanic trenches (which often surround the continents), corresponding to the existence of underlying upstream/ downstream convective currents in the mantle of Ampferer.

The major aspects of the plate tectonics theory were based on:

- the recognition of oceanic rifting in the mid-ocean ridges (Dietz 1961; Hess 1962, 1965; Talwani et al. 1965),
- the distribution of sediments in the oceans on both sides of the mid-ocean ridges (Ewing and Ewing 1967),
- the confirmation of Wegener's continental drift with the identification of the continental margins, mainly in the Atlantic Ocean, by the additional use of paleomagnetic data (Bullard et al. 1965; LePichon 1968),
- the distinction of the lithosphere and asthenosphere, together with the upper mantle's flow (McKenzie 1967, 1968), which allowed the acceptance of the existence and movement of the plates,
- the resulting seismicity on the Earth's sphere (McKenzie and Parker 1967),
- the distinction of the new mega-tectonic faults—boundaries of the tectonic plates (Wilson 1965; Morgan 1968),
- the affirmation of the opening rate of the oceans using paleomagnetic data (Vine and Matthews 1963; Cox et al. 1967; Heirtzler et al. 1968),
- the correlation of seismicity in the Benioff zones along the subduction zones and the other plate boundaries (Isacks et al. 1968),
- the explanation of the orogenic chains with the movement of the plates in previous geological times (Dewey and Bird 1970),
- the movement of the currents of the asthenosphere as a driving force of the lithospheric plates motion, (McKenzie 1967, 1968),
- the defining of the 12 largest plates (Morgan 1972; Turcotte and Oxburg 1972; Lepichon et al. 1973),
- the analysis of the stresses/driving forces controlling the movements in the asthenosphere and the plate boundaries (Elsasser 1969, 1971; Forsyth and Uyeda 1975; Chapple and Tullis 1977).

4.2 Orogenic Mechanism—Orogenic Arc

As previously mentioned, orogeny occurs along the convergent zones of the plates. In all the cases where a converging motion occurs along a boundary between two plates of any nature, an elongated doming is created in the crust, which, as it emerges from the surface of the sea, forms an elongated mountain. It is the natural result of the deformation of the lithosphere, which under lateral compression, moves vertically either downwards or upwards. Since upward resistance from the air is clearly smaller than the resistance from the underlying rocks in the downward movement, the upward movement of part of the lithosphere along the contact zone is dominant, with a simultaneous micro-mesoscopic deformation of the rocks, which in turn constitutes the orogenic phenomenon.

In the overwhelming majority, the convergent zones affect the passive margins of the continents, transforming them into active margins (Dewey 1969), with the tectonic contacts of the two plates occurring offshore and thus orogenic chains are often created in the form of long islands–island arcs (Karig 1970, 1974; Sugimura and Uyeda 1973). At the same time along the plate boundaries a deep trench is created from the subduction of one plate under the other. The melting of the subducted lithosphere occurring at more than 100 km depth behind the convergence front of the two plates leads to the formation of a volcanic arc, which is usually parallel to the trench and the island arc. These phenomena are interrelated and caused by the particular kinematics of the two plates. All together, they constitute a constantly occurring geometry, which is characterized as an orogenic arc and includes the following segments, as exemplified in the actual Hellenic arc (Angelier 1979; Papanikolaou and Dermitzakis 1981a, b; Papanikolaou 1986b, 1993) (Fig. 4.1):

i. **trench** (No 2, in Fig. 4.1): It is a deep elongate basin (4,500–5,000 m deep in the Ionian Sea along the Hellenic trench, and 10,000–11,000 m deep in the Mariana Trench in the Western Pacific), which defines the front of the relative movement/contact of the two converging plates. The subducting plate is oceanic and the advancing one is continental.
ii. **island arc** (No 3, in Fig. 4.1): This is a mountain chain forming elongate islands that can be traced parallel to the trench and is created by the deformation of the geological formations of the advancing plate (usually sedimentary) with folds, overthrusts, thrusts and metamorphic phenomena in greater depths. The thick-

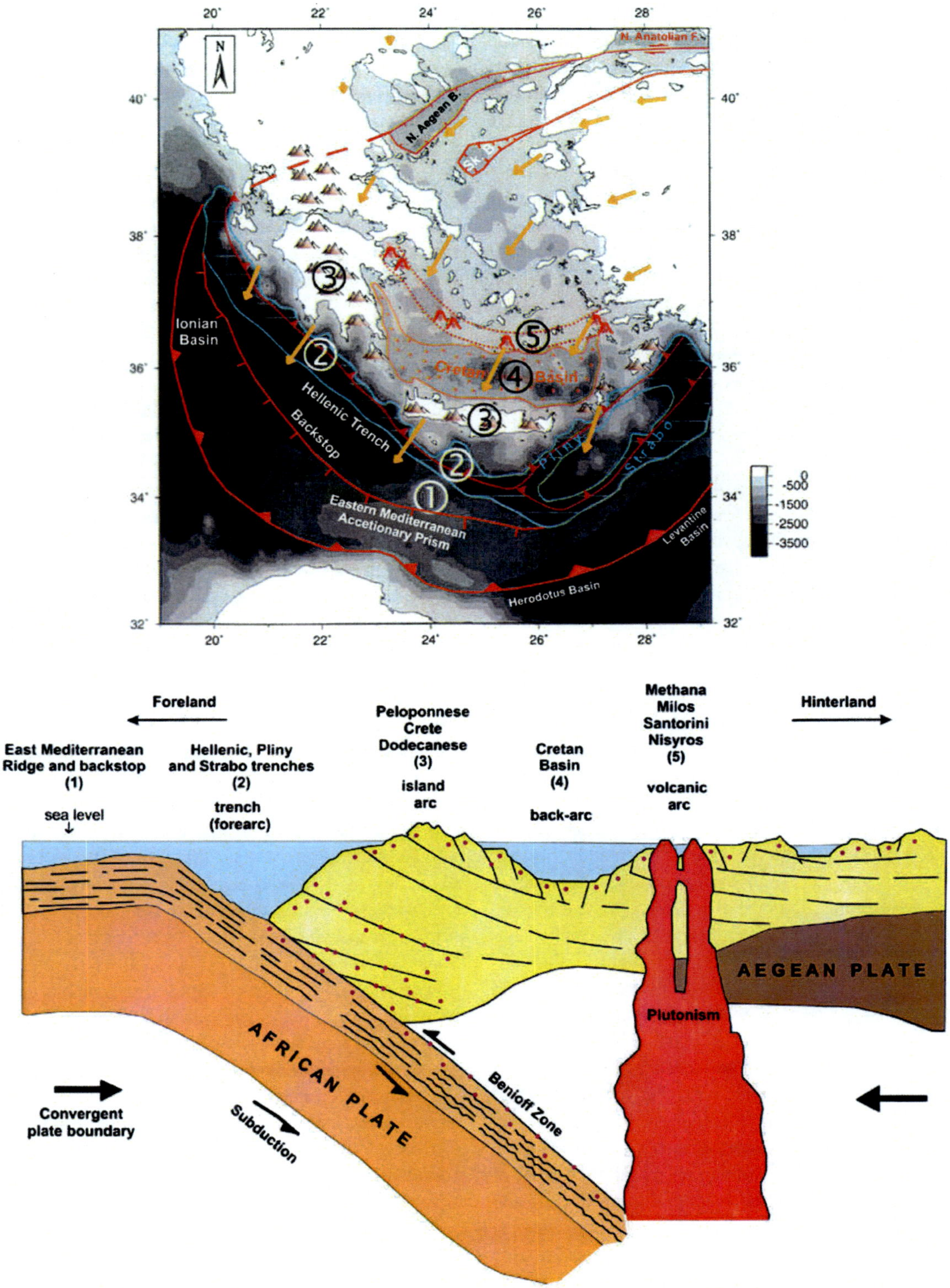

Fig. 4.1 a Simplified tectonic map showing the actual geometry of the Hellenic arc (based on Papanikolaou 1986b, 1993; Papanikolaou and Sideris 2007). b Schematic representation in a transverse section of an orogenic arc, adapted to the actualistic geometry of the Hellenic arc (based on Papanikolaou and Dermitzakis 1981a, b; Papanikolaou 1986b)

ness of the continental crust is increased with the addition of segments of the subducted lithosphere. The internal tectonic movements among the tectonic units form the so-called orogenic wedge structure. Peloponnese, Crete and the Dodekanese islands represent the island arc in the Hellenic arc today.

iii. **back-arc basin** (No 4, in Fig. 4.1): This is a generally more shallow basin than the trench, which can be seen behind the island arc. In this area, a decompression of the rocks by the creation of extensional structures is present. The continental crust becomes thinner due to the extension and it can be converted into a crust of basic composition when the back-arc basin turns into a Marginal Sea/Basin, like the Sea of Japan. In several cases the back-arc basin is missing depending on the relation between the convergence and the subduction rates. The Cretan Basin represents the actual back-arc basin of the Hellenic arc.

iv. **volcanic arc** (No 5, in Fig. 4.1): It is located always in a significant distance of a few hundred kilometers away from the trench on the advancing plate, and it sometimes coincides with the back-arc basin, or behind it, while often it occurs further ahead. Its position depends on the presence of the back-arc basin, as well as on the angle and rate of subduction. A high angle and high subduction rate brings the volcanic arc forward, near the trench, while a low angle and low subduction rate places it towards the back-arc basin and behind the arc. The Aegean volcanic arc occurs today at the internal northern margin of the Cretan back-arc basin.

The area in front of the trench over the subducting plate (No 1, in Fig. 4.1), where orogenic deformation has not occurred yet, is characterized by the direction of the tectonic transport as **foreland**, while the area behind the volcanic arc over the advancing plate, where orogenic deformation has been completed, as **hinterland**. A key point is that this geometry of the orogenic arc is not static but dynamic and remains diachronous as a geometry, while the participating rock formations change over time. Thus, new segments of the subducted plate enter in the trench structure and get subducted, but later on they may emerge again at the island arc or even further back in the hinterland, with varying degrees of deformation and metamorphism. That is, each area of the subducted plate ends up on the uplifted front of the advancing plate after a short or long, shallow or deep and simple or complicated journey above basal detachments. This transition implies the effect of several geodynamic phenomena, both at shallow and deep level, which differentiate the intact rock of the subducting plate from the same rock when it reaches the advancing plate, up to the hinterland.

The continuous movement of the plates creates the so-called orogenic arc migration, according to which, rocks of different paleogeographic regions of the subducting plate gradually participate in the orogenic arc. Thus, the horizontally deposited sediments in the oceanic basins, continental slopes, platforms, margins, etc. of the subducting plate are transformed to eroded, folded rocks in the orogenic arc at the front of the advancing plate. In fact, it is not the orogenic arc that migrates, but the rocks of the subducted plate, which gradually enter the orogenic zone and are underthrusted beneath the upper plate, which remains almost stable on the active margin of the continent («sous-chiarriage» of Aubouin 1974).

4.3 Mechanisms of Tectonic Detachments—Creation of the Tectonic Units

The whole tectonic process in the orogenic arc is based on the **decollement** phenomenon of an upper section of the subducted lithosphere, usually 2–5 km thick, which is detached from the rest of the lithosphere, either at shallow depth within the upper crust or deeper at a few tens of kilometers within the subduction zone. The remaining lithospheric slab continues to be subducted into the Earth's interior as it is depicted in seismic tomographs. The shallow detached fragment is accreted to the front of the advancing plate, where through compressive brittle or ductile deformation is transformed into a new portion of the continental crust of the upper plate, forming part of the **orogenic wedge**. This is the **thin skin tectonics mechanism**, which creates a zone of compressive deformation, known as **fold and thrust belt** along the arc front. Anisotropy in the upper crust of the subducting plate, due to alternating segments of neritic and abyssopelagic sedimentation, may lead to the formation of **tectonic units**, resulting from two types of detachments (Papanikolaou 1986b, 2015) (Fig. 4.2a): (1) a subhorizontal detachment (d1), of the upper stratigraphic sequence, detached from its basement, usually on a pronounced anisotropic surface, such as e.g. the unconformity of the Mesozoic Alpine sedimentary sequences over the underlying deformed and/or metamorphosed Paleozoic formations of the basement, and (2) a highly inclined detachment surface (d2), developed along the transitional boundaries of the individual basins and ridges, which had been usually formed at an early paleogeographic stage, through normal faults during the early rifting phase. This is the well-known phenomenon of **inversion tectonics**, in which the original normal faults of the rifting phase are re-activated and transformed into reverse faults/thrusts. In the cases of continental lithospheric fragments—tectono-stratigraphic terranes, a second deeper detachment zone (d3) is often observed within the continental crust, developed approximately at the transition of the brittle upper crust with the lower ductile crust, i.e. approximately at the base of the seismogenic layer, at a depth of 12–15 km (Papanikolaou 2015; Burchfiel et al. 2018). In this case the decollement is usually carried out at a greater depth, along the subduction zone, and is accompanied by tectonometamorphic phenomena. The continental fragment is integrated/accreted into the base of the upper plate and gradually slides under the island arc or even more internally (Fig. 4.2b).

The above complicated detachment mechanism at the front of the arc shapes the tectonic units observed at the upper plate, originating from the fragments of the lower plate and, thus, new continental crust is incorporated at the front of the advancing plate. Different types of tectonic units result, depending on the depth of the detachment within the subducting plate and the depth of accretion at the advancing plate, as they will be described later in the case of the Hellenides. Thus, we can distinguish: (1) units of shallow detachment and shallow rapid accretion, producing non

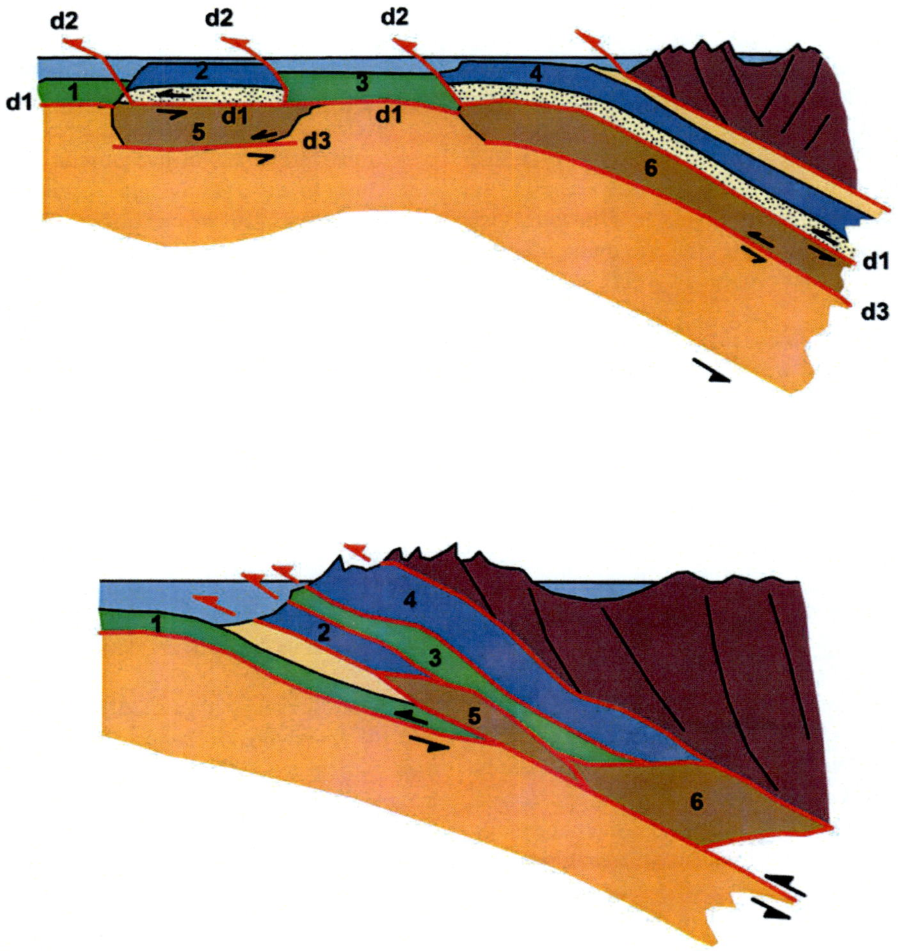

Fig. 4.2 **a** Tectonic detachment developing in the upper part of the crust from the rest of the lithosphere (d1) upon the subduction of the lower plate. The detached segment is integrated into the front of the base of the upper plate, while the rest of the slab enters at great depths, where it is detected in the seismic tomographs. At the same time, lateral detachments (d2) may also occur at the upper crust of the lower plate, at the weak transitional zones between basins and ridges, which had been previously created by synsedimentary normal faults. In the case of continental tectono-stratigraphic terranes the detachment can occurr deeper (d3), at the boundary of the brittle/plastic deformation in the crust, at the base of the seismogenic layer. **b** Intergration of the detached segments of the subducted plate through surfaces d1, d2, and d3 at the front and the base of the advancing plate. This is an accretion mechanism of the continental terranes of the subducted plate (Gondwana–Africa) in the advancing plate (Europe), whose continental crust is consequently growing. Numbers 1–6 refer to the pre-accreted units to the arc of figure (**a**) and their new position on the advancing plate. 1 and 3 correspond to abysso-pelagic units. 2 and 4 correspond to carbonate platforms. 5 and 6 correspond to pre-Alpine continental crustal fragments

metamorphic Alpine units (e.g. Ionian, Tripolis, Pindos, Parnassos of the so-called non metamorphic Hellenides), (2) units of shallow detachment and deep late accretion, leading to metamorphic Alpine units (e.g. Mani, Olympus, Cyclades of the so-called metamorphic Hellenides), (3) units of deep detachment and deep delayed accretion, leading to metamorphic pre-Alpine units (e.g. Arna, Flambouro).

In conclusion, three types of tectonic units are distinguished: the **metamorphic Hellenides,** the **non metamorphic Hellenides** and the **pre-Alpine units** (Papanikolaou 1980, 1986a) (see also Fig. 7.25). Apart from the creation of tectonic units from the detached fragments of the subducting plate this process also builds the new crust at the front of the advancing plate. In the retreating Hellenic subduction zone this process is facilitated and advanced by the increasing subduction angle with the roll back of the lower plate and produces the successive accretion of new lithosheric fragments at the base of the superficial nappes at the upper plate (Burchfiel et al. 2018). An alternative approach regarding the exhumation of the continental fragments following their subduction to depths of several tens of km through «caterpillar walk» mechanism was proposed by Tirel et al. (2013).

4.4 Shallow Geodynamic Phenomena in the Orogenic Arc

Along a transverse section of an orogenic arc and for a given time period we can distinguish three domains of different geodynamic regimes: (1) An outer external province, which develops on the subducting plate but away from the orogenic arc, i.e. outside the trench, which is characterized as a **pre-orogenic** area, where there is no effect of the orogenic arc. The main feature of this area is the uninterrupted sedimentation, deep or shallow, silicate or carbonate, etc., just as it was happening several million years before. (2) An intermediate province, comprising all the individual parts of the orogenic arc and undergoing all the orogenic geodynamic events, which is characterized as a **syn-orogenic** area. Here all the events, such as sedimentation, morphogenesis, volcanism, magmatism, metamorphism, seismicity etc., are under tectonic control. This area includes the edge of the subducting lithosphere and the front of the advancing plate. (3) An internal province, including the segments of the two plates that have already experienced the orogeny and now lie outside the influence of the orogenic arc, forming the stable new crustal part of the upper plate. This area is known as **post-orogenic** and is usually characterized by a mild continental physico-geographical picture. Active geodynamic phenomena are generally absent, although the previous ones are detectable and relatively recent.

If a chronological distinction of the individual events is desirable throughout the course of a paleogeographic unit from the pre-orogenic area of the subducting plate to the post-orogenic area of the advancing plate, then we can refer to rocks and structures of the pre-orogenic period, the syn-orogenic period, and the post-orogenic period. The major shallow geodynamic phenomena taking place within the orogenic arc will be analysed in the following.

(i) **Sedimentation**: The various environments, that can be found along a transverse section from one plate to the other through the orogenic arc, are characterized by their stratigraphic facies, which, in the case of the syn-orogenic area (or of the syn-orogenic period for each unit) is characterized by the presence of a specific type of clastic sediments, which are generally characterized as **tectono-sedimentary facies**, corresponding to the formations of flysch, wild flysch, molasse, etc. Thus, in each stratigraphic column of a unit of the subducting plate we can distinguish the lower segment, corresponding to the pre-orogenic period (which in the past was referred to as "geosynclinal sedimentation") and an upper segment, corresponding to the syn-orogenic sedimentation. The creation of the special tectono-sedimentary facies is due to the formation of the trench/fore-arc basin and the back-arc basin on either side of the island arc, which, due to the intense uplift, is supplying them with abundant clastic material from the intense erosion. The clastic material is mainly transported through rivers and is deposited either in the deltas or through the oceanographic circulation, forming turbidites at the bottom of the deep basins. Thus, **flysch** is deposited in the trench/fore-arc zone and **molasse** in the back-arc basin.

Flysch is usually deposited in stratigraphic continuity with the underlying biochemical pre-orogenic sediments and then participates in the orogenic deformation, forming an integral upper part of the stratigraphic column (Fig. 4.3). On the contrary, molasse is deposited in a stratigraphic unconformity—often important—with the underlying deformed Alpine sequences, which include both pre- and syn-orogenic sediments (Fig. 4.4). Thus, molasse is often observed to overly unconformably a folded older flysch. In addition, flysch belongs to a defined stratigraphic column of a tectonic unit, while molasse can overly various stratigraphic columns, that happened to form the basement of the back-arc basin during the deposition period of the particular molasse.

Some researchers like Aubouin (1974) use the term «tarditectonique» for molasse, trying to describe the fact that it participates in the last stage of the orogeny, in contrast to

Fig. 4.3 View of the contact of the Upper Eocene flysch with the underlying Upper Eocene pelagic limestones of the Ionian unit at a few km from the front of the Pindos nappe in Tzoumerka. The change of sedimentary facies is impressive and creates a strong relief in the area, due to the differential weathering and erosion, evident by an abrupt change of color, also due to the different vegetation

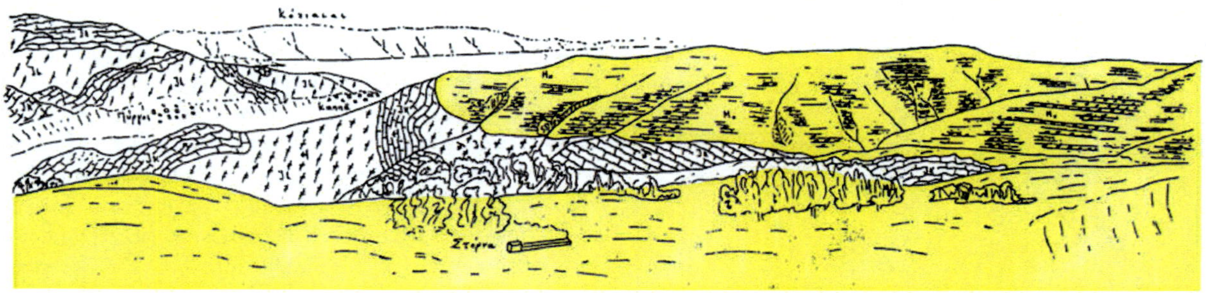

Fig. 4.4 A schematic panorama of the unconformable deposition of the sub-horizontal strata of the Oligocene—Lower Miocene molasse in the region of Kanalia—Pyrgos (Karditsa, central Greece), over the highly inclined to vertical strata of the Mesozoic limestones and cherts of the Western Thessaly unit at the western margin of the Meso-Hellenic trough in the area of Karditsa (from Papanikolaou and Sideris 1977, modified)

the **neotectonic** formations, which can be observed in the hinterland and are not genetically associated with the orogenic arc. Neotectonic formations include post-orogenic sediments above the Alpine formations. Based on the structure of the molasse in the Alps, Argand (1920) has distinguished, apart from the normal main folding of Eocene age, the early folding («plissements precurseurs») of Cretaceous age and the delayed folding («plissements tardifs») of Oligocene—Miocene age.

It should be noted that there are also cases of orogenic arcs without the presence of a back-arc basin, like the mountain ranges of the Andes, which give a most notable example. Additionally, there are some collisional mountain ranges, where instead of a back-arc basin, we have a continental plateau like Tibet north of the Himalayas. The creation of the back-arc basin and possibly of the marginal sea is related to the development of extensional stresses in the advancing plate behind the convergence/collision front

(Karig 1971). The extension associated with the creation of the back-arc basin or its opposite compression associated to the formation of the continental plateau depends on the relation between the convergence rate between the two plates and the subduction rate (Fig. 4.5) (Royden 1993).

Therefore, when the subduction rate is higher than the convergence rate, for example as is the case of the Hellenic arc, then tensile stresses are present and a back-arc basin is created, resulting in the so called **retreating plate boundaries**, while, on the other hand, when the subduction rate is less than the collision rate, compression stresses are present and a continental plateau is created, resulting in the so called **advancing plate boundaries**. The intermediate state, in which the two rates of convergence and subduction, are roughly the same, we refer to the **steady state plate boundaries**, with no back-arc basins or plateaus present. The retreating process of the Hellenic arc has been analysed since Eocene and the restoration model shows that the rate of

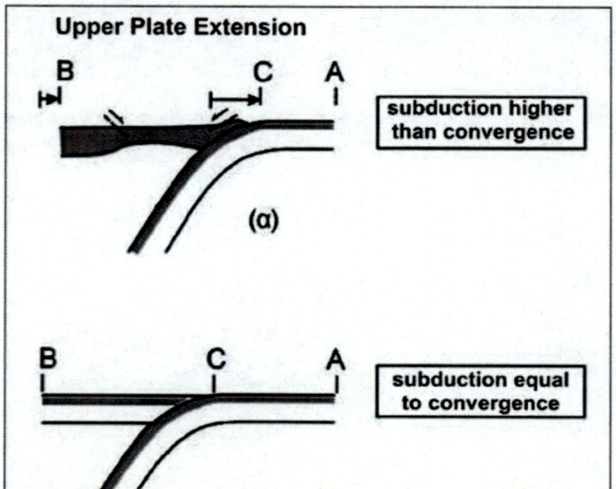

Fig. 4.5 The two cases of convergent zones with the convergence rate higher than the subduction rate, and vice versa. In the first case, extensional forces are created in the advancing plate that lead to the creation of a back-arc basin, while in the other, compression is present that leads to the creation of a plateau. In the intermediate scheme, the two rates are about the same and we have a steady state, with no back-arc basins or plateaus present (from Royden 1993)

the trench retreat was around 0.6 cm/yr during the first 30 Ma, from Late Eocene to the Middle Miocene, then it was accelerated to 1.7 cm/yr during Middle-Late Miocene and reached its maximum of 3.2 cm/yr during the last 5 Ma of the Plio-Quaternary (Brun et al. 2016). This procces of extension accommodated up to 600 km of trench retreat (Jolivet and Brun 2010; Jolivet et al. 2013).

Besides the typical flysch, there are some more specific cases of tectono-sedimentary facies, due to the proximity of the sedimentation area with intense tectonic activity. These are either: (i) olistolith and breccia-mylonite belts along steep subvertical cliffs, corresponding to large normal or strike-slip faults, or (ii) subhorizontal zones with breccias-conglomerates and olistoliths along the advancing front of the thrusts-nappes. The result is a chaotic image of the coarse clastic sedimentation, which is bibliographically known as **wild flysch**, usually comprising the upper part of the normal flysch deposits below the thrust surface of the overlying nappe (Masson 1976). A typical example in the Hellenides is the Tripolis wild flysch beneath the Pindos nappe (Fleury 1977, 1980; Lekkas 1978). It is impressive that the breccias due to friction "breches de friction" below the Pindos nappe in the Peloponnese have been described very early, already in 1909 by Negris.

Apart from the previous cases of flysch and molasse, there are intermediate stages of **flysch-molasse** formations, which correspond to deposits on the continental shelf of the advancing plate/island arc and are therefore unconformably deposited on their basement, while they are deformed under the compressive regime of the orogenic arc front (see also Sect. 5.1). The syn-orogenic sedimentation in the **foreland basin system** comprises an internal segment on the **orogenic wedge** developed above the advancing fold and thrust belt and especially the **wedge-top** depocentre, where flysch-molasse is deposited. This area is developed up to the compressive front of the wedge-top with the **foredeep** depocentre, where the maximum detritus is accumulated in the form of flysch and wildflysch (DeCelles and Gilles 1996) (Fig. 4.6). The outer part of the clastic sedimentation in the foreland basin may comprise the **forebulge** and **backbulge** depocentres, where the clastic material is minimum and wedges out towards the ocean. The forebulge area may be affected by bend faulting with normal faults (Ranero et al. 2005), creating small tectonic horsts, before entering the trench zone, as this is usually observed especially in cases of shallow platforms before their subduction (e.g. see transition to the flysch of the Tripolis and Parnassos platforms in Sects. 8.1.6 and 8.3.1).

The clastic material in the syn-orogenic sequences reveal the origin of the units participated at the island arc during the time of sedimentation. In general, the onset and the end of flysch sedimentation and the onset and the end of the molasse sedimentation allows us to date the tectonic

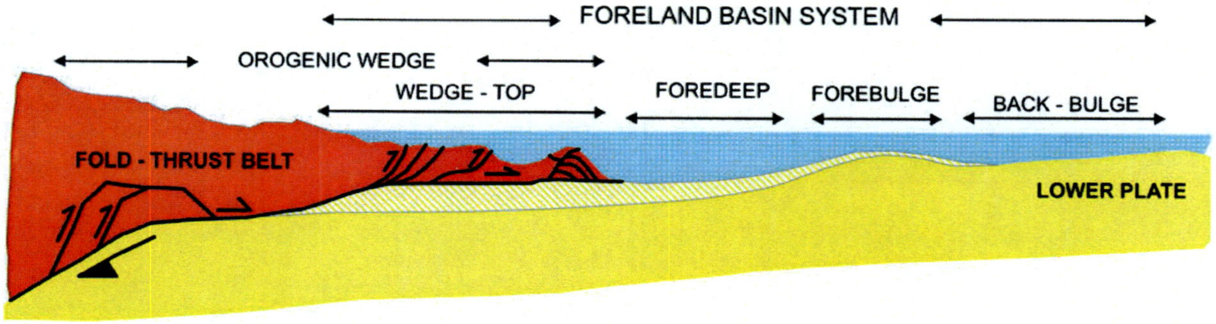

Fig. 4.6 Schematic cross section depicting the concept of the foreland basin system (from DeCelles and Giles 1996, modified)

phenomena along the orogenic arc and, at the same time, it can provide important time constraints in the palinspastic configuration of individual tectonic units in their pre-orogenic position (Papanikolaou 1986b, 1993).

The transition from the pre-orogenic, usually carbonate, sedimentation to the syn-orogenic clastic sedimentation of the flysch has some distinctive features, depending on whether the pre-orogenic paleogeography of the unit was neritic or pelagic. In the case of a pelagic unit (e.g. Ionian, Pindos) we have a characteristic gradual transition from biochemical pelagic to clastic pelagic sedimentation, while in the case of a neritic sedimentation (e.g. Tripolis, Parnassos) the transition is usually abrupt, often with the presence of unconformities and syn-sedimentary fault tectonics.

In the case of a pelagic pre-orogenic phase, we start with a calm environment with biochemical sedimentation, which at some geologic moment is interrupted by the arrival of clastic material in the form of turbiditic layers, which are deposited as interbeds within the carbonate sedimentation as well as intervals of marly limestones. There is no sudden change from biochemical to clastic sedimentation but a transitional period with alternative pulses of turbidity currents with deposition of clastic material and biochemical pelagic carbonate sedimentation. This creates an alternation between pelagic limestones and clastic material in the form of turbidites, which gradually show a decrease of the biochemical sedimentation and increase of the turbidites, until clastic sedimentation becomes dominant. The above transition from one sedimentation cycle to the other is characterized by the **transitional beds** («couches de passage») which may be several tens of meters thick, as described by Aubouin (1955) in the Pindos unit. In many occasions, the transitional beds are so thick that they can even be mapped as a separate stratigraphic formation.

On the contrary, on the shallow carbonate platform the transition to the flysch might take various forms. Sometimes we have normal stratigraphic boundaries, in which a change of facies is observed with marly transitional beds, a few meters thick, where pelagic foraminifers are found, which help to date them, like for example, Globotruncanes of the Upper Cretaceous in Parnassos and Globigerines of the Upper Eocene in Tripolis. In other times we have "abnormal" boundaries, with a disconformable or unconformable deposition of the flysch on the carbonate neritic basement, with presence of a paleorelief and syn-sedimentary tectonism, associated with ephemeral local emergences of the carbonate platform before the beggining of subsidence and the forthcoming flysch sedimentation (Richter and Mariolakos 1972, 1975; Papanikolaou and Lekkas 2001) (see also Figs. 8.22 and 8.64). These pre-flysch tectonic structures are related to the formation of bend faulting (Ranero et al. 2005) at the outer bulges observed in several cases in front of the trenches along convergent zones (DeCelles and Gilles 1996).

It is interesting to note that the drilling carried out in the Hellenic trench under the DSDP framework (Deep Sea Drilling Project), at a depth of 4,500–5,100 m, cut through 435 m of thick Pleistocene sediments overlying Lower Cretaceous neritic limestones, which seem to overthrust Upper Pliocene marls and marly limestones (Hsü and Ryan 1973). At the same time, the analysis of the samples from the DSDP drilling carried out in the Mediterranean showed the modern sedimentation of flysch, which is of marly composition in the Hellenic trench, while elsewhere, where large delta systems influence the clastic deposition, sandstones and locally conglomerates prevail (Stanley 1974).

ii. **Morphogenesis**: Apart from the creation of the trench and the back-arc basin, the morphology of the orogenic arc is characterized by the intense erosion and weathering of the rocks involved in the island arc. Here, the uplift of the mountains is proportional to the subduction angle and to the slip rate between the two plates. Typically, a mosaic of fault tectonic blocks is created, each of which moves relatively independently under the particular stress regime, resulting in the formation of tectonic grabens and horsts, with intensive deep erosion in the uplifted horsts and sediment deposition with creation of alluvial fans and cones in the subsiding

Fig. 4.7 **a** Characteristic raised coastlines by a few meters in southern Crete, showing tectonic elevation of the coastal block, resulted from a major earthquake Mw = 8.2 in 365 AD that uplifted western Crete up to 9 m. **b** In the same area, the view from the sea shows the newly elevated coastal zone of 2–3 m. The deep erosion of the canyon stops at the top of this recently elevated zone. Older morphological discontinuities observed on higher levels of the steep coast, show the successive stages of continuous uplift over the last tens of thousands of years

grabens. Marine terraces are formed along the coastal zones of uplifting tectonic blocks. Block tilting is often observed with the formation of stepped terraces, inclined planation surfaces and vertical motions of coastlines (Mariolakos and Papanikolaou 1981). Especially along the shores of the island arc, canyons are observed in elevated blocks, striving to balance the uplift with intense deep linear erosion, reaching the dynamic equilibrium level, usually near the present sea level. Coastline displacements are characteristic indicators of vertical motions, which can be often associated with specific earthquake events of the recent past (Fig. 4.7).

These phenomena are drastically degraded in the hinterland, where intense morphogenesis is restricted to a smaller number of active faults with lower slip rates. The morphology gradually acquires the image of mature relief.

iii. **Volcanism**: The volcanic phenomena are located along the volcanic arc area with either terrestrial or submarine volcanic structures (Figs. 4.8 and 4.9). Until 1987, research on active volcanos was limited to terrestrial outcrops along the Aegean volcanic arc. The acquisition of the oceanographic vessel "Aegeon" of the National Center for Marine Research (now Hellenic Center for Marine Research—HCMR), marked the launch of systematic oceanographic research in Greece. One of the first outcomes was the discovery of the Pafsanias submarine volcano in the Gulf of Epidaurus, north of the Methana peninsula (Papanikolaou et al. 1988, 1989; Pavlakis et al. 1989). In the following years several submarine volcanoes have been discovered and described along the Aegean volcanic arc (for a synthesis see Nomikou et al. 2013).

The volcanic rocks, as well as their plutonic counterparts in greater depths, are characterized by calc-alkaline chemical composition, especially when the subducted plate is oceanic, with andesites and rhyolites as the dominant rock types, as this was first demonstrated in the Mediterranean arcs by Ninkovitch and Hays (1972).

The volcanic rocks often intrude the molassic sediments of the back-arc basin, while sometimes they are found behind it towards the hinterland. In some cases, the volcanic arc develops in front of the back-arc basin, like the case of Japan, while there are cases of volcanic arcs without the existence of a back-arc basin, like the case of the Andes. The distance of the volcanic arc from the trench within the orogenic arc depends on the angle of the subduction zone and on the subduction rate, with main criterion the arrival of the subducted slab at temperatures above 700 °C, where melting occurs. Thus, a high subduction angle results into a narrow distance between the trench and the volcanoes, while a low angle leads to a longer distance. At the same time, high subduction rate results in a large trench—volcanoes distance, while a low rate leads to a short distance. Evidently, a high angle and high subduction rate leads to a volcanic arc proximal to the trench, while a low angle and low subduction rate would increase the distance between the volcanic arc and the trench. In the Hellenic molassic basins volcanism coexists in the North Aegean–Rhodope basin during the Late Eocene–Oligocene, but it is absent in the Meso-Hellenic Trough during the Oligocene–Middle Miocene (Papanikolaou 1993). In the current molassic basin of the Cretan Sea the volcanic arc is developed at the northern margin of the basin, towards the Cycladic plateau (see also the migration of the Hellenic volcanic arc in Fig. 10.8 of Chap. 10).

Dating of the orogenic calc-alkaline volcanism is usually achieved by both relative chronology geological methods, based on the age of the stratigraphic formations intruded by the

Fig. 4.8 Digital elevation model (DEM) of the island volcano of Nisyros, and view of the largest active crater Stefanos, of about 300 m in diameter and 30 m deep. The first geometry of the caldera is visible, created by the eruption of the first stratovolcano as well as the subsequent penetration and extrusion of the younger Prophitis Ilias lavas, which have created lava domes, forming the highest mountains of today and interrupting the circular structure of the previous caldera in the southwest (from Nomikou 2004)

volcanic rocks and the age of the transgressive sediments over the lavas, as well as by radiometric dating (e.g. Fytikas et al. 1976, 1984; Bellon et al. 1979; Pe-Piper and Piper 2002).

More detailed oceanographic research of the 2010s on the submarine volcanic centres has provided very important results both regarding the existence of submarine volcanoes and their volcanic history. Thus, a submarine volcanic chain of more than 22 volcanic craters or domes have been discovered along the NE-SW prolongation of the Kolumbo submarine volcano (Nomikou et al. 2012). For example, beneath the present Kolumbo volcanic crater, which has been built up during the 1650 AD disastrous eruption, with more than 70 casualties, 4 older eruptive phases have been detected within the Late Pleistocene/Holocene (Hubscher et al. 2015) (Fig. 4.10).

Additionally, detailed geological mapping both onshore and offshore of the Kammeni islands within the Santorini caldera, showed their submarine extension to a much wider area, with successive episodic lava flows and pyroclastic layers post dating the Minoan eruption 3.6 Ka ago, up to the recent eruptions of the twentieth century (Nomikou et al. 2014). A large break/landslide of the Minoan pyroclastic deposits along the northern rim of the caldera, where the present sea channel between Oea and Therasia occurs, created a huge post eruptive flooding of the previously formed caldera together with the disastrous tsunami generation (Nomikou et al. 2016).

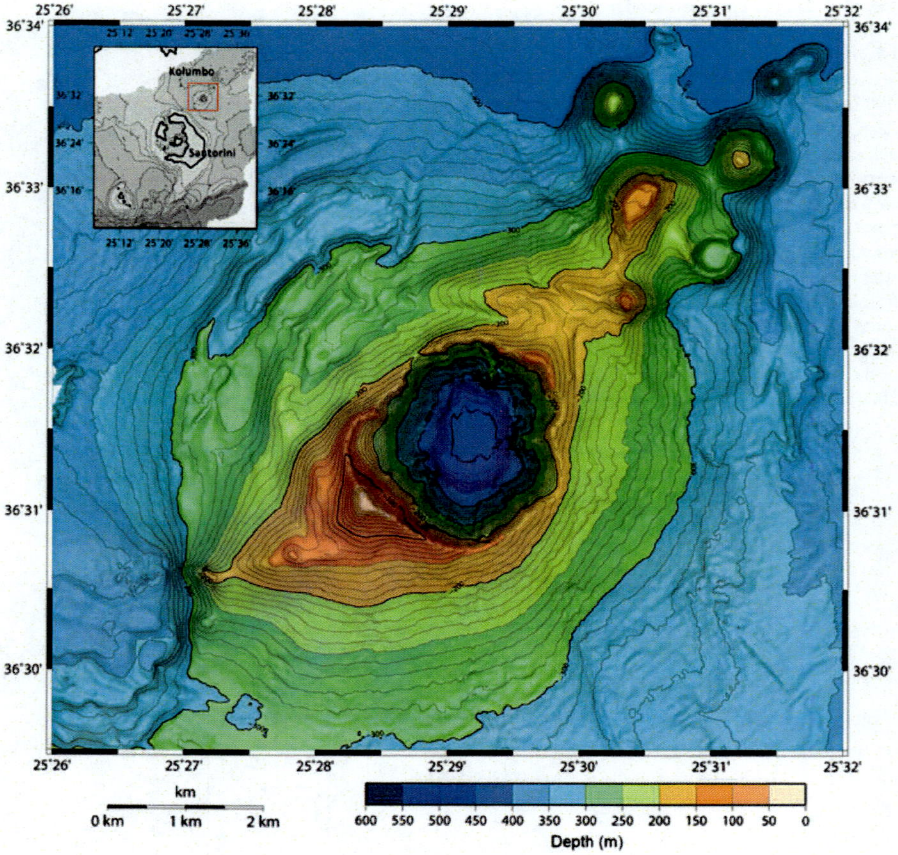

Fig. 4.9 Digital elevation model of the Kolumbo submarine volcano, northeast of Santorini. This is the only volcano whose last eruption in 1650 AD caused 70 fatalities. Its peak is now at a depth of about 15 m, while its crater base lies at a depth of 504 m (Nomikou et al. 2012), where hydrothermal vents with very important metalliferous deposits have been discovered (Kilias et al. 2013)

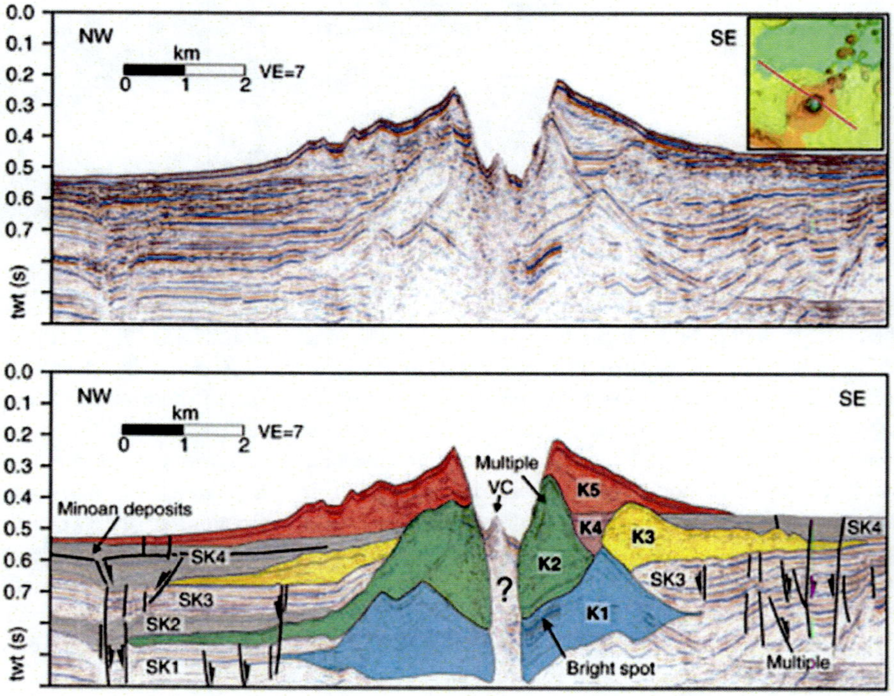

Fig. 4.10 NW–SE striking multichannel reflection seismic profile across the Kolumbo submarine volcano. Upper part shows seismic data, lower part shows interpretation. Grey shaded areas mark pyroclastic flows or mass-transport deposit. Coloured areas correspond to individual Kolumbo stratigraphic units/eruptions K1-K5, SK1-4 refer to intercalated units (from Hubscher et al. 2015)

iv. **Tectonic deformation**: The deformation observed in a transverse cross section of the orogenic arc is complex and multiform (see also Sects. 4.7 and 4.8). Therefore, both discontinuous deformation with fault surfaces and continuous plastic deformation with folding can be observed. The faults can be distinguished in: (i) an outer external zone of the arc, where compression prevails, comprising the trench and the outermost part of the island arc, where thrusts, overthrusts and reverse faults dominate (Fig. 4.11), and (ii) an internal zone, where extension prevails, comprising the inner part of the island arc and the back-arc basin, where normal faults are dominant (Fig. 4.12).

Folds are created mainly in the trench and island arc area, where compression prevails and their B axes are parallel to the trend of the arc and, naturally, transverse to the principal compressive stress $\sigma 1$, which is expressed by the direction of the plate convergence. In their overwhelming majority the folds are geometric, parallel, flexural slip folds with no significant changes of the thickness of the folded beds (Fig. 4.13). Gravity phenomena often occur when the lithology and inclination of the tectonic thrust surfaces and faults are favourable, combined with high liquidity. These phenomena generate landslides and slope instabilities in tectonically elevated areas, such as the Northern Peloponnese.

v. **Seismicity**: A zone of high seismicity is recorded from the trench area towards the interior of the arc, with a gradual weakening towards the hinterland (see also Fig. 3.6). Focal mechanisms reveal that reverse faults are activated in the trench area and at the front of the island arc (Kiratzi and Louvari 2003; Shaw et al. 2008), while in the inner part of the island arc in Crete and the back-arc basin normal faults are activated (Caputo et al. 2010; Mason et al. 2016; Lyon-Caen et al. 1988). At various locations of the arc and especially in the hinterland focal mechanisms of strike-slip faults can be observed (see also Fig. 10.6.9 in Chap. 10). In the Hellenic arc strike slip faulting can be observed at the northern boundary of the arc at Cephalonia and at the northern boundary of the Aegean microplate along the North Aegean Basin. It is interesting to note that the 1953 Mw = 7.3 earthquakes in the Ionian islands of Cephalonia and Zakynthos offered the opportunity for the first extraction of a focal mechanism in Greece and the determination of the direction of the seismic fault by Hodgson and Cock (1957). Seismicity in the arc is generally shallow, with the exception of the seismicity along the subduction zone/Benioff zone, which will be examined later in the deep geodynamic phenomena. Shallow seismicity is essentially limited to the upper crust of the advancing plate and is mainly manifested within the seismogenic layer at depths up to 10–18 km (Maggi et al. 2000). This depth depends on the thickness of the crust, the type of the stressfield and the corresponding faults, as well as the thermal flow in the crust. In rare cases of large earthquakes at sea, **tsunamis** are generated, like the 1956 Mw = 7.4 magnitude

Fig. 4.11 Characteristic overthrust of the Pindos nappe, made of Globotruncana bearing Upper Cretaceous pelagic limestones (3) over flysch (2) and overlying Eocene neritic limestones with Nummulites (1) of the Tripolis carbonate platform in Arcadia. The horizontal tectonic contact is associated with an overall tectonic transport of the nappe in the order of a few hundred kilometer

Fig. 4.12 Typical normal fault in the region of Pisia in Perachora, which generated a Mw = 6.7 magnitude earthquake in 1981, with a displacement of 0.80–1.00 m. This is the southern marginal fault that creates the North Corinth basin in the Alkyonides Gulf, with a fault throw of more than 500 m during the Pleistocene

earthquake of Amorgos, with sea waves up to 20 m (Galanopoulos 1957; Ambraseys 1960; Okal et al. 2009). Today we know that this phenomenon can also be caused by major submarine landslides, especially around steep volcanic slopes suffering gravitational collapse, triggered by strong surrounding earthquakes. The first description of tsunamis due to submarine landslide in Greece was reported in the Gulf of Corinth (Galanopoulos et al. 1966). However, in the case of Amorgos 7–9 m of recent fault scarps were traced in the submarine fault of Amorgos but no significant submarine landslides were observed (Nomikou et al. 2018).

4.5 Deep Geodynamic Phenomena in the Orogenic Arc

The subduction of segments of the lower plate at successive depths of 10, 20, 30 km and further deep is accompanied by a series of transformations, related to deep ductile deformation and metamorphism. All these processes contribute to the **tectono-metamorphic evolution** of each tectonic unit and shape its future stratigraphic column. The underground journey of each unit, from its disappearance at the trench in the subduction zone to its reappearance at some surficial position along the arc, can be distinguished in a descending curve down to its maximum subduction depth and an ascending curve from the maximum depth back to the surface. In general, the transition of a given unit from the subducting plate to the interior of the advancing plate can be accomplished by a number of different trajectories, such as (Fig. 4.14): (1–1′) with an almost superficial accretion to the front of the advancing plate in the island arc within the upper 5–10 km, (2–2′) with subduction to moderate depths of 20–30 km and rapid return at the surface of the island arc, often in the form of a tectonic window, through extensional detachments, (3–3′) with subduction to great depths of 50–60 km and return at the surface either in the back-arc basin or the hinterland, in the form of tectonic windows forming a metamorphic core complex, (4–4′-4″) with subduction to great depths and remaining in the lithosphere beneath the upper plate, with or without transition to the stage of anatexis and destruction of the original protolith. It should be

Fig. 4.13 Successive angular geometric folds observed in the Upper Cretaceous platy limestones of the Pindos unit in Agrafa. The general direction of the folding is N–S and shows E–W horizontal compression, perpendicular to the general direction of the Hellenic arc in western Greece

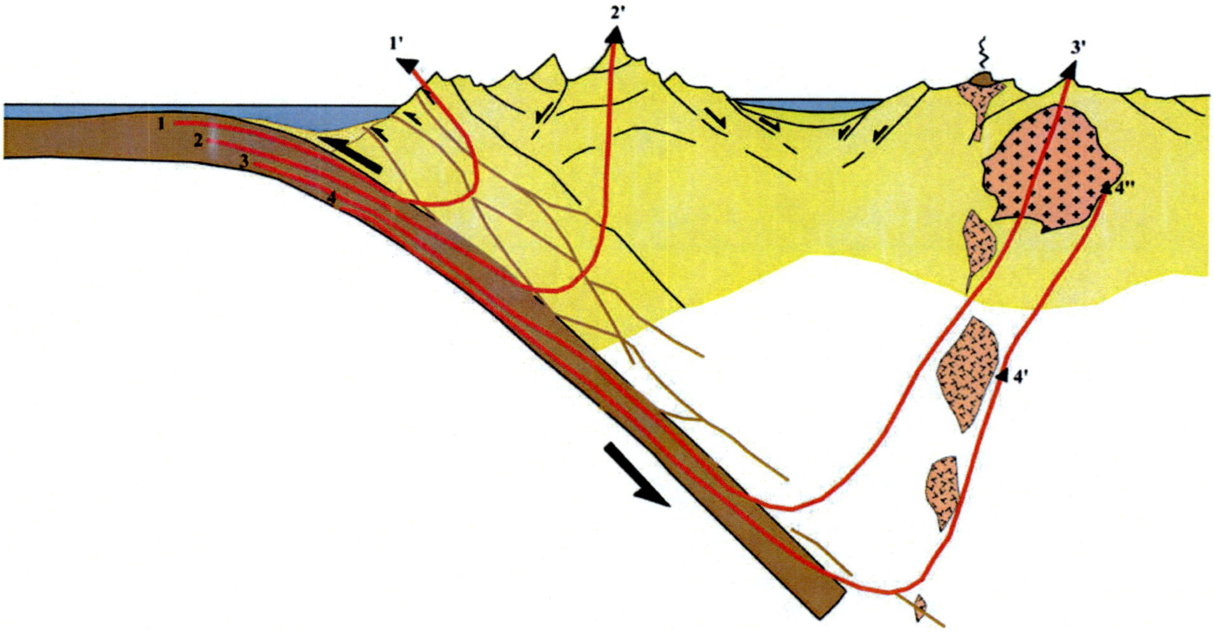

Fig. 4.14 Schematic representation of the possible paths/trajectories from the subducting to the advancing plate across the orogenic arc, depicted in four possible cases (explanation in the text)

noted that even in the case of a rapid uplift of a unit at the front of the upper plate there is still the possibility of its complete elimination (if it is not a very large unit) due to erosion. The above cases allow us to consider what percentage of the pre-orogenic paleogeographic organization of the subducting plate can be actually restored from the study of the outcrops in the orogenic arc at the advancing plate.

Deep deformation in the orogenic arc follows the general change of tectonic structures with depth, according to the changes of the deformation mechanisms, characterizing the so-called tectonic levels («etages tectoniques» or «niveaux structuraux») (Mattauer 1973). Deformation at depth is characterized by only sparse fault structures with a full dominance of plastic deformation with similar folds related to flow along new tectonic s-planes, along which cleavages or schistosities develop (Fig. 4.15).

The few fault structures found at depth are low angle thrusts and overthrusts, mainly along the subduction zone. This zone, starting from the trench is usually diping with 15°–45° below the upper plate, known as Benioff seismic zone (although it would be better named Wadati–Benioff zone, since Wadati had described it already in 1935 in Japan). This deep seismic zone can be detected up to 700 km depth (as in the Pacific arcs in Japan or Tonga or beneath the Andes) which is attributed to the rapid penetration of cold dense oceanic lithosphere in the mantle, where it continuous to be elastically deformed as in shallower depths. In the case of the present Hellenic arc **deep-focus seismicity** had been reported already in the 1940s by Gutenberg and Richter (1948, 1954), with a maximum depth of the Benioff zone at about 180–200 km. It is characteristic that all the major intermediate depth earthquakes of the Hellenic arc are confined within the Aegean microplate, bordered to the north by the arc boundary running along Preveza–North Aegean Basin, and to the east by the Asia Minor coastal zone (Papazachos and Comninakis 1982; Papazachos et al. 2000) (Fig. 4.16).

The deep-focus earthquakes show compressional focal mechanisms and correspond to reverse faulting, with a slip direction approximately perpendicular to the direction of the arc (McKenzie 1978; Jackson 1994; Taymaz et al. 1991). Additionally, some focal mechanisms corresponding to strike-slip faults are observed locally, probably accommodating the differential motions, due to the geometry of the arc curvature. A more detailed seismic analysis of the subduction zone along the western segment beneath Peloponnese showed the change of the dip angle from 17° to about 30–35° at about 70 km depth, indicating the roll back nature of the subduction zone (Suckale et al. 2009; Gesret et al. 2010; Sachpazi et al. 2016). Additionally, it showed a segmentation of the slab into dipping panels by along-dip faults of NE-SW orientation with systematic offsets to the deeper panels towards ESE, where the dip angles become higher (Sachpazi et al. 2016) (Fig. 4.17.). This disrupted geometry of the subducting slab may account for the usual thrusting interplate earthquake ruptures (e.g. the Mw 6.8 Methoni

Fig. 4.15 Recumbent almost isoclinal fold of km scale, facing south, within the marbles of the Mani unit from Central Crete (from Papanikolaou and Vassilakis 2010). This structure was formed at moderate depth of 10–15 km at the intermediate tectonic level

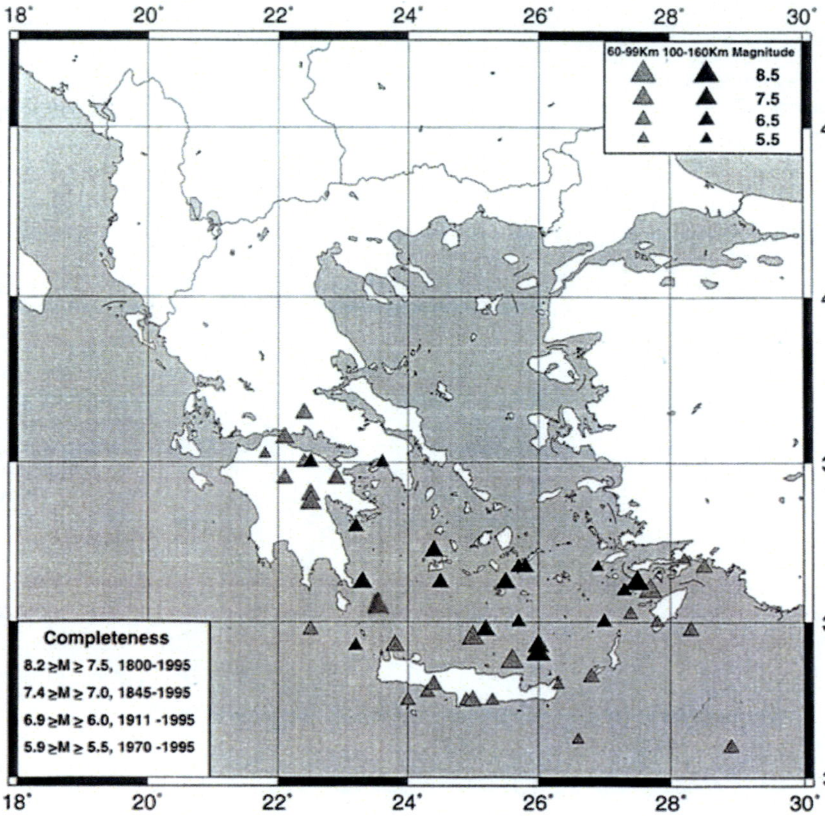

Fig. 4.16 Distribution map of intermediate depth earthquakes in the Hellenic arc (based on Papazachos and Comninakis 1982). The delimitation of the epicenters inside the Hellenic arc, below the Aegean microplate, where the Hellenic subduction zone is confined, is evident

2008, event) as well as for the intermediate slab earthquakes with right-lateral horizontal shear (e.g. the Mw 6.4 Achaia-Movri 2008, event).

The tectonic structures of the deformation at depth consist of similar, usually isoclinal folds, with flow parallel to the axial planes of the folds and transport direction towards their hinges. In cases where the subducted unit does not reach great depths, the B axes of the folds remain parallel to the general trend of the orogenic arc and to the B axes of the folds observed in the sedimentary rocks of the island arc at shallow level (e.g. the structures of the Helvetics in the Alps and of the Mani unit in the External Hellenides). On the other hand, when the subducted unit continues at greater depths, folds with transverse axes to the orogenic arc are observed, which may correspond in terms of kinematics to a-structures (Papanikolaou 1981, 1987; Rodgers 1984) (Fig. 4.18).

These deep-level structures are parallel to the slip direction of the subduction zone, by reorienting of the previous surface b-structures and with a syn-kinematic development of the fiber and phylloid minerals of the blueschist assemblages (mineralogic lineations parallel to the axes of the similar folds and intersection lineations). This structure is known as cross folding, characterizing the almost right angle between the deep-level folds (e.g. Cyclades) and the superficial folds (e.g. Pindos) formed during the same period (Late Eocene) in the arc. The arc parallel NW–SE structures (in green) observed in the metamorphic rocks of the Pelagonian and the Cyclades are later shallow structures (Miocene), superposed on the previous (Eocene–Oligocene) deep NE–SW structures (in red) (Fig. 4.18a). In any case, the dominant isoclinal folding of the Cycladic blueschists is characterized as extension—parallel folding with stretching lineations along the extension axis (Avigad et al. 2001).

In the case of units subducted to great depths, there are usually three or four **deformation phases** (Papanikolaou 1977, 1978) (Fig. 4.19): (i) an early syn-metamorphic deformational phase with transverse possibly a-kinematic structures, (ii) a younger syn-metamorphic phase with longitudinal b-kinematic structures, and (iii) a recent deformation phase, post-dating all the metamorphic events, including parallel b-kinematic type kink folds and various forms of brittle deformational fabric.

The underground journey of a given tectonic unit through time from its pre-orogenic position/period to its present

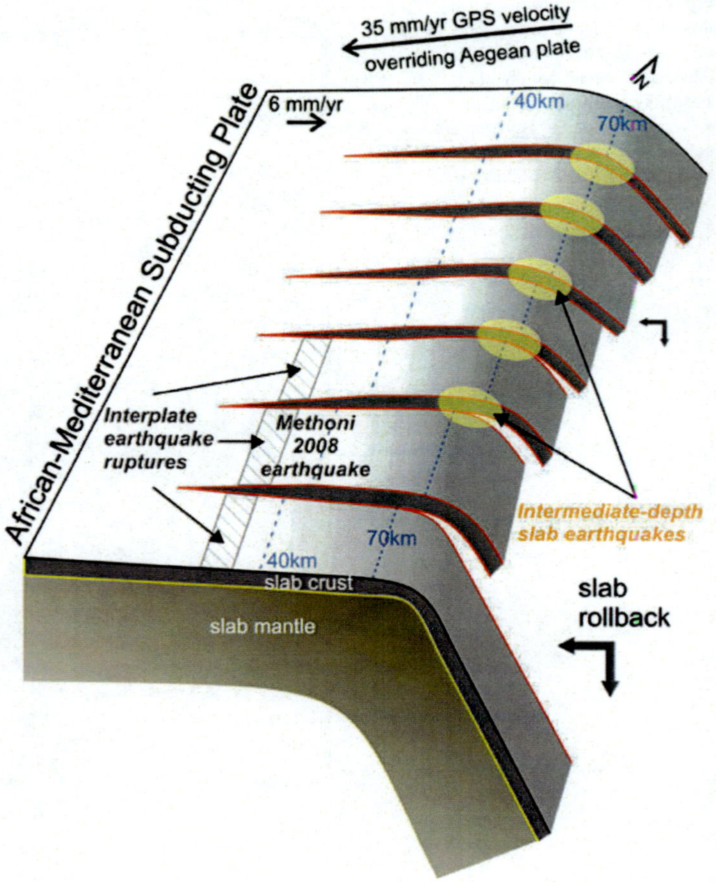

Fig. 4.17 Schematic diagram seen from the SE of the western Hellenic slab structure beneath Peloponnese and related seismicity (after Sachpazi et al. 2016)

location can be deciphered by the distinction of the number of deformation phases and their correlation with metamorphism, in combination with their kinematic interpretation (see also Fig. 4.26.).

Metamorphism in an orogenic arc area is characterized by a double metamorphic phase for each subducted unit, with a first subduction period along the Benioff zone, where the combination of high pressures and low temperatures forms the **blueschists** (Blake et al. 1969; Coleman 1972) (Fig. 4.20). A second period of uplifting/exhumation, usually in the same area with the rising magma due to the melting of the subducted lithosphere under the superficial volcanic arc, where the combination of low pressures and high temperatures forms the **greenschists** (Fig. 4.21a). This phenomenon of having two different metamorphic processes of P/T conditions in the same unit, but at different positions and different times, first described in Japan, produces inside the orogenic arc the so-called **paired metamorphic belts** of Miyashiro (1972, 1973).

Herein, it should be clarified that both deformation and metamorphic events are continuous and gradually the rocks evolve from one stage to the next, resulting in the so-called **progressive deformation** and **progressive metamorphism**. Therefore, the younger events occurring in smaller depths usually tend to obliterate the older ones that had been formed at greater depths, which can be retained only when the uplift is relatively rapid. Especially in cases where the unit passes near the ascending granitic magmatic bodies, high grade metamorphic gneissic rocks and amphibolites of the amphibolite metamorphic facies are created (Fig. 4.21b) and then migmatitic phenomena may follow during the initial stages of anatexis. The structures of this period are characterized as flow structures and the geometry of the previous tectonic planar flow is destroyed by flow folding, with curved fold axes and curved non-geometric axial surfaces of folds (Fig. 4.22).

Both deep-level deformation and metamorphism are dated by classic methods, when the age of the younger subducted formations can be determined (e.g. by finding non-destroyed fossils or by litho-stratigraphic correlations), as well as by radiochronology of the new metamorphic minerals and determination of the link between deformation phase and metamophic phase in each case.

The final metamorphic phase related with the greenschists often coincides with the intrusion of granites—granodiorites,

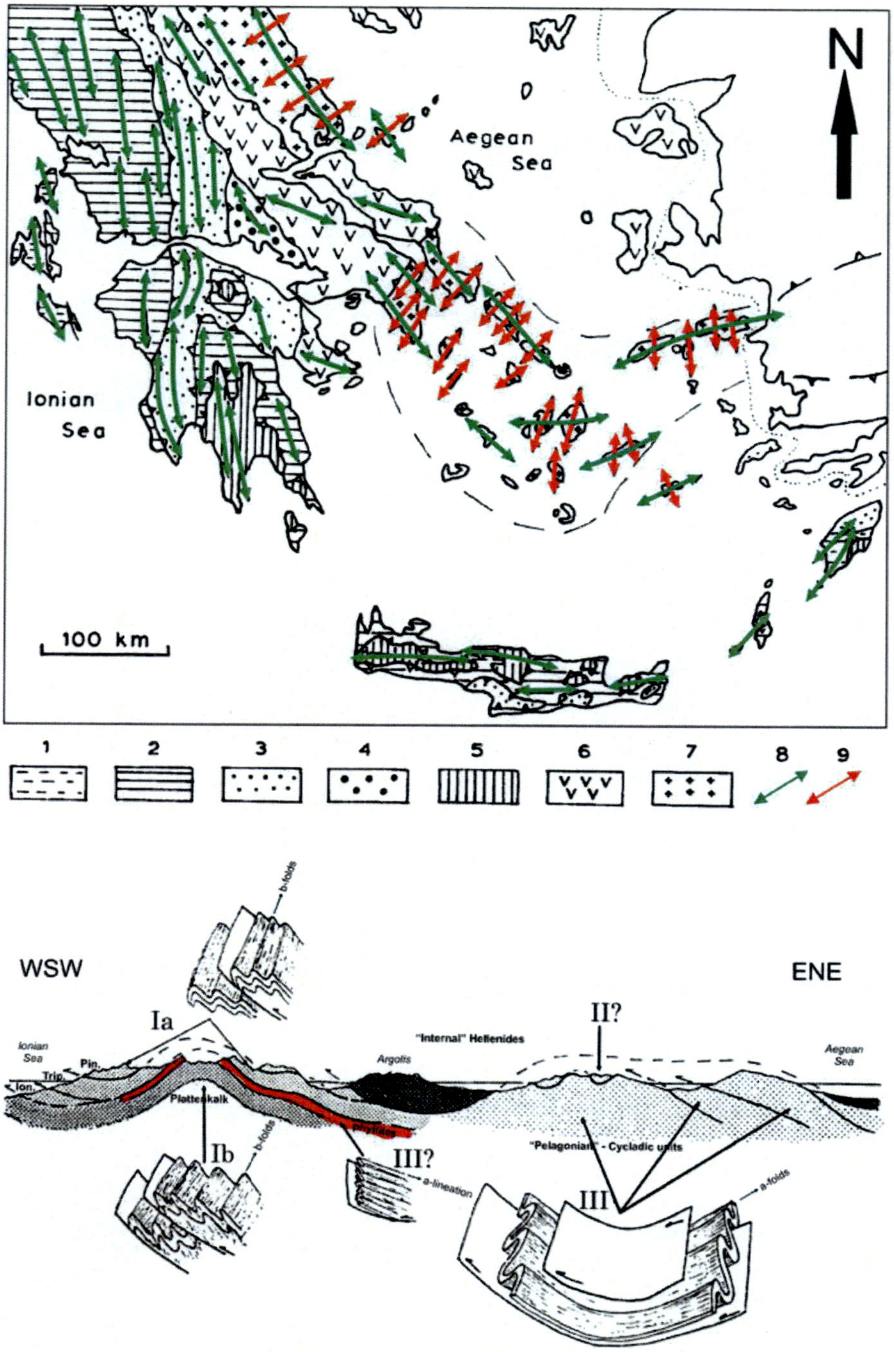

Fig. 4.18 **a** Simplified tectonic map of Greece showing the transverse to the arc deep-level folds of kinematic type a with their fold axes shown in red and the parallel to the arc superficial folds of type b with their fold axes shown in green (from Papanikolaou 1981). 1: Pre-Apulian, 2: Ionian and Gavrovo–Tripolis, 3: Pindos, 4: Parnassos, 5: Plattenkalk (Mani) and Phyllites (Arna), 6: Internal Hellenides, 7: Pelagonian and Cycladic units. **b** Schematic transverse cross section of the Hellenides through Peloponnese–Attica–Northern Cyclades, showing the structural style from a kinematic point of view in the various deep and superficial units (from Papanikolaou 1981). The classical parallel structures of the Hellenides type Ia correspond to B-structures within the non-metamorphosed units of Ionian, Pindos and Eastern Greece. The slightly metamorphosed Mani unit exhibits schistosity/cleavage along the axial planes of the folds of type Ib. In contrast, the metamorphosed Attica, Cyclades, and Arna units are characterized by transverse a-structures of type III corresponding to deep-level structures. In the slightly metamorphosed allocthonous nappe of Athens, we have an intermediate case of structures of type II, with coexistence of longitudinal and transverse structures

Fig. 4.19 Typical tectonic structures of the three major deformation phases from the metamorphic rocks of Andros, within the Northern Cyclades unit (from Papanikolaou 1977). **a** Isoclinal syn-metamorphic folds of phase A in blueschists. **b** Asymmetric microfolds of phase B in greenschists (Lcf), which deform the previous lineation (Lsf). **c** Conjugate system of kink folds (P1, P2) of phase C that deform all previous structures, comprising simple flexures post-dating the metamorphic events

resulting from the crystallization of calc-alkaline magmas in the area of the **magmatic arc**, which is formed underneath the volcanic arc. The ages obtained by the dating of these magmatic rocks correspond to the end of the tectono-metamorphic cycle acting transversely to the arc.

The overall **tectono-metamorphic—magmatic evolution** of some metamorphosed units of the Hellenides is provided in typical depth/temperature diagrams, which contain information regarding both the metamorphic assemblages and the tectonic microstructures (Jacobshagen 1986) (Fig. 4.23). In these diagrams the ages of the successive events are also displayed and thus, it is possible to estimate the uplift rate to the surface.

Thus, the Naxos metamorphic rocks seem to have been at a maximum depth of 50 km in the Eocene ($\sim$45 Ma), at temperatures of 450° C and under conditions of blueschist metamorphism, and then in the Late Oligocene–Early

Fig. 4.20 Blueschist assemblage from Syros, showing a major schistosity under the microscope. The glaucophane crystals can be distinguished together with phengite, quartz, zoisite, calcite, and opaque

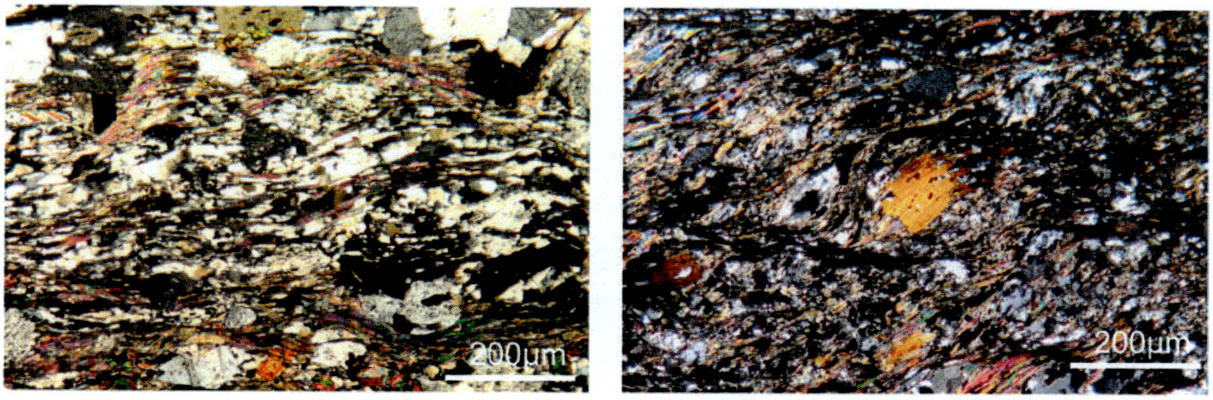

Fig. 4.21 **a** Greenschist assemblage under the microscope, coming from the Andros micaschists. The minerals albite, quartz, muscovite, and epidote are present. **b** Assemblage of amphibolites under the microscope, forming an initial schistosity, from the margin of the migmatite dome of Naxos and a subsequent greenschist assemblage forming a subsequent cleavage. The high temperature minerals plagioclase, quartz, white mica, and a large cyanite crystal can be observed, while the minerals chlorite, quartz, muscovite, and biotite have grown later

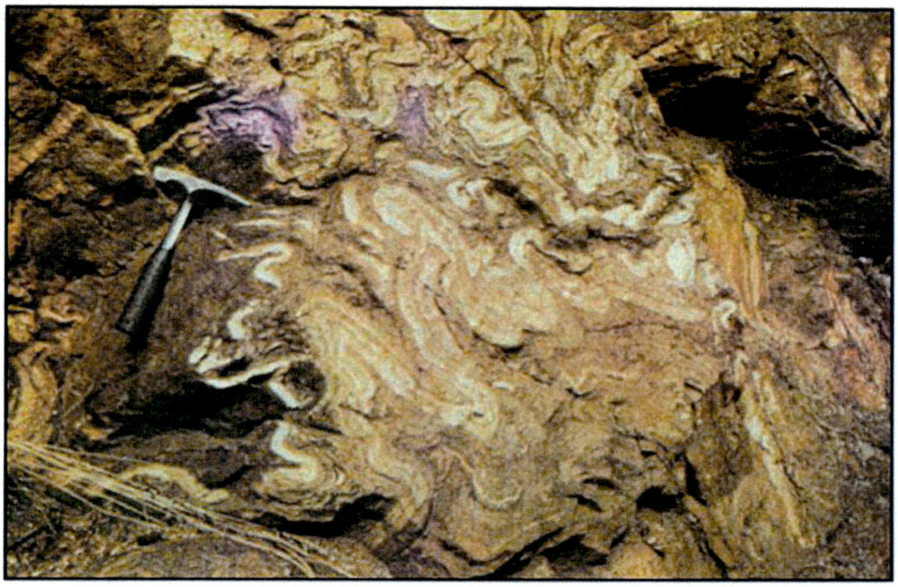

Fig. 4.22 Flow folding in migmatitic gneisses from the Sidironero nappe in Rhodope

Miocene (~25 Ma), they ascended to a depth of 20–25 km, with greenschist metamorphism and migmatization around magmatic intrusions, with a late position in the Late Miocene (~11 Ma) at 5–10 km, where the granodiorite intrusion of Western Naxos occurred. The last stages of uplift/exhumation of the metamorphic rocks at the surface are determined more precisely by the newly developed **thermochronology** methods, which focus on the tectonic contacts of extensional detachments (Thomson et al. 1998; Brichau et al. 2006; Hejl et al. 2008; Seward et al. 2009; Laurent et al. 2018).

An interesting formation, associated with deep-level deformation and metamorphism, is the **tectonic mélange** (or just mélange) (Bailey and McCallien 1950; Hsü 1968), which usually corresponds to wild flysch formation, acquiring deep tectonic structures and metamorphism during its subduction at several tens of kilometers depth (Ernst 1970). Therefore, the various lithologies appear in a chaotic complex resulting from a primary chaos during sedimentation of the wild flysch and then a secondary chaos through the ductile deformation and metamorphism with tectonic flow dominating everywhere.

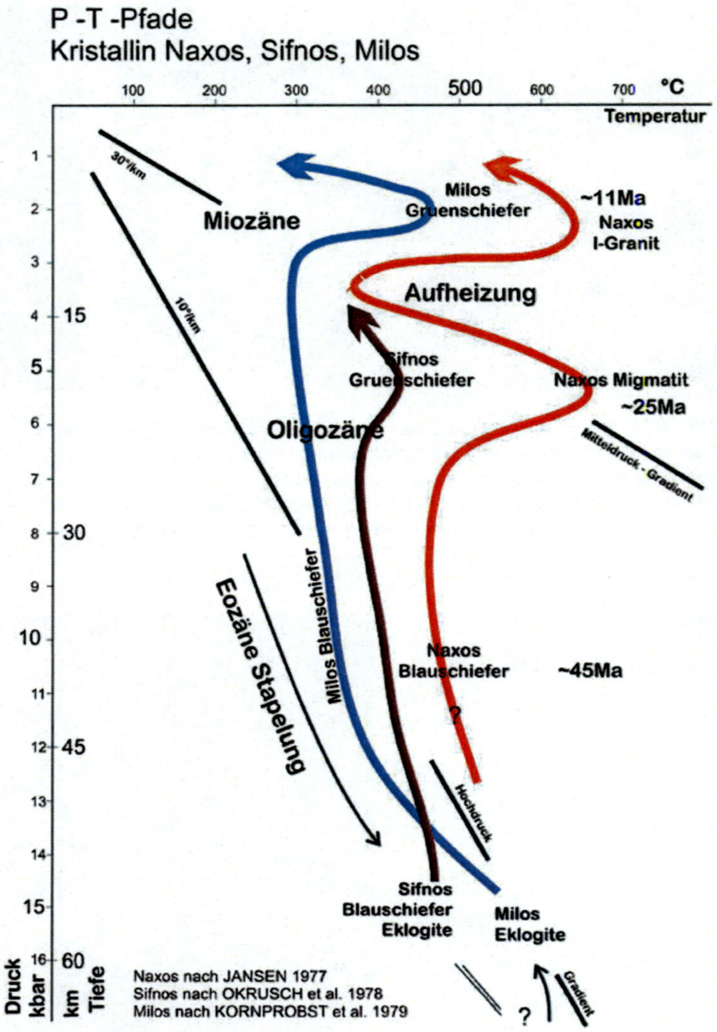

Fig. 4.23 Diagrams of pressure/temperature conditions during the successive metamorphic events of the Cycladic units in the arc, based on mineralogical assemblages and ages of metamorphic—magmatic events from the islands of Naxos, Sifnos and Milos (from Jacobshagen 1986, modified). The exhumation process of each unit is described by the characteristic phases that have been dated during its tectono-metamorphic evolution

In several cases, the term mélange is used when wild flysch is overlain by an ophiolite nappe. In this case it contains rocks of the ophiolite complex either as blocks or as matrix and is called **ophiolite mélange** (Gansser 1974). When additionally to ophiolitic rocks it contains other rocks, such as carbonate and silicate meta-sediments together with meta-clastics, a chaotic structure is born, characterized by colorful outcrops, known as **colored mélanges («mélanges colorés»)** (Fig. 4.24). In Greece typical cases of mélanges have been reported mainly by Mercier and Vergely (1972a, b) and Mountrakis and Soulios (1978) in the Almopia area, although the well-known "schist-sandstone-chert formation with ophiolites", of Malm age, generally corresponds to mélanges, but without metamorphism (Voreadis 1932; Renz 1955; Tataris 1975; Papanikolaou 1990). However, the term mélange may also be used (and has been used by many researchers) for other cases of chaotic outcrops in complex structures, such as for example in regions that experience intense diapir phenomena of evaporites.

Finally, the deep lithospheric tectonic processes have been studied more recently on the basis of **mantle flow**, as the major driving mechanism of plate movement within the subduction system, both beneath the subducting plate and within the mantle wedge between the two plates (Royden and Husson 2006; Faccenna et al. 2003; Evangelidis 2017). The lateral inhomogeneities due to the edges of subduction, relative to non subducted margins or relative to slower subduction systems may result in slab tears, where toroidal flow may affect the primary flow along the subduction axis (Govers and Wortel 2005; Guillaume et al. 2013; Jolivet et al. 2015). Additionally, in several cases, slab break and slab detachment may change the overall subduction parameters (Wortel and Spakman 2000), due to the major influence of the slab pull exerted to the subducting plate (Royden 1993).

Fig. 4.24 Multicolor ophiolitic mélange of Upper Jurassic age, observed under the Upper Cretaceous transgression in the Sub-Pelagonian unit from the region of St. Ioannis Mazarakis monastery in Beotia

4.6 The Issue of Continuity or Discontinuity of Folding in Orogenic Events—Stratigraphic Unconformities

Folding of the rocks is considered to be a distinctive feature of orogeny, and thus the classification of various orogenic events through time was based on the age of folding, occurring in a type locality, usually dated by the overlying unconformable sediments. Therefore, for example, the observation of the stratigraphic angular unconformity between the overlying Cenomanian strata and the underlying older folded Triassic-Jurassic formations was characterized as Austrian folding, since this unconformity was first described in Austria. Thus, the terminology established by Stille (1924, 1936) was globally used for folds of the Caledonian, Variscan and Alpine orogeny, whereby, for example, the Pyrenean folding imply that folding occurred in Late Eocene. In Greece, several periods of folding had been described along with their corresponding unconformities (Psarianos and Manolessos 1963; Maratos 1972) from the earliest Neo-Cimmeric folding (Voreadis 1938), the Austrian, with the characteristic Gosau transgressive layers, known from Austria (Mitzopoulos 1959), Pyrenean (Mitzopoulos and Trikkalinos 1937; Voreadis 1938), the Styrian (Trikkalinos 1939), as well as the younger folds and unconformities like the Attican, Wallachian and Pasadenian (Trikkalinos 1940).

Apart from the fact that this terminology adds nothing, since the classification of folding can be done directly through the usual geochronological scale, e.g. Early Miocene folding, without the use of new hard-to-remember terms, there arises also the main issue of whether or not the orogenic events globally or regionally occurred at the same period, with silent periods in between. This is the general question of whether the orogenic processes are continuous or result at specific time periods with intermediate periods of quiescence in between. The geotectonic frame of orogeny within the plate tectonics theory has definitely shown that plate convergence and hence orogenic deformation with folding is a continuous process and thus, the previous terminology of Stille has been abandoned. Moreover, if the table of orogenic events had been enriched with the data from the last fifty years of research, the number of folding episodes would have been multiplied and there would be a folding phase for almost all geological periods.

However, it is true that in some segments of the Tethyan orogenic system, intense orogenic phenomena of a certain period can be observed, while in adjacent segments there is no such observation. A typical example of this difference concerns the Cenomanian unconformity in the Hellenides, which is absent from the neighboring Taurides, where in turn the Maastrichtian unconformity occurs, which is absent in the Hellenides (Gutnic et al. 1979). This phenomenon was interpreted to be the result of the different direction and rate of the plate motion on either side of the existing right angle of the orogenic arc between the Hellenides and the Taurides. Thus, it is possible to have intense axial compression with folding on one branch but diagonal or lateral sliding on the other (e.g. pre-Cenomanian) and vice versa in another time period (e.g. pre-Maastrichtian). This phenomenon was called coin tectonics («Tectonique de coin», Ricou 1980), but it was based on the wrong assumption that the right angle of the Hellenic arc existed already in Late Cretaceous, whereas, as already explained, the Hellenic arc obtained its right angle curvature during the last 13 Ma period (LePichon and Angelier 1979, 1981).

The lateral differences on the timing of orogenic events along the Tethyan Alpine belt may well be explained by the existence of the tectono-stratigraphic terranes. The terranes during their accretion to the Eurasian margin produce different spatio-temporal orogenic phenomena and corresponding unconformities along Tethys, due to their lateral wedging out (Papanikolaou 2009). The eastward lateral transition of the terranes inside the Tethys Ocean is characteristic, when comparing the Hellenic terranes with the terranes in Asia Minor, occurring along the collisional zone between Arabia and Caucasus (Goncuoglu et al. 1997; Moix et al. 2008). Additionally, there is the problem of the "lost" elements either due to subduction—anatexis, or due to rapid erosion in the front of the island arc (see also Fig. 4.14), which may not allow the determination of folding of a certain period of time in a given orogenic segment. Finally, there is another important element concerning the apparently different timing of orogenic processes in adjacent segments of the orogenic system, which may be associated with: (i) the existing pre-orogenic paleogeographic organization and (ii) which crustal discontinuities were activated or created during the migration of the orogenic arc (see also Fig. 4.2). Thus, the front of the orogenic arc may change from a d2 detachment to the next, or stay at the same surface d2 for many millions of years. Therefore, each case will create a different orogenic timing in the adjacent parts of the orogenic system.

In Greece, we can distinguish: (1) **the orogenic unconformities** of the Alpine cycle, dividing the superjacent post-orogenic sediments from their Alpine basement during the Miocene–Quaternary and (2) **the intra-orogenic Alpine unconformities**, which include, apart from the major Cenomanian unconformity, the two older Jurassic unconformities of Malm and Lias.

4.7 Distribution of Stress Fields and Tectonic Structures in the Arc

The orogenic arc is generally characterized by the compression dominating along the convergent zone of the two plates. However, different stress fields do exist along individual sectors of the orogenic arc, generating different types of tectonic structures. Thus, in the trench area and the front of the island arc, compression produces extensive folding, with their axes parallel to the arc and almost perpendicular to the compressive stress, and reverse faults, thrusts and overthrusts, trending, again in the same direction. The above compressive structures are syngenetic, products of the same stress field, created on the upper tectonic level, forming the tectonic entity of a fold and thrust belt. The tectonic asymmetry of these compressive structures shows a tectonic transport from the advancing plate to the subducting one, and the axial planes of the asymmetric folds dip, like the thrust planes, towards the volcanic arc at the interior of the arc. On the contrary, in the area of the volcanic arc, the internal part of the island arc and the back-arc basin, the dominating extensional stress field creates normal faults, whose direction is perpendicular to the extensional stress and parallel to the directions of the previous compressive structures of the outer belt. The above arc structures alltogether, both of the compressive stresses in the front of the arc and the extensional stresses in the back arc area, are the so-called **arc parallel structures**, which are directly dependent on the existence and geometry of the arc.

The dominant tectonic structure in a cross section of the Hellenic arc is the subduction zone of the oceanic plate beneath the advancing continental plate with an angle varying between 10° and 50°, with the most common value being 35°. The geometry of the subduction zone determines the geometry of the shear zone along the compressive tectonic contact of the two plates and the resulting analysis of the normal and shear stress (Fig. 4.25). The horizontal compression zone with parallel shear along the tectonic contact of the two plates continues to a great depth of two hundred kilometers, as evidenced by the deep-focus earthquakes of the Benioff zone and their corresponding focal mechanisms, showing thrust faults with their slip parallel to the direction of convergence.

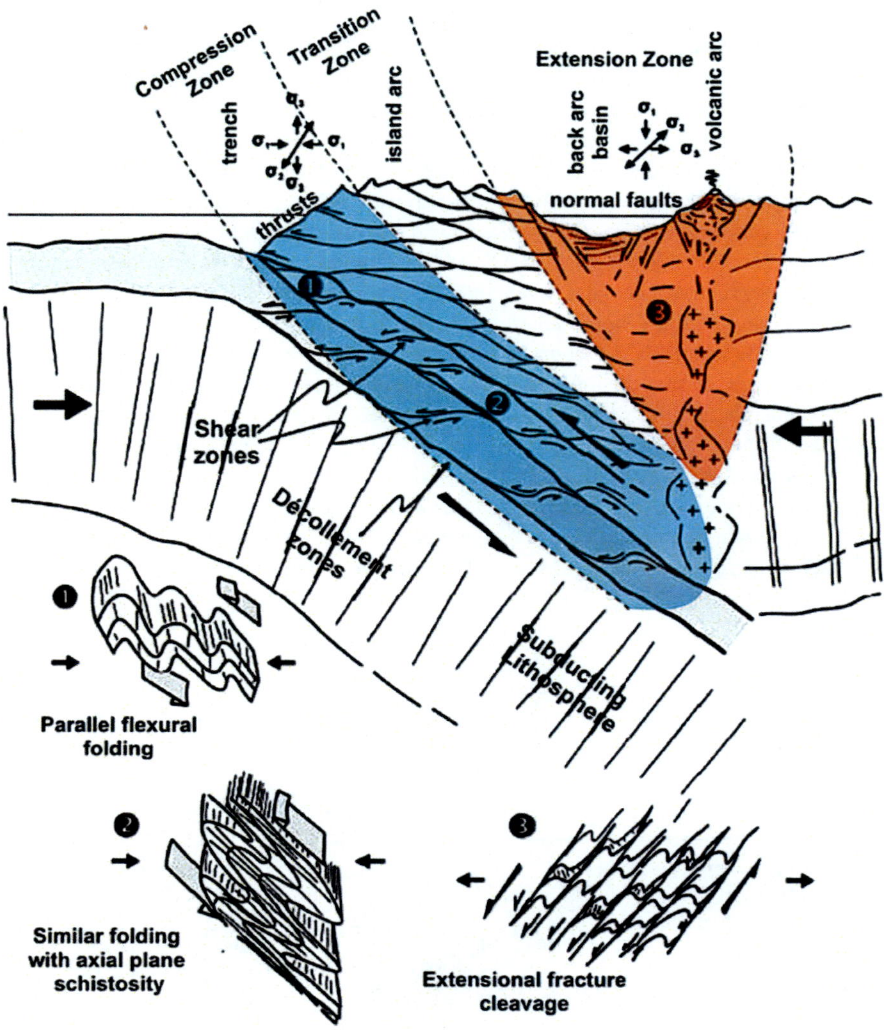

Fig. 4.25 Schematic cross-section of an orogenic arc with distinction of the stress fields and the corresponding tectonic structures at shallow and deep level (from Papanikolaou and Karotsieris 2005)

In the area of the back-arc basin and the volcanic arc, the extensional stress field extends to a depth of several tens of kilometers down to approximately the base of the crust. In the intermediate zone that develops between the two opposite deformation zones in the island arc we have a transition from compressive to tensile deformation. Thus, while the triaxial ellipsoidal stresses in the frontal compression zone have a principal stress σ1 horizontal and the minimum stress σ3 vertical, in the internal extensional zone the stress axes mutually change position so that σ1 becomes vertical while σ3 horizontal, with a direction perpendicular to the arc. The intermediate stress σ2 remains constant in a horizontal position, parallel to the arc in both deformation zones (Fig. 4.25).

The resulting tectonic structures change along the subduction zone, with parallel flexural—slip folds for example, at shallow depth, being replaced at greater depth by similar slip folds with flow and axial plane cleavage/schistosity on the newly formed s-structures (Mattauer 1973). In the tensile area behind the island arc the difference between surface and depth structures consists mainly in the development of non-penetrative large scale structures, in the form of normal faults and penetrative microstructures, in the form of fracture cleavage. Essentially, we have the succession of tectonic structures along with the tectonic levels, which are adapted to the geometry of the orogenic arc and not to a general theoretical scheme of homogeneous crust at the interior of a plate. On the contrary, there are two different expressions of the succession of tectonic features in the successive tectonic levels, under different stress regimes, compressive in the subduction zone and extensional in the volcanic arc and back-arc basin domains. In the interior of the Hellenic arc, several successive extensional structures have been described, with changes of the stress fields from the beginning of the Cenozoic to the present (Angelier 1979; Mercier 1979; Mountrakis 2006; Papanikolaou and Royden 2007; Burchfiel et al. 2008).

4.8 The Succession of Deformation Phases in the Arc

The number of deformation phases of a given unit depends on its final position in the orogenic arc. Therefore, if it is located in the front of the island arc, it may have undergone only one deformation phase (D1) during its transport from the frontal compression zone of the trench and the external part of the island arc (Fig. 4.26, 1–1′). This is the simplest of the cases where rocks are deformed through folding and reverse faulting with a general direction parallel to the arc.

If the unit is located in the internal part of the island arc or in the back-arc basin, then it may have undergone two deformation phases, the first of which occurred during its transport through the frontal compression zone and the second from the internal sector of the tensile stresses. Therefore, compressive structures with folds and thrusts of the first deformation phase (D1) can be distinguished, which will then be deformed by normal faults of the second phase (D2) (Fig. 4.26, 2–2′). In case when the tectonic unit has been subducted beneath the island arc, at depths of 10–20 km before it returns to the surface, it is possible to distinguish three deformation phases, resulting from the deep compressive structures of the first syn-metamorphic deformation phase (D1), followed by younger compression structures of the second phase (D2) in more shallow depths and finally by even younger shallow extensional structures of the third deformation phase (D3) (Fig. 4.26, 3–3′). Lastly, if the unit is located in the volcanic arc region or behind it, at least three deformation phases are expected, of which the first (D1) includes compression structures of usually isoclinal recumbent similar folds with axial plane schistosity and mineral lineation formed by the orientation of blueschist metamorphic facies minerals, followed by a second syn-metamorphic compressive deformation (D2) with close folds with cleavage and/or schistosity and lineation formed by the orientation of greenschist metamorphic facies minerals, and finally by a younger third post-metamorphic

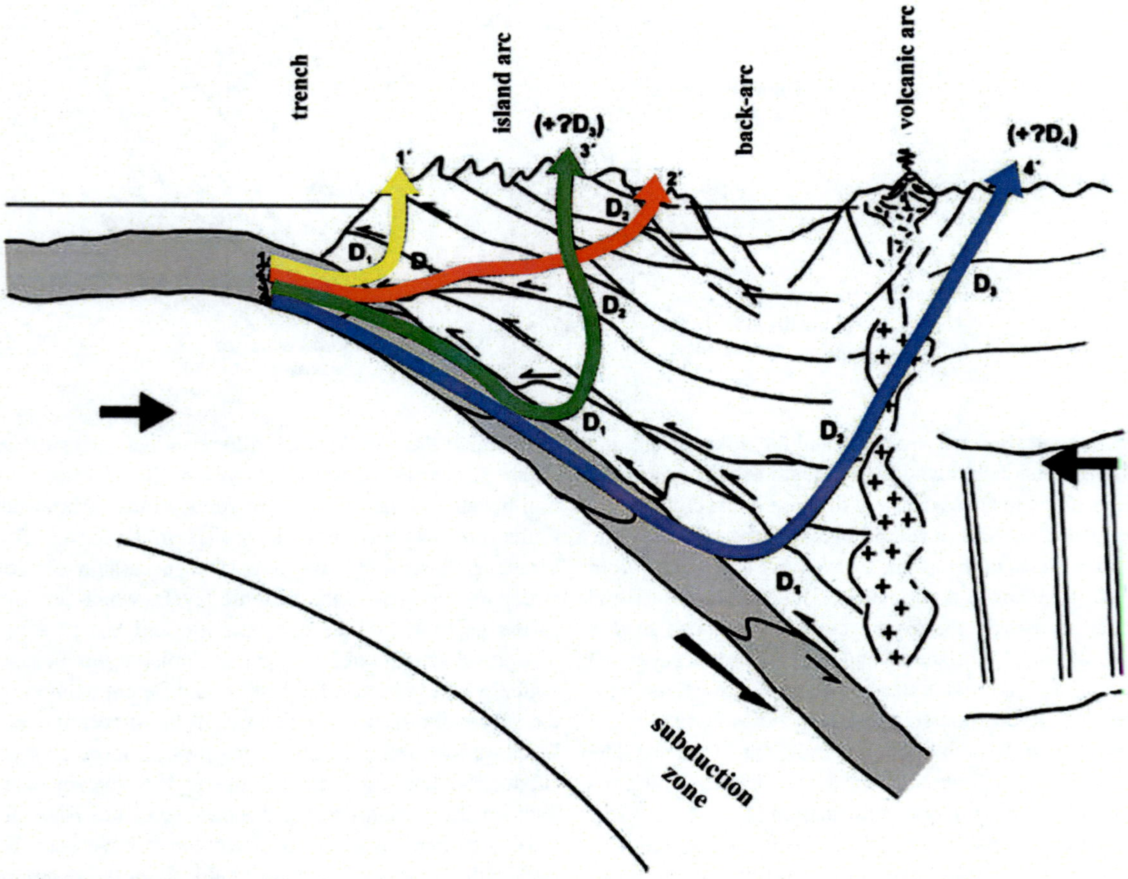

Fig. 4.26 Four possible scenarios for the creation of different deformation phases in the orogenic arc. The more internal and deeper a tectonic unit enters inside the arc, the higher the number and intensity of the deformation phases increase (from Papanikolaou and Karotsieris 2005)

deformation phase (D3) of extensional structures comprising fracture cleavages and normal faults (Fig. 4.26, 4–4′). This third deformation phase can often be distinguished in two different phases, a deep extensional phase with fracture cleavage, king folds, feather joints, etc., and a younger fourth shallow extensional deformation phase (D4) with normal faults and joints.

In conclusion, the rocks experience deformation in the successive segments of the arc characterized by the following:

1. In the trench and the front of the island arc, the deformation is relatively simple and geometric, consisting of a compressive deformation phase, including folds and thrusts with structures of the upper and intermediate tectonic levels.
2. In the axial area of the island arc, the deformation can be distinguished in two deformation phases, the first being a compressive phase, which has shaped the so-called Alpine tectonic fabric, with folding and thrusting, and a second phase of extensional brittle deformation with normal faults. In certain cases where exhumation of metamorphic units is observed along this belt in the form of tectonic windows, the first compressive phase may also involve deeper syn-metamorphic deformation.
3. In the internal sector of the island arc, as well as in the back-arc basin and the volcanic arc, the deformation varies depending on whether there is a metamorphic core complex or not.

 3.1. In non-metamorphic formations, the deformation may include a first compressive phase and one or two extensional phases, with successive tectonic structures of extensional detachments, low angle normal faults and normal faults.
 3.2. In a metamorphic core complex in the form of tectonic windows under the non metamorphosed rocks, deformation is complex, with one or even two syn-metamorphic compressional phases with isoclinal similar folds and axial plane schistosities, initially associated with blueschist HP/LT assemblages, and later on with greenschist HT/LP assemblages. Then, one or two extensional phases may follow, with an earlier deep level syn-metamorphic deformation dominated by crenulation or fracture cleavages, followed by flexural-slip structures of the intermediate layer and finally, extensional structures of the upper level, with normal faults and joints.

The above cases do not include the possible existence of older inherited structures from an earlier orogenic process (e.g. Variscan), which may have one or more deformation events by themselves. Additionally, transverse or diagonal strike-slip structures are not included, which may be developed in very specific zones, due to the presence of microplates or tectonic blocks with lateral asymmetries in the arc.

4.9 From Compression and Thrusting to Extension and Extensional Detachments—The Tectonic Windows—The Metamorphic Core Complexes

The synthetic view of the deformation created in the orogenic arcs is of particular interest, under the plate tectonics processes and the transition of each unit from the subducting to the advancing plate. Each unit starts with a compression in the trench area and the front of the island arc and continues in the interior of the arc through decollement from the subducted lithosphere at some depth, either on the upper 5–10 km, or deeper. During this transport shallow or deep structures are incorporated until the unit reaches the internal part of the arc, where extension prevails. The whole process can be distinguished in two stages: (1) a **building stage** of the Alpine tectonic fabric, which corresponds to the compressive phase with nappe emplacement and thrusting, representing the early deformation period, and (2) a **collapsing stage**, which corresponds to the extensional phase in the inner part of the arc and represents the late deformation period. The entire tectonic process is related to the increase of the crustal thickness below the island arc during the building stage and its thinning in the back-arc basin during the collapsing stage.

The reactivation of the tectonic surfaces created during the initial compression and building stage of the tectono-stratigraphic successions during the later extensional phase, may transform them into extensional detachments. An extensional detachment usually accomodates displacement in the order of several or decades of km. As a result in order to determine and locate such a structure you need to establish that certain nappes and/or tectono-stratigraphic horizons from the initial nappe pile and/or the initial tectono-stratigraphic sequence, are missing. This was shown in Peloponnese, Giona Mt and Crete (Papanikolaou and Royden 2007; Seidel et al. 2007; Papanikolaou et al. 2009; Papanikolaou and Vassilakis 2010). In several cases in the recent bibliography, extensional detachments have been “invented” in structures that constitute simple internal slides of a unit, due to disharmony, without substantial absence of tectono-stratigraphic horizons.

The tectonic fabric created at the front of the arc in the form of the fold and thrust belt comprises, the folds of the advancing plate and the folds of the subducting plate, separated by a transition zone engaged by the sediments of the tectonic wedge (Fig. 4.27). The folds here constitute typical

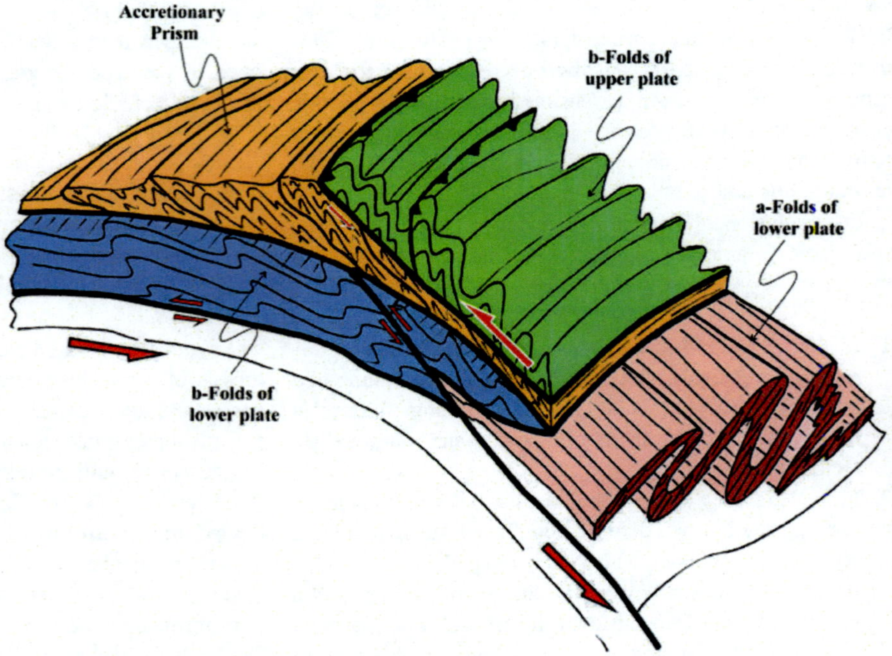

Fig. 4.27 Differentiation of the tectonic structures at the front of the arc, where we have the compressive stress field with the mega shear between the two plates and the simultaneous creation of shallow longitudinal and transverse deep structures, such as e.g. in western-central Peloponnese (from Papanikolaou and Lozios 2015)

b-structures of the upper and/or intermediate tectonic level, with parallel flexural slip folds as well as usual anisotropy phenomena on both plates. Therefore, folds develop approximately parallel to the orientation of the arc/trench and with axial planes dipping towards the inner part of the arc, showing a tectonic asymmetry toward the external front of the entire structure. Western Greece, from Epirus to the Peloponnese, is characterized by this general rather geometric structure.

Further deep, along the subduction zone, the fold geometry gradually changes to similar folds, with the development of axial plane schistosity at the lower tectonic level, (see also Fig. 4.18) until they are converted into similar folds of kinematic a-type, i.e. the folds' axes are re-orientated becoming parallel to the slip direction of the subduction zone. This means folding perpendicular to the arc and to the shallow folding. The deep process of the syn-metamorphic deformation is associated with a HP/LT environment, with blueschist assemblages, characterizing the descenting process, as this is observed in the metamorphic rocks of the Arna unit. The whole process of subduction and return at the surface/exhumation is maintained within a huge mega shear zone, several kilometers wide, between the two plates. In the case of the Cyclades, this shear zone of Late Eocene–Oligocene age is imprinted in the blueschist units of the Northern and Southern Cyclades, with underlying units of a lower metamorphic grade belonging to the external Hellenides platform, such as Olympus, Almyropotamos, Kerketeas, and Amorgos, and overlying non-metamorphosed units, such as Eastern Greece and the Cycladic nappe (Papanikolaou 1987) (Fig. 4.28). This particular mega shear blueschist zone of the Cyclades including asymmetric folds with opposite sense of shear is interpreted by Xypolias and Alsop (2014) as flow folding, created during the exhumation of subducted units to the surface.

In the accretionary prism area at the front of the arc, flysch is deposited in the trench along with olistoliths and large tectonic wedges/lenses of various rocks at the uppermost part forming this chaotic aspect of a wildflysh. These allochthonous transported elements have been detached from the contact zone of the two plates, brought to surface and through erosion they have slided in the trench basin. Therefore, we often find ophiolites, evaporites, exotic limestones and other exotic rocks, which form chaotic complexes of wild flysch or mélange due to the cannibalization of former clastic material. Tectonic units with spectacular outcrops of this type can be found in Crete, in the Upper Cretaceous Miamou unit and its counterpart, the Aderes unit in Argolis, as well as in the Eocene Laerma unit in Rhodes and its equivalent Eastern Kos unit in the Dodecanese (see corresponding Sects. in 8.2.6 and 8.1.12). The actualistic example of the Eastern Mediterranean accretionary prism, comprising Middle Miocene–Pleistocene clastic sediments, bordered by the backstop along the front of the Hellenic arc (see also Figs. 3.8, 3.9 and 3.10), gives an idea about the formation of the similar Late Cretaceous–Paleogene units.

Shear zones of cataclastic rocks are developed, between the internal thrusts of the upper plate and between the

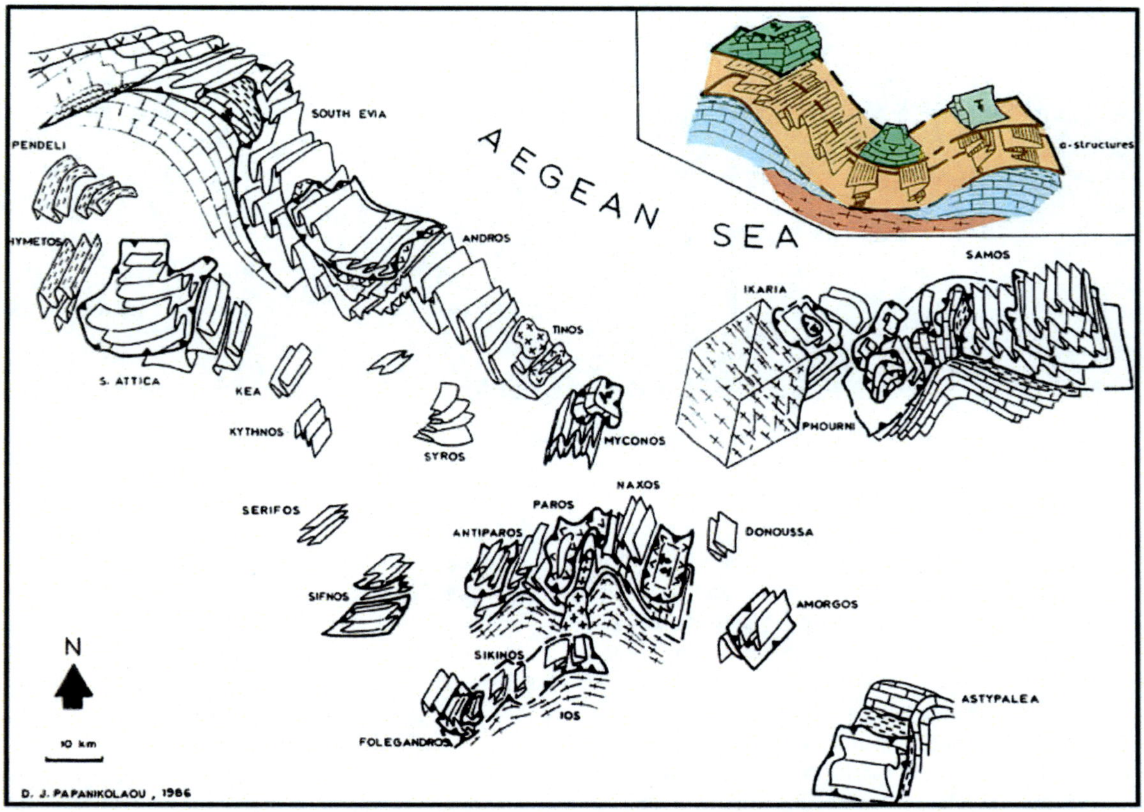

Fig. 4.28 Schematic 3D stereodiagram of the Cycladic blueschists tectonic structure, showing the megafolds croping out on the islands, representing deep a-structures transverse to the arc. These structures have been formed within the Oligocene megashear zone, created by the underlying low metamorphic grade external carbonate platform and the overlying non-metamorphosed units of the internal carbonate platform (from Papanikolaou 1987)

accretionary prism and the rocks of the two plates. These cataclastc rocks that reach deeper parts of the subduction zone are gradually transformed into mylonites, with tectonic cleavage and schistosity. Within these mylonite shear zones microlithons are created and kinematic indicators can be extracted in order to determine the shear sense (Gautier et al. 1993; Forster and Lister 1999; Jolivet et al. 2010, 2013; Roche et al. 2019).

In the collapsing stage occurring in the back-arc basin and the volcanic arc, the structures are developed in the interior of the upper plate on successive tectonic levels. Therefore, at the lower tectonic level, above the mega shear zone of the subduction, syn-metamorphic compressive structures may be observed, which re-fold the previous ones with approximately the same characteristics. Getting at shallower parts in the plate, at the beginning of the uplifting process, there is a transition from compression to extension and at the same time a change from HP/LT to HT/LP conditions, i.e. from blueschists to greenschists, a phenomenon known as **retrograde metamorphism**. Here the folds are more open and fold the previous axial planes of folds with the creation of new cleavages and lineations. Even higher, in the intermediate tectonic level, deformation ceases to be syn-metamorphic and extensional fracture cleavages are dominant, with transition to kink bands, which pass to shear zones of feather joints and finally to normal faults and extensional joints. The Cycladic units form characteristic examples of such tectono-metamorphic evolution (Papanikolaou 1977, 1978).

The above process of the evolution of the tectonic fabric from deep to shallow tectonic levels is not confined only to the micro- and meso-scopic scales, but occurs also in the megascopic scale. Thus, extensional shear zones, extensional detachments, low angle normal faults and tectonic gravity nappes in sedimentary basins are common structures developed during the collapsing stage at the interior of the arc. A characteristic extensional structure has been described in Parnon mt. along the East Peloponnese Detachment zone (Papanikolaou and Royden 2007) (Fig. 4.29).

Here, in the internal domain of the arc, fold and thrust structures can be observed, which are the result of an extensional field, unlike the conventional compressive structures of

Fig. 4.29 Extensional detachment in the form of a low angle normal fault, bringing in direct contact the non-metamorphic Tripolis unit in the hanging wall over the autochthonous metamorphic Mani unit in the footwall at Eastern Parnon Mt. The intermediate Arna unit and the lower section of Tripolis unit have been omitted, due to the extensional motion (from Papanikolaou and Royden 2007)

the arc and are characterized as **extensional folds** and **extensional nappes** (Wernicke 1981; Wernicke and Burchfiel 1982) (Fig. 4.30). That is to say, they are parasitic or drag folds, often created due to gravity over an underlying sliding surface in the event of a mass movement. This is a mega-landslide simulation, which is often justified by the listric form of the syngenetic normal faults. In Greece, structures of this type can be found in the deformed Miocene molasse of the Cyclades, occurring on top of the non-metamorphosed Cycladic unit, as well as in the margins of the molassic Cretan basin during the Middle-Late Miocene, as for example in the coastal zone of Northern Crete (Dermitzakis and Papanikolaou 1979; Papanikolaou and Vassilakis 2010) (Fig. 4.31). Similar events, with the creation of small tectonic gravity nappes during the Middle Miocene have also been described in the western margin of the molasse basin of Itea- Amfissa in North Giona mt (Papanikolaou et al. 2009; Gouliotis 2014) (see also Fig. 6.9).

Tectonic movements implying emplacement of nappes through gravity have been described in the Alps since the beginning of the previous century, with the emplacement of non metamorphosed Penninic units inside the Helvetic flysch formations. This structure was interpreted as a gravitationally triggered event described by the term **diverticulation** (Lugeon 1943). These are secondary gravity movements, occurring along the slopes of sedimentary basins, caused by the high uplift of the adjacent compressional mountain ranges. In the "Basin and Range" (Western USA) extensional detachments were described for the first time, which resulted in extensional tectonic nappes and related extensional folds (Wernicke 1981; Wernicke and Burchfiel 1982). It is important to note that extensional structures of collapsing type have been observed even in collisional ranges, such as the Himalayas (Burchfiel and Royden 1985), and have been established as a later phase of the orogeny, characterizing the **extensional collapse of orogens** by Dewey (1988).

Below the volcanic arc, magmatism intervenes in the entire process through the intrusion of granitic rocks (part of which reach at the surface), forming anticlinal folds, surrounded by syngenetic greenschist-amphibolite-migmatite rocks, creating **metamorphic core complexes** (Coney 1980). These metamorphic complexes, previously known as metamorphic cordilleras, lie on the axis of major uplift and extension. They are generally forming tectonic windows of extensional character, bordered by extensional detachments and are associated to the exhumation process of the blueschist rocks created during the preceding metamorphic phase (Platt 1993; Chemenda et al. 1995, 1996). The uplift and erosion of the metamorphic core complexes is related to the intrusive mechanisms of the granitic magma and their rapid unroofing is related to the rapid uplift. The metamorphic rocks of the Attica–Cyclades area are an example of a metamorphic core complex, where blueschists and eclogites metamorphosed during Eocene–Oligocene were brought to the surface already in the Miocene (Lister et al. 1984; Avigad and Garfunkel 1991; Avigad et al. 1997). It is characteristic that these late extensional detachments, separating the Cycladic Miocene molassic sediments from the underlying metamorphics have been used for the mineralisation in Mykonos and other similar cases (Menant et al. 2013).

In conclusion, a distinction should be made between the "cold" and "hot" tectonic windows, i.e. the tectonic windows appearing along the orogenic axis, usually below the island arc, with a maximum crustal thickness, not affected thermally by rising granitic magmas, as opposed to the tectonic windows associated with the rising of granitic dome structures in the metamorphic core complexes, with a reduced crustal thickness, usually in the volcanic arc area (Van Hinsbergen et al. 2005). Thus, the cold tectonic window

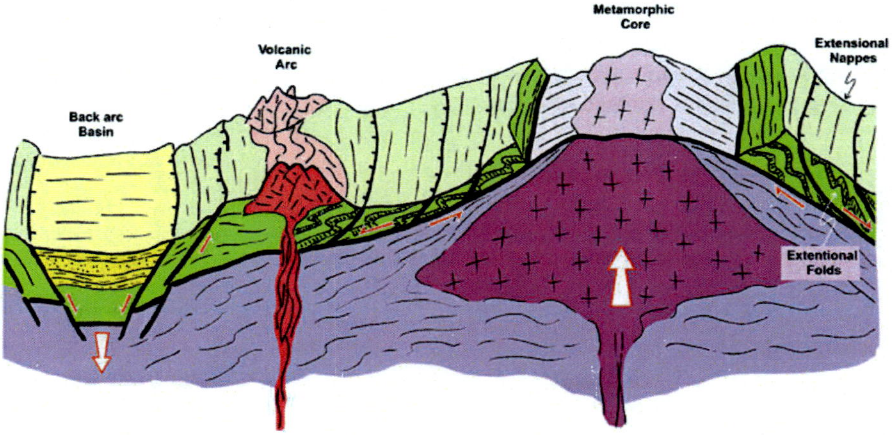

Fig. 4.30 Distribution of extensional structures in the back-arc basin and volcanic arc region, with the creation of tectonic windows and extensional detachments in metamorphic core complexes, e.g. in the Cyclades (from Papanikolaou and Lozios 2015)

Fig. 4.31 Typical outcrop of Mesozoic limestone (Ki) overlying the Middle-Upper Miocene clastic deposits of Crete on the Cretan Sea coasts, due to gravity sliding during the initial rifting phase of the Cretan basin

group in the Hellenides includes the cases of Taygetus, Parnon, Lefka Ori, Psiloritis, and Olympus(Godfriaux 1968; Thiebault 1982;, Bonneau 1973, Lekkas and Papanikolaou 1978; Fassoulas et al. 1994; Doutsos et al. 2000; Kilias et al. 2002; Papanikolaou and Royden 2007; Papanikolaou and Vassilakis 2010). These tectonic windows occur on the present island arc along the external tectono-metamorphic belt of the Hellenides, with the exception of Olympus, which is located north of the current Aegean microplate, away from the Oligo–Miocene Aegean granites. On the other hand, the tectonic windows of the Cyclades, Attica, Southern Evia and Samos, of the intermediate tectono-metamorphic belt, as well as the windows of Pangeon and the surrounding areas of the internal tectono-metamorphic belt, belong to the hot window group, with underlying Eocene-Miocene granites in the metamorphic core complexes (Lister et al. 1984; Papanikolaou 1987; Avigad and Garfunkel 1991; Dinter and Royden 1993; Gautier et al.1993; Warwenitz and Krohe 1998; Jolivet et al. 2010). Especially in the Cyclades area there have been several proposals for major detachments on top of the Cycladic blueschists units towards the North (e.g. Jolivet et al. 2010) or the West (e.g. Grasemann et al. 2012) or the South (e.g. Schneider et al. 2018). The overall existence of these late extensional structures during Miocene is considered as the main exhuming process for the creation of the metamorphic core complexes.

Finally, the development of strike-slip faults and block rotations marks the transition from the compression in the Hellenic trench and the Ionian islands at the front of the Hellenic arc to the extension in the Gulf of Corinth towards the axial zone of the Hellenides (Vassilakis et al. 2011).

References

Ambraseys, N. 1960. The seismic sea wave of July 9, 1956, in the Greek Archipelago. J. Geophys. Res., 65, 1257–1265.

Ampferer, O. 1906. Über das Bewegungsbild von Faltengebirgen. Jahr Geol. R.A., Wien.

Angelier, J. 1979. Néotectonique de I' arc Egéen. Soc. Geol. Nord, Publ., 3, 1–417.

Argand, E. 1920. Plissements precurseurs et plissements tardifs des chaines de montagne. Actes Soc. Helv. Sc. Nat. 31.

Argand, E. 1924. La tectonique de I'Asie. Congrès Géol. Intern., Bruxelles.

Aubouin, J. 1955. Les couches de passage au flysch dans l' E du Pinde meridionale (synclinal de Tirna-Perliango, Thessalie, Grece). C. R. somm. Soc. Geol. France, 137–141.

Aubouin, J. 1965. Géosynclines. Devel. Geotectonics, 1, 335 p., Elsevier.

Aubouin, J. 1974. Des tectoniques superposées et leur signification par rapport aux modèles géophysiques. L'exemple des Dinarides, paléotectonique, tectonique, tarditectonique, néotectonique. Bull. Soc. géol. France, XV, 426–460.

Avigad, D. & Garfunkel, Z. 1991. Uplift and exhumation of high-pressure metamorphic terrains: the example of the cycladic blueschist belt (Aegean Sea). Tectonophysics, 188, 357–372.

Avigad, D., Garfunkel, Z., Jolivet, L. & Azanon, J.M. 1997. Back arc extension and denudation of Mediterranean eclogites. Tectonics, 16, 924–941.

Avigad, D., Ziv, A. & Garfunkel, Z. 2001. Ductile and brittle shortening, extension-parallel folds and maintenance of crustal thickness in the central Aegean (Cyclades, Greece). Tectonics, 20 (2), 277–287.

Bailey, E. B. & McCallien, W. J. 1950. The Ankara melange and the Anatolian Thrust. Nature, 166, 938–943.

Bellon, H., Jarrige, J. & Sorel, D. 1979. Les activités magmatiques égéennes de l'Oligocène à nos jours et leurs cadres géodynamiques. Données nouvelles et synthèse. Rev. Géol. Dyn. Géogr. Phys., 21, 1, 41–55.

Benioff, H. 1955. Seismic evidence for crustal structure and velocity activity. Geol. Soc. Am. Sp. Papers, 62, 67–74.

Blake, M.C., Irwin, W.P. & Coleman, R.G. 1969. Blueschist facies metamorphism related to regional thrust faulting. Tectonophysics, 8, 237–246.

Bonneau, M. 1973. Sur les affinités ioniennes des «Calcaires en plaquettes» épimetamorphiques de la Crète, le charriage de la série de Gavrovo-Tripolitsa et la structure de I' arc égéen. C.R. Acad. Sc. Paris, 277, 2453–2456.

Brichau, S., Ring, U., Ketcham, R.A., Carter, A., Stockli, D. & Brunel, M. 2006. Constraining the long-term evolution of the slip rate for a major extensional fault system in the central Aegean, Greece, using thermochronology. Earth & Plan. Sci. Let., 241, 293–306.

Brun, J.P., Faccenna, C., Gueydan, F., Sokoutis, D., Philippon, M., Kydonakis, K. & Gorini, C. 2016. The two - stage Aegean extension, from localised to distributed, a result of slab rollback acceleration. Canadian Journal of Earth Sciences, 53, 11, 1142–1157.

Bullard, E., Everett, J. & Smith, A. 1965. The Fit of the Continents around the Atlantic. Philos. Trans., R. S. London, A258, 41–51.

Burchfiel, B.C., Nakov, R., Dumurtzanov, D., Papanikolaou, D., Tzankov, T., Serafimovski, T., King, R.W., Kotzev, V., Todosov, A. & Nurce, B. 2008. Evolution and Dynamics of the Cenozoic tectonics of the South Balkan extensional system. Geosphere, 4, 6, 917–938.

Burchfiel, B. C. & Royden, L. H. 1985. North – South extension within the convergent Himalayan region. Geology, 13, 679–682.

Burchfiel, B.C., Royden, L.H., Papanikolaou, D. & Pearce, F.D. 2018. Crustal development within a retreating subduction system: the Hellenides. Geosphere, 14, 2.

Caputo, R., Catalano, S., Monaco, C., Romagnoli, R., Tortorici, G. & Tortorici, L. 2010. Active faulting on the island of Crete (Greece). Geophys. J. Int., 183, 111–126.

Chapple, W. M. & Tullis, T.E. 1977. Evaluation of the forces that drive the plates. J. Geoph. Res., 82, 1967–1984.

Chemenda, A., Mattauer, M., Malavieille, J. & Bokun, A. N. 1995. A mechanism for syncollisional rock exhumation and associated normal faulting: results from physical modelling. Earth & Plan. Sci. Let., 132, 225–232.

Chemenda, A. I., Mattauer, M. & Bokun, A. N. 1996. Continental subduction and a mechanism for exhumation of high-pressure metamorphic rocks: new modelling and field data from Oman. Earth & Plan. Sci. Let., 143, 173–182.

Coleman, R.G. 1972. Blueschist metamorphism and plate tectonics. 24th Intern. Geol. Congress, sect. 2, 19–26.

Coney, P. 1980. Cordilleran metamorphic core complexes: An overview. Geol. Soc. Am. Mem., 153, 7–31.

Cox, A., Dalrymple, G. & Doell, R. 1967. Reversals of the Earth's Magnetic Field. Scient. Amer., 44–54.

Dana, J. 1873. On some results of the Earth's contraction from cooling, including a discussion of the Origin of Mountains. Amer. Jour. Science.

de Celles, P. G. & Gilles, K. N. 1996. Foreland basin systems. Basin Research, 8, 105–123.

Dermitzakis, M. & Papanikolaou, D. 1979. Paleogeography and Geodynamics of the Aegean Region during the Neogene. VII Int. Congress Medit. Neogene, Athens 1979, Ann. Géol. Pays Hellén., hors série IV, 245–289.

Dewey, J. 1969. Continental Margins: A Model for Conversion of Atlantic Type to Andean Type. Earth & Plan. Sc. Let., 6, 189–197.

Dewey, J. F. 1988. Extensional collapse of orogens. Tectonics 7, 1123–1139.

Dewey, J. F. & Bird, J. M. 1970. Mountain Belts and the New Global Tectonics. J. Geophys. Res., 75, 2615–2647.

Dewey, J. F. & Bird, J. M. 1971. Origin and Emplacement of the ophiolite suite: Appalachian ophiolites in Newfoundland. J. Geophys. Res., 76, 3179–3206.

Dietz, R. S. 1961. Continent and Ocean Basin Evolution by Spreading of the Sea Floor. Nature, 190, 854–857.

Dinter, D.A. & Royden, L.H. 1993. Late Cenozoic extension in NE Greece: Strymon Valley detachment and Rhodope Core Complex. Geology, 21, 45–48.

Doutsos, T., Koukouvelas, I., Poulimenos, G., Kokkalas, S., Xypolias, P. & Skourlis, K. 2000. An exhumation model for the south Peloponnesus, Greece. Intern. J. Earth Sci., 89, 350–365.

Elsasser, W.M. 1969. Convection and Stress Propagation in the Upper Mantle. In: The Application of Modern Physics to the Earth and Planetary Interiors, 223–246, Wiley.

Elsasser, W.M., 1971. Sea-floor spreading as thermal convection. J. Geoph. Res., 76, 1101–1112.

Ernst, W.G. 1970. Tectonic contact between the Franciscan melange and the Great Valley sequence. Crustal expression of a Late Mesozoic Benioff zone. J. Geophys. Res., 75, 886–901.

Evangelidis, C. P. 2017. Seismic anisotropy in the Hellenic subduction zone: Effects of slab segmentation and subslab mantle flow. Earth and Planetary Science Letters, 480, 97–106.

Ewing, M. & Ewing, J. 1967. Sediment distribution of the mid-ocean ridges with respect to spreading of the sea floor. Science, 156, 1590–1592.

Faccenna, C., Jolivet, L., Piromallo, C. & Morelli, A.,2003. Subduction and the depth of convection in the Mediterranean mantle. J. Geoph. Res.,108.

Fassoulas, C., Kilias, A. & Mountrakis, D. 1994. Postnappe stacking extension and exhumation of HP/LT rocks in the island of Crete, Greece. Tectonics, 13, 127–138.

Fleury, J.J. 1977. De Lamia à Messolonghi. La nappe de Pinde-Olonos et l'unité Meghdovas. Bull. Soc. Géol. France, 19, 1, 53–60.

Fleury, J.J. 1980. Les zones de Gavrovo-Tripolitza et du Pinde-Olonos (Grèce continentale et Péloponnèse du Nord). Evolution d'une plateforme et d' un bassin dans le cadre alpin. Publ. Soc. Géol. Nord, 4.

Forster, M. & Lister, G.S. 1999. Detachment faults in the Aegean core complex of Ios, Cyclades, Greece. In: Ring, U., Brandon, M.T., Lister, G.S., Willett, S.D. (Eds.), Exhumation Processes: Normal Faulting, Ductile Flow and Erosion. Sp. Publ., Geol. Soc., London, 305–324.

Forsyth, D. & Uyeda, S. 1975. On the relative importance of the driving forces of plate motion. Geoph. J. Roy. Astr. Soc., 43, 163–200.

Fytikas, M., Giuliani, O., Innocenti, F., Marinelli, G., Mazzuoli, R. 1976. Geochronological data on recent magmatism of Aegean sea. Tectonophysics, 31, 29–34.

Fytikas, M., Innocenti, F., Manetti, P., Mazzuoli, R., Pecerillo, A. & Villari, L. 1984. Tertiary to Quaternary evolution of volcanism in the Aegean region. Geol. Soc. London, Sp. Publ., 17, 687–699.

Galanopoulos, A. 1957. The seismic sea wave of 9th of July 1956. Proc. Acad. Athens, 32, 90–101 (in greek).

Galanopoulos, A. 1961. Elements of Seismology and Physics of the Earth's interior. 318 p., Athens (in greek).

Galanopoulos, A., Delimbasis, N. & Komninakis, P. 1966. Marine wave due to landslide without seismic triggering. Ann. Geol. Pays Hellen., 16, 93–110 (in greek).

Gansser, A. 1974. The ophiolitic melange, a world-wide problem on Tethyan examples. Eclogae Geol. Helv., 67, 479–507.

Gautier, P., Brun, J.P. & Jolivet, L. 1993. Structure and kinematics of upper Cenozoic extensional detachment on Naxos and Paros (Cyclades Islands, Greece). Tectonics, 12(5), 1180–1194.

Gesret, A., Laigle, M., Diaz, J., Sachpazi, M. & Hirn, A. 2010. The oceanic nature of the African slab subducted under Peloponnesus: Thin-layer resolution from multiscale analysis of teleseismic P- to S- converted waves. Geophys. J. Int., 183(2), 833–849.

Godfriaux, I. 1968. Etude géologique de la région de l'Olympe (Grèce). Ann. Géol. Pays Hellέη., 19, 1–281.

Goncuoglu, C., Dirik, K. & Kozlu, H. 1997. Pre-Alpine and Alpine terranes in Turkey: explanatory notes to the terrane map of Turkey. Ann. Géol. Pays Hellén., 37, 515–535.

Gouliotis, L. 2014. The tectonic structure of Giona Mt and its hydrogeological applications. PhD Thesis, University of Athens, 433 p (in greek).

Govers, R. & Wortel, M.J.R. 2005. Lithosphere tearing at STEP faults: Response to edges of subduction zones. Earth & Planet. Sci. Lett., 236, 505–523.

Grasemann, B., Schneider, D., Stockli, D. & Iglseder, C. 2012. Miocene bivergent crustal extension in the Aegean: evidence from the western Cyclades (Greece). Lithosphere, 4, 1, 23–39.

Guillaume, B., Husson, L., Funiciello, F., Faccenna, C. 2013. The dynamics of laterally variable subductions: Laboratory models applied to the Hellenides. Solid Earth, 4 (2), pp. 179–200.

Gutenberg, B. & Richter, C. 1948. Deep focus earthquakes in the Mediterranean region. Pure & Appl. Geoph., 12.

Gutenberg, B. & Richter, C. 1954. Seismicity of the Earth and Associated Phenomena. Princeton University Press.

Gutnic, M., Monod, O., Poisson, A. & Dumont, J.F. 1979. Géologie des Taurides Occidentales (Turquie). Mém. Soc. Géol. France, 137, 1–109.

Haug, E. 1900. Les geosynclinaux et les aires continentales; Contribution a l'etude des regressions et des transgressions marines. Bull. Soc. Geol. France, XXVIII.

Heirtzler, J.R., Dickson G.O., Herron, T.M., Pitman, W.C. & Le Pichon, X. 1968. Marine Magnetic Anomalies, Geomagnetic Field Reversals and Motions of the Ocean Floor and Continents. J. Geoph. Res., 73, 2119–2136.

Hejl, E., de Grave, J., Riedl, H., Weingartner, H., van den Haute, P. 2008. Fission-track thermochronology of the Middle Aegean Island Bridge; implications for Neogene geomorphology and palaeogeography. Zeit. Deuts. Ges. Geow., 159, 495–512.

Hess, H.H. 1962. History of Ocean Basins. Buddington Volume, Geol. Soc. Amer., 599–620.

Hess, H.H. 1965. Mid-oceanic ridges and tectonics of the sea-floor. In: Submarine Geology and Geophysics, 313–334, Butterworth, London.

Hodgson, J. & Cock, J. I. 1957. Direction of faulting in the Greek Earthquakes of August 9–13, 1953. Ann. Geol. Pays Hellen., 8, 29–47.

Hsü, K. 1968. Principles of melanges and their bearing on the Franciscan-Knoxville paradox. Bull. Geol. Soc. Am., 79, 1063–1074.

Hsü, K. & Ryan, W.B.F. 1973. Deep Sea drilling in the Hellenic Trench. Bull. Geol. Soc. Greece, X, 1, 81–89.

Hubscher, C., Ruhnau, M. & Nomikou, P. 2015. Volcano-tectonic evolution of the polygenetic Kolumbo submarine volcano/Santorini (Aegean Sea). J. volcanol. Geotherm. Res., 291, 101–111.

Isacks, B., Oliver, J. & Sykes, L. 1968. Seismology and the New Global Tectonics. J. Geoph. Res. 73, 5855–5899.

Jackson. J. 1994. Active tectonics of the Aegean region. Ann. Rev. Earth & Plan. Sci., 22, 239–271.

Jacobshagen, V. 1986. Geologie von Griechenland. Gebruder Borntraeger, 363 pp.

Jolivet, L. & Brun, J.P. 2010. Cenozoic geodynamic evolution of the Aegean. Int. J. Earth Sci., 99, 109–138.

Jolivet, L., Lecomte, E., Huet, B., Denele, Y., Lacompe, O., Labrousse, L., LE Pourhiet, L. & Mehl, C. 2010. The North Cycladic Detachment System. Earth & Plan. Sci. Let., 289, 87–104.

Jolivet, L., Faccenna, C., Huet, B., Labrousse, L., Le Pourhiet, L., Lacombe, O., Lecomte, E., Burov, E., Denele, Y., Brun, J.P., Philippon, M., Paul, A., Salaun, G., Karabulut, H., Piromallo, C., Monie, P., Gueydan, F., Okay, A.I., Oberhansli, R., Pourteau, A., Augier, R., Gadenne, L. & Driussi, O. 2013. Aegean tectonics: Strain localisation, slab tearing and trench retreat. Tectonophysics, 597–598, 1–33.

Jolivet, L., Menant, A., Sternai, P., Rabillard, A., Arbaret, L., Augier, R., Laurent, V., Beaudouin, A., Grasement, A., Huet, B., Labrousse, L. & le Pourhiet, L. 2015. The geological signature of a slab tear below the Aegean. Tectonophysics, 659, 166–182.

Karig, D.E. 1970. Ridges and basins of the Tonga-Kermadec island arc system. J. Geoph. Res., 75, 239–254.

Karig, D.E. 1971, Origin and development of marginal basins in the western Pacific. J. Geoph. Res., 76, 2542–2561.
Karig, D.E. 1974. Evolution of Arc Systems in the Western Pacific. Ann. Rev. Earth Plan. Sc., 2, 51–75.
Kauffmann, G. 1976. Perm und Trias im östlichen Mittelgriechenland und auf einigen ägäischen Inseln. Z. Deutsch. Geol. Ges., 127, 387–398.
Kay, M. 1947. Geosynclinal nomenclature and the craton. Bull. Amer. Assoc. Petrol. Geol. 31.
Kilias, A.A., Tranos, M.D., Orozco, M., Alonso-Chaves, F.M. & Soto, J.I. 2002. Extensional collapse of the Hellenides: a review. Rev. Soc. Geol. España, 15, 129–139.
Kilias, S., Nomikou, P., Papanikolaou, D., Polymenakou, P., Godelit sas, A., Argyraki, A., Carey, S., Gamaletsos, P., Mertzimekis, T., Stathopoulou, E., Goettlinger, J., Betzelou, K., Livanos, I., Chris takis, C., Croff-Bell, K. & Scoullos, M. 2013. New insights into hydrothermal vent processes in the unique shallow-submarine arc-volcano, Kolumbo (Santorini), Greece. Nat./Sci. Rep., 3, 2421.
Kiratzi, A. & Louvari, E. 2003. Focal mechanisms of shallow earthquakes in the Aegean Sea and the surrounding lands determined by waveform modelling: a new database. J. Geod., 36, 251–274.
Laurent, V., Lanari, P., Nair, I., Augier, R., Lahfid, A., & Jolivet, L. 2018. Exhumation of eclogites and blueschists (Cyclades, Greece): P-T evolution constrained by thermobarometry and thermodynamic modelling. Journal of Metamorphic Geology.
Lekkas, S. 1978. Contribution to the geological structure of the area SE of Tripolis. PhD Thesis, University of Athens, 192 p. (in greek).
Lekkas, S. & Papanikolaou, D. 1978. On the phyilite problem in Peloponnesus. Ann. Géol. Pays Hellén., 29/1, 395–410.
Le Pichon, X. 1968. Sea-Floor Spreading and Continental Drift. J. Geoph. Res., 73, 3661–3697.
Le Pichon, X. & Angelier, J. 1979. The Hellenic Arc and Trench system: a key to the neotectonic evolution of the Eastern Mediterranean area. Tectonophysics, 60, 1–42.
Le Pichon, X. & Angelier, J. 1981. The Aegean Sea. Phil. Trans. R. Soc. London, A300, 357–372.
Le Pichon, X., Francheteau, J. & Bonnin, J. 1973. Plate Tectonics. 300 p. Elsevier.
Lister, G.S., Banga, G. & Feenstra, A. 1984. Metamorphic core complexes of Cordilleran type in the Cyclades, Aegean Sea, Greece. Geology 12, 221–225.
Lugeon, M. 1902. Les grandes nappes de recouvrement des Alpes du Chablais et de la Suisse. Bull. Soc. Géol. France (4), 1, 723–825.
Lugeon, M. 1943. Une nouvelle hypothese tectonique: la diverticulation. Bull. Soc. Vaud. Sci. Nat., 62, 301–303.
Lyon-Caen, H., Armijo, R., Drakopoulos, J., Baskoutas, G. Delibasis, N. Gaulon, R., Kouskouna, V., Latoussakis, J., Makropoulos, K., Papadimitriou, P., Papanastasiou, D. & Pedoti, G. 1988. The 1986 Kalamata (South Peleoponnesus) Earthquake: detailed study of a normal fault, evidence for east-west extension in the Hellenic arc. Journal Geophysical Research, 93, 14,967–15,000.
Maggi, A., Jackson, J., McKenzie, D. P. & Priestley, K. 2000. Earthquake focal depths, effective elastic thickness and the strength of the continental lithosphere. Geology, 28, 6, 495–498.
Maratos, G. 1972. The geology of Greece. Athens, 189 p (in greek).
Mariolakos, I. & Papanikolaou, D. 1981. The Neogene Basins of the Aegean Arc from the Paleogeographic and the Geodynamic point of view. Intern. Symp. (H.EA.T.) Hell. Arc and Trench, Athens 1981, Proceedings I, 383–399.
Mason, J., Schneiderwind, S., Pallikarakis, A., Wiatr, T., Mechernich, S., Papanikolaou, I. & Reicherter, K. 2016. Fault structure and deformation rates at the Lastros-Sfaka Graben, Crete. Tectonophysics, 683, 216–232.
Masson, H. 1976. Un siecle de geologie des Prealpes: de la decouverte des nappes a la recherche de leur dynamique. Eclogae geol. Helv. 69/2, 527–575.
Mattauer, M. 1973. La deformation des materiaux de l' ecorse terrestre. 493p, Hermann.
McKenzie, D. 1967. The viscosity of the mantle. Geoph. J. Roy. Astron. Soc., 14, 297–305.
McKenzie, D. 1968. The geophysical importance of high temperature creep. In: The History of the Earth's Crust, 28–44, Princeton University Press.
McKenzie, D. 1978. Active tectonics of the Alpine - Himalayan belt: the Aegean sea and surrounding regions. Geophys.J.R. Astr. Soc., 55, (1), 217–254.
McKenzie, D. & Parker, R. 1967. The North Pacific: An Example of Tectonics on a Sphere. Nature, 216, 1276–1280.
Menant, A., Jolivet, L., Augier, R., & Skarpelis, N. 2013. The North Cycladic Detachment System and associated mineralization, Mykonos, Greece: Insights on the evolution of the Aegean domain. Tectonics, 32(3), 433–452.
Mercier, J.L 1979. Signification néotectonique de I'Arc Egéen. Une revue des idées. Rev. Géol. Dyn. Géogr. Phys., 21, 1, 5–15.
Mercier, J. & Vergely, P. 1972a. Les mélanges ophiolitiques de Macédoine (Grèce). Décrochements d' âge anté-crétacé supérieur. Zeit. deuts. Geol. Ges., 128, 469–489.
Mercier, J. & Vergely, P. 1972b. Les mélanges colorés (coloured mélanges) de la zone d'Almopias (Macédoine, Grèce). C.R. somm. Soc. Géol. France, 70–73.
Mitzopoulos, M. 1959. Erster Nachweis von Gosau schichten in Griechenland. Sitz. Oster. Akad. Wiss. 168.
Mitzopoulos, M. & Trikkalinos, J. 1937. Geologische Voruntersuchungen in Westthrazien. Proc. Acad. Athens, 12.
Miyashiro, A. 1972. Metamorphism and related magmatism in plate tectonics. Am. J. Sci., 272, 629–656.
Miyashiro, A. 1973. Paired and unpaired metamorphic belts. Tectonophysics, 17, 241–254.
Moix, P., Beccaletto, L., Kozur, H.W., Hochard, C., Rosselet, F.& Stampfli, G.M. 2008. A new classification of the Turkish terranes and sutures and its implication for the paleotectonic history of the region. Tectonophysics, 451, 7–39.
Morgan, W.J. 1968. Rises, Trenches, Great Faults and Crustal Blocks. J. Geoph. Res., 73, 1959–1982.
Morgan, W.J. 1972. Deep Mantle Convection Plumes and Plate Motions. Amer. Ass. Petr. Geol. Bull., 56, 203–213.
Mountrakis, D. 2006. Tertiary and Quaternary tectonics of Greece. In: Dilek, Y., Pavlides, S. (Eds.), Postcollisional Tectonics and Magmatism in the Mediterranean Region and Asia: Geol. Soc. Am. Sp. Paper, 409, 125–136.
Mountrakis, D. & Soulios, G. 1978. About an overthrusted schist-chert complex with ophiolites and the occurrence of ophiolite melanges in the area of Arnissa: their significance for the tecto-orogenic development of the Pelagonian zone. Bull. Geol. Soc. Greece, 13 (2), 18–33 (in greek).
Negris, Ph. 1909. Sur les brèches de friction dans les surfaces de charriage du Péloponnèse. C. R. Acad. Sci. Paris, 148, 1700–1702.
Ninkovitch, D. & Hays, J. 1972. Mediterranean island arcs and the origin of high potash volcanoes. Earth & Plan. Sci. Let., 16, 331–345.
Nomikou, P. 2004. Contribution to the geodynamics of Dodekanese: the submarine area of Kos – Nisyros islands. PhD Thesis, University of Athens, 470 p (in greek).
Nomikou, P., Carey, S., Papanikolaou, D., Croff Bell, K., Sakellariou, D., Alexandri, M. & Bejelou, K. 2012. Submarine volcanoes of the Kolumbo volcanic zone, NE of Santorini Caldera, Greece. Global & Planetary Change, 90–91, 135–151.

Nomikou, P., Papanikolaou, D., Alexandri, M., Sakellariou, D. & Rousakis, G. 2013. Submarine volcanoes along the Aegean Volcanic Arc. Tectonophysics, 597–598, 123–146.

Nomikou, P., Parks, M., Papanikolaou, D., Pyle, D., Mather, T., Carey, S., Watts, A., Paulatto, M., Kalnins, M., Livanos, I., Bejelou, K., Simou, E., Perros, I. 2014. The emergence and growth of a submarine volcano: The Kameni islands, Santorini (Greece). Geo. Res. J., Vol.1, 8–18.

Nomikou, P., Druitt, T.H., Hubscher, C., Mather, T.A., Paulatto, M., Kalnins, L.M., Kelfoun, K., Papanikolaou, D., Bejelou, K., Lampridou, D., Pyle, D.M., Carey, S., Watts, A.B., Weib, B. & Parks, M.M. 2016. Post-eruptive flooding of Santorini caldera and implications for tsunami generation. Nature Communications, 7:13332. https://doi.org/10.1038/ncomms13332.

Nomikou, P., Hubscher, C., Papanikolaou, D., Farangitakis, P., Runhau, M. & Lambridou, D. 2018. Expanding extension, subsidence and lateral segmentation within the Santorini- Amorgos basins during Quaternary: Implications for the 1956 Amorgos events, central-south Aegean Sea, Greece. Tectonophysics, 722, 138–153.

Okal, E., Synolakis, C., Uslu, B., Kalligeris, N. & Voukouvalas, E. 2009. The 1956 earthquake and tsunami in Amorgos, Greece. Geophys. J. Int., 178, 1533–1554.

Papanikolaou, D. 1977. The successive structural features of the last Alpine orogeny in Andros Island, Aegean Sea. VI Colloquium on the Geology of the Aegean Region, Proc. I, 311–319.

Papanikolaou, D. 1978. Contribution to the Geology of Aegean Sea. The island of Andros. Ann. Géol. Pays Hellén., 29/2, 477–553.

Papanikolaou, D. 1980. The Metamorphic Hellenides. 26th Int. Geol. Congress., Paris, 1980, Abstracts, I, 371.

Papanikolaou, D. 1981. Remarks on the kinematic interpretation of folds from some cases of the Western Swiss Alps and of the Hellenides. Ann. Géol. Pays Hellén., 30/2, 741–762.

Papanikolaou, D. 1986a. Late Cretaceous Paleogeography of the Metamorphic Hellenides. Geol. Geoph. Res., IGME, Special volume in honor of Prof. Papastamatiou, 315–328.

Papanikolaou, D. 1986b. Geology of Greece. Eptalofos Publications, 240 p. Athens (in greek).

Papanikolaou, D. 1987. Tectonic evolution of the Cycladic blueschist belt (Aegean Sea, Greece). In: Helgeson, H.C. (Ed.), Chemical Transport in Metasomatic Processes: NATO ASI Series, 429–450.

Papanikolaou, D. 1990. Probable geodynamic interpretation of the schist-chert formations of the Hellenides. Bull. Geol. Soc. Greece, 24, 135–148 (in greek).

Papanikolaou, D. 1993. Geotectonic evolution of the Aegean. Bull. Geol. Soc. Greece, 28, 33–48.

Papanikolaou, D. 2009. Timing of tectonic emplacement of the ophiolites and terrane paleogeography in the Hellenides. Lithos, 108, 262–280.

Papanikolaou, D. 2015. Geology of Greece. Patakis Publ. Co., 443 p., Athens (in greek).

Papanikolaou, D. & Dermitzakis, M. 1981a. Major changes from the last stage of the Hellenides to the actual Hellenic Arc and Trench system. Intern. Symp. on the Hellenic Arc and Trench (H.E.A.T.), Athens 1981, Proceedings, II, 57–73.

Papanikolaou, D. & Dermitzakis, M. 1981b. The Aegean Arc during Burdigalian and Messinian: a comparison. Riv. Ital. Paleont., 87, 1, 83–92 and Rapp. comm, int. Mer. Médit., 27, 8, 93–95.

Papanikolaou, D. & Karotsieris, Z. 2005. Tectonic Geology. University of Athens, 370 p (in greek).

Papanikolaou, D. & Lozios, S. 2015. Tectonic Geology. DaVinci Publications, 480 p, Athens (in greek).

Papanikolaou, D. & Royden, L. 2007. Disruption of the Hellenic arc: Late Miocene extensional detachment faults and steep Pliocene–Quaternary normal faults — or what happened at Corinth? Tectonics, 26.

Papanikolaou, D. & Sideris, C.H. 1977. Contribution to the knowledge of molasse in Greece. I: Preliminary research in Kanalia area of Karditsa. Ann. Géol. Pays Hellén., 28, 387–417 (in greek).

Papanikolaou, D. & Sideris C.H. 2007. Geology: The Science of the Earth. Patakis Publications, 291 p, Athens (in greek).

Papanikolaou, D. & Vassilakis, E. 2010. Thrust faults and extensional detachment faults in Cretan tectono-stratigraphy: implications for Middle Miocene extension. Tectonophysics, 488, 233–247.

Papanikolaou, D., Chronis, G., Pavlakis, P., Lykoussis, B., Roussakis, G. & Syskakis, D. 1989. Submarine Neotectonic Map of Saronic Gulf. Scale 1/100,000, Earthquake Planning & Protection Organization, National Centre Marine Research, University of Athens.

Papanikolaou, D., Gouliotis, L. & Triantaphyllou, M. 2009. The Itea–Amfissa detachment: a pre-Corinth rift Miocene extensional structure in central Greece. Geol. Soc., London, Sp. Publ., 311, 293–310.

Papanikolaou, D., Lykoussis, V., Chronis, G. & Pavlakis, P. 1988. A comparative study of neotectonic basins across the Hellenic arc: The Messiniakos, Argolikos, Saronikos and Southern Evoikos Gulfs. Basin Research, 1/3, 167–176.

Papanikolaou, I. & Lekkas, E. 2001. Synsedimentary tectonism during the transition from carbonate to clastic sedimentation in the Ionian unit (Messolongi area). Bull. Geol. Soc. Greece, 34, 191–198 (in greek).

Papazachos, B.C. & Comninakis, P.E. 1982. A catalogue of historical earthquakes in Greece and surrounding area for the period 479 B.C. – 1900 A.D. Publ. Geoph. Lab. Univ. Thessaloniki, 6, 1–24.

Papazachos, B.C., Karakostas, B., Papazachos, C.B. & Skordilis, E. 2000. The geometry of the Wadatti-Benioff zone and lithosperic kinematics in the Hellenic arc. Tectonophysics, 319, 275–300.

Pavlakis, P., Lykoussis, B., Papanikolaou, D. & Chronis, G. 1989. Discovery of a submarine volcanic centre in the West Saronic Gulf: The Paphsanias Volcano. Bull. Geol. Soc. Greece, 24, 59–70 (in greek).

Pe-Piper, G. & Piper, D. 2002. The Igneous Rocks of Greece. The Anatomy of an Orogen. Gebrueder Borntraeger, Berlin/Stuttgart.

Peyve, A. & Sinitzyn, V. 1950. Certains problemes fontamentaux de la doctrine des geosynclinaux. I zv. Akad. Nauk. SSSR, ser. Geologie.

Platt, J.P. 1993. Exhumation of high-pressure rocks: a review of concepts and processes. Terra Nova, 5, 119–133.

Psarianos, P. & Manolessos, N. 1963. Elements of Tectonic Geology. Athens, 152 p (in greek).

Ranero, C.R., Villasenor, A., Phipps Morgan, J. & Weinrebe, W. 2005. Relationship between bend-faulting at trenches and intermediate-depth seismicity. Geochemistry Geophysics Geosystems, 6, https://doi.org/10.1029/2005GC000997.

Renz, C. 1955. Die vorneogene Stratigraphie der normal sedimentären Formationen Griechenlands. I.G.S.R., 637 p., Athens.

Richter, D. & Mariolakos, I. 1972. Paläomorphologie und eozäne Verkastung der Gavrovo-Tripolis Zone auf dem Peloponnes (Griechenland). Bull. Geol. Soc. Greece, 9/2, 206–228.

Richter, D. & Mariolakos, I. 1975. Stratigraphische Untersuchungen an Kreide/Tertiär - Wende Im Gebiet von Delfi - Amfissa - Amfiklia (Parnass – Giona zone Griechenland). Ann. Géol Pays Hellén., 26, 417–434.

Ricou, L.E. 1980. La tectonique de coin et la genèse de l' arc égéen. Rev Géol. Dyn. Géogr. Phys..22, 147–155.

Roche, V., Jolivet, L., Papanikolaou, D., Bozkurt, E., Menant, A. & Rimmele, G. 2019. Slab fragmentation beneath the Aegean/Anatolia

transition zone: insights from the tectonic and metamorphic evolution of the Eastern Aegean region. Tectonophysics, 754, 101–129.
Rodgers, J. 1984. A geologic reconnaissance of the Cycladic Blueschist belt, Greece. Discussion. Bull. Geol. Soc. Amer., 95, 117–119.
Royden, L.H. 1993. Evolution of retreating subduction boundaries formed during continental collision. Tectonics, 12, 629–638.
Royden, L. & Husson, L. 2006. Trench motion, slab geometry and viscous stresses in subduction systems. Geoph. J. Intern., 167, 881–905.
Sachpazi, M., Laigle, M., Charalampakis, M., Diaz, J., Kissling, E., Gesret, A., Becel, A., Flueh, E., Miles, P., Hirn, A. 2016. Segmented Hellenic slab rollback driving Aegean deformation and seismicity. Geophysical Research Letters, 43(2), 651–658.
Schardt, H. 1893. Sur I' origine des Préalpes romandes. Arch. Sci. phys. nat., 30, 570–583.
Schardt, H. 1898. Les regions exotiques du versant N des Alpes suisses. Bull. Soc. Vaud. Sci. Nat., 34, 114–219.
Schneider, D., Grasemann, B., Lion, A., Soukis, K. & Draganits, E. 2018. Geodynamic significance of the Santorini Detachment System (Cyclades, Greece). Terra Nova, 2018; 1–9, DOI: https://doi.org/10.1111/ter.12357.
Seidel, M., Seidel, E, & Stoekhert, B. 2007. Tectono-sedimentary evolution of lower to middle Miocene half-graben basins related to an extensional detachment fault (western Crete, Greece). Terra Nova, 19(1), 39–47.
Seward, D., Vanderhaeghe, O., Siebenaller, L., Thomson, S., Hibsch, C., Zingg, A., Holzner, P., Ring, U. & Duchene, S. 2009. Cenozoic tectonic evolution of Naxos Island through a multi-faceted approach of fission-track analysis. Geol. Soc., London, Sp. Publ., 321, 179–196.
Shaw, B., Ambraseys, N., England, P., Floyd, M., Gorman, G., Higham, T., Jackson, J., Nocket, J., Pain, C. & Piggot, M. 2008. Eastern Mediterranean tectonics and tsunami hazard inferred from the AD 365 earthquake. Nature Geosciences, 268–276.
Spencer, E.W. 1977. Introduction of the structure of the Earth. McGraw Hill & Kogakusha P. Co. 640 p.
Stanley, D.J. 1974. Modern flysch sedimentation in a Mediterranean Island arc setting. In: Dott & Shaver editors, Modern and ancient geosynclinal sedimentation, Soc. Econ. Paleon. Miner. Publ. 19. Tulsa, Oklahoma.
Stille, H. 1924. Grundfragen der vergleichenden Tektonik. Berlin.
Stille, H. 1936. Wege und Ergebnisse der geologisch-tektonischen Forchungen. 25Jahre Kaiser Wilhelm - Gesellschaft zur Forderung der Wiss., 11, Die Naturw.
Suckale, J., Rondenay, S., Sachpazi, M., Charalampakis, M., Hosa, A. & Royden, L.H. 2009. High-resolution seismic imaging of the western Hellenic subduction zone using teleseismic scattered waves. Geoph. J. Intern., 178, 775–791.
Sugimura, A. & Uyeda, S. 1973. Island Arcs: Japan and its Environs. 246 p., Elsevier.
Talwani, M., Le Pichon, X. & Ewing, M. 1965. Crustal structure of the midocean ridges. 2. Computed model from gravity and seismic refraction data. J. Geoph. Res., 70, 341–352.
Tataris, A. 1975. Some questions on the «itinerary» of the younger (sh2) schist-psammite-chert complex and the relation Pelion – Olympus. Bull. Geol. Soc. Greece, 12, 95–112 (in greek).
Taymaz, T., Jackson, J.A. & McKenzie, D. 1991. Active tectonics of the north and central Aegean Sea. Geophysical Journal International, 106, 433–490.
Thiebault, F. 1982. L' évolution géodynamique des Hellenides externes en Péloponnèse méridional. Publ. Soc. Géol. Nord., 6, 574 p.
Thomson, S., Stoeckhert, B. & Brix, M. 1998. Thermochronology of the high-pressure metamorphic rocks of Crete, Greece: implications for the speed of tectonic processes. Geology, 26, 259–262.
Tirel, C., Brun, J.P., Burov, E., Wortel, M.J.R. & Lebedev, S. 2013. A plate tectonics oddity: caterpillar-walk exhumation of subducted continental crust. Geology, 41, 5, 555–558.
Trikkalinos, J. 1939. Uber die steirischen orogenen Bewegungen im Gebiete von Ferres. Proc. Acad. Athens, 14.
Trikkalinos, J. 1940. Uber die attischen und wallachissen orogenen Bewegungen in Attika. Proc. Acad. Athens, 15.
Turcotte, D.L. & Oxburg, E.R. 1972. Mantle convection and the new global tectonics. Ann. Rev. Fluid Mech., 4, 33–68.
Van Hinsbergen, D.J.J., Zachariasse, W.J., Wortel, M.J.R. & Meulenkamp, J.E. 2005. Underthrusting and exhumation: a comparison between the External Hellenides and the hot Cycladic and cold South Aegean core complexes (Greece). Tectonics, 24.
Vassilakis, E., Royden, L. & Papanikolaou, D. 2011. Kinematic links between subduction along the Hellenic trench and extension in the Gulf of Corinth, Greece: A multidisciplinary analysis. Earth & Plan. Sci. Let., 303(1–2), 108–120.
Vine F.J. & Matthews D.H. 1963. Magnetic Anomalies over Oceanic Ridges. Nature, 199, 947–949.
Voreadis, G. 1932. The schist-chert complex of Salamis and its mafic and ultramafic erruptions. Greek Geological Service, Publ. 19 (in greek).
Voreadis. G. 1938. Neocimmerian folding in Eastern Greece and the phases of the Alpine orogeny in Greece. Habilitation Thesis, Athens (in greek).
Wadati, K. 1935. On the activity of deep focus earthquakes in the Japan Island and neighborhood. Geophys. Mag., 8, 305–326.
Wawrzenitz, N. & Krohe, A. 1998. Exhumation and doming of the Thasos metamorphic core complex (S. Rhodope, Greece): structural and geochronological constraints. Tectonophysics, 285, 301–332.
Wegener, A. 1920. Die Entstehung der Kontinente und Ozeane. II Aufl.
Wernicke, B. 1981. Low-angle normal faults in the Basin and Range Province: Nappe tectonics in an extending orogen. Nature, 291, 645–648.
Wernicke, B. & Burchfiel, B.C. 1982. Modes of extensional tectonics. J. Struc. Geol., 4, 104–115.
Wilson, J.T. 1965. A New Class of Faults and their Bearing on Continental Drift. Nature, 207, 343–347.
Wortel, M.J.R. & Spakman, W. 2000. Subduction and slab detachment in the Mediterranean–Carpathian region. Science, 290, 1910–1917.
Xypolias, P. & Alsop, G. I. 2014. Regional flow perturbation folding within an exhumation channel: A case study from the Cycladic Blueschists. J. Struct. Geol., 62, 141–155.

Post-Alpine Formations in the Hellenic Region

5

5.1 General Characteristics—Ages—Geographical Distribution

The post-Alpine formations of the Hellenic region include strata that lie unconformably on the Alpine structure, belong to the interior and the hinterland of the present Hellenic arc and have been deposited during the Neogene and Quaternary. The timing of the closure of the Alpine orogenic cycle gradually progresses from the internal to the external part of the Hellenic arc. As a result, older post-Alpine formations are deposited in the interior of the arc and progressively younger formations towards the present day arc front.

The post-Alpine deposits are mostly terrestrial continental sediments, but also fluvial and lacustrine, that have been deposited in neotectonic grabens, with sedimentary sequences formed by the erosion of the surrounding Alpine formations, which crop out in the neotectonic horsts of continental Greece. Overall, a terrestrial phase is dominant throughout the post-orogenic evolution, without major changes in the physical geography of each region. However, there are also several sites where marine post-Alpine sediments can be found, mainly of coastal facies, which at present crop out on the mainland and occasionally at high altitudes, due to major neotectonic uplift movements. Such cases occur in Northern Peloponnese in the Trikala and Kalavryta areas at altitudes around 1200 m and in Central Crete, north of Anogia. Naturally, when we examine the Geology of Greece, we have to consider not only the present terrestrial area, but also the submarine area, which in many cases was, until recently, also land. In other words, we have to study the whole structure of the upper Hellenic crust, regardless of whether today it is above or below sea level. In the case of marine deposits on land the entire geologic—paleogeographic evolution becomes more complex. It can be separated in two phases with the distinction of a first subsiding phase, during which marine sedimentation was established for a certain geological period and then a second uplifting phase, with an interruption of the marine sedimentation, followed by the beginning of erosion of the uplifted post-Alpine sediments. In each case, the dating of the sediments will provide us with the necessary information to determine the geological history of each region. The basic information for each post-Alpine basin is the age of the base of the unconformable sedimentary sequence on the Alpine basement. In any case, before the beginning of the post-Alpine cycle, there is a long period of emersion and erosion of the Alpine unit(s) of the region until the beginning of the post-Alpine sedimentation. This erosion period may be relatively short or much longer, depending on whether molassic sedimentation was simultaneously occurring in the area or not.

The neotectonic horsts, naturally, form higher elevated areas, mainly in the island arc region, and are subject to erosion. They are affected by intense erosional processes that form highly incised linear gorges and canyons that transfer clastic material to the base level of the sedimentation basins/grabens (Fig. 5.1). These basins may be above sea level, forming terrestrial basins, which depending on the hydrogeological conditions of their basement may result into purely terrestrial and fluvial basins (permeable formations), or into lakes with lacustrine deposits (impermeable formations). When the basins are below sea level, we have marine neotectonic basins/grabens, with various sedimentary facies, mostly coastal ones.

Neotectonic structures are overprinting pre-Alpine structures and control newly established sedimentation basins. From the distribution map of the post-Alpine sediments in Greece (Fig. 5.2) it is evident that the main marine basins are of Upper Miocene–Pliocene age and can be observed mainly in the Peloponnese (only Pliocene and Pleistocene) and in Crete. Additionally, Upper Miocene marine basins can be observed in coastal areas of the North Aegean. On the contrary, in the Eastern Aegean island region (Samos, Chios, Kos, etc.) there are predominantly continental lacustrine basins of Upper Miocene–Pliocene age, while the rest of continental Greece is dominated by terrestrial basins. Of course, several basins have a complex multiphase evolution, where the terrestrial phase may succeed the lacustrine or

D. I. Papanikolaou, *The Geology of Greece*, Regional Geology Reviews,
https://doi.org/10.1007/978-3-030-60731-9_5

Fig. 5.1 Characteristic outcrop of an active fault in the coastal zone of Northern Peloponnese, in the Psathopirgos area, which uplifts the southern fault block of Northern Peloponnese. This uplift creates a canyon from deep linear erosion, perpendicular to the fault plane. The northern fault block of the Gulf of Corinth subsides and marine sedimentation has been established. The Psathopirgos fault is a marginal fault of the Corinth basin that controls the dynamic equilibrium of the two neotectonic fault blocks. An uplift of 0.7–0.8 mm/year was constrained in its immediate footwall by 234U-230Th coral dating (Houghton et al. 2003)

marine coastal phase and vice versa. Of particular interest are the post-Alpine basins that have been developed over previous molassic basins, such as the Western Thessaly and Western Macedonia Plio-Quaternary continental basins, established over the pre-existing marine Oligo-Miocene Meso-Hellenic Basin.

A typical terrestrial phase in Greece is the reddish clastic formations with fluvial deposits of Upper Miocene age, containing the well-known Pikermian mammal fauna (Hipparion mediterraneum, Tragocerus amaltheus, Gazella, Diceros pachygnathus, Mesopithecus pentelicus, Machairodus aphanistus etc.) (Gaudry 1867; Psarianos and Symeonidis 1975). These basins, apart from the abundance of fossil mammals that attract a lot of paleontological interest, do not contain deposits of economic importance, unlike the lacustrine basins, where lignite deposits can be found and form the main source of energy in Greece. Typical lignite bearing basins are located in Aliveri, of Miocene age, in Ptolemais and Vegora, of Upper Miocene–Pliocene age, in Megalopolis, mainly of Lower Pleistocene age, in the Philippi basin, mainly of Middle-Upper Pleistocene age, and other smaller ones (Koukouzas and Koukouzas 1995). In most cases, lignites have been formed at the margins of the basins, where marches were developed for long periods of time and where organic plant matter was accumulated (Christanis 2004) (Fig. 5.3). Typical fossils found in lacustrine post-Alpine sediments are (as for example in the Megara Basin, Theodoropoulos 1968) Vivipara, Melanopsis, Theodoxus, Melania, Planorbis, Congeria, Unio, et al. The post-Alpine continental basins of Greece usually comprise terrestrial alluvial and fluvial deposits interfingering with lacustrine formations. Around these environmental areas and especially at the coastal zones of the lakes there are often concentrations of mammals during the Late Miocene–Pliocene period all over the Aegeis land, bridging the Western Minor Asia areas with those of continental Greece. Thus, a large number of fossiliferous sites of Pikermian fauna have been studied in Eastern Greece and the Aegean islands (Symeonidis and Marcopoulou-Diacantoni 1977) (Fig. 5.4).

The marine post-Alpine basins are of greater economic importance, because they may contain hydrocarbons, like in the Upper Miocene basin of the North Aegean, west of Thassos Island (Lalechos and Savoyat 1977; Proedrou and Papakonstantinou 2004). The outcrops of marine post-Alpine sediments in Greece are relatively restricted,

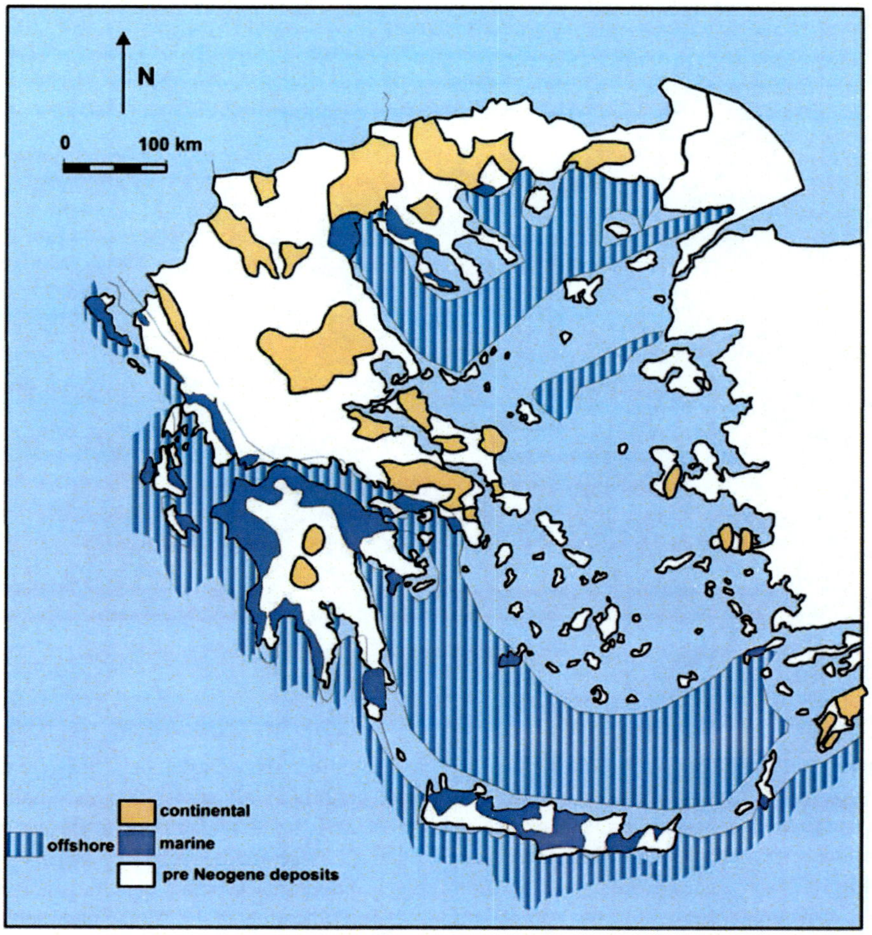

Fig. 5.2 Distribution of the post-Alpine formations in Greece. The arc boundaries inside the Aegean microplate show the control of the post-Alpine formations from the convergence of the plates, characterized by the uplift zone of the marine sediments in the front of the Hellenic arc. 1: Mainly terrestrial deposits of Miocene–Quaternary, both continental and lacustrine. 2. marine deposits of Upper Miocene–Quaternary, extended to the submarine area as well. 3: Alpine basement along with molassic deposits of Eocene–Miocene

including only some uplifted basins, whereas many Neogene marine basins are still active offshore. They comprise mainly Upper Miocene and Pliocene or even Lower Pleistocene sediments that have been deposited in basins of a few hundred meters depth, around 2–10 million years ago, which today, have been elevated up to 1.500 m above sea level due to neotectonic uplift (e.g. Northern Peloponnese). On the other hand other deep tectonic basins that were formed by subsidence during Miocene and/or Pliocene remain below sea level to date. Typical marine fossils in Greece are: (i) for the Late Miocene period various species of Ostrea, Glycymeris, Pycnodonta, Siderastraea, Porites, Tarbellastraea, Turritella, Chlamys, Cerithium, Terebralia, (ii) for the Pliocene period Pecten rhegiensis, Flabellipecten flabelliformis, Hinnites ercolianus, Amussium cristatum, Spondylus crassicosta and (iii) for the Tyrrhenian (Upper Pleistocene) Strombus bubonius, Conus testudinarius and Natica lactea (Psarianos and Symeonidis 1975).

The probability of finding oil on the mainland of Greece in post-Alpine marine basins today is relatively low, because their history was short and they have been highly deformed, with relatively small sediment thickness, that all together do not allow the maturation of the organic compounds to petroleum. Additionally, it should be noted that the potential oil easily escapes near surface conditions. On the other hand, the chances are high in several post-Alpine marine basins occurring at sea. The Messinian salinity crisis observed all over the Mediterranean basins, as described earlier, has

Fig. 5.3 Typical outcrop of horizontal lignite strata of Upper Miocene age, observed below 25 m of superjacent sterile sediments from the Mavropigi lignite mine in Ptolemais

favoured the creation of oil traps beneath the evaporites and thus, stratigraphy is considered more promising for gas and oil exploration within the Mediterranean basins. Whether oil reservoirs are developed or not depends on the tectonic structure and evolution of each basin, related to its position in the Mediterranean structure, either within the Alpine belt or outside it.

5.2 The Arc Parallel and the Arc Transverse —Oblique Basins

The post-orogenic basins are created by **marginal faults**, which have a different geometry and origin, depending on their position in the arc at each time period and their generic relation with the arc or other surrounding structures. Thus, neotectonic extensional arc parallel structures are observed across the arc, with succession of neotectonic horsts and grabens, from the island arc region to the back-arc basin and the volcanic arc region. A typical example of this arc parallel structure is given in the cross section from the Southwestern Peloponnese to the Central Aegean, which intersects the neotectonic horsts of Southwestern Messenia, Mani/Taygetus, Parnon, Argolis, Aegina, Attica and Southern Evia (Fig. 5.5) (Papanikolaou et al. 1988).

In between the afore mentioned neotectonic horsts, the Messiniakos, Laconikos, Argolikos, Western Saronikos, Eastern Saronikos, South Euboikos and Central Aegean neotectonic grabens/basins have been created along the same time period. In all cases, the marginal faults are normal NNW–SSE trending faults, with throws ranging from 1 to 7 km and with moderate to low angle diping planes. Some major structures separate the metamorphic from the non metamorphic units of the Alpine basement and correspond to extensional detachments with displacements of several km to several tens of km, whereas other faults represent medium to high angle normal faults with smaller displacements of 1–2 km. These are all extensional structures within the upper crust of the Hellenic arc, perfectly aligned with its dynamics, characterized by extension in the ENE–WSW direction. In a similar cross section on the eastern segment of the arc in the Dodecanese, the orientation of the marginal faults becomes ENE–WSW, parallel to the orientation of the Hellenic arc/trench system, and extensional stresses have a NNW–SSE orientation (Mariolakos and Papanikolaou 1981). It is evident that in between the two arc segments in the Cretan region, E-W basins are present, such as the characteristic Messara basin with N–S extension, where low-angle normal faults have ruptured the Late Miocene marine sediments overlying the upper nappes of the Cretan tectono-stratigraphy (Vassilakis 2006).

The previous examples show that the structure of the current Hellenic arc is dominant in the post-orogenic basins, without any major differentiation. However, the setting is different in other post-Alpine basins to the north, like for example the Plio-Quaternary basins of the Corinth and

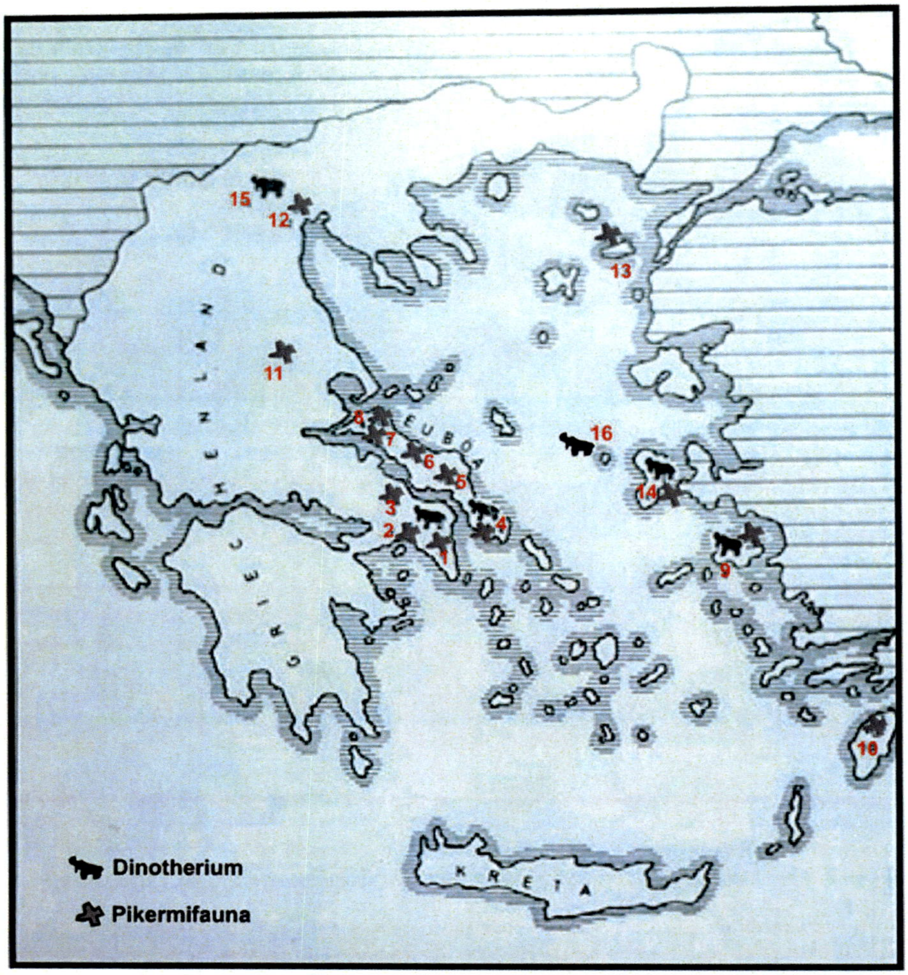

Fig. 5.4 Distribution of outcrops of Dinotherium and of the Pikermi fauna in Eastern Continental Greece and the Aegean islands (based on Symeonidis and Marcopoulou-Diacantoni 1977). Dinotherium has been found in localities: 1, 4, 9, 14, 15 and also in Psara Island 16 (Besenecker and Symeonidis 1974). Pikermian fauna in all the localities 1–15. 1: Pikermi, 2: Tour la Reine, Athens, 3: Tanagra, 4: Almyropotamos, 5: Triada, 6: Ahmet-Aga, 7: Rhovies, 8: Achladi, 9: Samos, 10: Rodos, 11: Alifaka, Thessaly, 12: Vathylakkos, Thessaloniki, 13: Imbros, 14: Chios, 15: Central Macedonia, 16: Psara

Maliac gulfs. Here the basins have a general E–W direction and disrupt the Alpine structures as well as the earlier structures of the molassic basins and even the earlier post-orogenic basins. These are arc transversal to arc oblique structures, which disrupt the previous arc parallel structures (Fig. 5.6) (Papanikolaou and Royden 2007).

A typical case of younger E–W basins that disrupt the previous activity of NW–SE basins is the Pleistocene Alkyonides basin in the Eastern Corinth Gulf, which has disrupted the Upper Miocene-Pliocene Megara basin (Mariolakos and Papanikolaou 1981). It is worth noting that these E–W marginal faults of the Alkyonides basin were activated during the 6.7 and 6.4 magnitude 1981 earthquakes (see also Fig. 4.12) that struck also Athens at a distance of 70 km (Mariolakos et al. 1981) (Fig. 5.7). The Megara Basin progressively became inactive and abandoned with the initiation of the younger E-W trending south Alkyonides fault system and the development of the Alkyonides Basin (Mariolakos and Papanikolaou 1981). Paleomagnetic studies also distinguish differential rotational domains showing that the Megara region belonged to the Beotia-Lokris block in the past, but now has been incorporated into the Peloponnese block (Mattei et al. 2004).

The transverse structures, such as the Gulf of Corinth which is the most prominent, are associated with the development of the Central Hellenic Shear Zone (CHSZ), which replaces the northern boundary of the Aegean microplate in continental Greece (Papanikolaou and Royden 2007), as the westward prolongation of the North Anatolian Fault and its equivalent faults in the North Aegean Sea. Interestingly, the post-orogenic basins north of the CHSZ, such as the Ptolemais and Elassona basins, keep up to the NW–SE orientation, like those of the southern part of the Hellenic arc, south of the CHSZ, becoming again arc parallel structures. It should be reminded that north of the CHSZ the geotectonic process is controlled by the collisional zone between the subducting Adria plate and the Northern Hellenides, with a rate of 8 mm/year. Thus, the geometric and

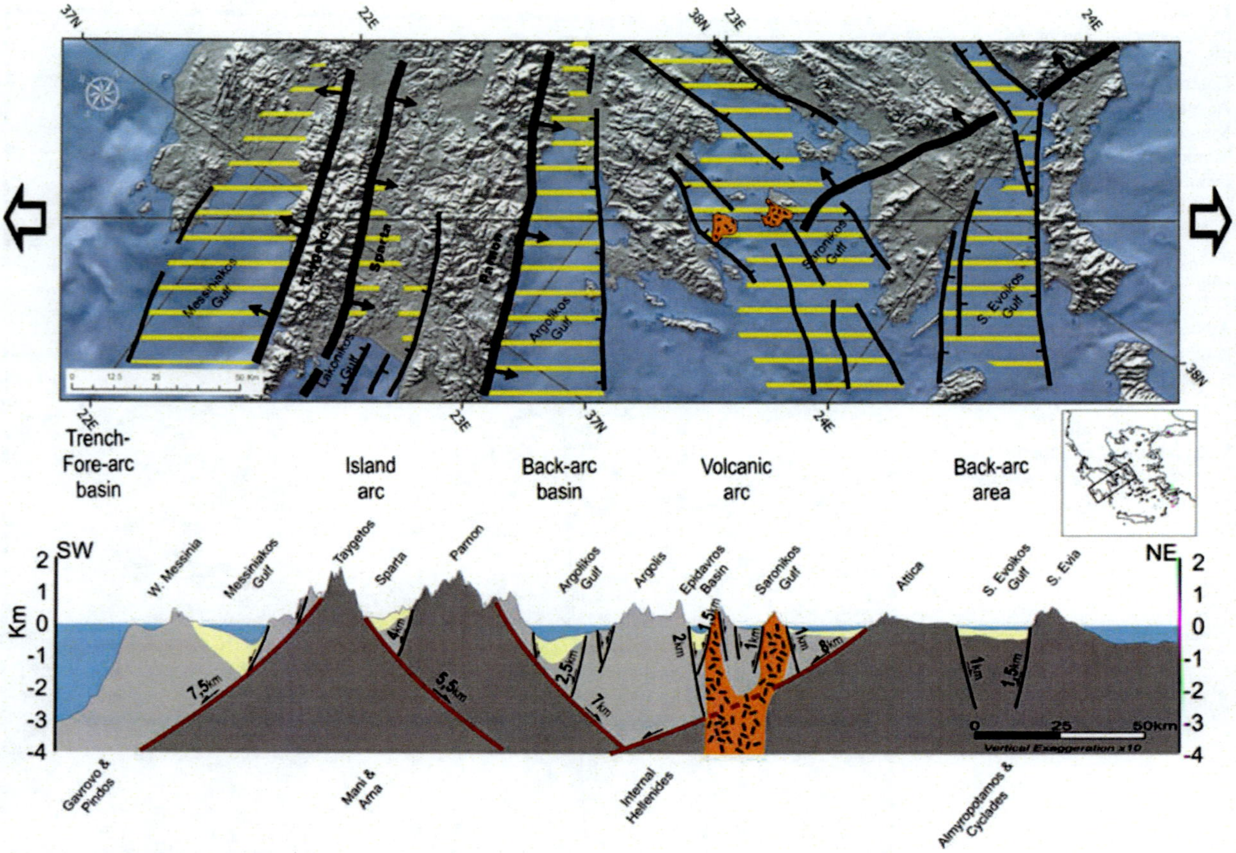

Fig. 5.5 Simplified neotectonic map and tectonic cross section, transverse to the Hellenic arc, from the Ionian to the Aegean Sea, which intersects the arc parallel neotectonic structures (from Papanikolaou 2010, modified from Papanikolaou et al. 1988). An alternation of neotectonic grabens and horsts of NW–SE trend is observed, with decreasing vertical displacements from the external part of the arc towards the internal. The post-Alpine sediments are shown in yellow. Dark gray corresponds to the metamorphic units occurring in the form of tectonic windows, whereas light gray corresponds to the non-metamophosed Alpine units. Red lines represent major extensional detachments and black lines major normal faults. The Plio-Quaternary volcanic extrusions are depicted in orange

dynamic-kinematic characteristics in the north are similar to those of the Hellenic arc to the south of the CHSZ, but of much lower intensity, probably due to the absence of a retreating oceanic subduction zone.

5.3 The Messinian Salinity Crisis

The Miocene–Pliocene marine sediments in Greece and throughout the Mediterranean, incorporate evaporites in the Upper Miocene and more especially in the Tortonian and mainly in the Messinian (Drooger 1973; Hsü et al. 1978; Roveri et al. 2014). Primary observations on outcrops of Upper Miocene marine sequences were confirmed by geophysical investigations and drilling in the submarine basins throughout the Mediterranean (Fig. 5.8). The overall Messinian salinity crisis developed in three main stages, each of them characterized by different paleo-environmental conditions (Roveri et al. 2014). During the first stage, evaporites precipitated in shallow sub-basins but the process was magnified and reached its peak in the second stage, when evaporite precipitation shifted to the deepest depocentres, whereas the third stage was characterized by large-scale environmental fluctuations in a Mediterranean basin transformed into a brackish water lake.

In Crete Messinian evaporites are traced in post Alpine sedimentary basins in between the middle Miocene–Tortonian sequences and the overlying Pliocene marls (e.g. Meulenkamp 1979; Meulenkamp et al. 1979) (Fig. 5.9). The famous Knossos Palace of the Minoan civilization (approximately 3000–1400 B.C.) situated 8 km south of Heraklion is actually built on the Messinian evaporites and most of the handcrafts are made of alabaster.

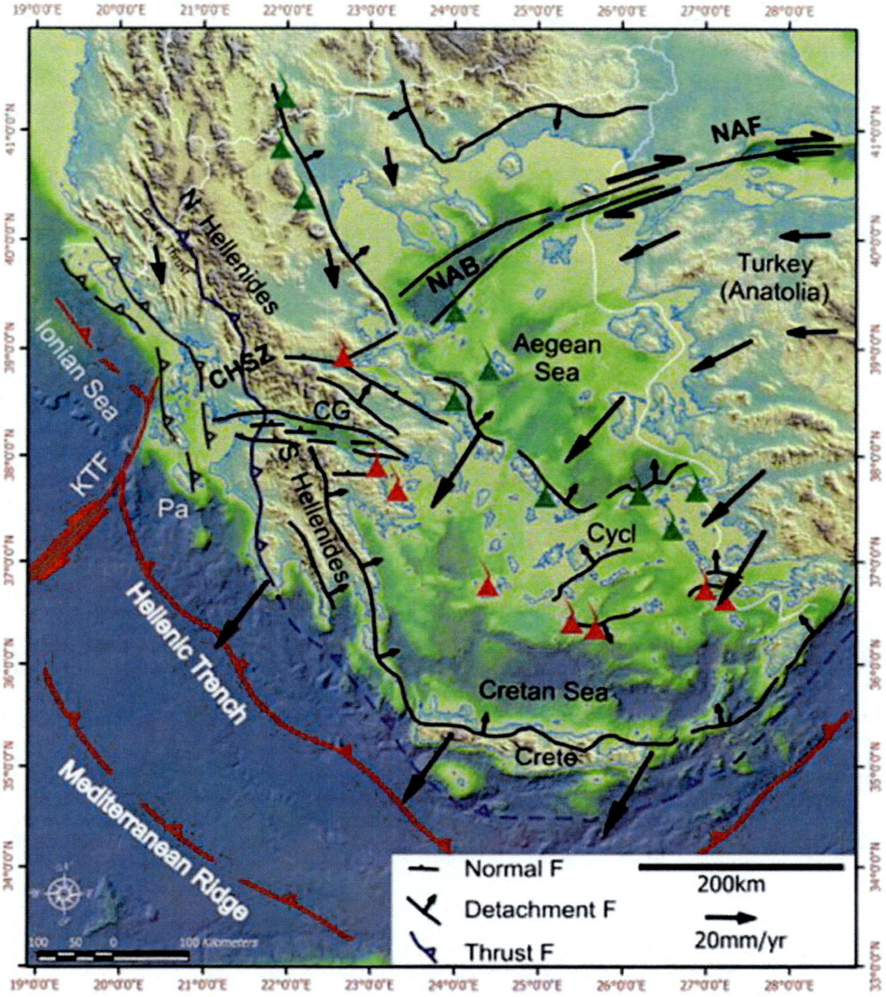

Fig. 5.6 Map of the major extensional structures of the Hellenic arc, showing the disruption of the previous arc parallel structures, of NW–SE orientation, from the new transverse structures, of E–W orientation, in the area of the Central Hellenic Shear Zone (based on Papanikolaou and Royden 2007 and Vassilakis et al. 2011)

Thus, it is expected that upper Miocene evaporites are likely to be found in Neogene marine stratigraphic sequences of the Mediterranean. The deformation of the evaporites together with their diapirism create suitable structures for hydrocarbon traps (anticline structures, faults, etc.). In addition, marine sediments are regarded as favourable formations for hydrocarbon. As a result, Upper-Miocene marine sedimentary sequences are generally considered as petroleum-promising. This was confirmed in the 1970s in the Prinos area, west of Thassos Island in the Northern Aegean Sea (Lalechos and Savoyat 1977; Proedrou and Papakostantinou 2004), where oil is extracted over the last 40 years.

The Mediterranean basin was isolated from its neighboring oceans, the Indian Ocean since the Middle Miocene and the Atlantic Ocean (Gibraltar) during long intervals of the Tortonian and especially the Messinian. At that time, the amount of evaporated water of the Mediterranean exceeded the fresh water input from the rivers of Africa (e.g. Nile) and Europe (Rhone, Po, etc.). Thus, up to 3–4 km thick evaporites were deposited, towards the Mediterranean sea bottom. Overall, two models had been proposed by sedimentologists: (i) the evaporation rate was enormous so that the entire Mediterranean was dried and only four large salt lakes remained, isolated from one another, mainly in the Balearic, Tyrrhenian, Ionian, and Levantine deep basins, as well as some smaller seasonal basins/lagoons, or (ii) the evaporation was significant, but chemical sedimentation with evaporite deposition occured everywhere, regardless the sea depth. Independently of the specific model of deposition of the evaporites, the main causes are of tectonic origin and more specifically, the collision of Arabia with Southeastern Asia Minor during the Middle/Late Miocene, the uplift of a

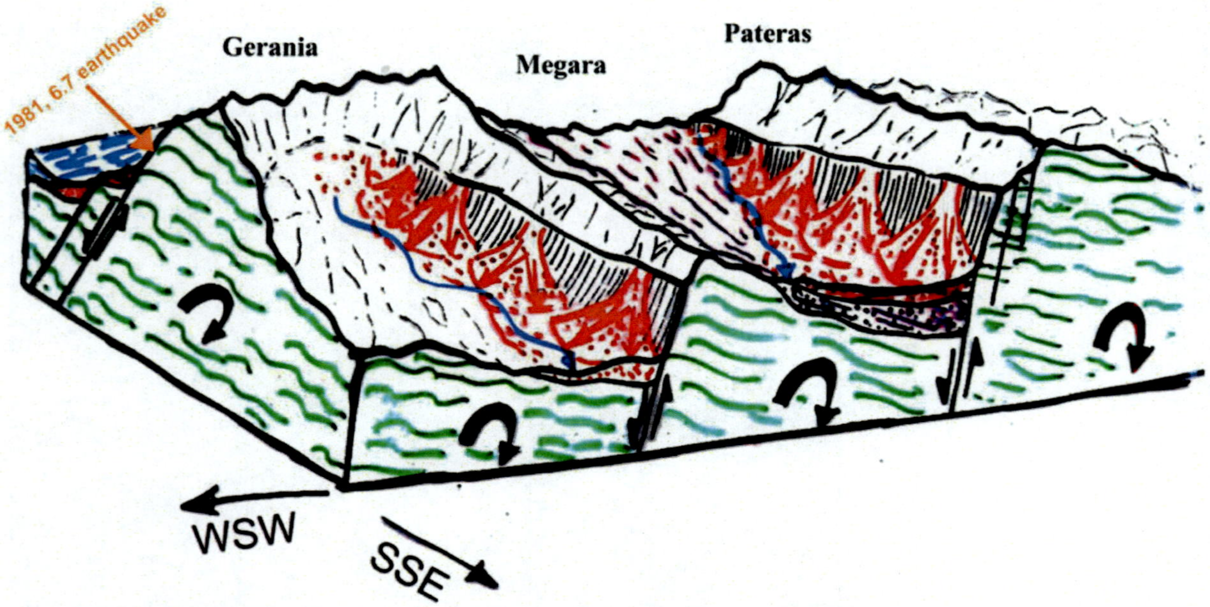

Fig. 5.7 Schematic 3D stereogram of the neotectonic structure of the Megara basin (from Mariolakos and Papanikolaou 1981). The two NW–SE and E–W fault sets can be observed to rotate the fault blocks, with characteristic dip of the bedding and morphological peculiarities (morphological slopes, drainage network etc.). The older NW–SE system has created the half-graben of the Megara basin tilted to the NE, while the younger E–W system interrupted its activity and formed the Alkyonides basin, tilted to the south. The 1981 earthquake events were part of this recent activity of the E–W faults forming the northern slopes of Gerania Mt

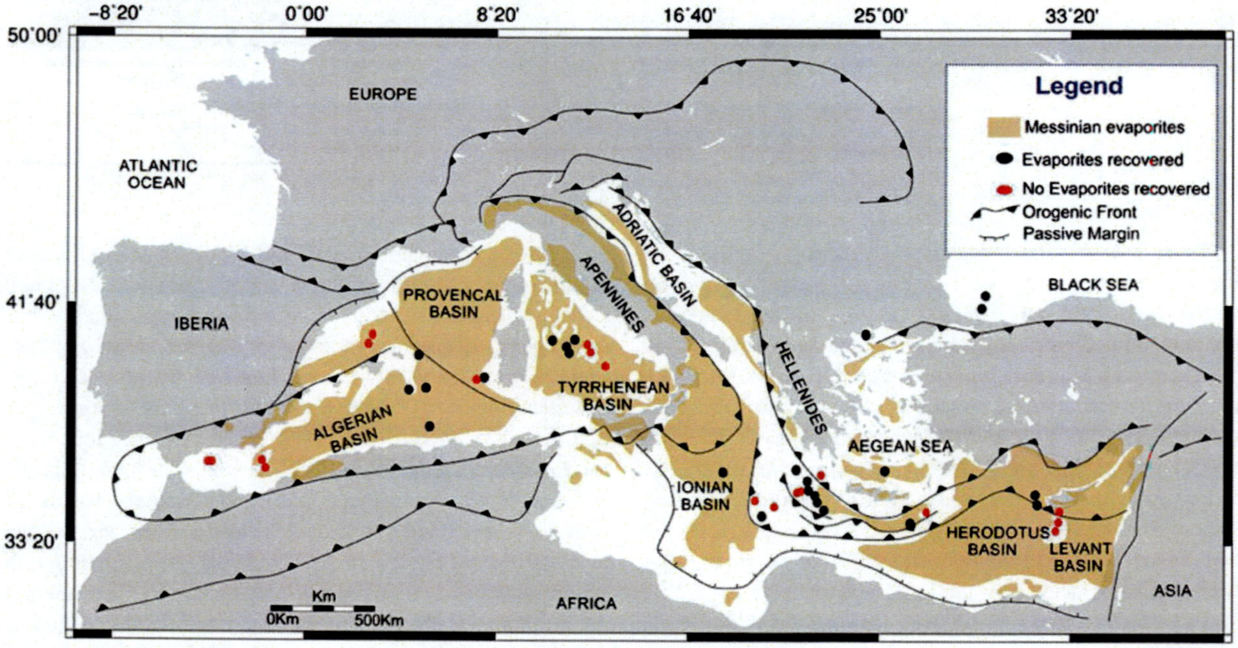

Fig. 5.8 The extension of the Messinian evaporites in the Mediterranean (after Roveri et al. 2014, modified)

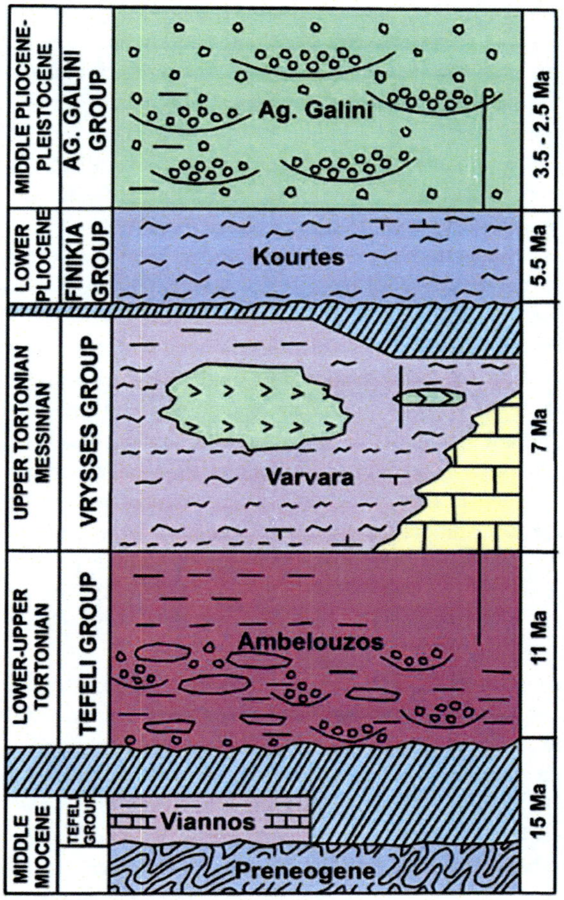

Fig. 5.9 Schematic stratigraphic column of the post-Alpine formations of the Southwestern Heraklion basin in Crete (from Meulenkamp 1979, modified). Gypsum deposits of Messinian age are observed in the upper section of the Varvara formation of the Vrysses group

natural dam from the rising Alps to the ParaTethys waters (Pannonian basin and Alpine molasse) and, finally, the elevation of the Bettic Cordillera and Rif around Gibraltar (Hsü et al. 1978). It is remarkable that despite the narrow and shallow strait of Gibraltar (1–1.5 km wide and 150 m deep as of today) after the Miocene–Pliocene boundary 5.4 Ma ago, the same process did not re-appear in the Mediterranean, except in isolated cases on a local scale. The Miocene/Pliocene boundary is one of the well defined stratigraphic boundaries and it is characterized by the deposition of marly limestones in the Mediterranean, of the so-called Trubi facies (from Sicily), over the Messinian evaporites. In Western Greece, the Messinian evaporites often coexist with the Triassic evaporites, occurring at the base of the stratigraphic column of the Ionian nappe, due to Plio-Quaternary tectonics (e.g. Zakynthos) (Underhill 1989).

5.4 The Subsidence of the Aegean During the Quaternary

The Central Aegean region is bordered to the north by the North Aegean basin and to the south by the Cretan basin. The intermediate area is primarily a shallow region-platform, with the exception of the Skyros basin and the small Southern Evia–Northern Cyclades and Chios–Ikaria basins. It is characteristic that this central Aegean area has been paleogeographically considered as a unique terrestrial area, named **Aegeis** (Philippson 1901), which connected the present day Central Eastern continental Greece with Western Asia Minor. The continental deposits of Middle/Upper Miocene–Pliocene observed on both sides of the Aegean (Attica-Beotia-Evia in the west, Chios-Samos-Kos in the east) but also in the centre of the Aegean (e.g. Psara) demonstrate the existence of Aegeis until its recent collapse in the Pleistocene, when migrations of mammals, including hominids, terminated due to subsidence (Papanikolaou et al. 2019b). Otherwise, the existence of Aegeis facilitated the migration of the mammals especially during Late Miocene, like those of the Upper Miocene Pikermi fauna, which is found in many outcrops on both sides of the Aegean Sea (Attica, Beotia, Southern Evia, Samos) (Mitzopoulos 1961) (see also Fig. 5.4). It is indicative that in this area there are generally no marine sediments of Neogene–Quaternary, with a few exceptions, such as: (1) the Messinian strata in Skyros (Grekoff et al. 1967), which are probably associated with a marine channel towards the North Aegean basin, (2) the "Lower-Pliocene" strata of Ikaria (Ktenas 1927), which had probably some connection to the southern Cyclades and the Cretan basin, (3) the Upper Pliocene strata of Rafina (Christodoulou 1958), signifying the first entry of the sea in the present South Evoikos gulf. On the contrary, in Northern Attica and Beotia (Oropos, Tanagra) and in Evia (Kymi, Agia Anna, Kerasia), as well as in Chios and Samos, there are Neogene sediments of lacustrine facies with fossil mammals, despite the fact that their physical profiles are observed today along the coastal zones adjacent to the Aegean Sea. At the same time, the distribution of the paleoflora assemblages (Velitzelos et al. 2014) leads to interesting conclusions, which, compared to the present distribution of flora and endemic species in the Aegean (Arianoutsou-Faragitaki et al. 2003) can provide us with a well-documented paleogeographic image during the late Cenozoic (see also Sect. 11.3).

The above observations show that the Central Aegean region, former Aegeis, has subsided fairly recently, mainly during the Quaternary. The occurrence of Cyclades is typical, showing the general Aegean subsidence with the former high mountains forming the islands today whereas the former flat coastal lowlands are now situated below sea level, forming the Cycladic plateau. Submarine research, shows either lack of sediments or a minimal sediment thickness over the Alpine basement all over the Central Aegean platform, in agreement with onshore findings (Papanikolaou et al. 2015, 2019a) (Fig. 5.10). Therefore, even if the direct dating of the base of the transgressive sediments over the platform is not feasible, we can use their thickness and sedimentation rate, determined by shallow coring, so as to conclude that the maximum age of the sediments is Middle Pleistocene and in some cases, Upper Pleistocene. The study of the sea level changes and shelf break prograding sequences during the last 400 Ka in the Aegean margins concluded that subsidence rates range between 1.46–1.88 m/Ka for the isotopic stages 2, 6, 8, 10 and probably 12 (Middle–Late Pleistocene) implying that during these glacial periods the intermediate land bridges between the two Aegean margins were gradually subsided (Lykousis 2009).

The Aegean subsidence, observed throughout the interior of the arc and its hinterland, occurred due to the extension caused by the roll back of the retreating Hellenic subduction zone (Royden 1993; Jolivet and Faccenna 2000; Royden and Papanikolaou 2011; Jolivet et al. 2013).

5.5 Climate Change During the Quaternary–Marine Terraces

The Quaternary covers the last 2.6 million years and is characterized by the occurrence of climate changes, with distinction between glacial and interglacial-warm periods (e.g. Klebelsberg 1949). Until the 1990s the general perception of the Quaternary was that we could perceive four glacial periods and four interglacial ones, of which the last interglacial-warm period is the present, called—and distinguished from all the previous ones—as the **Holocene**. The above subdivision was based on research from the Alps, where the widespread occurrence of glacial deposits had helped to map successive stratigraphic formations in a continental environment. However, the corresponding stratigraphic analysis in other European regions (e.g. the Urals) and in other continents showed that there was no agreement on this subdivision. At the same time, stratigraphic analysis has gradually included during the last decades additional results from research in lakes, glaciers and ice caps, such as in Greenland, and especially many cores of marine sediments both on the continental shelf and at the ocean bottom. Together with the improvement of sediment dating techniques, it has become apparent that the number of climatic alternations in the last million years was much higher, reaching 36 cycles. The use of oxygen isotopes has enabled the determination of paleo-temperatures in sediments (Chappell and Schakleton 1986) and therefore, along with the increasingly new technologies, the determination of the climatic cycles for most of the Quaternary was rendered possible (Shackleton 1978, 2000) (Fig. 5.11). Deep ice cores in Antarctica provide information over the climatic history of the past 800,000 years and future projects plan to extend the observations up to 1,500,000 years ago.

The diagram shows that there are extreme and intermediate cases of climate change and that the duration of each cycle was differentiated from the Lower to the Middle Pleistocene at approximately 900 Ka, with the dominance of 100.000-year cycles. It should be noted that the existence of 40.000-years cycles in climate change was first determined by Milankovitch (1936), based on astronomical data. Today we know that the cycles causing climate changes are astronomically induced, depending on the combination of three different astronomical motions of the Earth in respect to the Sun (Eccentricity with 100 Kyr period, Obliquity with 41 Kyr period and Precession with 21 Kyr period). The main cause is the fluctuation of the solar radiation on the planet in relation to these three individual movements.

In geology, climate change is more relevant to its indirect effect on the global **sea level changes**. Thus, instead of glacial and interglacial periods, the terms **low stand** and **high stand** sea level are used respectively. It is clear that low stand corresponds to glacial periods, when the water is bound to the ice of the poles and the glaciers, and high stand to warm periods, when the ice melts. Therefore, there are also the sea level change curves corresponding to the paleo-temperature change curves (Fig. 5.11). The above changes were assessed especially for the Eastern Mediterranean region and in more detail for the last 20 Ka in the Greek region, by the «Scientific Committee of the Study of the Climate Change Impacts» of the Bank of Greece, with the aim of forecasting the changes in the next 50–100 years, which concluded that an uplift of the sea level by 40–60 cm may be expected (Zerefos et al. 2011).

It is noteworthy that the latest sea level change from the previous glacial period of Wurm (Marine isotope Stage/MIS 2) to date is an uplift of 125 m, which represents the so-called **Holocene transgression**, which was described by Negris (1903) for the entire Mediterranean region (see also Sect. 11.3). It is essentially the greatest possible sea level change, which can be observed in extreme climate changes. In the previous similar change, which has been used as the boundary between the Middle and Upper Pleistocene, 126 Ka ago, the sea level was approximately the same as the present one and perhaps 6–7 m higher. The estimated paleoshoreline during the Wurm period at –125 m shows a series of significant changes in the physico-geographical configuration of Greece. Some typical examples are the

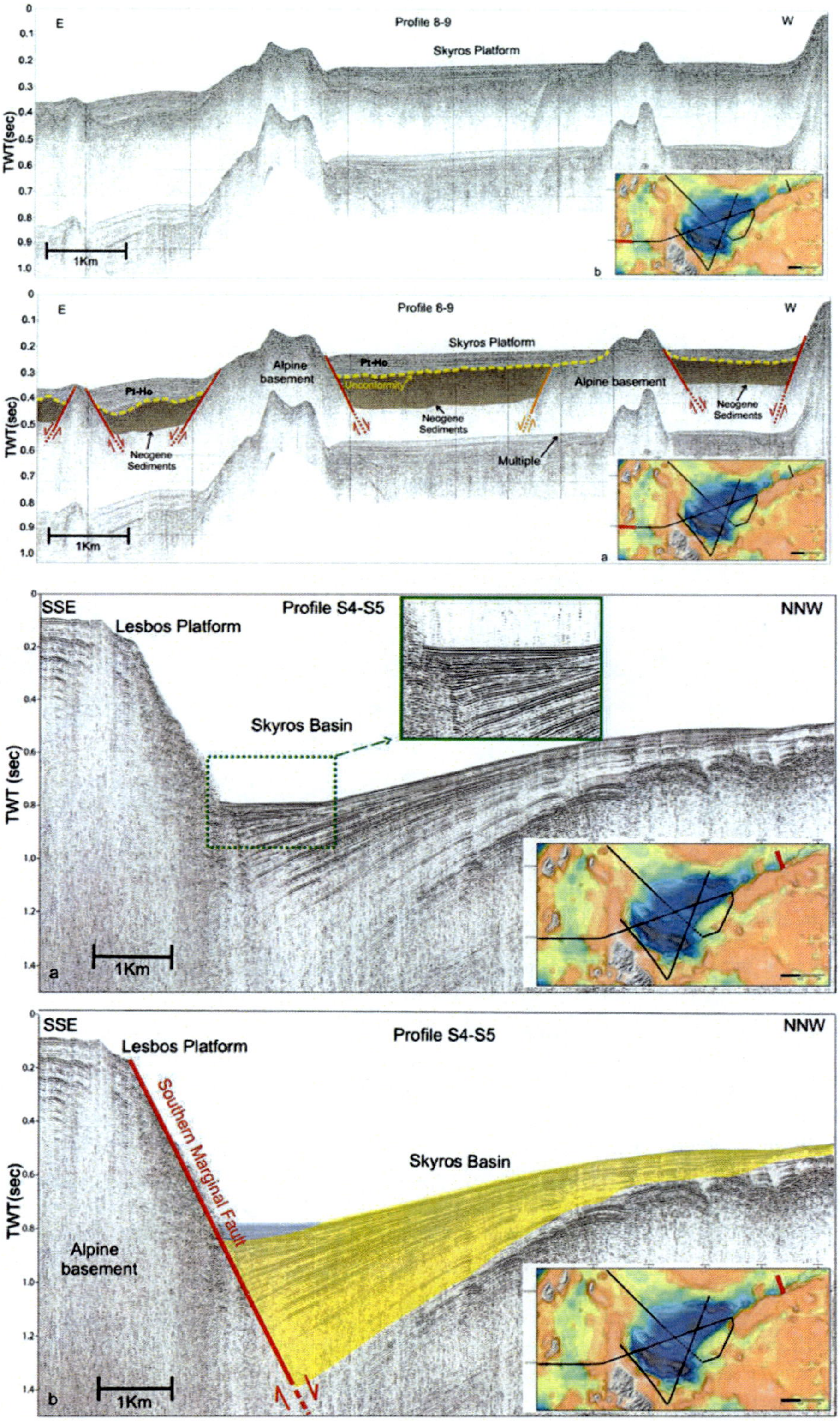

Fig. 5.10 Characteristic litho-seismic profiles of the Central Aegean over the Skyros–Northern Sporades platform (from Papanikolaou et al. 2015, 2019a), showing the minimal thickness of the recent sediments up to a few tens of meters, of Middle-Upper Pleistocene age, over the Alpine basement of the former Aegeis. **a** Lithoseismic profile from the Skyros–Northern Sporades platform, showing the minimum sediment thickness of only 10-30 m above the Alpine basement. **b** Lithoseismic profile transverse to the principal marginal fault of the Skyros basin to the NW of Lesvos, showing an increasing sediment thickness towards the fault plane on the hanging wall up to 600 m, due to syn-sedimentary tectonism/growth faulting. Over the Limnos platform to the north, the thickness of the same sedimentary sequence is limited to only 50 m. On the contrary, on the foot wall to the south the sediment thickness over the Alpine basement is limited only to a few meters

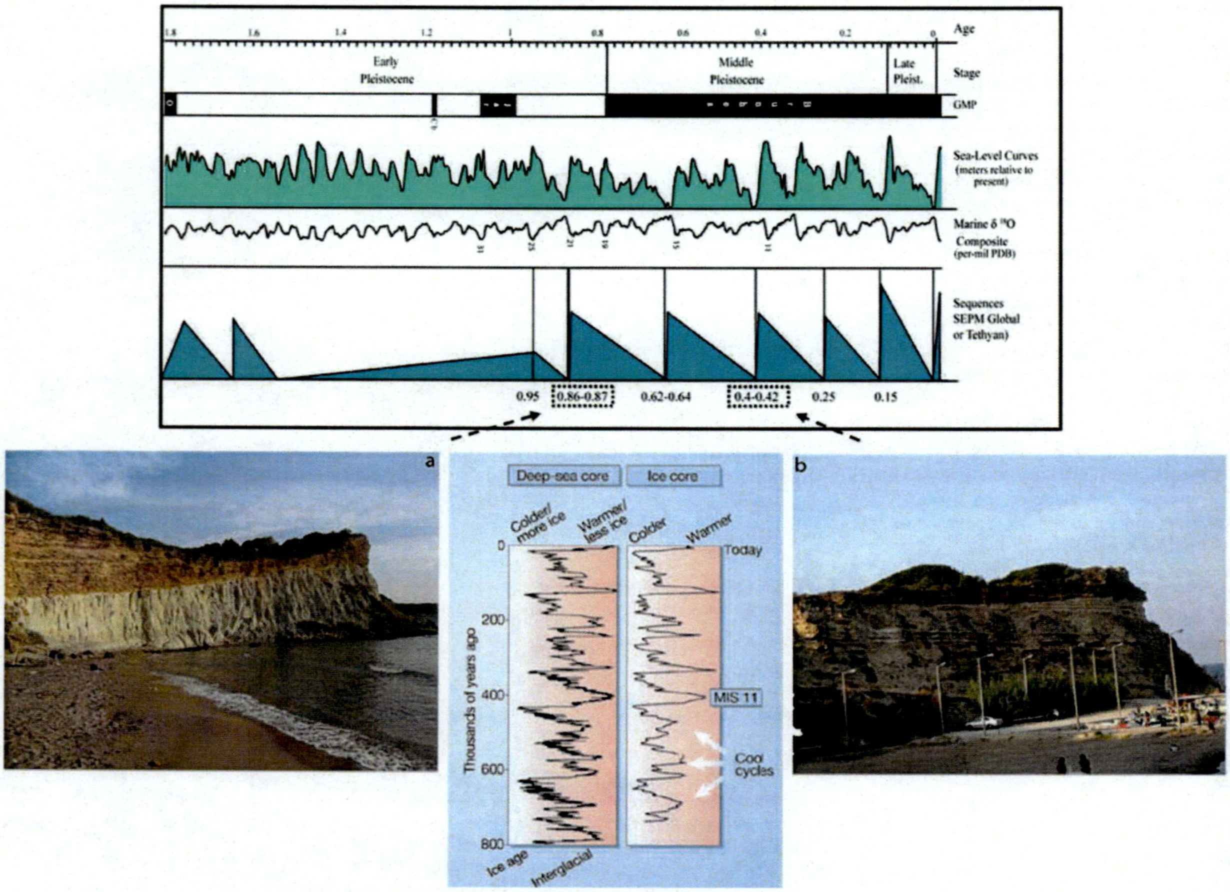

Fig. 5.11 The cycles of climate change during the Middle-Late Pleistocene and the corresponding periods of low and high sea level (from Woelbroeck et al. 2002). On top, above the curve of the sea level fluctuations, the magneto-stratigraphic scale is given. In the displayed pictures a and b two climatically induced stratigraphic unconformities are given from Zakynthos. **a** The typical disconformity of Lower–Middle Pleistocene (0.9 Ma) at cape Gerakas in Eastern Zakynthos (from Papanikolaou 2008). **b** The slightly angular unconformity of the Middle-Pleistocene (0.4 Ma) in the region of Gaidaros in Northern Central Zakynthos (from Papanikolaou et al. 2010). Between the two pictures a and b a detailed diagram of the climatic changes is given based both on deep sea cores and ice cores

unification of the Cyclades islands into one land, the transition of the Gulfs of Corinth and Northern Evoikos to lakes, the emergence of the Southern Evoikos Gulf, etc. (see also Sect. 11.3).

From a stratigraphic point of view, the Quaternary marine sequences have been divided into the Calabrian, corresponding to the Lower Pleistocene, and the Sicilian and Tyrrhenian, which have been mainly used to describe marine terraces of the Middle and Upper Pleistocene respectively. It is very important to note that the Quaternary marine sediments on the mainland are only observed in areas of large recent tectonic uplift, where the rise of the sea level due to tectonic uplift overcomes the positive eustatic movements due to climate changes (Flemming 1978). Thus, Pleistocene marine sediments are traced along uplifted coastal zones, such as in the Ionian islands and the Kyparissiakos Gulf along the frontal zone of the Hellenic arc and in uplifted blocks, such as the Northern Peloponnese, Southern Crete and Karpathos (Negris 1928; Keraudren 1970, 1979) (Fig. 5.12).

It is noteworthy that in the bibliography, the terms Tyrrhenian and Sicilian have created several problems, because they did not correspond to a certain age. The progress of research forced many researchers in the 1950s and 1960s to use intermediate terms such as Eutyrrhenian, Tyrrhenian I, Tyrrhenian II, etc. (Psarianos and Symeonidis 1975). Today, each marine terrace is attempted to be dated with an absolute dating and to be correlated with one of the thirty-six global sea level changes (MIS 1–36). The preservation of the term Tyrrhenian can be limited to Upper Pleistocene terraces and of the term Sicilian to Middle Pleistocene terraces, although their number might be much larger, like in the case of Corinth-Kiato, where there are

Fig. 5.12 A characteristic Pleistocene marine terrace in Southwestern Karpathos, where a few tens of meters of uplift can be observed, with an unconformable deposition of a horizontal thin cover of marine Upper Pleistocene sediments on the underlying tilted to the north sediments of Upper Miocene–Pliocene age. An intermediate period of erosion has peneplained the terrace during the Early–Middle Pleistocene

more than 11 marine terraces in the last 650 Ka (Armijo et al. 1996).

In 2009, the International Commission on Stratigraphy shifted the Pliocene/Quaternary boundary from 1.8 to 2.6 Ma. Basically, they thought that the former last stage of the Pliocene, the **Gelasian**, should be better added to the beggining of Quaternary. This change has important implications to literature, past publications and geological maps. In conclusion, it is now necessary to clarify which terminology is being used, by having also in mind that all of the pre-2009 references were considering the Pliocene/ Pleistocene boundary at 1.8 Ma.

References

Arianoutsou-Faragitaki, M., Giannitsaros, A. & Koubli-Sovantzi, L. 2003. Terrestrial ecosystems of Greece. University of Athens, Dept. of Ecology , 316p (in greek).

Armijo, R., Meyer, B., King, G., Rigo, A. & Papanastasiou, D. 1996. Quaternary evolution of the Corinth Rift and its implications for the late Cenozoic evolution of the Aegean. Geoph. J. Intern., 12, 11–53.

Besenecker, H. & Symeonidis, N. 1974. Der erste Saugertierfund aus dem Neogen der griechischen Insel Psara (Ostagais). Ann. Geol. Pays Hellen., 26, 109–117.

Chappell, J. & Shackleton, N. J. 1986. Oxygen isotopes and sea level. Nature, 324, 137–140.

Christanis, K. 2004. Coal facies studies in Greece. Int. J. Coal Geol., 58, 99–106.

Christodoulou, G. 1958. On the Pliocene foraminifera of Rafina. Bull. Geol. Soc. Greece, 3, 24–30 (in greek).

Drooger, C. W. (Ed.) 1973. Messinian events in the Mediterranean. North Holland Publ. Co., Amsterdam, 272 p.

Flemming, N.C. 1978. Holocene eustatic changes & coastal tectonics in the northeast Mediterranean. Implications for models of crustal consumption. Phil. Trans. R. S. London, A, 289, 405–457.

Gaudry, A. 1867. Animaux fossiles et geologie de l' Attique. 474 p, Paris.

Grekoff, N., Guernet, G. & Lorenz, C. 1967. Existence du Miocene marin au centre de la Mer Egee dans l'ile de Skyros (Grece). C. R. Acad. Sci., Paris, 265, 1276–1277.

Houghton, S.L., Roberts, G.P., Papanikolaou, I.D., Mcarthur, J.M. & Glimour, M. A. 2003. New 234U-230Th coral dates from the western Gulf of Corinth: Implication for extensional tectonics. Geophysical Research Letters, 29, https://doi.org/10.1029/2003GLO18112.

Hsü, K., Montadert, L, Bernoulli. D. Cita, M.B., Erickson, A., Garisson, R.E. Kidd, R.B., Melieres, F., Müller, C. & Wright, R. 1978. History of the Mediterranean salinity crisis. Init. Rep. D.S.D. P .XIII, 1, 1053–1078.

Jolivet, L. & Faccenna, C. 2000. Mediterranean extension and the Africa - Eurasia collision. Tectonics, 19, 6, 1095–1106.

Jolivet, L., Faccenna, C., Huet, B., Labrousse, L., Le Pourhiet, L., Lacombe, O., Lecomte, E., Burov, E., Denele, Y., Brun, J.P., Philippon, M., Paul, A., Salaun, G., Karabulut, H., Piromallo, C., Monie, P., Gueydan, F., Okay, A.I., Oberhansli, R., Pourteau, A., Augier, R., Gadenne, L. & Driussi, O., 2013. Aegean tectonics: Strain localisation, slab tearing and trench retreat. Tectonophysics, 597–598, 1–33.

Keraudren, B. 1970. Les formations quaternaires marines de la Grece. Dissertation Paris, 382 p.

Keraudren, B. 1979. Le Plio-Pleistocene marin et oligohalin en Grece: stratigraphie et paleogeographie. R. Geol. Dyn. Geogr. Phys., 21, 1, 17–28.

Klebelsberg, R. 1949. Handbuch der Gletscherkunde und Glazialgeologie, 2. Pleistozane Vergletscherung auf der Balkanhalbinsel. 737–740, Wien.

Koukouzas, C. & Koukouzas, N. 1995. Coals in Greece: distribution, quality and reserves. Coal Geology, Geol. Soc. London, Sp. Publ., 82, 171–180.

Ktenas, C. 1927. Decouverte du Pliocene inferieur marin dans l' ile de Nikaria. C.R. Acad. Sci., Paris, 184, 756–758.

Lalechos, N. & Savoyat, E. 1977. La sedimentation Neogene dans le Fosse Nord Egeen. VI Colloquium on the Geology of the Aegean Region, Athens 1977, Proceedings II, 591–603.

Lykousis, V. 2009. Sea-level changes and shelf break prograding sequences during the last 400ka in the Aegean margins: subsidence rates and palaeogeographic implications. Continental Shelf Research, 29(16), 2037–2044.

Mariolakos, I. & Papanikolaou, D. 1981. The Neogene Basins of the Aegean Arc from the Paleogeographic and the Geodynamic point of view. Intern.Symp.(H.EA.T.) Hell. Arc and Trench, Athens 1981, Proceedings I, 383–399.

Mariolakos, I. Papanikolaou, D., Symeonidis, N., Karotsieris, Z. & Lekkas, S. 1981. The deformation of the area around the eastern Korinthian Gulf, affected by the earthquakes of February–March 1981. International Symposium H.E.A.T. (Hellenic Arc & Trench System), Athens 1981, Proceedings I, 400–420.

Mattei, M., D'agostino, N., Zananiri, I., Kondopoulou, D., Pavlides, S. & Spatharas, V. 2004. Tectonic evolution of fault-bounded continental blocks: comparison of paleomagnetic and GPS data in the Corinth and Megara basins (Greece). J. Geophys. Res., 109, B02106.

Meulenkamp, J., 1979. The Aegean and the Messinian salinity crisis. VI. Coll. Geol. Aegean Region, Athens, 1977, Proc. Ill, 1253–1263.

Meulenkamp, J. in collab. with Dermitzakis, M., Georgiadou-Dikeoulia. E., Jonkers, H.A. & Böger, H., 1979. Field Guide to the Neogene of Crete. Publ. Dept. Geol. & Paleont. Univ. Athens, 1–32.

Milankovitch, M. 1936. Durch ferne Welten und Zeiten. Koehler und Amalang, Leipzig.

Mitzopoulos, M. 1961. Uber das Vorkommen von Elefanten in der Agais. Proc. Acad. Athens, 36.

Negris, Ph. 1903. Regression et transgression de la mer depuis l' epoque glaciere jusqu' a nos jours. Rev. Univ. Mines, 3, 1–33, 2 cartes.

Negris, Ph. 1928. Les terraces marines de la Grece. 1st Rep. Comm. Plioc. & Pleistoc. Terraces, IGU, 18–26, Oxford.

Papanikolaou, D. 2010. Major paleogeographic, tectonic and geodynamic changes from the last stage of the Hellenides to the actual Hellenic arc and trench system. Bull. Geol. Soc. Greece, 43, 71–85.

Papanikolaou, D. & Royden, L. 2007. Disruption of the Hellenic arc: Late Miocene extensional detachment faults and steep Pliocene–Quaternary normal faults—or what happened at Corinth? Tectonics, 26.

Papanikolaou, D., Lykoussis, V., Chronis, G. & Pavlakis, P. 1988. A comparative study of neotectonic basins across the Hellenic arc: The Messiniakos, Argolikos, Saronikos and Southern Evoikos Gulfs. Basin Research, 1/3, 167–176.

Papanikolaou, D., Nomikou, P., Rousakis, G., Livanos, I. & Papanikolaou, I. 2015. Active Tectonics and Seismic Hazard in Skyros Basin, North Aegean Sea, Greece. 6th INQUA Meeting on: Active Tectonics, Paleoseismology and Archaeoseismology, Fucino, Italy, 19–24 April 2015, Proc. 348–351.

Papanikolaou, D., Nomikou, P., Papanikolaou, I., Lampridou, D., Rousakis, G. & Alexandri, S. 2019a. Active tectonics and sesmic hazard in Skyros Basin, North Aegean Sea, Greece. Marine Geology, 407, 94–110.

Papanikolaou, D., Nomikou, P., Papanikolaou, I., Triantaphyllou, M., Galanidou, N., Dimiza, M. & Tzedakis, C. 2019b. Pleistocene palaeogeography and Palaeolithic archaeology. The tectonically active North Aegean Sea case study. 15th Congress Geol. Soc. Greece, Special Publications, 7, 82–83.

Papanikolaou, M. D. 2008. Quaternary palaeoenvironmental analysis of offshore and onshore marine sequences in the Eastern Mediterranean. PhD Thesis, University of Cambridge.

Papanikolaou, M., Papanikolaou, D. & Triantaphyllou, M. 2010. Post-Alpine Late Pliocene – Middle Pleistocene uplifted marine sequences in Zakynthos Island. Bull. Geol. Soc. Greece, 43/1, 475–485.

Philippson, A. 1901. Beiträge zur Kenntnis der griechischen Inselwelt. Peterm. Milt. Erganzunheft,134, 1–172, Gotha.

Proedrou, P. & Papakonstantinou, C.M. 2004. Prinos Basin – A model for Oil Exploration. Bull. Geol. Soc. Greece, 34, 327–333.

Psarianos, P. & Symeonidis, N. 1975. Stratigraphy. Athens, 186 p (in greek).

Roveri M., Flecker, R., Krijgsman,W., Lofi, J., Lugli, S., Manzi, V., Sierro, F., Bertini, A., Camerlenghi, A., de Lange, G., Govers, R., Hilgen, F., Hubscher, C., Meijer, P. & Stoica, M. 2014. The Messinian salinity crisis: past and future of a great challenge for marine sciences. Marine Geology, 352, 25–58.

Royden, L.H., 1993. Evolution of retreating subduction boundaries formed during continental collision. Tectonics, 12, 629–638.

Royden, L.H. & Papanikolaou, D.J., 2011. Slab segmentation and late Cenozoic disruption of the Hellenic arc. Geochemistry, Geophysics, Geosystems, 12, Q03010.

Shackleton, N.J. 1978. Ice-age paleo-oceanography of the Mediterranean. Nature, 276, 667–668.

Shackleton, N. J. 2000. The 100,000-year ice-age cycle identified and found to lag temperature, carbon dioxide and orbital eccentricity. Science, 289, 1897–1902.

Symeonidis, N. & Marcopoulou-Diakantoni, A. 1977. La faune pikermienne et le Neogene. Bull. Soc. Geol. Fr., 19, 1, 111–115.

Theodoropoulos, D. 1968. Stratigraphie und Tektonik des Isthmus von Megara (Griechenland). Erlanger Geol. Abh., 73, 23p.

Underhill, J. 1989. Late Cenozoic defromation of the Hellenide foreland, western Greece. Bull. Geol. Soc. Am., 101, 613–634.

Vassilakis, E. 2006. Study of the tectonic structure of the Messara Basin, Central Crete with the use of remote techniques and geographical information systems. PhD Thesis, University of Athens, 557 p (in greek).

Vassilakis, E., Royden, L. & Papanikolaou, D. 2011. Kinematic links between subduction along the Hellenic trench and extension in the Gulf of Corinth, Greece: A multidisciplinary analysis. Earth & Plan. Sci. Let., 303(1–2), 108–120.

Velitzelos, D., Bouchal, J.M. & Denk, T. 2014. Review of the Cenozoic floras and vegetation of Greece. Rev. Paleobotany & Palynology, 204, 56–117.

Woelbroeck, C.L., Labeyrie, E., Michel, J.C., Duplessy, J.G., McManus, K., Lambeck, E. Balbon, & M. Labracherie 2002. Sea-level and deep-water temperature changes derived from benthic foraminifera isotopic records. Quat. Sci. Rev., 21, 295–305.

Zerefos, C,H. and 21 collaborators, 2011. The climate of Eastern Mediterranean and Greece: past, present and future. In: «The environmental, economic and social consequences of climate change in Greece». Committee on the study of Consequences of Climate Change, Bank of Greece. 135p, Athens.

6 Molasse Formations in the Hellenides

6.1 Distinction of Flysch, Molasse, Flysch-Molasse

In the syn-orogenic region of the arc there are two sedimentation basins, in the trench/fore-arc basin and the back-arc basin, separated by the mountainous zone along the island arc/orogenic wedge, that is eroded, offering a lot of sediment material into both basins. Therefore, the clastic material derived from the mountainous chain is deposited in the stratigraphic sequences of the flysch facies in the trench and of the molasse facies in the back-arc basin. Both sequences include rock formations of deep clastic sedimentation, with turbidites, according to the Bouma sedimentary characteristics. The differentiation of their petrographic composition, depends on the lithologies of the eroded geological formations, the supply of the deltaic deposits of the island arc and its overall dynamic equilibrium, especially as far as its uplift rate. Nevertheless, there are always intermediate sea ways/channels connecting the two deep basins on both sides of the island arc, similarly to the present channels between Crete and Peloponnese on the one hand, and Crete, Karpathos and Rhodes on the other. Therefore, the distinction of the marine clastic syn-orogenic sedimentation is not so clear in all cases, at least from the physico-geographical point of view. Of course, when taking into account that flysch is deposited in continuity with the previous pre-orogenic biochemical sedimentation, unlike molasse, which is unconformably deposited over the Alpine formations, then the distinction becomes clear. Beyond the difference in the deposition characteristics, another significant difference relates to the deformation characteristics. In particular the flysch is highly deformed by a compressional stressfield, resulting into reverse faulting and extensive folding features. On the other hand the molasse is characterized by low deformation, mostly from extensional processes, involving simply normal syn-sedimentary faulting, tilted normal fault blocks and internal stratigraphic unconformities. In some cases, there are also intermediate phases of syn-orogenic clastic sequences, which do not exhibit the typical characteristics of either the flysch or the molasse and have been deposited at different segments of the fore land basins (DeCelles and Gilles 1996). Thus, unconformable deposition of clastic sediments may occur in the wedge-top area at the front of the advancing nappes, whereas disconformable deposition may occur in the fore deep and the fore bulge (see also Fig. 4.6). Compressional deformation dominates in the wedge top and part of the fore deep whereas extensional structures may occur at the fore bulge depozone. In these cases it is possible to use the term flysch-molasse, showing the mixed characteristics, as for example a clastic sequence which is unconformable on the Alpine basement but has been deformed with compressive structures. These cases represent transitional conditions, during the passage of a unit from the lower to the upper plate and from compression to extension. Schematically, we could distinguish the flysch-molasse domain in the internal slope of the trench towards the island arc (Fig. 6.1), where it may actually be deposited unconformably over the previously tectonised Alpine sediments, while it is still under the compressive stress field of the arc.

6.2 The Molasse Basins

The molasse formations of Greece can be distinguished in three large molasse basins, which were active during different periods and in different locations, following the southward migration of the Hellenic arc (Papanikolaou 1986, 1993) (Fig. 6.2). In particular, these are: (i) the Rhodope–Northern Aegean molasse, in the core of the arc during the Eocene–Oligocene, (ii) the Meso-Hellenic Trough molasse, in the middle of continental Greece during the Oligocene–Middle Miocene and its possible extension to the Cyclades, (iii) the Cretan Sea molasse, is the actualistic expression of the back-arc molasse basin of the Hellenic arc, initiated during the Middle/Upper Miocene, with only a few marginal outcrops in emersion. The aforementioned molasse basins exclude the Epirus flysch-molasse syncline. It is interesting to

D. I. Papanikolaou, *The Geology of Greece*, Regional Geology Reviews,
https://doi.org/10.1007/978-3-030-60731-9_6

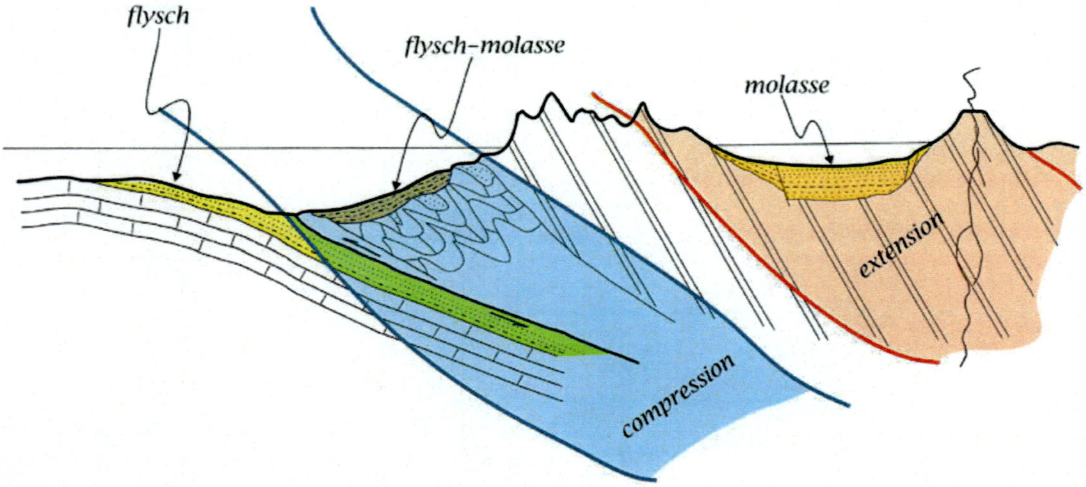

Fig. 6.1 Schematic representation of the areas of deposition of the synorogenic formations of flysch, flysch-molasse and molasse in the orogenic arc

Fig. 6.2 The main molasse basins of the Hellenic Arc. 1: Es-Ol, Rhodope–Northern Aegean (Middle Eocene–Oligocene), 2: Ol-Mi, Meso-Hellenic Trough, Tavas–Kale (Upper Eocene–Middle Miocene), 3: Mi, Epirus–Acarnania, Paramithia, Rhodes (Upper Oligocene–Middle Miocene), 4: Ol-Mi, Cyclades (Lower Miocene), 5: Ms-Pl, Cretan Sea (Middle Miocene–Quaternary)

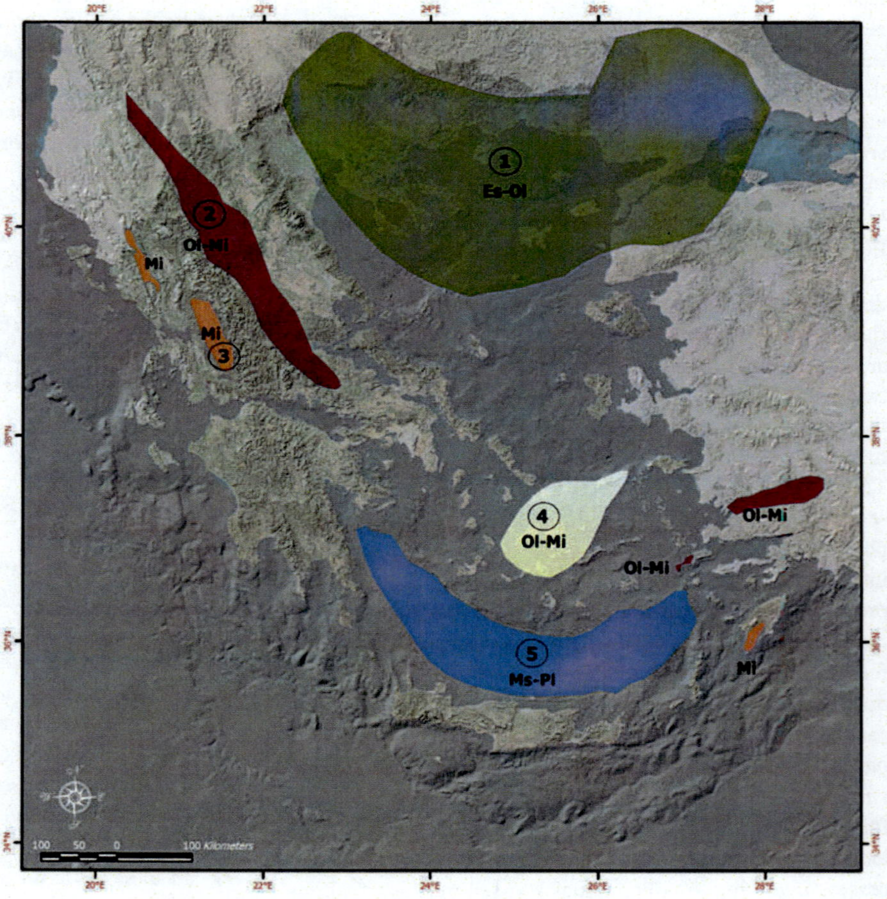

note that after the Middle Miocene no other molassic basins were created in the Northern Hellenides, north of the CHSZ.

6.2.1 The Rhodope–North Aegean Molasse

The Rhodope–North Aegean molasse was deposited unconformable on the metamorphic rocks of Rhodope, as well as in the area of the former Vardar/Axios zone. It was also extending to the Northern Aegean and especially the islands of Samothraki, Limnos and Aghios Efstratios. The basal sediments are of Lower–Middle Eocene age and the sedimentary sequences reach up to the Oligocene/Miocene boundary (Kopp 1966). It is remarkable that volcanic activity is observed in this region of the molasse basin, especially during the Oligocene, as in the case of Alexandroupolis, where there are sections with alternations of sediments and volcanic formations of more than one kilometer thickness. Today there are relatively few outcrops, either because they are underwater in the North Aegean, or because they have been covered by the post-Alpine sediments of Upper Miocene–Pleistocene (Serres, Axios), or because they have been eroded. There are no systematic studies about the general stratigraphic sequence of the Rhodope–North Aegean molasse neither a uniform terminology, since the stratigraphy changes considerably from region to region. This is due to the existence of sub-basins, bordered by syn-sedimentary strike-slip faults, which have affected the stratigraphic sequences of the molasse formations with unconformities, extruding wedges, and facies changes (Papanikolaou and Triantaphyllou 2010). A major unconformity is observed in the Alexandroupolis area, dated in the Upper Eocene. The unconformity separates the underlying clastic sediments of the Kirki Formation, characterized by abundant material from the erosion of mafic rocks, from the overlying Avandas formation comprising neritic limestones with Nummulites (Fig. 6.3) and the clastic sediments of the Pylaea formation that follow upwards. Towards the top of the molassic stratigraphic column in the Pythion area, characteristic Congeria bearing marly limestones occur, followed by cross bedded sandy horizons of upper Oligocene–Lower Miocene age (Fig. 6.4). This upper sedimentary facies of shallow coastal environment indicates the uplift of the basin and its filling up with clastic sediments, reflecting wider paleogeographic/geodynamic changes in the basin.

Significant outcrops of molasse formations occur in Limnos and Aghios Efstratios islands, with strong presence of volcanic rocks. In Limnos a very thick sedimentary sequence exceeding 2200 m extending from the Late Eocene up to the Oligocene/Miocene boundary (from about 38 to 23 Ma) has been reported (Papanikolaou and Triantaphyllou 2019). This Limnos sequence can be correlated with the Alexandroupolis outcrops of the upper unconformable sequence Avandas–Pylaea. It is characteristic that olistholites of the Nummulitic limestones of the Avandas formation occur at the base of the Linmos sequence.

In Eastern Thrace, at Turkey, several models of basin formation have been proposed, with an initial fore arc basin, which subsequently evolved into a back-arc basin (Gorur and Okay 1996). The Eocene stratigraphic evolution of the Eastern Thrace Basin (Siyako and Huvaz 2007) differs from the Oligocene deposition of thick clastic sequences, interpreted by some scientists as the result of post-collisional processes (e.g. D'Atri et al. 2012). In the Greek sector a supra-detachment origin of the basin has been proposed (Kilias et al. 2013). Research for hydrocarbons in these molasse sediments conducted by the Public Petroleum Company of Greece did not show any results, while in the eastern part of the Thrace basin, in Turkey, small deposits of oil and gas have been found, which are now being exploited.

6.2.2 The Meso-Hellenic Trough

The molasse sediments of the Meso-Hellenic Trough crop out in Karditsa, Trikala, Kalambaka, Grevena, and Kastoria up to the Greek-Albanian borders and continue inside Albania, with residual occurrences almost up to the Adriatic.

The basement of the Meso-Hellenic Trough crops out mainly along its western margin in the Pindos chain, made of the Western Thessaly, Pindos and Pindos ophiolites nappes, together with the peculiar flysch-molasse formation of Krania, which can be observed only within a "paleogulf" of a few tens of kilometers width. At the base of Krania, some transgressive ophiolitic conglomerates can be found, which evolve upwards to typical flysch litho-facies with cyclothemes. The whole formation is intensively folded, in contrast to the overlying Eptachori molasse, which is only slightly tilted. The eastern margin of the basin is croping out only in a few localities, because it is mostly faulted as the overall geometry of the basin is a tilted monoclinic to the east-northeast sequence of approximately 5 km thickness (Brunn 1956; Desprairies 1979). Above the Krania formation but also along hundreds of kilometers in the eastern slopes of Northern Pindos, one can observe the unconformable deposition of a thick sequence of clastic sediments, which are mainly marls, sandstones, and conglomerates. It should be emphasized that along various transverse profiles of the basin the stratigraphic formations do not remain stable but depend on the location of the profile along the basin, as there are several deltaic prisms with lateral thickness variations and wedges (Desprairies 1979). A major point of difference is the provenance of the clastic material either from the Pindos cordillera of the western margin of the Meso-Hellenic basin, representing the former island arc or

Fig. 6.3 Characteristic outcrop of the Upper Eocene Avandas neritic limestones with Nummulites at the northern exit of Avandas village

Fig. 6.4 Clastic horizons with cross bedding at the top of the Oligocene molasse in Western Thrace, in the Pythion area

the Pelagonian cordilliera, which is the eastern margin of the basin, representing the hinterland. The Pindos mountain chain, made of the ophiolite nappe, emplaced over the Pindos unit during the Late Eocene, provides clastic material from sedimentary rocks (limestones, radiolarites, flysch) and ophiolites, while the Pelagonian margin supplies with clastic material from metamorphic/magmatic rocks (gneisses, marbles, granites, mica schists). Consequently, there is no uniform stratigraphic column of the basin but each region develops its own. For example, abundant granitic material at the region of Kastoria is traced in the Miocene, due to the large outcrops of Paleozoic granites in the area.

Brunn (1956) first studied the Meso-Hellenic basin and distinguished various formations (Fig. 6.5). Above the peculiar flysch-molasse formation of Krania comes the Eptachorion formation, which comprises mainly marls of Oligocene age. Above the Eptachorion formation the Pentalofos formation of Upper Oligocene–Aquitanian age is placed, which comprises large masses of conglomerate and sandstone beds building up the Meteora hills, famous for their monasteries (Fig. 6.6).

The very thick Tsotyli formation, comprising marls and sandstones of Lower Miocene age is developed above the Pentalofos formation. Finally, the Ondria formation, which is 20–50 m thick, comprises sandstones and marly limestones of Middle to Upper Miocene age closing the stratigraphic sequence on top. It crops out at the northern part of the basin, south of Kastoria and it contains thick shell lamellibranchs, corals and other fossils of shallow marine environment. Thus, this formation marks the final uplift of the basin in a process of marine regression during the Middle Miocene, which terminates the former deep clastic sedimentation. It is characteristic that the top strata of the molasse sequence can be observed in the horizon from miles away, as they dip slightly to the ENE, forming a huge slightly inclined planation surface (Fig. 6.7). The altitudes of the area range from 600 to 900 m and therefore, the basin has been uplifted since Late Miocene by about one kilometer.

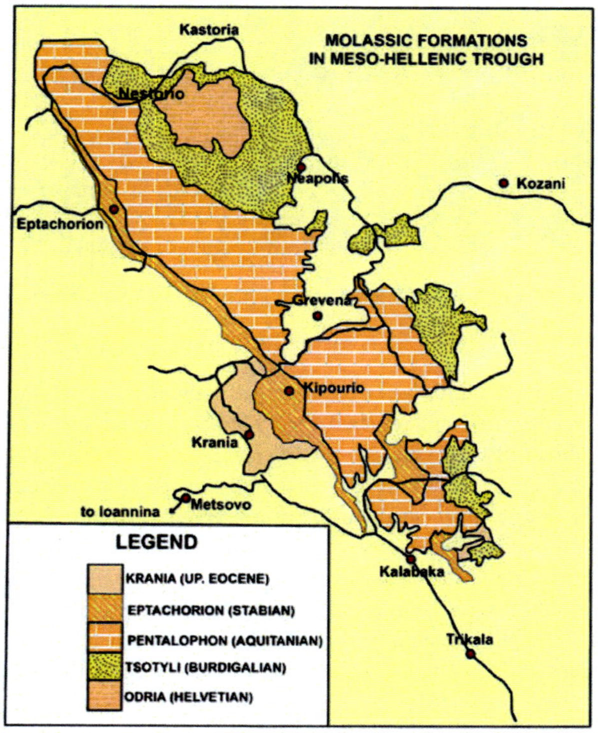

Fig. 6.5 Molasse formations of the Meso-Hellenic Trough by Brunn (1956), simplified by Papanikolaou and Sideris (1977)

Desprairies (1979) studied the Meso-Hellenic Basin from the sedimentological point of view and identified a series of deltaic fans, which supplied the basin mainly from the east with an overall thickness of about 5 km. This can be easily verified by examining the Meteora conglomerates, in which the overwhelming majority of the rounded pebbles and boulders are metamorphic and magmatic rocks of Pelagonian origin. It is noteworthy that while the bedding of the molasse generally dip to the east, their supplying source was also from the east rather than the west, as presupposed based on the dip of the beds. This general geometry is disrupted locally by some strike slip fault zones (Vamvaka et al. 2006), but overall the western margin of the basin appears with all the stratigraphic horizons in an approximately monoclinic sequence, while the eastern margin is tectonically disturbed and several stratigraphic horizons are absent, especially the lower ones. The Pelagonian margin of the basin has been considered as an «indentor» during its sedimentary evolution, providing with crystalline detritus the Lower Miocene conglomerates (Ferriere et al. 2011). Therefore, we have an asymmetric basin, with a general tilt towards the east, to the internal part of the arc.

From the geodynamic point of view the Mesohellenic basin has been regarded as a back arc molassic basin of the Hellenides during Oligocene–Middle Miocene (Papanikolaou 1986; Papanikolaou et al. 1988). Due to the position of the basin on top of the evolving nappes of the external Hellenides it was also characterized as a «piggy-back» basin (Vamvaka et al. 2006; Ferriere et al. 2011). Finally, it should be also noted that in Greece the potential for oil and gas has been examined and the possible source rocks (e.g. Eptachorion formation) and reservoirs (e.g. Lower member of Pentalofos formation) have been analysed (Kondopoulos et al. 1999) without any result up to the present. On the contrary, in Albania oil deposits have been found in formations similar to those of the Meso-Hellenic Trough.

Fig. 6.6 View of the Meteora conglomerates from the top of the Koziakas Mt. These outcrops of Pentalofos formation form a high relief area above the town of Kalabaka, due to the differential erosion of the conglomerates with respect to the adjacent marls

Fig. 6.7 View of the top strata of the Middle–Upper Miocene Ondria formation of the Meso-Hellenic molasse in the Nestorio–Damaskinia area, which are diping with 5–10° to the east

6.2.3 The Cycladic Molasse

The Cycladic molasse differs from the other molasses because it is strongly tectonised either on the top of the Cycladic non-metamorphic nappe, or in direct tectonic contact with the underlying metamorphic formations of the Cyclades, or even on top of the upper Paleozoic granite-gneisses of Paros (Papanikolaou 1980, 1996). The unconformity of the molasse over the Alpine basement has been traced in very few locations like in Paros, where it is observed above the ophiolite nappe and its age is at least partially lower Miocene (Dermitzakis and Papanikolaou 1980) (Fig. 6.8).

Folds and thrusts can be observed in the molassic clastic formations, as well as cataclasites at its base, where it comes in direct contact with the metamorphics, along extensional detachments, diping at very low angles or even horizontal. This means that the deformation observed corresponds to extensional folds and gravity sliding within the top structure of the metamorphic core complex of the Cyclades (see also Fig. 4.30). Opal-bearing silicified sediments are often present, possibly due to neighboring volcanic activity in the Miocene. More recently, new data have been presented: (i) based on geochronological analysis of clasts from the Miocene molassic sediments of Mykonos and Paros showing the back arc extesion in the Aegean (Sanchez-Gomez et al.

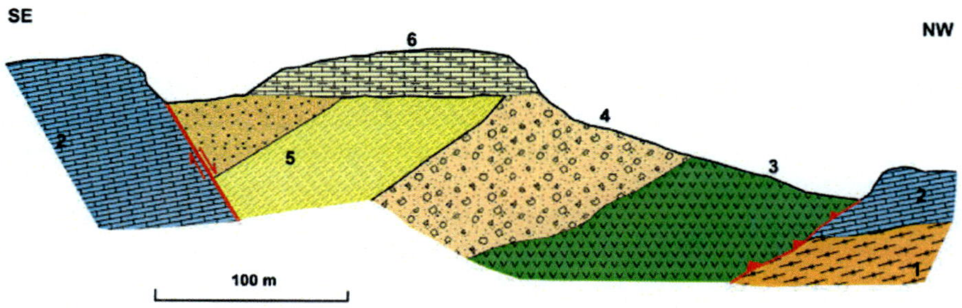

Fig. 6.8 The transgression of the Cycladic molasse of Burdigalian age, over the ophiolite nappe in the west of Naousa, Paros (from Dermitzakis and Papanikolaou 1980). 1: orthogneiss, 2: Marathi unit marbles (Southern Cyclades), 3: ophiolites, 4: sandstones-conglomerates of the base of the molasse, 5: molasse marls and clays, 6: travertines (unconformable Pliocene?)

2002) and (ii) successive evolutionary stages have been described from Paros, based on (U-Th)/He thermochronometry within the Miocene supra detachment basin (Bargnesi et al. 2013).

Some outcrops on the islands of **Kos** and **Rhodes** in the Dodecanese correspond to molassic deposits of Lower Miocene age. They lie unconformably on the metamorphic rocks of the relatively autochthonous Kefalos unit of western Kos (Papanikolaou and Nomikou 1998) and on the Laerma wild flysch of southern Rhodes (Mutti et al. 1970). The Western Kos basin at the Kefalos peninsula may correspond to the eastern extension of the Cycladic molasse basin whereas the Southern Rhodes basin might be homologous to the flysch-molasse basin of the Epirus-Akarnania syncline.

6.2.4 The Epirus–Akarnania Syncline

The Upper Oligocene–Lower Miocene molasse of the Epirus syncline (known also as Epirus–Acarnania syncline) in western Greece (Aubouin et al. 1977) lies on top of a large syncline of N-S orientation, which is observed on the several km thick flysch covering both the Ionian and Gavrovo units. Along the syncline axis the distinction of the molasse from the underlying Upper Eocene–Lower Oligocene flysch is not clearly visible beyond a significant lithological differentiation. On the contrary, at the margins of the syncline, the molasse appears in unconformity both over the flysch and over the pre-flysch basement, showing an impressive stratigraphic gap, especially in some localities, where it unconformably overlaps the Upper Triassic–Liassic limestones of the neritic Pantokrator facies of the Ionian unit (Richter 1978). Nevertheless, it participates in the syncline structure, but with generally gentle inclinations. The Radovitsi sequence, of Aquitanian age, lies unconformably over the underlying flysch, while the Polydorus sequence, of Burdigalian age, is also unconformable locally even on the Lias and with much more intense molassic character (IGRS and IFP 1966; Richter 1976, 1978).

To the north, at the height of **Paramythia** up to the Greek-Albanian borders, a similar but much smaller syncline structure with molassic sediments of the same age is present. It is remarkable that the Paramithia outcrops have been disrupted by left-lateral strike-slip fault zones with several km offset, transverse to the direction of the arc. Paleoseismicity studies of the major Souli Fault determined some seismic events during the Late Pleistocene–Holocene, proving that it is still active (Boccaletti et al. 1997).

6.2.5 The Cretan Basin

The molasse of the Cretan basin, which represents the actual back-arc basin, was first studied by Jongsma et al. (1977) who carried out a systematic oceanographic research. The age of the base of this molasse is considered to be Tortonian, although there are indications, mainly from Crete, that the rifting of the basin had already begun during the Middle Miocene (Dermitzakis and Papanikolaou 1979; Seidel et al. 2007; Papanikolaou and Vassilakis 2010). Similar conclusions are drawn from the northern margin of the basin in Anafi (Soukis and Papanikolaou 2004). The sedimentary thickness in the basin reaches 1.5–2 km, while its depth ranges from 1.5 to 2 km. This means that the basin has undergone a total subsidence of at least 3–4 km. Considering that the thickness of the crust under the basin is only 16–18 km (Makris 1973, 2010; Bohnhoff et al. 2001), it is evident that it represents a thinned continental crust with only half of its average crustal thickness. In addition to the geophysical research, a double drilling was conducted in the basin as part of the Deep Sea Drilling Project, (DSDP), which was made at a sea bottom depth of 1.845 m and penetrated 343 m of sediments, reaching the top of the Messinian evaporites (Hsü et al. 1978). The Quaternary (1.8 Ma) was determined with a thickness of 131 m of marls with sapropelic and volcanic ash horizons, the Upper Pliocene with a thickness of 155 m of marls and sapropels, the Lower Pliocene with a thickness of only 22 m of marls–

marly limestones, and the Messinian with a thickness of more than 32 m of evaporites (gypsum with carbonate breccias). Sedimentation rates vary significantly during the Plio–Pleistocene, with a minimum average of 0.9 cm/Ka in the Lower Pliocene, a maximum of 9.6 cm/Ka in Upper Pliocene, and a high 7.7 cm/Ka in the Pleistocene.

Outside of the margins of the present basin there are a few uplifted outcrops in the area of the northern coasts of Crete and in the Argolis peninsula (Spetses, Kranidi), and more to the north, in the Itea–Amfissa basin (Papanikolaou et al. 2009), where the sediments have been dated as Middle Miocene. Additionally, the lower Miocene molasse sediments of Western Kos might represent an initial phase of the basin. Finally, the Upper Miocene–Pliocene sediments of Southern Milos (Papanikolaou et al 1990), including some gypsum, probably represent uplifted sediments of the basin along its northern margin on the Cycladic plateau.

6.3 The Flysch–Molasse Basins

Recently, the entire succession of flysch–flysch-molasse–molasse was identified in the mountainous region of Giona Mt in Central Greece, where the Pindos and Parnassos units crop out, with flysch in the Upper Eocene, flysch-molasse in the Oligocene–Lower Miocene and molasse in the Middle Miocene (Gouliotis 2014). The **Giona flysch-molasse** includes sequences of clastic sediments of tens to a few hundreds meters thick, which lie unconformably on the Alpine basement and have been folded and ruptured by high-angle strike-slip faults. The **Itea–Amfissa** molasse basin was created as a supra-detachment basin in the Eastern Giona area (Papanikolaou et al. 2009; Gouliotis 2014). The deposition of the molassic sediments occurred in the Lower–Middle Miocene on the hanging wall of the Eastern Giona low-angle fault above the Paleocene red lutites of the Parnassos unit. It is remarkable that a small gravitational nappe, made of neritic limestones of Upper Jurassic–Lower Cretaceous age, comprising the Platyvouna peak (2.350 m), has been tectonically emplaced over the molasse (Fig. 6.9).

The stratigraphy of the Itea–Amfissa molasse basin includes (Gouliotis 2014): (a) a lower sequence comprising marls, sandstones and conglomerates, of Upper Oligocene–Langhian age. The sequence is tectonised and the clastic material sources from the Internal Hellenides, with pebbles of ophiolitic, granitic, and metamorphic rocks, (b) a plethora of small tectonic gravitational nappes and olistholites of Mesozoic limestones of the Parnassos unit, which have

Fig. 6.9 View of Northern Giona mt. from the east, showing the gravitational nappe of Platyvouna, consisting of Mesozoic neritic limestones (1), over the Lower–Middle Miocene molasse (2), which unconformably overlies the red Paleocene lutites of Parnassos (3), which in turn stratigraphically overlie the Upper Cretaceous pelagic limestones (4). The overall structure lies on the hanging wall of the Eastern Giona low angle normal fault

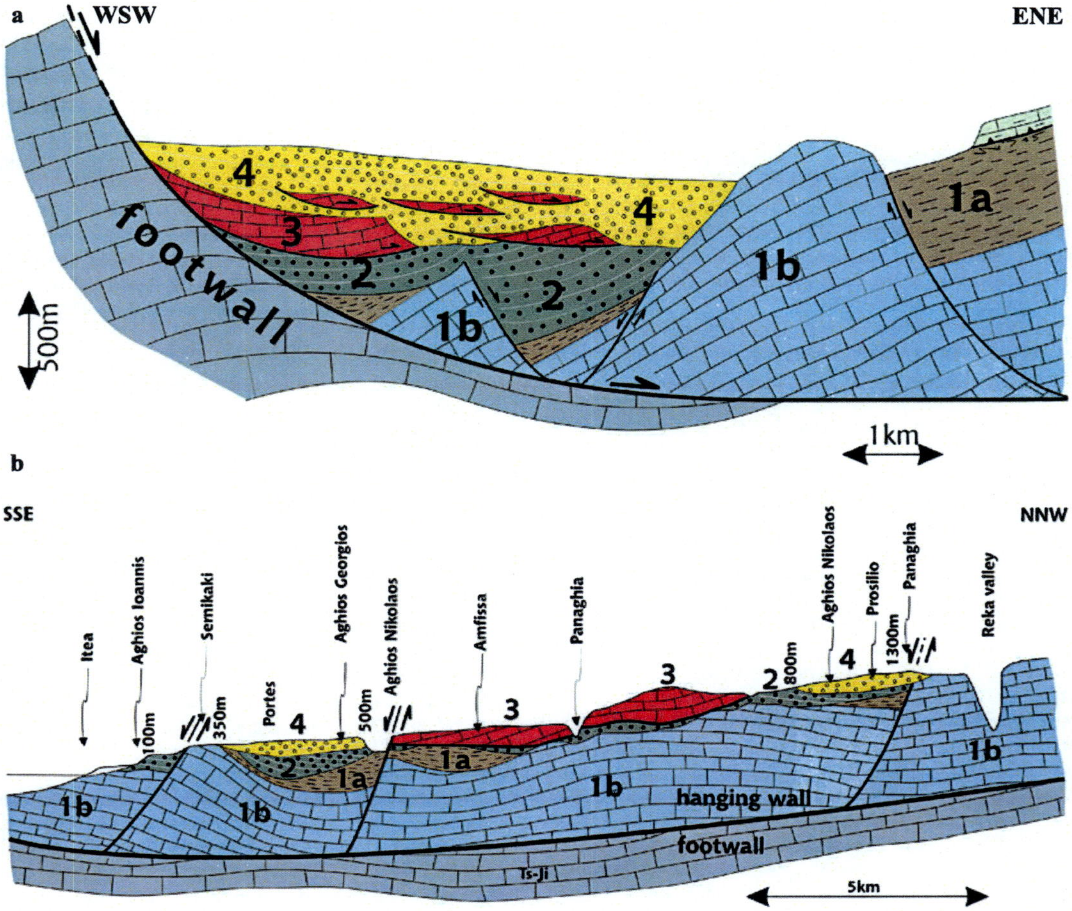

Fig. 6.10 Schematic representation of the tectonostratigraphic structure of the Itea–Amfissa molasse basin over the hanging wall of the extensional detachment of Eastern Giona (from Papanikolaou et al. 2009): **(a)** in a transverse section, and **(b)** in a longitudinal Sect. 1a: Upper Eocene–Lower Oligocene flysch, 1b: Upper Cretaceous limestones, 2: Upper Oligocene–Lower Miocene flysch-molasse, 3: olistoliths of Mesozoic limestones during the Middle Miocene, 4: carbonate breccia-conglomerates of Upper Miocene

slided into the upper strata of the molasse basin from the uplifted footwall of the Central Giona detachment during the Langhian–Serravallian, (c) an upper sequence of carbonate conglomerates and breccias, derived from the limestones of the Parnassos unit. These coarse clastic deposits overlap the previous relief of the slided blocks on top of the older molassic formations and form a planation surface, known as the Aghia Efthymia surface, probably during the Tortonian (Fig. 6.10). The top of the molassic formations is observed at sea level in the coastal area of Galaxidi - Itea, then it reaches altitudes of 250–400 m in Aghia Efthymia, and finally to the north, it is observed at 850–1200 m altitude in the area of Prosilio (see longitudinal schematic section of Fig. 6.10). This shows a gradual uplift of the top of the Itea-Amfissa basin of more than 1.2 km within a distance of 20 km since the Tortonian.

Some other special cases of restricted molassic outcrops occur: (1) in western Crete, where Middle-Upper Miocene clastic sediments have been deposited in the supra-detachment basin of **Topolia**, above the tectonic contact separating the Western Crete metamorphics from the overlying non metamorphosed Tripolis and Pindos/Ethia nappes (Seidel et al. 2007; Papanikolaou and Vassilakis 2010) (see also Fig. 8.12). (2) In central-eastern Crete, where Middle-Upper Miocene thick clastic sequences with large olistholites of Mesozoic limestones, usually supplied from the Tripolis unit, occur above the supra-detachment basin in the **Kalamafka** region (VanHinsbergen and Meulenkamp 2006).

References

Aubouin, J., Desprairies, A. & Terry, J. 1977. Le synclinal d' Epire – Akarnanie, la nappe du Pinde-Olonos & la nappe ophiolitique. Bull. Soc. Geol. France, 19, 20-27.

Bargnesi, E., Stockli, D., Mancktelow, N. & Soukis, K. 2013. Miocene core complex development and coeval supradetachment basin evolution of Paros, Greece, insights from (U-Th)/He thermochronometry. Tectonophysics, 595-596, 165-182.

Boccaletti, M., Caputo, R., Mountrakis, D., Pavlides, S. & Zouros, N. 1997. Paleoseismicity of the Souli Fault, Epirus, Western Greece. J. Geodynamics, 24/1-4, 117-127.

Bohnhoff, M., Makris, J., Papanikolaou, D. & Stavrakakis, G. 2001. Crustal investigation of the Hellenic subduction zone using wide aperture seismic data. Tectonophysics, 343, 239-262.

Brunn, J. 1956. Contribution à l'étude Géologique du Pinde Septentrional et d'une partie de la Macédoine Occidental. Ann. Géol. Pays Hellén., 7, 1 -358.

D'atri, A., Zuffa, G.G, Cavazza, W., Okay, A. DI Vincenzo,G. 2012. Detrital supply from subduction/accretion complexes to the Eocene-Oligocene post-collisional Thrace Basin, NW Turkey and NE Greece. Sediment. Geol., 143-144, 117-129.

De Celles, P. G. & Gilles, K. N. 1996. Foreland basin systems. Basin Research, 8, 105-123.

Dermitzakis, M. & Papanikolaou, D. 1979. Paleogeography and Geodynamics of the Aegean Region during the Neogene. VII Int. Congress Medit. Neogene, Athens 1979, Ann. Géol. Pays Hellén., hors série IV, 245–289.

Dermitzakis, M. & Papanikolaou, D. with contr. of S. Theodoridis and R. Mirkou 1980. The Molasse of Paros Island, Aegean Sea. Ann. Naturhist. Mus. Wien, 83, 59–71.

Desprairies, A. 1979. Etude sédimentologique de formations à caractère flysch et molasse, Macédoine, Epire (Grèce). Mem. Soc. Géol. France, 136, 1-80.

Ferriere, J., Chanier, F., Reynaud, Z.-Y., Pavlopoulos, A., Ditznjong, P., Migiros, G., Confand, I., Stais, A. & Bailleuil, J. 2011. Tectonic control of the Meteora conglomeratic formation (Mesohellenic basin, Greece). Bull. Soc. Geol. France, 182, 437-450.

Gorur, N. & Okay, A., 1996. A fore-arc origin for the Thrace Basin, NW Turkey. Geol. Rund., 85, 662–668.

Gouliotis, L. 2014. The tectonic structure of Giona Mt and its hydrogeological applications. PhD Thesis, University of Athens, 433 p (in greek).

Hsü, K., Montadert, L, Bernoulli. D. Cita, M.B., Erickson, A., Garisson, R.E. Kidd, R.B., Melieres, F., Müller, C. & Wright, R. 1978a. History of the Mediterranean salinity crisis. Init. Rep. D.S.D. P .XIII, 1, 1053–1078.

Hsü, K., Montadert, L, Bernoulli. D. Bizon, G., Cita, M.B., Erickson, A., Fabricius, F., Garisson, R.E. Kidd, R.B., Melieres, F., Müller, C. & Wright, R. 1978b. Site 378: Cretan Basin. D.S.D.P., XLII, 1, 321–341.

I.G.S.R. & I.F.P. 1966. Etude géologique de l'Epire. Technip., 306 p.

Jongsma, D., Wismann, G., Hinz, K. & Garde, S. 1977. Seismic studies in the Cretan Sea. 2. The Southern Aegean Sea: An extensional marginal basin without seafloor spreading? Meteor. Forsch. Ergebnisse, C, 27, 3-30.

Kilias, A., Falalakis, G., Sfeikos, A., Papadimitriou, E., Vambaka, A. & Gkarlaouni, C. 2013. The Thrace basin in the Rhodope province of NE Greece – A tertiary supradetachment basin and its geodynamic implications. Tectonophysics, 595–596, 90–105.

Kondopoulos, N., Fokianou, T., Zelilidis, A., Alexiadis, C. & Rigakis, N. 1999. Hydrocarbon potential of the middle Eocene-middle Miocene Mesohellenic piggy-back basin (central Greece): a case study. Mar. Pet. Geol., 16, 811-824.

Kopp, K.O. 1966. Geologie Thrakiens III: Das Tertiär zwischen Rhodope und Evros. Ann. Géol. Pays Hellén.,16, 315-362.

Makris, J. 1973. Some geophysical aspects of the evolution of the Hellenides. Bull. Geol. Soc. Greece, X, 1, 206-213.

Makris, J. 2010. Geophysical studies and tectonism of the Hellenides. Bull. Geol. Soc. Greece, 43, 32–45.

Mutti, E., Orobelli, G. & Pozzi, P., 1970. Geological studies on the Dodekanese Islands (Aegean Sea). IX. Geological map of the Island of Rhodes (Greece). Explanatory notes. Ann. Géol. Pays Hellén., 22, 77-226.

Papanikolaou, D. 1980. Contribution to the geology of the Aegean Sea. The Island of Paros. Ann. Geol. Pays Hellén. 30/1, 65–96.

Papanikolaou, D. 1986. Geology of Greece. Eptalofos Publications, 240 p. Athens (in greek).

Papanikolaou, D. 1993. Geotectonic evolution of the Aegean. Bull. Geol. Soc. Greece, 28, 33–48.

Papanikolaou, D. 1996. Paros Island. Geological Map of Greece at scale 1/50,000. IGME.

Papanikolaou, D. & Nomikou, P. 1998. The Paleozoic of Kos: a low grade metamorphic unit of the basement of the external Hellenides terrane. IGCP 276, Newsletter 6, Sp. Publ. Geol. Soc. Greece, 3, 155–166.

Papanikolaou, D. & Sideris, Ch. 1977. Contribution to the knowledge of molasse in Greece. I: Preliminary research in Kanalia area of Karditsa. Ann. Géol.Pays Hellén., 28, 387-417 (in greek).

Papanikolaou, D. & Triantaphyllou, M. 2010. Tectonostratigraphic observations in the western Thrace Basin in Greece and correlations with the eastern part in Turkey. Geol. Balc., 39/1-2, 293-294.

Papanikolaou, D. & Triantaphyllou, M. 2019. New stratigraphic data of the Limnos volcano-sedimentary sequence and correlations with the Thrace Basin. 15th Intern. Congress Geol. Soc. Greece, Special Publications, 7, 74-75.

Papanikolaou, D. & Vassilakis, E. 2010. Thrust faults and extensional detachment faults in Cretan tectono-stratigraphy: implications for Middle Miocene extension. Tectonophysics, 488, 233–247.

Papanikolaou, D., Gouliotis, L. & Triantaphyllou, M. 2009. The Itea–Amfissa detachment: a pre-Corinth rift Miocene extensional structure in central Greece. Geol. Soc.,London, Sp. Publ., 311, 293–310.

Papanikolaou, D., Lekkas, E., Mariolakos, I. & Mirkou, M. 1988. Contribution to the geodymanic evolution of the Mesohellenic Trough. 3rd Congress Geol. Soc. Greece, May 1986, Bull. Geol. Soc. Greece, 20/1, 17–36 (in greek).

Papanikolaou, D., Lekkas, E. & Syskakis, D. 1990. Tectonic analysis of the Milos geothermal field. Bull. Geol. Soc. Greece, 24, 27–46.

Richter, D., 1976. Die Flyschzonen Griechenlands III. Flysch sowie spät—und postorogene serien in West-Griechenland zwischen Albanien und dem Golf von Patras. N. Jb. Geol. Paläont. Abh., Teil. I, 151, 1, 73–100, Teil 2, 151, 2, 224–252.

Richter, D. with contrib. of Mariolakos, I. & Risch, H., 1978. The Main Flysch Stages of the Hellenides. In: Alps, Apennines, Hellenides, CLOSS et al editors, 434-438.

Sanchez-Gomez, M., Avigad, D. & Heimman, A. 2002. Geochronology of clasts in allochthonous Miocene sedimentary sequence on Mykonos and Paros Islands: implications for backarc extension in the Aegean Sea. J. Geol. Soc. London, 159, 45-60.

Seidel, M., Seidel, E, & Stoekhert, B. 2007. Tectono-sedimentary evolution of lower to middle Miocene half-graben basins related to

an extensional detachment fault (western Crete, Greece). Terra Nova, 19(1), 39–47.
Siyako, M. & Huvaz, O. 2007. Eocene stratigraphic evolution of the Thrace Basin, Turkey. Sed. Geol. 198, 75-91.
Soukis, K. & Papanikolaou, D. 2004. Contrasting geometry between Alpine and Late- to Post-Alpine tectonic structures in Anafi Island (Cyclades). Bull. Geol. Soc. Greece, 36, 1688-1696.
Vamvaka, A., Kilias, A., Mountrakis, D. & Papaoikonomou, J. 2006. Geometry and structural evolution of the Mesohellenic trough (Greece): a new approach. Sp. Publ. Geol. Soc. London, 260, 521-538.
van Hinsbergen, D.J.J. & Meulenkamp, J. 2006. Neogene supradetachment basin development on Crete (Greece) during exhumation of the South Aegean core complex. Basin Research, 18, 103-124.

7 Alpine and Pre-Alpine Formations of the Hellenides

7.1 Research History

Geological research on the structure of the Hellenic mountains began systematically during the nineteenth century, immediately after the liberation of Southern Greece from the Othoman Empire, mainly with the famous French Scientific Expedition of Peloponnese (Expedition Scientifique de Moree 1833), which comprised the French geologists E. Boblaye and Th. Virlet (1933) who worked during 1929–1930, following the, invitation of the first governor of modern Greece, Ioannis Kapodistrias. Previously, the only scientific information regarding the geology of the region was included in travel memoirs of various wanderers—explorers. Later on, other scientific expeditions followed by Austrians (Bittner et al. 1880) and Germans (Fiedler 1841; Bucking 1880; Lepsius 1893), who studied various areas and documented some basic observations. It should be noted that the first detailed geological map of Greece is the map of Attica by Lepsius (1893), at 1/25.000 scale, which has been widely used and remained virtually unchanged to date (despite the lack of tectonic structures).

At the end of the nineteenth century, the German geologist A. Philippson began his detailed life-long research in Greece, with several months of fieldwork in the Greek mountains of Peloponnese (1887–1889), Central Greece (1890), Epirus, Thessaly, and Zakynthos Island (1893), the Greek Aegean islands, mainly the Cyclades and Northern Sporades (1896). His research lead to the publication of monumental monographs for each part of Greece that he studied, which are accompanied by geological maps usually at 1/300,000 scale. His most important outcome was the publication of his first synthesis at the end of the nineteenth century, «La tectonique de l'Egeide» (1898), where the first complete picture of the Geology of Greece is depicted and isopic-tectonic zones are firstly been distinguished (Fig. 7.1). His work was later presented bibliographically enriched in his three major works, «Beitrage zur Kenntnis der griechischen Inselwelt» (1901), «Beiträge zur Morphologie Griechenlands» (1930) and «Die griechischen Landschaften» (1956–1959, which was published after his death).

Immediately after Philippson's appearance, another great researcher, the Swiss C. Renz, began his research in Greece, and published a plethora of new data, mainly stratigraphic, and conducted his research and fieldwork for about fifty years (1903–1955). From his entire work three major compositional publications stand out: «Die Tektonik der griechischen Gebirge» (1940) (Fig. 7.2), «Beiträge zur Stratigraphie und Palaontologie des ostmediterranen Junkpalaozoikums und dessen Einordnung im griechischen Gebirgsystem» (1945, in collaboration with M. Reichel), and his classic final piece «Die Vorneogene Stratigraphie der normalsedimentären Formationen Griechenlands» (1955, published one year after his death).

During World War I, various geologists worked mainly in Northern Greece, both French (Arambourg, Bourcart, Piveteau) and Germans (Kossmat), who presented the general characteristics of the geological structure. The synthetic work of Kossmat (1924) «Geologie der Zentral Balkanhalbinsel» is the main source of knowledge for the whole Balkan region.

More progressive perceptions about the tectonic structure of Greece, based on the new interpretations of tectonic nappes, were expressed by Blumenthal (1933) and especially Kober (1929,1931,1933), whose revolutionary—at the time-views about tectonic windows of metamorphic Alpine nappes («Metamorphiden») were finally confirmed in the 1970s (Fig. 7.3).

Among the Greek geologists, two international figures appear prior to world war II: (1) F. Negris (publications during 1901–1928), with the discovery of Triassic in the metamorphic formations of Attica–Cyclades, included in the «Roches cristallophyliennes et Tectonique de la Grèce», and with the description of the Pindos nappe in South-Western Peloponnese already in 1906, and (2) K. Ktenas (publications during 1906–1935), conducted significant research on the

D. I. Papanikolaou, *The Geology of Greece*, Regional Geology Reviews,
https://doi.org/10.1007/978-3-030-60731-9_7

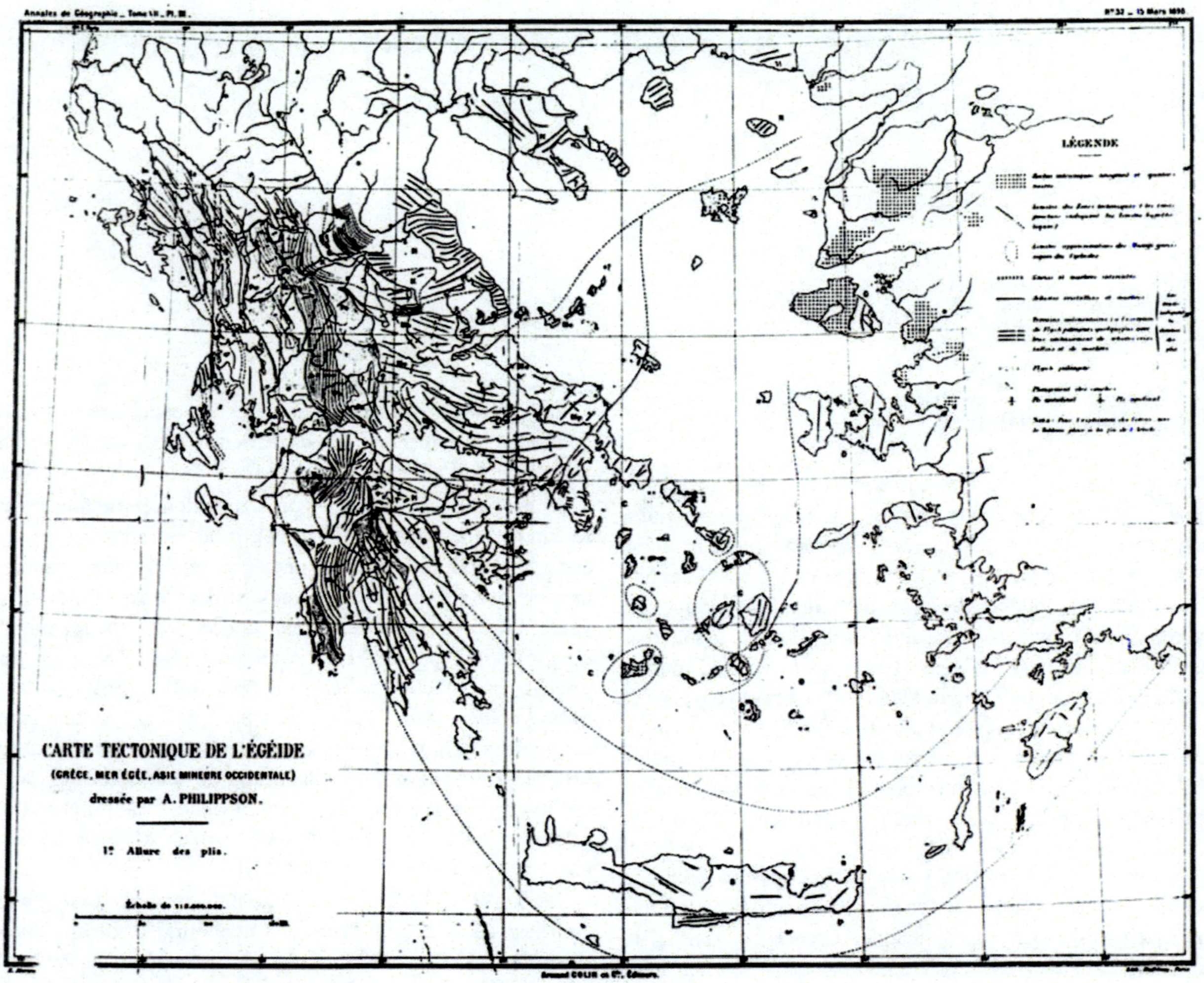

Fig. 7.1 Geotectonic map of Aegeis by Philippson (1898), which is essentially the first synthetic map of the Hellenic region

blueschists of the Cyclades, the finding of Lower Paleozoic in Chios, the discovery of Upper Paleozoic in Southern Peloponnese, and with important synthetic monographs, like: (i) the one presented in the 13th International Geological Congress in Liege (1922), entitled «Les plissements d'âge primaire dans la région centrale de la mer Egée». In this publication he presented a completely original geotectonic map of the Central Aegean region, where he tried to distinguish the Paleozoic folds from the Alpine ones and to classify the outcrops of the metamorphic rocks according to their metamorphic grade (M1 gneiss-amphibolites, M2 mica schists, M3 phyllites) (Fig. 7.4). The participation of five Greek geologists, three of which presented scientific communications (G. Georgalas, K. Ktenas, G. Voreadis) at the 13th International Geological Congress, held in Belgium in 1922, is characteristic for that period. (ii) The first synthesis on the lavas of Santorini, entitled «Le groupe d' iles du Santorin: Contribution a l'etude des laves tertiaires et quaternaires de la Mer Egee.» published as a Memoir of the Academy of Athens (1935), including all the recent data obtained during the volcanic eruptions in Nea Kammeni Island of the previous years (1925–1928). It should be noted that the Santorini eruption of 1925 was filmed by G. Georgalas, initiating the pioneer use of video for studying volcanic eruptions and geological processes worldwide (the documentary film is deposited at the Geology-Mineralogy Laboratory at the Agricultural University of Athens).

After world war II research is blooming and numerous foreign geologists study in Greece, while Greek geologists are beginning to make substantial contributions. This is how a plethora of new data are presented by French (Aubouin, Brunn, Celet, Dercourt, Mercier, Godfriaux, Ferrière, Angelier, Bonneau, Clement, Thiébault, Vergely etc.), Germans (Freyberg, Jacobshagen, Dürr, Altherr, Seidel, Creutzburg,

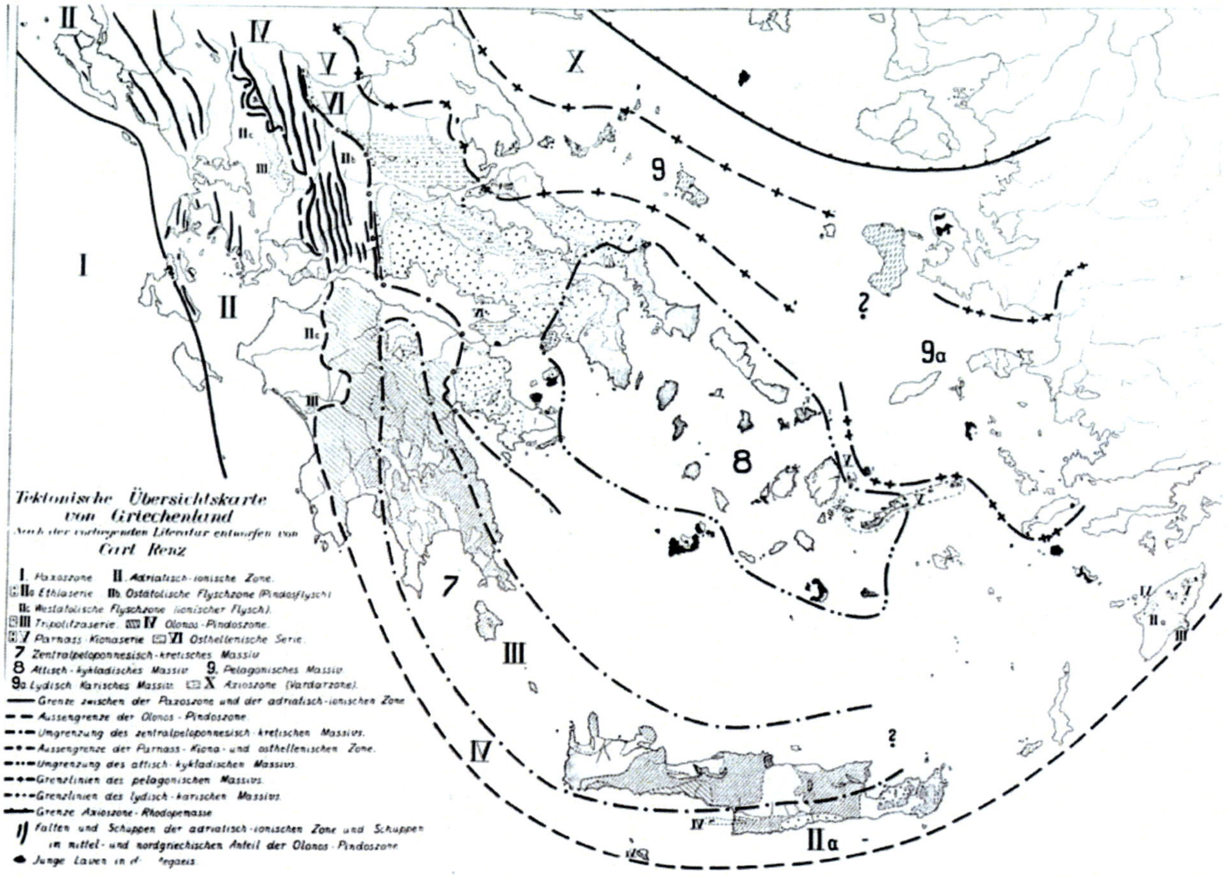

Fig. 7.2 Geotectonic map of Greece (Southern) by Renz (1940)

Richter, Kockel etc.), British (Smith, McKenzie, Dixon, Robertson, Jackson, etc.), as well as researchers from other countries (Bernoulli, Moores, Capedri, Schuiling, Paepe, Riedl etc.). The creation of I.G.M.E. (Institute of Geology and Mineral Exploration) since 1949, in the place of the former «Geological Office», marks the systematic geological research in Greece, under the guidance of Professors I. Papastamatiou, G. Marinos, and K. Zachos, who managed to gradually create the necessary scientific infrastructure for further development, through geological mapping at scale 1/50,000 and other surveys. Since 1960 and for several years, the tectonic model of the Hellenides, formulated by Aubouin (1959) was dominant (Fig. 7.5) and was made world-famous through his book «Geosynclines» (1965), which is the latest improved development of the theory of the geosyncline, although it was published while the new plate tectonics theory had already started to be developed.

The first application of the plate tectonics theory in Greece and the Mediterranean was based on active tectonics and mainly on seismological data, by McKenzie (1970, 1972), Papazachos and Comninakis (1971), Galanopoulos (1972). Applications of the new theory in the Tethyan Alpine system and the Hellenides emerge in the beginning of the 1970s (Smith 1971; Dercourt 1972; Dewey et al. 1973; Dimitrievic 1974). Since the end of the 1970s the current period of geological knowledge of Greece begins, based on the plate tectonics theory. A turning point in the evolution of geology in Greece is the first major international conference held in Athens in 1977 (VI Colloquium on the Geology of the Aegean Region), with the participation of almost all the foreign geoscientists working in Greece at the time, as well as several Greek scientists.

Synthetic monographs of the Geology of Greece appear already since the nineteenth century up to the 1970s by several geologists based on the old theory of the geosynclines (Cordella 1878; Mitsopoulos 1890; Negris 1901, 1915; Georgalas 1934; Renz 1955; Philippson 1956–59; Paraskevaidis 1959; Maratos 1972). During the 1980s new monographs, based on the plate tectonics theory are presented (Dercourt et al. 1980; Papanikolaou 1986b;

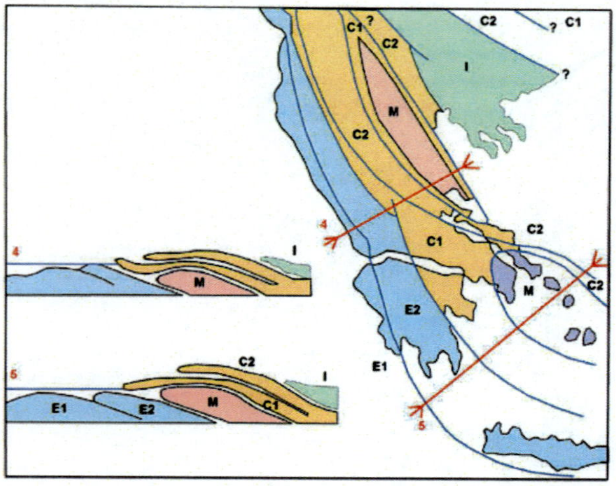

Fig. 7.3 Schematic geotectonic map of Greece with two transverse sections, 4 and 5, from the wider synthesis of the entire Alpine system by Kober (1931). The sections show the new, for that time, concept of metamorphic tectonic nappes (Metamorphiden—M), over the non-metamorphic tectonic nappes of the External units (Externiden—E1, E2), in the form of tectonic windows below the nappes of the central units (Zentraliden—C1, C2), which in turn are underlain below the nappes of the internal units (Interniden, I).

Jacobshagen 1986; Mountrakis 1986, 2010). The present monograph on the Geology of Greece is based on the evolution of the theory of plate tectonics with the younger theory of the tectono-stratigraphic terranes.

7.2 Distinction of the Tectono-Stratigraphic Terranes

As early as the 1980s, the plate tectonics theory was extended with the tectono-stratigraphic terrane theory (Coney et al. 1980; Howell 1980; Zen 1983; Jones et al. 1983; Schermer et al. 1984), which is essentially smaller-scale geotectonics, with India as a typical example. The continental fragment of India was rifted and detached from Eastern Africa within the Gondwana dispersion in the early Mesozoic, and after drifting within the Tethys Ocean, it was accreted to the sourthern margin of Eurasia, creating the collisional mountain range of the Himalayas since Oligocene–Miocene. At the same time as India was drifted northwards, the pre-existing Tethys Ocean was closing in front whereas the new Indian Ocean was opening in the back. This Kinematic pattern offers the possibility of distinguishing discrete paleogeographic and

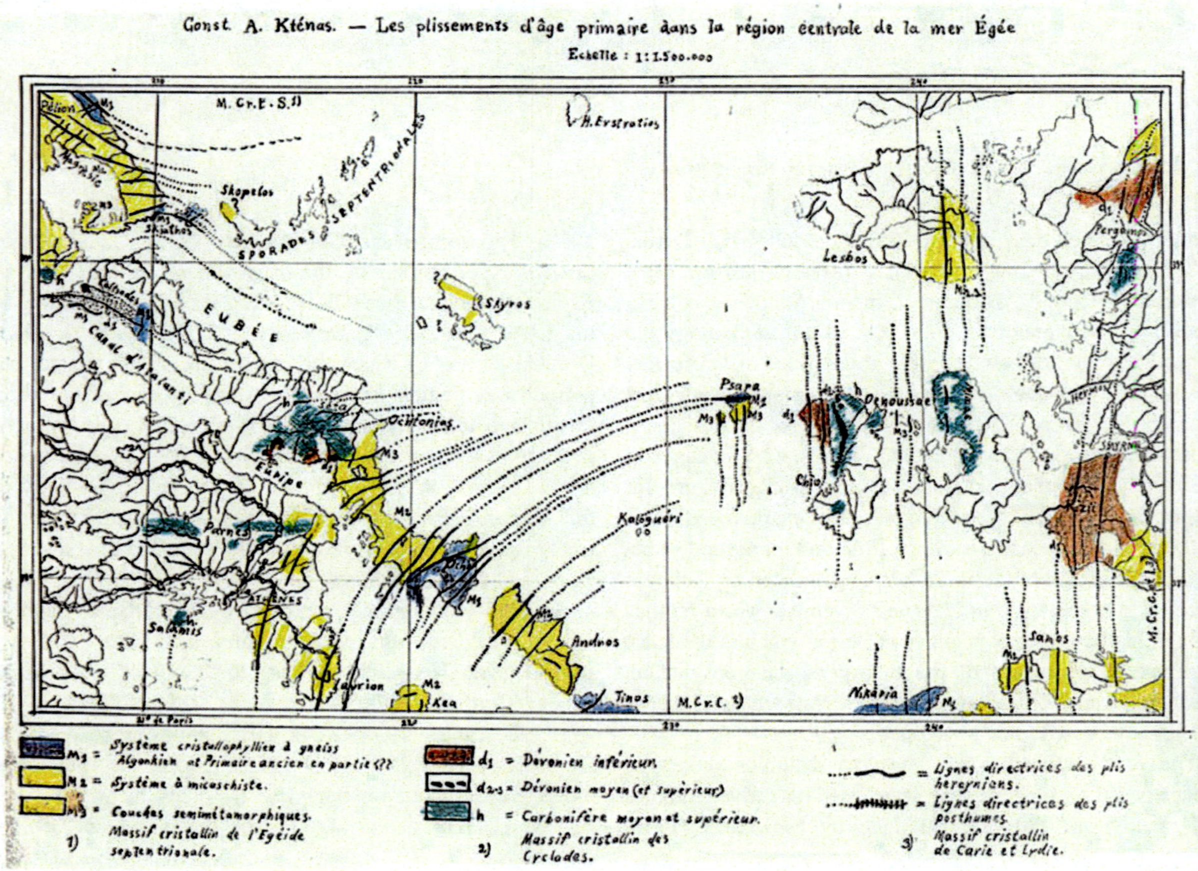

Fig. 7.4 The tectonic map of Ktenas (1923)

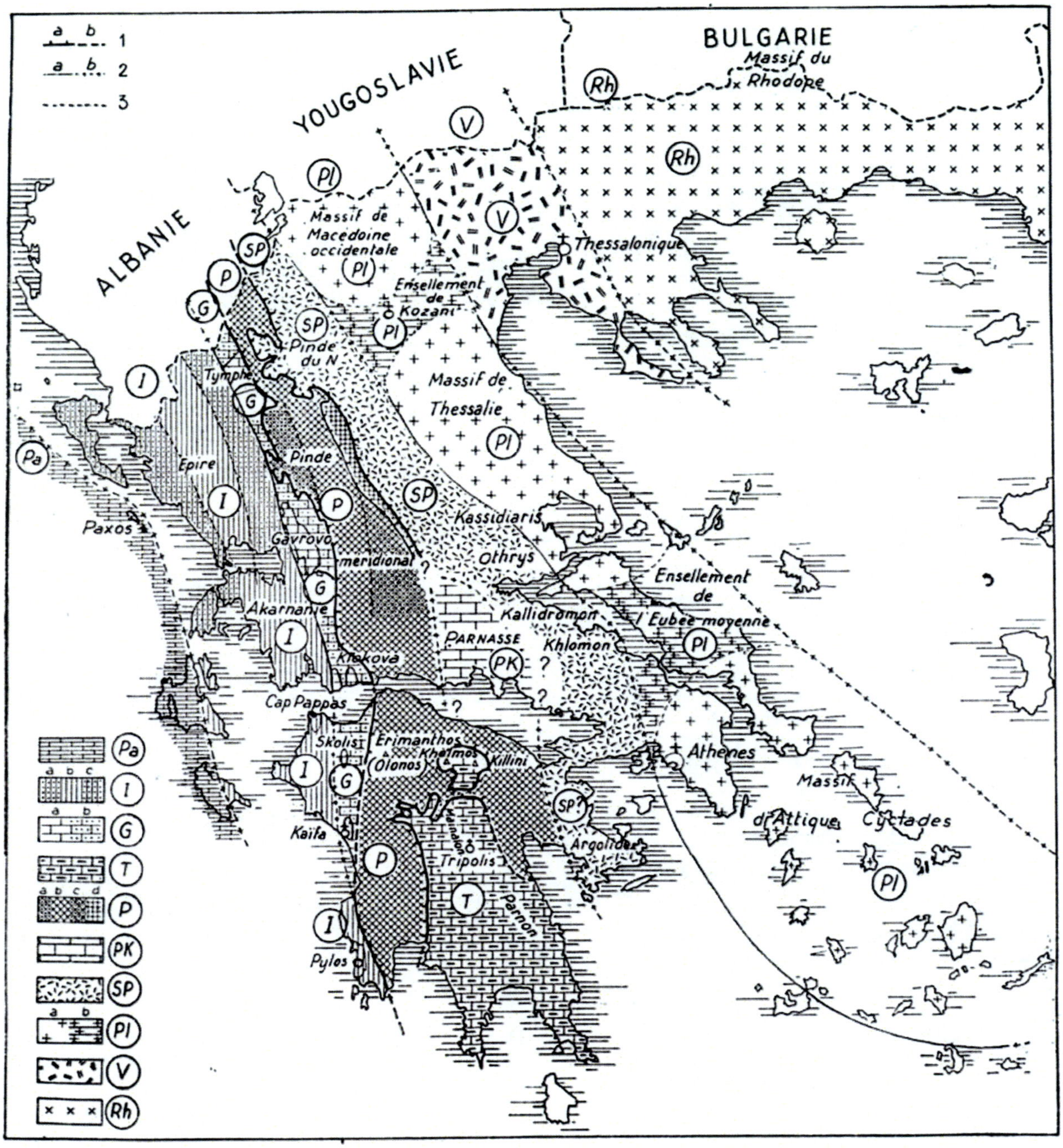

Fig. 7.5 Map of the "isopic zones of the Hellenides and their tectonic relations" (according to Aubouin 1959)

paleogeodynamic stages for the evolution of each tectono-stratigraphic terrane.

In the Mediterranean, the presence of intermediate crystalline massifs of pre-Alpine continental crust between different ophiolite sutures of the Tethyan Alpine system was well known (Papanikolaou and Sassi1989) (Fig. 7.6). The origin and evolution of these massifs were the subject of IGCP, Project No 276, "Palaeozoic geodynamic domains and their Alpidic evolution in the Tethys" (1987–1995). The results of this project included the documentation of the tectono-stratigraphic terranes throughout the Mediterranean, by issuing a special volume in «Annales Geologiques des Pays

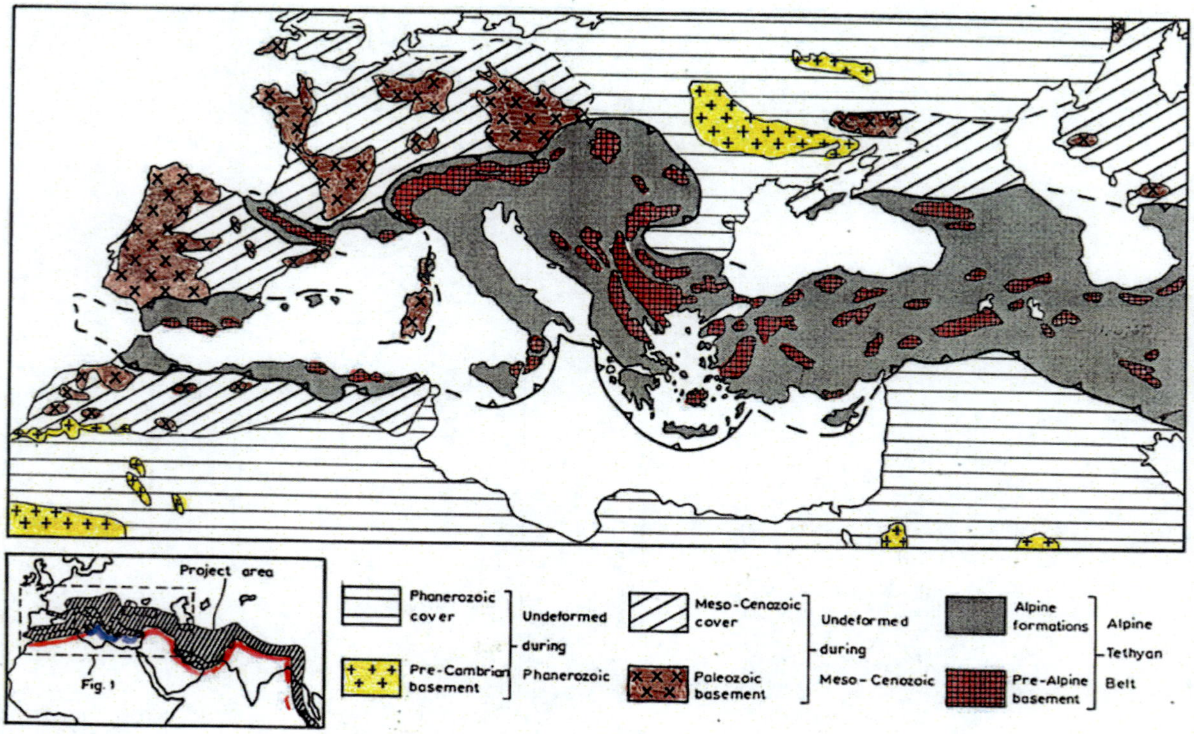

Fig. 7.6 Map of Europe and the Mediterranean with distinction of: (i) pre-Cambrian–Lower Paleozoic segments of continental crust with Phanerozoic sedimentary cover for the Pre-Cambrian and post-Silurian for the Caledonian orogeny. (ii) Upper Paleozoic segments with Paleozoic basement and Meso-Cenozoic sedimentary cover, resulted from the Variscan orogeny and (iii) Mesozoic–Early Cenozoic Tethyan Alpine Belt. Outcrops of pre-Alpine basement rocks observed within the Alpine Tethyan Belt justify the presence of the terranes (from Papanikolaou and Sassi 1989, based on the 1987 proposal to IGCP)

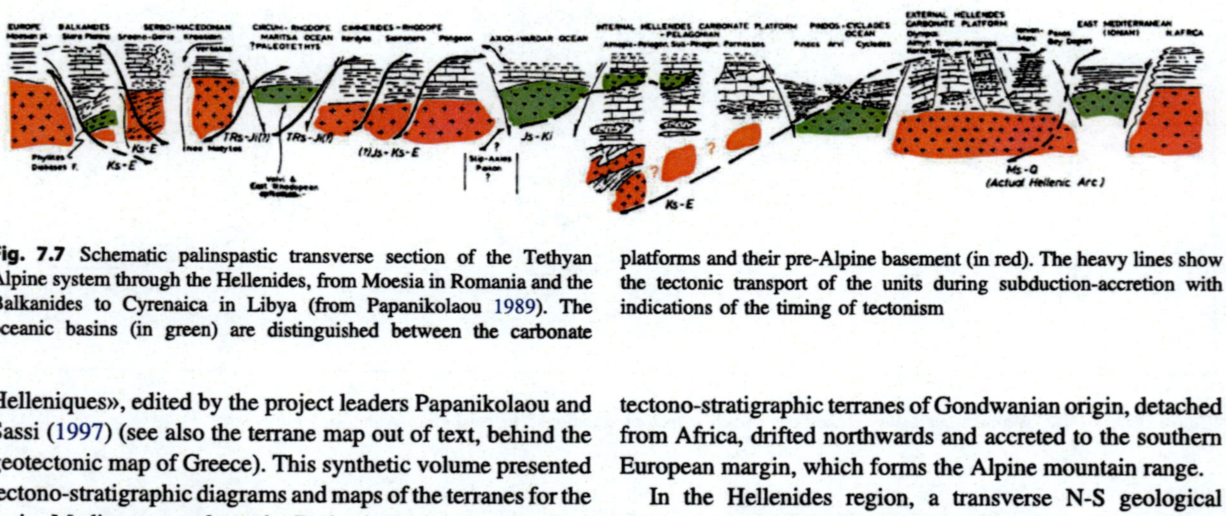

Fig. 7.7 Schematic palinspastic transverse section of the Tethyan Alpine system through the Hellenides, from Moesia in Romania and the Balkanides to Cyrenaica in Libya (from Papanikolaou 1989). The oceanic basins (in green) are distinguished between the carbonate platforms and their pre-Alpine basement (in red). The heavy lines show the tectonic transport of the units during subduction-accretion with indications of the timing of tectonism

Helleniques», edited by the project leaders Papanikolaou and Sassi (1997) (see also the terrane map out of text, behind the geotectonic map of Greece). This synthetic volume presented tectono-stratigraphic diagrams and maps of the terranes for the entire Mediterranean from the Betics in the west to the Caucasus in the east. Only a few outcrops of the pre-Alpine formations of Tethys along the northern margin of the Tethyan system are considered to be part of the Eurasian margin at the end of the Paleozoic. The remaining outcrops are considered tectono-stratigraphic terranes of Gondwanian origin, detached from Africa, drifted northwards and accreted to the southern European margin, which forms the Alpine mountain range.

In the Hellenides region, a transverse N-S geological section had been published for the first time (contrary to the E-W usual direction) extending from the Rhodope area to the Hellenic trench south of Crete, within the IGCP No 5 (Papanikolaou and Skarpelis 1980). Based on this section a new revised palinspastic extended section from the Moesian

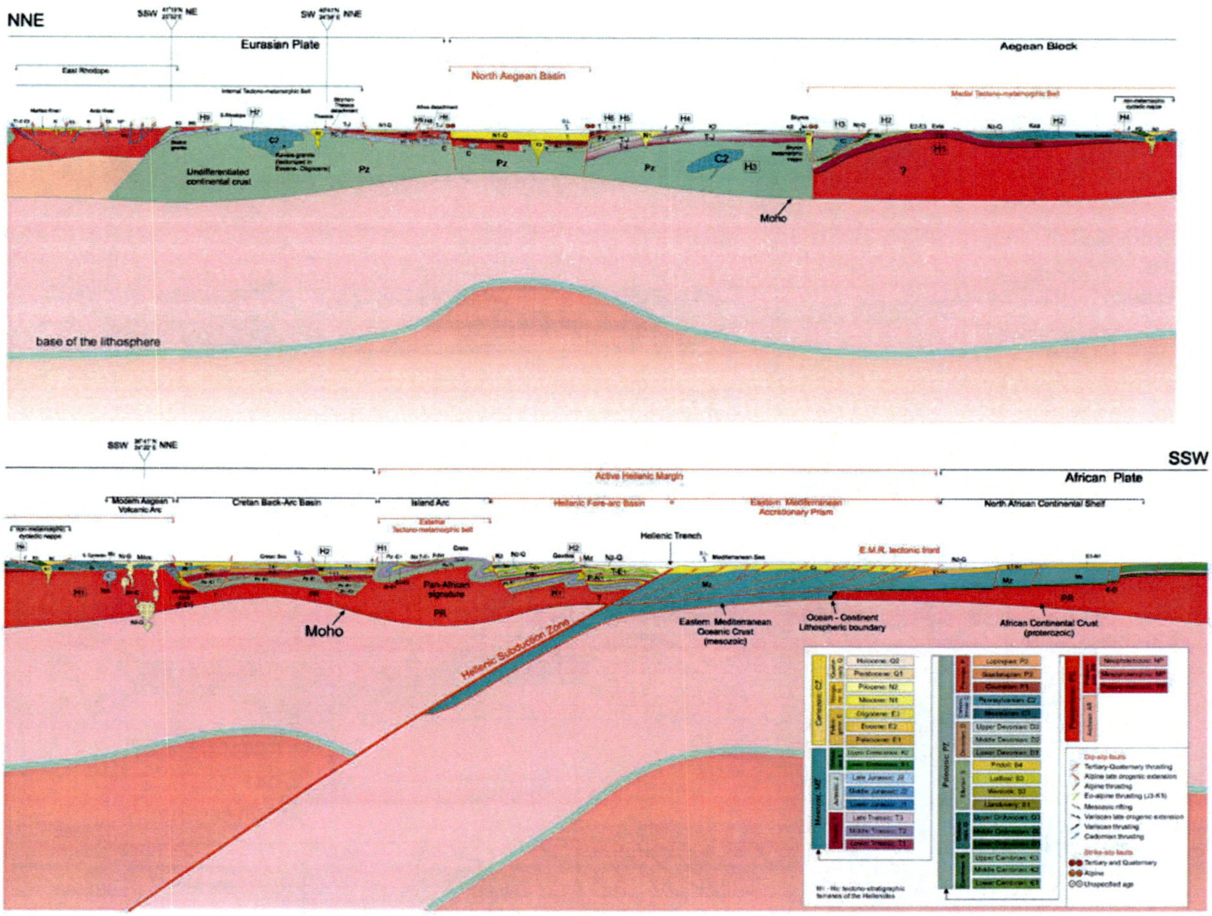

Fig. 7.8 Geological cross section through the Tethyan Alpine System in the Eastern Mediterranean, Geotraverse VII of the TRANSMED Atlas (from Papanikolaou et al. 2004)

platform to the Cyrenaica peninsula in Africa was presented within IGCP No 276 (Papanikolaou 1989) (Fig. 7.7). The succession of the Hellenic continental terranes and the intermediate Tethyan oceanic basins were identified along the section. Later on, this geological section was further elaborated and presented at the International Geological Congress of Florence in 2004 as Transect VII (Papanikolaou et al. 2004) (Fig. 7.8), in the framework of eight geotraverses (I-VIII) of the Mediterranean. These profiles were published in the «TransMed Atlas, the Mediterranean Region from Crust to Mantle» (Cavazza et al. editors, 2004, Springer). The geological section shows: (i) the Hellenic tectono-stratigraphic terranes, (ii) the three tectono-metamorphic belts, (iii) the Hellenic subduction and the active orogenic arc and (iv) the thickness of the crust, which is generally thin, with values between 16 and 18 km in the Cretan basin and the Northern Aegean basin, and 28–30 km in between the Aegean plate. To the north the thickness increases up to 40 km in Rhodope and Bulgaria. Herein, it should be noted that, if the section had crossed continental Greece, then the crustal thickness would be 40–50 km under the Pindos mountain range.

To the south of the Hellenic trench the allochthonous accretionary prism of the submarine mountain chain of the Eastern Mediterranean overlies the northern margin of the African plate, where several kilometers thick sediments of almost the whole Phanerozoic crop out. Therefore, the oceanic crust of the Eastern Mediterranean, which represents the last remnant of Tethys, characterized as the future H0 terrane, is already subducted for its most part, beneath the Hellenic arc, without any oceanic remnant remaining intact today at the seabottom (see also Sect. 3.5).

Nine terranes have been distinguished in the Hellenides, five of which are continental (H1, H3, H5, H7, H9) and four oceanic (H2, H4, H6, H8). Another final terrane of the Eastern Mediterranean Ocean (H0) could be added, which is the last remnant of the Tethys, during the forthcoming collision of the Hellenides with Africa. The main characteristic of the Tethyan

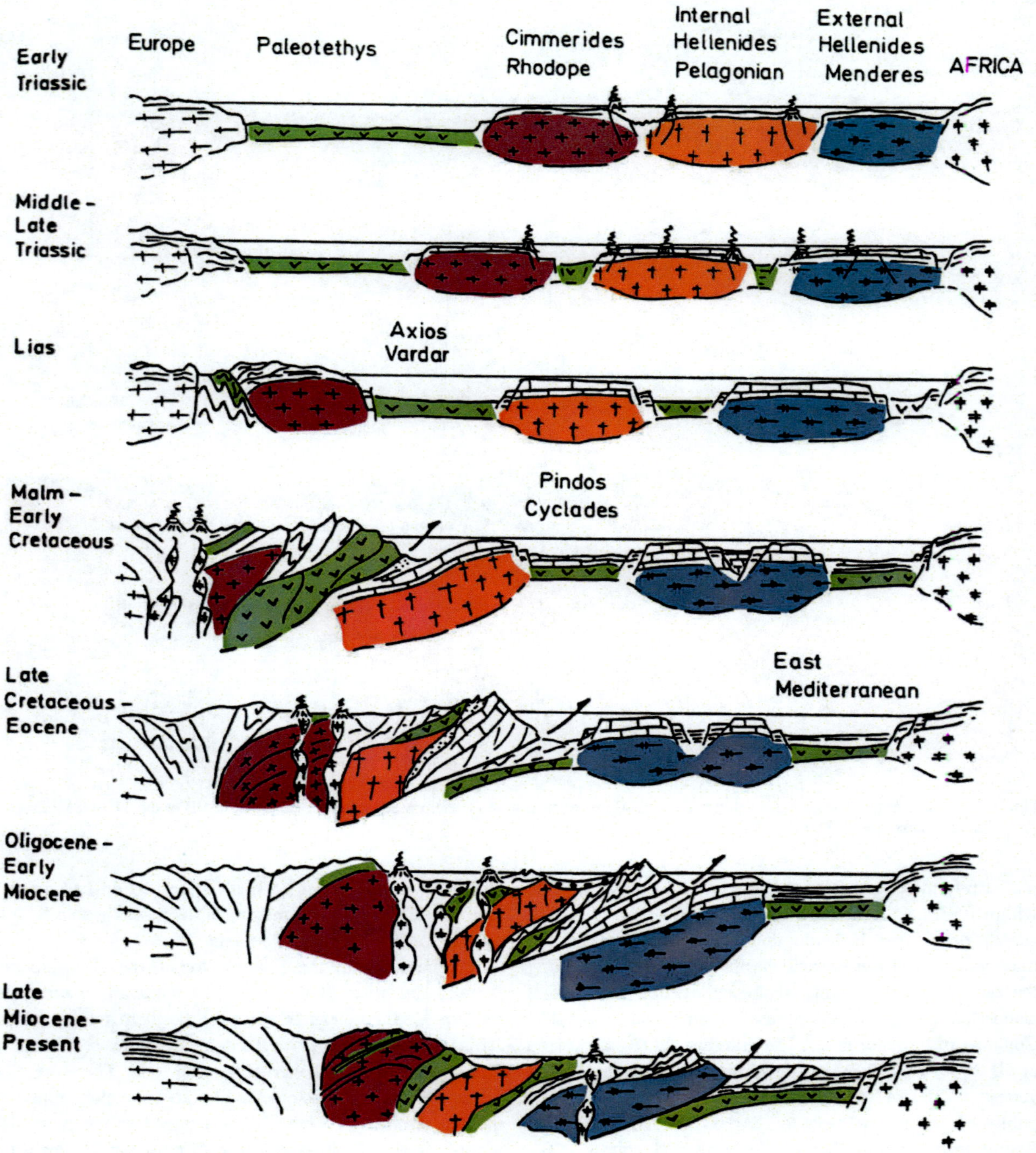

Fig. 7.9 Schematic tectonic profiles across the Hellenides, showing the general pattern of terrane drift of the continental terranes H7, H5, H3, and H1 from the Gondwana to the European margin from Triassic to Late Miocene. Parallel rifting, opening and subsequent subduction of the oceanic basins H6, H4, H2, and H0 occurred from the Triassic to present, when the final subduction phase of the last basin (East Mediterranean H0) beneath the Hellenic arc takes place (from Papanikolaou 1989)

paleogeography was that shallow water carbonate platforms had been developed on top of the continental terranes, while abyssopelagic sediments were deposited on top of the ophiolites inside the intermediate oceanic basins. The general drift movement of the Hellenic terranes was constantly occurring from south to north, with a gradual rifting from

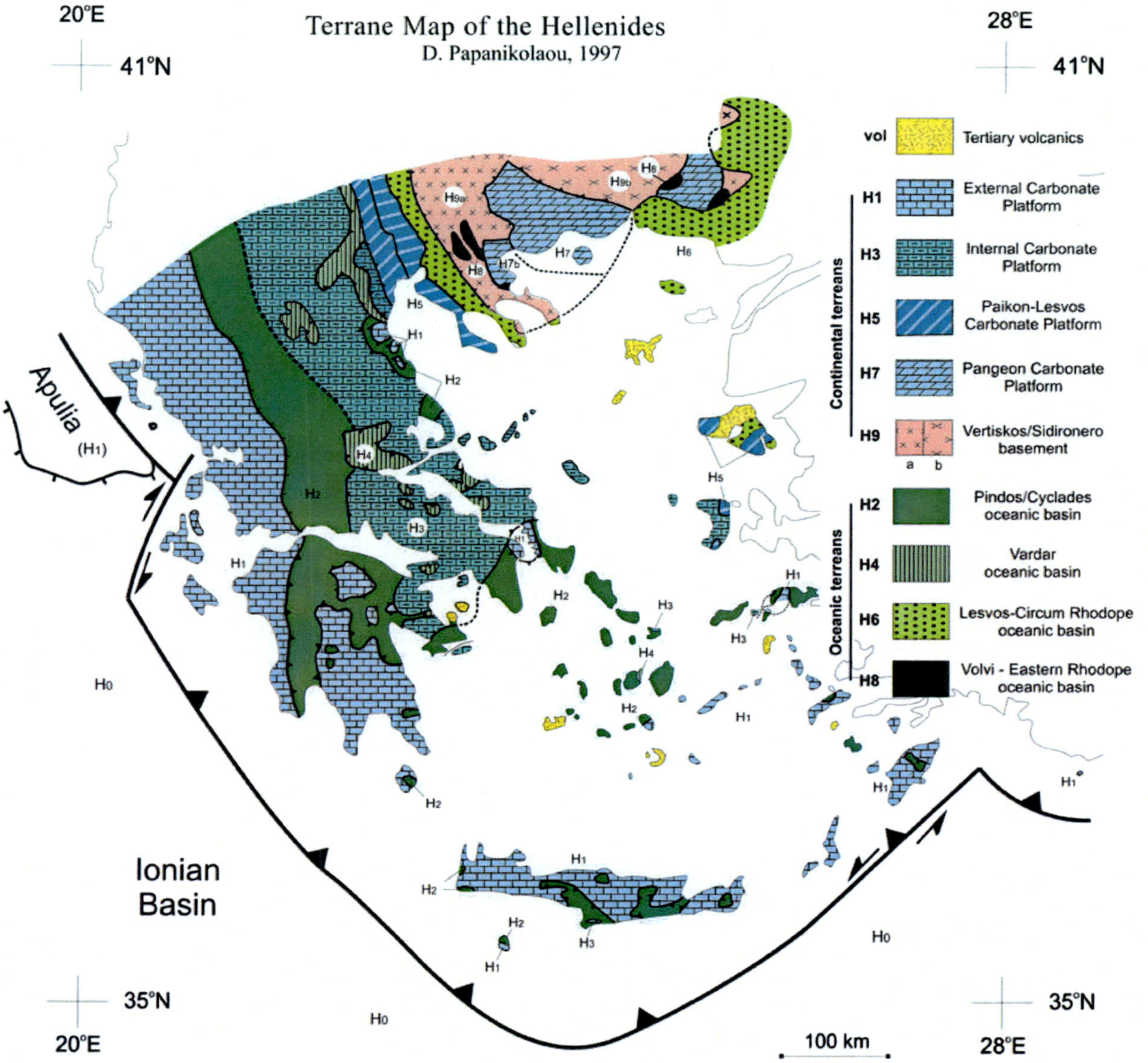

Fig. 7.10 Geotectonic map of the tectonostratigraphic terranes of Greece, according to Papanikolaou (1989, 1997, 2013)

Africa during the Triassic and a gradual arrival and accretion to Europe from the Jurassic to present (Fig. 7.9).

The Terrane map of the Hellenides (Fig. 7.10) shows the entire layout of the nine terranes, of which only four participate in the building of continental mainland Greece. The remaining five terranes make up the Northern Aegean region, the Central–Eastern Macedonia and Western Thrace.

The overall representation of the Hellenides paleogeography is impossible and thus meaningless, because it was constantly changing, with the drifting of the continental terranes bearing also their carbonate platforms and a continuous opening and closure of the intermediate oceanic basins. Thus, the overall paleogeographic representation (Fig. 7.11) serves only as a general sketch of the cumulative paleogeography, which has never existed as a whole, but it gives an overall diachronic dimension of the Hellenides. The scheme is based on the re-opening of the oceanic basins in their suture zones using the minimum possible dimensions, given their disappearance along the Hellenic subduction zone. In contrast, the dimensions of the continental terranes—platforms are more easily estimated because they generally remain in the crustal level and their competence during the orogenic deformation permits the preservation of their original volume.

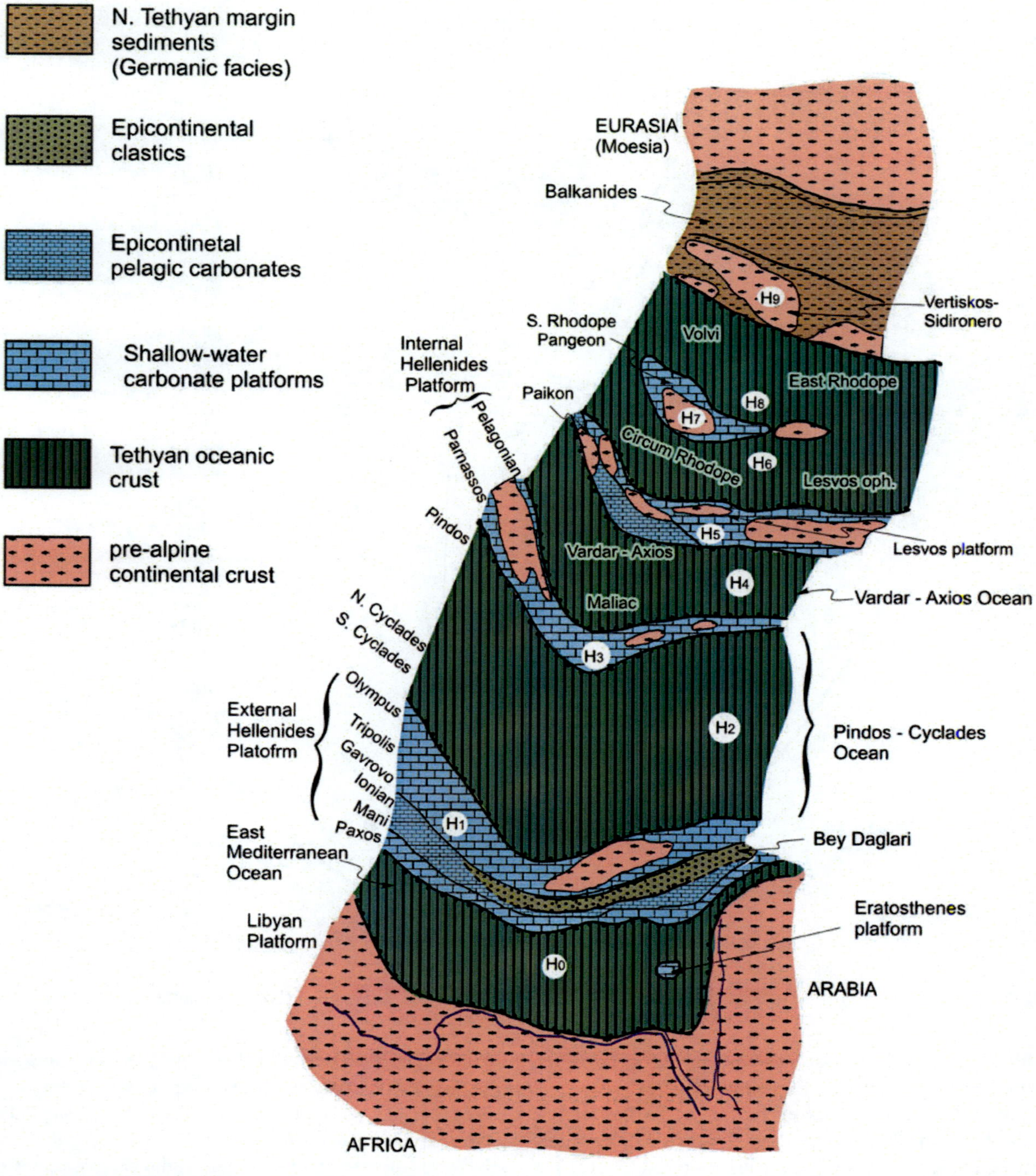

Fig. 7.11 Schematic representation of the tectono-stratigraphic terranes of the Hellenides in the Tethyan paleogeography, indicating the minimum extension of each oceanic terrane. This sketch has never existed in its entirety, mainly because while some basins were closing to the north, some others simultaneously were opening in the south (from Papanikolaou 1989, 1997, 2013)

7.3 Geodynamic—Paleogeographic Stages of the Terranes and Tectono-Stratigraphic Models of the Hellenides

One of the main outcomes of the application of the theory of tectono-stratigraphic terranes in the Hellenides propose that their stratigraphic structure can be summarized in only two types of stratigraphic columns, which arise from the overall terrane development and evolution in three stages (Papanikolaou 2013). The three stages of geodynamic and paleogeographic evolution will be analysed below, followed by the resulting two models of stratigraphic sequences. As it will be shown after the terrane analysis of the Hellenides, the two stratigraphic models can describe every unit by simply changing some ages (see also Figs. 8.2 and 8.3).

7.3.1 The Geodynamic—Paleogeographic Stages of the Terranes

The overall history of each continental tectono-stratigraphic terrane of the Hellenides can be divided in three stages (Papanikolaou 1989, 1997, 2013) (Fig. 7.12): (1) **the rifting stage**, which initiates with the creation of the new diverging plate boundary between the new crustal element/terrane and the African continent in the Northern Gondwana region during the Permian–Triassic, (2) **the drifting stage**, which represents the autonomous motion of the continental terrane within the Tethyan domain, which is related to the closure of pre-existing oceanic basins of Tethys northwards and the contemporaneous opening of new basins of Tethys southwards during the Late Triassic–Eocene, and (3) **the accretion stage** in the

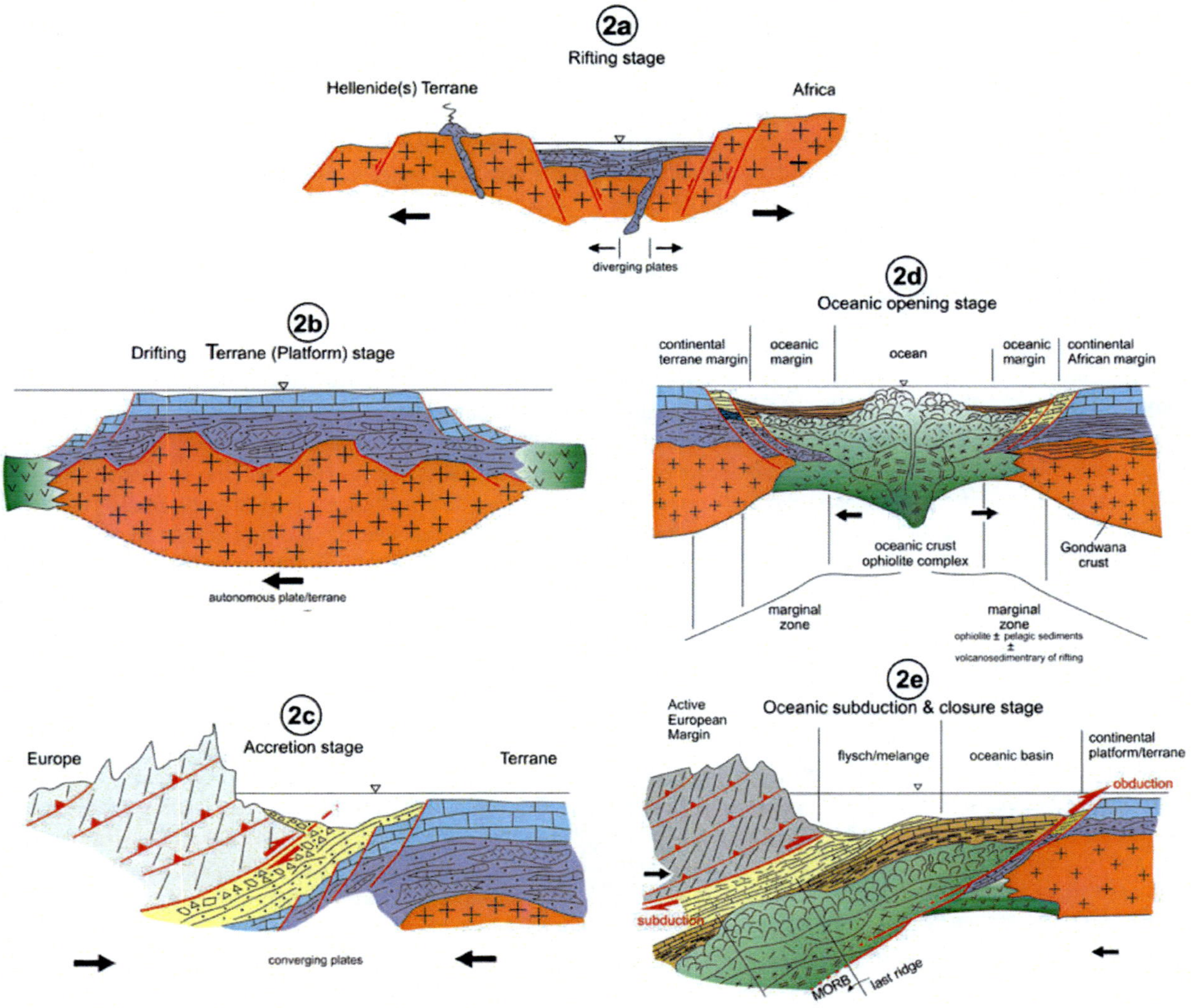

Fig. 7.12 Schematic representation of the three paleogeographic–geodynamic stages of the Hellenic terranes in Tethys, from their creation and detachment from Africa to their integration in Europe (according to Papanikolaou 2013)

active Eurasian margin and the European plate during the Early Jurassic—present.

Accordingly, oceanic tectono-stratigraphic terranes follow similar stages: (1) **the rifting stage**, during which the collapsed blocks within the rift will gradually form the two continental margins, one towards the terrane in the north and another towards Africa in the south, (2) **the oceanic opening stage**, during which ophiolites form and related abyssopelagic sediments will be deposited, (3) **the subduction and oceanic closure stage**, during which the oceanic basin will be sutured, following the subduction and accretion to the Eurasian margin, with eventual ophiolite obduction over the continental margin(s).

(1) Rifting stage

In the Permo–Triassic, extension with normal faulting is observed along the northern margin of Africa, creating a diverging plate boundary, which at its initial stages forms a continental basin with clastic sedimentation. The type of clastic sediments depends on the vicinity of the marginal faults of the basin and the new topography of the rift valley. Sedimentation may start as continental, with or without lacustrine deposits including coal and then continues as paralic with shelf deposits and finally it gradually deepens, until it becomes pelagic. Pelagic sequences may unconformably overlap the previous continental and coastal sedimentary facies in a transgressive setting. Evaporites may also be deposited during this stage, following the actualistic example of the Dead Sea rift. As soon as the rifting continues, crustal extension produces thinning of the crust and updoming of the isothermal contours, resulting in the creation and extrusion of alkaline lavas. Thus, the former graben/basin with clastic sedimentation is transformed to a more complex volcano-sedimentary environment with a varied tectono-stratigraphy over short distances, that is to say without stable stratigraphic horizons in the basin. The large variety of volcanic lavas and pyroclastics, as well as alterations in a variety of volcanic formations, creates a characteristic colorful complex, constantly changing, depending on the internal deformation of the rotating blocks. At the same time, some peculiar stratigraphic facies are created, such as the Hallstatt facies, widely observed in the Alps, of nodular marly limestones with ammonites, usually of Middle Triassic age, like the typical occurrence of Epidavros (Frech 1907). The above events in the area of initial divergence correspond to the intra-condinental rifting model of Cloos (1939) from the Rhine graben and mainly to the actualistic case study of the East African Rift (McKenzie et al. 1970; Tazieff et al. 1972).

The duration of the rifting stage can last from several millions of years up to a few tens of millions of years, depending on the extension rate. Indicatively, the actual example of the Red Sea rift lasted about 15 Myrs, from the initiation of extension during the Early Miocene to the extrusion of the mafic lavas in the Late Miocene–Pliocene. The duration of the rift stage in the Hellenides may be estimated by the age determination of the base of the volcano-sedimentary complex and of its top horizon below the shallow water carbonate sedimentation of the platform. Usually, this period includes parts of the Upper Permian and the Lower–Middle Triassic with a duration of about 10–15 Ma, as documented by a series of publications on several units of the Hellenides (Ktenas 1924; Kauffmann 1976; Ferriere 1976, 1982; Papanikolaou 1979b, 1989, Ardaens et al. 1979, Pamic 1984, Pe-Piper 1982, 1998; Migiros and Tselepidis 1990; Dimitriadis and Asvesta 1993; Robertson 2007; Himmerkus et al. 2009; Asvesta and Dimitriadis 2010; Palinkas et al. 2010).

(2) Drifting stage

As soon as the separation of the continental terrane from the African plate is established, there is a re-organization of the isostatic equilibrium and the continental block is floating above the mantle with its top almost reaching the sea level. Thus, shallow water carbonate platform sedimentation is established and continues during the drifting of the terrane towards the European plate, between the oceanic basins of Tethys, i.e. the older one closing north and the younger one opening south of the terrane. The isostatic balance is quite steady, as indicated by the constant deposition of the same facies of the shallow water carbonate platform over tens of millions of years. The permanent occurrence of fossils of organisms that use light to transform their energy needs (e.g. algae) shows that the deposition depth does not exceed 200 m, at about the boundary of the euphotic zone. The biostratigraphic markers are characteristic fossils, belonging to organisms that live in well determined special environments, such as tidal zones, where stromatolites are deposited, together with reefs with corals, gastropods, and related mollusks. These stratigraphic facies persist over tens of millions of years (e.g. Pantokrator), even though the thickness of the carbonate rocks of the platform may exceed 3–4 km. This observation demonstrates that the isostatic equilibrium of the terrane is not affected by the additional overburden of 3–4 km of carbonate sediments.

The timing of the onset of the drifting stage can be determined by the age of the first beds of the shallow water carbonate platform, which in the stratigraphic column corresponds to the transition from the volcano-sedimentary complex to the carbonate platform. The intercalation of some volcanic tuff layers at the base of the carbonate platform (Fig. 7.13) marks the end of the previous rift volcanism until its final elimination. Laterally, towards the margins of the

Fig. 7.13 Characteristic outcrop of the base of the shallow water carbonate platform of Tripolis in the Karnian, over the Middle Triassic volcanics of the Tyros beds in the Molai region. Thin intercalations of tuffs can be observed between the stromatolites of the carbonate sediments, which gradually fade towards the upper horizons

terrane with the oceanic basins, a change in the volcanism is observed, with a transition from the alkaline to the tholeitic domain.

Based on the thickness of the sediments of the carbonate platform and the duration of the sedimentation, an average sedimentation rate of 5–15 m/Ma can be calculated. From a stratigraphic point of view stage 2 marks a substantial change from stage 1, as there is a transition from the complicated volcano-sedimentary complex to the rather monotonous well-stratified sedimentation of the carbonate platform. At the same time, the boundary between the two facies is a mechanical anisotropic surface, and the subsequent compression event in the orogenic arc easily leads to a tectonic decollement. Thus the transition between the two facies is often not undisturbed, but tectonised.

During the drifting stage of the terrane within Tethys, additional extension may be exerted with the creation of an epicontinental basin but without separation of the crust of the two horsts of the platform. The extension in this case is limited to about 5–10%, and can be absorbed by the elastic behavior of the terrane and its crustal thinning. This additional extension is not accompanied by volcanism and is observed in the Hellenides usually during the end of Lias, with the creation of the pelagic sedimentation in the Ionian and Mani units (external carbonate platform), as well as in the Sub-Pelagonian and Beotia units (internal carbonate platform). The same phenomenon can be observed in other regions of Tethys during the Late Lias. From a stratigraphic point of view, the drifting stage corresponds to the shallow water carbonate platform sedimentation stage. From a tectonic point of view the detached terrane forms an independent plate or microplate, together with half of each of the two adjacent oceanic basins, i.e. up to the two mid-ocean ridges.

(3) Accretion stage

The end of the drifting stage coincides with the arrival of the continental terrane at the active European margin and the resulting continental subduction (Molnar and Gray 1979, Chemeda et al. 1996, DeFranco et al. 2008) and microcollision. The arrival of the continental terrane follows the previous subduction zone of the crust of the northern oceanic basin. After the initial microcollision of the terrane with its carbonate platform to the European margin, a tectonic decollement of the terrane from the underlying lithosphere follows and it is accreted to the front of the crust of the advancing European plate. The depth of decollement is approximately 10–15 km and corresponds to the upper portion above the ductile/ brittle crustal boundary (Molnar and Gray 1979; Burchfiel et al. 2018). The rest of the lithosphere under the decollement is subducted to great depths where it is identified by seismic tomography.

The change of the stratigraphy in the transition from stage 2 to stage 3 is dramatic, because the subduction produces a short gradual deepening in the trench, while the arrival of the clastic detrital material from the adjacent island arc installs deep clastic sedimentation of flysch. Thus, the stratigraphic column of each unit is completed with the third tectono-stratigraphic formation of flysch, usually deposited at 3–6 km of depth in the trench along the dynamic boundary of the two plates, the European to the north and the terrane to the south. The topmost beds of flysch become more chaotic, forming the so-called wild flysch with coarse clastics, olistoliths, olisthostromes etc., indicating the geological structure of the intensively uplifting and eroding island arc during the sedimentation period.

In the transition from the pre-orogenic biochemical sedimentation to the syn-orogenic clastic sedimentation of flysch, transitional beds of pelagic limestones with marly intercalations are observed at the first stage which soon after, become typical flysch deposits with turbiditic layers and cyclothemes. In several cases, the subsidence of the platform to the foreland basin is interrupted by local ephemeral movements that create micro-emersions and micro-unconformities in the fore-bulge depocentre, with formation of paleorelief and/or bauxite deposits or zones with iron-phosphorous hard ground, which constitute stratigraphic condensation horizons.

After the terrane has been detached and accreted to the advancing plate, and after being completely separated from the subducting slab, there comes the effect of isostatic buoyancy inside the upper plate. Thus, it is uplifted to the surface in the form of tectonic window, created by large low angle normal faults and extensional detachments.

The timing of the arrival in the trench area is determined by the age of the transitional beds of flysch, while the last phase of the beginning of the subduction is determined by the age of the topmost beds of wildflysch, before the final termination of sedimentation. The dating of the post-docking period in the upper plate is conducted through the dating of the first molasse sediments in the internal part of the arc, which seal all the structures, or by the intrusion of the volcanic rocks of the arc, which intersect all the formations (including also the syn-orogenic ones).

Granite intrusions in the interior of the arc create the final image of the new continental crust of the upper plate, which is compacted and gradually grows. The depth along the subduction zone before the decollement of a unit can be determined by the metamorphic assemblages (if any), while the uplifting is determined by the age of the first post-orogenic deposits. Thus, in the case of the Mani unit, the whole subduction and uplifting process in the tectonic windows of Peloponnese and Crete lasts about 10–15 million years, from the Late Oligocene (age of the flysch) to the Middle Miocene (age of the first post- Alpine sediments) (see Sect. 8.1.2).

(4) Oceanic opening stage

Immediately after the rifting stage and the separation of the continental terrane the rifting basin gradually grows into a new oceanic basin, where magmatic material with tholeitic affinities intrudes, replacing the previous alkaline volcanics. Thus the opening of the new ocean develops, with ophiolites and abyssopelagic sediments, usually siliceous, which are in lateral transition to the volcano-sedimentary complexes of the previous stage found in the two margins. The stratigraphy of the oceanic basin is continuously changing with different formations observed in various parts of the basin. Therefore, we have ophiolite complexes along the axis of the basin around the mid-ocean ridge, while abyssopelagic sediments can be observed at the abyssal basins and at the margins, where they overlie the volcano-sedimentary formations of the previous stage. With the opening of the ocean there is a symmetrical spread of the abyssopelagic sediments on either side of the mid-ocean ridge, which date the underlying ophiolites.

Depending on the drifting rate of the two continents and of the intermediate terranes, together with the opening rate of the oceanic basins, there may be an unconstrained expansion of the basin or termination of the oceanic basin opening, or even the beginning of a local intra-oceanic subduction zone. These phenomena do not affect the adjacent large geotectonic units/plates and usually disappear beneath the European margin together with the subsequent subduction of the whole oceanic basin. Indicative evidence of such intra-oceanic tectonic movements may be sometimes obtained through petrological and geochemical analyses, showing specific environments for the formation of ophiolites and associated mafic rocks, formed outside the mid-ocean ridge, such as the supra subduction zone ophiolites (SSZ), or the oceanic island arc ophiolites, etc. (Beccaluva et al. 1984; Saccani and Fotiades 2004, Dilek et al. 2005, 2008; Robertson et al. 2013). Another complication may be observed due to the creation of marginal seas adjacent to the subduction zones in the area of back-arc basins, as seems to be the case of the Upper Jurassic mafic rocks of Peonia unit in the Guevgeli area (Bebien and Mercier 1977).

(5) Oceanic subduction and closure stage

With the arrival of the oceanic basin in the trench area, a third stage begins, with the creation of deep clastic sedimentation inside the trench, of flysch or mélange type, which is characterized by greater depths of 5–8 km, in comparison to the former abyssal plain with depths of 3–5 km. Thus, the stratigraphic columns are completed with the flysch and wild flysch/mélange formations and at the same time, the presence of olistoliths within the flysch may indicate the petrological and stratigraphic composition of the island arc.

The duration of oceanic subduction depends on the dimensions of the ocean and the subduction rate. An indicative element is the duration of the flysch sedimentation, in relation to the average subduction rate. Thus, in the case of the Pindos, flysch deposition started in the Late Maastrichtian (65 Ma), as determined by the Globotruncanes found in the transitional beds, and ended sometime during the Middle to Upper Eocene (45 Ma) which is the age of the upper horizons of flysch (Aubouin 1959, Fleury 1980). Therefore, it lasted for about 20 million years, and by considering an average subduction rate of 50–60 mm/year, results to a width of about 1000-1200 km of the subducted Pindos oceanic basin.

Various models have been proposed for the ophiolite obduction process since the 1970s (Dewey and Bird 1970, 1971; Coleman 1971; Dewey 1976). The leading model, based on field observations, is the detachment and back-overthrusting towards the next platform to the south, which is underthrusted beneath the ophiolites, before they are completely subducted and smashed between Eurasia to the north and the terrane to the south. In the case of the Hellenides, the above model can be documented for all of the four oceanic basins, which were obducted to the south, over the younger carbonate platforms (Papanikolaou 2009, 2013) (see also Fig. 8.130).

It is interesting to note that during the oceanic crust subduction at great depths, of several tens of kilometers, blueschist metamorphic belts are created, such as the famous Franciscan Mélange in California (Ernst 1970; Blake et al. 1969). In Greece, a similar case is observed in the Cyclades blueschist belt (Blake et al. 1981), which have been exhumed at the surface after their detachment from the subduction zone and their uplifting together with the underlying margins of the continental terranes (Papanikolaou 1987, 2009; Avigad and Garfunkel 1991; Platt 1993; Avigad et al. 1997; Jolivet et al. 2003; Brun and Faccena 2008; Jolivet and Brun 2010; Ring et al. 2010).

7.3.2 The Two Tectono-Stratigraphic Models

The **stratigraphic synthesis** of the above evolution stages of the continental terranes with their carbonate platforms and the oceanic basins respectively leads to the existence of **two tectono-stratigraphic terrane models**, one for platforms and one for basins (Papanikolaou 2013) (Fig. 7.14).

The two models comprise their corresponding crusts, continental crust for the continental terranes with the overlying carbonate platforms and oceanic crust with ophiolites and abyssopelagic sediments for the oceanic terranes. The major difference is that the continental crust is pre-Alpine of Paleozoic or even PreCambrian age, underlying the Alpine sedimentation of the carbonate platform, whereas the oceanic crust is of Mesozoic age and is located in a lateral relation to the largest segment of the Tethyan abyssopelagic sediments. Therefore, in the case of the oceanic basins we must realize that there is no crustal basement under the ophiolites and the sediments of the basin. This apparent vacuum is essentially covered by the asthenospheric mantle, which naturally does not exist after the closure of the oceanic basin, and the resulting tectonic detachment, occurring both between the ophiolites and the associated sediments, and also between the ophiolites and the remaining subducted slab of the mantle lithosphere.

It is noteworthy that the two tectono-stratigraphic models share the same logic of the trilogy: (i) volcano-sedimentary complex, (ii) platform or basin, (iii) flysch, following the three stages of geodynamic–paleogeographic evolution described previously. That is, apart from minor differences in the volcano-sedimentary formations and the flysch of the two stratigraphic column types, the main difference is the development of a shallow carbonate platform over the continental terrane, which due to isostasy stands high at shallow sea level, unlike the abyssopelagic sedimentation adjacent and over the ophiolites inside the basin, which opens, subsides and becomes stable at a considerable depth. In the field, the stratigraphic trilogy of the continental terranes with the carbonate platform is obvious from the lithological difference of the carbonates both with the underlying volcano-sedimentary complex and the overlying flysch (Figs. 7.15 and 7.16).

On the contrary, the stratigraphic trilogy in the oceanic basins is not usually obvious in the field, due to the similar image of the volcano-sedimentary formations with the abyssopelagic sequences and the flysch.

The drifting of the continental terranes away from Africa and the simultaneous subsidence—opening of the oceanic basins are marked and dated by the transition from the volcano-sedimentary formations to the sediments of the carbonate platform on the one hand and of the basin on the other. On the contrary, the approaching of both types of terranes to the arc is marked and dated by the transitional beds to the flysch. The final integration of both the continental and the oceanic terranes in the Eurasian plate—continent is determined by the molasse and subsequent post-orogenic deposits, which unconformably overlie the tectonic contacts between the tectonic units and the terranes.

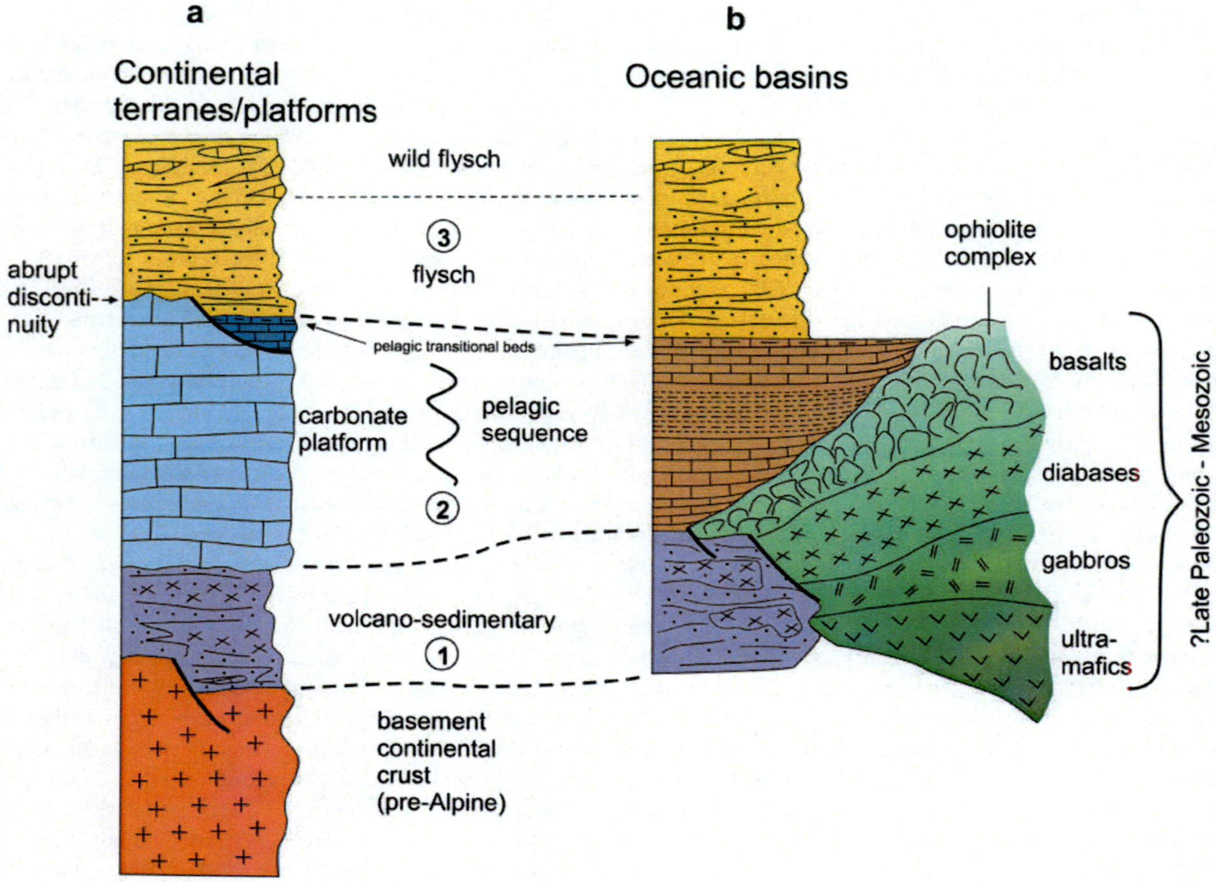

Fig. 7.14 The two stratigraphic columns corresponding to the two types of tectono-stratigraphic terranes of the Hellenides (according to Papanikolaou 2013). 1: volcano-sedimentary complexes of the rifting stage 2: shallow-water carbonates on the continental terranes during the drifting stage and parallel abysso-pelagic sequences of the oceanic basins during the oceanic opening stage 3: flysch-melange deposits of the subduction-accretion stage

7.4 Pre-Alpine Formations in Greece

The region of Greece is part of the Alpine cycle in the Tethys system and therefore it is expected to consist mainly of Mesozoic–Cenozoic formations. The paleogeographic area of the Hellenides does not include the two margins of Tethys, neither the Eurasian nor the African one, unlike e.g. the Alps, where the Helvetics are units of the Eurasian margin north of the development of the Tethys Ocean. Thus, while in the Alps (and the rest of the northern branch orogenic systems, such as the Carpathians, the Balkanides, the Pontides, etc.) outcrops of the Eurasian pre-Alpine basement participate in their tectonic structure, this is not expected in Greece. If all the units of the Hellenides had been developed in the Tethys Ocean, neither pre-Alpine basement nor shallow water carbonate platforms should be involved. Thus, only Mesozoic abyssopelagic sediments with ophiolites should be found, without any Paleozoic formation, as indeed this is the case of the Pindos and Maliac oceanic basins. However, the Hellenides also contain pre-Alpine outcrops of the "intermediate mountains" («zwinchengebirge»), such as the Pelagonian and Rhodope, which probably represent pre-Alpine continental crust of Gondwanian origin, overlain by large outcrops of shallow water carbonate platforms (Papanikolaou 1989, 2013). Thus, pre-Alpine continental crust is expected to form the basement below the Mesozoic carbonate platforms of the Hellenides either in normal stratigraphic position or detached. The same structure is expected to all the other similar carbonate platforms of the neighboring Tethyan belts, as shown in the Terrane map of the Mediterranean (see annotated Mediterranean terrane map out of text). The nature of the pre-Alpine basement may be more or less metamorphic Paleozoic sedimentary sequences

Fig. 7.15 View of the external carbonate platform of Tripolis over the underlying volcano-sedimentary complex of the Tyros beds of Permo-Triassic age within the H1 terrane, at the Tyros type locality

Fig. 7.16 View of the internal carbonate platform of H3 over the volcano-sedimentary complex of Upper Paleozoic–Lower Triassic age of the Chios autochthon

or deeper metamorphic and magmatic rocks. This depends on the pre-Mesozoic paleogeographic conditions in the margin of Gondwana and mainly on the erosion grade of the Variscan or earlier mountain ranges. Thus, a small grade of pre-Mesozoic erosion allows the preservation of non- or slightly metamorphosed Paleozoic sequences, whereas intense and long period of pre-Mesozoic erosion leads to outcrops of middle–lower crust of metamorphic and acid igneous rocks. Based on this distinction, the Variscan structures in Greece will be examined in more detail in the following subchapters.

7.4.1 Variscan Sequences in Greece

Pre-Alpine non-crystalline rocks are traced only in the islands of the Eastern Aegean, Chios (Ktenas 1921, 1923; Besenecker et al. 1968), and Kos (Desio 1931; Papanikolaou and Nomikou 1998), where fossils of the Ordovician, Silurian, Devonian, and Lower Carboniferous have been reported. From these fossil findings only Kos Island is considered a pre-Alpine unit today, since it turned out that in Northwestern Chios the lower Paleozoic fossils are restricted in blocks–olistoliths, together with blocks of mafic volcanic

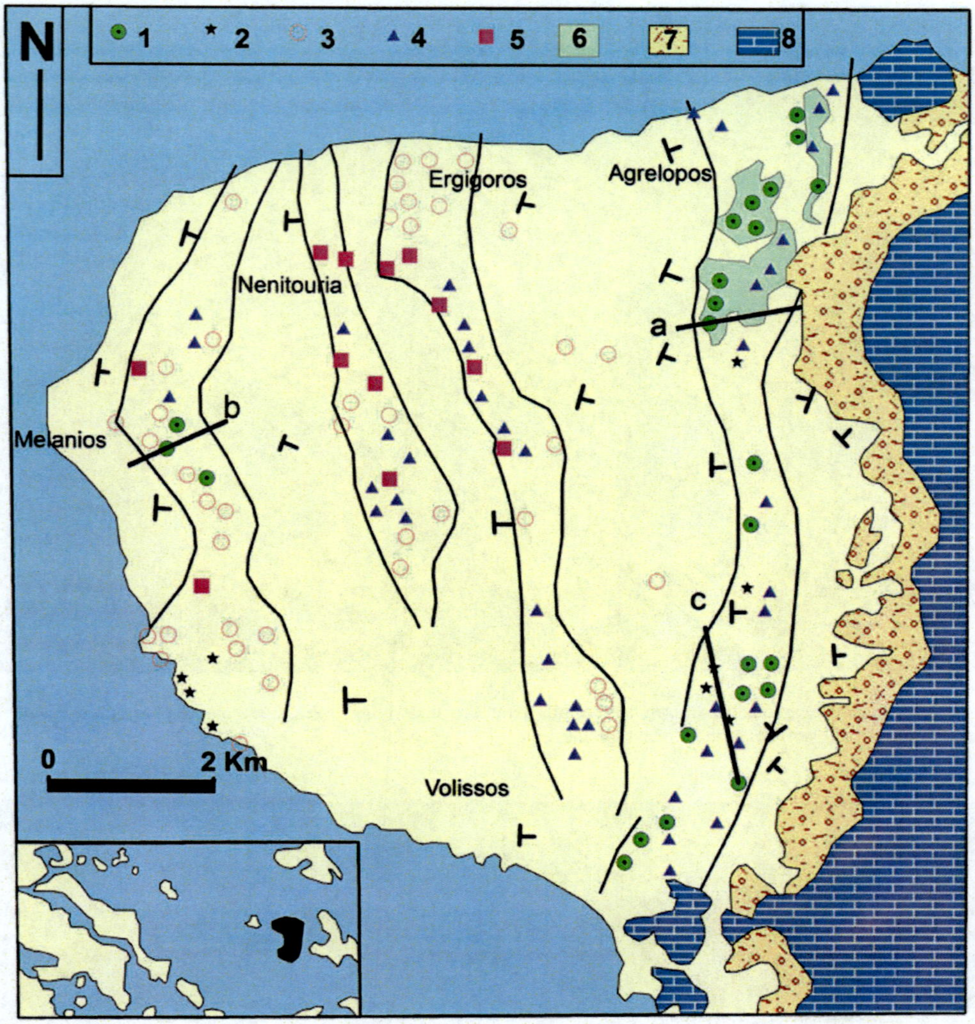

Fig. 7.17 Simplified map of Northwestern Chios, where fossiliferous rocks of the Lower Paleozoic are found in a chaotic flysch type complex, comprising four olisthostromes of Permian age (from Papanikolaou and Sideris 1983). 1: Silurian neritic limestones, 2: Devonian neritic limestones, 3: Lower Carboniferous neritic limestones, 4: basic volcanic rocks, 5: Devonian shales and lydites, 6: olisthostromatic matrix, 7: debris, 8: Triassic carbonate platform

rocks within a wild flysch formation (Figs. 7.17 and 7.18), whose age may range between Upper Carboniferous–Scythian (Papanikolaou and Sideris 1983, 1992).

In Chios Island it is interesting that the clastic formation has a thickness of several kilometers and includes at least four large olisthostromes with olistoliths, whose composition changes from the lower to the upper one. Therefore, while in the lower horizon the olistoliths are mainly neritic limestones of Carboniferous age, the upper horizon is dominated by neritic limestones of the Silurian with the famous tetracorallia (Rugosa) (Ktenas 1923; Besenecker et al. 1968) (Fig. 7.19). This differentiation indicates a progressive erosion of the adjacent basement during Late Paleozoic, with gradual deepening of the erosion from the upper to lower stratigraphic horizons. However, the origin of the olistholiths indicates that there are at least two sequences/units with different paleo-environmental character, a mainly neritic one with fossiliferous limestones and an abyssopelagic one, with lydites and mafic igneous rocks. It could be interpreted either as a Permian accretionary prism, associated with an orogenic arc process during the end of the Variscan cycle or as the result of rifting at the onset of the Alpine cycle. In Erythraea Peninsula on the opposite coast of Asia Minor, Ktenas (1925) reported extensive outcrops of the same formations like Chios, during his investigations there, aiming to correlate the two neighboring areas.

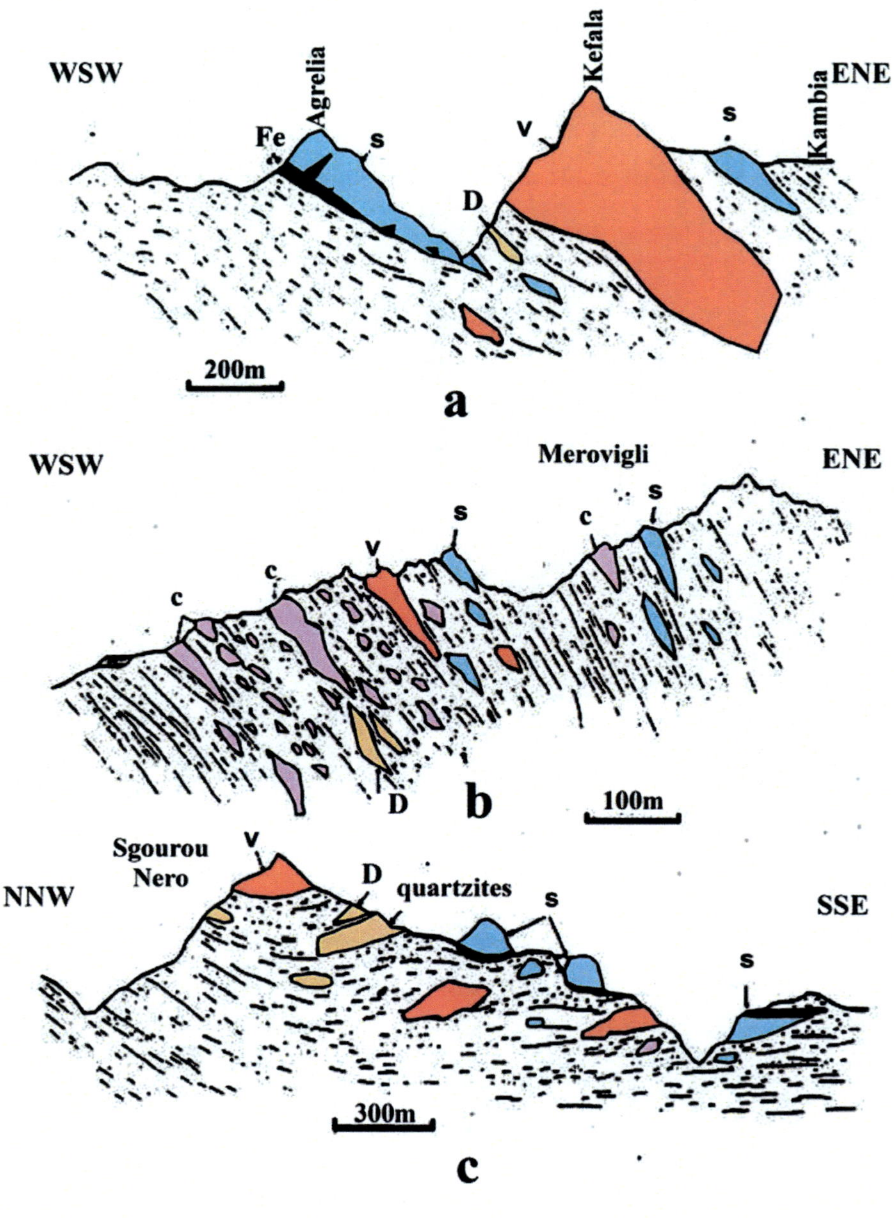

Fig. 7.18 Three sections (**a**, **b**, **c**) through the olisthostromes of Northwest Chios. The locations of the cross sections are marked on the map of Fig. 7.17. S: Silurian, D: Devonian, C: Lower Carboniferous, V: Volcanic rocks (from Papanikolaou & Sideris, 1983)

Therefore, the only case of fossil-bearing Paleozoic unit is **Kos** Island, where a Paleozoic slightly metamorphic sedimentary sequence crops out with some preserved fossils of Ordovician and Carboniferous age (Desio 1931). The typical fossils of the Ordovician include Fenestella corniculum and Orthis noctilo Sharpe, and of the Carboniferous Halia cylindrica, various crinoidea and brachiopods. The Paleozoic Kos unit is found only on **Mt. Dikeos**, in the form of a tectonic window under the nappes of the Alpine non metamorphosed units of Zia and Prophitis Ilias, which can be correlated with the Tripolis and Pindos units respectively (Papanikolaou and Nomikou 1998) (Fig. 7.20).

It is noteworthy that the Paleozoic Dikeos sequence is inverted, as shown by the exposure of the crystalline limestones of the Ordovician and the Carboniferous (Fig. 7.21).

Additionally, the Eastern Kos unit is also observed in tectonic contact with the Dikeos unit, on the eastern slopes of Mt. Dikeos, which is a chaotic clastic formation of Eocene age, determined from the fossils found in the olistholites (see also Fig. 8.39). On the western slopes of Mt. Dikeos the Miocene monzonite of Kos appears to intrude the Paleozoic rocks, but remains under the tectonic contact of the superjacent Tripolis nappe. This places the final evolutionary stage of the structure of Kos Island into the Miocene.

Fig. 7.19 View from the north of the Lower Paleozoic olistholite blocks (Pz) within the upper olisthostrome of the Permian wildflysch of NW Chios, underlying the Triassic platform (Tr)

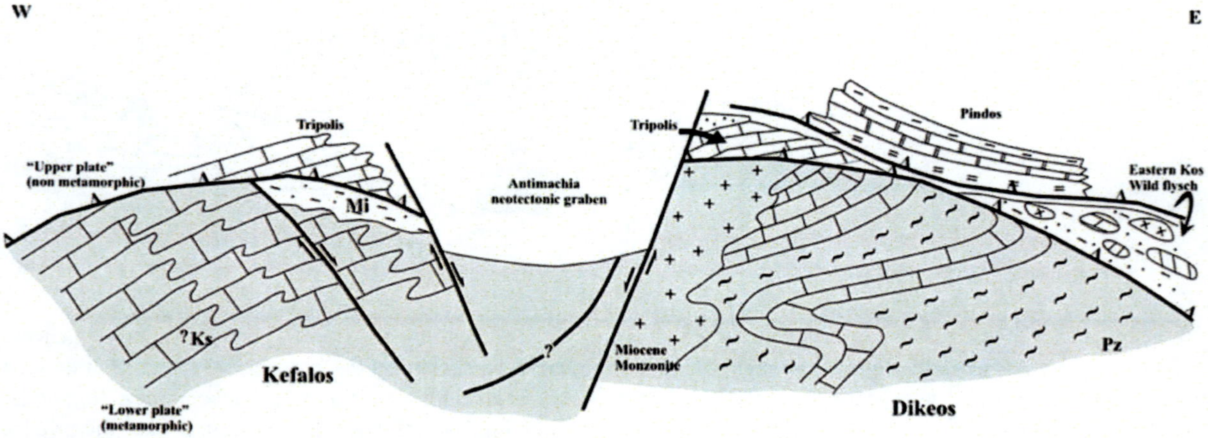

Fig. 7.20 Schematic geological E-W cross section along Kos Island (from Papanikolaou and Nomikou 1998). The metamorphic basement is shown in grey, comprising the Dikeos Paleozoic sequence in the east and the Kefalos Mesozoic (partly Cretaceous) crystalline limestones in the west

Similar conclusions can be drawn from the geology of Western Kos on the **Kefalos** peninsula, where the Tripolis unit can be observed as tectonic Klippen over a Lower Miocene clastic molasse type formation (Fig. 7.22). The molassic sediments lie unconformably directly over a metamorphic basement, consisting of crystalline neritic limestones of Upper Cretaceous age with traces of rudists.

Fig. 7.21 View of Mt. Dikeos along Southern Kos, where the inverted stratigraphic sequence of the Paleozoic can be observed, with the Carboniferous marbles underlying the Ordovician, observed on top of the mountain (from Papanikolaou and Nomikou 1998)

Fig. 7.22 View of the tectonic klippen of the Tripolis limestones over the Lower Miocene molasse of Western Kos in Kefalos

The relation between the metamorphosed Mesozoic formations of Kefalos with the Paleozoic of Dikeos is unknown. The disappearance of the Paleozoic of Dikeos to the west under the neotectonic graben of Antimachia allows the consideration of a possible continuity of the Paleozoic to the west, under the Mesozoic of Kefalos. In this case, correlations could be made with the Taurides platform further east in Asia Minor, where Paleozoic sequences are observed, continuing with or without unconformities to the Mesozoic (Papanikolaou and Demirtasli 1987).

Finally, it is interesting to note that the Upper Paleozoic - mainly Permian- is the base of the Alpine sequences and not their pre-Alpine basement, since there is no evidence of any significant unconformity between the Triassic and the Upper Paleozoic. In Mt. Knimis in Atalanti, it was proven that instead of a transgression of the Triassic on the Upper Paleozoic, we have a thrust of the Middle Triassic on the Upper Triassic (Sideris 1981). On the other hand, stratigraphic continuity is observed almost everywhere between the Permian and the Triassic with identification of Scythian strata (also known as Verfenian) in between (Renz and Reichel 1945; Marinos and Reichel 1958; Lekkas and Papanikolaou 1978; Papanikolaou 1979b; Thiebault and Kozur 1979; Baud and Papanikolaou 1981; Baud et al. 1991).

7.4.2 Variscan Metamorphism and Magmatism

Unlike the Variscan sedimentary sequences in Greece there are several units with metamorphic and magmatic rocks, which may be pre-Alpine, although this can not be confirmed for all of them. This uncertainty occurs mainly due to dating constraints of the metamorphic and magmatic rocks that have to be documented only through radiochronological data. Even though such dating techniques and data aquisition have been improved over the last decade several key informations on dating are still misssing. Nevertheless, it is quite clear that all the pre-Alpine metamorphosed units in Greece participate in the Alpine structure and have experienced all the deep level geodynamic phenomena during their involvement in the orogenic arcs of the Hellenides (Papanikolaou et al. 1982). Thus, even if sedimentation or metamorphism or deformation (or part of them) are pre-Alpine events, the unit involved is geotectonically considered Alpine, and this means that Alpine events of metamorphism, deformation, and magmatism have (also) overprinted past processes. Therefore, based on this perspective all the tectonic units will be examined together, pre-Alpine, possibly pre- Alpine and Alpine.

As we shall see in the detailed description of the tectonic units of the Hellenides in Sect. 8, the major pre-Alpine units are: (i) The Arna, (ii) The Sitia, (iii) The Southern Cyclades basement (e.g. Ios), (iv) The Asteroussia, (v) The Kastoria, (vi) The Flambouro, (vii) The Kerdylia, (viii) The Sidironero, (ix) The Vertiskos. Nevertheless, some small occurrences of metamorphic rocks remain outside this geotectonic subdivision, such as the Psara and Antipsara islands west of Chios, where Ktenas had distinguished two units of metamorphic rocks (Ktenas 1921, 1923). The lower unit comprising low grade phyllites and metagraywackes has a N-S tectonic trend whereas the upper unit comprising gneisses, schists and marbles has an E-W tectonic trend and its metamorphic degree is much higher (M3 versus M1 in his tectonic map, see also Fig. 7.4). Recent detrical zircon ages have indicated Permian age of their clastic sedimentation (Meinhold and Frei 2008). Thus these metamorphic tectonic units are differentiated from the neighboring non metamorphosed Permian units of Chios. On the contrary, Permian metamorphic rocks occur at the Lesvos autochthon unit (see section 8.5.1) and thus, interpretations involving the Cimmerides of Northwestern Minor Asia are under consideration.

The main outcome from the geochronologic data on metamorphism and magmatism of the pre-Alpine basement in the Hellenides is the extensive to exclusive presence of Upper Paleozoic metamorphic events and granitic intrusions, with a typical age of Carboniferous around 300 Ma (Dimitrievic 1972; Papanikolaou et al. 1982; Pe-Piper and Piper 2002). More recently, these Variscan ages have been confirmed, but older ones have also been reported (Anders et al. 2007; Himmerkus et al. 2006, 2007, 2009; Kroner and Sengor 1990; Reischmann 1998; Vavassis et al. 2000). However, the existence of an older orogenic cycle (e.g. Caledonian) can not be documented. Nevertheless, recent geochronological data from the Arna unit showed the presence of metalutites and quartzites, originating from supply of clastic material of Cambrian fans, within the paleogeographical organization of Northern Africa (Kydonakis et al. 2014).

7.5 Distinction of Internal and External Hellenides

The beginning of the major orogenic events of the Alpine cycle started in Late Jurassic–Early Cretaceous in the Hellenides, that is common also in the rest of the Tethyan system (Argand 1924; Trumpy 1980). These early orogenic phenomena occurred in the present interior of the Hellenic arc towards its core. These **paleo-Alpine** or **eo-Alpine**, as they are characterized, phenomena were fossilified below the Cenomanian transgression, which occurred throughout the entire formerly tectonised region (Brunn 1960; Jacobshagen et al. 1976). Then, the Alpine cycle continued, with shallow water carbonate sedimentation established over the paleotectonised region during the Late Cretaceous, and finally closed with the flysch sedimentation in the Eocene. It is with this second main Alpine orogenic phase during the Late Eocene–Oligocene, that the Alpine cycle ended for the paleo-Alpine units, but not for the rest external Hellenides. Thus, the **Internal Hellenides** are characterized by two orogenic periods separated by the Cenomanian unconformity. In contrast, the **External Hellenides**, which occur today at the circumference of the Hellenic arc, are characterized by continuous stratigraphic columns, tectonised only once, during Eocene–Miocene, i.e. during the main Alpine orogeny (Brunn 1960; Jacobshagen et al. 1976; Celet and Ferriere 1978; Jacobshagen 1980; Papanikolaou 1986b).

This peculiarity of the Internal Hellenides, concerning their participation in two orogenic arcs, the first during the Late Jurassic–Early Cretaceous and the second one during the Eocene–Oligocene, has not been fully analyzed and geotectonically explained, although similar events have been described all along the Tethyan Alpine system (e.g. in the Alps, Trumpy 1980). Generally, the Internal Hellenides are characterized by the tectonic emplacement of the H4 Axios ophiolites over the H3 internal carbonate platform, accompanied by metamorphism, deformation, magmatism and volcanism during the Late Jurassic–Early Cretaceous (Papanikolaou 2013). At the same time, pre-orogenic sedimentation continues in the external Hellenides. The general occurrence of the External Hellenides in the outer zone of the present arc in relation to the internal position towards the core of the arc of the Internal Hellenides, creates an increasing complexity from the periphery to the core. Thus, Western Greece has simpler and younger structures, while in Eastern Greece we have complex, older and polyphasic structures, with simultaneous occurrence of deep geodynamic events and ophiolites.

Large displacements of the tectonic nappes have occurred within the Hellenic arc during the main Alpine tectonic phase, throughout Eocene–Oligocene. Their general tectonic transport was from the inner to the outer arc. These displacements have created tectonic windows and tectonic klippen (Fig. 7.23), such as: (i) tectonic windows of the External Hellenides below the Internal ones, after long periods of erosion, such as for example, in the Olympus and the Cyclades and (ii) tectonic klippen of internal units over the external ones, like e.g. the non metamorphosed Cyclades unit and the Vatos unit in Crete. The additional presence of some small tectonic units/klippen, such as the Makrotantalo and the Anafi in the Cyclades and the Asteroussia in Crete, which have been tectonised during the Late Cretaceous, i.e. between the two large orogenic events, adds complexity to the entire process. These units tectonised during the Late Cretaceous have been considered as early external units in the overall tectonic scheme (Fig. 7.23). Several tectonic windows exist also within the External Hellenides of metamorphosed units (Mani, Arna, Western Crete) beneath non metamorphosed units of more internal paleogeographic origin (Tripolis). Additionally, tectonic windows of internal units are observed within the Internal Hellenides (Rhodope).

Today, however, it is known that apart from the Upper Cretaceous unconformity, there are also two Jurassic unconformities in Malm and Lias. These unconformities are observed in the internal part of the Internal Hellenides, sometimes observed below the Upper Cretaceous unconformity (see Sects. 4.6, 8.3.5, 8.3.8, 8.5.1, 8.5.2 and 8.6.3). The Liassic unconformity is related with the Cimmerides orogeny, while the Malm unconformity is included in the early stages of the paleo-Alpine orogeny of the internal units, without further distinction.

7.6 Distinction of Metamorphic and Non-Metamorphic Hellenides–Tectono-Metamorphic Belts

Until the 1970s most of the metamorphic rocks of the Hellenides were considered as pre-Alpine metamorphic massifs—basement of the Hellenides (Philippson 1901; Deprat 1904; Ktenas 1923; Renz 1940, 1955; Trikkalinos 1954, 1955, 1960; Brunn 1956; Aubouin 1959; Dercourt 1964), despite the opposing views expressed by various researchers, such as Negris (1915), Kober (1929), and more recently Marinos and Petrascheck (1956), Marinos (1961). A typical example of the most conservative views regarding the age of the metamorphic rocks of the so-called "Attic-Cycladic Crystalline Massif" is the following conclusion of Trikkalinos (1960) (academician-professor), in his very critical review on publications of G. Marinos, who supported the Permo-Mesozoic age of the metamorphics and Alpine structure:

> …extensive research on the above rock formations, which form the basement of Greece and Western Asia Minor, lead into the following conclusion: In the fore mentioned region there was an extended mainland during the Precambrian, which consisted of older metamorphosed rocks. Later, tectonic movements allowed the gradual transgression of the Paleozoic sea from the east to the west and the unconformable deposition of sedimentary strata of this younger period on part of the crystalline basement of Greece. This explains the presence of Paleozoic in Western Asia Minor since the Cambrian, in the Rhodope massif since the Sillurian, and in the rest of Greece in some regions since the Devonian and in others since the Carboniferous.

Similar views also prevailed in the neighboring countries of the Balkan Peninsula until relatively recently (e.g. Boncev 1966, 1986; Arsovski et al. 1977, Zagorcev 1998).

However, numerous fossils of Mesozoic–Cenozoic were discovered sporadically with characteristic outcrops both in Mt. Olympus (Godfriaux 1962, 1968) and in the area of the

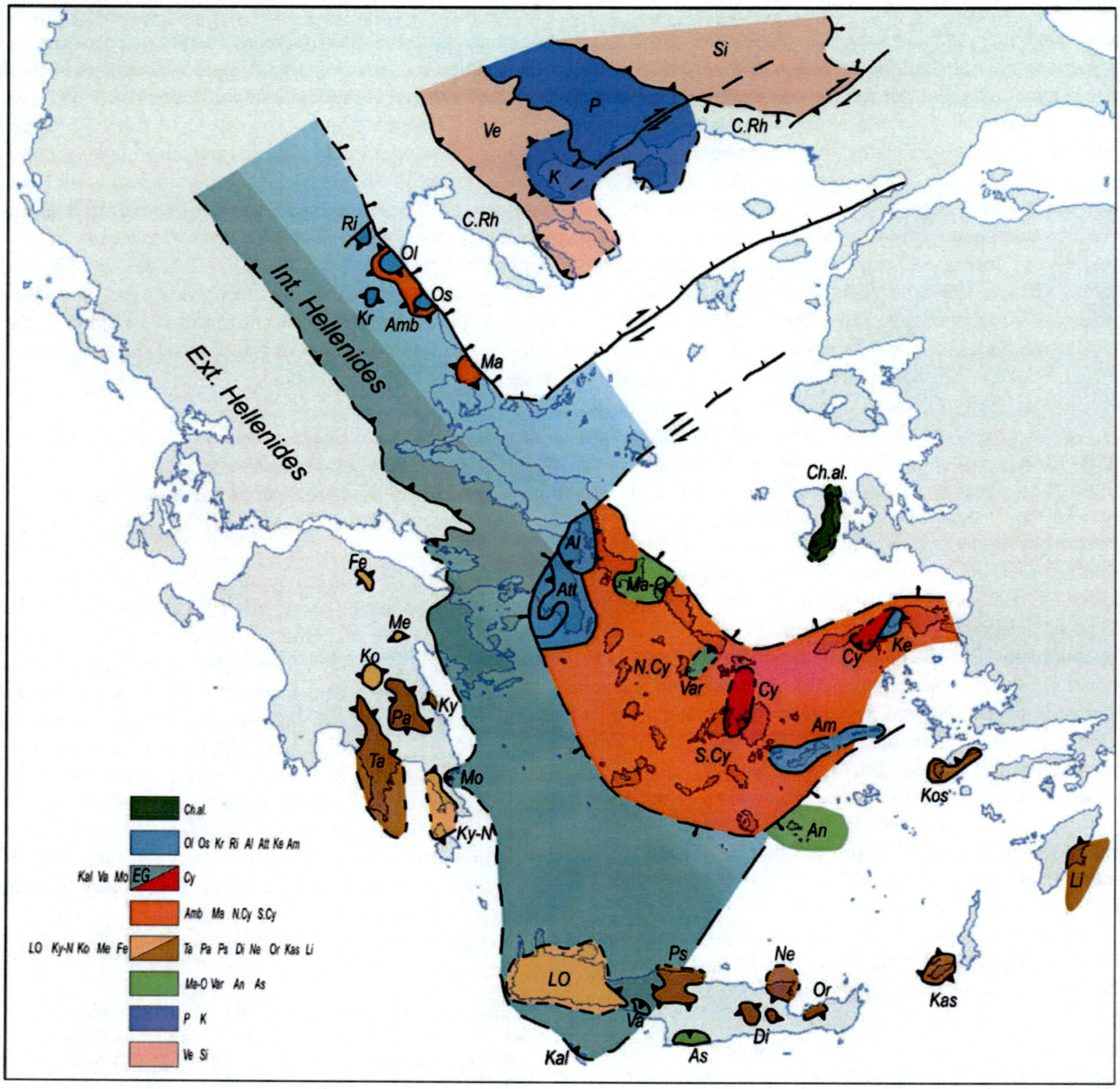

Fig. 7.23 Schematic tectonic map of the main tectonic windows and tectonic klippen in Greece, based on the distinction between Internal and External units (from Papanikolaou 1986, modified). Minor similar tectonic structures within the external and internal units are also highlighted. – Tectonic windows of external under the internal units: Ol Olympus, Os Ossa, Ma Makrynitsa, Al Almyropotamos, Ke Kerketeas, Riz Rizomata, Kr Krania, Am Amorgos. – Tectonic klippen of internal over the external units: Cy Cyclades non metamorphic, Kal Kalypso, Va Vatos. – Tectonic windows of external under external units: Ta Taygetos, Pa Parnon, Fe Feneos, Me Merkouri, Ko Kollines, Ky Kyparissi, Ky-N Kythera-Neapoli, L.O. Lefka Ori, Ps Psiloritis, Di Dikti (Western and Eastern), Ne Neapolis–Elounta, Or Orno, Kas Kassos, Li Lindos, Kos. – Tectonic klippen of early external (Late Cretaceous) over external units: As Asteroussia, An Anafi, Ma-O Makrotantalo–Ochi, Var Vari. – Tectonic windows of internal under internal units: P Pangeon, K Kerdylia, CH al Chios Allocthon, and Rhodope–Serbo-Macedonian in its entirety under C.Rh Circum Rhodope

former Attic-Cycladic Massif, and especially in Almyropotamos unit in southern Evia, which is homologous to the Olympus, (Argyriadis 1967; Katsikatsos 1969). Thus, in the early 1980s the overall image shows that all over the former Attic-Cycladic Massif, rocks of the Alpine cycle actually dominate (Fig. 7.24). Thus, besides the fossils found in the low grade metamorphic formations (Thera, Amorgos, Anafi) and the relics of the non metamorphic Cycladic nappe, several

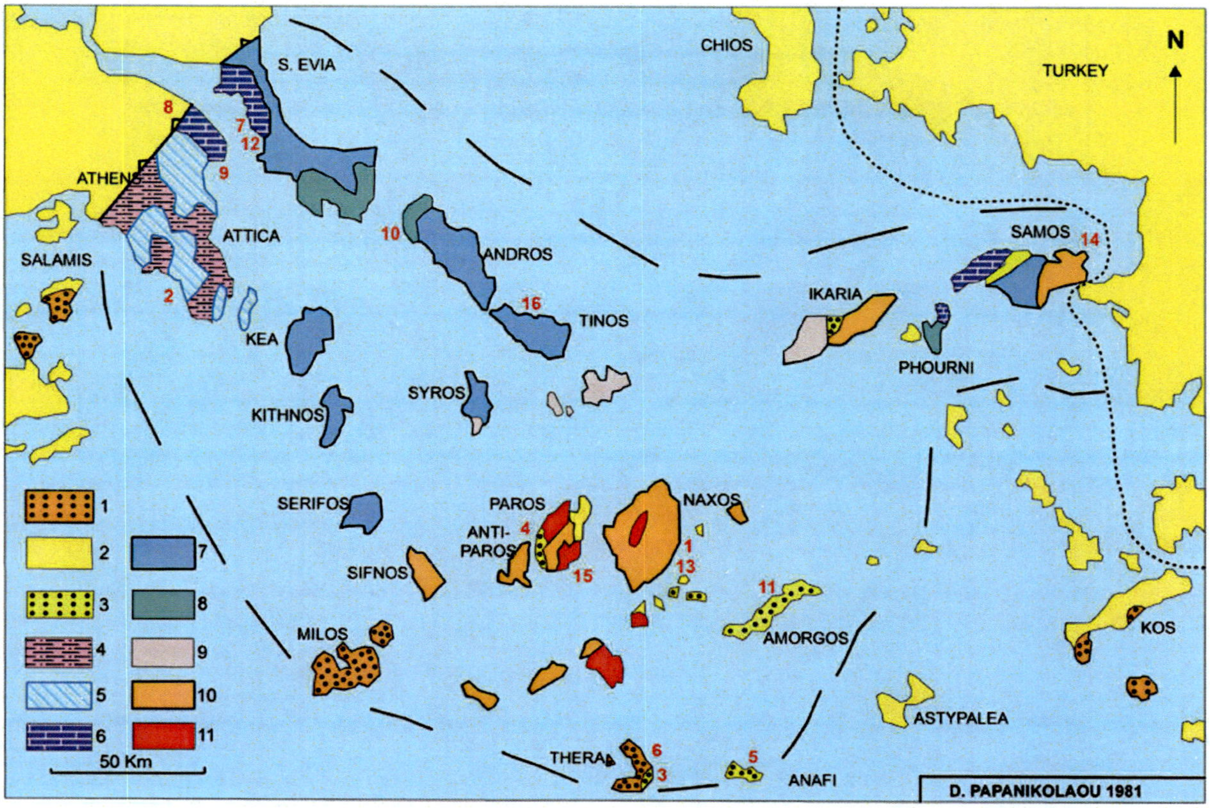

Fig. 7.24 Map of the former Attic-Cycladic Massif, showing the individual tectonic units and their fossil-bearing locations (according to Papanikolaou 1986c, 1988a). 1: Plio–Quaternary volcanics, 2: non-metamorphic Hellenides, 3: low grade metamorphic units (Amorgos, Anafi, Thira, Dryos of Paros, Mesaria of Ikaria), 4: Allochthon of Attica, 5: Autochthon of Attica, 6: Almyropotamos and Kerketeas, 7: Northern Cyclades, 8: Makrotantalo–Ochi, and Fourni, 9: schistose granites, 10: Southern Cyclades, 11: pre-Alpine rocks of the Southern Cyclades. Fossiliferous sites are shown in red numbers. (1) Triassic, Negris (1915), (2) Triassic, Marinos and Petrascheck (1956), (3) Triassic, Papastamatiou (1958), (4) Permian, Anastopoulos (1963), (5) Eocene, Melidonis (1963), (6) Eocene, Tataris (1965), (7) Triassic–Upper Cretaceous, Katsikatsos (1969) (8), Cretaceous, Papadeas (1973), (9) Cretaceous, Argyriadis (1967), (10) Permian, Papanikolaou (1976), (11) Triassic–Eocene, Durr et al. (1978), (12) Eocene, Dubois and Bignot (1979), (13) Triassic, Durr and Flugel (1979), (14) Cretaceous, Papanikolaou (1979a), (15) Permian, Papanikolaou (1980a), (16) Triassic, Melidonis (1980).

Permian to Eocene fossiliferous sites of great stratigraphic value were discovered such as: (i) Permian marbles in Antiparos (Anastopoulos 1963), in Makrotandalon/ Andros (Papanikolaou 1976), in Dryos/ Paros (Papanikolaou 1980). (ii) Triassic marbles in Naxos (Durr & Flugel 1979), Tinos (Melidonis 1980). (iii) Upper Cretaceous in emery bearing marbles in Vourliotes/Samos (Papanikolaou 1979a). (iv) Eocene at the contact marbles/flysch of Almyropotamos/ S. Evia (Dubois & Bignot 1979).

A similar story occurred in the metamorphics of Peloponnese–Crete, where it was also proved that the previously considered Lower Paleozoic rocks of the metamorphic basement belong to Alpine formations and can be correlated with the non-metamorphic rocks of the Ionian unit (Fytrolakis 1972; Epting et al. 1972; Bonneau 1973, Bizon and Thiebault 1974).

The discovery of the Mesozoic and/or Cenozoic fossils in the metamorphics of Attica-Cylcades, Olympus, and Peloponnese–Crete, was complementary with the observation-confirmation of the existence of large tectonic nappes in the region of the Hellenic arc, which allowed the revision of the existing views on the palinspastic restoration of the Hellenides (Bernoulli and Laubscher 1972). Thus, several metamorphic tectonic units proved to be Alpine and certainly all the units have participated in the Alpine Tectonics (Papanikolaou 1979b, 1981). At the same time, it became evident that the paleogeographical region of the Hellenides was not limited only to the classical non-metamorphic Hellenides, but included also the **Metamorphic Hellenides,** comprising a number of other units, which were subducted and metamorphosed during their journey through the orogenic arc (Papanikolaou 1980b,1986a) (Fig. 7.25).

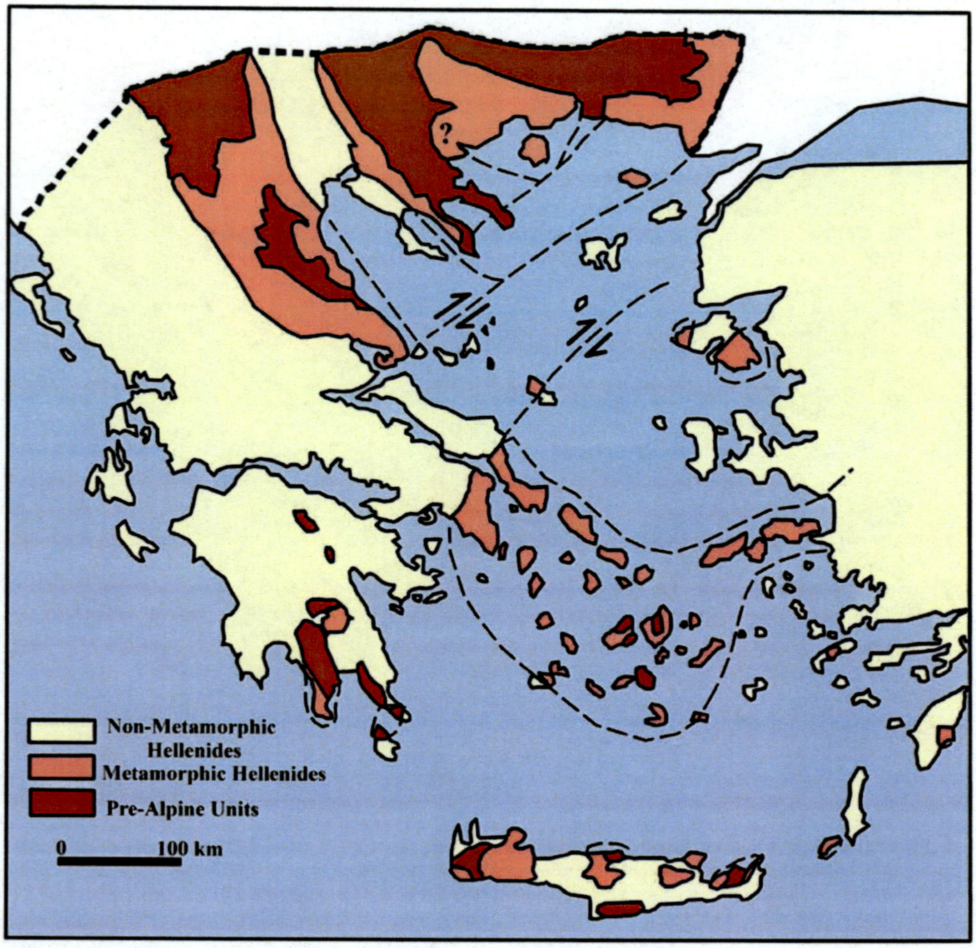

Fig. 7.25 Simplified map of Greece that separates the metamorphic from the non-metamorphic Hellenides and the pre-Alpine units (from Papanikolaou 1986c, modified)

Thus, the overall paleogeographic organization of the Hellenides, needs to be modified and incorporate the metamorphic Hellenides in their initial paleogeographic position in between the scheme of the non-metamorphic classical Hellenides (e.g. in the classic scheme of Aubouin 1959) (see Sect. 9.3).

Although the paleogeographic representation of the non metamorphic Hellenides has become generally accepted, this is not the case for the whole entity of both metamorphic and non-metamorphic Hellenides, because: (i) only a few efforts have been made for the placement of all units, and (ii) there are several unresolved issues in terms of the stratigraphy, the paleoenvironment and the overall geological development of the metamorphic Hellenides, that prevent the extraction of only one scenario for each unit.

Regardless of the above issues, the study of the metamorphic units of Greece, both Alpine and pre-Alpine, is facilitated by the distinction of three parallel **tectono-metamorphic belts** in the Hellenic arc (Papanikolaou 1984) (Fig. 7.26). This new distinction replaced the older one, referring to the various metamorphic/crystalline massifs, which considered all the metamorphic rocks as pre-Alpine, with only minor Alpine effect (e.g. Brunn 1956).

The three tectono-metamorphic belts of the Hellenic arc are (Papanikolaou 1984): (i) the **internal** tectono-metamorphic belt of **Rhodope** (sensu lato) in the core of the arc, which includes pre-Alpine rocks but also metamorphic Hellenides, with tectono-metamorphic events during the Mesozoic and a final tectono-magmatic –tectono-volcanic cycle during the Early Tertiary, (ii) the **medial** tectono-metamorphic belt of the

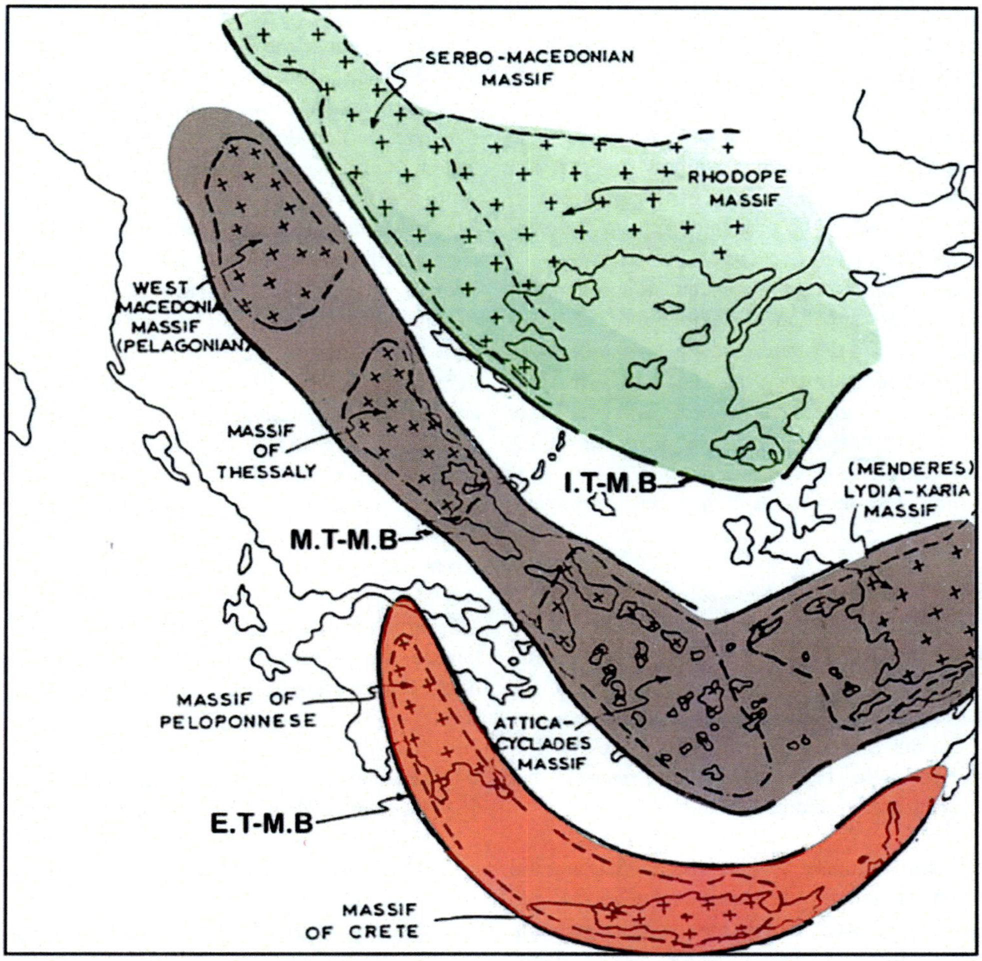

Fig. 7.26 The three tectono-metamorphic belts of the Hellenic arc, external (E.T-M.B), medial (M.T-M.B) and internal (I.T-M.B) (from Papanikolaou 1984, 1986b). Each belt consists of several Alpine and/or pre-Alpine metamorphic units, which form a composite deep level tectono-metamorphic superunit. The boundaries of the former "Pelagonian crystalline massifs" are also included in the medial belt

Pelagonian (sensu lato), which includes both pre-Alpine (former massifs of Western Macedonia, Eastern Thessaly, Attica-Cyclades, Lydia-Caria), but also several units of metamorphic Hellenides, especially in the Cyclades region, with tectono-metamorphic events in the Early Cenozoic and a final stage of magmatism/volcanism in the Late Cenozoic, (iii) the **external** tectono-metamorphic belt of **Peloponnese–Crete**, which includes mainly metamorphic Hellenides together with some pre-Alpine units, with tectono-metamorphic events during the Oligo–Miocene and without magmatic/volcanic events, since it is located within the present island arc, in more external position than the contemporary Aegean volcanic arc. The occurrence of PreCambrian and/or Paleozoic rocks within the three metamorphic belts of the Hellenides have been discussed within IGCP No 22 (PreCambrian in younger Fold Belts) (Papanikolaou 1988a,b).

7.7 Criteria of Distinguishing the Tectonic Units

Several stratigraphic sequences—formations share the same characteristics despite their spread in different geographic locations. The latter lead to the classification of the Alpine formations into geotectonic units. This classification is based on: (i) stratigraphic criteria, that is mainly the general character of a stratigraphic column (neritic, pelagic, transitional for a certain period, etc.) which represented a certain succession (or stability) of paleogeographic environments. The distinction of the formations, based on their lithology–stratigraphy–paleogeography resulted in the so-called **isopic zones** (Philippson 1898; Renz 1940; Aubouin 1959), which define the paleoenvironment of the Hellenides. The isopic zones are characterized by a particular stratigraphic column, irrespective of its geodynamic characteristics, that is basically a division of the pre-orogenic paleogeography, which later on takes part in the orogeny. (ii) Tectonic criteria, the style and intensity of deformation, the existence of tectonic nappes with overthrusts and tectonic windows/klippen etc. In this case, the division concerns the **geotectonic zones**, which are clearly related to the orogenic stage of the evolution of the Hellenides (Aubouin 1959).

The interdependent relation between lithology and deformation, with the distinction of flexible–rigid, ductile–brittle rocks in each stratigraphic sequence—isopic zone resulted in a certain tectonic style, so that the concepts of isopic and geotectonic zones have been mixed. Thus, sometimes the two concepts co-exist in the same stratigraphic sequence, while in other cases they do not.

The isopic and geotectonic zones do not always correlate, because:

(i) The general tectonic orientation is often diagonal in relation to the pre-existing isopic orientation or the pre-existing tectonic one, as shown by the parallel sections of Western continental Greece (Fig. 7.27). Here, the major tectonic line/front, separating the Internal from the External Hellenides, is not parallel to one particular unit, but successively transects the underlying tectonic structures.

Thus, the tectonic boundary between the internal and the external units is not parallel to the isopic or the geotectonic zoning of the external units, but it cuts successively the underlying tectonic contacts, separating the nappes. Thus, for example in Northern Pindos area, in the cross section of Grevena–Konitsa, the Northern Pindos ophiolite nappe has been emplaced directly on the Upper Eocene flysch of the Pindos unit, which is detached from its underlying Mesozoic sequence and thrusted on the Ionian unit. On the eastern slopes of Northern Pindos Mt the Oligocene molasse sediments of the Meso-Hellenic Trough usually cover the underlying nappe of the Eastern Greece unit, which has been thrusted on the Northern Pindos ophiolites. In Central Pindos, at the section of Trikala–Arta, the nappe of Western Thessaly unit is observed over the Pindos unit, (which here comprises its complete stratigraphic column), which is thrusted directly on top of the Ionian unit. In the next cross section at Central Pindos Mt, Karditsa- Amfilochia, the Pindos unit is tectonically emplaced over the Gavrovo unit, which is thrusted further west on the Ionian unit. Finally, at Southern Pindos Mt, the corresponding section of Livadia–Mesolonghi shows the Eastern Greece unit, on the Beotia unit, and the latter on the Parnassos unit, which is thrusted over the Pindos unit, through the intermediate thrust sheets of the transitional Vardousia unit, and then the Pindos unit thrusted on top of the Gavrovo unit and finally the Gavrovo unit on the Ionian unit. These series of E-W trending cross-sections from the Southern Central Greece up to the Albanian border, demonstrate the successive elimination of tectonic units from the Eastern Greece and the Beotian unit up to the Ionian unit. Thus, the intermediate units of Parnassos, Vardousia, Pindos (except for the Eocene flysch) and Gavrovo have successively been "buried" below the more internal nappes. It is characteristic that the Ionian unit remains always at the outer, more external part of the nappe structure. The way the Ionian unit is separated from the internal Hellenides differs significantly from southwards to northwards. In particular, there are five units in between the Ionian and the internal Hellenides in the south, but only two towards the northern part of Greece. In conclusion, the nappe structure of western Greece is characterised by the diagonal relation between the successive nappes, which cover successively more external units from south to north and at the same time the tectonic displacement increases. This structure can be related to a sinistral rotation of western Greece during the development of the external fold and thrust belt in the Late Eocene–Early Miocene.

(ii) It creates a sense of "cylindrism", i.e. of a parallel arrangement of the paleogeographic environments together with a parallel tectonic zonation (see also Fig. 9.2). This is contrasting the actualistic models and is not acceptable for long distances, due to the different stress release along the various arcuate segments of the convergent plate boundaries, where normal slip may be gradually substituted by strike slip motion within a length of a few hundred km along the belt.

In general, the concept of «tectonic zonation», either isopic or geotectonic, is currently avoided, (preserved mainly for bibliographical reasons) because it implies the notion of cylindrism, degrading the existing irregularities of the organization of a real physico-geographic region and the importance of transverse or diagonal geotectonic processes. Thus, the term **tectonic unit** is used, which, while preserving all the common pre- and syn- orogenic characteristics, it simultaneously allows a non-cylindrical geological interpretation. Another important issue is that the isopic zone begins from the interpretation that various outcrops, are considered to belong to a certain uniform paleogeographic area, degrading their geodynamic characteristics, which may be non-consistent with this interpretation. On the contrary, a tectonic unit is based on the overall characteristics of the rocks, both pre- and syn- orogenic, and so, using palinspastic processes, the effects of the orogenic geodynamic phenomena can be gradually substracted until we reach the

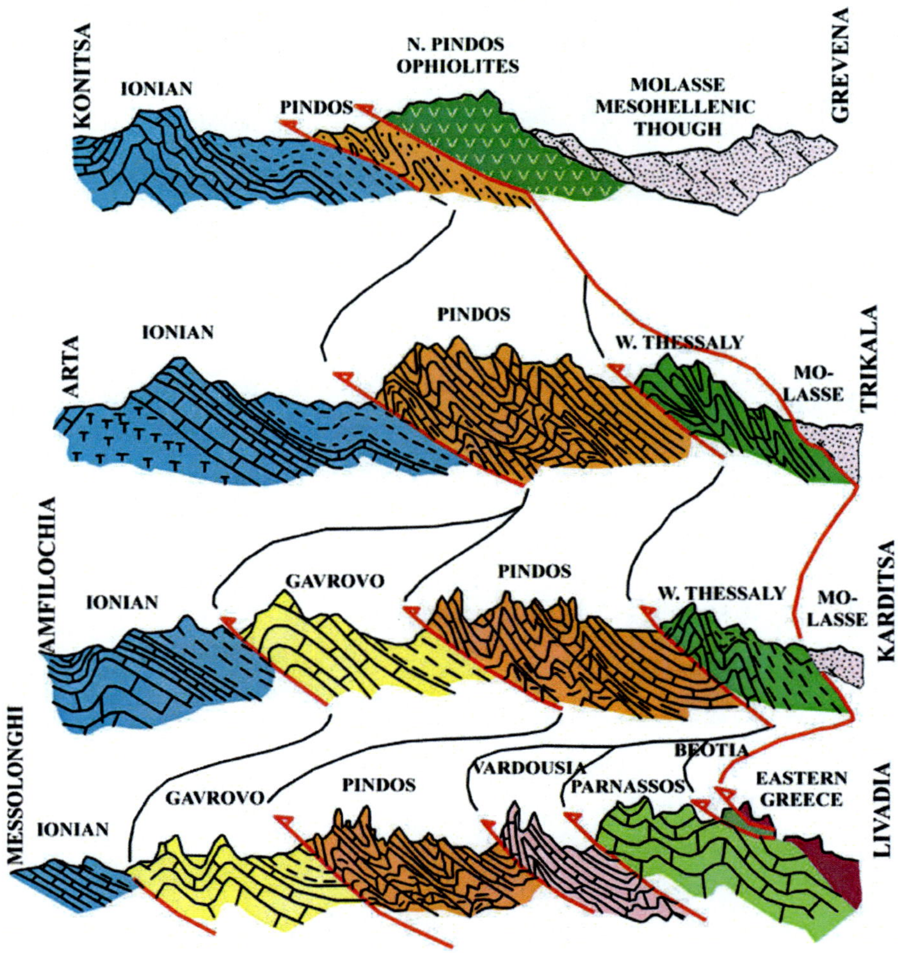

Fig. 7.27 Schematic tectonic sections of continental Greece, showing the diagonal relation of the frontal thrust of the internal units (e.g. Eastern Greece unit) with the tectonic nappes of the external units (from Papanikolaou 1986b)

pre-orogenic state. Thus, two tectonic units may be considered as belonging to one isopic zone, such as e.g. Ionian and Mani. In this case, they differ in the type of orogenic evolution (involving metamorphism (Mani) or not (Ionian)), while their pre-orogenic evolution was similar. In contrast, their integration directly into the same zone (Ionian) degrades their completely different orogenic evolution, by simulating completely different rocks and processes. On the contrary, the blueschists of the Cyclades, which are perceived as one unit by some researchers, include more tectonic units, with the Northern and Southern Cyclades units showing substantial differences. Thus, the Northern Cyclades unit has pelagic features with volcano-sedimentary sequences, while the Southern Cyclades unit comprises a metamorphosed shallow water carbonate platform with meta-bauxites. Hence, the pre-orogenic evolution of the two Cycladic units was quite different, although the intense and uniform syn-orogenic tectono-metamorphic evolution, especially in the subduction zone during the Paleogene unified them, in the form of blueschists. The chronological data of the common tectono-metamorphic-magmatic cycle of the Cycladic units are characteristic, as summarized by Schliestedt et al. (1987), in both Northern and Southern Cyclades, showing: (i) the first blueschist metamorphism in the Middle Eocene–Early Oligocene (Fig. 7.28), (ii) the second greenshchist metamorphism in the Early–Middle Miocene (Fig. 7.29), and (iii) the third contact metamorphism around the intrusive granites of the magmatic/volcanic arc in the Middle-Late Miocene (Fig. 7.30).

Consequently, it is evident that the inclusion of the orogenic geodynamic phenomena in the definition of a tectonic unit is of great importance and especially of the deep level phenomena, which until recently were completely neglected.

Until the late 1970s, only seven or eight geotectonic zones have been distinguished in Greece (Aubouin et al. 1961, 1976, 1979; Jacobshagen et al. 1978). Today, we can distinguish nine tectono-stratigraphic terranes (5 continental and 4 oceanic), and about fifty tectonic units as they appear in the map of geotectonic units out of text. Even though

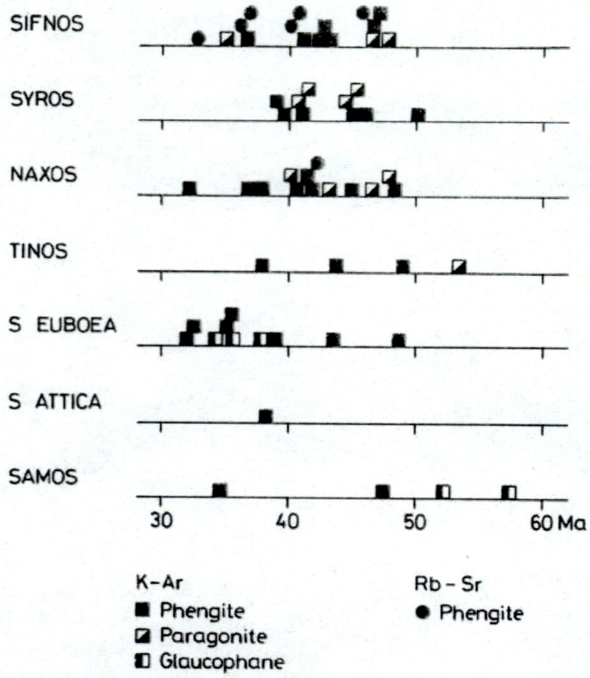

Fig. 7.28 Geochronological data of the blueschist metamorphic events of the Cyclades and adjacent metamorphic areas in the Eocene (55–35 Ma) (from Schliestedt et al. 1987)

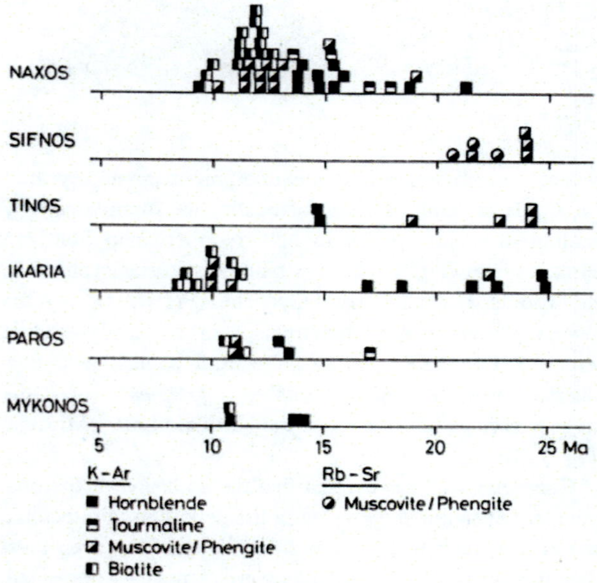

Fig. 7.29 Geochronological data of the greenschist metamorphic events of the Cyclades and adjacent islands in the Miocene (25–8 Ma) (from Schliestedt et al. 1987)

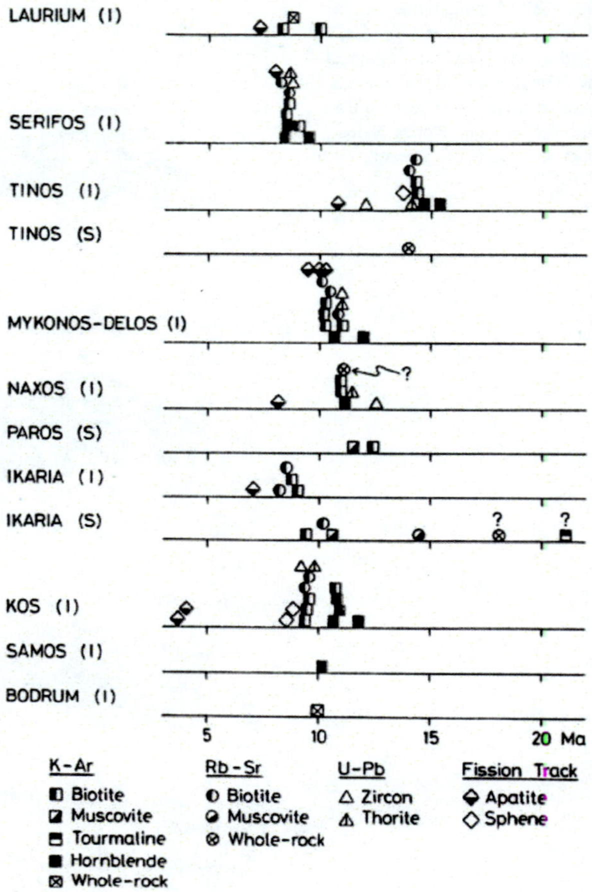

Fig. 7.30 Geochronological data of the granitoids and related contact metamorphic events of the Cyclades and adjacent metamorphic areas in the Middle–Late Miocene (15–8 Ma) (from Schliestedt et al. 1987)

some of these units have limited outcrops of only a few square kilometers, they provide us with valuable data for unravelling the complex geological history of the Hellenides. The description of the tectonic units of the Hellenides will start from the most external (Paxos) to the most internal ones (Rhodopean units), following a gradual unfolding of the structure of the Hellenic fold and thrust belt. This process is based on their probable paleogeographic organization, which however, may include several alternative scenarios. Additionally, there are some "ephemeral" units, related to the orogenic processes (e.g. accretionary prisms) that have existed only for a certain period. Consequently, the units to be described in the next chapter belong to an overall diachronic complex structure, that has never existed simultaneously.

References

Anastopoulos, I. 1963. Geological structure of Antiparos Island and neighboring islets. Geol. Geophys. Res., IGME, 7, 4, 102–234 (in greek).

Anders, B., Reischmann, T. & Kostopoulos, D., 2007. Zircon geochronology of basement rocks from the Pelagonian Zone, Greece: constraints on the pre-Alpine evolution of the westernmost Internal Hellenides. Intern. J. Earth Sci., 96, 639–661.

Ardaens, R., Colin, J.P. & Kozur, H. 1979. Sur la présence du Scythien supérieur fossilifère dans la chaine du Vardoussia (Grèce continentale). Conséquences paléogéographiques.C. R. somm. Soc. Géol. France, 132–135.

Argand, E. 1924. La tectonique de l'Asie. Congrès Géol. Intern., Bruxelles.

Argyriadis, I. 1967. Sur le problème des relations structurales entre formations métamorphiques en Attique et Eubée. C. R. Acad. Sc. Paris, 264, 438–441.

Arsovski, M., Dumurdzanov, N., Hristov, S., Ivanov, T., Ivanova, V., Petkovski, P. & Stojanov, R. 1979. Correlation of the Pre-Cambrian complexes of the Pelagonian massif, Vardar zone and Serbo-Macedonian massif. VI Coll. Geol. Aegean Region, Athens1977, Proc. II, 549–557.

Asvesta, A. & Dimitriadis, S., 2010. Facies architecture of a Triassic rift-related silicic volcano-sedimentary succession in the Tethyan realm, Peonias subzone, Vardar (Axios) Zone, northern Greece; regional implications. J. Volc. Geoth. Res., 193, 245–269.

Aubouin, J. 1959. Contribution à I' étude géologique de la Grèce septentrionale: Les confins de l'Epire et de la Thessalie. Ann. Géol. Pays Hellén.,10, 1–483.

Aubouin, J. 1965. Géosynclines. Devel. Geotectonics, 1, 335 p., Elsevier.

Aubouin, J., Brunn, J.H., Celet, P., Dercourt, J.. Godfriaux, I. & Mercier, J. 1961. Esquisse de la géologie de la Grèce. Mem. Soc. Géol. France, (hon. Prof. Fallot), 583–610.

Aubouin, J., Bonneau, M., Davidson, J., Leboulenger, P., Matesco, S.& Zambetakis, A. 1976. Esquisse structurale de l arc égéen extême: des Dinarides aux Taurides. Bull. Soc. Géol. France, XVIII, 2, 327–336.

Aubouin, J., Desprairies, A. & Terry, J. 1977. Le synclinal d' Epire—Akarnanie, la nappe du Pinde-Olonos & la nappe ophiolitique. Bull. Soc. Geol. France, 19, 20–27.

Aubouin, J., Le Pichon, X., Winterer, E. & Bonneau, M. 1979. Les Hellenides dans I' optique de la tectonique des plaques. VI Coll. Geol. Aegean Region, Athens 1977, Proc., 3, 1333–1354.

Avigad, D. & Garfunkel, Z., 1991. Uplift and exhumation of high-pressure metamorphic terrains: the example of the cycladic blueschist belt (Aegean Sea). Tectonophysics, 188, 357–372.

Avigad, D., Garfunkel, Z., Jolivet, L. & AZANON, J.M., 1997. Back arc extension and denudation of Mediterranean eclogites. Tectonics, 16, 924–941.

Baud, A. & Papanikolaou, D., 1981. Olistholites and flysch facies at the Permo-Triassic transition from Attica, Eastern Greece. IGCP Project 106, Sarajevo, 33–34.

Baud, A., Jenny, C., Papanikolaou, D., Sideris, C., & Stampfli, G., 1991. New observations on Permian stratigraphy in Greece and geodynamic interpretation. Bull. Geol. Soc. Greece, XXV, 187–206.

Bebien, J. & Mercier, J., 1977. Le cadre structural de l'association ophiolites-migmatites-granites de Guevgueli. (Macédoine, Grèce)—une croûte de bassin interarc ? Bull. Soc. Geol. France, 19, 927–934.

Beccaluva, L., Ohnenstetter, D., Ohnensteter, M. & PAUPY, A., 1984. Two magmatic series with island arc affinities within the Vourinos ophiolite. Contr. Miner. Petr., 85, 253–271.

Bernoulli, D. & H. Laubscher 1972. The palinspastic problem of the Hellenides. Eclogae geol. Helv., 24, 347-372.

Besenecker, H., Dürr, S., Herget, G., Jacobshagen, V., Kauffmann, G., Lüdtke, G., Roth, W. & Tietze, K.W. 1968. Geologie von Chios (Ägäis). Geologica et Paleontologica, 2, 121–150.

Bittner, A., Neumayr, M, & Teller, F. 1880. Uberblick uber die geologischen Verhaltnisse eines Theiles des Agaischen Kustenlander. Denkschr. D. Kais. D. Wiss. Math.-nat. Kl., 40, 379–415.

Blake, M.C., Irwin, W.P. & Coleman, R.G. 1969. Blueschist facies metamorphism related to regional thrust faulting. Tectonophysics, 8, 237–246.

Blake, M.C., Bonneau, M., Geyssant, J., Kienast, J.R., Lepvrier, C., Maluski, H. & Papanikolaou, D., 1981. A geologic reconnaissance of the Cycladic Blueschist Belt, Greece. Bull. Geol. Soc. Amer., 92, 247–254.

Blumenthal, M. 1933. Zur Kenntnis des Querprofils des zentrales und nordlichen Peloponnes. N. Jb Min., 703, 449–514.

Boblaye, P. & Virlet, Th. 1933. Expédition scientifique de Morée. 2e section des Sciences Physiques. Paris.

Boncev, E. 1966. Revue general de la structure geologique de la Bulgarie. Bull. Inst. Geol. Sofia, 15, 5–24.

Boncev, E. 1986. Apercu general sur la tectonique des Balkans. Geol. Balc., 16.2, 3–32.

Bonneau, M. 1973. Sur les affinités ioniennes des «Calcaires en plaquettes» épimetamorphiques de la Crète, le charriage de la série de Gavrovo-Tripolitsa et la structure de I' arc égéen. C.R. Acad. Sc. Paris, 277, 2453–2456.

Brun, J.P. & Faccenna, C., 2008. Exhumation of high-pressure rocks driven by slab rollback. Earth & Plan. Sci. Let., 272, 1–7.

Brunn, J. 1956. Contribution à l'étude Géologique du Pinde Septentrional et d'une partie de la Macédoine Occidental. Ann. Géol. Pays Hellén., 7, 1–358.

Brunn, J.H. 1960. Les zones helléniques internes et leur extension. Réflexions sur I' orogenèse alpine. Bull. Soc. Géol. France, (7), II, 470–486.

Burchfiel, B.C., Royden, L.H., Papanikolaou, D. & Pearce, F.D. 2018. Crustal development within a retreating subduction system: the Hellenides. Geosphere, 14, 2.

Bizon, G. & Thiebault, F. 1974. Données nouvelles sur l' age des marbres et quartzites du Taygète (Péloponnèse méridional, Grèce). C.R. Acad. Sc. Paris, 278, 9–12.

Bucking, H. 1880. Uber die Lagerungsverhaltnisse der alteren Schichten in Attika. Sitzungber. Der Kgl, Preuss. Akad. D. Wiss. zu Berlin, 39, 935–950.

Cavazza, W. et al, 2004. The TRANSMED Atlas: geological-geophysical fabric of the Mediterranean region. Episodes, 27, 4, 244–254.

Celet, P. & Ferriere, J. 1978. Les Hellénides internes: Le Pélagonien. Eclogae geol. Helv.,71/3, 467–495.

Chemenda, A. I., Mattauer, M. & Bokun, A. N., 1996. Continental subduction and a mechanism for exhumation of high-pressure metamorphic rocks: new modelling and field data from Oman. Earth & Plan. Sci. Let., 143, 173–182.

Cloos, H. 1939. Hebung, Spaltung, Vulkanismus. Geol. Rdsch. XXX.

Coleman, H. 1971. Plate tectonic emplacement of upper mantle peridotites along continental edges. J. Geoph. Res., 76, 1212–1222.

Coney, P. J., Jones, D. L. & Monger, J. W. H. 1980. Cordillieran suspect terranes. Nature, 288, 329–333.

Cordella, A. 1878. La Grece sous le rapport geologique et mineralogique. 188 p. Paris.

De Franco, R., Govers, R. & Wortel, M. J. R. 2008. Dynamics of continental collision: influence of the plate contact. Geophys. J. Int.,174, 1101–1120.

Deprat, J. 1904. Etude geologique et petrographique de l' ile d' Eubee. 232p, Besancon.

Dercourt, J. 1964. Contribution à l' étude géologique d'un secteur du Péloponnèse septentrional. Ann. Géol. Pays Hellén.,15, 1–418.

Dercourt, J. 1972. The Canadian Cordilliera, the Hellenides and the sea-floor spreading Theory. Canad. Jour. Earth Sci, 9, 709–743.

Dercourt, J., Makris, J. & Melidonis, N., 1980. Grece. Introduction a la geologie generale. 26th International Geological Congress, Livret-Guide, G12, 159p.

Desio, A. 1931. Le isole italiane dell' Egeo. Mem. Carta geol. Ital., 24, 534 p.

Dewey, J. F., 1976. Ophiolite obduction. Tectonophysics, 31(1-2), 93–120.

Dewey, J. F. & Bird, J. M. 1970. Mountain Belts and the New Global Tectonics. J. Geophys. Res., 75, 2615–2647.

Dewey, J. F. & Bird, J. M. 1971. Origin and Emplacement of the ophiolite suite: Appalachian ophiolites in Newfoundland. J. Geophys. Res., 76, 3179–3206.

Dewey, J. F., Pitmann, W. C., Ryan, W. B. F. & Bonnin, J. 1973 .Plate tectonics and the evolution of the Alpine system. Bull. Geol. Soc. Amer., 84, 3137–3180.

Dilek, Y., Shallo, M. & Furnes, H. 2005. Rift–drift, seafloor spreading, and subduction tectonics of Albanian ophiolites. Inter. Geol. Rev., 47, 147–176.

Dilek, Y., Furnes, H. & SHALLO, M. 2008. Geochemistry of the Jurassic Mirdita Ophiolite (Albania) and the MORB to SSZ evolution of a marginal basin oceanic crust. Lithos, 100, 174–209.

Dimitriadis, S. & Asvesta, A., 1993. Sedimentation and magmatism related to the Triassic rifting and later events in the Vardar–Axios Zone. Bul. Geol. Soc. Greece, 28, 149–168.

Dimitrijević, M. 1972. Hercynian Metamorphism in the Central Part ofthe Balkan Peninsula. Z. Deutsch. Geol. Ges., 123, 329–335.

Dimitrievic, M. D. 1974. The Dinarides: a model based on the new global tectonics. In: Jancovic, S. (Ed.), Metallogeny and concepts of the geotectonic development of Yugoslavia. Faculty of Mining and Geology, University of Belgrade, 141–178.

Dubois, R. & Bignot, G. 1979. Presence d'un «hard-ground» nummulitique au sommet de la serie cretace d' Almyropotamos (Eubee meridionale, Grece). Consequences. C. R. Acad. Sci., Paris, 289, 993–995.

Dürr, S. & Flugel, E. 1979. Contribution à la stratigraphie du cristallindes Cyclades: Mise en évidence du Trias Supérieur dans les marbresde Naxos (Grèce). Rapp.Comm. int. Mer Médit, 25/26,2a, 31–32.

Dürr, ST., Altherr, R., Keller, J., Okrusch, M.& Seidel, E. 1978. The median Aegean crystalline belt: Stratigraphy, Structure, Metamor-phism, Magmatism. In: Alps, Apennines, Hellenides, 455–477.

Epting, M., Kudrass, R. & Schafer, A. 1972. Stratigraphie et position des séries métamorphiques aux Talea Ori/Crète. Z. Deutsch. Geol. Ges., 123, 365–370.

Ernst, W.G. 1970. Tectonic contact between the Franciscan melange and the Great Valley sequence. Crustal expression of a Late Mesozoic Benioff zone. J. Geophys. Res., 75, 886–901.

Ferriere, J. 1976. Sur la signification des séries du massif de l'Othris (Grèce continentale centrale): la zone isopique maliaque. Ann. Soc. Géol. Nord, 96/2 , 121–134.

Ferriere, J. 1982. Paléogéographies et tectoniques superposées dans les Hellénides internes: les massifs de l'Othrys et de Pélion. Soc. Géol. Nord, Publ. 8, 970 p

Fiedler, G. 1841. Reise dursch alle Teile des Konigreiches Griechenland. 565 S., Leipzig.

Fleury, J.J. 1980. Les zones de Gavrovo-Tripolitza et du Pinde-Olonos (Grèce continentale et Péloponnèse du Nord). Evolution d'une plateforme et d' un bassin dans le cadre alpin. Publ. Soc. Géol. Nord, 4.

Frech, F. 1907. Die hallstatter Kalke bei Epidauros (Argolis) und ihre Cephalopoden. N. J. B. Min., Festband, 1–32.

Fytrolakis, N. 1972. The effect of some orogenetic mouvements and the formation of gypsum in Eastern Crete (Sitia district). Bull. Geol. Soc. Greece, 9, 81–100 (in greek).

Galanopoulos, A. 1972. Plate tectonics in the area of Greece as reflected in the deep focus seismisity. Bull. Geol. Soc. Greece, IX, 2, 266–285.

Georgalas, G. 1934. Geology of Greece. Great Hellenic Encyclopaedia, vol. 10 (in greek).

Godfriaux, I. 1962. Sur la presence de l'Eocene dans les calcaires superieurs de l ' Olympe (Thessalie septentrionale, Grece). Bull. Soc. Geol. France, 4, 49–54.

Godfriaux, I. 1968. Etude géologique de la région de l'Olympe (Grèce). Ann. Géol. Pays Hellén., 19, 1–281.

Himmerkus, F., Reischmann, T. & Kostopoulos, D. 2006. Late Proterozoic and Silurian basement units within the Serbo-Macedonian Massif, northern Greece: the significance of terrane accretion in the Hellenides. Geol. Soc., London, Sp. Publ., 260, 35–50.

Himmerkus, F., Anders, B., Reischmann, T .& Kostopoulos, D. 2007. Gondwana-derived terranes in the northern Hellenides. Geol. Soc. Am. Mem., 200, 379–390.

Himmerkus, F., Reischmann, T. & Kostopoulos, D. 2009. Triassic rift-related meta-granites in the Internal Hellenides, Greece. Geol. Mag., 146, 252–265.

Howell, D.G. 1980. Mesozoic accretion of exotic terranes along the New Zealand segment of Gondwanaland. Geology, 8, 487–491.

Jacobshagen, V. 1980. Die eozäne Orogenese in the Ägäis. Berl. Geow. Abh., 21–33.

Jacobshagen, V. 1986. Geologie von Griechenland. Gebruder Borntraeger, 363 pp.

Jacobshagen, V., Risch, H. & Roeder, D. 1976. Die eohellenische Phase. Definition & Interpretation. Z. Deut. Geol. Ges., 127, 133–145.

Jacobshagen, V., Dürr, ST., Kockel F., Kopp, K.O., Kowalcyk, G. with contr. BERCKHEMER H., BÜTTNER, D. 1978. Structure and Geodynamic Evolution of the Aegean Region. In: Alps, Apennines, Hellenides, 537–564, Stuttgart.

Jolivet, L. & Brun, J.P. 2010. Cenozoic geodynamic evolution of the Aegean. Int. J. Earth Sci., 99, 109–138.

Jolivet, L., Faccenna, C., Goffe, B., Burov, E. & Agard, P. 2003. Subduction tectonics and exhumation of high-pressure metamorphic rocks in the Mediterranean orogens. Am. J. Sci., 303, 353–409.

Jones, D.L. Howell, D.G., Coney, P.J. & Monger, J.W.H. 1983. Recognition, character and analysis of tectonostratigraphic terranes in western North America. In: Hashimoto & Uyeda ed. Accretionary tectonics in the Circum-Pacific regions: Advances in Earth & Planetary Sciences, Terra Science Publ. Co., Tokyo, 21–35.

Katsikatsos, G. 1969. The age of the metamorphic system of Southern Evia and its stratigraphic succession. Proc. Acad. Athens, 44, 223–238 (in greek).

Kauffmann, G. 1976. Perm und Trias im östlichen Mittelgriechenland und auf einigen ägäischen Inseln. Z. Deutsch. Geol. Ges., 127, 387–398.

Kober, L. 1929. Beitrage zur Geologie von Attika. Sitz. Akad. Wiss. Wien, 138, 299–327.

Kober, L. 1933. Die Orogentheorie. Berlin.

Kober, L. 1931. Das Alpine Europa. Verlang von Gebrüder Borntraeyer, Berlin.

Kossmat, F. 1924. Geologie der Zentral Balkanhalbinsel. In: Die Kriegsschanplätz 1914–1918 geologisch dargestelt, 13, 198 p., Berlin.

Kroner, A. & Sengor, A.M.C., 1990.Archean and Proterozoic ancestry in late Precambrian to early Paleozoic crustal elements of southern

Turkey as revealed by single-zircon dating. Geology 18, 1186–1190.

Ktenas, C. 1921a. Sur la découverte du Dévonien à l' île de Chios. C. R.somm. Soc.Géol. France, p. 131.

Ktenas, K. 1921b. Sur le Carbonifere de l' ile de Chios. C. R. somm. Soc. Geol. France, p. 146.

Ktenas, K. 1923. Sur la decouverte d' un horizon a Productus cora a l'ile de Chios. C. R. somm. Soc. Geol. France, p. 206.

Ktenas, C. 1924. Formations primaires semimétamorphiques au Péloponnèse central. C.R. somm. Soc. Géol. France, 61–63.

Ktenas, K. 1925. Contribution to the geology of Northern Erythraea (Minor Asia). Scientific Annals of Physical and Mathematic Faculty of Athens University, 1, 57 p (in greek).

Kydonakis, K., D. Kostopoulos, M. Poujol, J. P. Brun, D. Papanikolaou& J. L. Paquette, 2014. The dispersal of the Gondwana Super-fan System in the eastern Mediterranean: New insights from detrital zircon geochronology. Gondwana Res., 25, 1230–1241.

Lekkas, S. & Papanikolaou, D. 1978. On the phyilite problem in Peloponnesus. Ann. Géol. Pays Hellén., 29/1, 395–410.

Lepsius, R. 1893. Geologie von Attika. 196 S, Berlin.

McKenzie, D.P. 1970. Plate tectonics in the Mediterranean Region. Nature, 226, 239–243.

McKenzie, D.P. 1972. Active tectonics of the Mediterranean Region. Geoph. J.R. Astron.Soc., 30, 109–185.

McKenzie, D., Davies, D. & Molnar, P. 1970. Plate tectonics of the Red Sea and East Africa. Nature, 226, 243-248.

Meinhold, G. & Frei, D. 2008. Detrital zircon ages from the islands of Inousses and Psara, Aegean Sea, Greece: constraints on depositional age and provenance. Geol. Mag., 145, 886–89

Migiros, G. & Tselepidis, V. 1990. Der erste Nachweis von Hallstater kalken in der Nord-Pindos Decke (NW Griechenland). N. Jahr. Geol. Paläont., H4, 248–256.

Maratos, G. 1972. The geology of Greece. Athens, 189 p (in greek).

Marinos, G. 1961. Paleontologic and stratigraphic research in Eastern-continental Greece. Bull. Geol. Soc. Greece, 4/1, 14–28 (in greek).

Marinos, G., & Petrascheck, W.E. 1956. Laurium. Geol. Gephys. Res., IGME, IV, 1–247 (in greek).

Marinos, G. & Reichel, M., 1958. The fossiliferous Permian in Eastern continental Greece and Evia. Special Studies Geol. Greece, IGME, 8, 1–16 (in greek).

Melidonis, N. 1963. The geology of Anafi Island. Geol. Geophys. Res., IGME, 8, 61–308 (in greek).

Melidonis, N. 1980. The geological structure and the metalliferous occurrences of Tinos Island. Special Studies Geol. Greece, IGME, 13, 1–80 (in greek).

Mitzopoulos, K. 1890. Geological configuration of the greek terrain. Promitheus, 77 p (in greek).

Molnar, P. & Gray, D. 1979. Subduction of continental lithosphere: Some constraints and uncertainties. Geology, 7, 58–62.

Mountrakis, D. 1986. The Geology of Greece. Thessaloniki (in Greek).

Mountrakis, D. 2010. Geology and geotectonic development of Greece. University Studio Press, 373 p, Thessaloniki (in greek).

Negris, Ph. 1901. Plissements et dislocations de l'ecorse terrestre en Grece. 209 p, Athenes. NEGRIS, PH. 1915. Roches cristallophylliennes et tectonique de la Grèce. 123 p. Athènes.

Palinkas, L., Kovacs, S., Molnar, F., Haas, J., Jozsa, S. Dosztaly, L., Gulacsi, Z.G.K., Kover, S., Ozswart, P., Mikes, T., Halamic, J., Hrvatovic, H., Sudar, M., Jovanovic, D., Djeric, N., Migiros, G., Papanikolaou, D. & Tselepidis, V. 2010. Triassic rift-type basalts and related deep-water sediments in the western ophiolite belt of the Hellenides—Dinarides (from Orthrys Mts., Greece to Darno Hill, NE Hungary). Geol. Balc., 39/1-2, 285–286.

Pamic, J.J. 1984. Triassic magmatism of the Dinarides in Yugoslavia. Tectonophysics, 109, 273–307.

Papadeas, G., 1973. Zur geologie der Kristallinen Gesteine von Marathon. Bull. Geol. Soc. Greece, X, 2, 13–64.

Papanikolaou, D. 1976. The age of the metamorphics in Andros Island. Proc. Acad. Athens, 51, 292–301 (in greek).

Papanikolaou, D. 1979a. Unités tectoniques et phases de déformation dans I' île de Samos, Mer Egée, Grèce. Bull. Soc. Géol. France, (7), XXI, 6, 745-752.

Papanikolaou, D. 1979b. Stratigraphy and structure of the Paleozoic rocks in Greece: an introduction. In: Sassi, F.P. (Ed.), IGCP Project No. 5 "Correlation of Prevariscan andVariscan Events of the Alpine–Mediterranean Mountain Belt, 93–102.

Papanikolaou, D. 1980a. Contribution to the geology of the Aegean Sea. The island of Paros. Ann. Géol. Pays Hellén. 30/1, 65–96.

Papanikolaou, D. with contribution by N. Skarpelis, 1980b. Geotraverse southern Rhodope – Crete (Preliminary results). In: Sassi F. P. (ed.), I.G.C.P. No 5, Newsletter, 2, 41–48.

Papanikolaou, D. 1980c. The Metamorphic Hellenides. 26th Int. Geol. Congress., Paris, 1980, Abstracts, I, 371.

Papanikolaou, D. 1981. Some problems concerning correlations within the metamorphic belts of the Pelagonian and the Rhodope. In: Sassi F.P. (editor), IGCP Project No 5, Newsletter; 3, 119–123.

Papanikolaou, D. 1984. The three metamorphic belts of the Hellenides: a review and a kinematic interpretation. Geol. Soc. London, Spec. Publ., 17, 551–561.

Papanikolaou, D. 1986a. Late Cretaceous Paleogeography of the Metamorphic Hellenides. Geol. Geoph. Res., IGME, Special volume in honor of Prof. Papastamatiou, 315–328.

Papanikolaou, D. 1986b. Geology of Greece. Eptalofos Publications, 240 p. Athens (in greek).

Papanikolaou, D., 1987. Tectonic evolution of the Cycladic blueschist belt (Aegean Sea, Greece). In: Helgeson, H.C. (Ed.), Chemical Transport in Metasomatic Processes: NATO ASI Series, 429–450.

Papanikolaou, D. 1988a. Precambrian in the Hellenides (Pelagonian, Cyclades, Peloponnesus–Crete). In: Zoubek, V., Cogne, J., Kozhoukharov, D., Krautner, H. (Eds.), IGCP No 22, «Precambrian in Younger Fold Belts», 821–840, Wiley and Sons, New York.

Papanikolaou, D. 1988b. The Rhodope Massif in Greece. In: Zoubek, V., Cogne, J., Kozhoukharov, D., Krautner, H. (Eds.), IGCP No 22, "Precambrian in younger fold belts", 765–778, Willey & sons, New York.

Papanikolaou, D. 1989. Are the medial crystalline massifs of the eastern Mediterranean drifted Gondwanian fragments? In: Papanikolaou, D., Sassi, F.P. (Eds.), Sp. Publ. Geol. Soc. Greece, 1,63–90 and IGCP 276, Newsletter, No 1, Athens.

Papanikolaou, D. 1997. The tectonostratigraphic terranes of the Hellenides. Ann. Géol. Pays Hellén., 37, 495–514.

Papanikolaou, D. 2009. Timing of tectonic emplacement of the ophiolites and terrane paleogeography in the Hellenides. Lithos, 108, 262–280.

Papanikolaou, D. 2013. Tectonostratigraphic models of the Alpine terranes and subduction history of the Hellenides. Tectonophysics, 595–596, 1-24.

Papanikolaou, D. & Demirtasli, E. 1987. Geological correlations between the Alpine segments of the Hellenides–Balkanides and Taurides–Pontides. In: Flügel, H.W., Sassi, F.P., Grecula, P. (Eds.), IGCP No 5, «Pre-Variscan and Variscan Events in the Alpine–Mediterranean Mountain Belts.» Alfa Publishers, Bratislava, 387–396.

Papanikolaou, D. & Nomikou, P. 1998. The Paleozoic of Kos: a low grade metamorphic unit of the basement of the external Hellenides terrane. IGCP 276, Newsletter 6, Sp. Publ. Geol. Soc. Greece, 3, 155–166.

Papanikolaou, D. & Sassi, F.P. 1989. Palaeozoic Geodynamic Domains and their Alpidic evolution in the Tethys: A brief outline of the

IGCP No 276 proposal. IGCP 276, Newsletter 1, Sp. Publ. Geol. Soc. Greece, 1, 7–10.
Papanikolaou, D. & Sideris, Ch. 1983. Le Paléozoique de l' autochtone de Chios: Une formation à blocs de type wildflysch d' âge Permien (pro parte). C. R. Acad. Sc .Paris, 297, 603–606.
Papanikolaou, D & Sideris, Ch. 1992. Introduction to the geology of Chios island—fieldguide. 6th Congress of the Geological Society of Greece on «The Geology of the Aegean» and IGCP 276, 1992 FieldMeeting, Athens, 15 p.
Papanikolaou, D., Sassi, F.P. & Skarpelis, N. 1982. Outlines of the Pre-Alpine Metamorphisms in Greece. In: Sassi & Varga (editors), I.G.C.P. No 5, Newsletter, 4, 56–62 and Ann. Géol. Pays Hellén., 31/1, 16–31.
Papanikolaou, D., Bargathi, H., Dabovski, C., Dimitriu, R., El-Hawat, A., Ioane, D., Kranis, H., Obeidi, A., Oaie, C., Seghedi, A. & Zagorchev, I. 2004. TRANSMED Transect VII: East European Craton–Scythian Platform–Dobrogea–Balkanides–Rhodope Massif–Hellenides–East Mediterranean–Cyrenaica. In: Cavazza, W., Roure, F., Spakman, W., Stampfli, G., Ziegler, P. (Eds.), The TRANSMED Atlas: the Mediterranean Region from Crust to Mantle. Springer-Verlag, Heidelberg.
Papastamatiou, J. 1958. On the age of the crystalline limestones of Thera Island. Bull. Geol. Soc. Greece, Ill, 1, 104–113 (in greek).
Papazachos, B.C. & Comninakis, P.E. 1971. Geophysical and tectonic features of the Aegean arc. J. Geophys. Res., 76, 9817–8539.
Paraskevaidis, I. 1959. Simple and short Geology of Greece. Introduction to the map of Geology and Mineral Wealth of Greece at scale 1/500,000. IGME, Athens (in greek).
Pe-Piper, G. 1982. Geochemistry, tectonic setting and metamorphism of mid-Triassic volcanic rocks of Greece. Tectonophysics, 85, 253–272.
Pe-Piper, G. 1998. The nature of Triassic extension-related magmatism in Greece: evidence from Nd and Pb isotope geochemistry. Geol. Mag., 135, 331–348.
Pe-Piper, G.& Piper, D. 2002.The Igneous Rocks of Greece.The Anatomy of an Orogen. Gebrueder Borntraeger, Berlin/Stuttgart.
Philippson, A. 1898. La tectonique de l' Egéide. Ann. de Géographie, 112–141.
Philippson, A. 1901. Beiträge zur Kenntnis der griechischen Inselwelt. Peterm. Milt. Erganzunheft,134, 1–172, Gotha.
Philippson, A. 1956-59. Die Griechischen Landschaften. Volumes I-V, V. Klostermann, Frankfurt.
Platt, J.P. 1993. Exhumation of high-pressure rocks: a review of concepts and processes. Terra Nova, 5, 119–133.
Reischmann, T. 1998. Pre-Alpine origin of tectonic units from the metamorphic complex of Naxos, Greece, identified by simple zircon Pb/Pb dating. Bull. Geol. Soc. Greece, 32, 101–111.
Renz, C. 1940. Die Tektonik der griechischen Gebirge. Pragm. Akad. Athinon, 8.
Renz, C. 1955. Die vorneogene Stratigraphie der normal sedimentären Formationen Griechenlands. I.G.S.R., 637 p., Athens.
Renz, C. & Reichel, M. 1945. Beiträge zur Stratigraphie und Paläontologie des Ostmediterranen Jungpaläozoikums und dessen Einordnung im griechischen Gebirgssystem. Eclogae geol. Helv., 38, 2, 211–313.
Ring, U., Glodny, J., Will, T. & Thomson, S. 2010. The Hellenic subduction system: high pressure metamorphism, exhumation, normal faulting, and large-scale extension. Ann. Rev. Earth & Plan. Sci., 38, 45–76.
Robertson, A.H.F. 2007. Overview of tectonic settings related to the rifting and opening of Mesozoic ocean basins in the Eastern Tethys: Oman, Himalayas and Eastern Mediterranean regions. Geol. Soc., London, Sp. Publ., 282, 325–388.
Robertson, A.H.F., Trivic, B., Deric, N. & Bucur, I. I. 2013. Tectonic development of the Vardar ocean and its margins: Evidence from the Republic of Macedonia and Greek Macedonia. Tectonophysics, 595–596, 25–54.
Saccani, E. & Photiades, A. 2004. Mid-ocean ridge and supra-subduction affinities in the Pindos ophiolites (Greece): implications for magma genesis in a forearc setting. Lithos, 73,229–253.
Schermer, E. R., Howell, D.E. & Jones, D. L. 1984. The origin of allochthonous terranes: perspectives on the growth and shaping of continents. Ann. Rev. Earth & Plan. Sci., 12, 107–131.
Schliestedt, M., Altherr, R. & Matthews, A. 1987. Evolution of the Cycladic crystalline complex: petrology, isotope geochemistry and geochronology. In: Helgeson, H.G. (Ed.), Chemical Transport in Metasomatic Processes, Reidel Publishers, Dordrecht, 389–428.
Sideris, Ch. 1981. A new perception of the Atalanti «Paleozoic». Ann. Géol. Pays Hellén., 30/2, 637-646 (in greek).
Smith, A.G. 1971. Alpine deformation and the oceanic areas of the Tethys, Mediterranean and Atlantic. Geol. Soc. Am. Bull., 82, 2039–2070.
Tataris, A. 1965. About the presence of Eocene in the semimetamorphic basement of Thera Island. Bull. Geol. Soc. Greece, 6, 232–238 (in greek).
Tazieff, H., Varet, J., Barberi, F. & Giglia, G. 1972. Tectonic significance of the Afar (or Danakil) Depression. Nature, 235, 144–147.
Thiebault, F. & Kozur, H. 1979. Precisions sur l'age de la formation de Tyros (Paleozoique superieur—Carnien) & la base de la serie de Gavrovo-Tripolitza (Carnien) (Peloponnese meridional, Grece). C. R. Acad. Sc. Paris, 288, 23–26.
Trikkalinos, J. 1954. Über die paläogeographische Bedeutung der Kykladen—Masse für die tektonische Entwicklung des östlichen Teiles Griechenlands. Pragm. Akad. Athinon, 18, 2, 1–48.
Trikkalinos, J. K. 1955. Uber das Alter des metamorphen Gesteine Attikas. Ann. Geol. Pays Hellen., 6, 193–198.
Trikkalinos, I. 1960. Contribution to the investigation of the Tectonic structure of Greece. Some observations on studies recently performed in Attica, Eastern Orthrys and Northern Evia. Ann. Géol. Pays Hellén., 11, 297–312 (in greek).
Trumpy, R. 1980. An outline of the Geology of Switzerland, Wepf and Co, Basel.
Vavassis, I., De Bono, A., Stampfli, G.M., Giorgis, D., Valloton, A., Amelin, Y. 2000. U–Pb and Ar–Ar geochronological data from the Pelagonian basement in Evia (Greece): geodynamic implications for the evolution of Paleotethys. Schweizerische Mineralogische und Petrographische Mitteilungen, 80, 21–43.
Zagorcev, I. 1998. Rhodope controversies. Episodes, 21, 159–166.
Zen, e-an. 1983. Exotic terranes in the New England Appalachians - limits, candidates and ages; A speculative essay. Geol. Soc. Am. Mem., 158, 55–82.

8 Description of the Tectonic Units

The Alpine structure of Greece comprises the description of the tectono-stratigraphic terranes and the geotectonic units within each terrane. The description follows the paleogeographic organization of the tectono-stratigraphic terranes of the Hellenides from the external (H1) to the internal (H9) parts of the chain. The description of the geotectonic units within each terrane similarily follows its position from the external towards the internal margin of the basin or the platform respectively. The general integration of the tectonic units of the Hellenides within the terranes has been described in detail in a series of publications (Papanikolaou 1989b, 1997, 2009, 2013; Papanikolaou et al. 2004a). Additionally, it has been included in the tectonostratigraphic diagrams of IGCP 276 final volume, together with the terrane maps (Papanikolaou and Ebner 1997). A brief description of the terrane stratigraphy and the integration of the main geotectonic units in each terrane of the Hellenides is given in the tectono-stratigraphic diagrams of the nine terranes in Fig. 8.1.

The succession of the terranes and tectonic units according to their probable paleogeographic location is relatively easy in the beginning but gets gradually harder while progressing towards the interior of the arc. At the same time, the relative position of the units within the terranes is not always well established and in some cases it is merely just probable. New data may lead us to a different structure/organization. This is entirely related to the presence of the metamorphic Hellenides and the palinspastic restoration techniques we can use. The lack of accurate dating of the protoliths and/or of the tectono-metamorphic events within the orogenic arc make the final options complicated and in some cases, different integrations may be expected in the future.

The nine tectono-stratigraphic terranes of the Hellenides are distinguished in two groups, according to their overall continental or oceanic signature and their Tethyan history can be outlined by the two tectono-stratigraphic models, presented previously in Sect. 7.13 and shown in Fig. 7.14. Thus, the major paleogeographic stages of the continental terranes with their pre-Alpine basement and their Alpine shallow water carbonate platforms are illustrated in the diagrams of Fig. 8.2. In these diagrams, the transition from the volcano–sedimentary complexes of the rifting stage to the carbonate platform stage is given for each terrane. A younger age of this transition is observed from the internal towards the external terranes, from the Late Paleozoic (H7) to the Carnian (H1). These systematically younger ages indicate the successive onset of the terrane drift from the northern to the southern terranes. Similarily, younger ages are observed in the transition from the platform stage to the synorogenic flysch stage from the Lias (H7?, H5) to the Miocene (H1). These systematically younger ages demonstrate the successive docking of the terranes in the active European margin and the orogenic migration of the Hellenides towards the south.

The same process is observed in the corresponding tectono-strartigraphic diagrams of the oceanic basins (Fig. 8.3). The transition from the volcano-sedimentary complexes of the rifting stage to the pelagic/abyssal sequences of the basins becomes younger from the Late Paleozoic (H8?) to the Carnian/Norian (H2). Similarily, the transition from the pelagic sedimentation to the flysch of the oceanic basins becomes younger from the Lias (H8? H6) to the Maastrichtian (H2).

The following description of the tectonic units is accompanied by the geotectonic map of Greece, given in annex out of text, together with a part of the tectono-stratigraphic terrane map of the Eastern Mediterranean in the back page. This geotectonic map is a revised and completed version of previous maps (Papanikolaou 1986c, 1989b) and shows the overall extension of the geotectonic units of the Hellenides. The two maps are available also in highly reduced size at the end of the book.

D. I. Papanikolaou, *The Geology of Greece*, Regional Geology Reviews,
https://doi.org/10.1007/978-3-030-60731-9_8

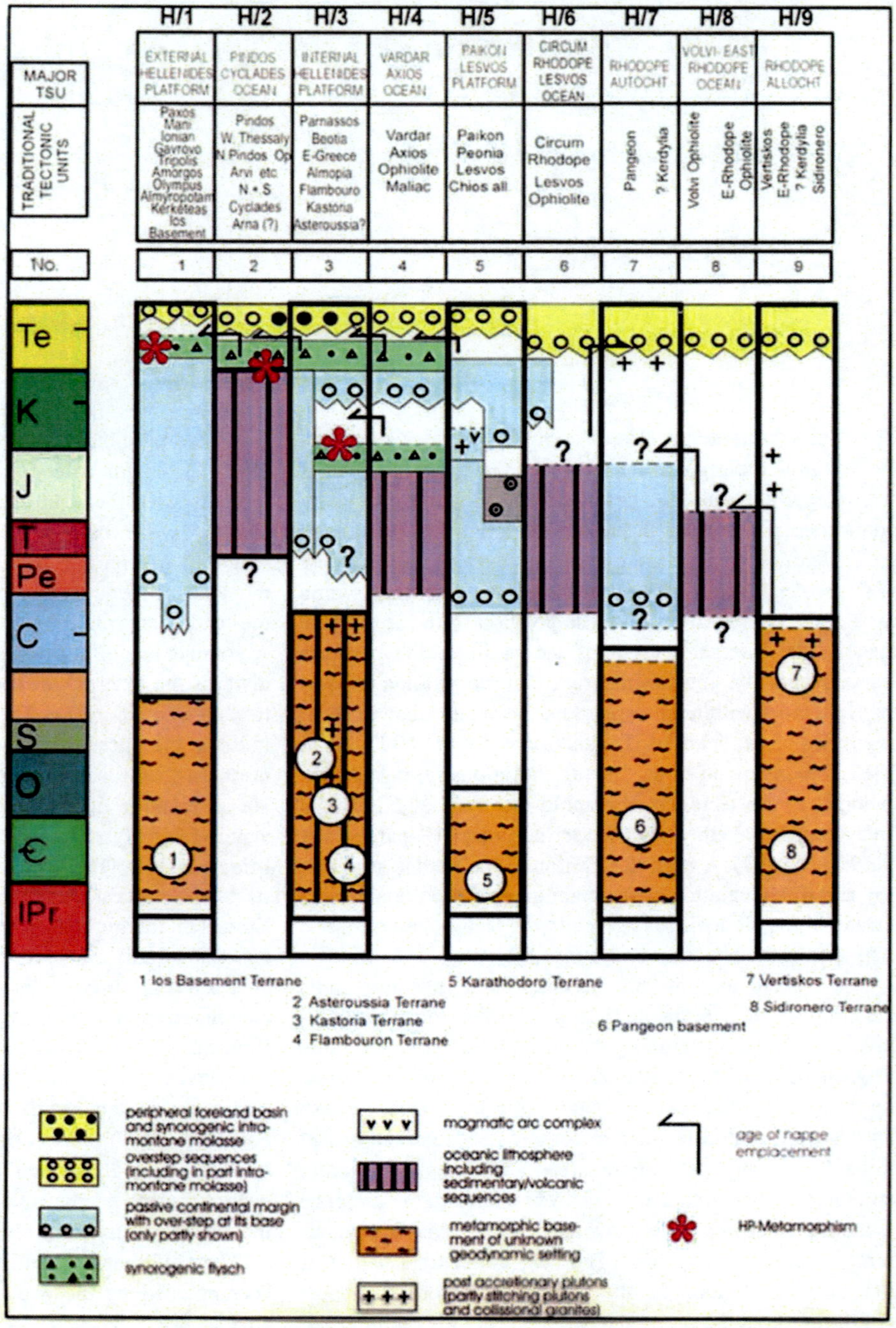

Fig. 8.1 Diagram of stratigraphic correlation charts of the Hellenic Tectono-stratigraphic Terranes and integration of the geotectonic units of Greece within them (from Papanikolaou 1997, within the final volume of IGCP 276)

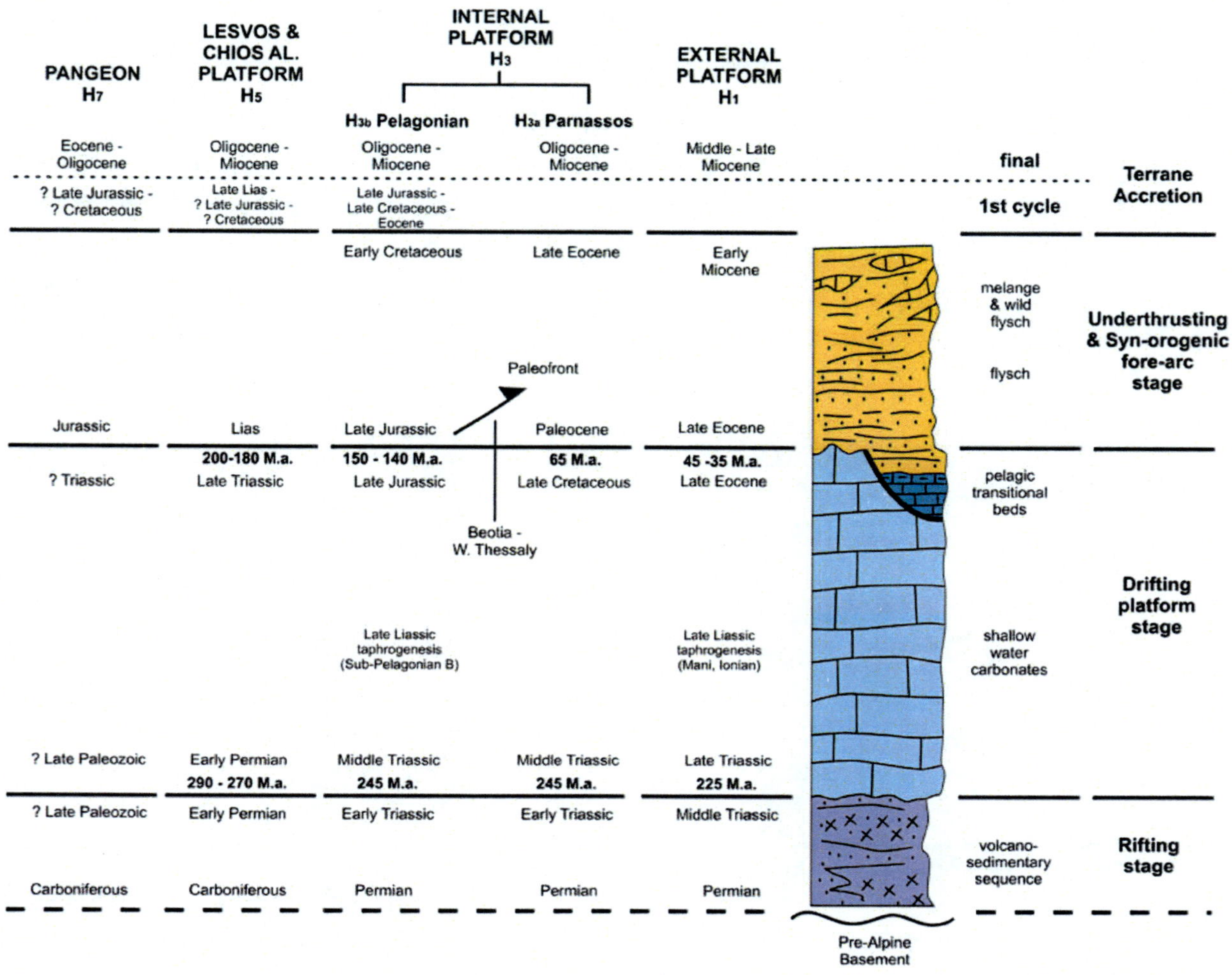

Fig. 8.2 Stratigraphic columns of the continental terranes H1, H3, H5 and H7, showing the timing of the three paleogeographic stages for each terrane (from Papanikolaou 2013). Thus, the shallow water carbonate sedimentation in the External Platform H1 lasts from the Carnian to the Late Eocene, whereas the volcano-sedimentary facies of the rift period comprises the Permian–Middle Triassic. The transition from the carbonate platform to the flysch occurs in the Late Eocene

8.1 The External Platform of the Hellenides—H1

8.1.1 The Paxos (or Pre-Apulian)–Kastellorizo Unit

The Paxos unit was distinguished by Renz (1940, 1955) as the most external unit of the Hellenides. It originates from the western margin of the external platform of the Hellenides. Offshore geophysical research (Kokinou et al. 2005) has concluded that the Paxos unit continues westwards until the present East Mediterranean Ocean in the Ionian Sea (H0). The Ionian oceanic crust has started to be subducted under the Paxos and the other more internal units of the external platform (H1) since Late Miocene–Early Pliocene. This is indicated by the timing of the final thrusting of the Ionian unit over the Paxos unit together with its compressive deformation and uplift (e.g. Underhill 1989). The platform carbonates overlie Triassic evaporites and Paleozoic continental crust. The top of the sequence comprises Messinian evaporites and the post orogenic deposits comprise an unconformably overlying non-deformed Upper Pliocene–Quaternary sedimentary cover. The unit can be observed in the Paxos islands (type locality), Lefkada, Cephalonia and Zakynthos (Renz 1955). It was also called Pre-Apulian (Aubouin 1959), due to the fact that it is not a clear ridge in Greece, but also a slope, characterized by several horizons of microbreccia limestones. It becomes a typical shallow water carbonate platform in the Apulia peninsula in Southern Italy (Fig. 8.4).

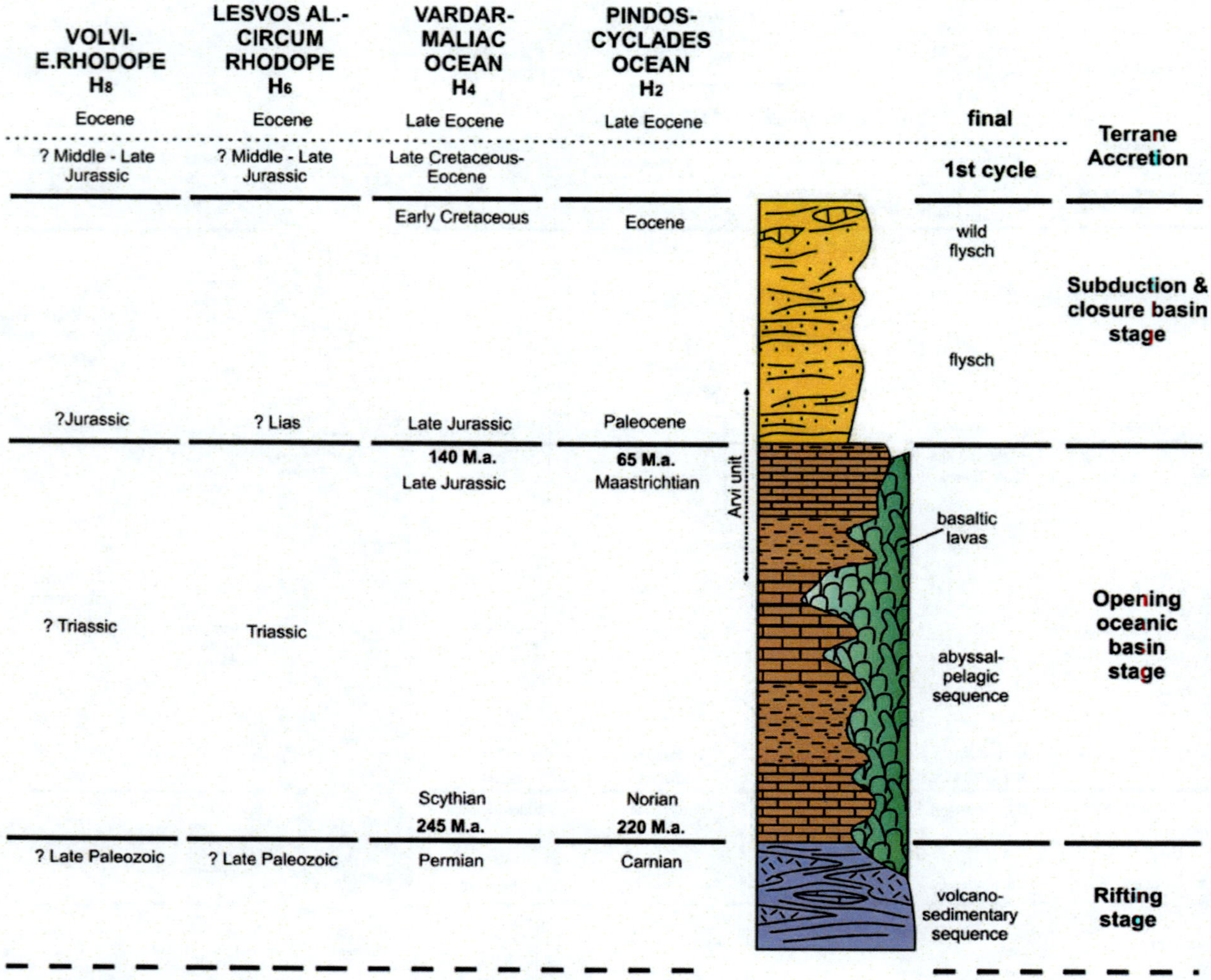

Fig. 8.3 The stratigraphic columns of the oceanic basins in relation to the three paleogeographical geodynamic stages (from Papanikolaou 2013). Thus, Pindos basin H2 begins its pelagic sedimentation in the Norian and comes to an end in the Maastrichtian. Previously, during the Middle Triassic–Carnian, it was in the volcano-sedimentary rifting stage, whereas after the Maastrichtian–Danian, it was in the synorogenic flysch sedimentation

The lateral transition between the Paxos unit and the Apulian platfrom can be generally accepted, because of the obvious stratigraphic correlations and the similar tectonic evolution. In the Northern Ionian Sea, there is no evidence of a large tectonic contact differentiating the two areas. Nevertheless, west of the Corfu Island, the tectonic front of the Northern Hellenides probably occurs along a deep channel of about 1 km depth (Monopolis and Bruneton 1982; Del Ben et al. 2015). However, the overthrusting formations belong to the Ionian and not the Paxos unit, whose continuity to the north is not clear. In any case, Apulia is the foreland of both the Hellenides and the Appenines, with opposite sense of tectonic transport; westwards for the Hellenides and eastwards for the Apennines. More recently, oceanographic studies in the North Ionian and the South Adriatic area have shown the existence of the **South Apoulian basin** in the area west of Corfu (Del Ben et al. 2015). This is a separate epicontinental basin developed between the Apulian and the Paxos carbonate platforms, whose subsidence seems to have occurred during the Late Jurassic (?)—Early Cretaceous. Thus, the link between the western part of the External Hellenides platform and Apulia platform is much more complicated than previously assumed. It seems that the previous continuous Triassic-Liassic carbonate platform, extending from Apulia to Olympus, has suffered several distinct rifting episodes in the Late Lias (e.g. in the Ionian unit) as well as later in the Cretaceous. Generally, there are several references where the

Fig. 8.4 Outcrop of the upper horizons, made of marly limestones of Upper Miocene–Pliocene age, of the Apulian carbonate platform in Otranto. The horizontal position of the platform and its very recent uplift during the Late Pliocene–Pleistocene is remarkable

External carbonate platform of the Hellenides is attributed to the wider **Adria plate**, including also the Apulian platform (e.g. Channel et al. 1979; Brun et al. 2016).

Stratigraphically, it is a neritic carbonate sequence from the Jurassic to the Miocene, with minor disconformities in the Paleogene, like in Zakynthos. The Jurassic formations appear only in the western Lefkada Island, and for this reason a schematic stratigraphic column and a simplified geological map of Lefkada are presented in Figs. 8.5 and 8.6 (Bornovas 1964).

In the Aquitanian, marly formations and turbiditic limestones are observed, which substitute the typical flysch sedimentation, found in all the other geotectonic units. In the Burdigalian and the Middle Miocene clays and marls are prevailing, without distinctive sandstone horizons. The above lithological peculiarities are the reason why the Paxos unit has been considered as the only one lacking typical flysch (Aubouin 1959).

The tectonism of the Paxos unit is of Miocene age, and in Zakynthos and Cephalonia islands in particular, it is dated in the Miocene/Pliocene boundary (Underhill 1989). In Zakynthos Island, the stratigraphic column ends up with Messinian gypsum, above which the Triassic gypsum of the Ionian unit, croping out in the Skopos peninsula, is overthrusted. In Zakynthos Island, an almost complete stratigraphic profile of the Miocene can be observed, several hundred meters thick (Dermitzakis 1978). It is important to note that the intra-Pliocene chronostratigraphic gap between the tectonism of the Miocene–Lower Pliocene beds (Messinian evaporites overlain by the Lower Pliocene Trubi facies and superjacent marls) and the unconformable/disconformable non-deformed post orogenic beds of the Upper Pliocene–Middle Pleistocene is only 1–2 million years (Papanikolaou et al. 2010) (Fig. 8.7).

From a geotectonic point of view, Paxos unit is the foreland of the Hellenides. This is shown in the geotectonic map of Greece by the westernmost location of the Ionian frontal thrust over the Paxos relative autochthon unit. This frontal Ionian thrust can be followed from the Western Lefkada Island to the Eastern Cephalonia and to Eastern Zakynthos islands. Of course, we know that west of the above Ionian islands, the actual front of the advancing system of the Hellenic fold and thrust belt lies along the Hellenic Trench and even further to the west in the backstop, above the subducting East Mediterranean oceanic crust. Thus, the tectonic emplacement of the Ionian nappe, which occurred in the Paxos unit in Early Pliocene, may continue today, several tens of kilometers to the west, since plate convergence is still active. Thus, the Paxos unit is the foreland of the Hellenides fold and thrust belt but only relatively autochthon, since all the Hellenides are now allochthonous above the advancing plate boundary.

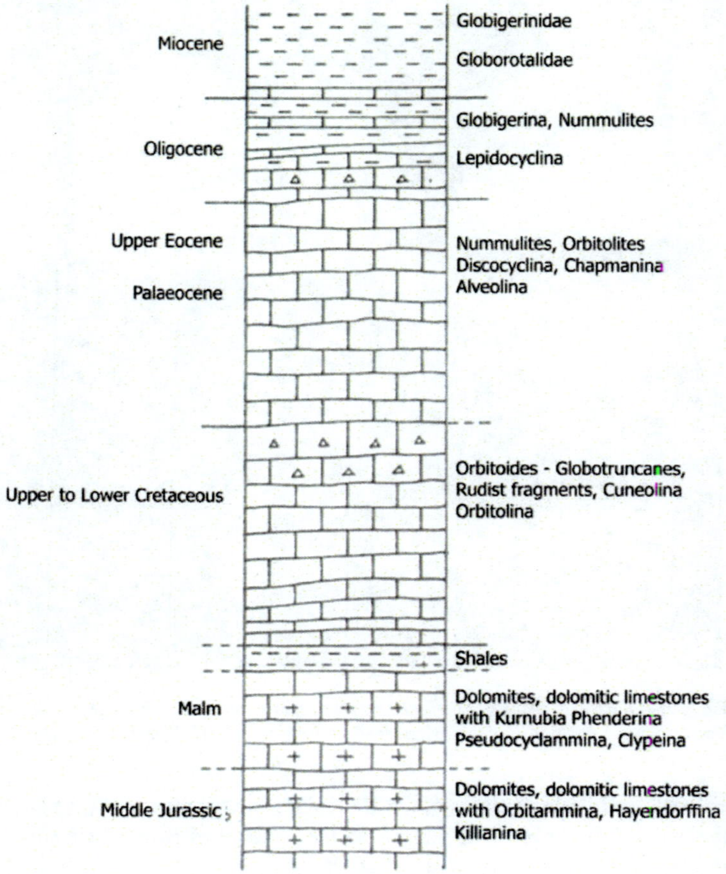

Fig. 8.5 Schematic stratigraphic column of the Paxos unit, from the data of the geological map of Lefkada Island (from Bornovas 1964)

The largest outcrop of the Paxos unit is observed In the Cephalonia Island, where two tectonic sub-units can be distinguished, separated by a thrust; the Paliki sub-unit to the west and the Ainos sub-unit to the east. These two units comprise different sedimentary/stratigraphic facies, with a neritic carbonate platform in the Paliki sub-unit, different from the slope facies of the Ainos sub-unit. It is characteristic that the intermediate NNW-SSE thrust passes through the Argostoli Gulf and is characterized as seismically active. This was shown by the Mw = 6.1 and Mw = 6.0 magnitude 2014 earthquakes, whose focal mechanisms showed N-S strike slip and E-W thrust faulting respectively (Valkaniotis et al. 2014). Along the southern coast of the Paliki peninsula the modern deformation is characterized by growth folding, observed by a N-S anticline folding the Pliocene-Middle Pleistocene marine sediments (Papanikolaou and Triantaphyllou 2013) (see also Fig. 11.16).

On the island of **Kastellorizo** in the Dodekanese islands there is another tectonic unit similar to the Paxos unit. The Kastellorizo unit is located at the eastern continuation of the Hellenic arc and trench system, immediately east of the Rhodes deep basin and before the beggining of the Taurus and Cyprus arcs. The Kastellorizo unit corresponds to the Bey Daglari unit of Southwestern Asia Minor, which is the deepest tectonic unit of the Taurides–Pontides system (Bernoulli et al. 1974; Brunn et al. 1976; Gutnic et al. 1979; Papanikolaou and Demirtasli 1987). Thus, similarly to the Paxos unit, Kastellorizo and Bey Daglari units are the relatively autochthonous foreland on the Lycian nappes (which correspond to the external Hellenides), thrusted from the north–northwest. These autochthonous units constitute a huge tectonic window between the Lycian nappes and the Antalya nappes emplaced from the east-southeast, which do not have a counterpart in the Hellenic arc (Fig. 8.8).

Systematic research with geological mapping have shown that the top stratigraphic horizons of the Kastellorizo unit are observed in **Ro** islet, to the west of Kastellorizo, which comprises limestones with algae and corals of Middle-Upper Miocene age (Tsaila-Monopolis and Galeos 1984; Galeos 1986). Therefore, the earlier view that Kastellorizo could correspond to one of the external units of Paxos, Ionian or Tripolis, based on the presence of limestones with Nummulites of Eocene age (Christodoulou 1972) is excluded. At the same time, the new data show that the pre-orogenic

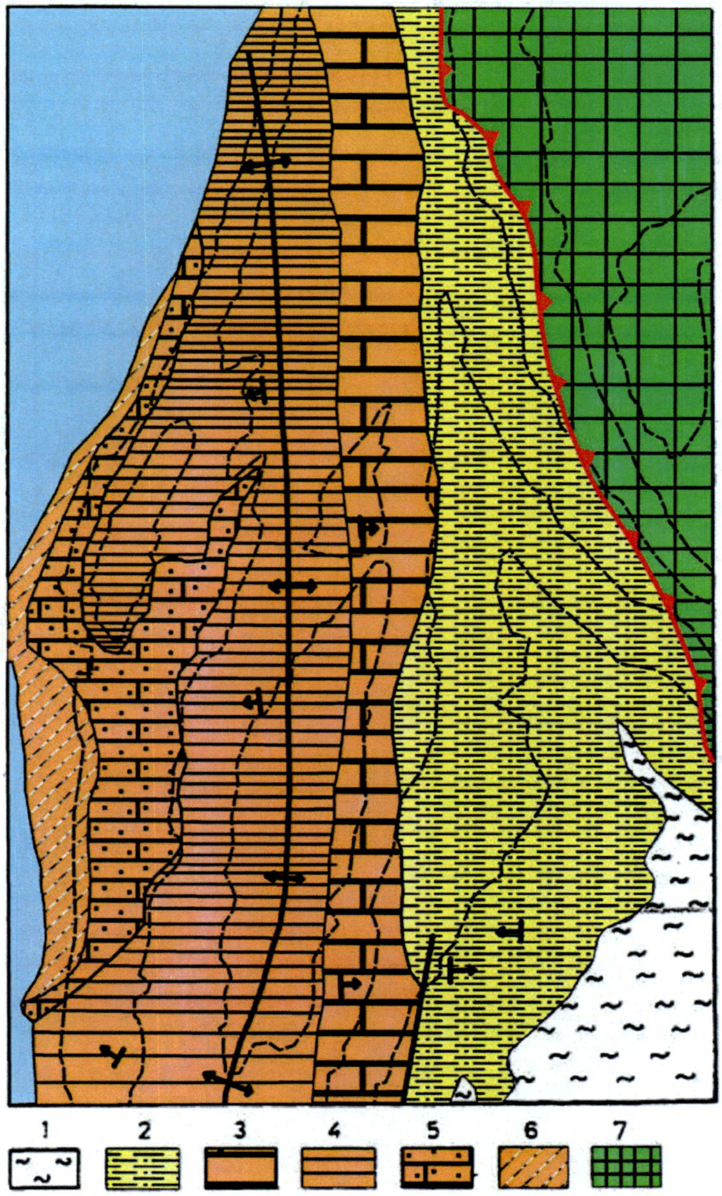

Fig. 8.6 Simplified geological map of Southwestern Lefkada Island, where the Paxos unit can be observed below the Ionian nappe (from Bornovas 1964). 1. Alluvial 2: Burdigalian–Tortonian, compact blue, green-brown marls, with intercalations of breccia-limestones, 800 m thick. 3: Paleocene–Aquitanian, bedded limestones, microbrecciated, micronodular, with echinoderm fragments in alternations with pelagic limestones and cherts. They develop upwards into marly platy limestones, 250 m thick. Fossils: Alveolina sp, Discocyclina archiaci, Gypsina globula, Orbitolites complanatus, etc. 4. Upper Cretaceous, bedded limestones with microbreccia and echinoderm fragments and molluscs, alternating with pelagic limestones. They develop upwards into thick-bedded limestones with rudist fragments and to oolithic limestones, 200 m thick. Fossils: Orbitolina concara, Globotruncana lapparenti, Orbitoides media etc. 5: Lower Cretaceous, bedded limestones, micronodular, with sparse chert intercalations, 100 m thick, 6: Upper Jurassic, ammonite-bearing limestones and black bituminous shales, 40 m thick. 7: Ionian unit formations, mainly Pantokrator limestones of Upper Triassic–Liassic

carbonate sedimentation lasted in Kastellorizo, until the Late Miocene i.e. 10–15 million years later than the end of the carbonate sedimentation in the Paxos unit, which is observed at the Oligocene/Miocene boundary. Thus, the age of tectonism of the Kastellorizo unit coincides with that of Bey Daglari, where Upper Miocene–Pliocene flysch appears on top of the carbonate platform, showing tectonism in the Pliocene (Gutnic et al. 1979). This Pliocene age of deformation shows a tectonic synchronization of the Kastellorizo unit with the Paxos unit in Western Greece in the Ionian Sea on the two sides of the Hellenic arc.

8.1.2 The Mani Unit (Metamorphic Ionian)

The next more internal unit, from the paleogeographic point of view, in the external platform of the Hellenides is not the Ionian (although it is observed to be thrusted on the Paxos unit in the islands of Lefkada, Cephalonia and Zakynthos since the Late Miocene-Early Pliocene) but Mani unit, which is croping out in tectonic windows in central–southern Peloponnese and Crete. This conclusion resulted from the discovery of fossils in the relative autochthon metamorphosed unit of Peloponnese and Crete and its stratigraphic

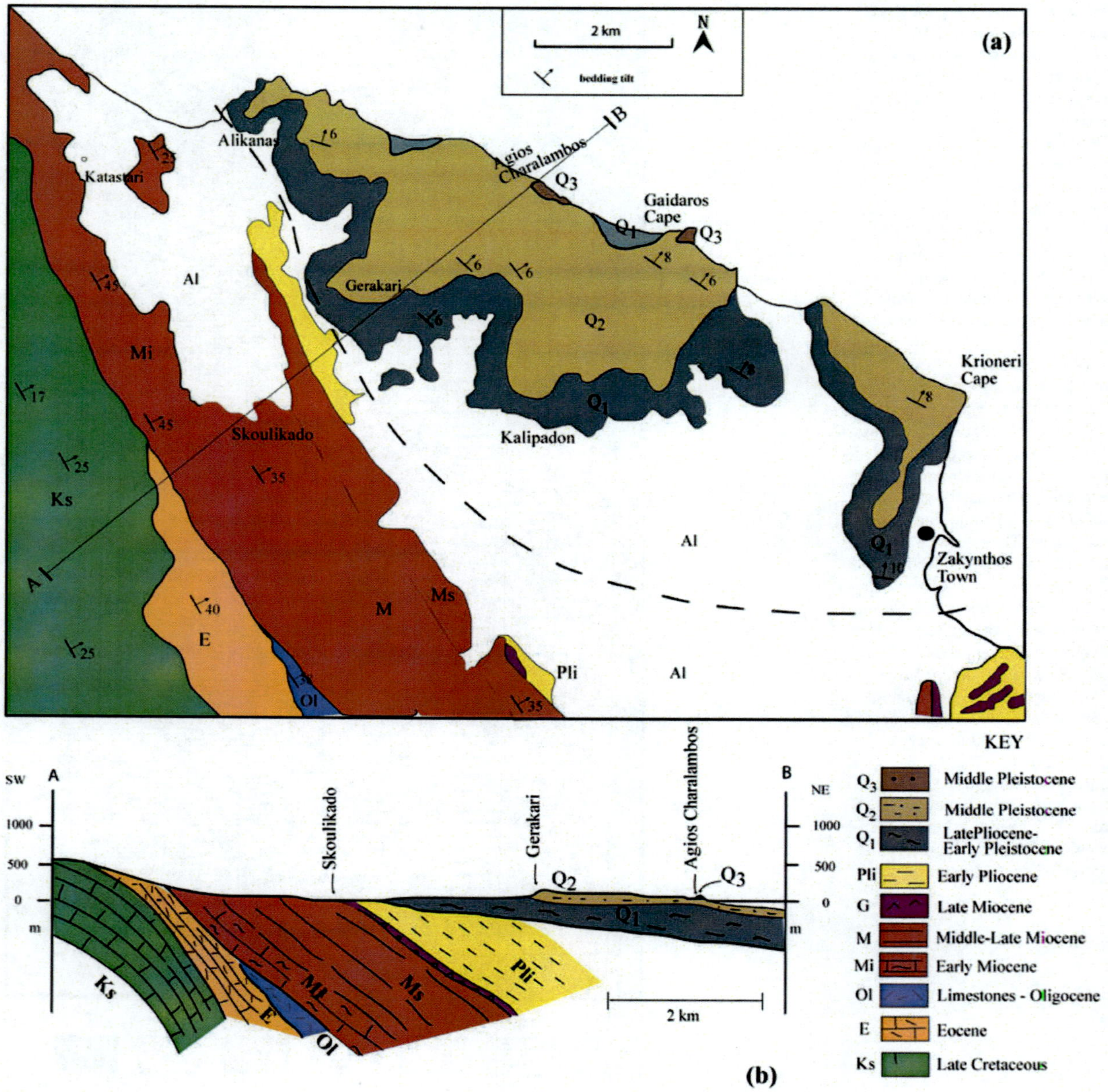

Fig. 8.7 Geological map of the central Eastern Zakynthos Island, showing the unconformity of the Upper Pliocene–Pleistocene beds dipping with 5–7° to the NNE, above the Miocene–Lower Pliocene sequence of the top of the Paxos unit, dipping with 30–40° to the NE (from Papanikolaou et al. 2010)

correlation to the Ionian unit (Fytrolakis 1972, 1980; Bonneau 1973a; Thiebault 1977, 1982) (Fig. 8.9a).

In fact, in a cross section of the Peloponnese, the nappe structure comprises the Ionian unit beneath the Gavrovo-Tripolis unit in the west, whereas in its central and eastern part the Tripolis unit is observed under the Pindos unit and above the Mani unit (formerly known also as Plattenkalk) (see also Fig. 8.33). Thus, both the Ionian and Mani units occur below the Gavrovo and Tripolis units at a distance of approximately 60–70 km. Their major differences are the tectono-metamorphic characteristics and the tectonic position, since Mani appears in metamorphic tectonic windows at the central-eastern zone of the Peloponnese. If we remove the tectonically emplaced, non metamorphosed nappes of Pindos, Tripolis and Ionian back to their original position towards the Aegean Sea, then Mani

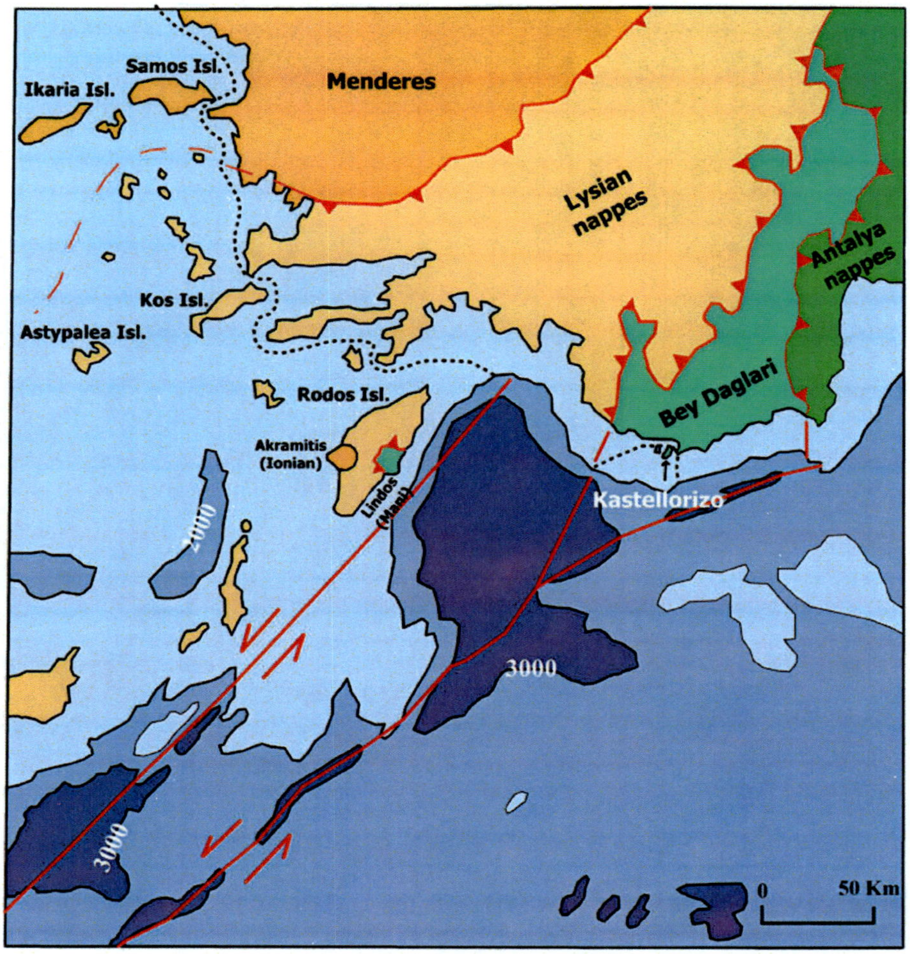

Fig. 8.8 The position of the Kastellorizo unit as a continuation of the Bey Daglari unit in the Southwestern Asia Minor. The closest units of the Hellenides are those of the Akramytis and Lindos units in Rhodes, corresponding to the Ionian and Mani units respectively

unit should be paleogeographically located between the Paxos and Ionian units (see also Fig. 9.6). Thus, several scientists use the term «metamorphic Ionian» for the Mani unit. A key structure of the entire kinematics is the Oligocene Mani flysch, which is slightly metamorphosed, in the form of phyllites, with isoclinal folding and axial plane schistosity (Fig. 8.9b). Thus, the beginning of subduction of the Mani unit under the more internal platform units and the resulting tectono-metamorphic evolution is post-Oligocene.

In the bibliography of the 1970s the term **Plattenkalk** was used for Mani unit, derived from the German term describing platy limestones. However, this term, is only a lithological characterization widely used also for other formations and units (e.g. Pindos, Amorgos, etc.). Additionally, it does not represent the entire stratigraphic sequence of the unit and does not refer to a type locality, as required by modern international practices. Thus, the term Mani unit is used, as proposed by Fytrolakis (1980), referring to the typical area of the Mani peninsula, made almost exclusively from this particular unit. In Crete, Bonneau (1973a, 1976) had used the term **Ida** zone, while Epting et al. (1972) and Jacobshagen (1979, 1980) had used the term **Talea Ori**. Finally, in Rhodes Island, the Mani unit corresponds to the **Lindos** unit (Mutti et al. 1970; Aubouin et al. 1976).

The stratigraphic column of the Mani unit is similar to that of the Ionian, as described by Renz (1955). Thus, the Pantokrator formation is observed at the base (Upper Triassic–Liassic), in the form of solid white marbles, followed by some schists and siliceous crystalline limestones, corresponding to the "Posidonia shales", continuing upwards to the Vigla formation, which corresponds to the «Plattenkalk» sensu stricto (platy crystalline limestones with cherts), and closing with the clastic-breccia crystalline limestones (Senonian) and the multicolored cipolline marbles at the transitional beds towards the phyllites of the slightly metamorphosed flysch. The first fossils of the Mani unit were found in eastern Crete, in the area of Kalavros by Fytrolakis (1972), in the transitional beds from the crystalline limestones to the flysch, comprising Globigerines of the Upper Eocene (see also Fig. 8.38). Today, we know

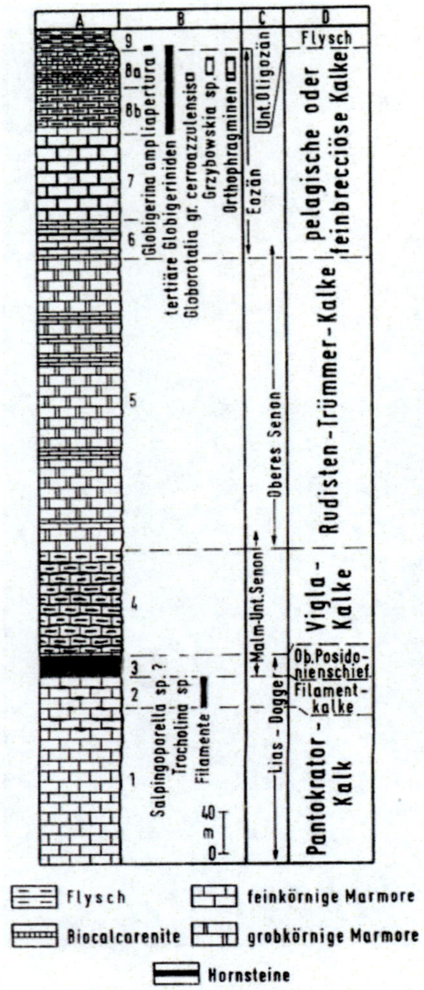

Fig. 8.9 **a** Stratigraphic column of the Mani unit and correlation with the Ionian unit (from Jacobshagen 1986 based on Thiebault 1977). h: flysch, i: bioendocalcarenite, j: fine crystalline marbles (former pelagic limestones), k: coarse crystalline marbles, l: siliceous horizons. **b** Isoclinal recumbent fold inside the phyllites of the Oligocene meta-flysch of the Mani unit from the region of Alika in the Mani peninsula

almost the entire stratigraphic column, documented by fossils (see Thiebault 1977, 1982).

In Crete, to the west of Heraklion, an occurrence of Upper Triassic carbonates unconformably overlying the Upper Paleozoic (Permian) and Scythian formations, known as **Fodele** and **Sisses** formations respectively, was reported (Epting et al. 1972). These formations comprise black neritic limestones with Productus of the Permian, characterizing a shallow water carbonate platform and marly shale sediments with some gypsum occurrences in the Scythian, separated by a stratigraphic unconformity from the overlying Upper Triassic Pantokrator limestones. The Northern Crete tectonic structure is an impressive kilometric scale isoclinal fold, where a general reversal of the above Upper Paleozoic–Scythian formations is observed (Fig. 8.10). It is noteworthy that the above Late Paleozoic–Scythian continental margin environment is contrasting the volcano-sedimentary rifting sequences of the Tyros beds, observed below the more internal units of the External platform (e.g. Western Crete, Tripolis) (Papanikolaou 1979b).

The metamorphic conditions of the Mani unit are difficult to be estimated due to the carbonate lithology. Blumör et al. (1994) have described the presence of Fe–Mg carpholite in the Kastania Phyllites from Taugetus mt, which they attribute to the base of the Mani unit, leading to P-T estimates around 7–8 kbars and 310–360 °C. However, the Kastania phyllites are probably belonging to a thrust sheet of the Arna unit and not to the basement of the Mani unit. A key point is

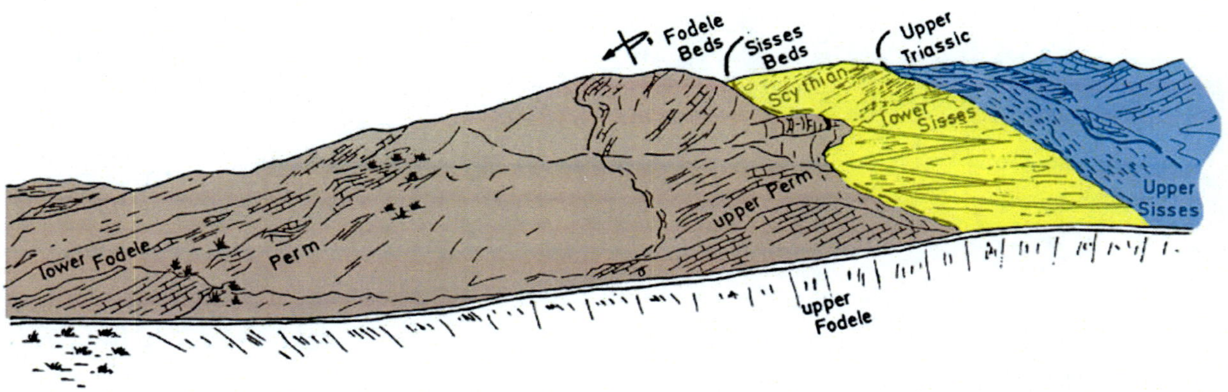

Fig. 8.10 Panoramatic view looking to the east, of the inverse Permo-Triassic sequence, occurring at the base of the Mani unit in the northern slopes of the Talea Ori, at the 106th km of the road Rethymno–Heraklio (from Papanikolaou 1988c)

their similarity to the Arna metamorphic rocks, contrary to the Upper Paleozoic–Scythian formations from Crete, occuring in the Fodele and Sisses area, beneath the Triassic marbles of the Mani unit (Epting et al. 1972). In conclusion, Mani unit is characterized by a relatively low grade of metamorphism, intense limestone crystallinity, and an intense tectonic structure with tectonic flow along the isoclinal folds in all scales (see also Fig. 4.15). Its outcrops correspond to tectonic windows within the external Hellenides both in Peloponnese and in Crete (see also Fig. 7.23).

8.1.3 The Western Crete–Trypali Unit

The presence of a distinct tectonic unit comprising crystalline limestones to marbles of neritic facies with fossils of the Late Triassic–Liassic, above the Mani unit in Crete, was known as the **Trypali** unit (Creutzburg and Seidel 1975; Kopp and Ott 1977). The particularity of the Western Crete ourcrops was previously mentioned by both Creutzburg (1958), who distinguished the «Madara Kalke» formation and Tataris and Christodoulou (1965), who described the "Triassic transgressive system of Western Crete", when the deeper metamorphic rocks were still considered as pre-Permian basement. Thus, in Crete, the Trypali unit, comprising Upper Triassic–Liassic marbles of Pantokrator facies limestones is tectonically emplaced upon the Mani unit. According to Karakitsios (1979) and Bonneau (1984) this unit is a tectonic duplication of the lower part of the Mani unit, whereas according to Hall et al. (1984) it is a unit made of supra-tectonic breccias, in the form of a mélange, due to horizontal sliding between Mani and the overlying more internal units.

The distinction of the **Western Crete** unit was made later on by Papanikolaou (1988c), by incorporating the Trypali section of Upper Triassic–Jurassic age in its upper part together with the underlying volcano-sedimentary sequences of Upper Paleozoic–Middle Triassic age (resembling the "Tyros beds" beneath the Tripolis carbonate platform) (Fig. 8.11). At the base of the unit, mica schists with quartzites and thin marble horizons are observed, in which Carboniferous fossils have been identified (Krahl et al. 1983) (Fig. 8.12).

Large outcrops of gypsum are observed in the Middle Triassic, like those along the western coasts of Chrysoskalitissa, resembling the Triassic gypsum of the Ionian unit in Western continental Greece (Papanikolaou 1979b). The distinction of the unit was based on the observation of the stratigraphic continuity between the carbonate Triassic–Liassic rocks asociated to gypsum-raywackes and the Permo-Triassic volcano-sedimentary formations. At the same time, geological mapping of the metamorphic rocks in Western Crete showed the distinction of the Arna unit from the tectonically underlying volcano-sedimentary formations of Western Crete, as opposed to the overlying volcano-sedimentary formations of the Ravdoucha-Tyros beds of the base of the Tripolis nappe. All these metamorphic formations were mixed together in the Cretan bibliography as the so-called "phyllites-quartzites" (Papanikolaou and Vassilakis 2010) (Fig. 8.13).

The absence of the stratigraphically overlying formations (Dogger-Eocene) in the Trypali unit raises some questions, although it is generally assumed that the upper formations, probably of pelagic facies, have been detached from the underlying massive Pantokrator limestones and have been left behind, during the Tertiary tectonic emplacement of the nappes. This assumption was supported by Krahl et al. (1983) who have reported the occurrence of younger schistosed formations of Jurassic-Cretaceous age above the Liassic of the Trypali unit in central Crete. Outcrops similar to the Trypali Unit have been described also in the southern

Fig. 8.11 Characteristic outcrop of the multicolored volcano-sedimentary Triassic "Tyros beds" type, in the Western Crete unit

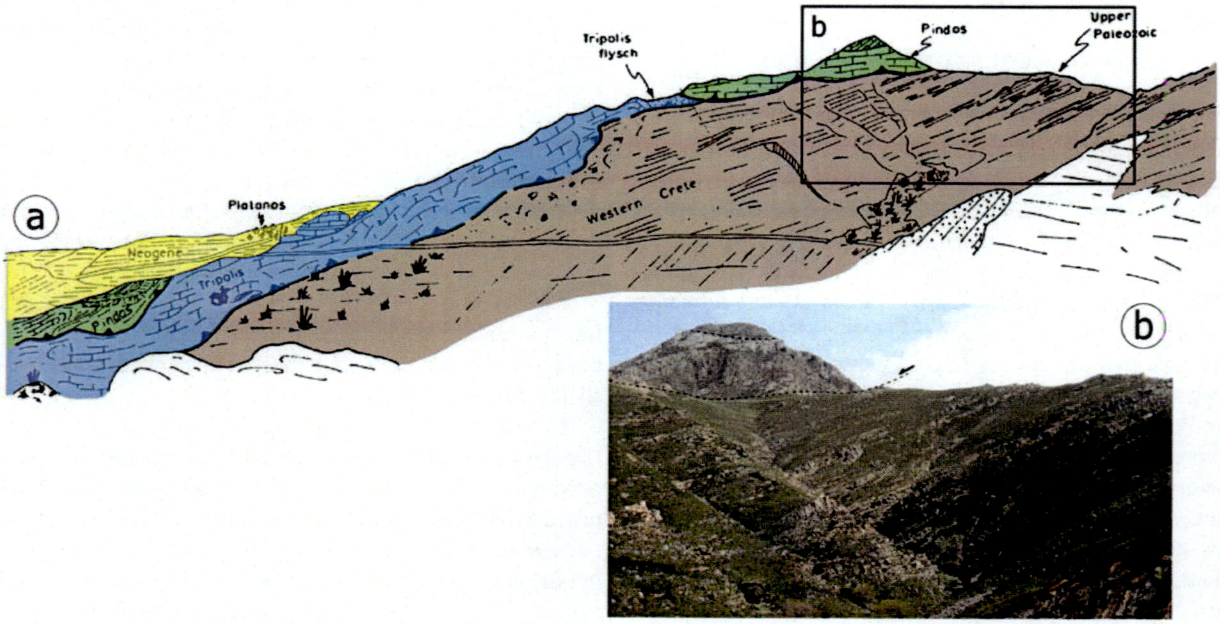

Fig. 8.12 **a** Panoramatic sketch looking eastwards of the lower section of the Western Crete unit, with its Carboniferous horizons under a sub-horizontal fault—extensional detachment, separating the tectonic klippen of the Tripolis and Pindos (Ethia) units at the top of the mountain, in the area south of Platanos (from Papanikolaou 1988c). **b** Detail of the previous panorama, showing the moderately inclined to the north Carboniferous horizons below the sub-horizontal tectonic klippen of the Tripolis and Pindos/Ethia units

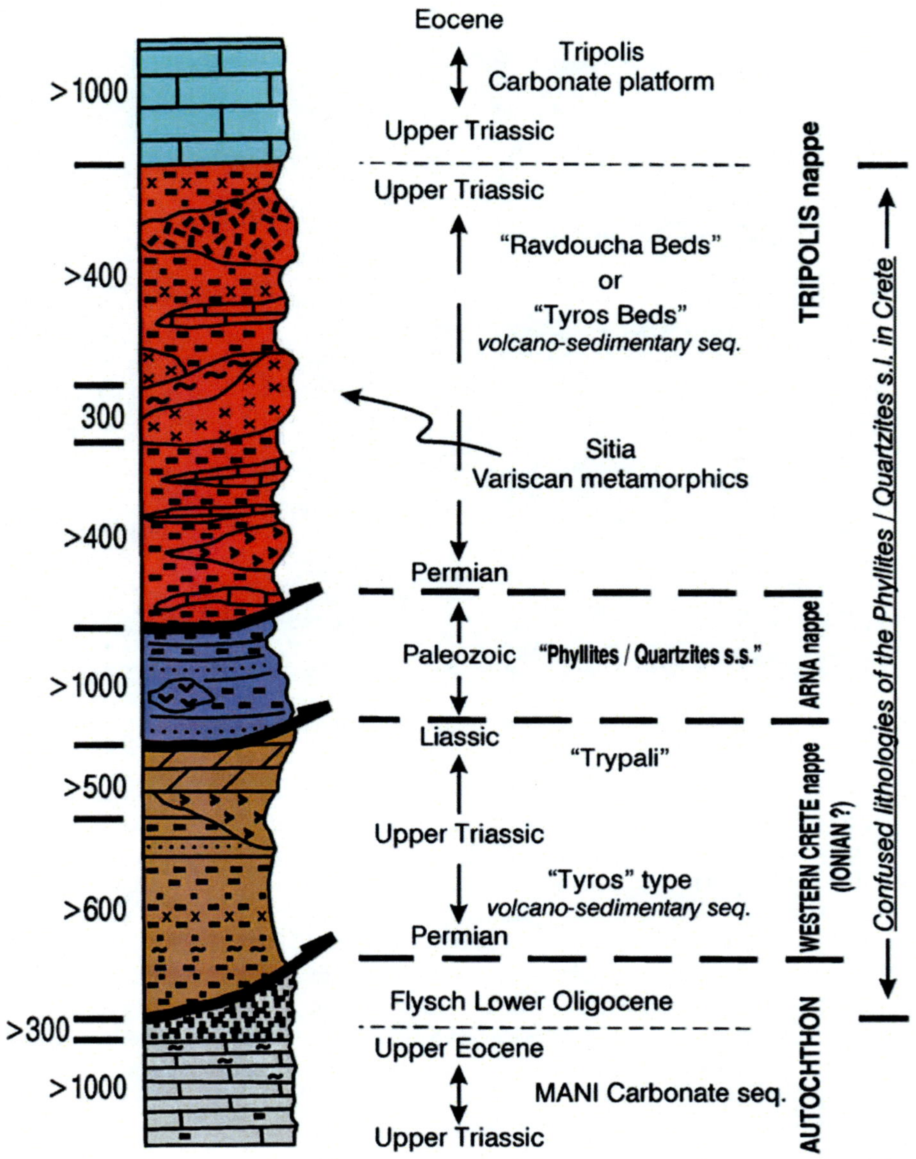

Fig. 8.13 Tectono-stratigraphic column of Crete, resulted by analyzing the units and sequences between the relatively autochthonous metamorphic carbonate sequence of the Mani unit and the base of the non metamorphosed carbonate platform of the Tripolis unit. The nappe of Western Crete between the Mani and Arna units is distinguished, comprising both terms of the «Tyros type» Permo-Triassic volcano-sedimentary complex and the «Trypali» Triassic-Jurassic carbonates (from Papanikolaou and Vassilakis 2010)

Peloponnese, at the areas of Parnon Mt and Gythio, named **Kosmas-Gythio Unit** (Skourtsos and Lekkas 2004). These are crystalline neritic limestones of Upper Triassic–Jurassic age, overlying the metamorphic rocks of the Arna and Mani units

8.1.4 The Ionian Unit

The Ionian unit is traditionally limited to the Ionian Islands, Epirus, Western Sterea and Northwestern Peloponnese (Renz 1940, 1955). From then on, we have to cross the entire arc and reach its eastern edge at the Dodecanese islands, and especially Karpathos and Rhodes, to observe Ionian type units. Such is the **Akramitis** unit, in SW Rhodes Island, where a non-metamorphosed stratigraphic column, comprising the Upper Jurassic–Oligocene is observed, similar to the Ionian unit (Mutti et al. 1970) (Fig. 8.14).

It is interesting to note that in Rhodes Island, the Akramitis unit, resembling to the Ionian, occurs at the southwestern side, while the Lindos unit, resembling to the Mani unit, is observed at the southeastern side (Fig. 8.15). Between the two units, the distinct tectonic unit of **Laerma**, comprising a peculiar chaotic melange formation of Oligocene wild flysch type appears (Mutti et al. 1970). The tectonic peculiarity of the Ionian unit in Rhodes lies in the fact

Fig. 8.14 Outcrop of the Upper Cretaceous pelagic platy limestones of the Akramitis unit (Ionian) in the southwestern Rhodes Island

that it appears behind the Mani unit, in more internal position, contrary to the structure of western continental Greece. There, the Ionian is located in more external position, next to the Hellenides tectonic front, while the Mani unit appears 70–80 km further back in tectonic windows in the Peloponnese and Crete. Additionally, while the Mani unit in Peloponnese and Crete is tectonically overlain by units of metamorphic rocks (Western Crete, Arna), non-metamorphic units are directly overlying the Lindos unit in Rhodes.

It should be noted that all the external units of the Mani, Ionian, Akramitis, Lindos, Karpathos, etc. have similar stratigraphic columns and belong to a wide paleogeographic area of the Hellenides, characterized by the Liassic rifting and the transition from shallow water carbonate platform to pelagic sedimentation of an epicontinental sea. Their differences are found in their present tectonic position and their post-Eocene geodynamic evolution within the Hellenic arc, expressed by the different grade of deformation and metamorphism.

The main feature of the Ionian unit is that its stratigraphic column can be distinguished into three periods, divided by two unique "boundaries": (i) the first change is from neritic to pelagic biochemical sedimentation, during the Late Liassic in the pre-orogenic period, and (ii) the second change is from pelagic carbonate sedimentation to synorogenic clastic sedimentation of flysch during the Late Eocene. In the Ionian unit, typical flysch can be observed, with characteristic transitional beds, several tens of meters thick, during the Late Eocene up to Early Oligocene (Triantaphyllou 2013) (it varies from region to region within the unit). Syn-sedimentary tectonism during the onset of flysch sedimentation has been described in the Retsina region of Aetolia-Acarnania (Papanikolaou and Lekkas 2001). Here, flysch deposition occurs in tectonic grabens, contemporaneously with carbonate pelagic sedimentation in the adjacent tectonic horsts. This phenomenon is well known from the carbonate platforms of Tripolis and Parnassos (Richter and Mariolakos 1972, 1974, 1975), but it seems that it occurred also here, due to the immediate proximity of the particular area of the internal Ionian basin with the neritic carbonate platform of Gavrovo.

The change from the neritic to the pelagic sedimentation stage is chronologically defined accurately at the end of the Lias (Renz 1955). The lower pre-Liassic sequence, comprises neritic limestones, thickly bedded to unbedded, usually dolomitic, with algae, lamellibranches, etc., and gypsum at its base (below the Carnian). The upper sequence comprises pelagic, mainly carbonate, sediments, with or without chert intercalations, with clastic breccia limestones mainly at the upper horizons (Senonian). The succession of the stratigraphic horizons and the tectonic structure of the successive thrusts and folds is observed in the characteristic geological map from Epirus (Fig. 8.16). Transitional features between these two paleogeographic stages are

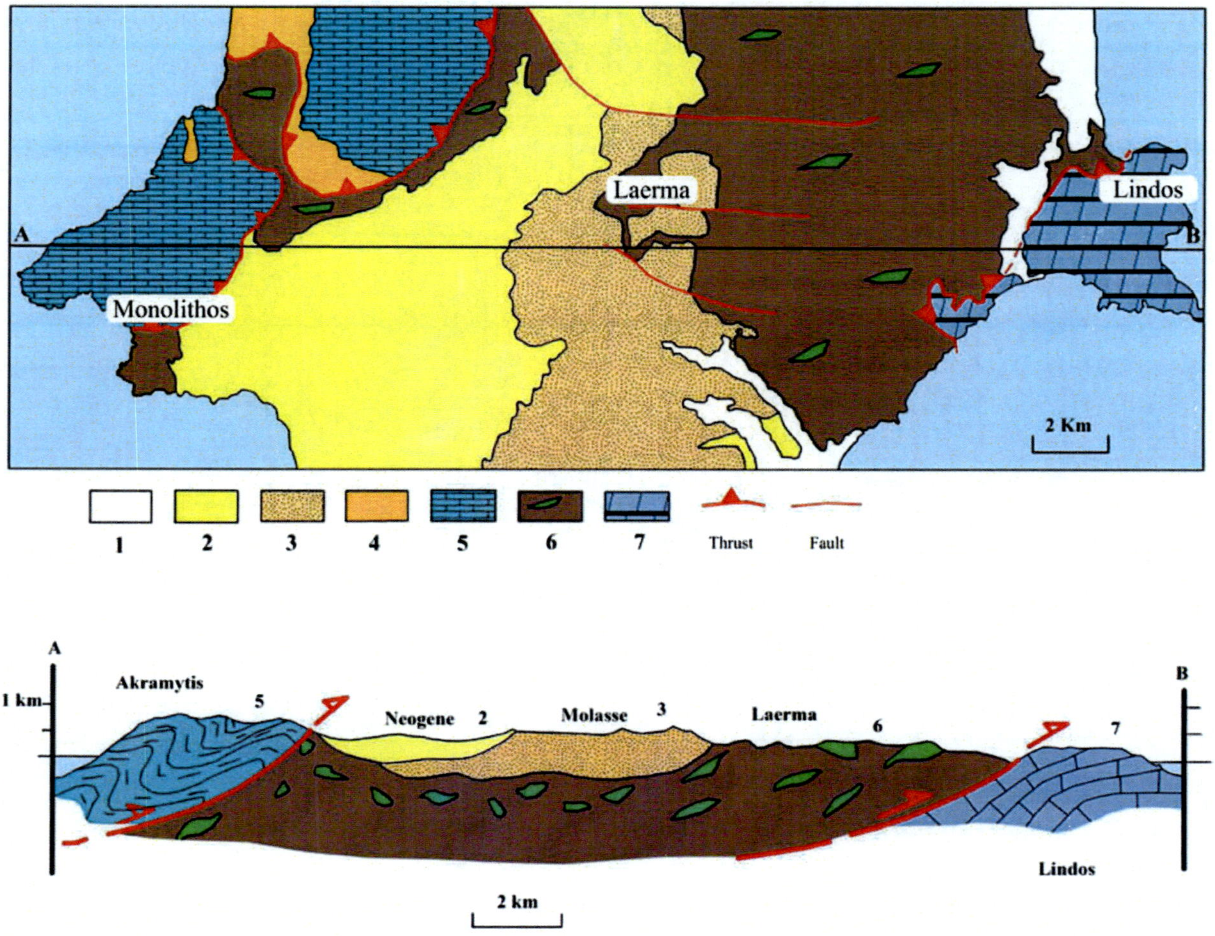

Fig. 8.15 a Simplified and modified geological map of the central section of Rhodes Island, showing the non-metamorphosed Akramitis unit to the west and the metamorphosed Lindos unit to the east (according to Mutti et al. 1970). **b** Geological cross section A-B, with an E-W orientation, showing the presence of the intermediate Laerma unit, of Oligocene age, between the Akramitis/Ionian and Lindos/Mani units. 1: Alluvial, 2: Upper Miocene–Pliocene, 3: Upper Oligocene–Aquitanian molasse, 4: Katavias flysch (Akramitis unit), Lower Oligocene, 5: pelagic limestones with cherts of the Upper Jurassic–Eocene of the Akramitis unit, 6: flysch–mélange formation, Lower Oligocene, of the Laerma unit, 7: crystalline limestones to marbles of the Lindos unit, partly of Cenomanian age

expressed through some rocks of specific facies, observed between the two sequences, such as e.g. red-violet nodular limestones with ammonites (ammonitico rosso facies) or clastic formations, such as the Posidonia shales, as well as siliceous limestones of Dogger–Early Malm (Renz 1955).

Thus, from the paleogeographic point of view the external carbonate platform was from the Carnian to the Late Lias a continuous shallow water carbonate sedimentary environment, comprising the Paxos, Mani, Western Crete, Ionian, Gavrovo, Tripolis and other more internal units up to the Olympus. Since Late Lias-Dogger, the paleogeographic region of the Ionian, Western Crete and Mani units deepens in the subsiding tectonic graben, while Paxos and Gavrovo tectonic horsts on both sides of the graben continue their sedimentation with shallow water carbonates.

The rifting mechanism comprised some syn-sedimentary normal faults, which gradually produced a subsidence of a segment of the previously uniform platform. This change resulted into paleogeographical consequences on the various tectonic segments involving block tilting, local emersions, etc. This spatial diversity resulted in a diversity of sedimentary facies during the transition period. Thus, some areas were uplifted and emerged for some time and then they subsided again. Some other areas that were deeper, were uplifted but without emersion, and then subsided even deeper than before. Thus, between the two main facies, the

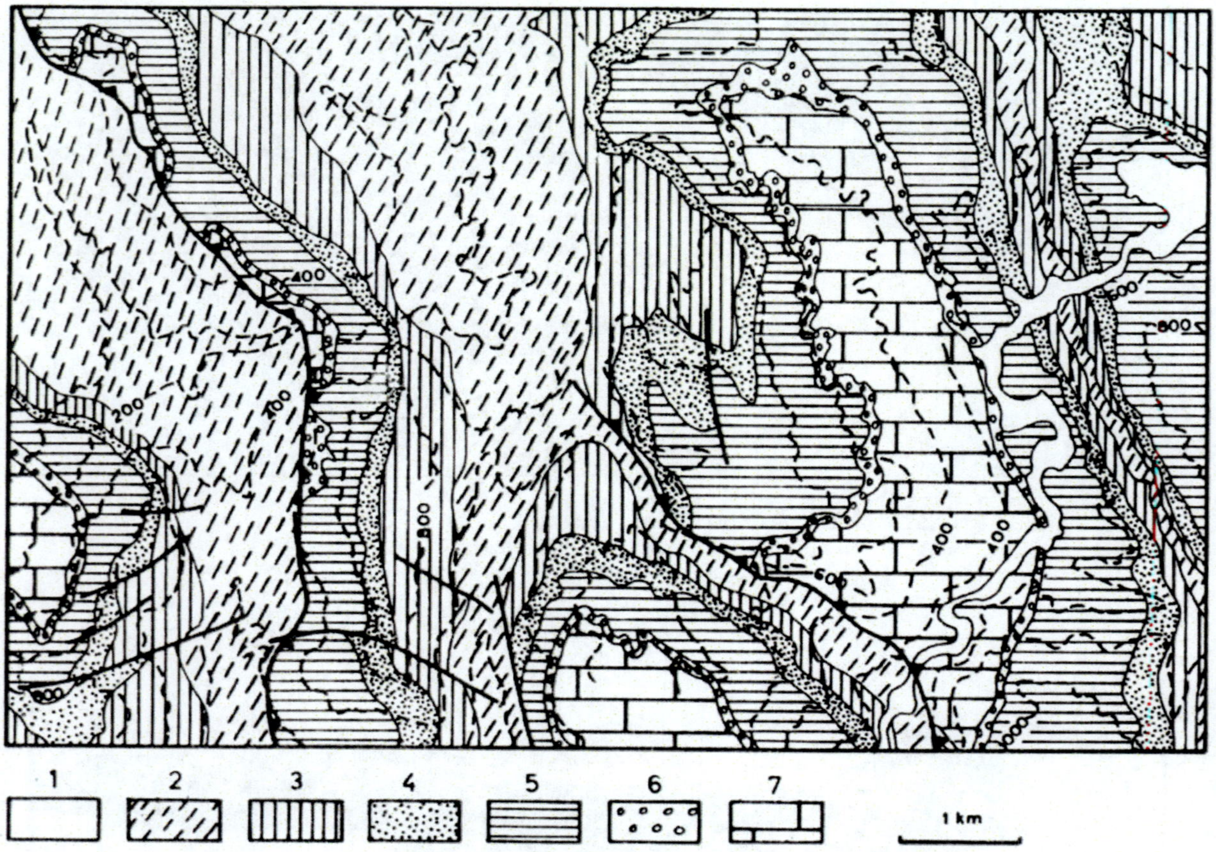

Fig. 8.16 Geological map of the Xiromero area, where the Ionian unit crops out (from the Filiates sheet at scale 1/50,000, Perrier and Koukouzas 1967). 1: Alluvial, 2: undivided flysch, 3: thin layered pelagic limestones with Globigerines and mircobreccia horizons with Nummulites, Alveolines, and chert intercalations, 4: microbreccia limestones, compact, with rudist fragments, with Orbitoides, 5: pelagic limestones with actinozoa and intercalations of cherts–Calpionelles of Tithonian age are found in the lower and Globotruncanes in the upper horizons, 6: shales with Posidonies, with siliceous interlayers of Dogger, 7: thick bedded, compact, fine-grained limestones, with calc-algae of Lower–Middle Lias

neritic and the pelagic, we have a variety of stratigraphic formations or even stratigraphic gaps in certain areas, with the pelagic sequence unconformably overlying the neritic one. There are cases where the stratigraphic hiatus includes almost the entire Jurassic, as in the Perdika region, where platy limestones with Calpionelles are observed overlying directly thick-bedded limestones with Megalodon and Gyroporelles. Elsewhere, the stratigraphic hiatus is shorter, while stratigraphic continuity can be observed in other locations. The relatively short total thickness of the Dogger formations shows that they represent condensation horizons, deposited on steep faulted slopes between the horsts and grabens during the collapsing period.

The Late-Eocene–Oligocene compression tectonics of the arc reactivated some of the older extensional structures of the Middle-Jurassic rifting of the Ionian unit. This is a characteristic case of inversion tectonics that created favourable conditions for petroleum reservoirs (Karakitsios 1995).

The stratigraphic column of the Ionian remains essentially the same, since its definition by Renz (1955), who established various stratigraphic names, from several type localities, for some characteristic Ionian formations, such as:

- Foustapidima limestones, for the black sub-lithographic limestones, overlying the evaporites. They are of Carnian age and contain Cardita.
- Dolomites "Haupt – Dolomit", similar to rock formations of the same age of the Eastern Alps (Lower Norian).
- Pantokrator limestones, with Gyroporella algae, Paleodasycladus, Thecosmilia corals, Stylophyllopsis, Phyllocoenia, Coccophyllum, and other neritic fossils, such as Spiriferina, Terebratulla, Rhynconella, Kokinckina, of the Upper Triassic–Lias.

- Posidonia shales (mainly Posidonomya bronni).
- Ammonitico rosso facies, with various ammonites, Harpoceras, Hildoceras, Lytoceras, Leioceras etc., of Dogger–Lower Malm.
- Vigla Limestones, of Malm–Lower Senonian age, with Calpionelles at the base and Globotruncanes on top, characterized by chert intercalations within the platy pelagic limestones.
- Breccia limestones, of Senonian–Eocene age, in alternation with pelagic limestones and benthic fauna from re-sedimentation (rudist fragments).

A detailed stratigraphic analysis particularly of the Cenozoic formations and tectonic maps of the Ionian unit are included in the reports for oil exploration, carried out by IFP and IGSR (1966) and BP (1971). These reports distinguished the Ionian unit in three zones, external, axial and internal.

8.1.5 The Gavrovo–Pylos Unit

The name of this unit is somewhat confusing because there is also the term Gavrovo–Tripolis, which is widely used in the literature. In fact, the term Gavrovo–Tripolis is a composite interpretative term, assuming that the Gavrovo outcrops in western Greece belong to the same unit as the Tripolis outcrops in central Peloponnese. Although there is strong evidence that Gavrovo and Tripolis units are parts of the same paleogeographic region of the external carbonate platform, their tectonic position and evolution is different. Thus, the Gavrovo and Pylos outcrops are exactly the same, both in terms of stratigraphy (with minor Eocene unconformities and bauxite outcrops) and tectonic position in the front of the Pindos nappe. On the contrary, the Tripolis unit appears further back in the arc, in tectonic windows under the Pindos nappe. In any case, the Gavrovo outcrops in their type locality in Epirus find their counterpart re-emerging in the Klokova and Varassova mts in western Sterea Hellas across Patra, and then again in Pylos in southwestern Peloponnese (Kiskyras 1962, 1972). Thus, the Gavrovo unit may belong to the same carbonate platform as the Tripolis unit with continuous sedimentation from the Late Triassic to the Late Eocene, but it certainly represents the western segment of the platform whereas the Tripolis represents the central segment. Their differences are: (i) a different tectonic position, (ii) the Permian-Triassic base of the stratigraphic column and the tectonic basement of the Gavrovo-Pylos units are unknown, as opposed to the Tripolis carbonates, which are stratigraphically underlain by the Permo-Triassic Tyros beds and tectonically underlain by the Arna and Mani units. iii) the Oligocene–Miocene flysch on the top of the Gavrovo stratigraphic sequence, which seems to have a post-Late Eocene common evolution with the Ionian unit. (iv) the Gavrovo–Pylos unit shows mild tectonism and no metamorphism, while Tripolis shows an intense tectonic structure, with imbrications due to decollements and very low grade metamorphism at its base. Therefore, it is clear that the paleogeographic regions of the Gavrovo-Pylos, and Tripolis were neighboring regions of the external platform during Triassic–Middle Eocene, but from Late Eocene onwards they were tectonically differentiated.

The problem of the uniform flysch, covering both the Ionian unit to the west and the Gavrovo unit to the east, has significant implications concerning the existence of a lateral transition between the Ionian and Gavrovo units in the area between Messolongi and Varassova (Fig. 8.17).

The first possibility is to have a lateral transition between the Ionian and Gavrovo units within 5 km distance. In this case the flysch is indeed common for both units. This is supported by the observation of gradual transition to the flysch, both from the neritic Upper Cretaceous–Eocene limestones of the Gavrovo unit on the mountains of Klokova and Varassova, and from the Eocene pelagic limestones of the Ionian unit in Messolongi–Retsina. The second possibility is to have a buried tectonic thrust of the Gavrovo unit over the Ionian, probably representing an inversion tectonic structure from an initial Late Liassic synsedimentary normal fault bordering the platform from the basin. This thrust could have occurred during the deposition period of the older flysch horizon a, during the Latest Eocene–Early Oligocene, which is present only above the Ionian limestones. On the contrary, the upper flysch formations (horizons b-e) of Late Oligocene–Early Miocene age cover both units.

The Gavrovo–Pylos unit is characterized by continuous neritic carbonate sedimentation from the Late Triassic until the Late Eocene, whereby local unconformities are observed, with a characteristic unconformable deposition of flysch over the Upper Cretaceous limestones on its type locality at the Gavrovo mountain (Fig. 8.18).

This unconformable relation can be observed in several localities in the geological map of the Gavrovo area (Fig. 8.19), where the flysch comes in direct contact over the Upper Cretaceous limestones, without the interference of Eocene limestones. The overall structure also includes a system of longitudinal NNW-SSE trending normal faults affecting the limestone/flysch boundary.

In other locations an unconformable deposition of Upper Eocene limestones is observed over a paleorelief with bauxite material, developed over Middle Eocene limestones, like north of the Riza village of Eastern Klokova and in many localities of Pylos (Kiskyras 1962).

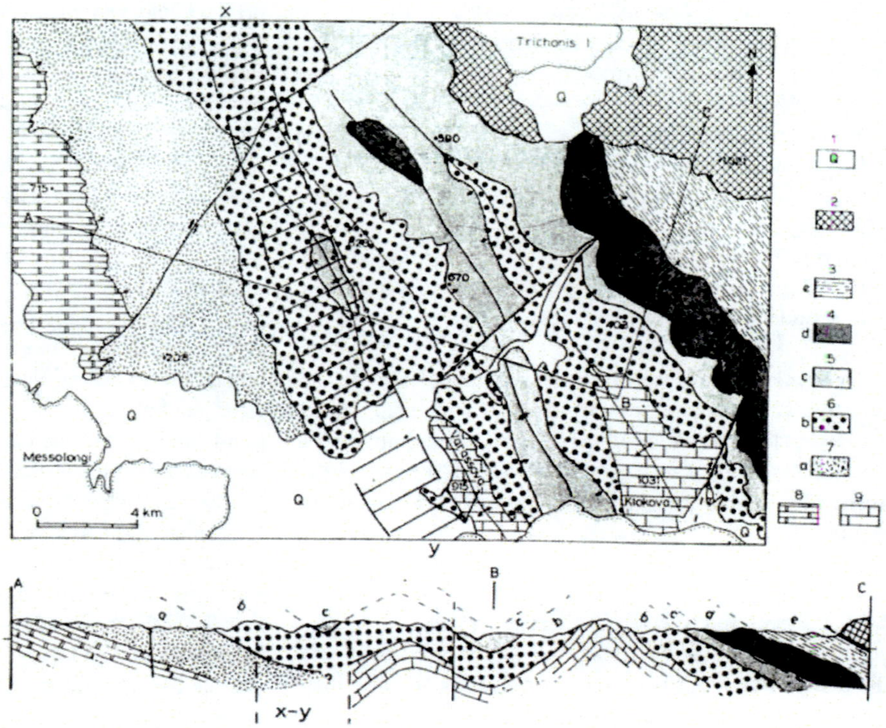

Fig. 8.17 Geological map of the area east of Messolonghi (from British Petroleum Co 1971, simplified and modified) and geological cross section, including the marginal zone between the Gavrovo and Ionian units. The x-y zone shows the possible location of the Gavrovo unit overthrust on the Ionian unit, under the common formation b of the flysch (from Papanikolaou 1986a). 1: quaternary deposits, 2: pelagic sequence of the Pindos nappe, 3-7: a, b, c, d, and e formations of the common flysch of the Ionian–Gavrovo units, 8: Upper Cretaceous–Eocene pelagic limestones with cherts of the Ionian unit, 9: Upper Cretaceous–Eocene neritic limestones of the Gavrovo unit

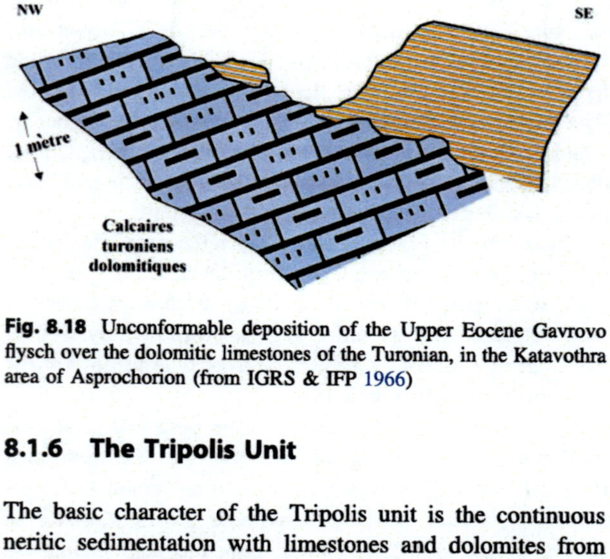

Fig. 8.18 Unconformable deposition of the Upper Eocene Gavrovo flysch over the dolomitic limestones of the Turonian, in the Katavothra area of Asprochorion (from IGRS & IFP 1966)

8.1.6 The Tripolis Unit

The basic character of the Tripolis unit is the continuous neritic sedimentation with limestones and dolomites from the Late Triassic to the Late Eocene, with characteristic black bituminous thick-bedded limestones, known as « Tripolitza kalk» by Philippson (1892). That is, during the previous sedimentation period the sea was very shallow (10–200 m deep) and never acquired pelagic sedimentation. This shallow water carbonate platform, is characterized by algae, corals, molluscs etc., although more complex paleoenvironments were also developed, since there were areas in direct contact with the open sea and others closed, shallow areas forming lagoons with deposition of characteristic stromatolitic horizons (Fig. 8.20).

The nature of this unit allows us to address a major and general problem, ragarding the issue of the sedimentation rate. Within one million years we may have the deposition of 10 m or 50 m thick neritic limestones. Generally, the development of reefs on a carbonate platform with corals, algae, etc., produces a high sedimentation rate. Thus, this area, characterized by a steady shallow sea depth from the Triassic to the Eocene, was actually subsiding, because it is impossible for carbonate sediments 3,000–4,000 m thick to fit into basins 50, 100, or even 200 m deep. Therefore, the subsidence rate of the platform should be compensated by the sedimentation rate, which implies dynamic isostatic equilibrium.

The total stratigraphic column of the Tripolis unit is very simple in terms of lithology, made of a very thick sequence of neritic limestones from the Upper Triassic to the Upper Eocene, as this is shown in geological maps (Fig. 8.21). What naturally changes is the bio-characteristics/fossils. Thus, we can distinguish: Diplopora algae–Megalodon in the Triassic, corals–Clypeines in the Jurassic, Nerinees–rudists in the Cretaceous and finally Nummulites in the Eocene. The lithological monotony with limestones of the same facies, makes mapping really difficult. We need to be able to distinguish stratigraphic horizons and this requires experience both in the field and in paleontological

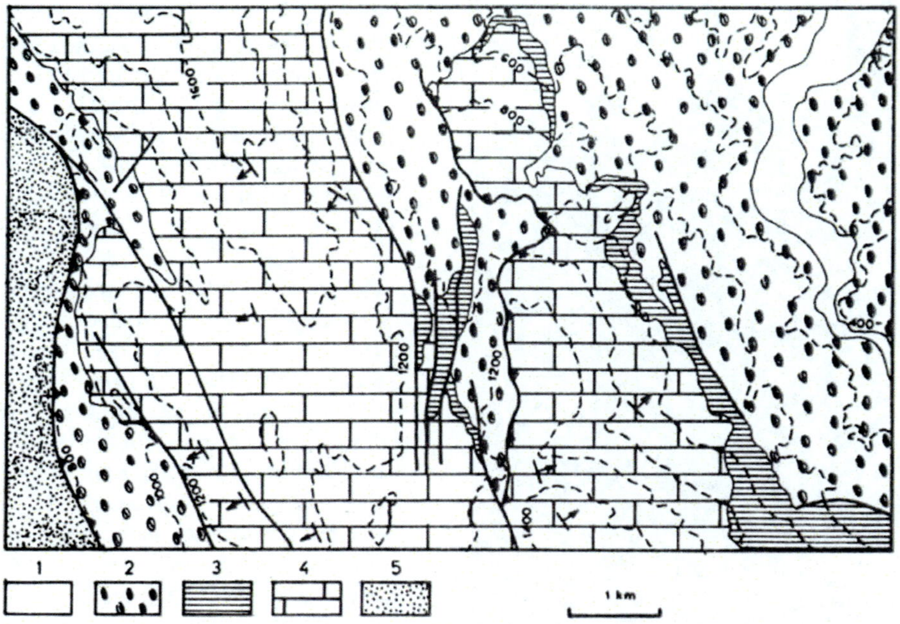

Fig. 8.19 Geological map of the Makrynoros area, where the Gavrovo unit can be observed (type locality, Raptopoulon sheet at scale 1/50,000, by Savoyat et al. 1970). 1: alluvial, 2: flysch, alternations of blue marls and sandstones, platy calcarenites, and polymictic conglomerates, of Eocene–Oligocene age, 3: Eocene limestones, black, sub-lithographic, breccias, reefal, with Asterodiscus, Discocyclina, Microcodium, etc., 4: Cretaceous limestones, undivided, with dolomites, sub-lithographic, breccia limestones, oolithic, with Rudistes, Nerinees, Miliolidae, Orbitoides, 5: Upper Eocene–Oligocene flysch of the Ionian unit

determinations (macro–micro-). An additional problem is the dolomitization, which usually destroys the fossils. On top flysch is present, above the Upper Eocene limestones.

If we focus on the Eocene section of the column, we often observe an unconformity inside the Eocene limestones (Fig. 8.22). In this unconformity small outcrops of bauxites can be found, economically insignificant. The event that created this interruption of the sedimentation process with uplifting, emergence and erosion during middle-late Eocene: (i) removed part of the previously deposited Eocene or even Upper Cretaceous limestones, (ii) created the bauxite deposits and then (iii) subsided and was covered by the transgressive Upper Eocene limestones, followed by (iv) the flysch (Fig. 8.22a). The above process lasted longer in some areas of this huge carbonate platform and shorter or not at all in other areas, depending on the block fault kinematics. Thus, in many locations flysch can be observed directly overlying the Eocene or the Upper Cretaceous limestones, whereas in others there is a continuous stratigraphic succession from the limestones to the flysch with some thin marly transitional beds (Fig. 8.22b). The unconformable deposition of flysch on a paleorelief is usually observed on the footwall of a synsedimentary fault, separated from the continuous conformable transition between the limestones and the flysch observed in the hangingwall. The paleofault surfaces are often characterized by the deposition of hard ground (Richter and Mariolakos 1972, 1975).

We will now examine in more detail: (i) the top and (ii) the base of the stratigraphic column of the Tripolis unit:

(i) As it is well known, the Tripolis unit is currently located under the Pindos tectonic nappe. In the upper flysch horizons immediately under the tectonic contact with the nappe a unique tectono-sedimentary formation of wild flysch is observed. Besides the usual cyclothems with pelites and sandstones, it also contains massive clastic material in the form of conglomerates, breccias and olistholiths. The majority of the blocks originate from the various straigraphic formations of the Pindos nappe (DeWewer 1976; Lekkas 1979) and from other tectonically higher units, coming from the interior of the arc. The creation of the wild flysch is important, as this tectono-sedimentary formation contains the last sediments, deposited inside the flysch sedimentary basins, shortly before being covered by the forthcoming Pindos nappe. The latter during its slow advance, was pushing various blocks on its front, which were falling in the wild flysch and were enclosed inside it, thus dating the Pindos tectonic emplacement. The overall tectono-sedimentary process has been

Fig. 8.20 Characteristic outcrop of neritic limestones with stromatolites of the Upper Triassic at the base of the Tripolis platform from the Rodopou peninsula in Western Crete

described in detail from the famous nappes of the PreAlps of Western Switzerland (e.g. Masson 1976).

(ii) Regarding the base of the Tripolis column the change of views from the 1960s to the 1980s is interesting. Initially, it was believed that beneath the Triassic limestones there was a Paleozoic metamorphic basement, based on the report by Ktenas (1924a) of Carboniferous–Permian fossils found within the underlying formations of the **Tyros Beds**. These beds are dominated by slightly metamorphosed shales and sandstones, comprising also andesitic lavas, pyroclastics and pelagic limestones. Thus, it was proposed that the Tripolis limestones represented transgressive formations above the various metamorphosed rocks, such as phyllites, crystalline limestones, lavas, shales, mica schists and marbles in the deeper horizons, generally representing the Variscan basement (e.g. Dercourt 1964). These views were abandoned when a gradual transition from the lower formations of the Tripolis limestones to the underlying Permian-Middle Triassic volcano-sedimentary complex of the Tyros Beds was described (Lekkas and Papanikolaou 1978). The age of the volcano-sedimentary complex comprises according to new determinations the Carboniferous–Permian up to the Middle Triassic (Thiebault and Kozur 1979). Thus, the Upper Paleozoic–Triassic age of the formations underlying the carbonate platform date the lower horizons of the Tripolis column and not its basement. The Tyros beds are a slightly metamorphosed complex, characterized by tectonic cleavage and recrystallization, with only a few new minerals (Papanikolaou and Skarpelis 1987). The same beds can also be found in Crete (Bonneau and Karakitsios 1979), where they are also known as **Ravdoucha beds** (Sannemann and Seidel 1976; Kopp and Ott 1977) described from the Tripolis base in the Rodopou peninsula. In fact, these beds were deposited during the rifting stage of the H1 terrane, when the Tripolis unit was still together with the other H1 units in the Gondwanian–African margin. The age of the onset of the drifting stage of H1 and the beginning of the shallow water carbonate sedimentation is Carnian, based on dating of algae from the deeper carbonate forizons of the platform and while there is still interlayering of volcanic tuffs within the stromatolites (Fig. 8.23).

The Carnian age of the transition from the rifting to the drifting stage of H1 and the establishment of the shallow carbonate platform is characteristic of H1, as opposed to that of the internal platform of the Hellenides H3, where the transition takes place much earlier, in the Late Scythian–Anisian (see also Fig. 8.2).

The metamorphosed rocks, found below the Tyros beds of the Tripolis unit, belong to the metamorphic tectonic unit of Arna, while the marbles, previously considered as the oldest formations of the Paleozoic basement, are proved today to be of Triassic–Eocene age, belonging to the Mani unit.

In the Dodecanese, there are sporadic outcrops of units with similar characteristics as those of the Tripolis unit, such as for example the **Archangelos** unit in Rhodes (Desio 1931), where the corresponding Tyros Beds were also reported (Lekkas et al. 2001), as well as the **Zia** unit in Kos (Papanikolaou and Nomikou 1998). In some cases, the Tripolis unit is highly

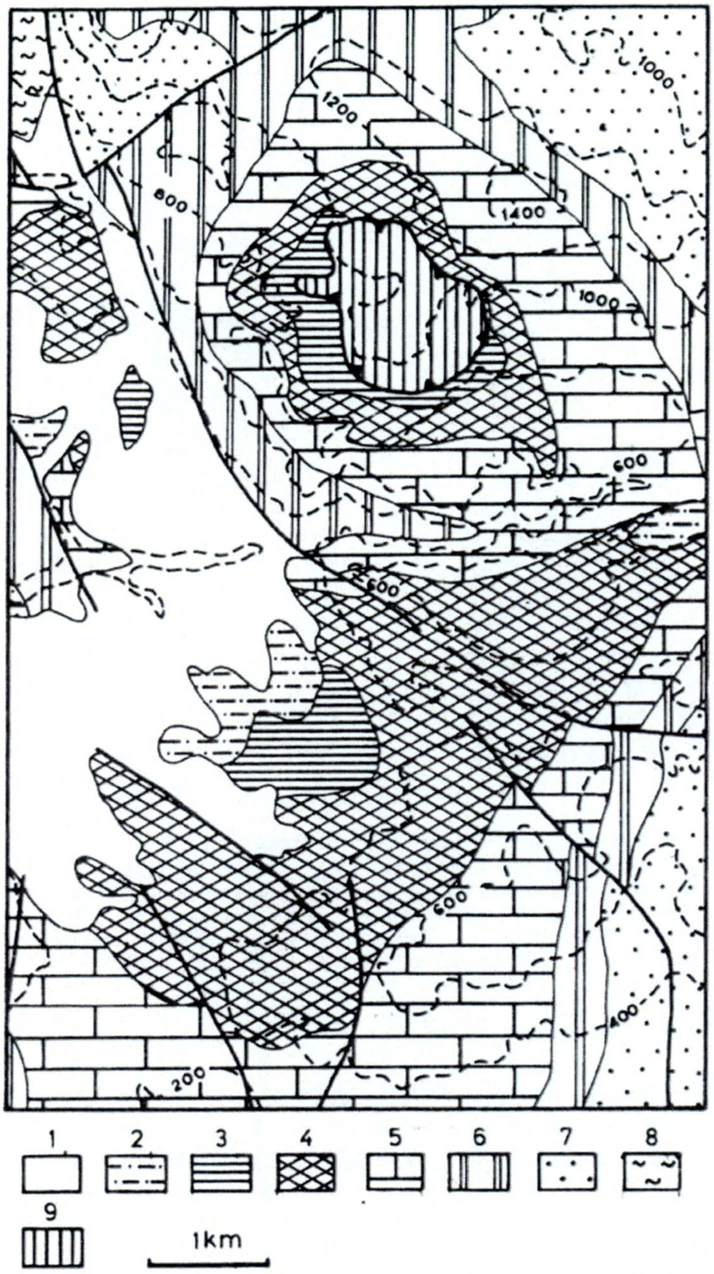

Fig. 8.21 Geological map of the Western Taygetus Mt, where the Tripolis unit can be observed with almost all the horizons of the carbonate platform (from the Kardamyli sheet, at scale 1/50,000, Psonis and Latsoudas 1982). 1: Quaternary, 2: Pliocene with marls and sandstones, 3: flysch, 4: bituminous black limestones with Nummulites, 5: gray to black limestones with Rudists, 6: dolomites and thick-bedded limestones with Orbitolina, 7: limestones and dolomites of Late Triassic–Jurassic, 8: shales with intercalations of crystalline limestones with conodonts of Triassic age and tuffs and lavas of andesitic composition (Tyros beds), 9: pelagic platy limestones with Globotruncanes of the Pindos unit

deformed as in Astypalaea Island (Marnelis and Bonneau 1979). Additionally, the **Prophitis Ilias** unit in Santorini has characteristics of the Tripolis unit with age determinations of Upper Triassic for the Carbonate platform (Papastamatiou 1958) and Eocene for the flysch (Tataris 1965). More recently, a Pelagonian origin of the Prophitis Ilias unit has been proposed (Schneider et al. 2018) because: (i) Jurassic-Cretaceous fossils have not been found on top of the Triassic carbonates and (ii) an unconformable stratigraphic contact of the flysch over the limestones was suggested. However, the stratigraphic or tectonic nature of the contact of the Eocene flysch is in question and unconformable Eocene flysch on the Triassic limestones is not compatible with the structure and geological history of the Pelagonian domain.

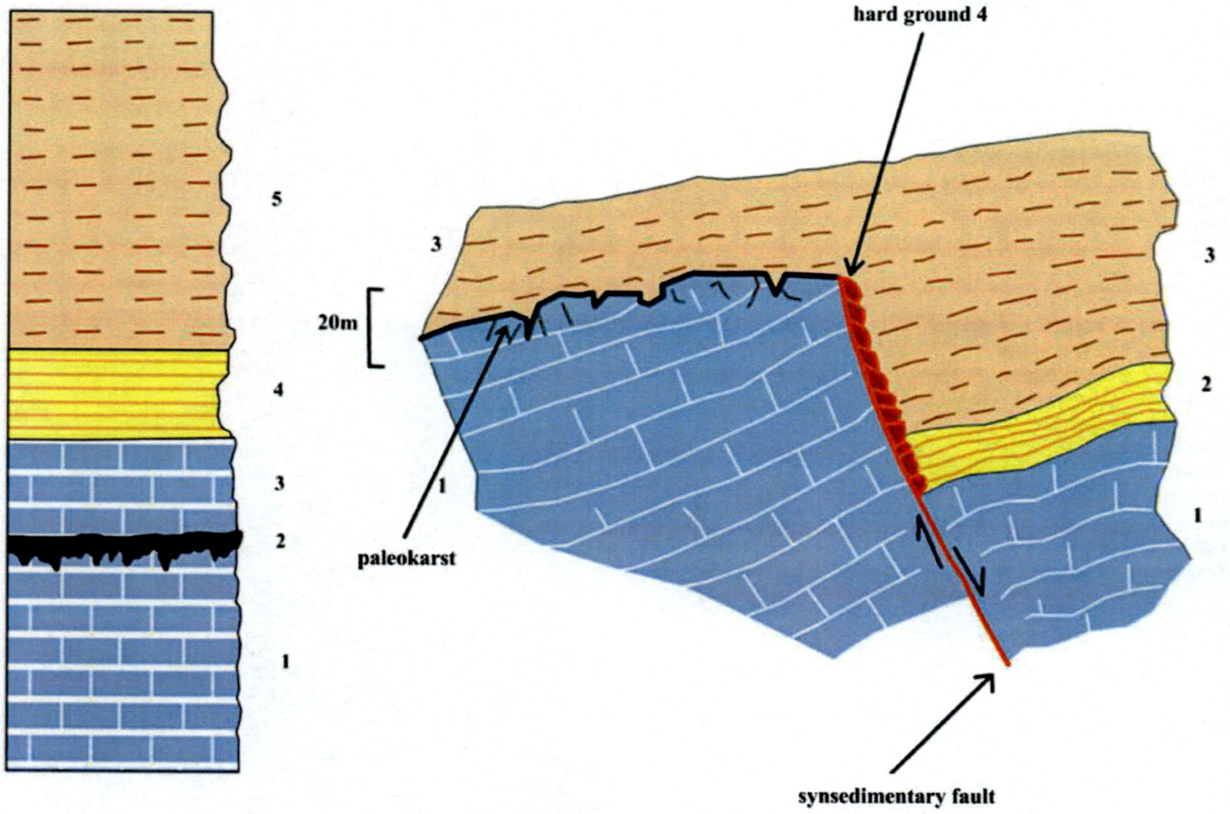

Fig. 8.22 Two cases of transition from the Tripolis limestones to flysch. **a** gradual transition from the limestones (3) to the flysch (5) through transitional marly beds (4), with a possible prior unconformity of the Eocene limestones (3) and bauxite deposition (2) over the Upper Eocene limestones paleorelief (1). **b** unconformable deposition of the flysch (3) over the Eocene or older limestones (1). Paleofault surfaces of syn-sedimentary faults with phosphate-iron encrustations (hard ground) can be observed (4), separating the uplifted fault block with the unconformity at the footwall from the subsided block in the hangingwall, where a stratigraphic continuity is observed, with the presence of transitional marly beds with Globigerines (2)

8.1.7 The Amorgos Unit

The Amorgos unit lies at the southeastern edge of the Cycladic metamorphics and has been studied by Renz (1955), Durr et al. (1978), Fytrolakis and Papanikolaou (1981), Minoux (1981), Rosenboum et al. (2007) and Chatzaras et al. (2011). This unit comprises the entire Amorgos island (excluding Nikouria islet) and the smaller islands of Keros, Antikeros, Levitha and Kindaros.

Its stratigraphic column, according to Fytrolakis and Papanikolaou (1981), is quite unusual resembling the Ionian unit (the lower part) and the Tripolis unit (the upper part) (Figs. 8.24 and 8.25). At its base, there are dolomites and limestones of the Upper Triassic with Megalodon (Kryoneri formation, mr1). In the deeper dolomite horizons thin intercalations of violet-colored tuffs are observed, similar to those found at the base of the Tripolis carbonate platform directly above the Tyros beds. Above them there is a schistosed–slightly metamorphic complex of clastic rocks, with some tuffs in various locations, and with large olistoliths of Jurassic fossil-bearing limestones, as well as massive conglomerates (Katapola formation, sch1). Above them slightly crystalline platy limestones are observed, with cherts and Silex (especially at the middle part), whose age should be Upper Jurassic to Lower Cretaceous (Chozoviotissa formation, mr2). These pelagic facies platy limestones closely resemble the Vigla formation of the Ionian unit and even more the respective formation of the Mani unit (Plattenkalk), due to its isoclinal folding and slight metamorphism. Above them follows a clastic formation of phyllites with small thickness of 10–20 m (Potamos formation, sch2) and afterwards a very thick formation of crystalline neritic limestones, in which occurrences of meta-bauxites have been exploited until the 1950s (Krikela formation, mr3). The stratigraphic sequence ends with typical flysch of Upper Eocene age (Tholaria formation, sch3).

Fossils are present in the Upper Triassic Kryoneri formation, in the Jurassic Katapola formation, in the Upper

Fig. 8.23 Transitional beds of the volcano-sedimentary Tyros/Ravdoucha beds in Central Crete towards the base of the Tripolis carbonate platform in the Carnian

Cretaceous Krikela formation and in the Upper Eocene flysch of the Tholaria formation (Fytrolakis and Papanikolaou 1981). This stratigraphic column shows features that can be found in the Lycian nappes in Asia Minor and not in continental Greece (Bernoulli et al. 1974). It is characteristic that during the Late Jurassic–Early Cretaceous, the same pelagic sedimentation of the Vigla limestone facies, of the Ionian unit can be observed in the Chozoviotissa formation of Amorgos. On the contrary, during Late Cretaceous–Eocene, neritic characteristics with carbonate rocks similar to those of the Tripolis unit can be observed in the Krikela formation. At the lower part of the column, the transition from the carbonate platform sedimentation to the pelagic sedimentation is taking place in Dogger, as in the case of the Ionian unit. However, a major difference with the Ionian Unit concerns the absence of the transitional facies of ammonitico rosso, posidonian shales, and siliceous pelagic limestones, which are substituted in Amorgos Island by a much more important coarse clastic formation (Katapola formation), indicating strong syn-sedimentary tectonism, with remarkable emersion and erosion.

The Amorgos unit is slightly metamorphosed under HP/LT conditions, at about 6 Kbars and 320 °C (Minoux 1981). The structure is purely Cycladic, with two deformation phases, with isoclinal folds etc., but not as prominent as in the typical Cyclades (Fytrolakis and Papanikolaou 1981). In any case, it comprises a continuous stratigraphic column, tectonised after the Late Eocene and it certainly belongs to the External Carbonate Platform H1. It should be noted that when the Cycladic units underwent blueschist metamorphism at depths of 40–60 km in the subduction zone, Amorgos was still at the surface, with deposition of Nummulite-bearing neritic limestones and flysch.

Within the thrust sheets of the Katapola region two small thrust sheets of HP/LT metamorphic rocks have been found (Rosenbaum et al. 2007): (i) a thrust sheet of a metabasite unit, with parageneses showing 500–600 °C and more than 13 kbars and (ii) a thrust sheet of a conglomerate, with parageneses of 300–450 °C and 10–14 kbars. The same outcrops have been studied later also by Chatzaras et al. (2011) who have concluded that the blueschists were primarily tectonically emplaced above the Amorgos carbonate platform, similarly to the Cycladic blueschists above the Almyropotamos unit. This conclusion confronted the previous proposal of Rosenbaum et al. (2007) that the blueschists were the basal formations beneath the Amorgos carbonate platform.

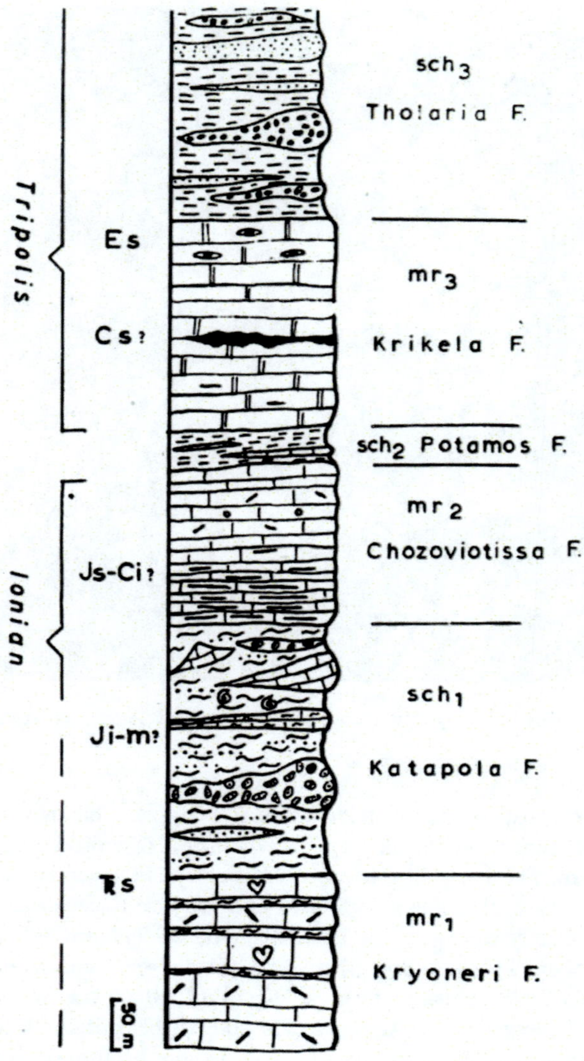

Fig. 8.24 Schematic stratigraphic column of the Amorgos unit (from Fytrolakis and Papanikolaou 1981)

8.1.8 The Olympus–Almyropotamos–Kerketeas Units

These are three homologous units, with different outcrop sites, but in approximately the same tectonic position, in the internal boundary of the medial tectono-metamorphic belt, in the form of tectonic windows. They are also homologous based on their litho-stratigraphy, their macro- and micro-tectonic structure, and their metamorphic grade. They appear in Olympus and Ossa (Godfriaux 1968; Derycke and Godfriaux 1978; Katsikatsos et al. 1982; Schmitt 1983; Schermer 1990), Almyropotamos and Marathon (Argyriadis 1967; Katsikatsos 1969, 1979; Katsikatsos et al. 1976a, b; Lozios 1993), as well as in Kerketeas (Western Samos) and Korakas (Fourni Islands) (Papanikolaou 1979a, 1986a, b; Roche et al. 2019).

They are characterized by: (i) a stratigraphic column exclusively made of slightly metamorphosed limestones and dolomites from the Triassic up to the Eocene, with neritic carbonate sedimentation (Godfriaux 1962, 1968; Katsikatsos 1969) (Fig. 8.26), and (ii) possible occurrence of unconformities, which have been reported in Almyropotamos (Katsikatsos 1979) and Olympus (Schmitt 1983). Thus, the Eocene has been documented with Nummulites below the flysch and above rudists bearing limestones of the Upper Cretaceous, which probably lie unconformably over the Upper Triassic–Liassic limestones with Megalodon and algae. This implies the existence of a possible stratigraphic hiatus of the Late Jurassic–Early Cretaceous period, which has not yet been identified with fossils. Especially at the Almyropotamos unit, there is an unconformable deposition of flysch over the Eocene and Upper Cretaceous limestones (Dubois and Bignot 1979), just like in the Gavrovo and Tripolis (Fig. 8.27). At Kerketeas, meta-bauxite occurrences have been reported within the thick carbonate sequence (Mposkos 1978), possibly linked to the aforementioned unconformities.

These three units are the deepest –tectonically- known formations in the region of the "former Pelagonian". Towards the interior of the arc they do not reappear, because they are buried under the Internal Hellenides nappes, while eastwards along the medial belt, next to Samos Island, the same Kerketeas marbles re-emerge in the so-called Meander massif (Menderes). The Mesozoic nappe of the Northern Menderes is possibly relative to the Kerketeas unit, which is geotectonically equivalent to the Almyropotamos and Olympus units (Papanikolaou and Demirtasli 1987; Ring et al. 1999). Thus, we have the successive tectonic structures of: (i) the Olympus and Ossa window, (ii) the Almyropotamos and Marathon window, (iii) the western Samos and Korakas/Fourni window, and then (iv) a huge tectonic window, which comprises the entire Menderes massif, where Mesozoic marbles are observed peripherally all along its northern border, whereas at its core there is a pre-Alpine crystalline complex with granites, gneisses, etc. Above all these tectonic windows, the Cycladic blueschist units can be observed, such as in the typical outcrops of the Samos Island (Figs. 8.28 and 8.29).

The Olympus–Ossa tectonic window was later extended with two more significant outcrops in the **Rizomata** region in Pieria (Fig. 8.30) (Kilias and Mountrakis 1985) and the **Krania** region in Elassona (Kilias et al. 1991b).

The tectonic structures of the main outcrops in Olympus and Krania form tectonic windows that have been interpreted as the result of multiphase extensional tectonics with low angle normal faults, followed by high angle normal faults

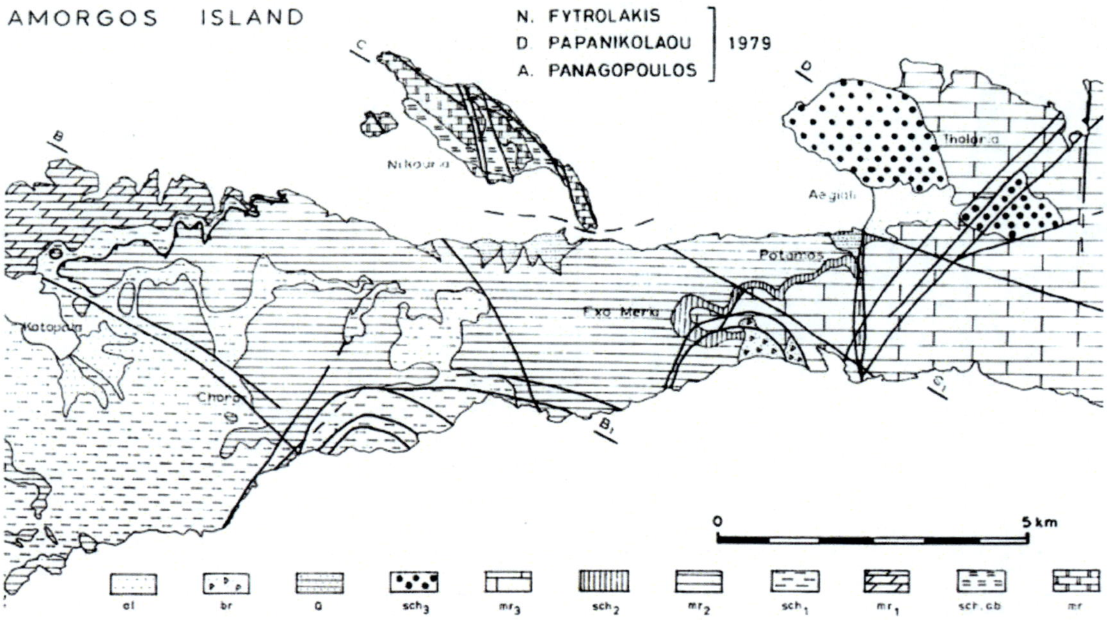

Fig. 8.25 Geological map of the central part of Amorgos island, where all the stratigraphic horizons are present: mr1 Kryoneri formation, sch1 Katapola formation, mr2 Chozoviotissa formation, sch2 Potamos formation, mr3 Krikela formation, sch3 Thollaria flysch. Sch-ab and mr are amphibolites and marbles of Nikouria, while al, br, and Q are alluvial, debris, and sandstones-conglomerates of marine terraces, respectively

after the initial creation of the tectonic nappe pile (Kilias 1996) (Fig. 8.31).

The metamorphism of these units is generally difficult to be determined due to the carbonate nature of their rock formations. However, in the Almyropotamos and Kerketeas units pressures and temperatures showing a moderate subduction at a depth of 20–30 km have been suggested, with pressure values of 7–10 Kbars and temperatures of 350–400 °C (Shaked et al. 2000; Chen 1995).

8.1.9 The Attica Unit

This unit is the relatively autochthonous, tectonic unit of Attica, overlain by other tectonic units, both metamorphic and non metamorphic, such as the Eastern Greece, Lavrion, and Almyropotamos units. The non-metamorphic nappe of the Eastern Greece unit is observed along the southern slopes of the Aigaleo, Poikilo and Parnitha mountains, north of a tectonic contact of ENE-WSW orientation. This contact is an extensional detachment fault which may be followed in Southern Evia, between Aliveri and Oktonia peninsula and further to the northeast in Skyros Island, where it becomes a right-lateral strike slip fault (Papanikolaou and Royden 2007). The normal throw component of the tectonic contact is several kilometers, if we consider the absence of intermediate formations between the non-metamorphosed rocks of the Eastern Greece unit and the underlying metamorphics of Attica. The contact of the Attica unit with the Almyropotamos unit is oriented NW-SE in the Marathon area, where the superposition of the Upper Cretaceous marbles of the Almyropotamos unit over the formations of the Attica unit is observed (Katsikatsos 1979; Lozios 1993). On the contrary, the nappe of the Lavrion–Athens Allochthon is observed in scattered outcrops, both inside the Athens basin (Papanikolaou et al. 2004b) and in the Lavrion area (Marinos and Petrascheck 1956).

The Attica unit is metamorphosed and intensively deformed, with primary ductile structures in a NE-SW orientation and younger ones in a NW-SE orientation (Mariolakos 1971; Mariolakos and Papanikolaou 1973). It consists of a thick marble sequence, usually dolomitic, overlain by mica, amphibolitic, etc. schists, with fine horizons of intermediate pelagic marbles. Within the mica schists, mafic-ultramafic metamorphic rocks are observed. The stratigraphic column remains relatively the same as determined by Lepsius (1893), in the Hymettus mt area, with: (a) Vari Schists at the base, (b) Pirnari Dolomites, (c) Lower Marble, (d) Kessariani Schists, (e) Upper Marble.

The age of the marbles is partially Upper Triassic- Lower Jurassic (Marinos and Petrascheck 1956), based on algae, corals, and lamellibranches, which have been sporadically found in the so-called Lower Marble. A more detailed stratigraphic analysis has not yet been conducted and there

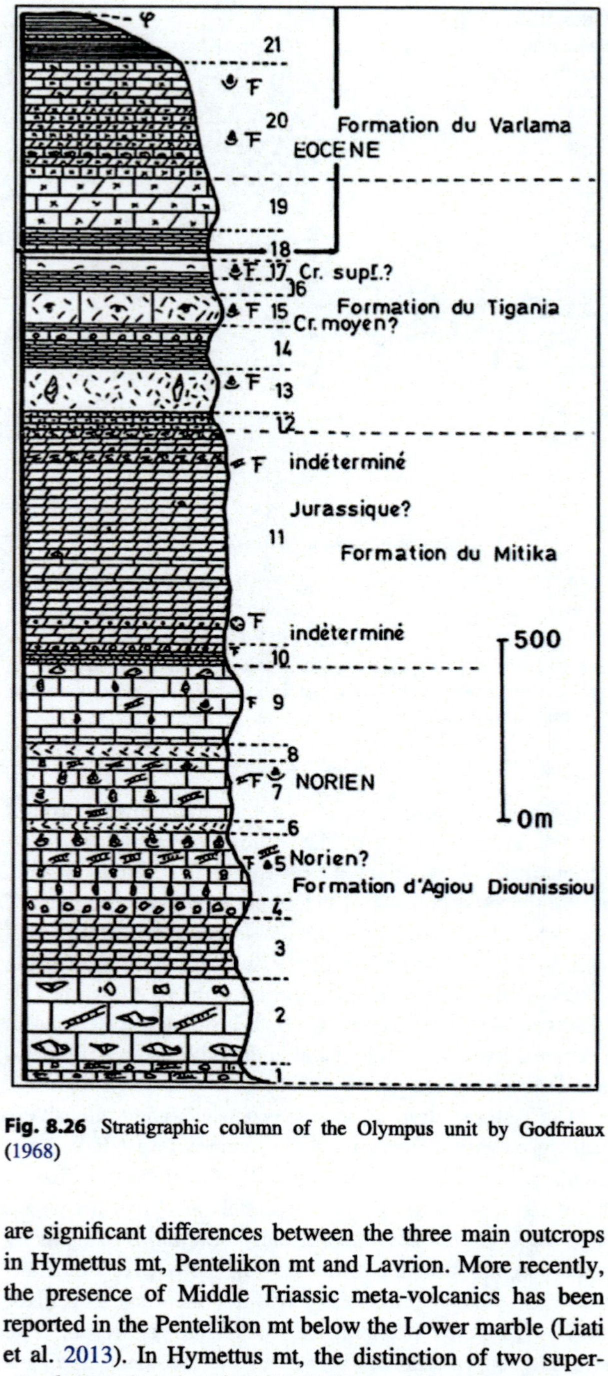

Fig. 8.26 Stratigraphic column of the Olympus unit by Godfriaux (1968)

are significant differences between the three main outcrops in Hymettus mt, Pentelikon mt and Lavrion. More recently, the presence of Middle Triassic meta-volcanics has been reported in the Pentelikon mt below the Lower marble (Liati et al. 2013). In Hymettus mt, the distinction of two superposed tectonic sub-units with Triassic carbonate sedimentation has been suggested (Lekkas and Lozios 2000) (Fig. 8.32).

It is evident that the intergration of the Attica unit in the Hellenides is problematic, due to the fact that the age of the uppermost beds of its stratigraphic column is not known. Thus, it could belong either to the external platform H1 or to the internal H3; that is to say, it could be analogous either to the Olympus–Almyropotamos–Kerketeas units or to the Almopia unit. In the first case the Kessariani Schists should represent the Tertiary metaflysch, with some tectonic duplication of the stratigraphic column between the Lower and the Upper marble, whereas in the second case the Kessariani Schists should represent the so-called «schist-chert formation» of Upper Jurassic-Lower Cretaceous age and the overlying «Upper Marbles» should represent the Upper Cretaceous transgression.

Nevertheless, its tectonic position inside the large composite Attica–Cyclades tectonic window, its structural characteristics with early folding and lineations in the NE-SW direction, similar to the Northern Cyclades unit and the Almyropotamos overthrust in particular, do not allow it to be regarded as an internal unit. On the contrary, its position in the external platform is more favorable in amore external position than the Olympus–Almyropotamos–Kerketeas units, probably between the paleogeographic regions of Amorgos and Tripolis.

8.1.10 The Arna Unit (Phyllites–Quartzites)

The Arna unit includes the medium grade metamorphic rocks of blueschist type of the external tectono-metamorphic belt of Peloponnese–Crete, which are tectonically interlayered between the Mani and Tripolis units (Skarpelis 1982; Papanikolaou 1984c; Papanikolaou and Skarpelis 1987) (Fig. 8.33).

The blueschist facies metamorphism is of Oligocene–Early Miocene age (Seidel et al. 1982) and was formed at pressures of about 7 Kbars and temperatures of 350 °C (Skarpelis 1982), without having undergone a subsequent retrograde greenschist facies. More recent data have shown pressure values up to 9–14 Kbars and temperatures between 320 and 500 °C from various outcrops in Northern and Southern Peloponnese (Trotet et al. 2006; Jolivet et al. 1996). The large differences of the estimated P/T conditions of the blueschist metamorphism have been attributed to the different rates of the exhumation process of the unit towards the surface, without any thermal effect from ascending magmas.

Intense deformation comprising at least three phases, with the first two characterized by ductile isoclinal folding, are observed with the early structures oriented ENE-WSW transverse to the arc, probably representing a-kinematic type structures (Papanikolaou 1981a, 1984c; Papanikolaou and Skarpelis 1987) (Fig. 8.34).

In the Peloponnese, Arna unit has been confused with the overlying Permian–Triassic Tyros beds or with the underlying Oligocene flysch of the Mani unit, despite the fact that

Fig. 8.27 Outcrop of the top of the Eocene marbles of the Almyropotamos unit under the phyllites of the meta-flysch in the Koskina region. Nummulites have been found along the contact at the base of the flysch

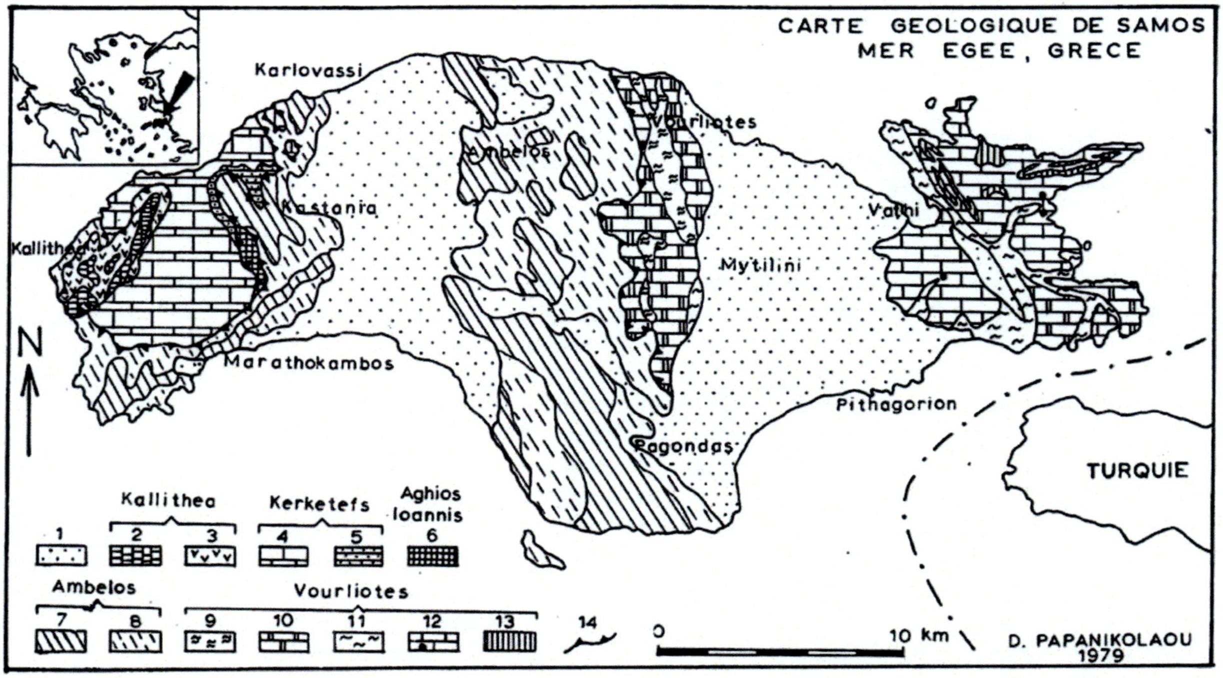

Fig. 8.28 Schematic geological map of Samos Island, showing the relatively autochthonous Kerketeas unit, under the tectonic nappes of the blueschist bearing Ag. Ioannis, Ampelos and Vourliotes units, as well as the non-metamorphic Kallithea nappe (from Papanikolaou 1979a) (see also the schematic tectonic section of Samos in Fig. 8.76). 1: sedimentary deposits and volcanics of Neogene, 2: Upper Triassic–Jurassic limestones, 3: spilites, diabases, radiolarites, and pelagic limestones of the Middle Triassic, 4: Kerketeas marbles, 5: Kerketeas phyllites (metaflysch), 6: metamorphic mafic igneous rocks, 7: Ampelos marbles, 8: Ampelos schists, 9: lower Vourliotes schists, 10: lower Vourliotes marbles, 11: intermediate Vourliotes schists, 12: upper Vourliotes marbles of Upper Cretaceous, 13: upper Vourliotes schists (meta-flysch)

Fig. 8.29 View of the tectonic window of Kerketeas (Ke) in Western Samos Island below the blueschists bearing Ambelos nappe (Amb)

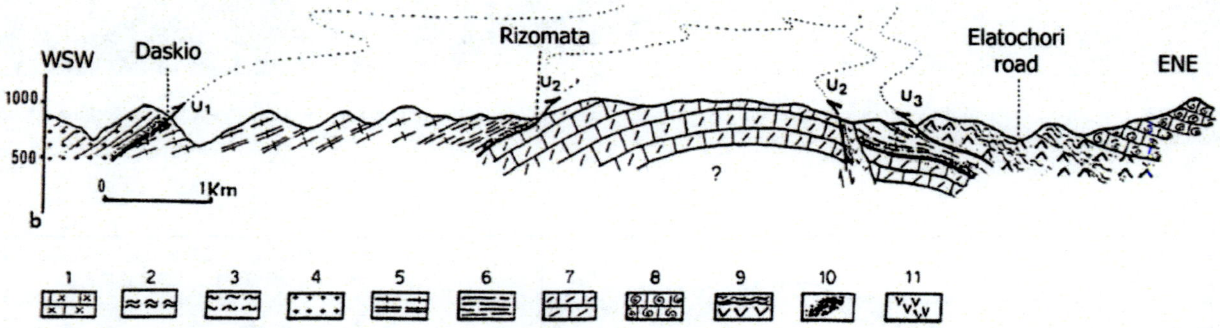

Fig. 8.30 Geological cross section of the Rizomata tectonic window in Pieria (modified from Kilias and Mountrakis 1985). 1: Triassic–Jurassic marbles, 2: garnet mica schists, 3: amphibolitic schists, 4: gneissic granite, 5: paragneisses and amphibolites, 6: glaucophane schists, 7: limestones of the Olympus–Ossa window, 8: Upper Cretaceous limestones, 9: calc-phyllites and metabasites, 10: mylonites, 11: post-Alpine volcanics

those two formations have a different lithology, metamorphic grade and deformation style (Lekkas and Papanikolaou 1978; Papanikolaou and Skarpelis 1987).

The Arna unit is often bibliographically referred to as Phyllite–Quartzite unit, due to the fact that it is usually mixed with the Tyros beds, which contain phyllites. This name is inadequate (like the Plattenkalk case) because: (i) they are not phyllites from a petrological point of view, (ii) it does not cover all the characteristic formations, such as the meta-basalts, and (iii) it does not refer to a specific type locality, so that comparisons could be made. Arna, on the other hand, is a type locality in the slopes of Eastern Taygetus Mt, with characteristic lithologies of meta-basalts, meta-tuffs, meta-conglomerates, meta-pelites and quartzites. In other locations of Northern Taygetus Mt. ultra-mafic rocks are also observed as well as marbles in the form of olistoliths (Skarpelis 1982) (Fig. 8.35).

The main problem is the confusion of the Arna unit, which contains the true metamorphic blueschist rocks, with other metamorphic formations, either of a lower grade, or just schistosed, which occur between the underlying Mani marbles and the overlying Tripolis limestones (see also Fig. 8.13). In the case of Crete, the "phyllites-quartzites" may include (Papanikolaou and Vassilakis 2010): (i) the phyllitic and meta-sandstone rocks of the Oligocene metamorphosed Mani flysch, (ii) the Permian-Triassic phyllites, carbonates, and meta-tuffs of the Tyros type of the Western Crete unit (together with the raywackes and the gypsums under the carbonates of the Trypali unit), (iii) the metamorphosed blueschists of the Arna unit, with the presence of all the characteristic formations but with particularly impressive outcrops of quartzites (Fig. 8.36), (iv) the Permian-Triassic of the volcano-sedimentary complex of the Ravdoucha area—equivalent of the Tyros beds—in which

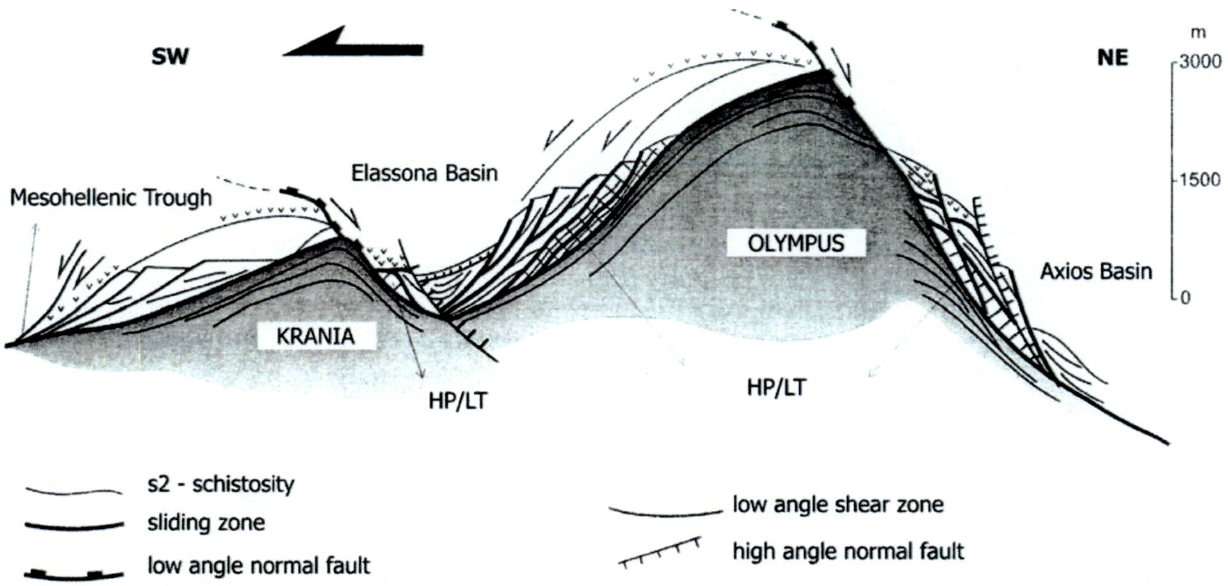

Fig. 8.31 Tectonic schematic representation depicting the collapse of the tectonic nappes and the creation of the Olympus and Krania tectonic windows through successive extensional structures (modified from Kilias 1996)

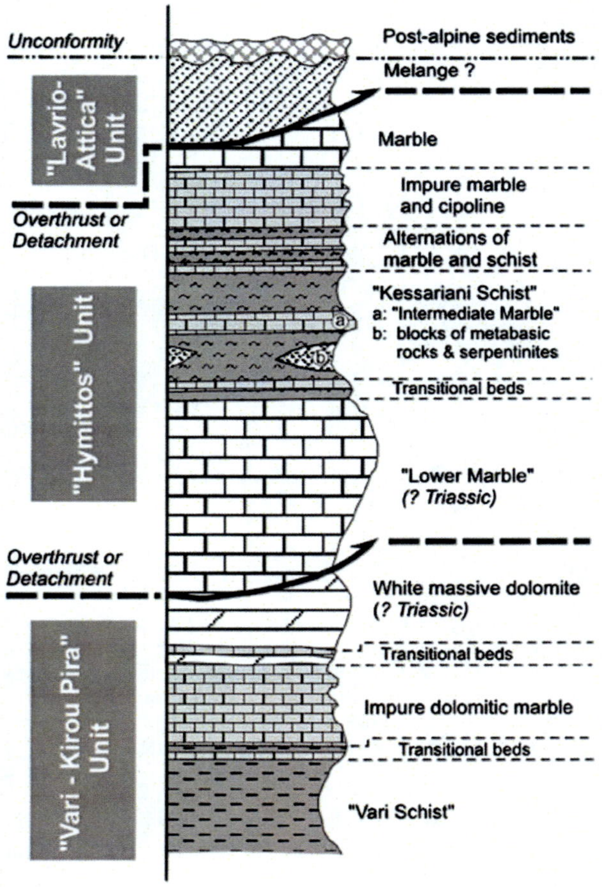

Fig. 8.32 Schematic lithostratigraphic column of the metamorphic formations of Hymettus mt and distinction of two possible tectonic sub-units of Vari–Kyrou Pyrra and Hymettus (from Lekkas and Lozios 2000)

outcrops of gypsum can be also observed, at the base of the Tripolis platform as well as (vi) the pre-Alpine metamorphosed rocks of the Sitia unit. Unique cataclastic rocks known as corngeule are observed in several locations, due to hydraulic fracturing along the contact of the Tyros beds with the underlying metamorphics of the Arna unit (Papanikolaou 1988c) (Fig. 8.37), resembling similar outcrops of the nappes in the Prealpes, in western Switzerland (Masson 1972).

The age of the Arna unit is not well defined, mainly due to the lack of fossils. Several references occuring in the bibliography of fossil findings, usually of Permian-Triassic age, are not related to the Arna unit, but to the phyllites, either of the overlying Tyros beds at the base of Tripolis, or to the underlying counterparts in the Western Crete unit. The only available ages are: (1) the Upper Paleozoic age of the Arna granite in Kythira (unpublished data by Papanikolaou and Danamos, which have been reported as personnal communication by Stampfli et al. 2003, verified later by Xypolias et al. 2006), and (2) the Cambrian age, found in the meta-clastics of Arna in the tectonic window of Feneos in Zarouchla of the Northern Peloponnese (Kydonakis et al. 2014b). Therefore, the age of the Arna unit is now generally considered as Paleozoic and the unit represents a detached tectonic segment of the African basement of the H1 terrane, below the External platform of the Hellenides H1 (Burchfiel et al. 2018) (see also Fig. 9.6). This conclusion confirms the second hypothesis of Jacobshagen (1979) concerning the position of the Arna unit within the Hellenides.

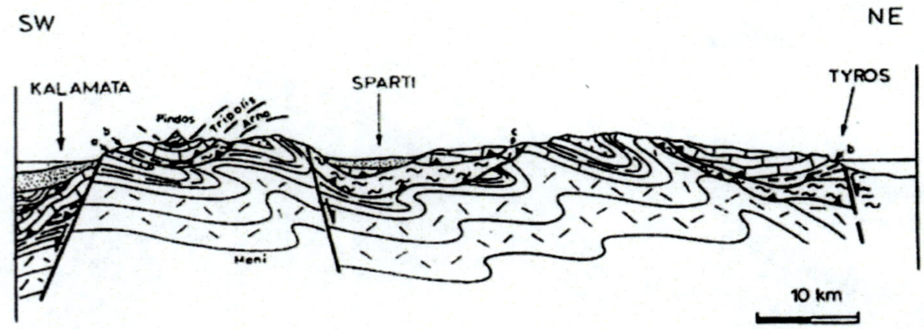

Fig. 8.33 Schematic geological cross section of Southern Peloponnese, showing the tectonic intercalation of the Arna unit between the underlying low metamorphic grade Mani unit with the Oligocene meta-flysch at the top and the overlying non-metamorphic Tripolis unit with the Upper Paleozoic–Triassic Tyros beds at its base (from Papanikolaou and Skarpelis 1987)

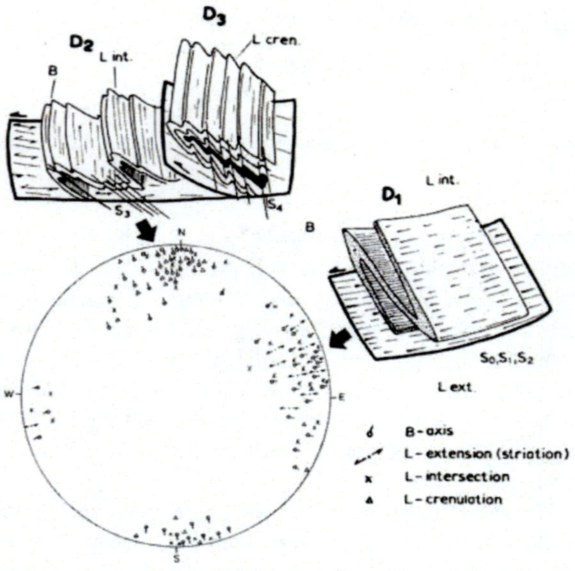

Fig. 8.34 Stereographic projection of the meso- and micro-scopic structures (fold axes and lineations) of the Arna unit, mainly from the Taygetus region and schematic representation of the geometry of the structures of the three deformation phases (D1, D2 and D3) (from Papanikolaou and Skarpelis 1987)

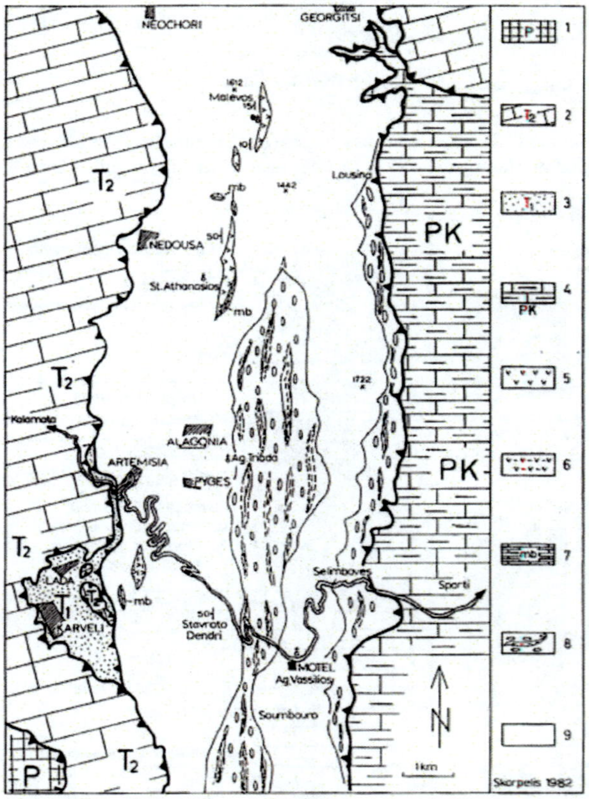

Fig. 8.35 Geological map of Northern Taygetus Mt with petrographical distinction of the Arna horizons (based on Skarpelis 1982, from Papanikolaou and Skarpelis 1987). 1: Pindos limestones, 2: Tripolis limestones, 3: Permian–Triassic phyllites and carbonate intercalations of the Tyros beds, 4: Mani marbles, 5–9: Arna's lithologies, 5: meta-basalts, 6: serpentines, 7: marbles, 8: meta-conglomerates, 9:meta-pelites

8.1.11 The Sitia Unit

In Eastern Crete, at the Sitia region, there are large outcrops of high grade metamorphic rocks (amphibolites, mica schists, granitic dykes, pegmatites, etc.), mixed with the low grade volcano-sedimentary Permian-Triassic Tyros beds. Radiometric dating reported an age of about 300 Ma for the high grade metamorphism of these rocks, hence belonging to the Variscan cycle (Wachendorf et al. 1975; Seidel et al. 1977, 1981). In the area of Kalavros, there is an E-W syncline structure with the Sitia Variscan metamorphics at its core, lying above the Permo-Triassic Tyros beds, dated by Papastamatiou and Reichel (1956) (Fig. 8.38). Immediately below the Tyros beds the Oligocene metaflysch of the Mani unit crops out at the top of the autochthon metamorphosed carbonates, where Fytrolakis (1972) first discovered the

Fig. 8.36 Characteristic outcrop of the isoclinally folded Arna quartzites in Western Crete near Ravdoucha

Fig. 8.37 Outcrop of cataclastic rocks of the corngeule type along the contact of the Tyros beds—Arna metamorphics in the Plakalona region of Western Crete (from Papanikolaou 1988c)

Globigerines and proved that it is a tectonic window and not a pre-Alpine basement. Thus, in this area of Eastern Crete we observe successively older rocks from base to top with the Mani Oligocene metaflysch, overlain by the Permian Tyros Beds, overlain by the medium-high grade Variscan metamorphics of Sitia. The Sitia outcrops in Eastern Crete are generally interlayered within the Permo-Triassic Tyros beds, with several Permo-Triassic fossiliferous sites (Krahl et al. 1986) stratigraphically overlain by the Upper Triassic–Eocene Tripolis carbonate platform. It is one of a few locations of the external tectono-metamorphic belt where pre-Alpine rocks with pre-Alpine metamorphism can be found at the base of a stratigraphic column (Papanikolaou et al. 1982).

The issue of the Sitia unit is whether it is: (i) the pre-Alpine basement of the Tripolis unit, or (ii) olistoliths inside the Permian-Triassic Tyros beds, or (iii) tectonic wedges, which have been emplaced together with the arrival of the tectonic nappes during the Alpine orogeny in Eocene–Oligocene This

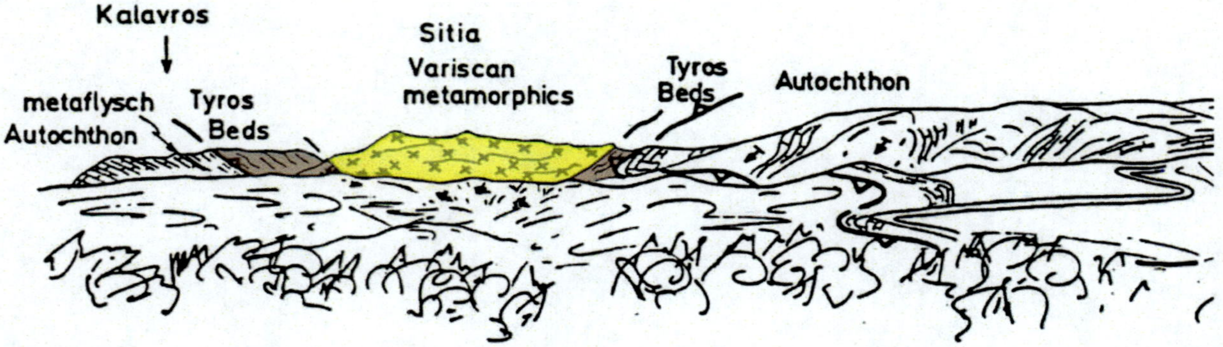

Fig. 8.38 Panoramatic view towards the east of the syncline along the northern coastal zone of Eastern Crete, showing the Variscan metamorphics of Sitia, above the Permian Tyros beds and the top of the Mani autochthon with the Oligocene metaflysch (from Papanikolaou 1988c)

last hypothesis leads to a probable different origin of the pre-Alpine Sitia unit, not belonging to the Tripolis column in Crete. However, in some localities of the Sitia region Variscan detritus has been observed (Kopp and Wernado 1983) and especially at the Vai region at the easternmost Cretan coast, there are some breccia and conglomerate horizons of high grade metamorphic rocks of the Sitia type inside the Permian–Triassic Tyros beds (Papanikolaou 1988c). These observations indicate that the large fragments of the Sitia metamorphics are large-scale olistoliths–extensional nappes, embedded during the rifting stage at the northern margin of Africa (Papanikolaou 1988c; Stampfli et al. 2003). This is endorced by the new data on the age of the Sitia metamorphics, which include the Cambrian (Romano et al. 2004, 2006). Similar outcrops of pre-Alpine basement rocks beneath the external carbonate platform have been reported also from the Dodekanese islands of **Kalymnos, Telendos, Leros and Lipsos**, with characteristic tectono-metamorphic succession of a Variscan medium grade barrovian event, overprinted by a low grade Alpine high pressure metamorphism (Franz et al. 2005; Roche et al. 2019).

8.1.12 The Laerma Rhodes and Eastern Kos Units

Apart from the usual tectonic units, which include a stratigraphic sequence, representing a specific paleogeographic region of the Hellenides, there are also some units that include ephemeral tectono-sedimentary formations, created during the orogenic phase in the arc, without a pre-existant paleogeographic setting in the Hellenides. Such units include, among others, the Laerma unit in Rhodes Island (Mutti et al. 1970; Papanikolaou et al. 1995) and the Eastern Kos unit in Kos (Papanikolaou and Nomikou 1998). The lithology of these units corresponds to a wild flysch–mélange formation, which, besides the various members of rhythmic clastic sedimentation, also contain chaotic horizons, with a strong presence of large scale olistoliths of various rocks, either carbonates, or mafic volcanics, or siliceous radiolarites etc. (Fig. 8.39). The age of both melange formations is Upper Eocene, based on the presence of the Nummulite-bearing limestones in between the olistoliths. It should be noted that the olistoliths are of different sedimentological facies and originate from different palaeogeographic–tectonic units.

The entire image justifies the characterization of these units as result of chaotic clastic sedimentation, in the context of an accretionary prism during the Late Eocene at the front of the advancing Hellenic fold and thrust belt. Obviously, the accretionary prism was detached from the ephemeral basement and followed a main decollement surface at the front of the arc. The unifying element for these two identical units of Rhodes and Kos is their location above the metamorphosed autochthon of the two islands. In the case of Rhodes Island it is the metamorphosed Mani unit (local Lindos unit), while in Kos Island it is the metamorphosed Paleozoic basement of Mt. Dikeos (Fig. 8.40). Of course, in Western Kos at the Kefalos Peninsula, there are also the Upper Cretaceous marbles, which may overlap the Paleozoic metamorphics of Dikeos Mt. In this case, they could be considered as equivalent to the higher horizons of the Mani unit. In both islands, these unique units occur beneath the non metamorphosed nappes of Akramitis (Ionian), Archangelos (Tripolis), Profitis Ilias (Pindos) in Rhodes Island and Zia (Tripolis) and Profitis Ilias (Pindos) in Kos Island. That is, the two clastic units functioned as an intermediate shear zone between the subducted metamorphosed units and the upper non-metamorphosed nappes. The interpretation of the tectonic contacts above the relative autochthon metamorphics at the base of the overlying clastic formations as extensional detachments is also open, although several complications may arise regarding their position at the frontal zone of the Hellenic arc. The Eastern Kos wildflysch is similar to the Karabortlen flysch of the Lycian nappes (Bernoulli et al. 1974).

Fig. 8.39 View of the chaotic formation of the Eastern Kos unit. The olistoliths are limestones of various sedimentological facies and mafic volcanic rocks

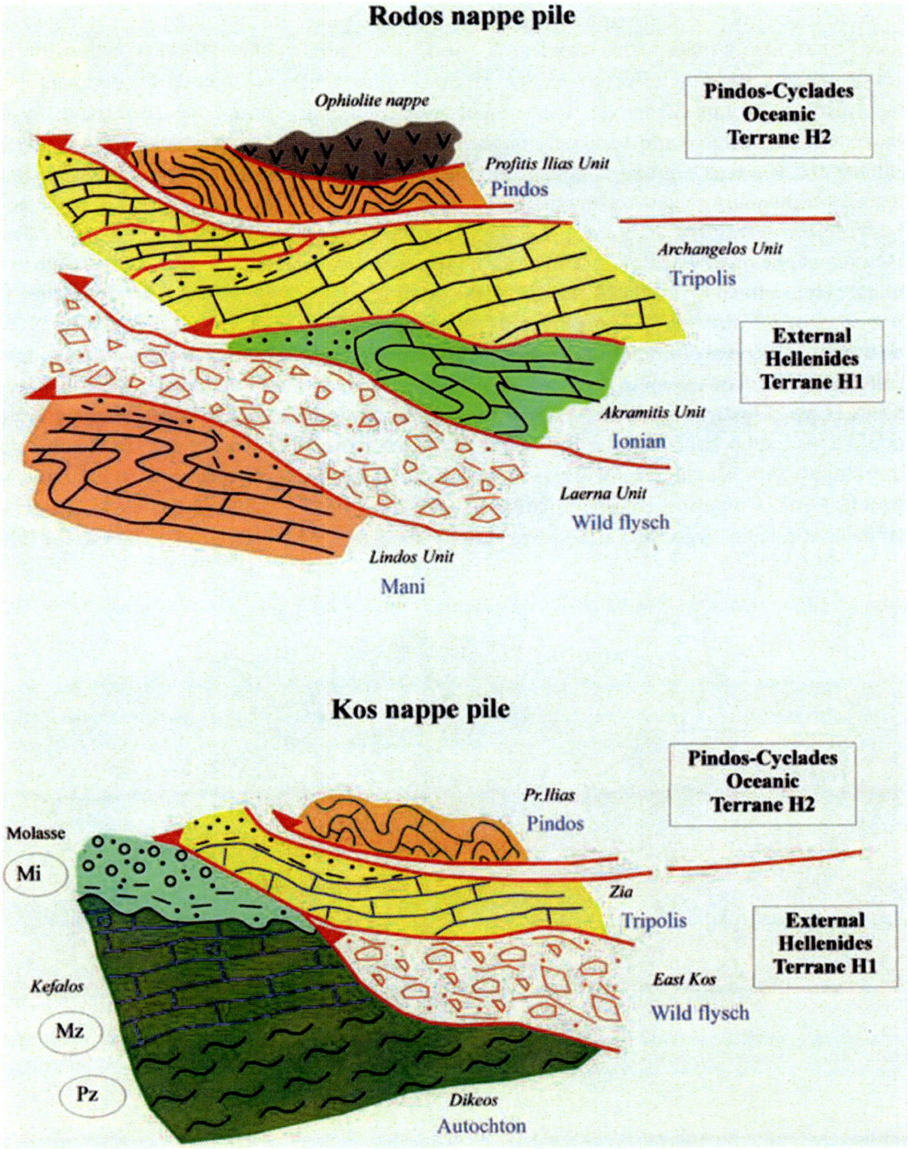

Fig. 8.40 Schematic representation of the tectonic nappe piles in Rhodes and Kos islands, where the position of the chaotic mélanges of the Laerma and Eastern Kos units is shown, in between the metamorphic rocks of the relatively autochthon units and the non metamorphosed upper nappes

8.2 The Pindos–Cyclades Ocean—H2

8.2.1 The Pindos Unit

The Pindos unit was first described by Philippson (1892) as «Olonos-Pindos» zone with a distinct pelagic stratigraphic sequence. It is the most typical nappe of the Hellenides as well as the most obvious in the field, hence it was the first nappe to be described since 1903 by Cayeux and 1906 by Negris and Ktenas (Cayeux 1903; Negris 1906a, b; Ktenas 1906). Thus, while the tectonic nappes were conceived by the geological reasoning only in the early 20th century (in 1898 the "Exotic terranes" in the pre-Alps were described by Schardt and in 1902 the first large synthesis on the tectonic nappes of the Western Alps was composed by Lugeon in Switzerland), these researchers identified Pindos as a nappe, because it was easy to observe (in the Peloponnese or elsewhere) the Nummulites bearing neritic Eocene limestones of the Tripolis unit underneath the pelagic Triassic limestones with Halobies, or the Upper Cretaceous limestones with Globotruncanes of the Pindos unit (see also Fig. 4.11). Moreover, the sub-horizontal tectonic contact of the nappe could be clearly observed along tens of kilometers, whereas a strong contrast of the pelagic over the neritic sedimentary facies was evident (Fig. 8.41).

Despite the fact that the Pindos unit was the first classical tectonic nappe recognised in the Hellenides, the general view on its displacement up to the 1970s was of the order of only a few tens of kilometers. It was only after the revolutionary discovery of Eocene Nummulitic crystalline limestones in the tectonic window of Olympus (Godfriaux 1962, 1968) to realise that the displacement of this nappe was enormous (Bernoulli and Laubscher 1972). Additionally, light crystalline limestones with Megalodon of Triassic age were found, as well as algae, Hippurites, and Nummulites (Godfriaux 1968) whereas the facies (bio- and litho-) are exactly the same as those observed in the Gavrovo and Tripolis (Fleury 1980). This discovery showed that the Olympus unit is the eastward continuation of the carbonate platform of Gavrovo-Tripolis below the Pindos and the other more internal nappes. Thus, from the Makrynoros region, where the front of the Pindos nappe is actually observed over the Oligocene Gavrovo wild flysch, until Olympus, the Gavrovo-Tripolis-Olympus platform is buried beneath the Pindos and all other nappes (Bernoulli and Laubscher 1972; Aubouin et al. 1979) (Fig. 8.42). Thus, the roots of the Pindos nappe must be traced somewhere in the present day Aegean Sea, east of the Olympus Mt and the overall displacement is of the order of a few hundred km.

Another criterion for the magnitude of the displacement of the Pindos nappe refers to the duration of its emplacement over the more external units. Thus, the onset of the transport of the Pindos nappe starts during its decollement from its basement and the adjacent units, indicated by the end of its flysch sedimentation in the Middle Eocene, at about 50 Ma. Since then, it started to slide over the neighboring more external units within the fore-arc basin areas, where flysch sedimentation was taking place. Beneath the Pindos overthrust the successive flysch formations were aquiring the characteristics of wild flysch with breccias, olistholites etc. derived from the Pindos nappe. The end of the Pindos journey occurred when it reached the front of the Pindos nappe, over the wild flysch of the Epirus–Acarnania syncline, where the age of the upper strata are dated Lower-Middle Miocene (Burdigalian–Langhian) at about 15 Ma (IFP & IGSR 1966; BP 1970). This means that the Pindos nappe was travelling for about 35 million years, from the Middle Eocene to the Middle Miocene, throughout the main Alpine orogenic period of the Hellenides.

Several small tectonic units in the form of tectonic wedges are observed at the base of the Pindos nappe. These small units comprise particular tectono-stratigraphic formations not belonging neither to the Pindos nor to the underlying Tripolis

Fig. 8.41 A classical outcrop of the Pindos (Ethia) nappe over the Eocene Tripolis flysch, overlying the Tripolis limestones, with their characteristic Nummulite bearing horizons, from Kastelli, Central Crete

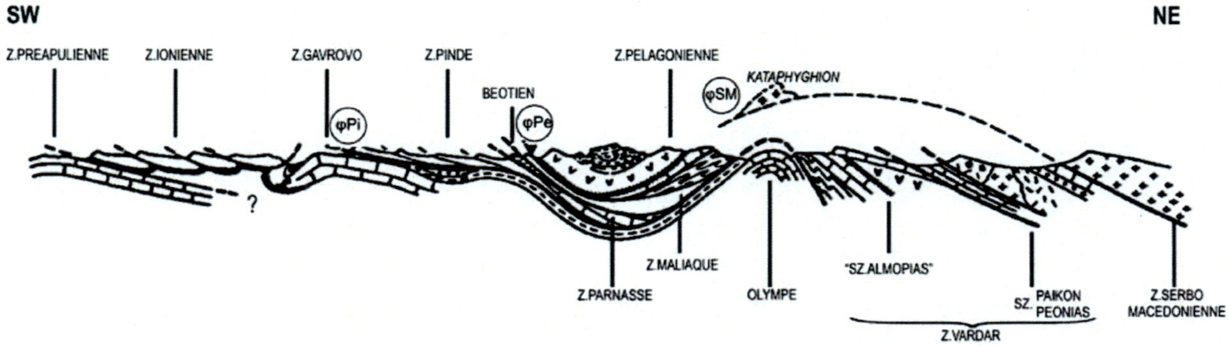

Fig. 8.42 Schematic cross section of the Hellenides, where the major allochthony of the Pindos nappe in relation to the tectonic window of the Olympus unit can be observed (from Aubouin 1977)

Fig. 8.43 Typical outcrop of abyssal-pelagic sediments of radiolarites, cherts and diabasic tuffs from the Jurassic of the Pindos Unit in the Agrafa area

unit. Their paleogeographic position within the external Hellenides is dubius, based on their stratigraphy, which comprises: (i) a shallow water carbonate platform facies of the Upper Triassic- Jurassic over a volcano-sedimentary complex of the Lower–Middle Triassic and unconformable wild flysch of the Eocene in the **Megdovas unit** (Fleury 1977) and (ii) intermediate facies with radiolarites, micro-breccia limestones and pelagic limestones of ammonitico rosso facies of Upper Cretaceous age, grading to Paleocene–Eocene flysch in the **Agridaki unit** (Karotsieris 1981).

In the Pindos stratigraphic sequence, silicate or carbonate pelagic sedimentation is observed with radiolarites and other siliceous rocks and pelagic limestones throughout the duration of biochemical sedimentation from the Late Triassic to the Late Cretaceous (Philippson 1892, 1959; Renz 1955; Fleury 1980).

The location of the Pindos unit inside the Pindos–Cyclades Ocean (H2) is questionable and under discussion. The lateral transition of the Pindos towards the internal platform in the Parnassos region (H3) is documented (Celet 1962, 1979; Papastamatiou & Tataris 1963; Gouliotis 2014) by : (i) the Vardousia tectonic thrusts, where the stratigraphic column is characterized by the presence of carbonate mirco-breccia slope facies, as well as by other transitional facies (Penteoria Unit, etc.) and (ii) the same timing of transition to the flysch in the Maastrichtian-Danian. Based on the above lateral transition the Pindos unit should correspond to the internal segment of the Pindos–Cyclades oceanic basin towards the margin of the Internal Carbonate Platform (Papanikolaou 1986a). However, the presence of the allochthonous, Upper Cretaceous oceanic basalts bearing Arvi unit above the Pindos (or Ethia) and the presence of the ophiolitic outcrops of Kerassia - Milia as tectonic wedges along the Vardousia tectonic front, complicate this simple interpretation. In any case, Pindos does not have significant outcrops of mafic volcanic rocks, except of limited outcrops of spilites and tuffs in the Jurassic (Pe-Piper and Piper 1984) (Fig. 8.43). It is therefore probable that the Pindos paleogeographic area was a parallel deep trough, not directly related to the ophiolite suture of the Pindos-Cyclades oceanic basin.

Fig. 8.44 Typical outcrop of platy pelagic limestones with silicate intercalations of the Pindos unit, observed on both sides of a gorge in the Agrafa region

The upper part of the stratigraphic column of the Pindos unit has a several km thick flysch formation with characteristic transitional beds from the underlying pelagic limestones of significant thickness 20–50 m, dated from their Globotruncana bearing fauna in the Maastrichtian-Danian (Aubouin 1955; Fleury 1980). Flysch sedimentation comprises the Paleocene–Middle Eocene period. Below the flysch, alternations of silicate and carbonate pelagic sediments are observed (Fig. 8.44). The stratigraphy generally comprises Mesozoic radiolarites and pelagic limestones with many variations and lateral transitions from region to region.

The stratigraphic column described by Fleury (1980) is a representation of an average situation from central continental Greece (Fig. 8.45). Under the flysch, a formation of pelagic limestones is located, which include Silex or chert nodules with Globotruncanes of Cenomanian–Maastrichtian age. Below this Upper Cretaceous formation we have a clastic formation, which was named by some scientists «first flysch». This is a misleading term, because its characteristics are not those of a flysch, neither from the sedimentological nor from the tectonic point of view. It is a clastic event within the Cretaceous pelagic sedimentation of the Pindos basin, not followed by folding and emersion but by carbonate sedimentation up to the Tertiary flysch. Below the clastic interval radiolarites and related siliceous sediments with some pelagic carbonate intervals occur. Pure radiolarites are observed mainly in Dogger–Malm, where some mafic rocks, mainly tuffs, are also observed. In the Lower Cretaceous radiolarites alternating with pelagic limestones with Calpionelles occur. Below the radiolarites of Dogger siliceous pelagic limestones are observed, dated Upper Triassic (Norian) and especially Lias. Below these formations there is another Upper Triassic clastic formation, which was again erroneously called «Triassic flysch» , with famous Halobia bearing marly limestones («filaments») dated as Carnian. In several outcrops of the lower part of the Pindos stratigraphic sequence, mafic volcanic rocks are present, probably of Middle Triassic age, forming the base of decollement of the nappe from its original paleogeographic position.

The clastic Cretaceous formation («first flysch»), deserves special attention because for the first time we find a 50–100 m thick clastic sequence with turbidites inside the pelagic biochemical section of a stratigraphic column (Maillot 1979) (Fig. 8.46). During this period, more internally than the Pindos unit the early Alpine orogeny is present, involving the tectonic emplacement of a large segment of the Axios/Vardar ocean. This early orogenic phase, yielded clastic detritus with abundant ophiolite material, which reached the Pindos basin, forming a flysch type sedimentary interval.

However, this orogenic process did not continue to reach the more external Pindos unit. Instead it was rendered

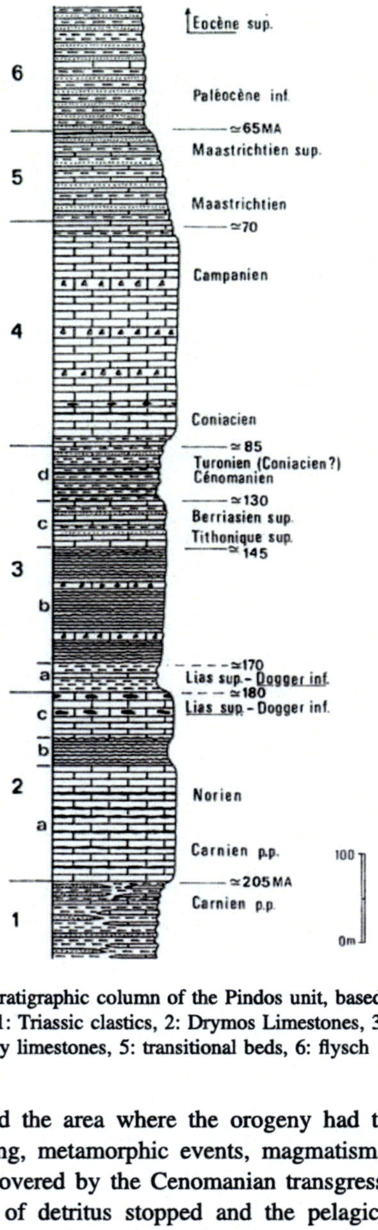

Fig. 8.45 Stratigraphic column of the Pindos unit, based on Fleury's data (1980). 1: Triassic clastics, 2: Drymos Limestones, 3: radiolarites series, 4: platy limestones, 5: transitional beds, 6: flysch

inactive and the area where the orogeny had taken place (with folding, metamorphic events, magmatism, etc.) was subsided, covered by the Cenomanian transgression. Thus, the arrival of detritus stopped and the pelagic carbonate sedimentation with Globotruncana bearing limestones was restored in the Pindos basin until the end of the Cretaceous. The Pindos unit belongs to the External Hellenides, as shown by its continuous stratigraphic column, although the presence of the Cretaceous clastic formation shows that it was close to the Internal Hellenides at the time of the early Alpine orogeny, receiving detritus from the emerging chains.

Regarding the internal tectonic structure of the Pindos unit it is expected to be intensively folded and thrusted, because its stratigraphy is characterized by several alternations of limestones and radiolarites with deviating rigid-plastic lithologies, producing disharmonic phenomena. Thus, besides its far travelled nappe nature it is characterized by tight folds with inverted limbs and overthrusts. Its main structural characteristic is its imbricate structure, with monoclinal sequences and periodical repetitions in the same stratigraphic order, as this can be observed in the geological maps (Fig. 8.47). It is characteristic that the Pindos imbrications are colorfully toned in the field from the repetition of red–crimson (radiolarites), white–yellow (pelagic limestones) and yellow–brown (flysch). The main horizons of tectonic detachment in the stratigraphic column of the Pindos unit are: (i) its base, with the clastic Triassic sediments (Negris 1909; Aubouin 1959), (ii) the base of the Upper Cretaceous limestones directly above the first flysch, creating the **Pindos Arcadic nappe** in Peloponnese (Dercourt 1964; Kiskyras 1972; Lekkas 1978), and (iii) the base of the flysch, appearing as a separate nappe in the Northern Pindos mt (Brunn 1956; Aubouin 1959).

The base of the stratigraphic column includes, in addition to the Carnian clastic sediments, some occurrences of mafic volcanic rocks, usually diabases, probably of Middle Triassic age, which all together represent the volcano-sedimentary complex of the initial rifting phase, before the opening of the oceanic basin of H2 (see also Fig. 8.3). This was supported by geochemical analysis of the mafic rocks from the base of the Ethia unit in Central Southern Crete (Robert and Bonneau 1982). It is characteristic for H2 that the onset of the pelagic–abyssal sedimentation in the context of the opening of the oceanic basin occurs in the Late Triassic, immediately after the Carnian, when the clastic sedimentation is terminated and the pure biochemical carbonate and silicate pelagic sedimentation is established. On the contrary, the transition from the first rifting phase to the opening of the oceanic basin in the Axios region of H4 takes place in the Scythian, i.e. about 15 million years earlier than in H2 (see also Fig. 8.3).

In Crete, besides the classical Pindos unit, the **Ethia** unit is also present (Renz 1955; Creutzburg 1958; Creutzburg and Papastamatiou 1969). Its type locality is in the Asteroussia mts and its difference from the Pindos, is mainly due to its late flysch sedimentation in the Late Eocene, instead of the Late Cretaceous-Paleocene. Additionally, during the Paleocene–Eocene carbonate sedimentation with alternating pelagic and neritic facies is observed in Ethia, whereas in classical Pindos during this period we have the flysch sedimentation. The above difference between Pindos and Ethia shows an uplifting of the Ethia sea bottom prior to the flysch sedimentation. This is probably due to the different paleogeographic position within the Pindos-Cyclades Ocean, where the Pindos unit s.s. corresponds to the inner half of the Pindos basin towards the internal Parnassos platform, while Ethia unit corresponds to the external half of the basin towards the external Tripolis platform (Papanikolaou 2013).

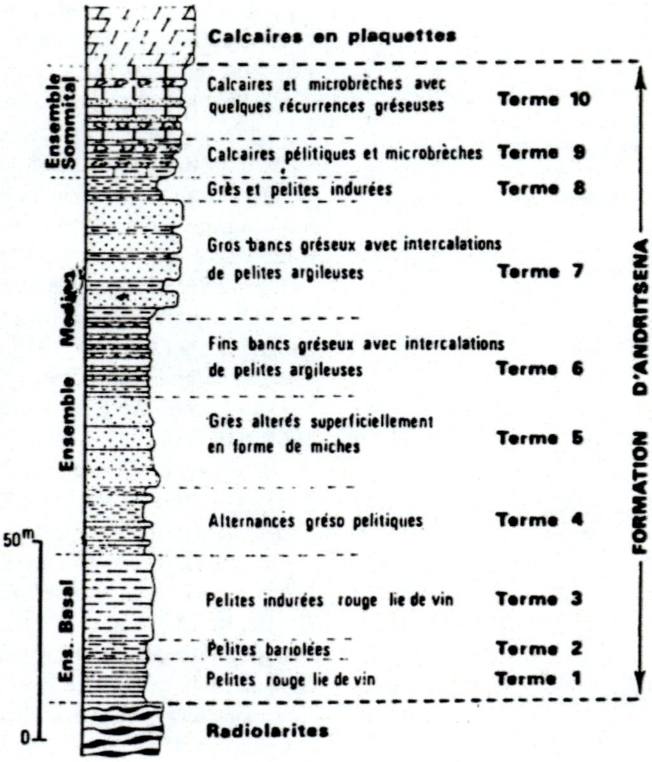

Fig. 8.46 Detailed lithological description of the clastic sediments of the Lower Cretaceous–Cenomanian ("first flysch") of the Pindos unit, which are interlayered between the underlying radiolarites and the overlying platy limestones, from the Andritsaina area (from Maillot, 1979)

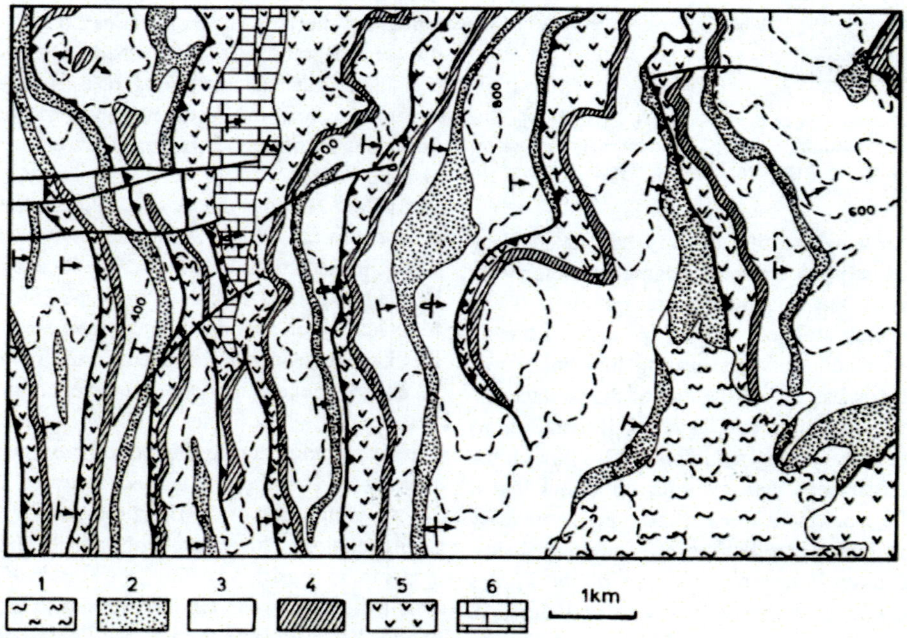

Fig. 8.47 Geological map of the Kefalovrisi area, where imbrications of the Pindos unit can be observed (from the geological map of Figalia sheet, Lalechos 1973). 1: alluvium and scree, 2: Danian–Eocene flysch, 3: platy limestones of the Upper Cretaceous with Clobotruncana, Pseudocyclamina, etc., 350–400 m thick, 4: limestones and sandstones of the Cenomanian–Early Turonian with Globotruncana, Orbitolina, etc., 100 m thick, 5: radiolarites with sandstones, marls, and limestone alternations of the Dogger–Early Cretaceous, 250–300 m thick, 6: pelagic limestones with Halobia and cherts, sandstones and marls of the Upper Triassic- Lias

This different paleogeographic position may explain the time difference of about 20 million years observed both in the onset of the flysch sedimentation and the beggining of subduction/underthrusting of the two segments of the H2 basin.

In Eastern Crete another different variation of the Pindos unit is present, known as the **Mangassa** unit (Papastamatiou et al. 1959). This is also different from the Pindos unit s.s. with younger flysch and neritic sediments in the Paleocene-Eocene as well as from the Cretan Ethia unit. The main additional difference concerns the absence of the radiolarites and the clastic "first flysch" formation in the Early Cretaceous (Bonneau and Zambetakis 1975; Fytrolakis 1980).

The above differences in the stratigraphy and the tectonic evolution of the Pindos-type units in Crete, along with the other unique units of Crete (e.g. Western Crete–Trypali, Sitia, Asteroussia, Vatos) not appearing in continental Greece, led Wunderlich (1971) to propose the term "Minoides" for the Hellenides of Crete.

The Pindos type units extend also in the Dodekanese islands of Tilos, Rhodes (Prophitis Ilias unit) and Kos (Prophitis Ilias unit) (similar names of churches existing in different type locality sites/hills of the two islands, see also Fig. 8.40). These units correspond to the Koycegiz sequence of the Lycian nappes (Bernoulli et al. 1974) and to the Marina sequence of Leros and Kalymnos islands (Durr et al. 1978).

8.2.2 The Arvi Unit

The Arvi unit was named by Bonneau (1973b, 1976) to include the mafic volcanic rocks that appear in Central Southern Crete, associated to pelagic sediments with Globotruncanes of the Upper Cretaceous. These volcanic rocks had already been documented in 1964 by Tataris, who had described both the basaltic pillow lavas and their affiliation with the marly pelagic limestones, of crimson-red color, that largely cover them (Fig. 8.48). The stratigraphic column of this unit does not include other formations older than those of the basaltic lavas, while it is also complemented by a formation of Eocene flysch, which is normally overlying the Upper Cretaceous–Paleocene limestones.

The general character of the basaltic lavas as well as their geochemical characteristics indicate basalts of the mid-ocean ridge (Robert and Bonneau 1982). More recently, these basalts were considered, based on new geochemical analysis, as a result of an intra-oceanic sea mount (Palamakumbura et al. 2013).

It is interesting that the Arvi unit has been observed also in several other regions of continental Greece and not only in Crete, such as: (i) in the **Ermioni** area of Argolis and (ii) in the **Kerassia–Milia** zone in Central Greece (Papanikolaou 1989a, 2009; Clift and Robertson 1989; Bortolotti et al. 2009).

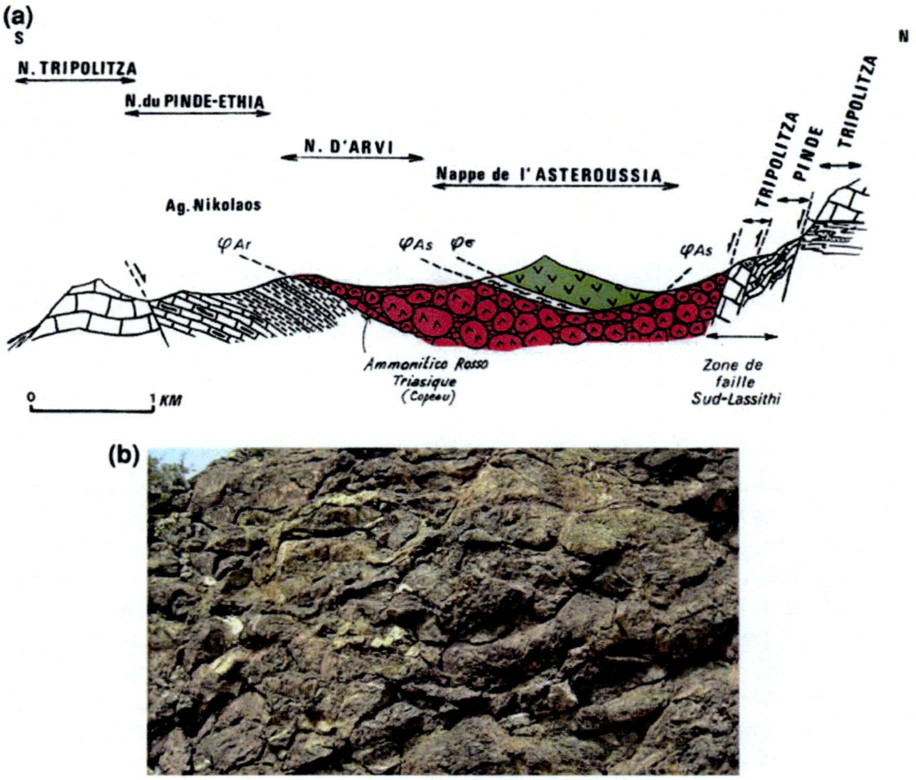

Fig. 8.48 a Geological cross section in the Vianos area, where the Arvi unit can be observed between the Pindos/Ethia and the Asteroussia units (from Bonneau 1973a). b Outcrop of the Upper Cretaceous basaltic pillow lavas of the Arvi unit in Crete

Fig. 8.49 Outcrops of pillow lavas of Upper Cretaceous age in the Kerassia region, inside the upper layers of the Pindos flysch, under the Vardousia nappes

(i) In Argolis, the outcrops of the Arvi basalts occur at the base of the Aderes clastic complex and they were known as hosts of metal-bearing mixed sulfides (Aranitis 1963), exploited in the 1960s and 1970s; in fact, the mines were visited by students of the Natural Sciences department of the University of Athens, guided by the late Prof. G. Paraskevopoulos. Another small outcrop of the Arvi formations occurs within the blocky flysch of the uppermost tectonic unit of central-east Peloponnese in the Parnon mt, known as **Glypia Unit** (Skourtsos et al. 2001). Besides the Upper Cretaceous basalts and pelagic limestones of Arvi type formations, several more lithologies occur within this flyschoid complex, generally resembling to the Aderes complex, such as neritic and pelagic limestones of Cretaceous age, Permian neritic limestones, radiolarites and ophiolitic rocks.

(ii) In Central continental Greece, the outcrops are lenticular, along a tectonic zone of several tens of kilometers length, within the Pindos flysch in the front of the Vardousia nappes (see also the geotectonic map out of text). In this zone, besides the outcrops of the typical Arvi basalts (Fig. 8.49), other mafic and ultra-mafic rocks are also present (gabbros, peridotites). This observation reinforces the view that the Arvi unit represents an ophiolitic suture and not just a mere volcanic episode of the Pindos-Cyclades Ocean.

In any case, Arvi is found as a nappe over the Pindos–Ethia units, indicating its more internal position, close to the internal margin of the ocean towards the Parnassos platform. At the same time, both in Crete and in Argolis, and to a lesser extent in Central Greece, it is located below or at the base of the chaotic clastic formations of the Miamou–Aderes accretionary prisms. This indicates the presence of a significant tectonic discontinuity with the superjacent more internal nappes.

It is interesting to note that in the internal Dinarides in the Bosnia and Herzegovina area there are outcrops of basaltic lavas and related intra-oceanic magmatic rocks, interfingering syngenetically with Upper Cretaceous pelagic limestones and related deep sediments (Ustaszewski et al. 2009). These outcrops resemble those of the Arvi unit in Greece (personnal observations during the Dinaric ophiolites fieldtrip in 2006). Thus, this peculiar oceanic unit has much more extended paleogeographic significance than it was assumed during its early recognition in Crete. The paleogeographic position of Arvi has been proposed to be at a more internal region than Pindos and/or Ethia units and before the Pelagonian microcontinent (Bonneau 1984; Palamakumbura et al. 2013). However, the relation between the Pindos and Parnassos units was not taken into account in the above proposals. The previously described stratigraphic, tectonic and paleogeographic differences between the Pindos and the Ethia nappes and the Arvi characteristics may be comprised

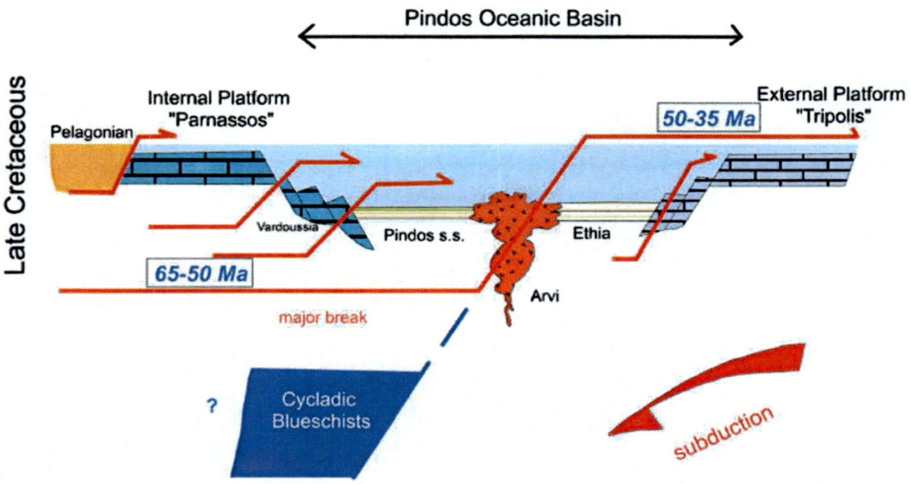

Fig. 8.50 Schematic diagram showing the probable position of the Arvi unit during the Late Cretaceous in between the Pindos internal branch and the Ethia external branch of the Pindos-Cyclades oceanic basin. The Arvi mafic rocks are considered as a middle oceanic ridge but they could also be considered as an oceanic sea mount. The Cycladic units might be also involved within this paleogeographic scheme on either side of the Arvi unit

in a paleogeographic organization, where the Arvi unit lies in between the two branches of the Pindos-Cyclades Ocean during its final opening stage (Fig. 8.50).

The above scheme explains both the tectonic position of the Arvi, above the Ethia unit and also the lateral transition between the Pindos basin and the Parnassos platform through the Vardoussia transitional facies. The paleogeographic break within the external Hellenides may explain, more generally, the 15–20 million years difference, between the flysch sedimentation and the onset of subduction of the Pindos and Ethia units (65–50 Ma versus 50–35 Ma) as well as the more general similar differences between the Internal Carbonate Platform of the Parnassos-Beotia etc. and the External Carbonate Platform of Tripolis. The same break could also involve the Cycladic units, before their subduction and HP/LT metamorphism in the Eocene.

8.2.3 The Northern Cyclades Unit

The Northern Cyclades unit is the typical blueschists unit of the Hellenides, croping out in the Northern Cyclades, Southern Evia, Pelion and around the Olympus tectonic window (Papanikolaou 1984c, 1986a, b, 1987). It comprises a lithostratigraphic column established in Andros Island, containing mainly pelagic marbles at its base and unique thick horizons of cipollinic marbles of gray, blue, greenish and violet colors towards the top, as well as meta-tuffs and meta-lavas of basic composition with a plethora of metaclastics, mainly siltstones (Papanikolaou, 1978d) (Fig. 8.51). The presentation of a specific litho-stratigraphic column is difficult, as it changes from one location to another.

Unfolding the tectonic structure of Andros Island, which comprises several large isoclinal folds over a length of 40 km, permits to correlate the local stratigraphic columns and determine a generalised stratigraphic sequence (Fig. 8.52). The marbles–cipollinic marbles are dominant in some areas and the meta-volcanics are dominant in other areas. The meta-volcanics comprise meta-tuffs or meta-lavas in the form of amphibolitic schists and massive amphibolites. Intercalations of meta-pelites and meta-sandstones are observed everywhere. Thus, the general character of the Northern Cyclades unit is a volcano-sedimentary pelagic sequence with pelagic limestones of presumably Mesozoic age that was metamorphosed in Early Tertiary.

This unit is known all along the medial tectono-metamorphic belt (Papanikolaou 1986b) (see also Fig. 8.54a), with different type locality names such as: the Styra (Southern Evia, Katsikatsos 1979), the Ampelakia (formerly Ossa, Godfriaux 1977; Katsikatsos et al. 1982; Kilias et al. 1991a), the Makrinitsa (Pelion, Ferriere 1979), the Ampelos (Samos, Papanikolaou 1979a). It crops out mainly in the Northern Cyclades, where the type locality of the mineral glaucophane occurs (in Syros, Hausmann 1845) together with the rest of the sodium mineral amphiboles (such as jadeite), showing high pressures/low temperatures. These metamorphic rocks croping out in Syros, Siphnos and Tinos islands have been described already since the beginning of the 20th century by Ktenas (1907a, b). Thus, the HP/LT metamorphosed blueschist rocks of the Northern Cyclades appear in Syros–Tinos–Andros–Gyaros–Southern Evia, and to a lesser degree in Kea, Kythnos, Serifos, where the HP/LT metamorphism has been overprinted by the

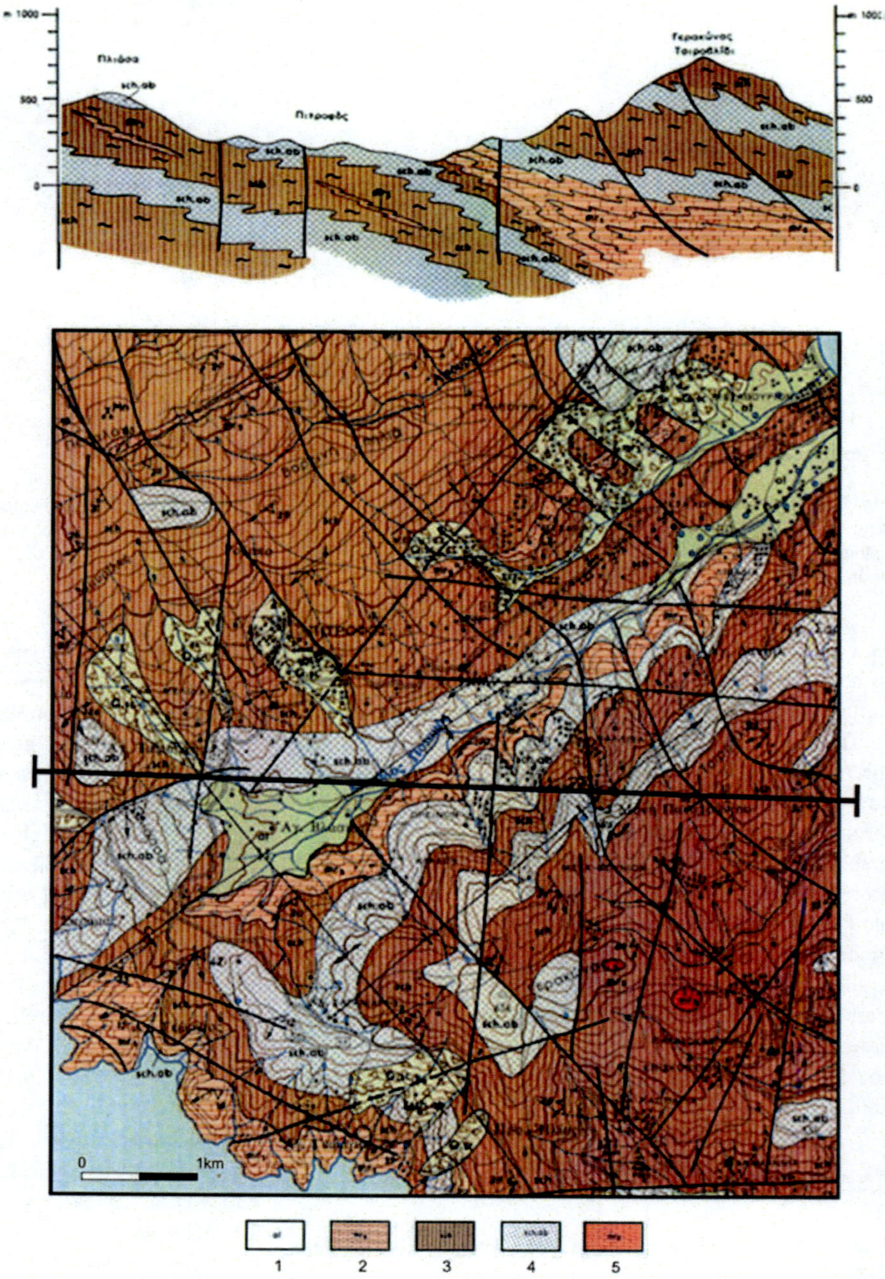

Fig. 8.51 Geological map of Central Andros Island and geological cross section AB, showing the core of a large, northwest-vergent isoclinal fold in the Northern Cyclades unit (simplified from the Gavrion–Andros–Piso Meria maps, Papanikolaou 1978a). 1: Quaternary, 2: Lower marble, 3: mica schists, 4: amphibolitic schists, 5: thin intermediate marbles

younger greenschist metamorphic event. The latter is often accompanied by the Miocene granites of the Aegean volcanic arc (Papanikolaou 1986b, 1987, Schliestedt et al. 1987). However, apart from this region, the Northern Cyclades unit reappears in Pelion (Ferriere 1979) (Fig. 8.53) and around the tectonic window of Mt Olympus to the north (Katsikatsos et al. 1982, 1986) as well as in Samos Island to the east (Papanikolaou 1979a), up to the Menderes region in Asia Minor (Durr 1975). Thus, the Northern Cyclades unit can be followed all along the medial metamorphic belt from the Olympus area in the north to the Menderes area in the east (Papanikolaou 1986b) (Fig. 8.54a).

In general, the Northern Cyclades unit is the most typical blueschist unit of the Aegean and its tectonic position is stable throughout the entire length of the medial tectono-metamorphic belt (Papanikolaou 1986b, 1987). It is overlying the Olympus–Almyropotamos–Kerketeas units and underying different units along the belt, such as the

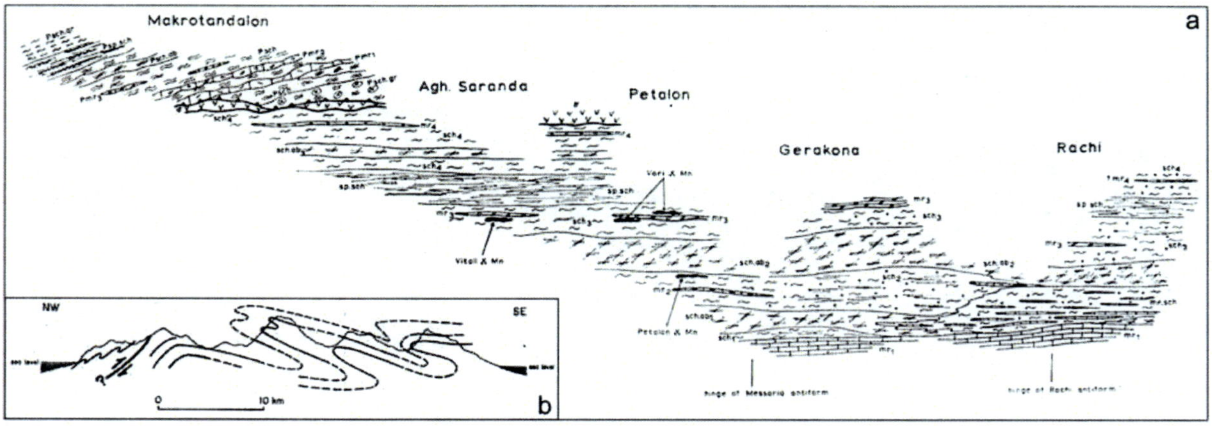

Fig. 8.52 Schematic representation of the stratigraphic distribution of the Northern Cyclades unit in Andros Island, after the unfolding of the tectonic structure, given in sketch b. The diagram shows the lateral facies transitions along 70 km of the unfolded structure. In the same figure the stratigraphic sequence of the tectonically overlying unit of Makrotantalon is included (from Papanikolaou 1978d)

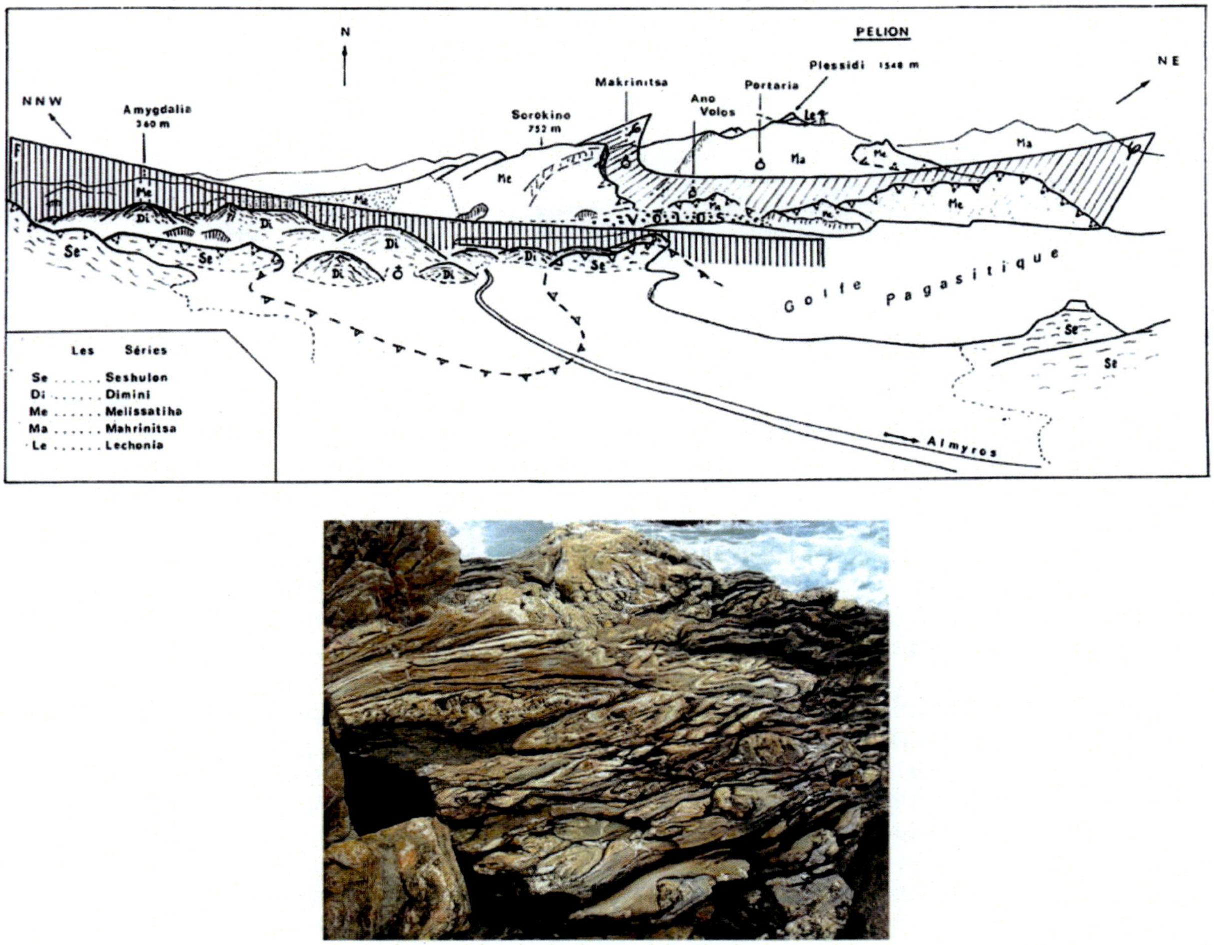

Fig. 8.53 **a** Panoramic view from the west of the tectonic window of the blueschists of Pelion (Makrinitsa unit), under the metamorphic rocks of the Flambouro unit (mainly gneisses) and the Almopia unit (mainly Pelagonian marbles) (from Ferriere 1977). **b** Outcrop of metamorphic rocks of the Northern Cyclades unit from Eastern Pelion, with characteristic plastic-flow deformation

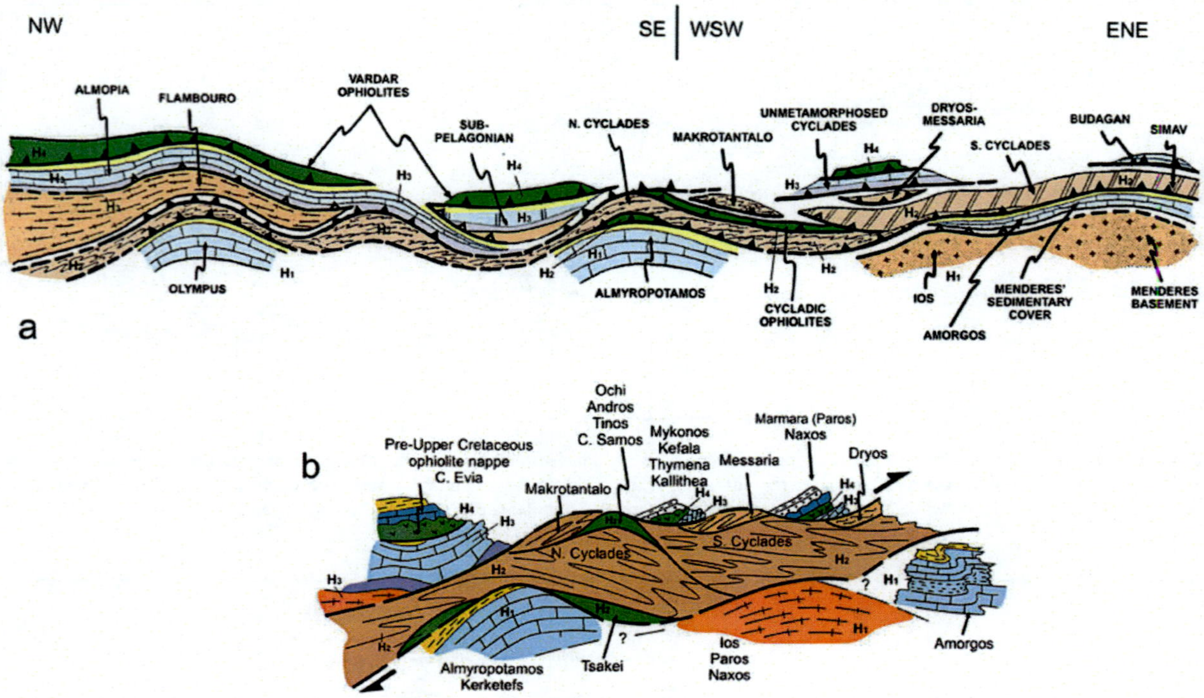

Fig. 8.54 **a** Schematic longitudinal section of the medial tectono-metamorphic belt, showing the stable presence of the Northern Cyclades unit, with various type-locality names throughout its length, from the Olympus area to Samos and Menderes (from Papanikolaou 1986b, 2013). **b** Schematic transverse section of the medial tectono-metamorphic belt showing the Cycladic megashear zone between the underlying units of the H1 and the overlying units of the H3 terrane (from Papanikolaou 1987, 2013)

Flambouro unit in Eastern Thessaly, the Mankrotantalon–Ochi unit in Southern Evia–Andros, the Southern Cyclades unit (Vourliotes) in Samos and the non-metamorphosed Cyclades unit in various locations (Fig. 8.54).

In other words, the tectonic basement of the Northern Cyclades blueschists is the internal segment of the external carbonate platform of the Hellenides H1, underthrusted during the Late Eocene–Oligocene beneath the Pindos–Cyclades oceanic domain H2. On the contrary, the overlying nappes of the Eastern Greece and the non-metamorphosed Cyclades unit belong to the internal platform of the Hellenides H3, characterized by the Upper Cretaceous transgression.

The ages of the stratigraphic formations of the unit are generally unknown due to the lack of fossils, which, have been destroyed by the medium metamorphic grade. An Upper Triassic–Liassic age, reported by Melidonis (1980) in Northern Tinos Island was originally believed to belong to the lower horizon of the marbles of Andros Island, but later it was suggested that it might represent a tectonic window of the underlying Almyropotamos unit (Avigad and Garfunkel 1989).

The pelagic nature of the unit, apart from the presence of the cipollinic marbles, is underligned also by the presence of Mn- and Fe-deposits (hematitic schists, etc.) associated with the meta-volcanic rocks. These metalliferous deposits are probably the result of fume-hydrothermal activity in deep basins with mafic volcanism. This contrasts the Southern Cyclades character, where extensive meta-bauxite horizons are present inside the neritic marbles, indicating a shallow-water environment with ephemeral emersions.

At the top of the column, there is a metamorphosed wild flysch, where large blocks of metamorphosed ophiolites are present, with impressive outcrops especially in Northern Syros Island (Bonneau et al. 1980). The age of the flysch is not well documented, although it should be Upper Cretaceous to Eocene, since the blueschist metamorphism of the unit is of Late Eocene age (Brocker and Franz 1998; Brocker and Pidgeon 2007; Zeffren et al. 2005) and the following retrograde greenschist metamorphism is of Upper Oligocene–Lower Miocene age (Schliestedt et al. 1987, see also Figs. 7.26, 7.27 and 7.28).

Another element that differentiates the Northern Cyclades unit from the Southern Cyclades unit is the tectonic trend, characterized by isoclinal folds and stretching lineations in the NE-SW direction, between N40° and 70°E, observed in all scales (micro-meso-macro-scopic scale) (Fig. 8.55). The same structural trend in the Southern Cyclades and in Ikaria

Fig. 8.55 Characteristic outcrop of alternations of cipollinic marbles and mica, amphibolite schists of the Northern Cyclades unit in Southern Evia (Styra). The repetition of the cipollinic marbles is due to isoclinal folding in the ENE-WSW direction. Successive morphological discontinuities are formed along the tectonic repetitions, due to the differential erosion between the marbles and the schists

and Samos islands is oriented N-S. The NE-SW trend is observed all along the medial tectono-metamorphic belt from the Olympus area up to Tinos Island.

The metamorphic grade varies, with gradual attenuation from the south to the north (Blake et al. 1981). According to Bonneau and Kienast (1982), 8 Kbars and 320 °C are present in Southern Evia, 10 Kbars and 480 °C in Andros, and 14 Kbars and 500 °C in Syros. In a detailed study of Gyaros, Katagas (1984) described the individual stages of progressive metamorphism, from the original blueschist facies to the retrograde greenschist facies. More recent data indicate a wide distribution of pressures and temperatures for both the Late Eocene blueschist event, with values of 12–20 Kbars/450–550 °C, as well as for the retrograde Miocene greenschist event, of 4–9 Kbars/400–500 °C, in Syros–Sifnos–Tinos islands (Altherr et al. 1982; Matthews and Schliestedt 1984; Trotet et al. 2001; Para et al. 2002).

8.2.4 The Makrotantalon–Ochi Unit

The Makrotantalon–Ochi unit appears in Northern Andros, in the Makrotantalon area, where it was described by Papanikolaou (1976, 1978a, d), as well as in areas of Southern Evia, and especially in Mount Ochi. This unit is characterized by blueschist metamorphism and the same tectonic structure, like the Northern Cyclades unit, but with a different litho-stratigraphy. In the outcrops of Northern Andros marble horizons occur (Fig. 8.56), where Permian fossils have been found (Mizzia, Pseudoschwagerina, Carinthiaphyllum, Fusulinidae etc.) (Papanikolaou 1976, 1978c, d). The fossils were found in reefal crystalline limestones, intercalated with meta-siltstones and meta-sandstones and superjacent mafic volcanic rocks, metamorphosed to amphibolites. The same Permian marbles have been observed in the outcrops of Ochi Mt, as at the top of the hill within the old castle above Karystos.

The amphibolites are possibly of Triassic age, due to the fact that they are overlying the Permian marbles and resemble other Triassic volcano-sedimentary formations in the Hellenides. Recently, a geochronological Triassic age of meta-volcanic rocks was identified in the outcrops of the unit in Southern Evia (Chatzaras et al. 2013). At the same time, there are many Mn deposits present, associated with this volcanic activity, similar to those in the underlying Northern Cyclades unit but in a different stratigraphic and tectonic position. At the base of the Makrotantalon unit typical mica schists with garnets and pyrites are observed, which tectonically overlie the upper formations of the underlying Northern Cyclades unit, through a discontinuous zone of ultramafic serpentinised rocks. Thus, the Makrotantalon unit is thrusted over the Northern Cyclades unit, with a complex

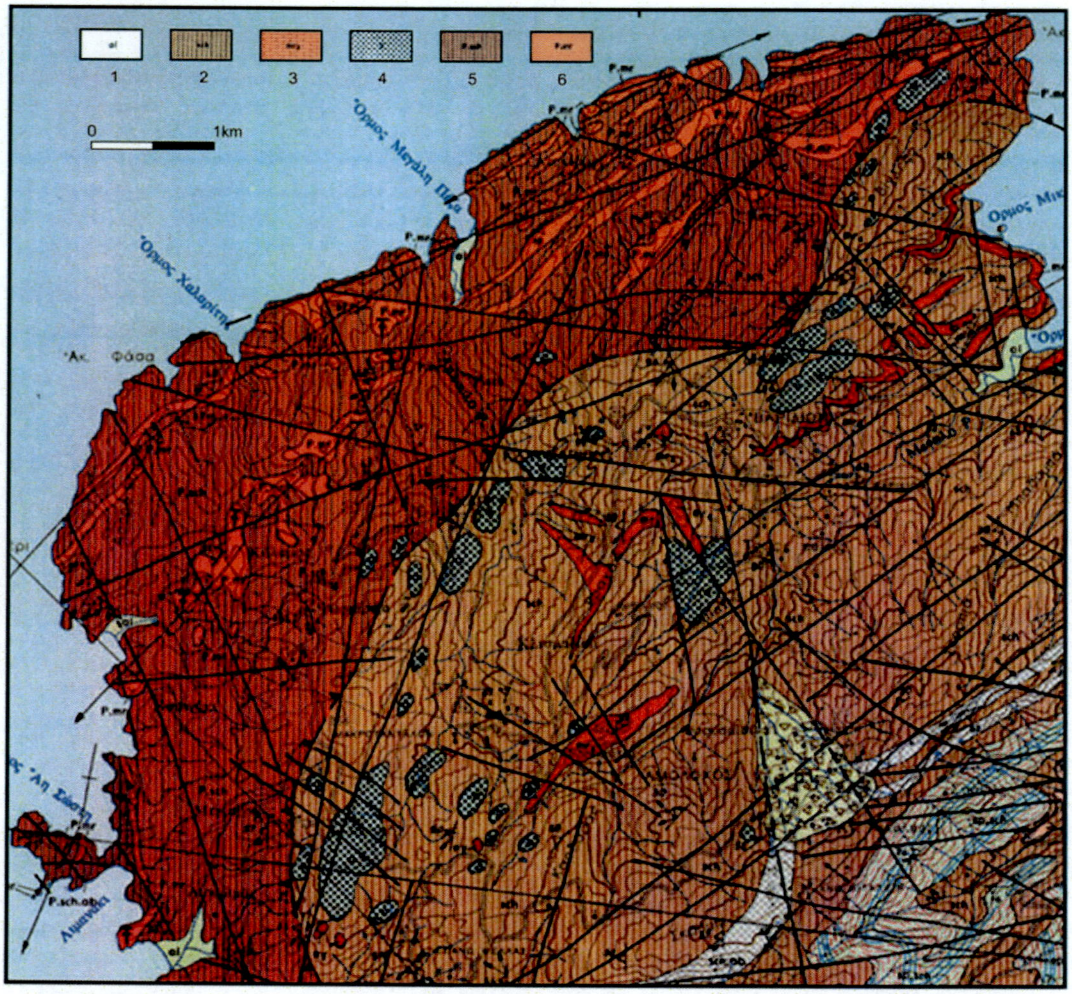

Fig. 8.56 Geological map of the Makrotantalon area in Northern Andros, showing the intercalations of the Permian marbles. Folded and schistosed ultra-mafic rocks are observed along the contact with the underlying Northern Cyclades unit (from the Gavrio–Andros–Piso Meria map, Papanikolaou 1978a). 1: Alluvium, 2: Upper schists of the Northern Cyclades, 3: Upper marbles of the Northern Cyclades, 4: serpentinised peridodites, 5: Markotantalon schists, 6: Makrotantalon marbles

tectonic contact in Northern Andros and Southern Evia. Especially in Andros Island, it seems that it is not a shallow thrust, but a very deep tectonic contact (at least 20 km deep), with contemporaneous metamorphism and isoclinal folding conditions and with presence of schistosed ophiolitic rocks (Papanikolaou 1978d, 1984c).

The metamorphism of the unit has the characteristics of blueschist metamorphism (Papanikolaou 1978d), which, more recently was determined at a pressure of 18 Kbars and a temperature of 550 °C in an initial phase, followed by a retrograde greenschist event with pressure values of 4–9 Kbars and temperatures of 450–550 °C (Huet et al. 2014). The age of the initial metamorphism is Upper Cretaceous, between 105 and 75 Ma, followed by a Late Eocene blueschist event between 45 and 35 Ma, with intermediate pressure values of more than 7 Kbars and temperatures above 300 °C. Finally, a retrograde greenschisht event, with a pressure drop but a rise in temperature, was dated between 25 and 20 Ma (Brocker and Franz 1998, 2006; Zeffren et al. 2005). The above dates suggest that this unit is an internal segment of H2 (more internal than the Pindos and Northern Cyclades units), which began to subduct during the Late Cretaceous and then followed the same path as the Northern Cyclades unit during the Eocene–Oligocene, with a final common exhumation process during the late Miocene.

Similar occurrences with the Makrotantalon unit can be observed in Northern and Southeastern Tinos Island, in the Tsiknia peninsula, where a metamorphic ophiolitic nappe with meta-sediments is overlying the Northern Cyclades unit. Its metamorphism has been dated also as Late

Cretaceous, between 95 and 70 Ma in meta-gabbro and phyllites (Brocker and Franz 1998; Zeffren et al. 2005).

8.2.5 The Pindos Ophiolitic Nappes–Northern Pindos–Central Greece–Crete–Dodecanese

Apart from the sedimentary pelagic sequences of the Pindos–Cyclades Ocean, there are also ophiolitic outcrops without syngenetic sediments, which belong to the Mesozoic oceanic crust of the H2 basin. These outcrops create a linear belt on a large-scale map, from the Greek-Albanian border to the north, along Western Macedonia, Western Thessaly, Central Sterea, Argolis, Crete, and the Dodecanese to the east of the arc, until we reach Karpathos and Rhodes (Fig. 8.57). Another smaller ophiolitic belt should be added in more internal position, in the form of a tectonic window, comprising the metamorphosed ophiolites of the H2 Ocean, which are tectonically integrated in the Cycladic blueschists during their Eocene subduction.

The ophiolitic outcrops of the H2 are found tectonically emplaced over the Eocene Pindos flysch as a rule and in the Northern Pindos mt they are covered unconformably by Oligocene molassic sediments of the Meso-Hellenic Trough (Brunn 1956; Aubouin et al. 1977; Papanikolaou 2009) (Fig. 8.58).

Therefore, the age of the tectonic emplacement of the H2 ophiolites is Late Eocene. We reach the same tectonic emplacement age also through the corresponding analysis of the position of the Cretan ophiolites, in the area of the southern slopes of Psiloritis mt. (Bonneau 1984; Papanikolaou 2009). Here the ophiolites constitute the uppermost tectonic nappe and due to a major extensional detachment fault, they are directly overlying the autochthon Mani unit (see also Fig. 8.101). In the hanging wall the ophiolites occur above the flysch formations of the Ethia unit and a few wedge-shaped outcrops of the Arvi lavas and of the Asteroussia metamorphics (Papanikolaou 2009) (Fig. 8.59).

In the Dodecanese region, the same structure is generally maintained, with the ophiolitic nappes found in the uppermost position over the Ionian, Tripolis and Pindos units (Hatzipanagiotou 1988). Both in Crete and the Dodekanese there are also geo-chronological datings of the rocks of the ophiolitic complex, which show Cretaceous ages (Delaloye et al. 1977; Koepke et al. 2002; Pe-Piper and Piper 2002). This is different from the Liassic ages determined by the metamorphic amphibolitic sole of the ophiolitic base in continental Greece (Spray and Roddick 1980; Pe-Piper and Piper 2002).

An important parameter for the Northern Pindos ophiolites is that inside the tectonic structure of the Eocene tectonic nappe, there are Triassic outcrops with lavas (Terry 1971; Migiros and Tselepidis 1990), in the context of a volcano-sedimentary formation, which probably correspond to the initial stage of the opening of the ocean. A description of the tectono-stratigraphy of the ophiolitic Pindos nappe by Jones and Robertson (1991) links the dismembered ophiolitic complex and a Mesozoic pelagic sequence of Pindos type. Finally, Late Cretaceous intra-oceanic magmatism has been reported also from the internal Dinarides in Bosnia and Herzegovina (Ustaszewski et al. 2009), probably belonging to the same oceanic area (H2).

8.2.6 The Miamou (Crete)–Aderes (Argolis) Unit

The Miamou unit was distinguished in Asteroussia mt by Bonneau (1976) as a separate formation of clastic rocks, in the form of a mélange, due to the chaotic presence of lenses–olistoliths of different sedimentary facies inside the flysch material. Such olistholites comprise Jurassic neritic limestones with coralls, mafic volcanic rocks and other limestones with pelagic fossils such as Globotruncanes of Upper Cretaceous age. The synthesis of the above data and field observations lead to the interpretation of a clastic wild flysch formation within an accretionary prism of the Paleocene–Eocene period. The tectonic position of the Miamou unit is above the Upper Eocene Ethia flysch and, wherever it is present, also above the Arvi unit. Thus, it has been probably formed in the front of the advancing internal nappes over the Pindos units during the Early Cenozoic.

A corresponding unit can be also observed in the Argolis peninsula, in the Aderes mountain range (Papanikolaou 1989a). This is also characterized by a chaotic structure, with the presence of olistoliths of ultra mafic and other mafic rocks, as well as granite, in addition to Upper Cretaceous pelagic limestones and formations of the Arvi unit occurring at the base of the nappe. This unit was formerly known in the bibliography as schist-sandstone complex of Ermioni (Marinos 1955; Aranitis 1963). Another small outcrop of similar composition and tectonic position occurs at the eastern slopes of the Parnon Mt, known as Glypia unit (Skourtsos et al. 2001).

The basic difference between the Laerma and the Eastern Kos units, previously examined in H1, with the Miamou and Aderes units involved in H2, is their different tectonic position and in lesser degree, the age and composition of the clasts. More specifically, the Laerma and Eastern Kos units

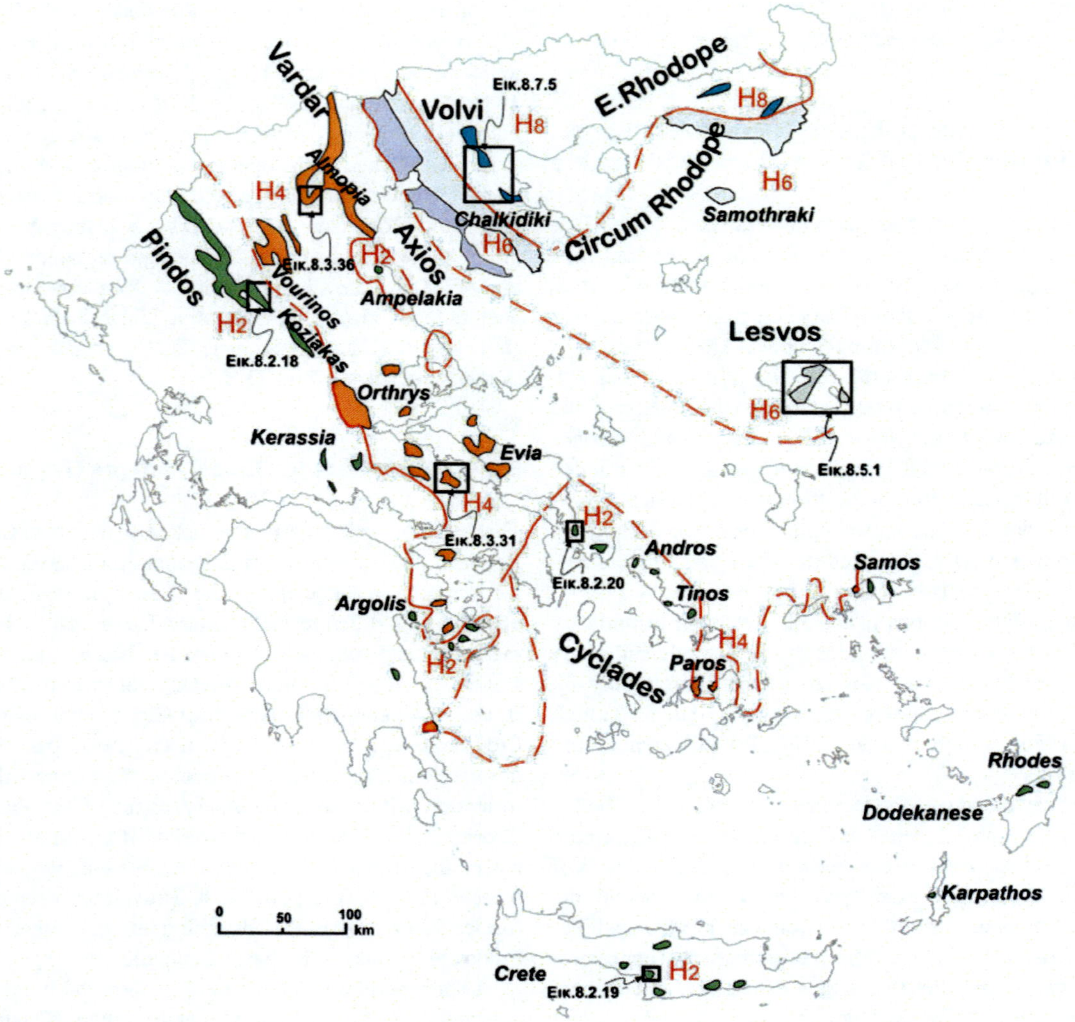

Fig. 8.57 Map of the major ophiolitic outcrops of the H2, H4, H6 and H8 basins (from Papanikolaou 2009). The map frames of the specific ophiolite outcrops on this map correspond to the following figures of the book: 8.2.18 = Fig. 8.58, 8.2.19 = Fig. 8.59, 8.2.20 = Fig. 8.60, 8.3.31 = Fig. 8.92, 8.3.36 = Fig. 8.97, 8.5.1 = Fig. 8.117, 8.7.5 = Fig. 8.129. Fig. 8.123 corresponds to the Ne part of Greece in E. Rhodope and Circum Rhodope

are located above metamorphic units of the External Hellenides platform and below the Pindos unit, while the Miamou and Aderes units are located above the Pindos unit, within the tectonic front of the internal units.

8.2.7 The Metamorphic Ophiolitic Nappes of the Cyclades

In the Cyclades region, the ophiolitic outcrops of the H2 have been tectonically emplaced during Late Eocene–Oligocene as shown by the underlying Almyropotamos Eocene flysch on top of the marbles and the overlying Eocene blueschists of the Northern Cyclades unit (Papanikolaou 2009) (Fig. 8.60). Similarly, there is a tectonic contact of the meta-basic rocks of the small, wedge shaped Aghios Ioannis unit over the relatively autochthon Kerketeas unit in Western Samos Island, occuring at the base of the Ampelos unit blueschists (equivalent to the Northern Cyclades unit) (Papanikolaou 1979a).

Apart from the ophiolites occurring at the base of the Northern Cyclades nappe, there are also ophiolites at the top, which comprise several occurrences in Southern Evia, Andros, Tinos, Syros and Samos islands. In Southern Evia the uppermost ophiolites are encountered in the complex tectonic contact between the underlying Northern Cyclades unit and the overlying Markotantalon–Ochi unit (Katzir et al. 2007). In Andros Island, they are observed along the

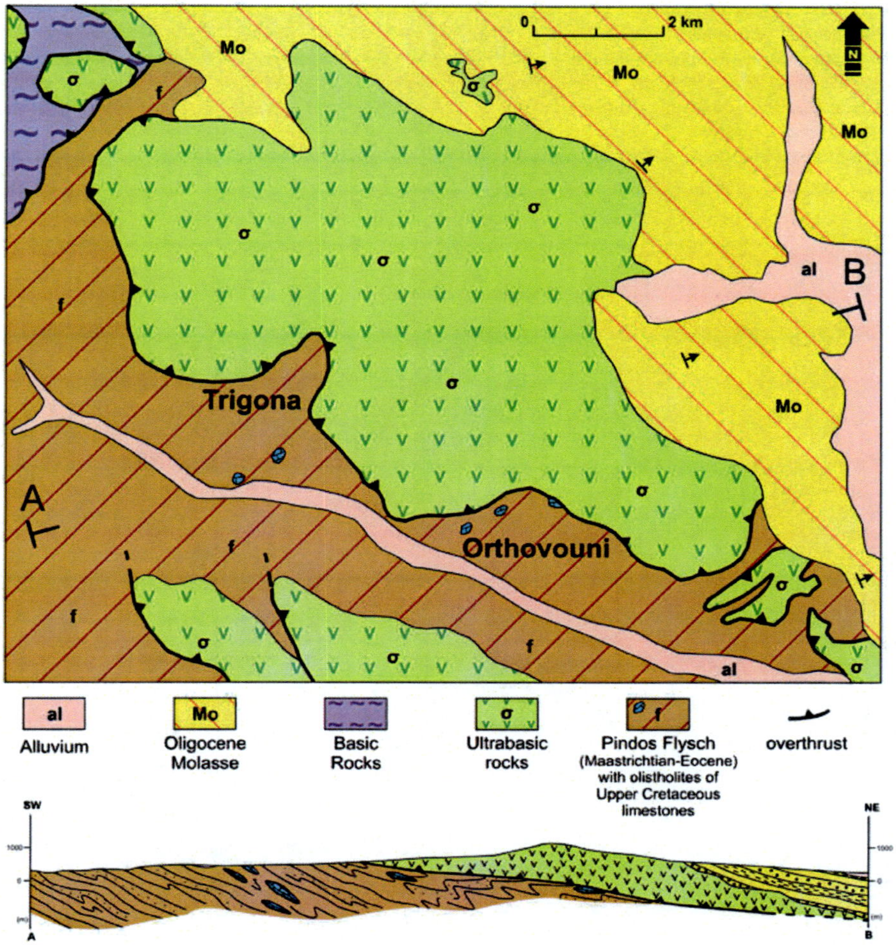

Fig. 8.58 Geological map and cross section of the Northern Pindos ophiolites (from Papanikolaou 2009). The ophiolites are observed tectonically emplaced on the Eocene Pindos flysch and are unconformably overlain by the Oligocene molasse

complex tectonic zone separating the Northern Cyclades unit from the overlying Makrotandalon nappe (Papanikolaou 1978a, d). Similarly, in Samos Island, they are located between the Northern Cyclades (Ampelos) and the Southern Cyclades (Vourliotes) nappes (Papanikolaou 1979a). In the case of Syros and Tinos islands, the ophiolites are found at the top of the blueschists (Bonneau et al. 1980).

The age of the Cycladic ophiolites ranges from the Jurassic to the Upper Cretaceous, as documented by the age determination of the meta-gabbros from the islands of Andros, Tinos, and Syros, at 80 Ma (Bulle et al. 2010), while the age of the blueschist metamorphism is Late Eocene–Oligocene (Philippon et al. 2012).

8.2.8 The Lavrion–Athens Allochthon Unit

The Allochthon unit of Lavrion overlies the relative autochthon unit of Attica and has been studied by Marinos and Petrascheck (1956) within their metallogenetic study of the ancient Lavrion mines, which have been re-exploited during the end of the 19th and most part of the 20th centuries. This unit gradually passes northwards to the complex Athens Allochthon unit, which is generally less metamorphosed (Papanikolaou 1986a; Papanikolaou et al. 2004b) (Fig. 8.61).

The outcrops of the unit in Athens are characterized by the difficulty to define a stratigraphic column, as there are many contradictions (Marinos et al. 1971, 1974; Tataris

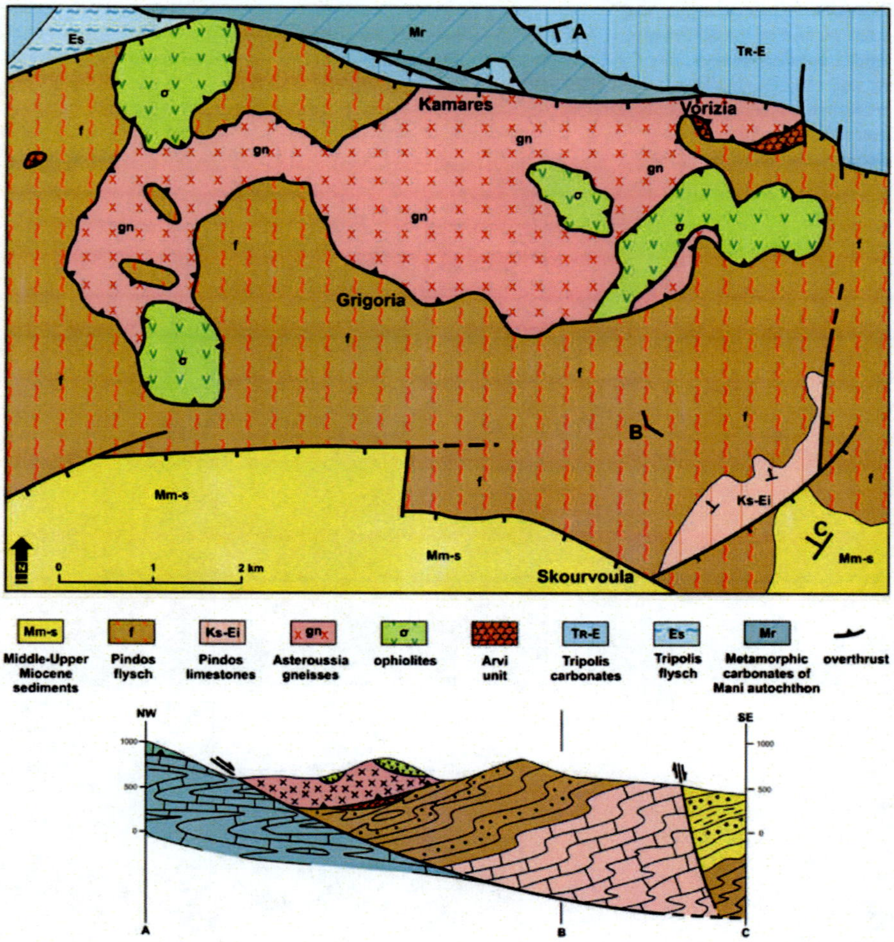

Fig. 8.59 Geological map and cross section of Southern Central Crete, showing the location of the ophiolitic nappe over the Arvi, Asteroussia, and Ethia units in the Late Eocene–Oligocene (from Papanikolaou 2009)

1972; Paraskevaidis and Chorianopoulou 1978). It contains a lot of large olistoliths within the shaly-sandy matrix with a variety of lithologies and ages and additionally it has been intensively deformed with cleavages and tectonic wedges, resulting in a mélange (Papanikolaou 1986a). The general picture is an arenaceous-pelitic slightly metamorphosed complex, partially of Cretaceous age, which contains blocks of a variety of ages and compositions especially at its base, such as in the area of Panepistimoupoli Zografou/Kessariani, with Triassic limestones, Jurassic limestones, basic igneous rocks, serpentinites etc. At the top of the unit Upper Cretaceous limestones can be observed both in Athens (Acropolis, Lycabettus, Turkovounia) and in Lavrion (Bertzeko area) (Lepsius 1893; Ktenas 1907c; Leleu and Neuman 1969). These limestones emerge from the phyllitic matrix due to differential erosion without having the uppermost stratigraphic position. More recent research has shown that the low grade metamorphic formations of the Athens plain are superjacent to the relatively autochthon unit of Attica with low angle normal faults (Papanikolaou et al. 2004b). The most characteristic lithologies in Lavrion are the sericite-chlorite schists, which include meta-basalts and meta-gabbros with parageneses of the blueschist type. From the tectonic point of view, transverse E-W structures (foliations and microfolds) are dominant, contrary to the autochthon unit which displays N-S to NNE-SSW structures (Marinos and Petrascheck 1956).

Geochemical research of the prasinites has shown that they originated from tholeitic type protoliths of the Mid-Ocean Ridge (Arikas et al. 2001), while their metamorphism shows blueschist type parageneses of 7–8 Kbars and 300–340 °C, followed by a retrograde greenschist event (Baltatzis 1996; Baziotis et al. 2009). It is evident that the metamorphism of the meta-mafic rocks is much higher than

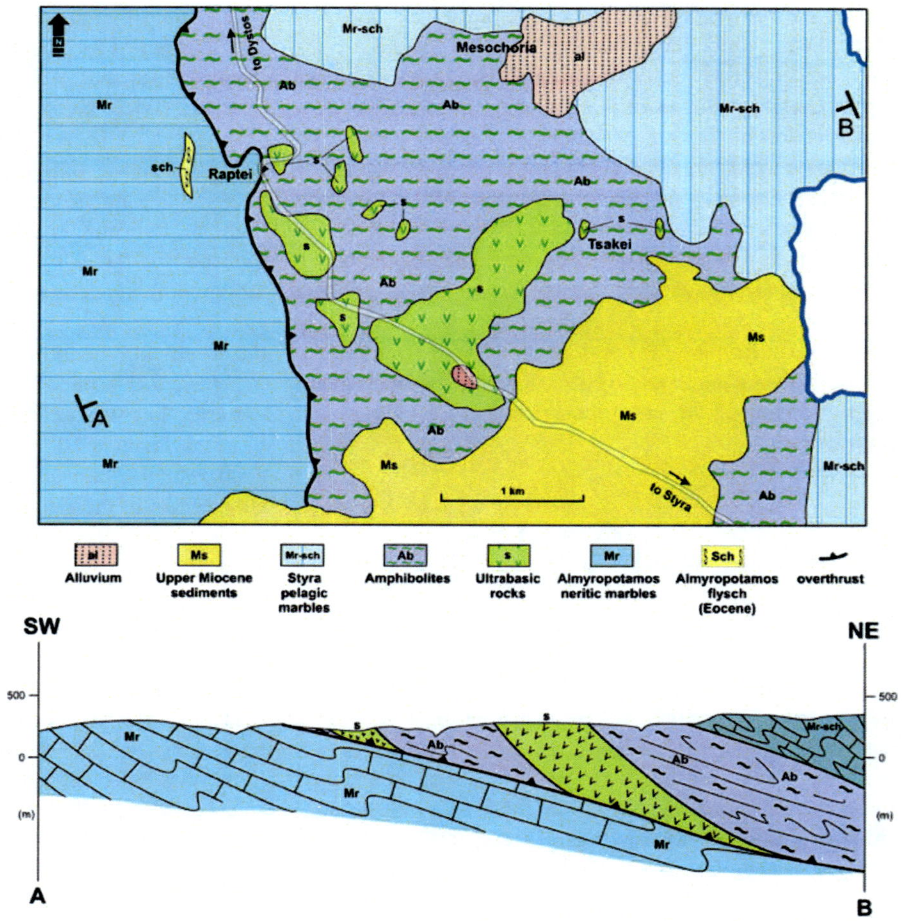

Fig. 8.60 Geological map and cross section of the ophiolites of Southern Evia, which are observed between the underlying Almyropotamos unit and the base of the Northern Cyclades (Styra) blueschists. The age of tectonic eplacement is Late Eocene–Oligocene (from Papanikolaou, 2009)

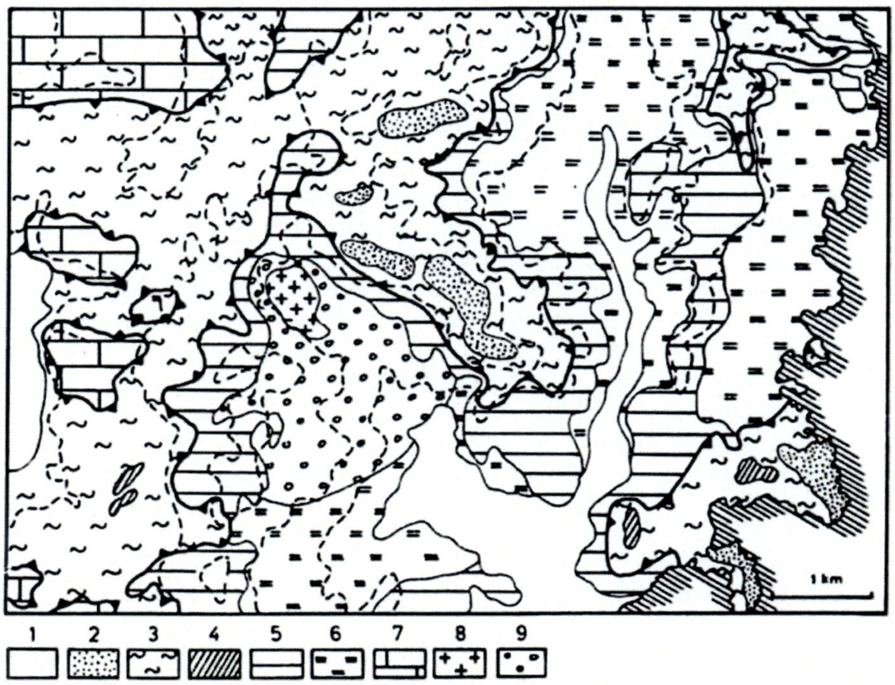

Fig. 8.61 Geological map of the Lavrion area, where the Lavrion nappe is observed (formations 2–4) above the relatively autochthon unit of Attica (formations 5–7) (modified from Marinos and Petrascheck 1956). 1: Quaternary, 2: crystalline limestones to marbles inside the Allochthon schists, 3: chloritic sericitic schists of the Lavrion unit, 4: prasinites inside the Allochthon, 5: Upper marble of the Autochthon, 6: Kessariani schists, 7: Lower marble, 8: Plaka granodiorite of Late Miocene, 9: hornfelses, as a result of contact metamorphism of the Kessariani schists with the Miocene granodiorite

that of the surrounding phyllites, thus creating a metamorphic hiatus, which confirms the olistolith nature of the meta-basites inside the clastic matrix. The presence of the Upper Cretaceous limestones inside the metamorphic chaotic formation of the Athens Allochthon shows a post-Late Cretaceous age of metamorphism, most probably during the Paleocene–Eocene. The Lavrion–Athens Allochthon unit generally resembles the Aderes unit, with the exception of the low grade metamorphism. Thus, they could belong to the same tectono-sedimentary formation, but in a different geodynamic position in the Hellenic arc during the Early Cenozoic. In this case, the Aderes unit would represent the front of the accretionary prism, the Athens Allochthon deeper parts along the subduction zone and the Lavrion unit even deeper.

8.3 The Internal Platform of the Hellenides—H3

8.3.1 The Parnassos Unit

The Parnassos unit is a neritic carbonate platform whose paleogeographic position was more internal than the Pindos basin (Renz 1940, 1955; Papastamatiou 1960; Celet 1962). It is similar to the Gavrovo and Tripolis units of the external carbonate platform but with a major difference in the transition to the flysch, Maastrichtian in Parnassos contrary to Priambonian in Gavrovo and Tripolis. The existence of transitional formations with slope characteristics in the **Vardousia** mt (Celet 1962, 1979) indicates the paleogeographic continuity between the Pindos basin and the Parnassos platform. However, the Parnassos unit is not present all along continental Greece. Today, the Parnassos unit forms an "almond" bordered between two significant E-W neotectonic zones, the Corinth rift from the south and the Sperchios river/valley in the north. The absence of the Parnassos unit north of the Sperchios valley and south of the Corinth Gulf cannot be justified by the neotectonic kinematics. Instead, these two neotectonic zones, may originate from the reactivation of older transform faults of the Tethyan realm (Aubouin and Dercourt 1975). The similarity of the High Karst unit in the Dinarides with the Parnassos platform, was considered as a possible re-appraisal of the Parnassos unit in the north beneath the internal nappes (Aubouin et al. 1970). Nevertheless, the High Karst unit could be homologous to the Parnassos unit, but paleogeographically independent, like another "almond" type isolated carbonate platform.

The stratigraphic column of the Parnassos unit is quite simple, like the Gavrovo one, consisting of a Mesozoic carbonate sequence that concludes with Paleocene-Eocene flysch (Figs. 8.62 and 8.63).

The transition from the carbonate sedimentation to the flysch occurs in the Maastrichtian/Danian boundary, just like in the Pindos unit. Two different transitional cases can be observed (Fig. 8.64).

In the final stages of the carbonate sedimentation, the Senonian neritic limestones with rudists are overlain by pelagic platy limestones with Globotruncanes in the Maastrichtian, indicating a gradual deepening of the basin. The transition to the flysch is marked by a thin formation of red-violet marly-limestones with Globigerines of Paleocene age. Thus, we have a classical transition for a ridge, based on the succession: neritic thick-bedded limestones—deepening—pelagic thin-bedded limestones—arrival of clastic material—flysch. In some locations though, an abnormal deposition of flysch is observed, related to syn-sedimentary tectonism, similar to the corresponding events of the Ionian in the Dogger and of the Gavrovo and Tripolis in the Eocene. The result is the development of particular facies of hardground along the fault surfaces, which is a reddish phosphorus-iron crust of several centimeters to decimeters thick, i.e. a concentrated pelagic sedimentation horizon (Fig. 8.65). Pelites, marly limestones and marls are present above the hard ground known as "red series" (Fig. 8.66) with a variable thickness of 5–50 m, which develops gradually, within 2–5 m, to the typical Eocene flysch. Exactly the same formation of red Paleocene pelites is also observed in the directly adjacent Western Thessaly–Beotia unit.

In conclusion, there are two transitional paths from the carbonate to the clastic sedimentation: the first case—the "orthodox" one—is conducted from the neritic limestones through the pelagic limestones with Globotruncanes of the Maastrichtian and a gradual development with transitional beds to the flysch, while in the second case hard ground formations (horizons of minimum thickness, but very wide stratigraphic range, with extremely slow sedimentation rate) are directly deposited over the Campanian or older limestones. The hard ground is known also as condensation horizons, that chrono-stratigraphically correspond to the sedimentation period of the platy limestones with Globotruncanes of the Maastrichtian and a portion of the transitional beds (Kalpakis 1979) (Fig. 8.67).

Thus, during the same time period when in some areas of the Parnassos platform transitional deposition of pelagic sediments (up to 100 m thick) was taking place, in some other areas only nodular phosphorus-iron crustal deposition occurred, 10 cm thick, which is an isochronous sediment, comprising the fossils of the Maastrichtian, Danian, and Paleocene. Therefore, we have another case of syn-sedimentary tectonism, which preceeds the deposition of flysch and the final tectonism of the unit (Richter and Mariolakos 1974, 1975). These synsedimentary tectonic structures occur in the fore bulge area, where bend faulting is observed, especially in cases of shallow water carbonate

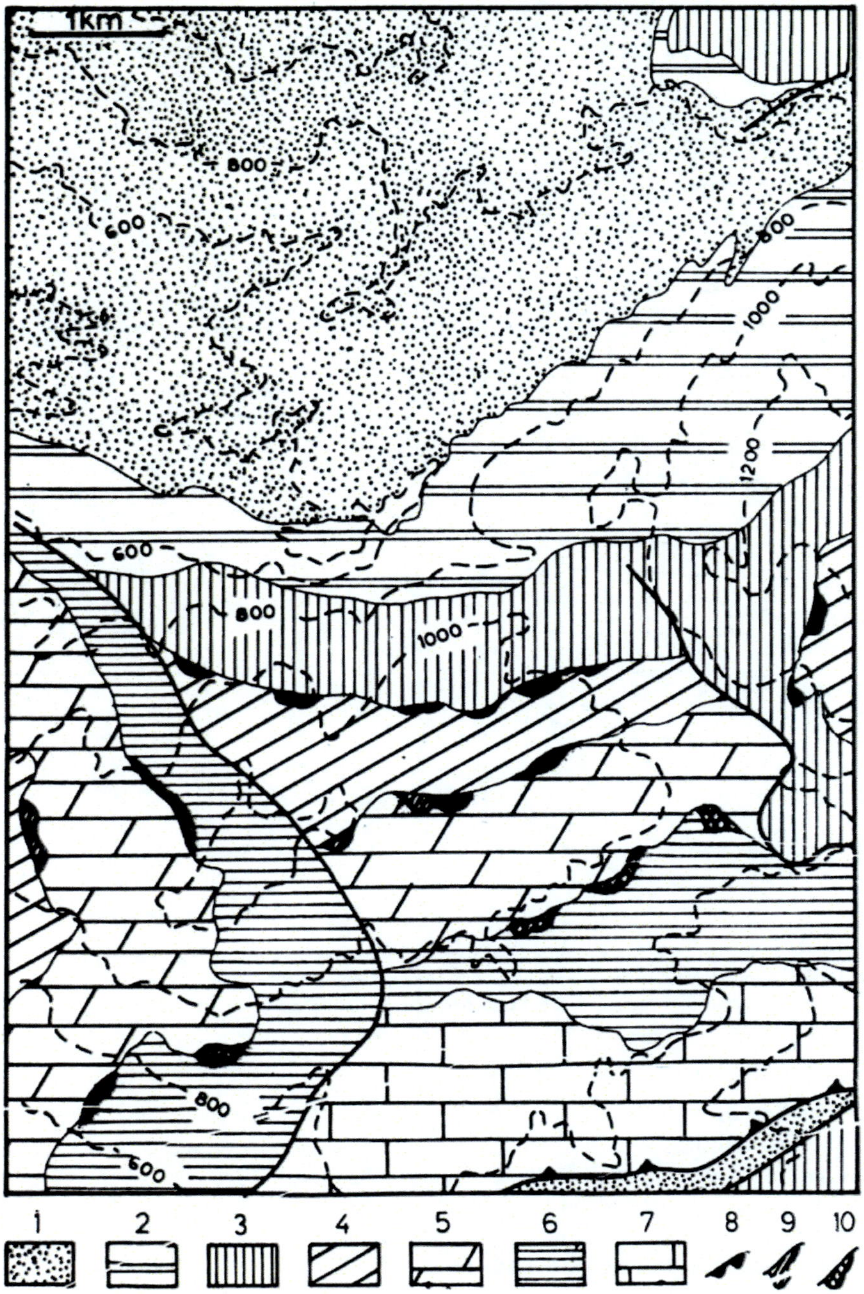

Fig. 8.62 Geological map of the Delphi area, where the entire stratigraphy of the Parnassos unit can be observed (from the Delphi sheet, Aronis et al. 1964). 1: flysch, 2: thin-bedded limestones of the Senonian–Paleocene with Globotruncana, Globigerina, etc., 50–70 m thick, 3: rudist-bearing limestones, gray to black, bituminous, ceiling of the upper b3 bauxite formation, Turonian–Senonian, with Rudistae, Miliolidae, Hippurites, Cuneolina, 80–100 m thick, 4: "intermediate" limestones, Tithonian–Cenomanian, in between the b2 and b3 bauxites, with gastropods, lamellibranches, molluscs, corals, and foraminifera Valvulinidae, Miliolidae, Trocholina, 400 m thick, 5: thick-bedded limestones, stiff, with Cladocoropsis mirabilis, Clypeina cf jurassica, Kurnubia jurassica, Pseudocyclamina etc., Upper Jurassic, ceiling of the lower bauxite layer b1, 300 m thick, 6: limestones of the Lower-Middle Jurassic, dark colored, bituminous, often oolithic, 200 m thick, 7: crystalline dolomites, white or gray, Upper Triassic, more than 600 m thick, 8: bauxites b3, 9: bauxites b2, 10: bauxites b1

platforms, where uplift and erosion is easy to result even through minor tectonic deformation.

With the exception of this pre-orogenic event in the Late Cretaceous, all the stratigraphic horizons of the shallow-water carbonate platform are present from the Late Triassic to the Late Cretaceous, either as limestones or dolomites with 1.700 m of total thickness, as it was calculated from the thickness of the stratigraphic formations shown in the geological maps. The distinction and mapping of the successive stratigraphic formations of the platform would be very difficult without the presence of the bauxite horizons. Their presence in the platform is due to several emergence events, followed by erosion, with creation of karstic cavities, filled with material either from the platform itself or from more internal paleo-tectonised areas with ophiolites and lateritization. The presence of a thin lignite

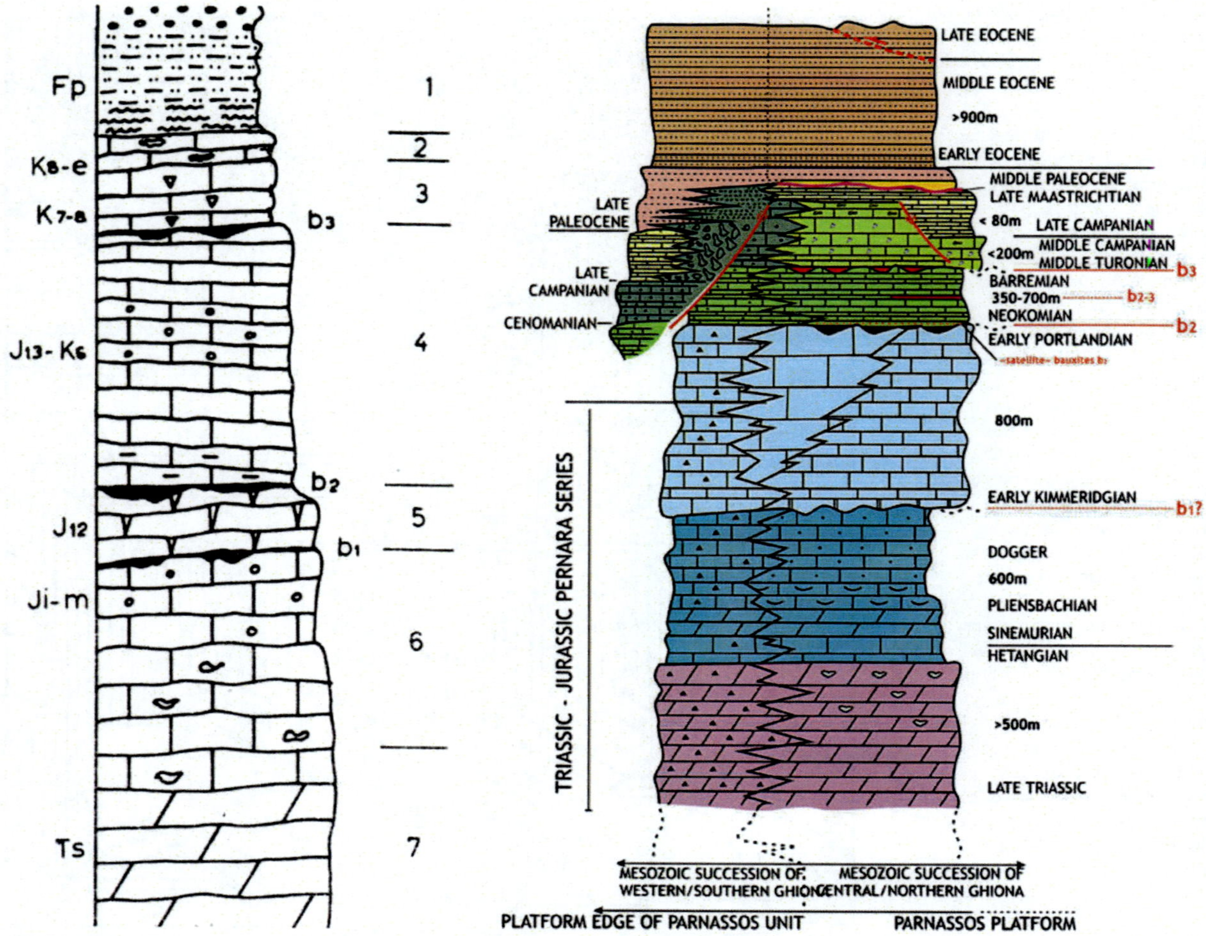

Fig. 8.63 **a** Stratigraphic column of the Parnassos unit (numbers 1–7 correspond to the legend of the map of Fig. 8.62, from Aronis et al. 1964). **b** Stratigraphic column of the Parnassos unit in Giona mt, distinguishing the Eastern Giona, which is similar to the Parnassos unit, from the Western Giona, which shows transitional facies towards the Vardousia unit (absence of bauxite horizons and presence of carbonate breccia facies) (from Gouliotis 2014)

layer at the base of the b3 bauxite in the Parnassos platform (Fig. 8.68), as well as thicker layers on top of the b3 bauxite, reported from some underground exploitations (Kalaitzidis et al. 2010) shows that for some period of time during the Cenomanian, a lagoon environment had been established over the uplifted and karstified limestones.

The Parnassos platform is the western margin of the internal platfrom of the Hellenides H3 and is also the closest to the Internal Hellenides, which had already been tectonized much earlier, in the Jurassic and the Early Cretaceous. Therefore, it was close to active orogenic events that affected it (unlike the other, more external areas) with short periods of uplift, emersion and erosion throughout the paleo-Alpine cycle. There are several such ephemeral events accompanied by bauxite deposition. However, the most extended and important ones are generally the following three (Fig. 8.69):

- b3: under the Upper Cretaceous, rudist-bearing limestones
- b2: in the Lower Cretaceous, usually within the Aptian
- b1: under the Kimmeridgian, in the Upper Jurassic, with Gladocoropsis bearing limestones.

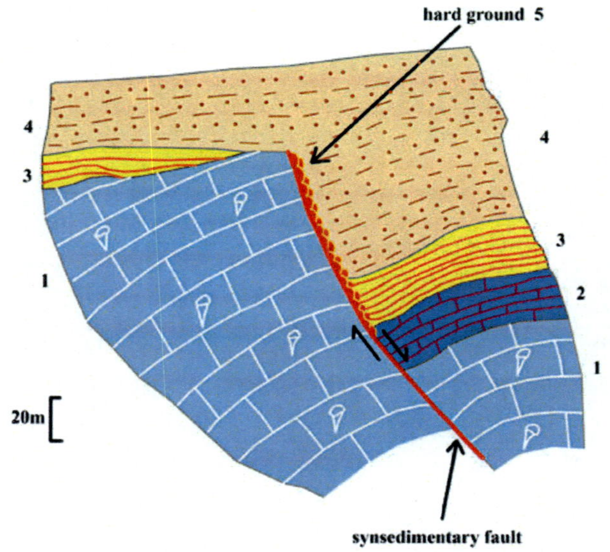

Fig. 8.64 Two cases of development from the Parnassos limestones to the flysch, based on the syn-sedimentary tectonism during the onset of the subsidence of the carbonate platform into the foreland basin. On the hanging wall, a gradual transition from rudist-bearing limestones (1), to pelagic limestones with Globotruncana (2) occurs, followed by red Paleocene pelites (3) and then by usual gray flysch (4). On the contrary, at the footwall, an irregular contact with underlying rudist-bearing limestones (1) and unconformably overlying red pelites of Paleocene age with Globigerines (3) and flysch (4). Condensation horizons of hard ground (5) are observed on the surface of the syn-sedimentary fault

The above three bauxite horizons are not necessarily present everywhere, and all events of emergence and erosion were not linked with bauxites. The most significant bauxite horizon is the b3 one, which occurs below the Upper Cretaceous (Cenomanian) transgression. This unconformity shows that the Parnassos unit is of an intermediate state between the internal and external units.

In Vardousia mt, transitional sedimentary sequences from the Parnassos platform to the Pindos basin occur throughout the Late Triassic–Late Cretaceous. At the lower thrust sheets of the Vardousia unit above the Pindos flysch some volcano-sedimentary formations occur below the Middle–Upper Triassic carbonate platform. These are clastic sediments with some pelagic limestones, of ammonitico rosso facies, as well as basic lavas and volcanic tuffs, whose age was determined by conodonts as Scythian–Anisian (Ardaens 1979). The front of the Vardousia nappe over the Pindos flysch is an impressive cliff of several hundred meters, along which a recumbent almost isoclinal fold of km scale is observed, immediately alove the basal thrust. Its sub-horizontal axial plane dips to the east and its inverted limb is thinned out, whereas its normal limb remains with its average stratigraphic thickness. The overall structure simulates the pattern of the fold-nappe, first described in the Swiss nappe of Morcles in the Western Alps (Lugeon 1902) (Fig. 8.70).

Fig. 8.65 **a** View of a syn-sedimentary fault, occurring along the contact of the Upper Cretaceous limestones with the red Paleocene pelites in the Distomon region. **b** Detail of the fault surface, where the hard ground crust is developed with a plethora of fossils

Fig. 8.66 Outcrop of the red Paleocene marly limestones–pelites, known as "red series", with Globigerines, from Eastern Giona mt

In the western outcrops of the Parnassos unit in the Southwestern Giona mt, an intersection of three transitional tectonic units occurs, in the form of tectonic wedges. These units are characterized by different sedimentary facies in various horizons. The most important tectonic wedge of the **Penteoria unit** shows a gradual deepening from the neritic Upper Triassic, with Megalodon, to the pelagic Jurassic, with ammonites (Papastamatiou and Tataris 1963; Gouliotis 2014) (Fig. 8.71). This early subsidence of the shallow water carbonate platform is observed at the transitional zone between the Parnassos platform and the Pindos Basin.

Generally, the Parnassos unit shows a complex tectonic contact to the west, with a relatively small displacement of a few tens of kilometers towards the Pindos unit. In some locations, the Parnassos limestones pass gradually to the flysch of the Mornos syncline, towards the transitional Vardousia units (Gouliotis 2014). On the other hand, towards the east, major tectonic klippen can be observed over the Parnassos unit from the Western Thessaly–Beotia nappe, such as for example, the Gerolekkas klippe (Celet 1979). It is noteworthy that along the northern slopes of the Parnassos mt in the Amfiklia region a low angle detachment fault brings in direct contact the Eastern Greece–Sub-Pelagonian units over the Parnassos unit (Kranis and Papanikolaou 2001). Similar large extensional detachment faults have been mapped along the Eastern Giona mt slopes, where they form the base of the supra-detachment basin of the Miocene Itea–Amfissa molasse (Gouliotis 2014) (see also Figs. 6.9 and 6.10). This tectonic event has been dated in the Middle Miocene (Papanikolaou et al. 2009).

8.3.2 The Western Thessaly–Beotia Unit

The main characteristic of the Western Thessaly–Beotia unit is the existence of an Upper Jurassic–Lower Cretaceous "flysch" established over various paleogeographic areas, both neritic and pelagic (Papanikolaou and Sideris 1979). Thus, in the stratigraphic columns of the Koziakas mt, Western Othrys mt, Northern Iti mt (in tectonic klippen), Beotia (from Livadia to Gerania mt) and partly in Argolis, there are Triassic–Jurassic sequences where during the Kimmeridgean, and more specifically, the Tithonian, there is deposition of terrigenous clastic material, creating a kind of "flysch". This formation resembles the "first flysch" of Pindos, only that here it starts much earlier in the Tithonian–

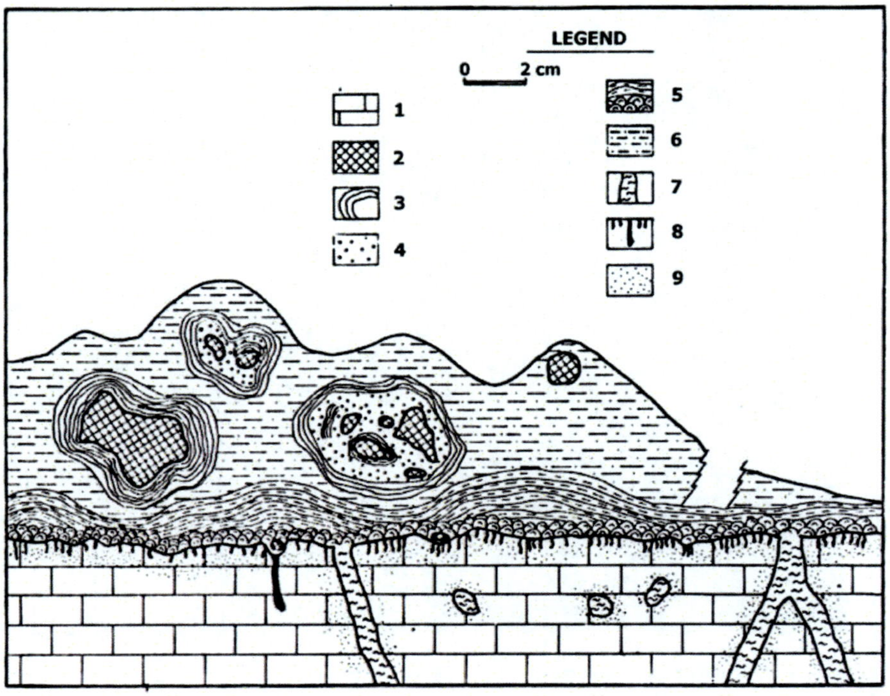

Fig. 8.67 Detail of the transition from the Cretaceous limestones to the red Paleocene pelites of the Parnassos unit, in the Osios Loukas region (from Kalpakis 1979). 1: Upper Cretaceous biomicrites, 2: fragments of underlying limestones, quartz and igneous or metamorphic rocks, 3: successive crustations of iron-oxides (gaetite), 4: the previous iron-oxides without internal structure, 5: Iron-phosphate crustations, upper part: stromatolites LLH type, lower part: stromatolites SH type, 6: red compact marls, 7: nests of P-Fe constituents and/or micrite, 8: bio-tunnels and micro-cracks, 9: infiltration by brownish oxides

Berriasian, while in Pindos it starts approximately in the Cenomanian (Fig. 8.72). However, the Tithonian–Berriasian "flysch" with turbidites, sandstones, pelites, clastic limestones, with Calpionelles as the typical fossil, are overlain in the Cenomanian–Turonian, and especially during the Campanian–Maastrichtian, by pelagic carbonate sedimentation with Globotruncanes. This Upper Cretaceous formation contains also characteristic slope facies with clastic breccia-limestones, with breccias coming from ophiolites and cherts, known as "Thymiama facies" (Aubouin 1959). The pelagic limestones are deposited until the Maastrichtian–Danian, when the typical Tertiary flysch begins. The transition from the Upper Cretaceous pelagic limestones to the flysch is unique, marked by the presence of red Paleocene pelites, exactly like in the Parnassos platform, but without the hard ground crusts and syn-sedimentary tectonism. This is probably due to the Upper Cretaceous facies of Western Thessaly–Beotia unit, which is already pelagic, before the onset of the flysch sedimentation.

The existence of the Tithonian–Lower Cretaceous "flysch" is therefore the unifying element of all these paleogeographic regions, which comprise the Western Thessaly–Beotia units, tectonically underlying the internal units (Fig. 8.73). Their particular features are:

(i) The existence of an Upper Jurassic–Lower Cretaceous clastic formation, meaning that for several millions of years it was a deep basin, filled with clastic material, originating from the more internal cordillera–island arc of the Paleo-Alpine orogeny.
(ii) The Western Thessaly–Beotia units are Early Tertiary nappes above the Pindos and Parnassos units coming from more internal paleogeographic areas.
(iii) The clastic material is dominated, by ophiolitic detritus and cherts. This is the indirect evidence that during the Jurassic/Cretaceous boundary, a segment of the Tethyan oceanic crust (from the H4 terrane) emerged through obduction and started to be eroded along the evolving orogenic arc.

The major significance of these units is that they are the innermost units with continuous stratigraphic columns from the Triassic to the Eocene. The observation of interbedded clastic sequences of the flysch type shows that during the Tithonian–Berriasian they occupied the external edges of the foreland basin of the orogenic arc. However, this area was not filled up and did not participate in the orogenic process of the adjacent island arc for unknown geotectonic reasons. Thus, in the Internal Hellenides all stratigraphic columns reach up to the Tithonian–Berriasian, forming the paleotectonised lower Triassic–Jurassic sequences. Their upper sequence comprises the unconformable Upper Cretaceous shallow carbonates and the Tertiary flysch, tectonised during the Late Eocene. The resulting composite Eastern Greece unit is overthrusted on the Western Thessaly–Beotia units.

Fig. 8.68 Thin lignite horizon observed at the base of the b3 bauxite, showing a lagoon environment during the Cenomanian, prior to the bauxite deposition and the subsequent transgression of the sea

Fig. 8.69 View of the two upper bauxite horizons, b2 and b3, in Northern Giona mt. The occurrence of the "intermediate" limestones of Lower Cretaceous age, observed between the two bauxites is characteristic

Fig. 8.70 The front of the recumbent isoclinal fold-nappe of the Vardousia basal thrust sheets over the Pindos flysch, seen from the west

It should be noted that the so-called "Beotian flysch" (Clement 1971; Celet and Clement 1971) and the "Beotian zone" (Celet et al. 1976), practically represent the «upper schist-chert formation» of the Eastern Greece unit (Voreadis 1932, 1938; Renz 1955; Tataris 1967a, b). The two similar tectono-stratigraphic formations represent equivalent geodynamic environments, but they concern different cases, because the Boetian flysch is an interstratified formation within the continuous sequence from the Triassic to the Eocene, whereas the schist-chert formations are the top of the paleo-tectonised units and are unconformably overlain by the Upper Cretaceous transgression (Papanikolaou 1990). Thus, any flysch occurrences of Upper Jurassic–Lower Cretaceous age not continuing into the Upper Cretaceous–Eocene sedimentation do not belong to the Western Thessaly–Beotia unit, but to the Sub- Pelagonian or the Maliac units within the broader composite Eastern Greece unit.

In Western Thessaly, it has been documented that (Fig. 8.74): (i) there is a lateral transition between the Pindos and Western Thessaly in the Tavropos region during the Cretaceous–Eocene (Papanikolaou and Lekkas 1979), (ii) an ophiolitic complex occurs in the Eastern Koziakas mt, stratigraphically related with the Malm radiolarites, approximately at the position of the "Upper Jurassic - Lower Cretaceous flysch" (Capedri et al. 1985), and (iii) a Triassic clastic formation was reported from the base of the Koziakas sequence, similar to the "Triassic flysch" of the Pindos unit (Lekkas 1986, 1987). More recent studies have indicated: (i) the presence of Triassic radiolaria bearing sediments, associated to the basic volcanic rocks above the Upper Jurassic radiolarites of the Koziakas sequence, forming an ophiolitic melange (Chiari et al. 2012). (ii) a Lower–Middle Jurassic age of the amphibolites occurring at the base of the overlying ultrabasic rocks, thrusted on top of the Upper Jurassic ophiolitic melange (Pomonis et al. 2002).

Comparing the sequences of the Beotia and Western Thessaly units, we can observe that the Western Thessaly has pelagic features, passes laterally towards the Pindos basin and has a syngenetic relation with the Jurassic ophiolites of the Eastern Koziakas mt, whereas the Beotia has neritic features and a lateral transition to the Parnassos platform. It is characteristic that in the Beotia sequence the bauxite horizon b1 occurs in the Late Jurassic, exactly like in the Parnassos platform, suggesting that the two units were part of the same H3 platform until the Late Jurassic. Additionally, the Koziakas mountain range, predominately structured by the Western Thessaly unit, was originally known as the "Ultra-Pindic Zone" (Aubouin 1959), implying the extension of the Pindos basin eastwards towards the ophiolites of the Internal Hellenides.

8.3.3 The Southern Cyclades Unit

The Southern Cyclades unit was distinguished from the Northern Cyclades unit by Papanikolaou (1980c) and

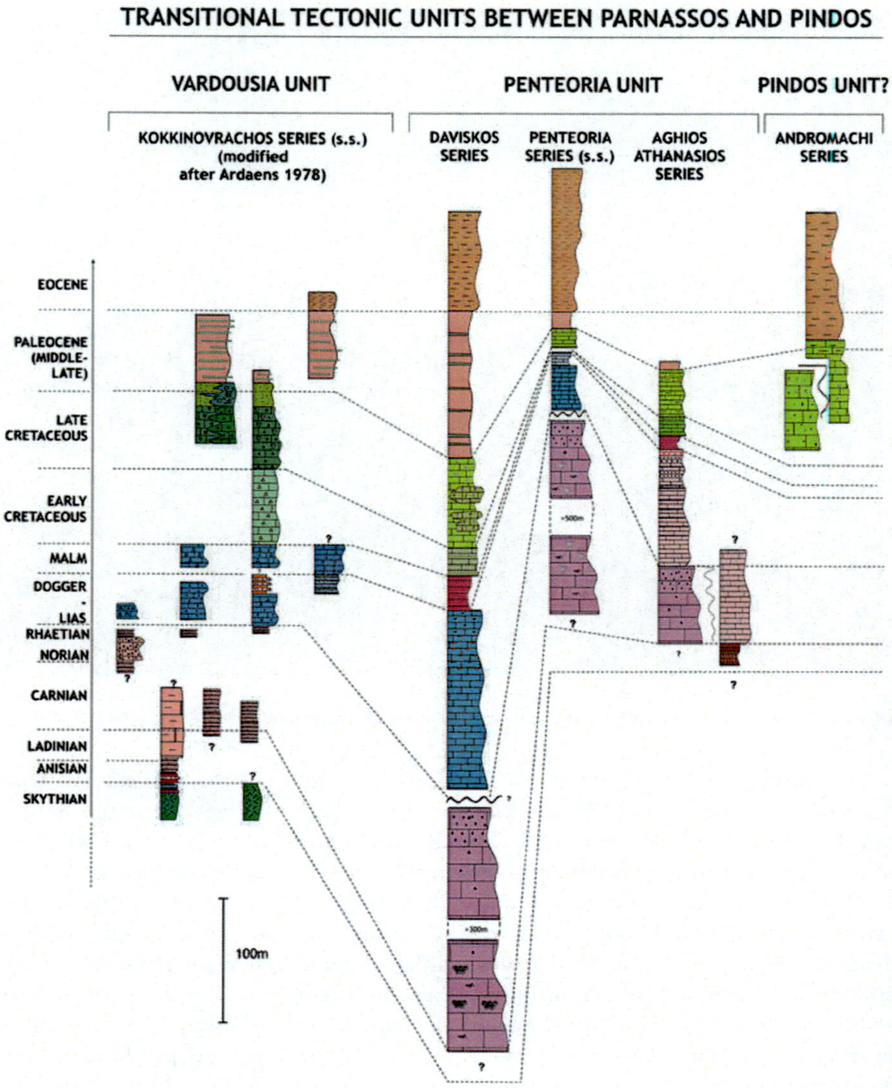

Fig. 8.71 The lateral transitional facies in the intermediate tectonic units between the Parnassos platform and the Pindos basin in the Southern Giona mt (from Gouliotis 2014)

includes the islands of Naxos, Paros (Marathi unit), Antiparos, Sifnos, Sikinos, Ios, Folegandros, Iraklia, as well as some other homologous units, whose stratigraphic columns have minor changes, like the relatively autochthon Ikaria unit (Papanikolaou 1978b), and the Vourliotes unit in Samos (Papanikolaou 1979a). The characteristic litho-stratigraphic column of the unit, like in Paros Island (Papanikolaou 1980a, 1996), comprises thick neritic marbles with emery deposits (meta-bauxites). At the base of the column, there are also mica and amphibolitic schists with minor interbeds of pelagic marbles with silex. This basal schist formation represents the metamorphosed volcano-sedimentary complex of Permian–Triassic age, occurring below the carbonate platform. Metamorphosed clastic formations of small thickness, probably representing flysch, are observed also at the top of the column, probably of Eocene age. These upper schsits are usually observed, at the core of recumbent synclines of consecutive isoclinal folds (Fig. 8.75). Thus, the stratigraphic column comprises a Permian–Triassic volcano-sedimentary complex at its base, followed by a thick sequence of Triassic to Upper Cretaceous marbles (with meta-bauxites) and a type of meta-flysch at the top (with the existence of wild flysch–mélange as an uppermost bed, with ophiolitic blocks etc.). A similar litho-stratigraphic distribution is also observed in Naxos Island, where the Triassic marbles (Negris 1915; Dürr and Flugel 1979), the emery deposits (Papavasiliou 1914; Papastamatiou 1951), as well as the isoclinal folds of marbles-schists with N-S orientation (Bonneau et al. 1978) are well known.

The Triassic has been dated mainly in Naxos Island (Negris 1915; Dürr and Flugel 1979), while the Upper Cretaceous in the Vourliotes unit (Southern Cyclades) of

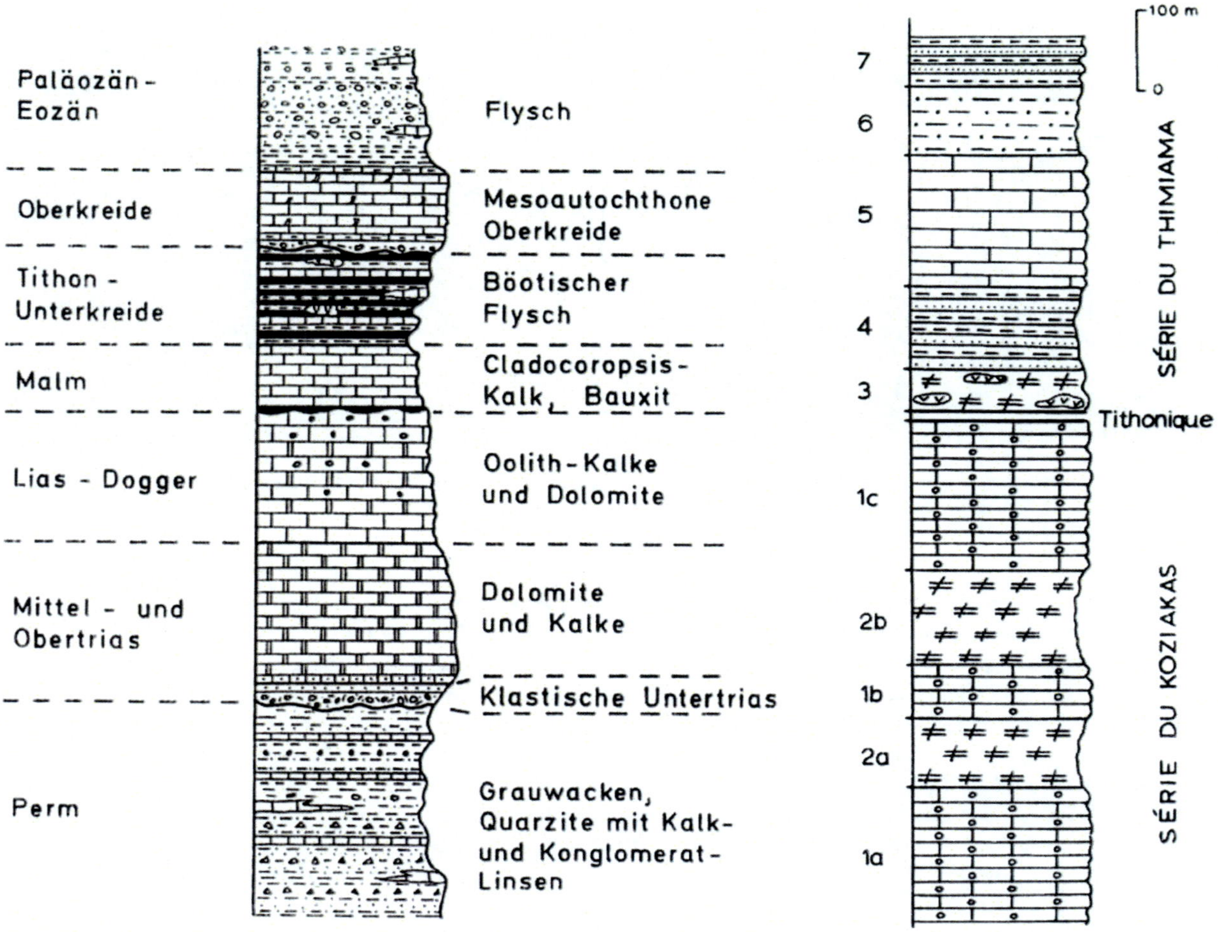

Fig. 8.72 a Stratigraphic column of the Beotia unit (from Dercourt et al. 1980). b Stratigraphic column of the Western Thessaly unit (from Papanikolaou and Sideris 1979). 1a-c: successive limestone horizons of Koziakas, of the Upper Triassic–Tithonian, 2a-b: intercalations of radiolarite horizons of Koziakas, 3: polygenetic ophiolitic conglomerate inside radiolaritic matrix, 4: Lower Cretaceous flysch, 5: Upper Cretaceous limestones of the "Thymiama facies", 6: red Paleocene pelites, 7: Tertiary flysch of Thymiama

Samos Island (Papanikolaou 1979a), where it is overthrusted on the Ampelos unit (Northern Cyclades) (Fig. 8.76). The stratigraphic column of the Southern Cyclades is different in Ikaria and Samos, because thick schists/amphibolites formations are observed between the Triassic and the emery bearing marbles.

Another characteristic feature of the Southern Cyclades unit is that in only a few cases (e.g. Folegandros, Sikinos, Ios) typical blueschist parageneses can been found, as this early HP/LT metamorphism has been overprinted by the retrograde greenschist metamorphism, accompanied by massive granitic intrusions. The migmatite of Naxos occurring at the central zone of the metamorphic core complex is a typical case (Fig. 8.77) (Jansen 1977) (as also in Paros, Ikaria, Mykonos, etc.) (see also Fig. 4.23).

The succession of the metamorphic facies in Naxos Island shows a Late Eocene blueschist phase, with pressure values of more than 10 Kbars and temperatures of 450–480 °C, followed by a retrograde greenschist event with pressure values of 5–7 Kbars and temperatures of 500–700 °C (Buick and Holland 1989; Avigad 1998; Duchene et al. 2006). The anatexis and migmatization phenomena are associated with the rise of granite bodies in the Late Oligocene–Early Miocene. The final tectono-magmatic event comprises granite intrusions with contact metamorphism in the Late Miocene. Similar conditions are observed in Sifnos and Milos islands (Jansen 1977; Okrush et al. 1978; Komprobst et al. 1979; Andriessen et al. 1979; Schliestedt et al. 1987; Wijbrans and McDougall 1988; Roche et al. 2016) (see also Figs. 4.23 and 7.29). In Ios Island, the Eocene blueschist

Fig. 8.73 **a** The Upper Cretaceous limestones (Ks) of the Beotia unit, overlying the Boetian flysch (Ki) in the Distomo region, in front of the Parnassos mountain range, made of the Mesozoic carbonate platform (Tr-J). **b** Outcrop of asymmetric folding with curved axial surfaces and disharmonic phenomena in the Lower Cretaceous Beotian flysch of the Distomo region

metamorphism M1 was characterized by a minimum pressure of 9–11 Kbars and a temperature of 350°–400 °C, while the Miocene greenschist metamorphism M2 occurred at a pressure of 5 Kbars and a temperature of 380°–420 °C (Van der Maar and Jansen 1983). On the contrary, very high pressures and temperatures, of blueschist to eclogitic facies, were recently described in the Southern Cyclades unit, from the islands of Sikinos and Folegandros, with 17–25 Kbars and 480–550 °C, followed by a retrograde greenschist event, with 4–7 Kbars and 440–470 °C (Augier et al. 2015).

Geochronological data have shown that all the granitic intrusions of the Cyclades are of Miocene age (Altherr et al. 1982; Wijbrans and McDougall 1988). Their latest products are the non tectonised, younger Upper Miocene granites, such as the granodiorite of Western Naxos, the granite of Serifos, the granites of Tinos, the granodiorite of Lavrion etc. Generally, the tectono-metamorphic and tectono-magmatic cycle in the Cycladic region is terminated during the Late Miocene (Soukis 2011). Of course, this last phase of deformation, metamorphism, and magmatism is not restricted to the Southern Cyclades unit, but to all the units occupying the maximum curve of the arc along the medial tectono-metamorphic belt and coincide with the position of the volcanic arc during Miocene (see also Figs. 7.28, 7.29, 7.30 and 10.8). However, the Miocene granites of the central zone of the Cyclades have been interpreted more recently as a localised thermal event, along a E-W to ENE-WSW zone, resulted from the transverse slab tear along the Aegean/Anatolian microplate boundary and the subsequent upwelling of magmas from the mantle flow above the subducting Aegean slab (e.g. Jolivet et al. 2013, 2015; Brun et al. 2016).

The differentiation of the Southern from the Northern Cyclades unit concerns: (i) lithostratigraphy and paleoenvironment, with pelagic sequence in the Northern Cyclades versus shallow water carbonate platform in the Southern Cyclades (ii) bauxites from ephemeral emersions of the carbonate platform in the Sourthern Cyclades versus Mn-bearing volcano-sedimentary formations in the Northern Cyclades, (iii) NE-SW to ENE-WSW orientation of the early synmetamorphic HP/LT tectonic structures in the Northern Cyclades versus N-S to N10°E tectonic trend in the Southern Cyclades. (iv) low grade metamorphic nappes of the Dryos–Mesaria units and/or the non-metamorphosed Cycladic nappe (composed of ophiolites, Cretaceous limestones and Lower Miocene molasse) are observed in tectonic klippen above the Southern Cyclades (e.g. Paros, Naxos) (Fig. 8.78), contrary to the medium grade HP/LT units of Makrotandalon–Ochi, observed above the Northern Cyclades units. (v) Pre-Alpine basement rocks occur only beneath the Southern Cyclades carbonate platform.

The occurrence of **the pre-Alpine basement in the Southern Cyclades**, which is lacking from the Northern Cyclades, is a difference expected from the crustal nature of the two units, as a pre-Alpine basement is expected below the carbonate platform (H3) contrary to the pelagic sequence of the Northern Cyclades which is associated with the ophiolites of the Pindos-Cyclades oceanic basin H2. It is remarkable that the Olympus–Almyropotamos–Kerketeas external carbonate platforms of H1 are tectonically overlain by the Northern Cyclades unit (see also Fig. 8.54).

The contact between the underlying gneiss and the Southern Cyclades nappe is a tectonic decollement, whose displacement cannot be determined (Papanikolaou 1980a) (see also Fig. 8.75). Thus, in Paros Island, different stratigraphic horizons are observed against the contact with

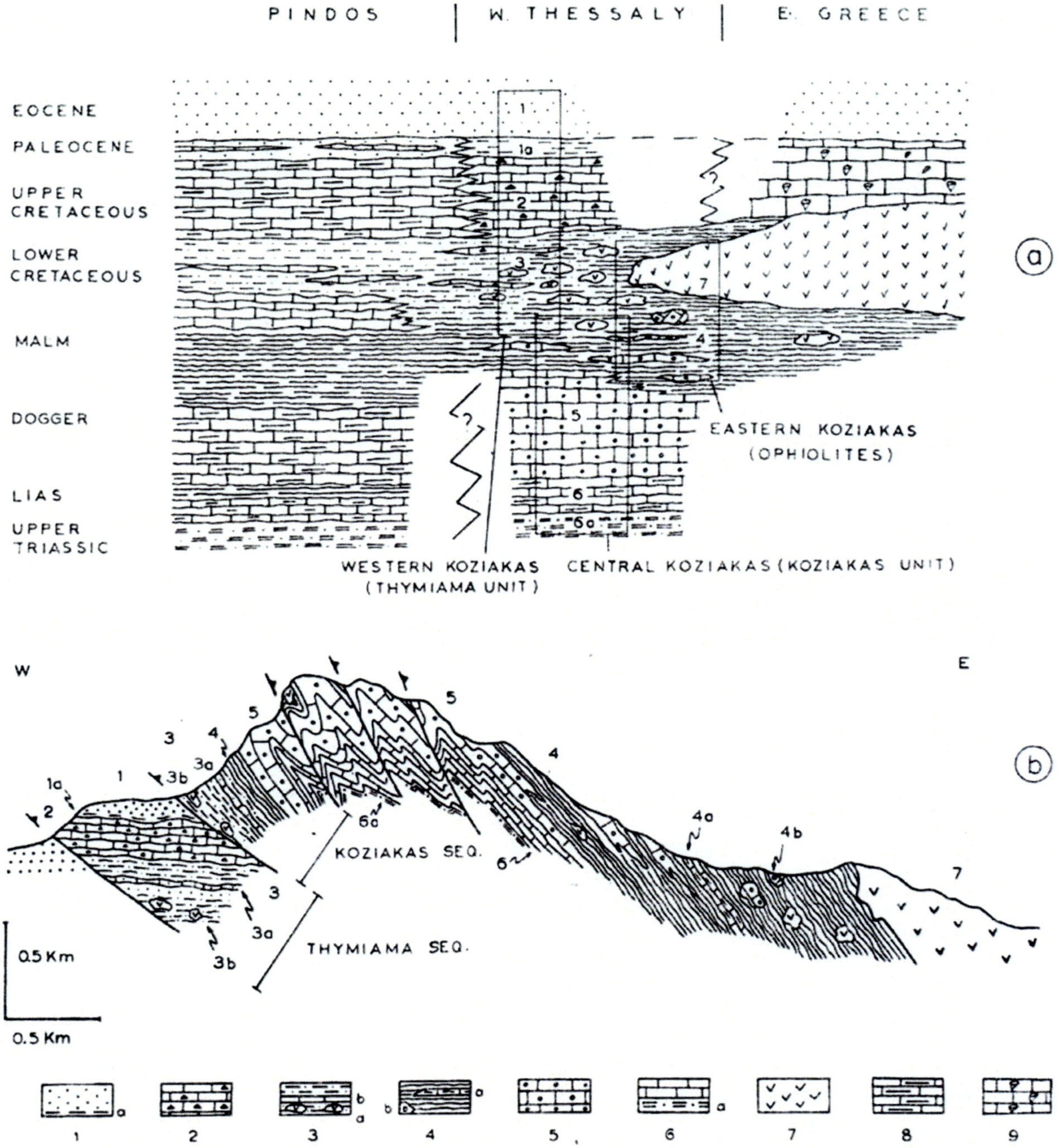

Fig. 8.74 **a** Schematic stratigraphic diagram showing the possible relations between the Pindos, Western Thessaly, and Eastern Greece units (from Papanikolaou and Lekkas 1979a, b). **b** General geologic cross section of the Koziakas mountain range (from Capedri et al. 1985). 1: Tertiary flysch, 1a: Paleocene red pelites, 2: Upper Cretaceous micro-breccia pelagic limestones «Thymiama facies», 3: Upper Jurassic–Lower Cretaceous clastic sequence, rich in ophiolite clasts («Beotian flysch»), 3a: limestones with calpionelles, 3b: radiolarites and red pelites, 4: radiolarites, 4a: intercalations of pelagic limestones, 5: oolite limestones of Dogger–Malm, 6: Upper Triassic–Liassic limestones, 6a: Triassic clastic formation, 7: ophiolites, 8: pelagic limestones with silex, 9: rudists bearing limestones

the underlying gneisses with varying thicknesses, depending on their position in the multiple isoclinal folding (Fig. 8.78).

Additionally, in Ios Island deep level extensional detachment faults with mylonites have been described, which have brought in contact the superjacent metamorphic

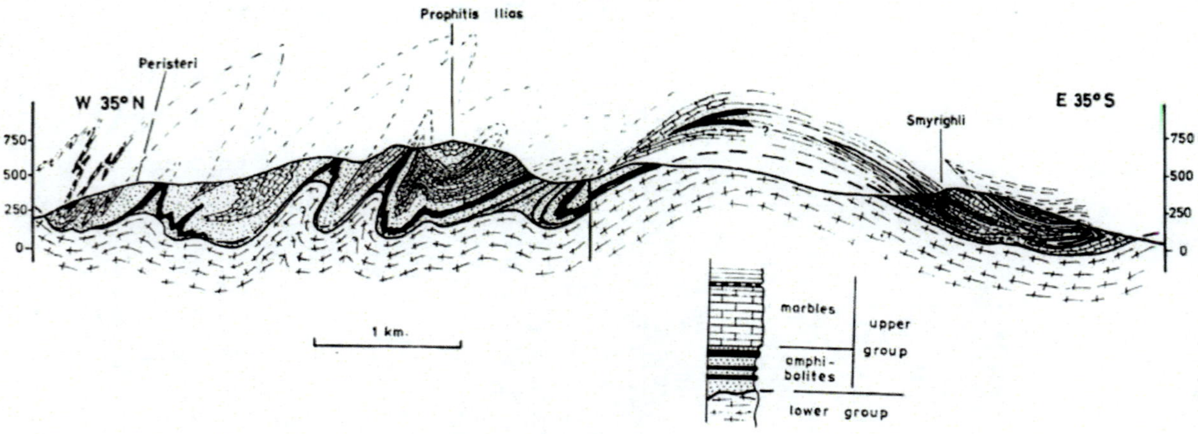

Fig. 8.75 Transverse geological section of Paros Island, showing the complex deformation of the Southern Cyclades unit as well as the disharmony with the underlying gneisses of the pre-Alpine basement (from Papanikolaou 1980a)

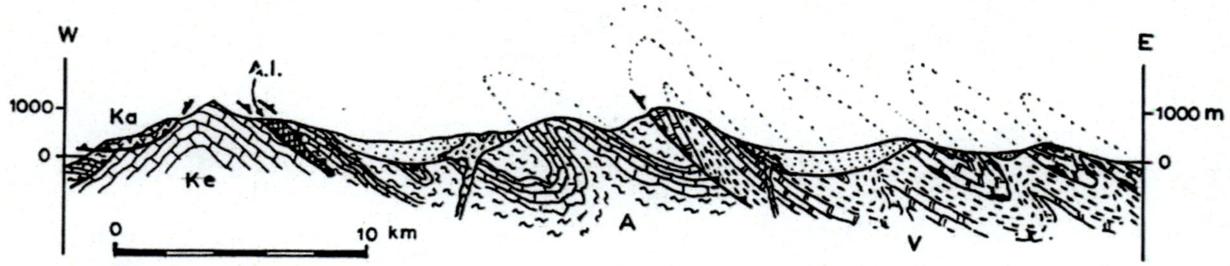

Fig. 8.76 Synthetic geological cross section of the island of Samos, which includes the units of Kallithea (Ka, Cycladic), Kerketeas (Ke), Ambelos (A, Northern Cyclades), Vourliotes (V, Southern Cyclades) and Aghios Ioannis (A.I., metamorphic mafic igneous rocks) (from Papanikolaou 1979a). The Late Miocene volcanics at the margins of the two Neogene basins of continental/lacustrine facies are also shown

rocks with the gneiss core (Forster and Lister 1999). Another tectonic contact several meters thick of shallow tectonic level, with cataclastic character has been observed, between the underlying garnet bearing mica schists of the Paleozoic basement with the superjacent marbles and blueschists of the Southern Cyclades unit (Fig. 8.79).

Consequently, the Southern Cyclades unit can be characterized either as para-autochthonous or as allochthonous, regarding their relation with the pre-Alpine basement. In the first case, the pre-Alpine basement should belong to the H3 basement, if the Southern Cyclades are indeed part of the front/ external margin of the internal carbonate platform and not in the internal margin of the external platform. In the second case the pre-Alpine basement could belong to the external platform of H1, below the Almyropotamos–Kerketeas–Amorgos units, which are the deeper units of the medial tectono-metamorphic belt of the Hellenides (Papanikolaou 1986b, 2013) (see also Sect. 8.3.9 and Fig. 8.54).

8.3.4 The Dryos–Messaria Unit

The Dryos unit crops out mainly in Southern Paros and Eastern Antiparos islands (Papanikolaou 1977, 1980a), and the Mesaria unit in Ikaria Island (Papanikolaou 1978b). Both units consist of low grade metamorphic rocks, showing a significant difference in metamorphic grade, compared to the underlying Southern Cyclades unit. Thus, from amphibolitic metamorphic facies with gneisses, amphibolites and mica schists, we descend to epizonal rocks of low greenschist facies, with phyllites, and greenschists with sericite, chlorite, albite, etc. The tectonism of the Dryos–Messaria units is intense, with isoclinal folding and thrusting, but not as complex as in the underlying Southern Cyclades unit. Their lithostratigraphic columns cannot be reconstructed due to their restricted outcrops, usually in peninsulas above late extensional detachment faults or small tectonic klippen. Permian age was determined in crystalline dark gray limestones with Gymnocodium and Staffella in the Dryos unit in

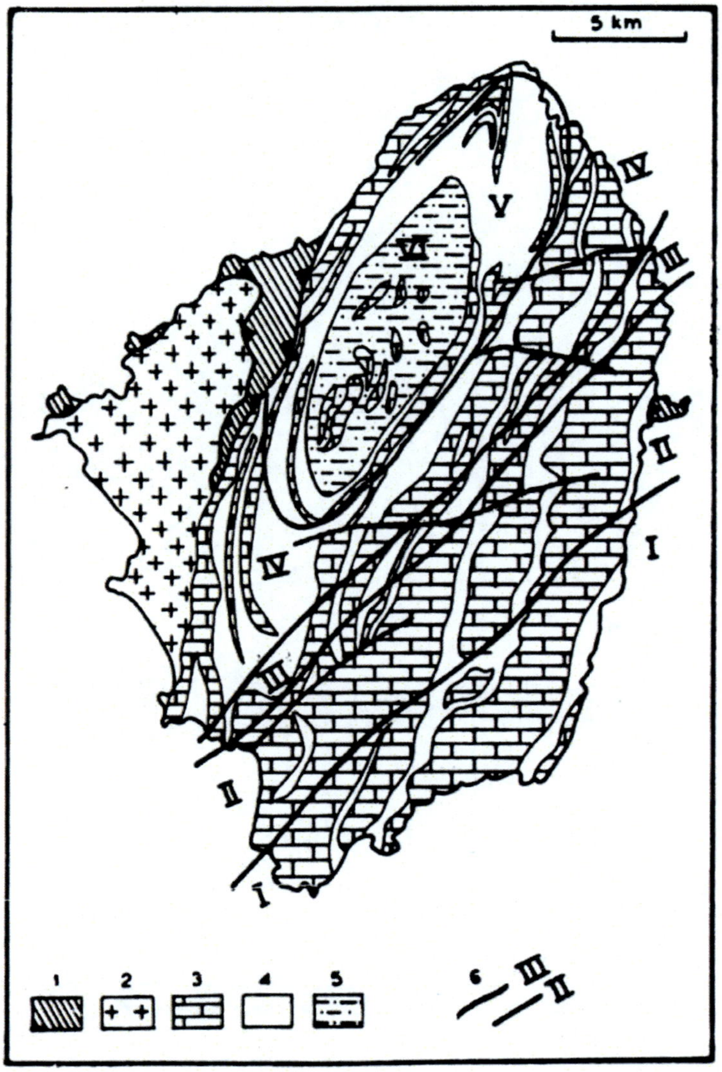

Fig. 8.77 Geological map of Naxos Island with metamorphic isograds, based on the parageneses around the migmatite dome (from Jansen and Schuiling 1976). 1: non-metamorphic nappe (Cycladic unit), 2: Upper Miocene granodiorite, 3: marbles with emery deposits of the Southern Cyclades, 4: amphibolites and mica schists of the Southern Cyclades, 5: migmatite, 6: metamorphic isograds with I: diaspore, II: chlorite–sericite, III: biotite–chloritoid, IV: kyanite, V: kyanite–sillimanite, VI: migmatite

SE Paros Island (Papanikolaou 1980a). Nevertheless, the presence of mafic igneous rocks is also important in the type locality of Dryos as well as in the tectonic klippen at the Keraki hill of central western Paros (Fig. 8.80a). On the contrary, in the southern outcrops of Paros Island in the area of Alyki and in SW Antiparos Island the presence of phyllites and calc-phyllites intercalated with crystalline limestones becomes more important. It is characteristic that both units are tectonically intercalated between the strongly metamorphosed rocks of the Southern Cyclades unit and the non-metamorphosed Cycladic unit (similar to the Eastern Greece unit). Especially in Central Ikaria, the Messaria unit is intercalated as a tectonic sheet of 200 m thickness between the underlying emery-bearing marbles of the Lower Ikaria unit (probably corresponding to the Southern Cyclades unit) and the non-metamorphic Triassic limestones of the Kefala Unit (local name of the Cycladic unit) (Papanikolaou 1978b; Kumerics et al. 2005) (Fig. 8.80b). These structures are generally of Miocene age and are associated with low angle normal detachment faults (Papanikolaou 1980a, 1987).

8.3.5 The Eastern Greece Unit

Paleo-tectonised units during the Late Jurassic–Early Cretaceous, unconformably overlain by the Upper Cretaceous transgression occur more internally than the Western Thessaly–Beotia units (Papanikolaou 1984b, 1986a,c). These are the internal Hellenides, where continuous stratigraphic

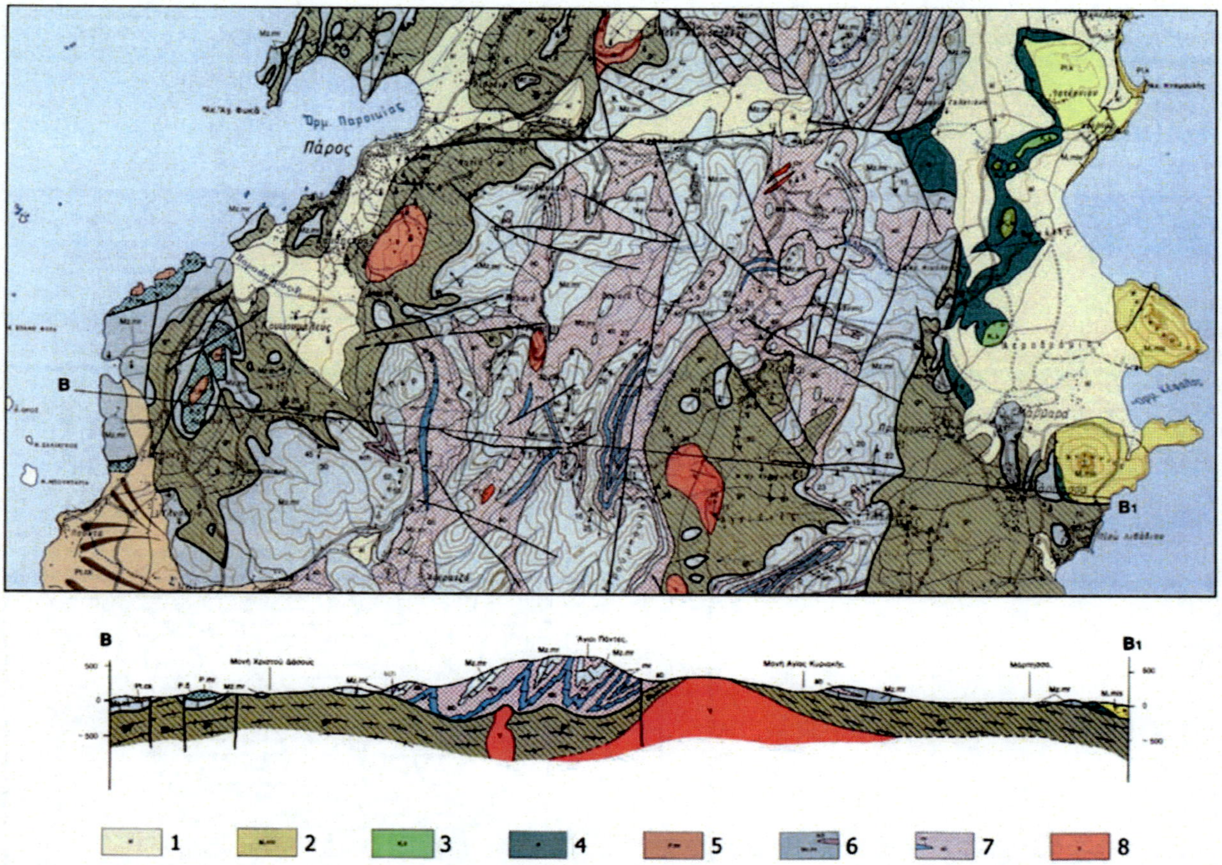

Fig. 8.78 Geological map of the central part of Paros Island and E-W geological section (from Papanikolaou 1996). The two surficial nappes of Dryos in the west and Marmara (non metamorphic Cycladic nappe) in the east form small tectonic klippen above the Southern Cyclades Unit and its pre-Alpine basement. 1: Quaternary, 2: Lower Miocene Cycladic molasse, 3: Upper Cretaceous limestones, transgressive over the ophiolites, 4: serpentinised ophiolites, 5: marbles, phyllites, and meta-diabases of Permian age of the Dryos unit, 6: marbles with emery of the Marathi unit (Southern Cyclades), 7: amphibolites with marbles and mica schists intercalations, of the Marathi unit, 8: granite and orthogneiss of the pre-Alpine basement of the Southern Cyclades

columns from the Triassic to the Tertiary do not exist but instead they comprise two stratigraphic sequences: (i) the lower Triassic–Jurassic sequences comprising various columns, both pelagic and neritic, metamorphosed and non-metamorphosed, with or without ophiolites. (ii) The sediments above the Late Cretaceous transgression (usually Cenomanian) unconformably overlying the paleo-tectonised units (Fig. 8.81). The transgressive sequence includes a conglomerate at the base, with abundant pebbles coming from ophiolite lithologies, a thick sequence of neritic limestones, with Rudists and sometimes some pelagic intercalations of limestones with Globotruncanes and some clastic turbidite formations and the Tertiary flysch on top, deposited during Danian–Eocene.

Thus, Eastern Greece unit is meaningful only during the Late Cretaceous–Eocene and only for its purely Alpine tectonism in the Eocene. On the contrary, for the pre- Late Cretaceous period this term is meaningless, since its main characteristic, the Late Cretaceous transgression, does not exist. Thus, in Dogger, for example, the rocks actually observed below the Late Cretaceous unconformity belonged to a variety of paleo-geographic areas, such as pieces of oceanic crust, carbonate platforms, basinal pelagic sediments, slope facies, etc. Thus, the transgression has buried an entire orogeny, involving a complex paleogeographic organization, with participation of the H3 and H4 terranes. Hence, the paleo-geographic area of Eastern Greece constitutes a «mega-unit» , observed only during the Cenomanian–Eocene. This unit comprised several older paleotectonised units, whose meaning ends in the Late Jurassic/Early Cretaceous. For example, the Maliac unit, represents an abyssal-pelagic basin of the H4, active during the Triassic–Jurassic period, which was folded and thrusted together with the ophiolite nappe during the Late Jurassic.

Fig. 8.79 The shallow cataclastic zone separating the marbles and blueschists of the Southern Cyclades unit above the garnet bearing mica schists of the pre-Alpine basement, along the northern side of the Ios Gulf

Consequently, it became a mountain range, subjected to erosion, until it was transgressed by a shallow sea during the Cenomanian.

The main tectonic characteristic of the internal units from the Eastern Greece unit and more internally is the distinction of two major periods/phases of tectonism: the paleo-Alpine tectonism during Late Jurassic–Early Cretaceous and the main Alpine tectonism during Eocene–Oligocene. The main Alpine tectonism folds the transgressive sequence of sediments and re-folds the underlying paleo-tectonised sediments. In other words, the internal Hellenides have been tectonised twice, whereas the external Hellenides only once.

The Eastern Greece area was not shallow everywhere during the Late Cretaceous, nor was it so calm as it might be considered at first glance. Instead, a synsedimentary tectonism during the Late Cretaceous, occurred with the creation of tectonic grabens–horsts. Thus, the shallow water carbonate platform sedimentation was interrupted by the formation of some tectonic grabens/basins, where turbiditic currents produced clastic formations of "flyschoid" type (Mariolakos and Papanikolaou 1982). Additionally, pre-Maastrichtian synsedimentary strike-slip faulting has been observed in some localities of Western Macedonia, such as the Nission Fault, south of Kaimaktsalan/Voras mt (Mercier and Vergely 1977). The generalized presence of rudist-bearing limestones is observed mainly of the Campanian–Maastrichtian, below the transitional marly limestones with Globotruncanes of the Maastrichtian–Danian and the Eocene flysch.

The Cenomanian unconformity has been used in several occasions during the Eocene tectonism as a surface of tectonic decollement. In several cases, the entire sedimentary sequence of the Upper Cretaceous–Eocene has slided, over the more external units, forming a separate tectonic nappe, as in **Vermion** mt (Brunn 1956; Pichon 1976) (see also Fig. 8.97).

In some locations in Northern continental Greece, older unconformable formations of Upper Jurassic age were documented overlying the ophiolites and underlying the Cenomanian sediments. These sediments are either pelagic with Calpionelles or neritic with algae, and they are found mainly in the Vourinos mt region (Brunn et al. 1970; Pichon 1976; Mavridis et al. 1979). Later studies doubted the existence of a Jurassic transgression, because the fossils found in these clastic sediments are transported together with the breccias and olistoliths, at the base of the Cenomanian sediments (Mariolakos and Papanikolaou 1982). On the contrary, a classical Tithonian unconformity over the pillow lavas of the Peonia ophiolites can be observed a few kilometers to the north of the Greece-former Yugoslavia borders, in the Demir Kapija area (Papanikolaou and Stojanov 1983). More recent studies have shown that in Northern Greece, there are two unconformities in the Internal Hellenides area; one in the Late Jurassic and another in the Late Cretaceous, which occur above different ophiolitic nappes (Galeos et al. 1994; Sharp and Robertson 2006; Robertson et al. 2013). The Late Jurassic transgression is a precedent tectonic phase of the Late Cretaceous transgression, observed in approximately the same units and the same terranes, with outcrops mainly

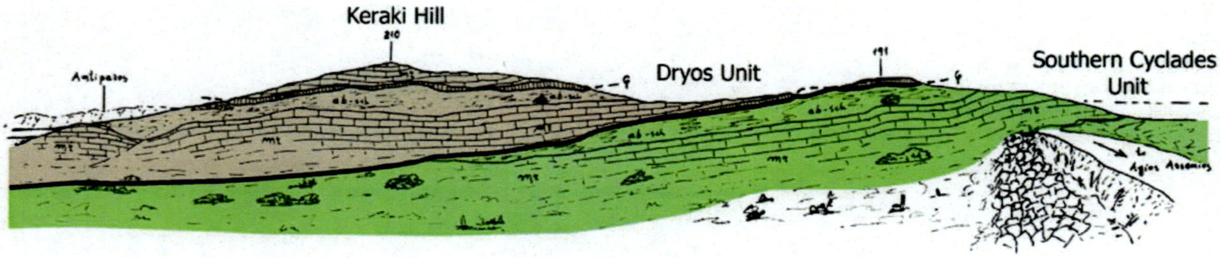

Fig. 8.80 a Panoramatic view from the east of the Keraki hill at central western Paros. The sub-horizontal tectonic klippe of the Dryos unit is observed above the marbles and amphibolites of the Southern Cyclades unit (based on Papanikolaou 1977)

Fig. 8.80 b Geological map and geological cross section of Central Ikaria (Evdilos area), where the low grade metamorphic nappe of Messaria (No 5, 6) is intercalated between the relatively autochthon Lower Ikaria unit (No 7, 8, 9), with medium grade amphibolitic metamorphic facies, and the upper non-metamorphosed Kefala unit (Cycladic unit, No 3, 4) (simplified from Papanikolaou 1978b). 1: Alluvium, 2: Western Ikaria Miocene granite, 3: Upper Triassic limestones and dolomites with Megalodon, 4: Middle Triassic mafic igneous rocks (diorite), 5: Messaria marbles, 6: Messaria phyllites, 7: alternations of marble and schists of the Ikaria autochthon, 8: marbles of the Ikaria autochthon with small emery outcrops, 9: gneisses of the Ikaria autochthon

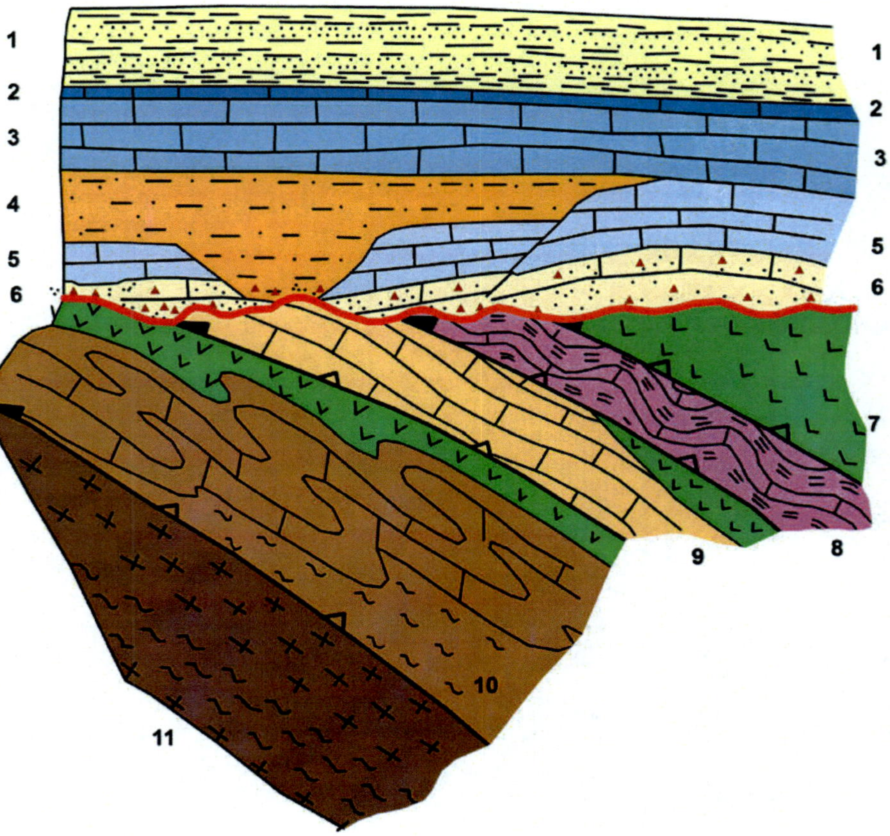

Fig. 8.81 Schematic representation of the complex geologic structure of the Eastern Greece unit (from Papanikolaou 1986c). 1-6: sediments of the Late Cretaceous transgression, 7-11: pre- Late Cretaceous tectonic units. 1: Danian–Eocene flysch, 2: pelagic limestones with Globotruncana of the Maastrichtian, 3: rudist-bearing limestones, mainly in the Campanian—Maastrichtian, 4: clastic turbidite formations of flyschoid features within synsedimentary grabens, mainly in the Cenomanian–Turonian, 5: neritic limestones, mainly of the Cenomanian–Turonian, 6: conglomerate and sandy limestones (Cenomanian). 7: Axios ophiolites (H4), 8: Maliac abyssal-pelagic unit, 9: Sub-Pelagonian unit (carbonate platform), 10: Almopia unit (metamorphosed carbonate platform), 11: Flambouron and Kastoria units (pre-Alpine crystalline basement)

in Northern Greece, such as the Axios ophiolite nappe (H4) on the metamorphosed Almopia unit (H3). It is remarkable that pre-Late Cretaceous paleo-Alpine metamorphism is observed also in some Upper Jurassic–Lower Cretaceous transgressive sequences (Mercier and Vergely 1984) (Fig. 8.82).

It should be noted that the Late Jurassic–Early Cretaceous transgression has also been observed in more internal units, such as the Circum-Rhodope unit (Maratos and Andronopoulos 1965; Kockel et al. 1977; Ivanova et al. 2015). However, a clear distinction should be made with the Liassic unconformity, observed in the Chios Allochthon, and generally in the Cimmerides, showing different features (Papanikolaou 2009) (see also Sect. 8.5). In the Greek literature, the term «Cimmerian continent» has been used for the Pelagonian orogeny (Mountrakis 1986), referring to the Cimmerides, although the tectonic activity was not in the Late Triassic–Liassic but in the Late Jurassic or even in the Early Cretaceous.

The number of the paleo-tectonised units in the Eastern Greece unit is large and the description for all of them impossible, as their stratigraphic column and other geodynamic features change from one unit to the other, or even laterally inside the same unit. Some characteristic units will be described in the following chapters such as: the Sub-Pelagonian, the Maliac, the Almopia, the Skyros, the Asteroussia, the Kastoria and the Flambouron. All these units occur below the Late Cretaceous transgression and therefore during the Late Cretaceous–Eocene period they belonged to the Eastern Greece unit (Fig. 8.81). Most of these paleo-tectonised units belong to the relatively autochthon internal platform H3 and its pre-Alpine basement, with the exception of the ophiolite nappes and the Maliac unit, which belong to the H4 Allochthon.

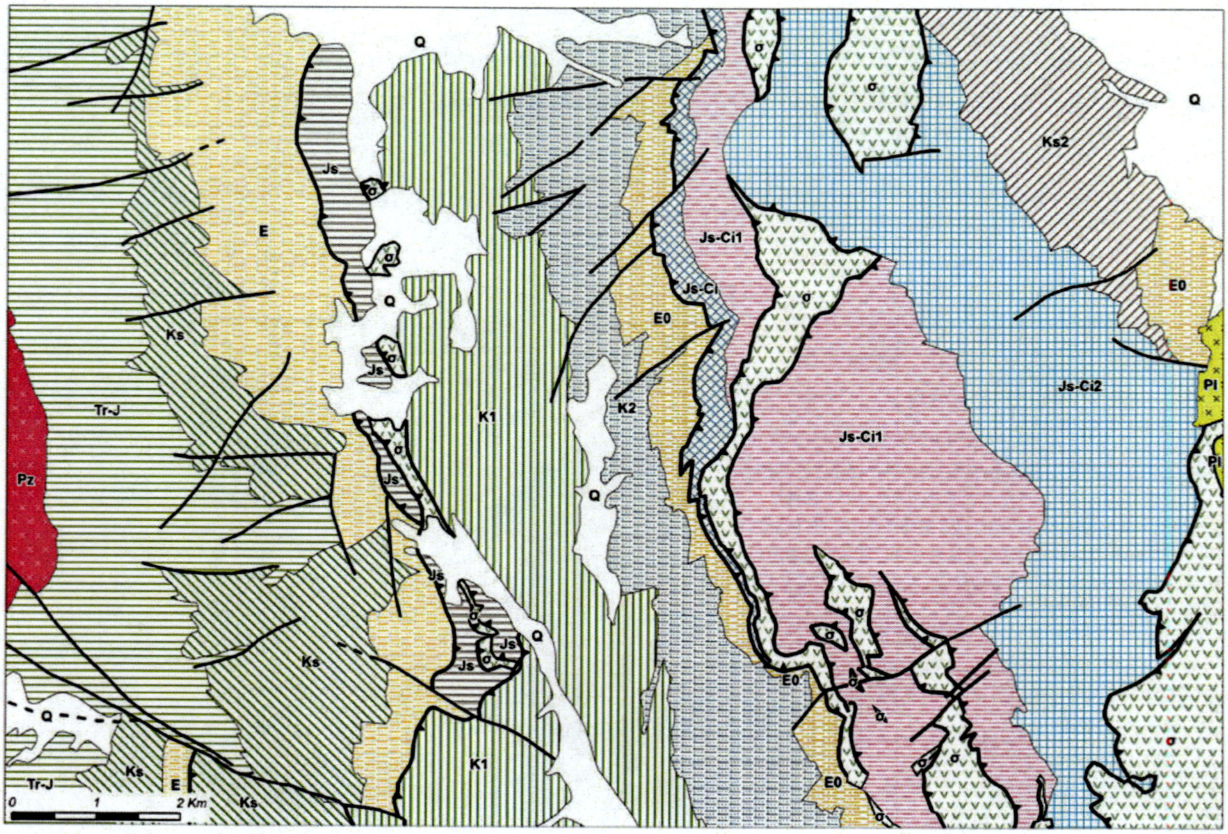

Fig. 8.82 Geological map of the Aridaia region, with transgressive volcano-sedimentary deposits of the Late Jurassic–Early Cretaceous over the Axios/Vardar ophiolites (H4) (modified from Mercier and Vergely 1984). Above the Upper Jurassic–Lower Cretaceous rocks the well known Upper Cretaceous unconformity can be observed on the tectonic imbrications of the Almopia ophiolites. The Upper Cretaceous unconformity directly covers the metamorphosed Almopia sequence (metamorphic Pelagonian) to the west, without the presence of Upper Jurassic–Lower Cretaceous formations. The Upper Jurassic formations are metamorphosed in the upper ophiolite imbrications to the east. Almopia unit H3. Pz: Paleozoic basement, TR-J: metamorphosed Triassic–Jurassic carbonate platform, Ks: Upper Cretaceous transgressive sediments, E: Eocene flysch. Almopia/ Axios Ophiolites H4. σ: ophiolites, Js: Upper Jurassic limestones and tuffs with Cladocoropsis, Upper Jurassic–Lower Cretaceous, JsCi1: volcanic extrusions, tuffs and breccia-conglomerates of the Upper Jurassic, JsCi2: meta-dolerites, meta-basalts, meta-rhyolites, and metamorphic breccia-conglomerate of the Upper Jurassic (?), K1: Aptian–Campanian limestones of neritic facies with rudists and of pelagic facies with Globotruncanes, K2: Campanian–Maastrichtian limestones, Ks2: Aptian–Albian limestones with Nerinea and Senonian–Maastricthian with Globotruncana, Eo: Upper Maastrichtian–Paleocene flysch

8.3.6 The Cycladic Unit (Non Metamorphic)

This unit crops out mainly in the Cyclades region, in the Central Aegean Sea and to the east up to Samos Island, mostly in restricted outcrops including all the non-metamorphic Alpine rocks of the Central Aegean Sea (Papanikolaou 1980b) (Fig. 8.83).

The main difference between the Cycladic nappe and the unit of Eastern Greece concerns the particular tectonic feature of a late tectonism, observed in the Cycladic nappe during the Middle–Late Miocene (mainly Tortonian). This late deformation is characterized by extensional tectonics and comprises extensional detachments and gravity sliding, involving the Alpine formations as well as the Miocene molasse of the Cyclades. Thus, the stratigraphic column of the Cycladic unit includes in addition to the formations of the Eastern Greece unit also the unconformably overlying Miocene molasse.

The Miocene tectonic event is determined by the presence of the unconformable Lower Miocene molassic formations (equivalent to those of the Meso-Hellenic molassic basin) above the Upper Cretaceous transgressive limestones and the older paleotectonised Triassico-Jurassic rocks, involving also the ophiolites, as observed especially in Paros Island (Papanikolaou 1980a). The molassic sediments are tectonised with folding and faulting and sometimes they form an allochthon thrust sheet, emplaced directly over the lower units, such as the gneissic pre-Alpine core of Paros Island, as

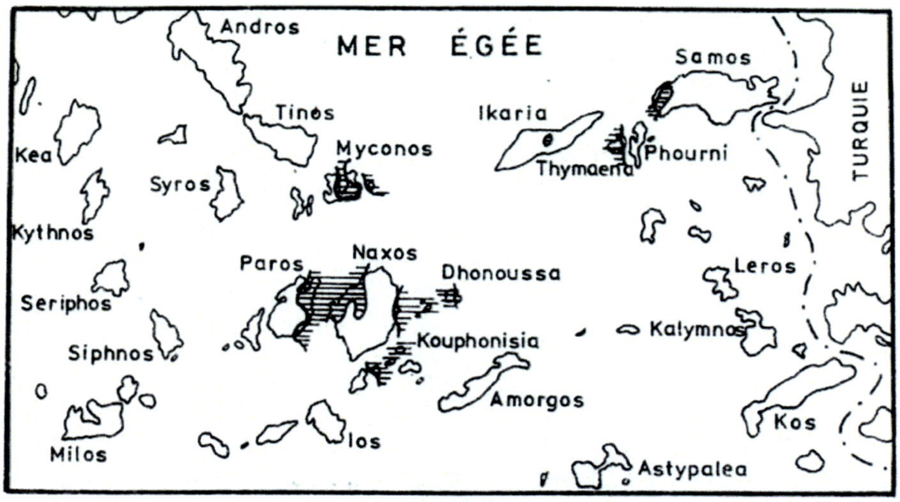

Fig. 8.83 The main outcrops of the non-metamorphic Cycladic nappe in the Central Aegean Sea (from Papanikolaou 1980b)

indicated in the geological map (Papanikolaou 1996) (see also Chap. 6). It is interesting to note that both in Thymaena (Fig. 8.84) and Mykonos islands the geometry of the sub-horizontal tectonic contacts suggests a tectonic transport of the Cycladic nappe from the south to the north, i.e. in contrast to the general asymmetry of the Hellenic arc (Dürr and Altherr 1979; Papanikolaou 1980b). The internal structures of the Cycladic nappe generally involve low angle normal faults that represent an extension during Miocene, as observed also in the neighboring Dodekanese Islands, to the south of Phourni (Roche et al. 2019).

The Cycladic nappe does not appear anywhere with its complete stratigraphic column. Various segments of the unit appear in several islands, where they are called with local, names, such as: **Marmara** unit in Paros Island (Papanikolaou 1980a), **Kallithea** unit in Samos Island (Theodoropoulos 1979; Papanikolaou 1979a), **Kefala** unit in Ikaria Island (Papanikolaou 1978b), **Thymaena** unit in Phourni islands (Papanikolaou 1980b). The Permian age has been determined in Naxos (Marks and Schuiling 1965) and Mykonos (Papastamatiou 1963), the Triassic age in Koufonisia, Mykonos, Ikaria, Thymaena and Samos (Cayeux 1911; Dürr et al. 1978; Papanikolaou 1978b, 1979a, 1980b), the Jurassic in Koufonisia, (Dürr et al. 1978), the Upper Cretaceous in Paros (Trikkalinos 1942; Papageorgakis 1968a; Papanikolaou 1980a) and the Lower Miocene in Paros, Naxos and Koufonisia (Negris and Boussac 1914; Papageorgakis 1968b; Angelier et al. 1978; Dermitzakis and Papanikolaou 1980). In some locations the Permian can be observed at the base of the nappe or the neritic Triassic, while in others the ophiolites, or the Cretaceous transgressive sediments, or the Lower Miocene molasse. In Paros Island the Miocene molasse is also observed in direct tectonic contact with the underlying gneisses of the basement with strong mylonitization (Papanikolaou 1980a, 1996). In Western Samos Island, the Kallithea/Cycladic unit lies directly above the Kerketeas autochthon with a low angle normal fault, which has "omitted" the several kilometers thick intermediate metamorphic rocks of the Ampelos and Vourliotes units (Fig. 8.85).

In Thymaena Island of the Fourni complex, and in the Kallithea unit of Western Samos Island there are important outcrops of basic lavas and tuffs, together with ammonitico rosso facies limestones and clastic sediments of Middle Triassic age, determined by conodonts (Papanikolaou 1979a, 1980b). These formations belong to the Triassic volcano-sedimentary complex underlying the internal carbonate platform H3 (Fig. 8.86).

8.3.7 The Sub-pelagonian Unit (Non Metamorphic Pelagonian Platform)

The term Sub-Pelagonian is unsuccessful, since it does not correspond to the paleo-geographic region that Aubouin (1959) had determined when it was first used. The Sub-Pelagonian unit had been defined as the slope area of the Pindos basin towards the Pelagonian ridge, along which massive ophiolite extrusions were taking place (Fig. 8.87).

The plate tectonics theory established a new view on the formation of the ophiolites along the mid-ocean ridges and thus, the granite type intrusion and extrusion views were abandoned. Additionally, the subsequent tectonic emplacement of the ophiolites and their obduction over the continental margins rendered invalid the term Sub-Pelagonian, both in terms of paleogeography and tectonics. Thus, in the 1970s it became evident that the Sub-Pelagonian ophiolites are actually a tectonic nappe emplaced upon the underlying Triassic–Jurassic rocks. Additionally, it came out that the

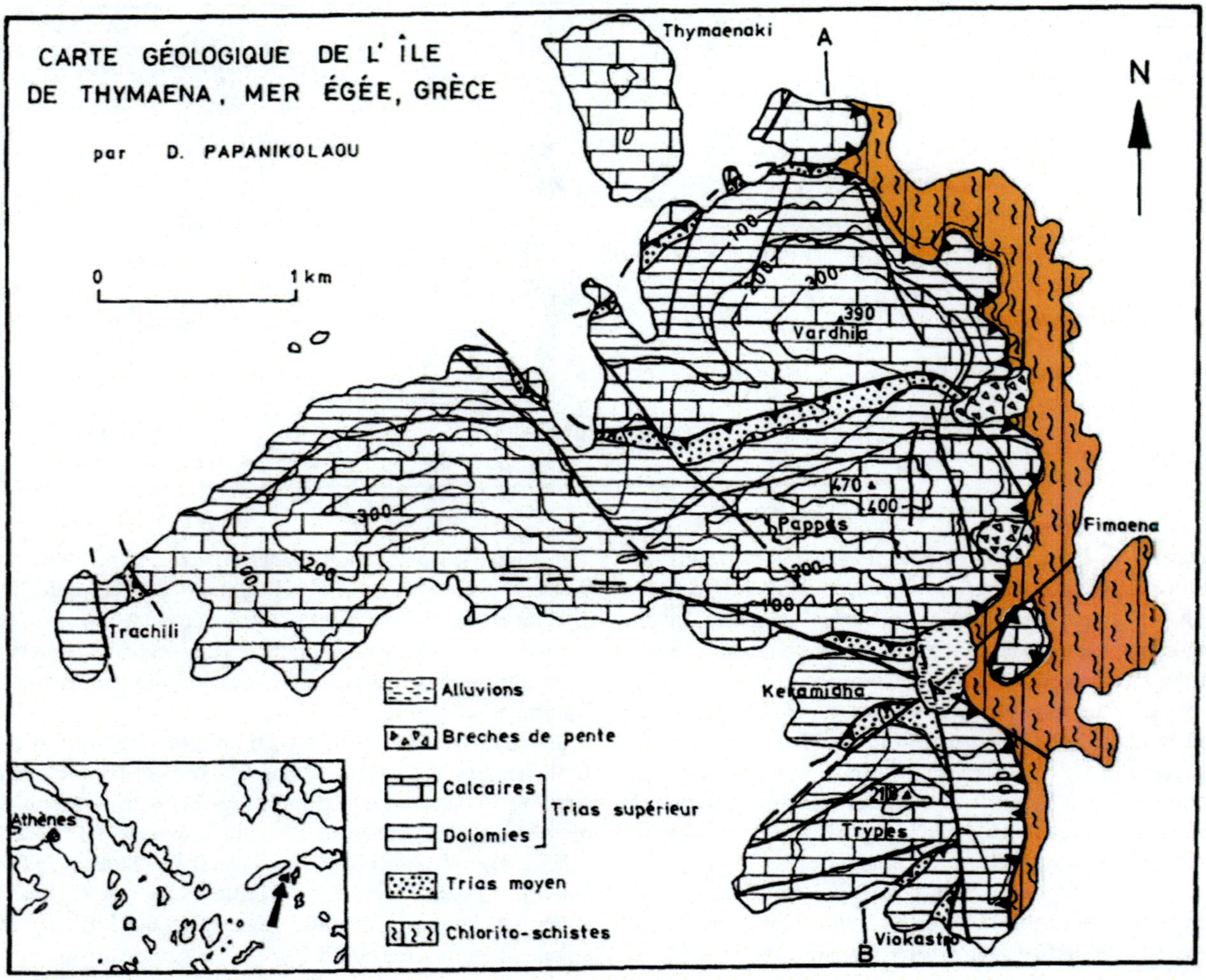

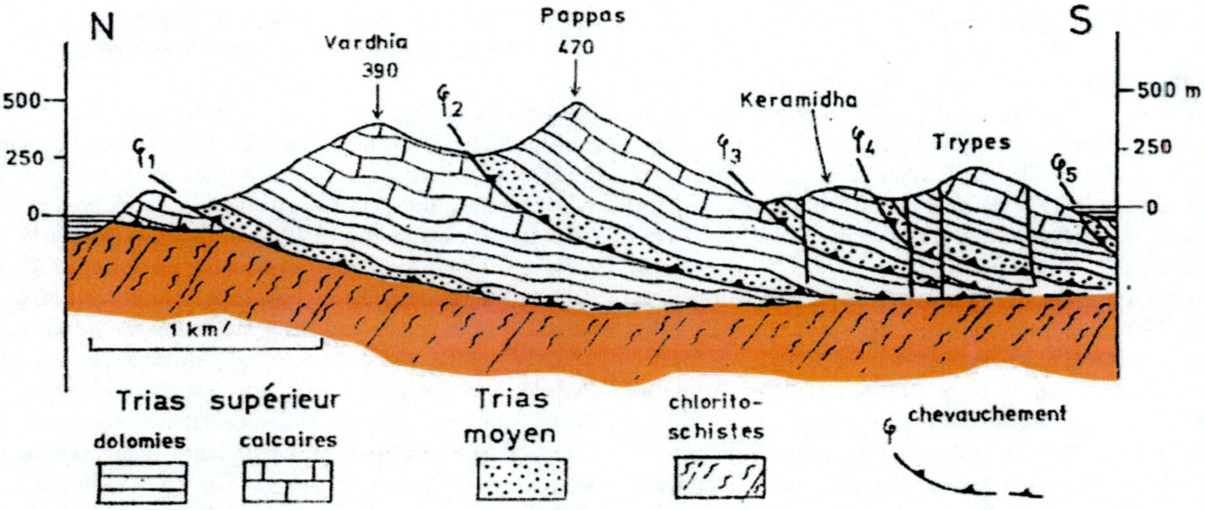

Fig. 8.84 a Geological map of Thymaena Island (Fourni island complex), and b geological cross section of Thymaena Island, showing the Cycladic nappe over the metamorphic basement and the imbrications within the Triassic formations above the sub-horizontal basal tectonic contact (from Papanikolaou 1980b)

Fig. 8.85 The low angle normal fault separating the Kallithea unit and the relatively autochthon Kerketeas unit near the Drakei village of western Samos Island

Fig. 8.86 Outcrop of red-pink colored limestones of ammonitico rosso facies of Middle Triassic age from the base of the Kallithea nappe in Western Samos Island

term "Pelagonian zone" included a multitude of metamorphic Alpine and pre-Alpine units. Hence, the term Sub-Pelagonian lost its initial meaning and it was proposed to be replaced by the term "non-metamorphic Pelagonian" (Celet and Ferriere 1978). Due to the fact that the term Pelagonian has been associated with the metamorphic rocks, both Alpine and pre-Alpine, the term Sub- Pelagonian was retained to describe the non-metamorphic rocks including the Triassic–Jurassic carbonate platform (Papanikolaou, 1986c).

The Sub-Pelagonian unit extends over large areas of the Eastern Greece unit, in its classic sense (Renz 1940, 1955), mainly comprising the neritic type rocks of a carbonate platform, together with the usually overlying **«schist–sandstone-chert formations»** initially known as "Schiefer-hornstein" formation, as in the Kallidromon mt. (Papastamatiou et al. 1962) (Fig. 8.88).

Until the 1970s (e.g. Tataris 1975), there was much confusion about the number of the schist-sandstone-chert (sh) formations as well as about their age and geodynamic significance. The investigation and analysis of characteristic localities of the sh formations showed that there was some confusion between the «schist-chert» formations and the «schist-sandstone-chert» formations. Thus, the « schiefer-hornstein»/schist-chert formations were deeper pelagic sequences, mainly observed over the subsiding carbonate platforms during the rifting stage, usually in the Late Lias–Dogger, just like the case of the external Ionian unit. On the contrary, the «schist-sandstone–chert» formations were relevant to the deep level clastic sedimentation of the flysch type, in the fore-arc setting of the orogenic arc, mainly during the Late Jurassic–Early Cretaceous (Papanikolaou 1990). This was a major problem in some cases, where the schist-sandstone-chert formation of the Upper Jurassic was

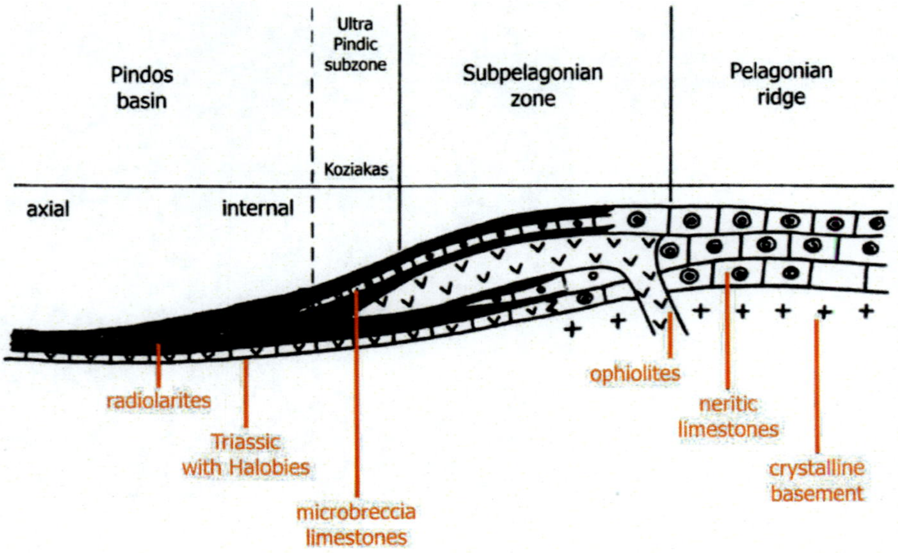

Fig. 8.87 Schematic representation of the paleogeographic region of the Sub-Pelagonian zone during the Jurassic–Cretaceous as an area of intrusion and extrusion of ophiolites on the slopes between the Pindos basin and the Pelagonian ridge (from Aubouin 1959)

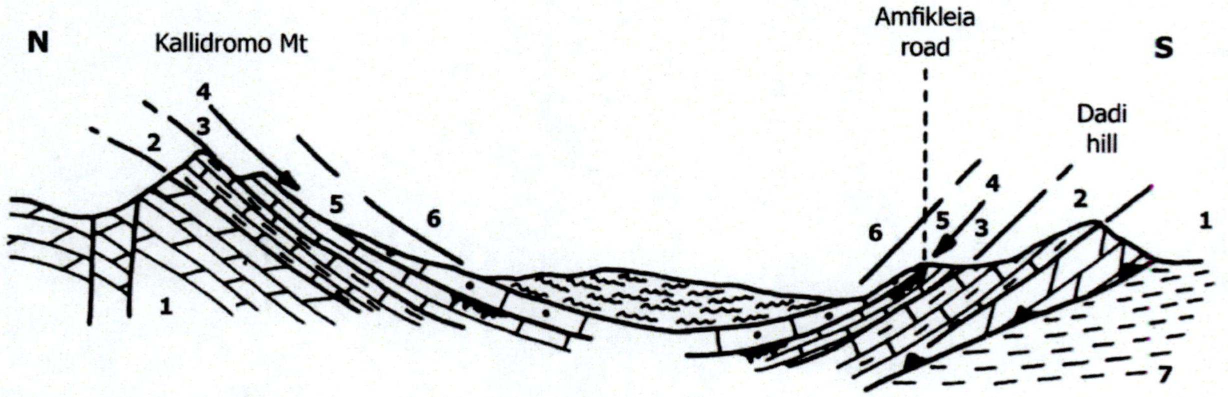

Fig. 8.88 Schematic cross section from Kallidromon mt to the footslopes of Parnassos mt, showing the tectonic superposition of the Sub-Pelagonian over the Parnassos unit and the Triassic-Jurassic formations of the Sub-Pelagonian unit (from Papastamatiou et al. 1962). This tectonic contact was originally considered as an Eocene thrust, but later on, it was characterized as a Miocene extensional detachment, which has omitted the Beotia unit (Kranis and Papanikolaou 2001). 1: Upper Triassic Dolomite, 2: Lower Jurassic limestone, 3: Middle Jurassic limestone, 4: Bauxite horizon (b1), 5: Cimmeridgean limestone with Cladocoropsis, 6: schist-hornstein formation with ophiolites, 7: Eocene Parnassos flysch

deposited above a schist-chert formation of the Upper Lias–Dogger. This explains the different ages previously reported for the sh formations, ranging from the Lias to the Lower Cretaceous (Philippson 1892, 1959; Renz 1940, 1955; Voreadis 1932, 1938; Tataris and Kallergis 1965; Tataris 1967a, b, 1972, 1975). It is characteristic that during the Late Jurassic, the Sub-Pelagonian sequences comprised in some cases a shallow water carbonate sedimentation and bauxite horizons (b1), while in others, deep pelagic sequences of shales and radiolarites («schiefer-hornstein») (Fig. 8.89). Thus, it was necessary to distinguish between the two Sub-Pelagonian types (Papanikolaou 1990): (1) the first sequence, of the Sub-Pelagonian type A, comprised carbonate platform sedimentation from the Triassic and bauxites (b1) until the Kimmeridgean and an overlying schist-sandstone-chert formation, i.e. flysch, in the Tithonian–Early Cretaceous. (2) another sequence, of the Sub-Pelagonian type B, comprised a carbonate platform of Triassic–Liassic age, followed by a transitional period of pelagic sedimentation towards the Dogger (Fig. 8.90), with predominance of silicate sedimentation, lasting up to the Kimmeridgean–Tithonian, where the flysch/melange («schist-sandstone-chert») was deposited.

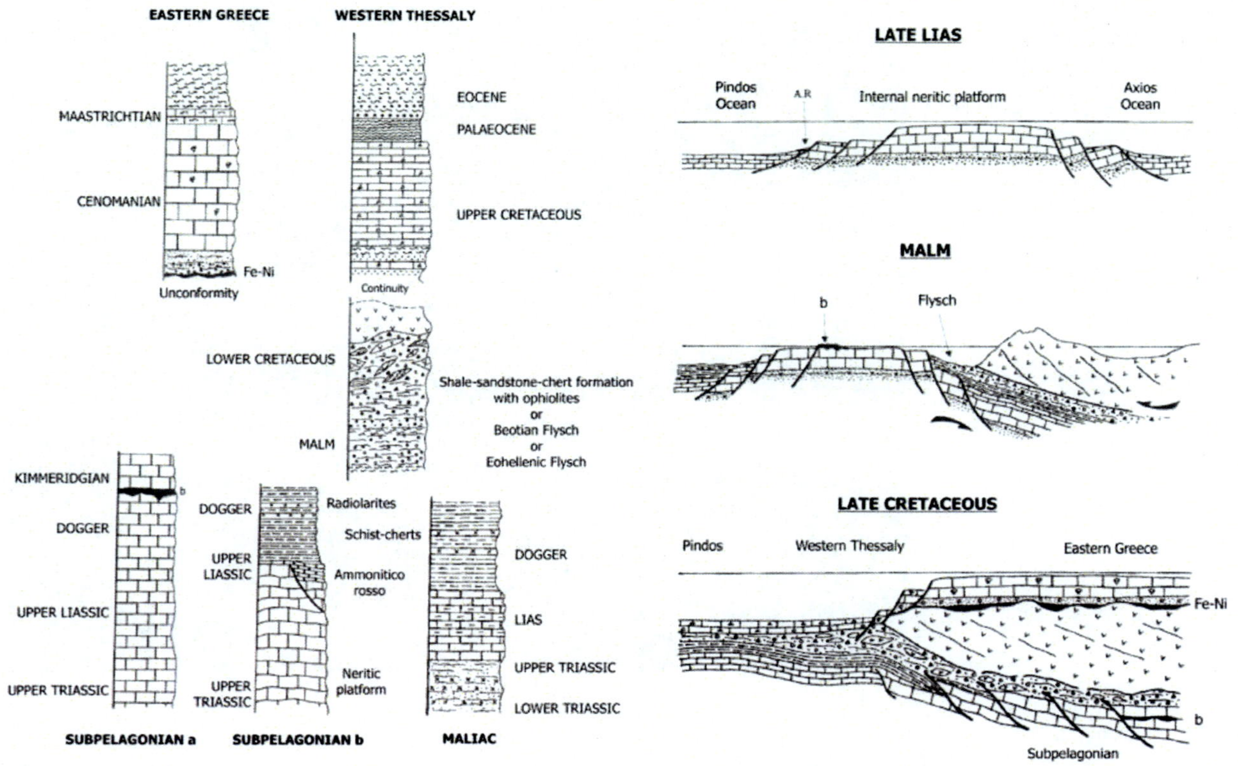

Fig. 8.89 **a** Distinction of the schist-sandstone-chert formations in the area between the Western Thessaly–Beotia and Eastern Greece units (from Papanikolaou 1990). The Sub-Pelagonian unit can be distinguished into two types: (A) a stratigraphic sequence with continuous carbonate sedimentation until the Malm with bauxites b1 and directly over them schist-sandstone-chert/flysch in the Late Malm, (B) a stratigraphic sequence of a Triassic- Liassic carbonate platform, overlain by a schist-chert formation in the Late Lias–Dogger and then by a schist-sandstone-chert/flysch in the Late Malm. The Maliac unit, is characterized by alternations of schist-chert formations and pelagic limestones, overlain by schist-sandstone-chert/flysch in the Late Malm. The Western Thessaly–Beotia unit has a schist-sandstone-chert/Beotian flysch in the Malm–Early Cretaceous followed by Late Cretaceous pelagic breccia limestones of the "Thymiama facies", in contrast to the other more internal units, which remain below the Late Cretaceous transgression. **b** The three stages of geological evolution from the Late Lias to the Late Cretaceous in the transitional area between the external and internal Hellenides, where «schist-sandstone-chert/flysch formations» were deposited (from Papanikolaou 1990)

In Beotia area, the rifting of the Sub-Pelagonian B was developed during the Dogger, based on dating of radiolarites above the carbonate platform (Danelian and Robertson 1995). At the top of the Sub-Pelagonian unit of both types of sequences A and B, the schist-sandstone-chert formation assumes the characteristics of a mélange, which is underlying the ophiolite nappe. This case corresponds to the older term "schist-sandstone-chert formation with ophiolites". This is observed not only in the Sub-Pelagonian units, but also in the Maliac unit, where ophiolite nappes occur between the Cenomanian transgression and the top of the Maliac sequence. In Argolis, ophiolite mélange of the Late Jurassic can be observed both upon the Sub-Pelagonian unit (local name Trapezona unit), and the Maliac unit (local name Epidavros unit) (Baumgartner 1985; Vrielynck 1982).

In several outcrops of the Sub-Pelagonian A, one or two bauxite-bearing horizons can be observed, mainly in the Kimmeridgean, below the limestones with Cladocoropsis mirabilis. These are similar phenomena with the Parnassos and Beotia units within the internal carbonate platform, before their differentiation and separation in the Late Jurassic. In the Late Paleozoic–Early Triassic outcrops known since the early 19th century (Renz 1908) several clastic formations can be observed, which include volcanic rocks and limestone olistoliths of Upper Paleozoic age (Attica, Salamis, Evia, Chios) (Baud and Papanikolaou 1981; Papanikolaou and Baud 1982; Baud et al. 1991; DeBono et al. 2001). In the Ladinian, red-pink or greenish nodular marly limestones of the ammonitico rosso facies (Hallstat facies) are often observed, associated with submarine volcanism, such as the unique outcrops of quartz keratophyre, in Parnitha mt (Ktenas 1909, 1924b). The Permian-Triassic formations of the volcano-sedimentary complex mentioned above correspond to the rifting stage of the H3 terrane. During the Late Triassic-

Fig. 8.90 Characteristic transition from the neritic carbonate sedimentation of the Upper Liassic to the pelagic sedimentation of thin-bedded pelagic limestones with cherts of Dogger (subvertical strata), from the Sub-Pelagonian B of the Aghios Ioannis Mazarakis region in Beotia

Early Jurassic, a neritic sedimentation of considerable thickness is dominant, including fossils of Megalodon, Gyroporella, Diplopora, Paleodasycladus etc., throughout the entire H3 internal carbonate platform.

In Northern Greece, in the Lakes Prespes area, extended outcrops of the Sub-Pelagonian unit can be found, showing significant stratigraphic peculiarities, like in the case of Krystallopigi, which resembles the Sub-Pelagonian B, with a deepening of the platform in the Late Lias- Dogger (Mountrakis 1983, 1984) (Fig. 8.91).

An interesting issue is the distinction of the paleo-tectonic structure of the Late Jurassic–Early Cretaceous from the Alpine structure of the Eocene (Papanikolaou 2009). This distinction is possible when the surface of the Cenomanian unconformity is kept almost intact, as in the Pavlos region in Beotia, where the Alpine tectonics are expressed with folding and thrusting observed in the Upper Cretaceous limestones and the Eocene flysch in the NE-SW direction, with tectonic asymmetry towards the NW (Fig. 8.92). On the contrary, the Sub-Pelagonian structures below the transgression, observed in the Triassic–Jurassic carbonates of the platform, the schist-sandstone-chert formation and the ophiolitic nappe, are characterized by folding and thrusting in ENE-WSW direction, with an asymmetry towards the SSE.

In the Ypaton region in Beotia, two phases of tectonic thrusting can be observed, with an initial paleo-Alpine phase including the tectonic emplacement of the ophiolites upon the Triassic–Jurassic carbonate platform of the Sub-Pelagonian and a younger Alpine thrusting of the Triassic–Jurassic platform over the same ophiolites, as a result of the Eocene tectonics (Fig. 8.93).

Two phases of thrusting have been reported during the Paleo-Alpine orogeny in some cases, like in Othrys mt (Smith et al. 1975; Smith and Woodcock 1976) and in the areas of Elikon mt and Gerania mt (Kaplanis et al. 2013), with a first thrusting of the ophiolites towards the east in the Late Jurassic, followed by a thrusting towards the west during the Berriasian.

8.3.8 The Almopia Unit (Metamorphic Pelagonian Platform)

The Almopia unit was considered as part of the so-called Axios/Vardar belt, as defined by Kossmat (1924). This belt was divided in several sub-units, Almopia representing the most external unit, characterized by a metamorphosed Triassic-Jurassic carbonate platform (Mercier 1966, 1968). Apart from the classical outcrops of the Almopia unit in the Edessa region, there are also similar outcrops of the former «metamorphic Pelagonian» or the former Pelagonian belt sensu stricto, in Western Macedonia (Vermio, Askio, Vourinos mts) and Eastern Thessaly (Ossa, Mavrovouni, Pelio mts). At the same time, the term Pelagonian zone had lost its initial meaning of a neritic ridge overlying a Paleozoic basement according to its definition by Aubouin (1959) because several different units and even different terranes

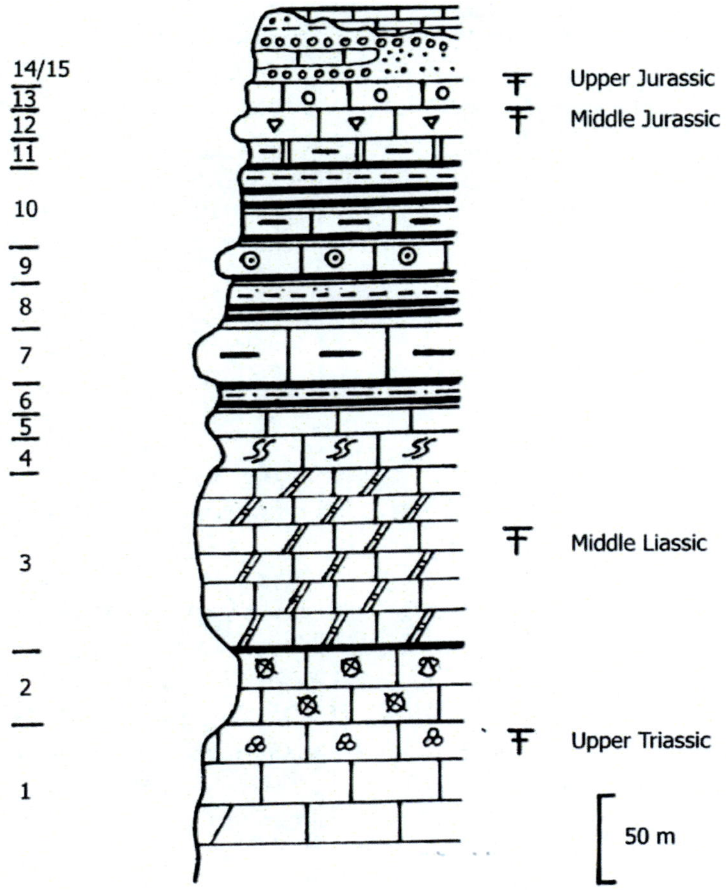

Fig. 8.91 Schematic stratigraphic column of the Sub-Pelagonian unit in Krystallopigi (from Mountrakis 1983). 1: dolomitic limestones of the Upper Triassic, 2: limestones with corals, 3: neritic limestones of the Middle Lias, 4: limestones with Litiotis algae, 5: platy limestones, 6: black cherts and pelites, 7: siliceous pelagic limestones, 8: radiolarites and marls, 9: neritic limestones of Lias, 10: alternations of cherts, pelites, limestones, 11: siliceous limestones with Posidonia, 12: breccia limestones of the Middle Jurassic, 13: clastic limestones of the Middle-Upper Jurassic, 14: turbiditic formation with detritus of ophiolites and Triassic-Jurassic limestones, 15: yellow rudist-bearing limestones

were participating in its structure (Papanikolaou 1981b, 1986b, 1988a, 1989b, 2013). The Almopia unit actually forms a tectonic nappe, emerging beneath the Axios/Vardar suture zone in the east, reaching up to the Meso Hellenic Basin in the west. It overlies the pre-Alpine basement units (Flambouro and Kastoria) on top of the nappe pile of the medial tectono-metamorphic belt (former Northern Pelagonian) (Papanikolaou 1988a) (Fig. 8.94).

The pre-Alpine basement units underlying the Almopia platform carbonates are distinguished into at least two units, the Flambouron unit in the east and the Kastoria unit to the west. In a deeper tectonic level, mainly at the Olympus tectonic window, the Northern Cyclades blueschists (local name of Ampelakia unit) appear below the pre-Alpine basement (Flambouron unit) and beneath them the relatively autochthon Olympus unit of the External Carbonate platform of the Hellenides.

The stratigraphy of the Almopia unit consists of:

(i) A lower formation of phyllitic rocks, crystalline limestones, often of the ammonitico rosso facies, meta-clastic rocks, lavas and tuffs, observed along the northern slopes of Askion mt, in the area of Namata village (Papanikolaou 1984a) (Figs. 8.95 and 8.96). It is a volcano-sedimentary complex, several hundred meters thick, similar to the Tyros beds of Tripolis, although here the limestonres of the ammonitico rosso facies forms a distinctive horizon. Its age has been determined, mainly on the basis of conodonts, as Lower–Middle Triassic in Namata/ Askion mt (Papanikolaou and Zambetakis-Lekkas 1980) and in the area of Kozani-Grevena (Mavridis and Matarangas 1979).

(ii) A thick formation of crystalline limestones to marbles. The basal horizons of the carbonate platform were dated by conodonts, as Middle-Upper Triassic (Papanikolaou and Zambetakis-Lekkas 1980). The Upper Triassic age has been also determined by algae (Mercier 1968). A Middle Jurassic age is assumed for the top of the platform carbonates, but without paleontological evidence. Generally, these marbles belong to the metamorphosed shallow water internal carbonate platform (H3).

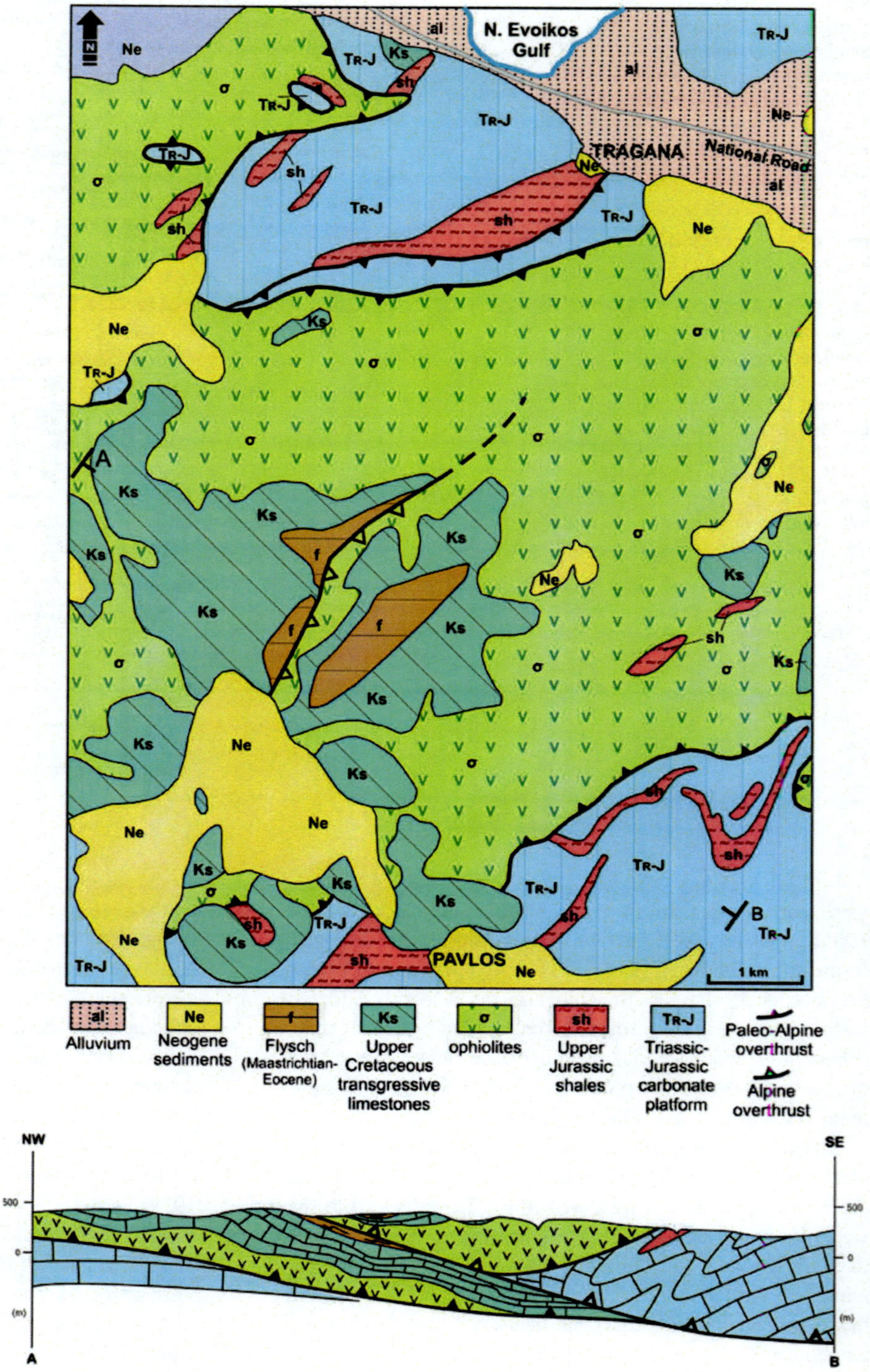

Fig. 8.92 Geological map and cross section of the Eastern Greece unit in Pavlos area in Beotia, showing the ENE-WSW paleo-Alpine structures below the Upper Cretaceous transgression and the NE-SW Alpine structures folding also the Eocene flysch (from Papanikolaou 2009)

(iii) An upper schist formation, forming a tectonic mélange (Mercier and Vergely 1972), with sandstones, cherts, shales, limestones and ophiolitic blocks, observed below the tectonic contact of the ophiolite nappe. It is remarkable that the ophiolites are intensively tectonised with isoclinal folds at the km scale, in the same style as the underlying sedimentary formations (Vergely 1976, 1984) (Fig. 8.97).

Fig. 8.93 View of the Ypaton region, where two phases of compressional tectonics can be observed, with a first phase of the ophiolites (σ) thrusting over the Jurassic limestones of the carbonate platform (Tr-J) and a second phase of the carbonate platform overthrusting the ophiolites. The steep cliff of the mountain is formed by the overthrusted carbonate platform above the younger thrust

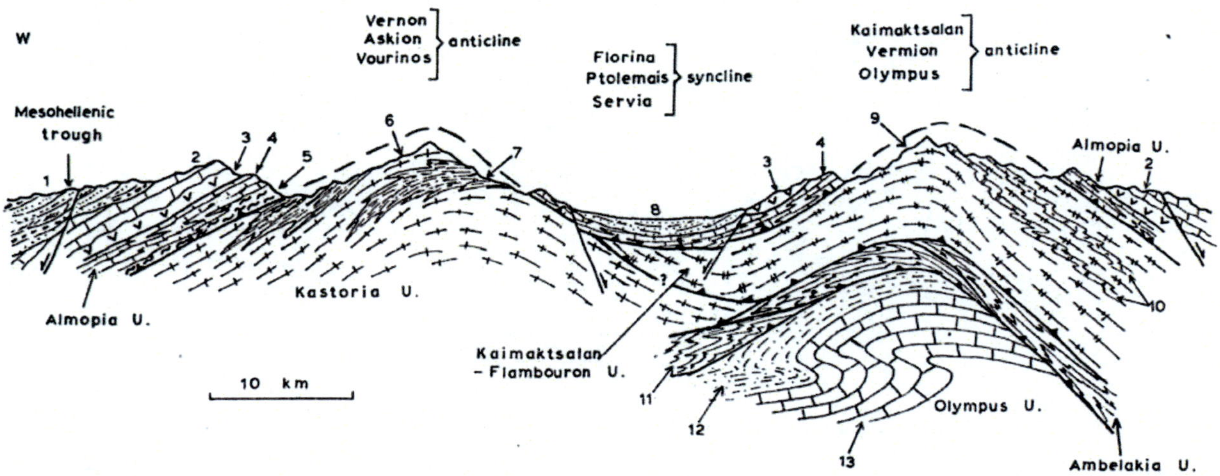

Fig. 8.94 Synthetic cross section of the northern part of the medial tectono-metamorphic belt (former Pelagonian) (from Papanikolaou 1988a). 1: MesoHellenic molasse, 2: Upper Cretaceous transgressive sediments, 3: Axios/Vardar ophiolite nappe, 4: Almopia marbles of Middle Triassic–Jurassic, 5: Almopia phyllites, marbles and meta-volcanics of Lower–Middle Triassic, 6: Kastoria granites and gneisses (Paleozoic), 7: Kastoria mica schists, 8: Post-Alpine sediments of Ptolemais basin, 9: Flambouron gneisses, granites, amphibolites and mica schists (Paleozoic), 10: Flambouron marbles, 11: Ambelakia blueschists (Northern Cyclades), 12: Olympus flysch (Eocene), 13: Olympus Triassic-Eocene crystalline limestones

(iv) The ophiolite nappe, which originates from the Axios/Vardar suture zone of H4 and can be observed mainly in the Edessa area and Vourinos mt.
(v) The sediments above the Late Cretaceous transgression in the Almopia unit, which are usually detached and tectonised during the Eocene orogenic phase.

The paleo-tectonic structures with isoclinal folding that deform the Almopia unit together with the ophiolite nappe in the Aghios Dimitrios area in Vermion mt, do not continue to the overlying non-metamorphosed sediments of the Upper Cretaceous transgression, implying that the metamorphism and the ductile deformation is pre-Upper Cretaceous in age (Vergely 1984). These tectonic structures are generally oriented ENE-WSW, with a tectonic transport towards SSE. Approximately the same tectonic structure and asymmetry can be observed in the paleo-tectonic structures of the non-metamorphosed Sub-Pelagonian unit and the ophiolite nappe at the Beotia area in Sterea (Papanikolaou 2009) (see also Fig. 8.92). In Vermion mt, the Alpine structures are restricted to the tectonic detachment of the Late Cretaceous–Eocene sedimentary sequence above the transgression and the creation of the surficial Vermion nappe over the non-detached transgressive sediments above the Almopia unit. Thus, a duplication of the Upper Cretaceous–Eocene sedimentary sequence is observed in Vermion mt.

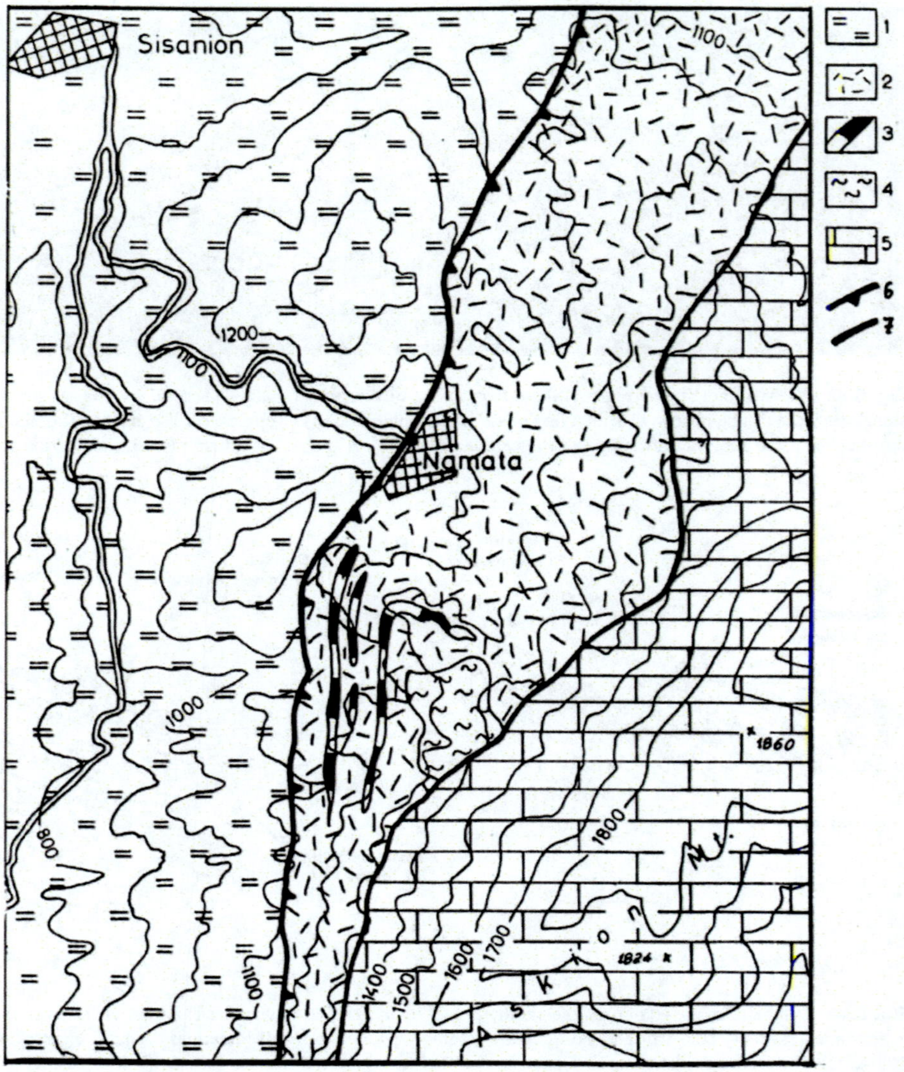

Fig. 8.95 Geological map of the Namata area at the northern slopes of Askio mt, where the Lower–Middle Triassic volcano-sedimentary formations of the Almopia unit tectonically overlie the Carboniferous granite-gneisses of the Kastoria unit (from Papanikolaou 1984a). 1: granite-gneisses of Kastoria, 2: Lower–Middle Triassic volcano-sedimentary formations of the Almopia unit, 3: crystalline limestones of ammonitico rosso facies, 4: mica schists and graphitic phyllites, 5: marbles of the Middle–Upper Triassic of the Almopia unit, 6: overthrust, 7: tectonic decollement

The outcrops of the Almopia unit are limited to the north of the Maliac tectonic zone, in Thessaly and Western Macedonia regions. In contrast, from Evia and Phtiotis to the south, the non-metamorphosed Sub-Pelagonian unit crops out. In Eastern Orthis mt and Northwestern Evia region some subunits can be observed with transitional characteristics regarding their metamorphic grade, showing a superposition of the metamorphosed subunits of the Almopia type over the non-metamorphosed subunits of the Sub-Pelagonian type (e.g. Ferriere et al. 2016).

The P/T conditions of the pre-Late Cretaceous metamorphism of Almopia unit has been determined with pressure values of more than 12 Kbars and temperatures of 450–500 °C regarding the initial blueschist facies, and 4–5 Kbars and 280–380 °C regarding the retrograde greenschist facies (Kilias et al. 2010).

8.3.9 The Ios–Southern Cyclades Basement

At the base of the Southern Cyclades unit, there are gneisses, granites, mica schists and amphibolites, which represent the pre-Alpine basement. First geochronological data from Ios Island showed, besides the Eocene metamorphism of the blueschists and the Miocene retrograde metamorphism of the greenschists, also an older metamorphic event of Paleozoic age (Kreuzer et al. 1978; Henzes-Kunst and Okrush 1978; Henzes-Kunst and Kreuzer 1982). Later on, new ages were determined, mainly from the schistosed granites of Ios, Paros, and Naxos islands, which confirmed the Paleozoic age about 300–320 Ma of the metamorphism and the acidic magmatism (Andriessen et al. 1987; Engel and Reischmann 1998) (Fig. 8.98).

Fig. 8.96 View of the northern slopes of Askio mt, where the volcano-sedimentary formations of the Lower-Middle Triassic (2) can be observed under the Triassic-Jurassic marbles of the Almopia unit (3) and over the Carboniferous granite-gneisses of the Kastoria unit (1) (from Papanikolaou 2013)

The contact between the pre-Alpine rocks of the basement and the base of the Southern Cyclades unit is tectonised through decollement and ductile shearing, like in Paros Island (see also Fig. 8.75). Generally, a large difference is observed of the deformation between the Paleozoic basement and the Mesozoic cover. In Ios Island, a thick cataclastic tectonic zone is observed between the top of the basement rocks and the base of the Southern Cyclades marbles and schists (see also Fig. 8.79). The outcrops of this tectonic contact in different islands indicate a tectonic detachment zone (Forster and Lister 1999; Huet et al. 2009) that passed through different tectonic layers of the crust, with transition from ductile to brittle deformation. More detailed mapping and chronostratigraphy in the islands of Ios and Sikinos have shown that besides the Carboniferous gneissic granites there are also Triassic granitoid bodies as well as Upper Neoproterozoic and Lower Paleozoic metasediments involved in the basement structure (Flansburg et al. 2019). Additionally, a parautochthonous nature of the Southern Cycladic blueschists with the pre-Alpine basement of the Southern Cyclades has been proposed (Poulaki et al. 2019).

It is interesting that younger granite intrusions of Miocene age resulting from the Neogene volcanic arc, are observed in Paros Island (Papanikolaou 1996), intruding the Paleozoic granite-gneisses. Their outcrops are producing characteristically different rough morphological relief in the landscape in contrast to the mild topography of the gneisses (Fig. 8.99).

These Miocene granitic occurrences show the large scale recent uplift of several kilometers, that occurred in the central Aegean region during the last 10 Ma, considering that the intrusion depth of the Miocene granites was about 5–8 km.

8.3.10 The Asteroussia Unit

Asteroussia is the mountain range on the southern part of Central Crete with an E-W general orographic orientation, where the type locality outcrops of the Asteroussia unit are observed. The Asteroussia unit is the uppermost tectonic unit in Crete, above the units of Mani, Western Crete, Arna, Tripolis, Pindos/Ethia, Arvi, Miamou, and Vatos (Fig. 8.100). It comprises medium-high grade metamorphic rocks and granites (Davis 1967; Bonneau 1972).

The Asteroussia unit is tectonically imbricated with ophiolites, observed either below or above the metamorphic rocks. A unique feature of the tectonic structure of Crete is that while in the Asteroussia mountains the unit is located together with the ophiolites in its initial topmost tectonic position over the Ethia, Arvi, and Miamou units (Fig. 8.101a), in other areas it comes to a direct tectonic contact with the lower unit/autochthon Mani unit, due to extensional detachment faults. A characteristic example is observed along the southern slopes of Psiloritis mt, where the extensional detachment shows total displacement of

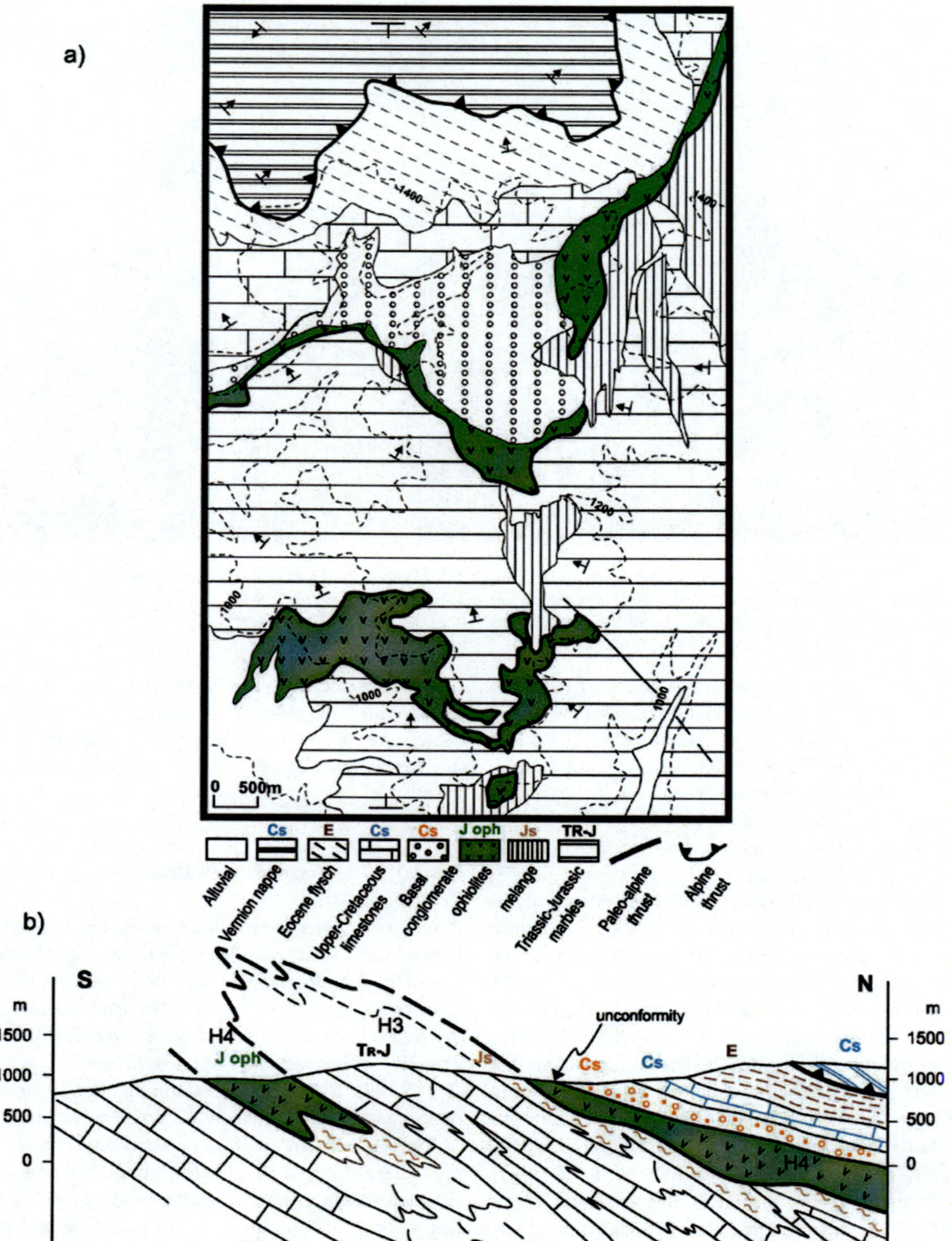

Fig. 8.97 Geological map (a) and N-S geological section (b) of the Aghios Dimitrios area in the southwestern Vermio mt. The upper horizons of the Almopia unit are observed, folded with large isoclinal folds together with the Axios ophiolite nappe, beneath the non-metamorphic Upper Cretaceous unconformable sequence. The epidermic Upper Cretaceous-Eocene nappe of Vermion unit is also observed in the northern area (from Papanikolaou 2009)

Fig. 8.98 Outcrop of granite-gneiss of Ios Island with aplite veins of Carboniferous age

Fig. 8.99 View of granitic outcrops in Northern Paros, where the Carboniferous granite-gneisses forming the mild relief, are penetrated by the intruding Miocene granites, producing an intense rough relief

about 25 km. The Messara supra-detachment basin and the detached parallel structure of the Asteroussia mt occur on its hanging wall (Papanikolaou and Vassilakis 2010) (Fig. 8.101b).

The metamorphic rocks of the Asteroussia unit are of medium–high grade, comprising amphibolites, gneisses, mica schists, and marbles, with very characteristic outcrops of thin alternations of amphibolites and marbles (Fig. 8.102). The P/T conditions of the amphibolite metamorphic facies, were estimated 4–5 Kbars and 700 °C (Koepke and Seidel 1984).

Geochronological data determined an Upper Cretaceous tectono-metamorphic age of about 70 Ma (Lippot and Baranyi 1976; Seidel et al. 1976, 1981). This age constitutes an unique event, due to the fact that the tectonism of the Hellenides shows two peaks, one in the Late Jurassic-Early Cretaceous (about 140–110 Ma) and a second one in the Late Eocene (about 45 Ma). The 70 Ma age does not correlate with any of the widely known geotectonic events in the Hellenides, while it is the main age of tectonism with characteristic transgression of sediments over the ophiolite nappes in the Lycian nappes of Asia Minor in the Taurides

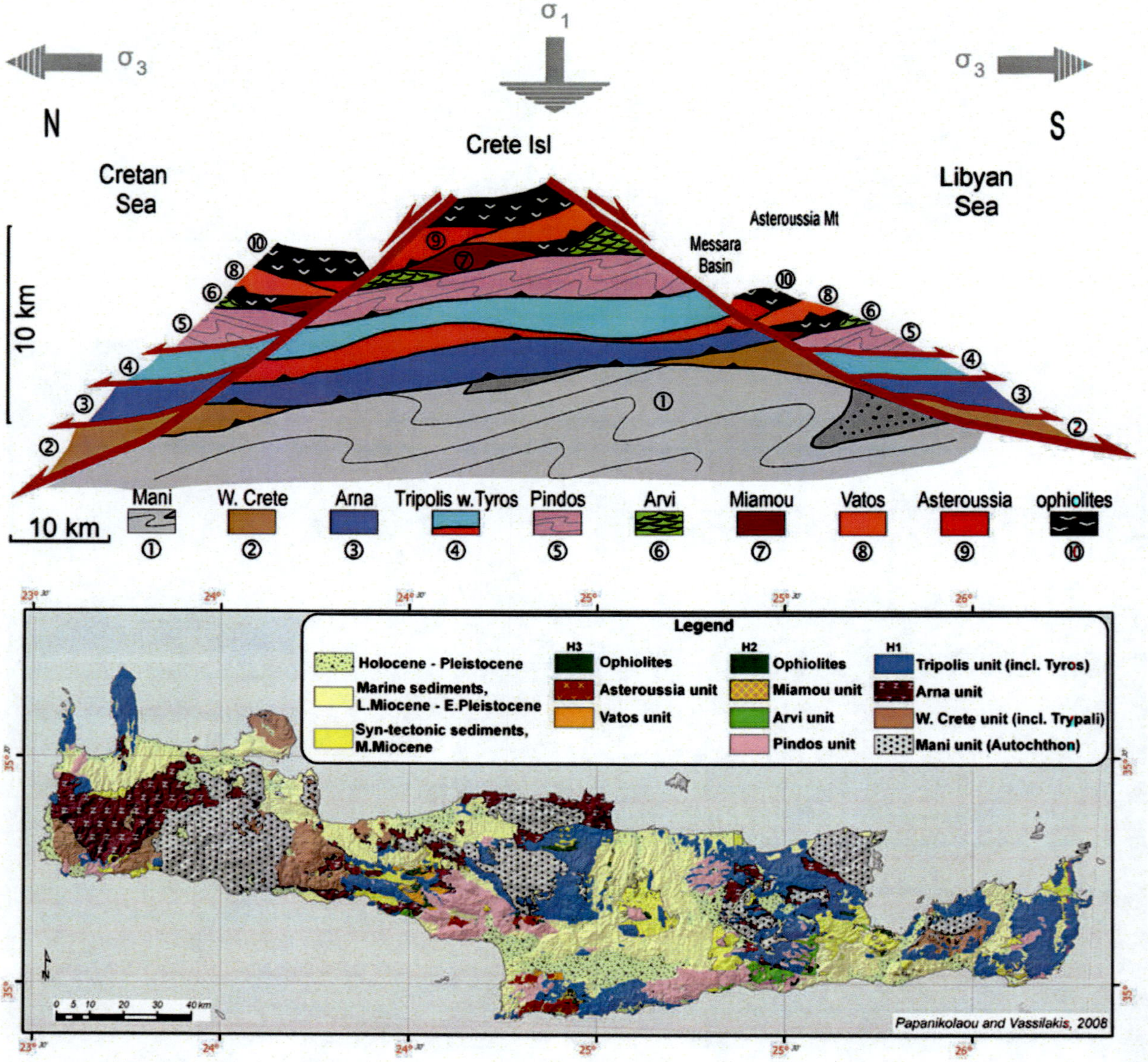

Fig. 8.100 **a** Schematic diagram of the tectonic nappe pile of Crete. The Asteroussia unit is located at the top of the tectonic nappe pile above the Vatos, Miamou, Arvi, and Ethia units, with a Late Eocene–Oligocene tectonic emplacement (from Papanikolaou and Vassilakis 2010). **b** Map of the geotectonic units of Crete (from Papanikolaou and Vassilakis 2010)

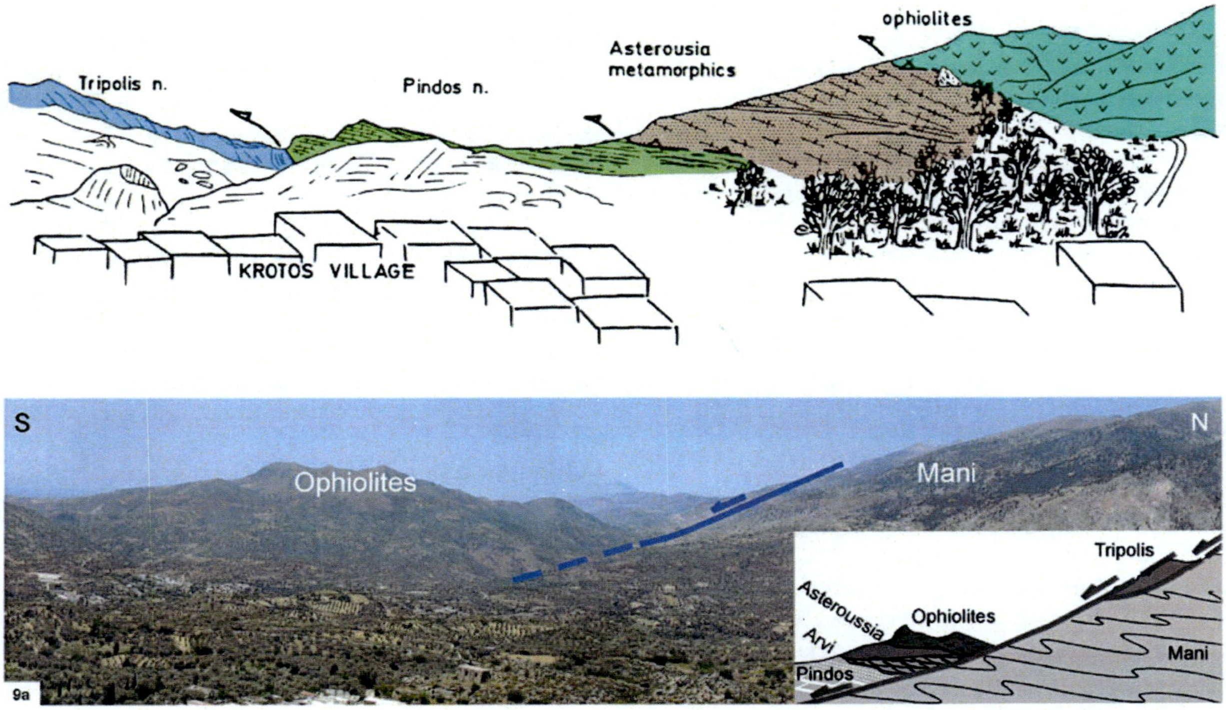

Fig. 8.101 a Panoramatic view of the upper part of the Cretan nappe pile in the Asteroussia mt, type locality of the Asteroussia unit (from Papanikolaou 1988c). b View of the large extensional detachment fault of Southern Crete, which brings the uppermost ophiolites and the metamorphics of the Asteroussia unit in contact with the relative autochthon Mani unit (from Papanikolaou and Vassilakis 2010)

Fig. 8.102 Characteristic outcrop of the Asteroussia unit lithologies, with thin alternations of amphibolites and marbles from the Asteroussia mountain range in Southern Crete

(Brunn et al. 1976; Gutnic et al. 1979; Ricou 1980; Papanikolaou and Demirtasli 1987; Rimmele et al. 2006). Of course, in the framework of a continuous orogenic Alpine cycle, within the plate tectonics convergence these intermediate tectono-metamorphic features are expected along laterally variated Alpine segments.

The Late Cretaceous tectono-metamorphic event combined with the nature of the pre-Alpine basement rocks of the Asteroussia unit, the characteristic HT/LP conditions of its amphibolitic metamorphism and its tectonic intercalation with ophiolites, favours the geodynamic positioning of the unit at the lower crust, near the magmatic arc, directly above the subduction zone of a Late Cretaceous orogenic arc.

8.3.11 The Anafi Units

It is of great interest that the Late Cretaceous metamorphic event of the Asteroussia unit was found also in some minor units of the Southern Cyclades, such as the Anafi and Nikouria units (Durr et al. 1978; Reinecke et al. 1982; Martha et al. 2016). In Anafi Island, a metamorphic nappe pile of amphibolitic schists, meta-ophiolite rocks, marbles, and granites is observed upon the relatively autochthon slightly schistosed Eocene flysch (Melidonis 1963) (Fig. 8.103).

The pre-flysch basement of the Eocene flysch of the Anafi unit does not crop out but based on its characteristics, it could be homologous to the flysch of the Profitis Ilias unit in the neighboring Santorini Island (Tataris 1965). The entire nappe pile of the Anafi Island has been disrupted by extensional detachment faults of Miocene age (Soukis and Papanikolaou 2004) (Figs. 8.104 and 8.105). These extensional detachment faults formed the northern margin of the Cretan back-arc basin, during the Middle–Late Miocene.

A very impressive aspect of the Late Cretaceous metamorphic nappes of the Anafi Island is their different paleo-geographic provenance and their common Late Cretaceous metamorphic event. Thus, the lower nappes above the autochthon exhibit oceanic abyssal-pelagic features whereas the upper nappe continental characteristics of shallow water carbonate platform with granitic intrusions. Nevertheless, the same age of metamorphism has been found also in the Makrotantalon–Ochi unit, composed of typical blueschists, which however, later in the Early Tertiary followed the same path in the orogenic arc as the Northern Cyclades unit (see also Sect. 8.2.4). On the contrary, the ophiolitic Anafi units have a pre-Late Cretaceous subduction phase, with a consecutive amphibolitic metamorphic phase and granitic intrusions in the Late Cretaceous, and finally, a tectonic emplacement over the non-metamorphic Eocene flysch.

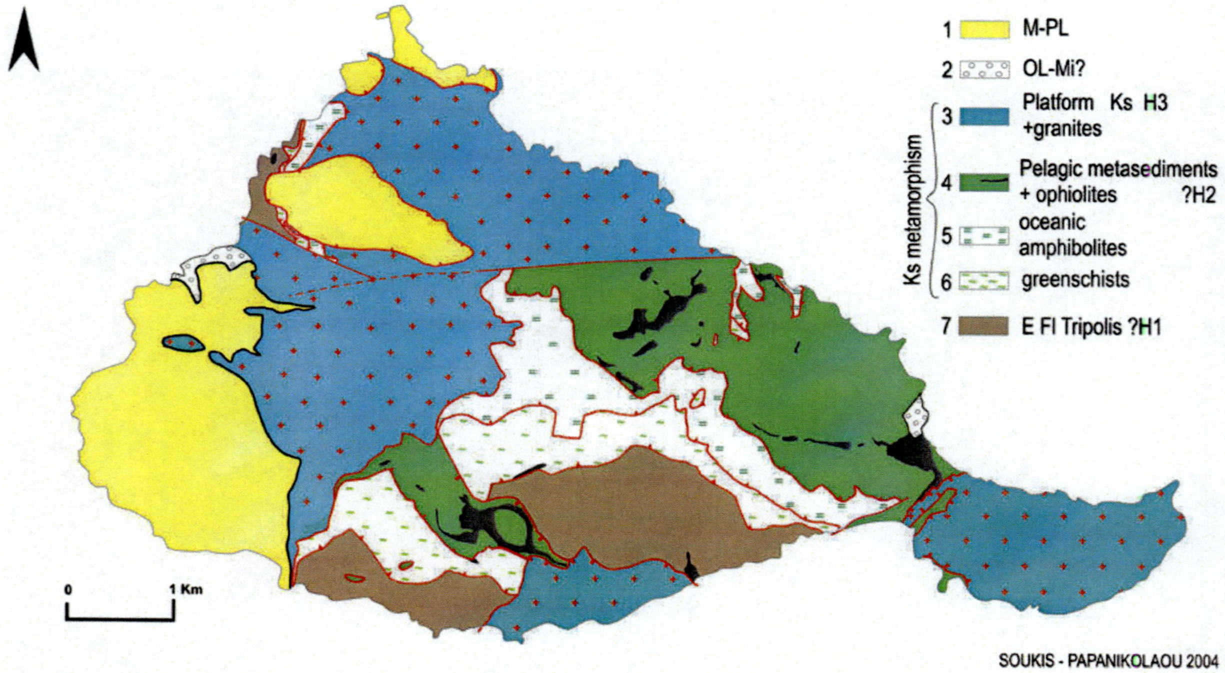

Fig. 8.103 Simplified geological map of the Anafi Island, based on Melidonis (1963), showing the Late Cretaceous metamorphic units of the H2 and H3 terranes, over the autochthon Eocene flysch of the H1 (from Soukis and Papanikolaou 2004, modified). 1: Neogene deposits of continental facies in western Anafi—including the detached Theologos beds. 2: molasse deposits of sandstones-conglomerates of Oligocene–Miocene (?) age. 3: marbles and intrusions of Late Cretaceous granites of the upper unit (H3). 4: pelagic meta-sediments and ophiolites (H2), 5: amphibolites (meta-gabbros) (H2), 6: greenschists (meta-diabases, meta-tuffs) (H2), 7: Eocene flysch of Tripolis (?) (H1)

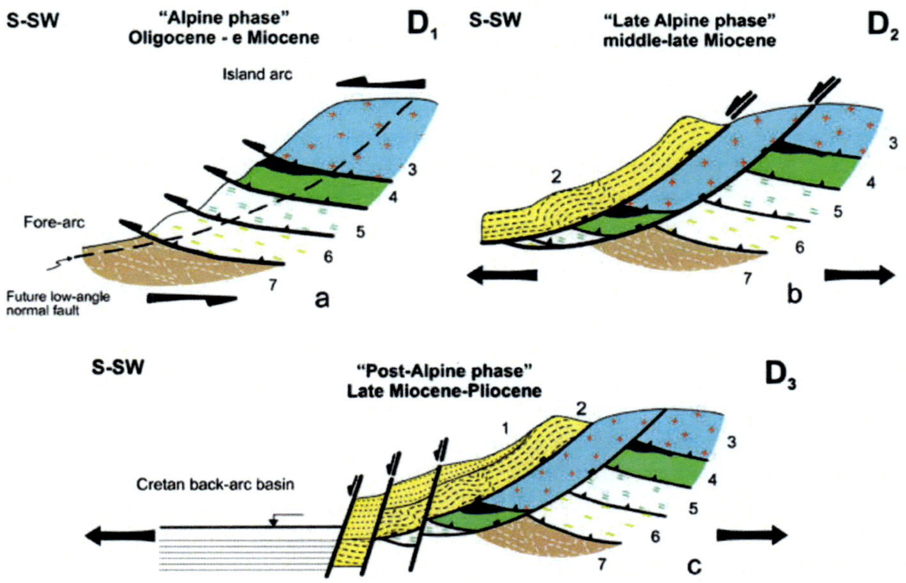

Fig. 8.104 Schematic representation of the compressive deformation phase D1, during the Oligocene, which resulted in the emplacement of the Anafi nappes (3, 4, 5 and 6) over the autochthon flysch (7). During the Miocene (2) the extensional deformation phase D2, disrupted the previous nappe pile with low angle normal faults. During the Late Miocene–Pliocene (1) the extensional deformation phase D3 formed the northern margin of the Cretan basin through normal faulting (from Soukis and Papanikolaou 2004). The numbers refer to Fig. 8.103

Fig. 8.105 Characteristic extensional low angle normal fault, bringing the upper tectonic unit of the marbles/granites (3) in direct contact with the relatively autochthon unit of the Eocene flysch (7) at cape Roukounas in Southern Anafi (from Soukis and Papanikolaou 2004). The three intermediate tectonic units of the metamorphosed pelagic–ophiolitic formations (4, 5 and 6) have been omitted. The numbers refer to Fig. 8.103

The difficulty of further analyzing the paleogeographic area of the Anafi units and its subsequent geotectonic evolution is due to the lack of dating the metasediments. The Late Cretaceous age of the tectono-metamorphism implies that the Anafi ophiolites may originate from the internal area of H2, with an initial subduction during the Late Cretaceous beneath the more internal carbonate platform H3. Later on all these units together suffered the HT/LP metamorphism, asssociated with the Upper Cretaceous granitic intrusions. The final tectonic transport over the most external non-metamorphic Upper Eocene flysch, possibly of the external platform H1 (? Tripolis Unit), resulted in the final Anafi nappe pile in the Late Eocene–Oligocene.

The Late Cretaceous metamorphic event is part of the orogenic organization of the Hellenides between the paleo-Alpine and the Alpine phase. It should be noted that similar outcrops are observed in the Nikouria islet to the north of Amorgos Island (Durr et al. 1978), as well as in Southern Syros Island, at the Vari region (Bonneau et al. 1980; Soukis and Stockli 2013) and the Akrotiri peninsula in Tinos Island (Patzak et al. 1994). New datings, including thermo-chronometric data, from Southern Syros and Western Tinos, showed that this upper greenschists bearing unit, consisting of Triassic granitic rocks, was metamorphosed during the Late Cretaceous (Soukis and Stockli 2013) and later, during the Miocene, was tectonically emplaced upon the relatively autochthon unit of the Northern Cyclades.

During the Miocene–Pliocene several hundred meters of a clastic sedimentary sequence of continental facies, known as Theologos Beds, were deposited in Anafi Island (Melidonis 1963). These sediments are paleogeographically related to the existence and evolution of the Aegeis median landmass (Philippson 1901), before the submergence of the area and the creation of the modern Aegean Sea.

8.3.12 The Kastoria Unit

In Northern Greece Godfriaux (1968) had distinguished the Flambouron unit, made of pre-Alpine basement rocks, above the Olympus autochthon and the blueschists bearing Ambelakia nappe. Later the Kastoria unit was distinguished from the Flambouro unit due to some unique features of its stratigraphy and tectonic position (Fig. 8.106) (Papanikolaou and Zambetakis-Lekkas 1980; Papanikolaou and Stojanov 1983; Papanikolaou et al. 1984). In the Northern Pelagonian region, Mountrakis (1983, 1984) distinguished the Vernon and Voras units, respectively. From the lithological point of view, the Kastoria unit can be distinguished in two groups: a lower group, made of granites, gneisses, mica schists etc., and an upper group, made of relatively low grade metamorphic meta-clastic rocks of greenschist facies, made of phyllites, quartzites, sericite-chlorite schists and meta-tuffs (Papanikolaou and Stojanov 1983; Papanikolaou 1988a). This low metamorphic grade upper group extends further north to the former Yugoslavia, where crystalline limestones have been dated with fossils, showing Upper Cambrian to Devonian ages (Papanikolaou and Stojanov 1983). In the Greek region there are no occurrences of crystalline limestones and therefore, the Greek members of the group (e.g. in the Aposkepou area) can only indirectly be dated as Lower Paleozoic.

Another feature of the Kastoria unit is the large granitic plutons that intruded in more than one phases (Kilias 1980; Katerinopoulos 1983). Thus, we have successive granitic intrusions, creating contact metamorphic aureols with hornfelses etc. (Kilias 1980). Therefore, we can distinguish older and younger granites, schistosed or not, which based on geochronological data belong to the Paleozoic, with Silurian as well as Carboniferous ages (Fig. 8.107) (Marakis 1970, 1972; Kilias 1980; Mountrakis 1984). During the Late Paleozoic, intrusions of basic dikes have been also reported within the Variscan orogenic cycle in the metamorphics of the Kastoria unit, both in Northern Greece and in Southern former Yugoslavia (Katerinopoulos 2008).

In Greece, the contact metamorphism around the Late Paleozoic granitic plutons is very intense and in most locations it overprints the regional low grade greenschist metamorphism. It should also be noted that in Southern Yugoslavia, around the Ochrida lake area, as well as in Eastern Albania, in the Episkopi area, the Triassic transgression is maintained at the base of the carbonate platform of the Sub-Pelagonian of H3. The Triassic unconformity is also observed in the metamorphosed Triassic–Jurassic carbonate platform of the Almopia unit, observed above the Kastoria/Perister unit, in the Kicevo region (Papanikolaou and Stojanov 1983).

8.3.13 The Flambouron Unit

The stratigraphic column of the Flambouron unit consists of a thick sequence of granites, gneisses, amphibolites, mica schists, etc., which pass upwards into a marble formation, overlain by alternations of mica schists, amphibolites, and marbles. The distinction of the Flambouron unit within the former Pelagonian zone was introduced by Godfriaux (1968), immediately after the discovery of the Olympus Nummulites and the understanding that the overlying granite-gneisses constitute a huge metamorphic basement nappe.

The marble formation is very thick in some locations and thin in others probably due to synmetamorphic flow. In the Vermio mt, in the Kastania area, and in the Voras mt (Kaimaktsalan), its thickness is great and it becomes even greater in Yugoslavia, whereas it decreases southwards. The age of the marbles is unknown although in Southern Yugoslavia, very close to the Greek borders, the age was determined as Riphean–Cambrian (transition from Pre-Cambrian to Cambrian) based on phytoplankton,

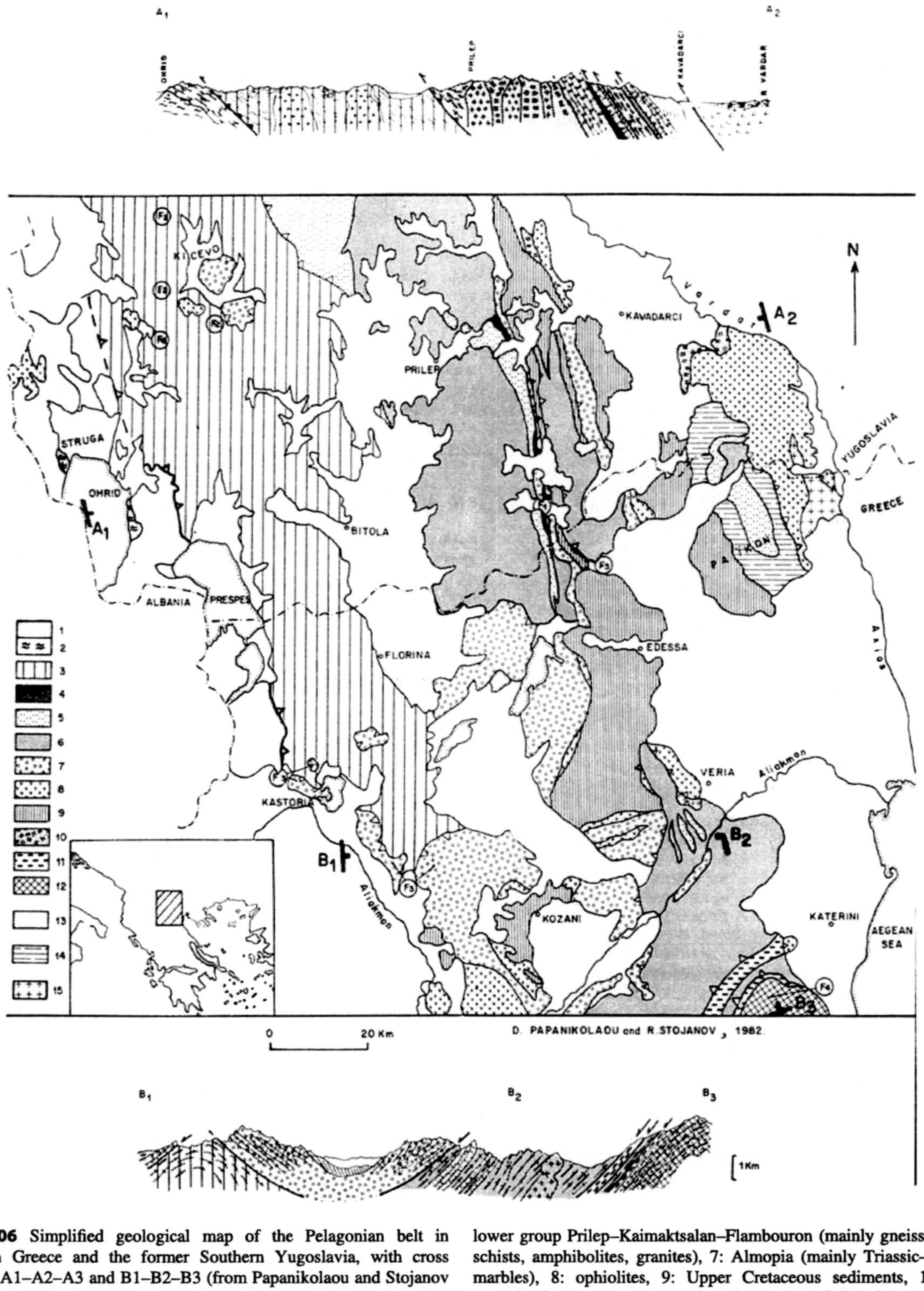

Fig. 8.106 Simplified geological map of the Pelagonian belt in Northern Greece and the former Southern Yugoslavia, with cross sections A1–A2–A3 and B1–B2–B3 (from Papanikolaou and Stojanov 1983). 1: Sub-Pelagonian Mesozoic sediments, 2: Sub-Pelagonian Paleozoic sediments, 3: Perister–Kastoria unit (partially Cambrian-Devonian), 4: Trojaci formation (Riphean–Cambrian?), 5: upper group, Prilep–Kaimaktsalan–Flambouron (mainly marbles), 6: lower group Prilep–Kaimaktsalan–Flambouron (mainly gneisses, mica schists, amphibolites, granites), 7: Almopia (mainly Triassic–Jurassic marbles), 8: ophiolites, 9: Upper Cretaceous sediments, 10: Late Jurassic–Cretaceous transgressive limestones of Peonia (e.g. Demir Kapija), 11: blueschists of Ampelakia, 12: Olympus autochthon, 13: molassic and post-Alpine formations, 14: Triassic–Jurassic of Paikon (limestones–rhyolites), 15: Upper Jurassic Fanos granite

Fig. 8.107 Granite-gneiss outcrop, in the form of augengneiss of Carboniferous age in the Kastoria unit, from the Namata–Sisani area

determined at the upper part of the marble formation (Arsovski et al. 1979). However, the results of this dating in rocks of medium–high grade metamorphism are highly questionable, because the organic material can not survive at such high temperatures (Papanikolaou 1981b). The above marbles, considered of lower Paleozoic age in S. Yugoslavia, were considered as Triassic in N. Greece region, as they were confused with the Triassic marbles of the Almopia unit, that can be found right next to each other (as the Almopia marbles occur also upon the gneisses and the schists). Thus, on the one hand, the fossiliferous Triassic marbles of Almopia, are tectonically emplaced upon the gneisses, mica schists, and granites of the Flambouron unit, and on the other hand, the marbles of the Flambouron unit, without any chronological data in the Greek region are considered as Riphean–Cambrian? in S. Yugoslavia.

The first geochronological data of the granites in the Flambouron unit indicated probable Carboniferous and Permian ages (Yarwood and Aftalion 1976). These ages were later confirmed by new geochronological datings (Reischmann et al. 2001; Koroneos et al. 2013), which concluded to the already known Variscan age of about 300 Ma.

Another feature of the Flambouron unit is the multiple metamorphic phases and not just a usual retrograde greenschist facies overprinting an initial blueschist phase, as it occurs in e.g. the underlying Ambelakia blueschists (Northern Cyclades unit). Therefore, we do not observe two successive metamorphic events within one orogenic cycle but more metamorphic events belonging to more cycles (Godfriaux 1968; Yarwood and Dixon 1977; Migiros 1983; Schermer et al. 1993; Mposkos and Perraki 2001; Kilias et al. 2010). Thus, the tectono-metamorphic history of Flambouron unit, (as well as of the Kastoria unit), comprises: (i) an old pre-Alpine Variscan metamorphic event of amphibolitic facies, (ii) an Eo-alpine, Early Cretaceous event of blueschist facies, which overprints the previous one, and (iii) a mild metamorphic event, which is due to the Late Eocene tectonism of the already metamorphosed rocks, when they were thrusted together with the rest of the internal units over the external units, like the relatively autochthon, Olympus unit. Characteristic synmetamorphic structures of isoclinal folding and associated intersection lineations occur in the Maastrichtian-Paleocene flysch above the Almopia unit, as in the Edessa Railroad Station (Mercier and Vergely 1977). These observations indicate strong deformation of the Post-Cenomanian sediments at deeper tectonic levels, during Eocene, which have not been further analysed up to present. The distinction of the same Eocene structures in the Paleozoic basement rocks of the Flambouron unit is very difficult.

8.4 The Axios/Vardar Ocean—H4

8.4.1 The Maliac Unit

The Axios/Vardar Ocean (H4) is mainly observed in the form of ophiolite complexes, tectonically emplaced upon the internal Hellenides platform H3, and the only sedimentary stratigraphic sequence of the H4 oceanic domain is the Maliac unit. The sedimentary history of the Maliac unit is limited from the Late Permian to the Early Cretaceous whereas during the Late Cretaceous–Eocene it is included in the wider Eastern Greece unit. Its general stratigraphic column is similar to the Pindos sequence, especially for the Upper Triassic–Jurassic period. Thus, radiolarites are dominant with intercalations of pelagic breccia-limestones, as well as volcanic tuffs and basaltic lavas of Norian–Rhaetian (Fig. 8.108) up to the Upper Jurassic, where pelagic siliceous limestones are also present (Ferriere 1976, 1979, 1982; Hynes et al. 1972; Smith et al. 1975) (Fig. 8.109).

The differences between the Maliac and the Pindos unit, besides the duration of the abyssal-pelagic sedimentation, is that the Maliac unit has less limestones than the Pindos unit and much more mafic volcanics, with a strong presence of submarine volcanism. Additionally, the stratigraphic column of the Maliac unit begins in the Permian with a volcano-sedimentary complex of the rifting stage, in which shallow neritic limestones can be observed whereas the initation of the abbysal-pelagic facies, showing the oceanization of the former rift, started already in the Scythian (Fig. 8.110). In contrast, the volcano-sedimentary complex of the Pindos unit, is dated in the Middle Triassic (?)—Carnian and the initiation of the abyssal-pelagic sedimentation postdates the Carnian (see also Fig. 8.3).

Thus, the Pindos unit has clearly pelagic features during the Late Triassic–Jurassic period, whereas the Maliac unit has pelagic-abyssal features, as it is deeper and closer to centres of mafic submarine volcanic activity, with a plethora of basalts and diabases. This can be clearly observed in the geological cross sections of the area east of Neochorion, in Central Otrhis mt, at the Loggistion unit (Fig. 8.111). The name of the Maliac unit derives from the Maliac gulf (Ferriere 1976), due to the typical outcrops all over the Othris mt, with unique Triassic radiolarites. Similar outcrops are present in other locations too, such as in Central Evia and especially in Argolis, where the pelagic **Epidavros** unit of Maliac affinities, is thrusted over the neritic Sub-Pelagonian type **Trapezona** unit, both covered by the Late Cretaceous unconformity (Vrielynck 1982).

In a recent review of the Maliac unit by Ferriere et al. (2016) there is a proposal to include in the Maliac oceanic basin, besides the well known outcrops of Othris, Evia and Argolis, also the outcrops of Central Macedonia, belonging to the units of Peonia and Paikon. Additionally, they consider that the Othris outcrops represent the western margin of the Maliac Ocean whereas the Paikon and Peonia outcrops represent the eastern margin, which during the Middle Jurassic convergence period became an active margin. The central area of the Maliac Ocean comprised Middle to Late Triassic oceanic lithosphere, associated with Middle Triassic–Middle Jurassic sediments. However, correlations with the other pre-Cretaceous ophiolite bearing outcrops/units of H4 (e.g. Vourinos, Vermion, Chalkidiki) are not proposed

Fig. 8.108 Outcrop of abyssal-pelagic sediments of Upper Triassic age from the Maliac unit of Central Evia

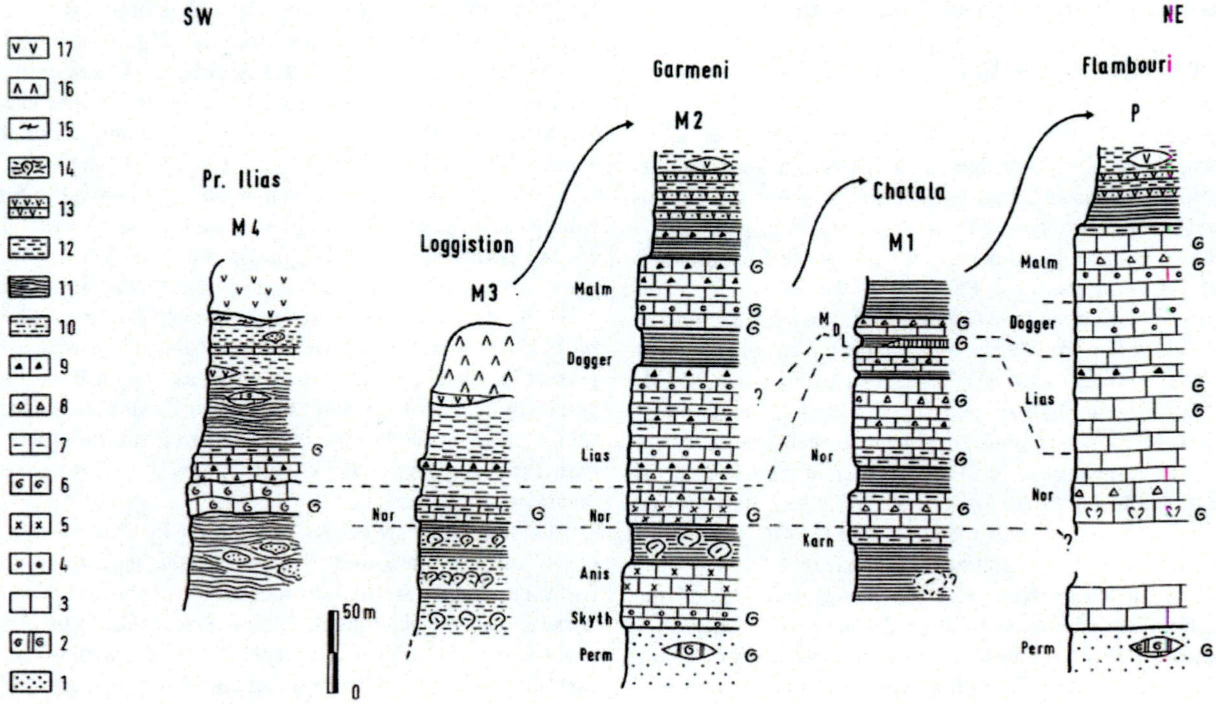

Fig. 8.109 Stratigraphic columns of the Maliac unit by Ferriere (1979). Four stratigraphic columns are described, which are incorporated into the Maliac unit (M1, M2, M3 and M4), as well as the relatively autochthon Sub-Pelagonian unit (P). A progressive deepening to the SW can be observed with abyssal-pelagic features in the upper units of Prof. Ilias and Loggistion, which pass laterally into pelagic features of the Garmeni and Chatala units before the carbonate platform of the relative autochthon Flambouri unit. 1: Permian shales-sandstones, 2: limestones with Fusulines, 3: neritic limestones, 4: oolithic limestones, 5: dolomitic limestones, 6: Hallstatt facies limestones with ammonites, 7: limestones with cherts, 8: breccia limestones, 9: microbreccia limestones, 10: sandstones, pelites, 11: cherts, radiolarites, 12: pelites, 13: chaotic mélange of Malm age with volcanics, 14: Triassic pillow lavas, 15: amphibolites, 16: pillow lavas and radiolarites, 17: peridotites and gabbros

and there is no coherent paleogeographic scheme proposed for the internal Hellenides pre-orogenic organization.

8.4.2 The Vatos Unit

The Vatos unit is known from Central Crete, where it was distinguished by Bonneau (1976, 1984). This unit belongs to the upper part of the Cretan nappe pile, overlying the Ethia and Arvi units (see also Fig. 8.100). However, it is often observed sitting tectonically directly over deeper units, such as the Tripolis unit, due to large low angle normal faults (Fig. 8.112).

The stratigraphic column is unique, with: (i) Permian limestones at its base, (ii) argillaceous–silicate rocks with a few carbonate interbeds at the Mesozoic sequence up to the Late Jurassic and (iii) ophiolitic rocks in the form of a mélange at the top. The unit is accompanied by members of an ophiolitic complex emplaced on top of the melange. Additionally it is characterized by a low grade blueschist-type metamorphism. Its classification in the oceanic basin of Axios (H4) is based on its paleo-Alpine tectonism in the Late Jurassic and especially on its subduction-type HP/LT metamorphism. The presence of non-metamorphosed Upper Cretaceous pelagic pinkish limestones with Globotruncanes (mainly of Campanian–Maastrichtian age) over the ophiolites verifies its internal origin (Fig. 8.113).

Flysch is not observed above the Upper Cretaceous limestones. Thus, the ophiolite rocks are observed between the underlying chaotic structure of the Upper Jurassic ophiolitic melange and the overlying pelagic Upper Cretaceous limestones. Hence, the ophiolites occur at the Lower Cretaceous part of the Vatos column. The Upper Cretaceous pelagic sequence observed on top of the Vatos ophiolites cannot be correlated with the Upper Cretaceous transgression in the Eastern Greece unit, where shallow water platform carbonates dominate. On the contrary, it simulates the limestones of the same age and sedimentary facies observed syngenetically related to the Arvi basalts.

Fig. 8.110 **a** Outcrop of thin-platy pinkish-crimson limestones with cherts of Scythian–Anisian age (Hallstat facies) in the volcano-sedimentary complex of the Maliac unit in Central Othris mt. **b** Characteristic outcrop of the Triassic radiolarites of the Maliac unit from Central Othris mt, in contrast to the Pindos unit, where the associated radiolarites are mainly of Middle-Upper Jurassic age

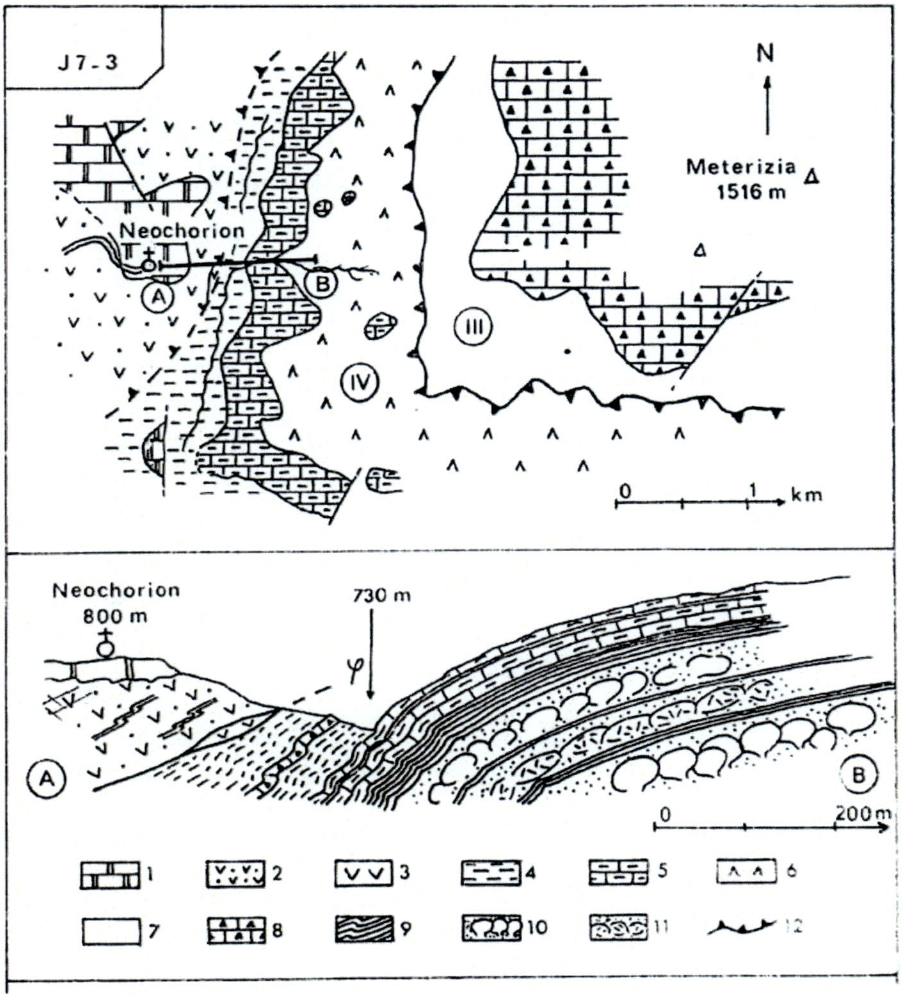

Fig. 8.111 Geological cross section A-B east of Neochorion by Ferriere (1977), showing the stratigraphic sequence of the Loggistion unit. In the accompanying map, the Loggistion unit is thrusted over the Garmeni unit in Central Othris (Meterizia summit). 1: transgressive limestones of the Upper Cretaceous, 2: basaltic lavas, 3: serpentinised peridotites, 4: silicate shales (Jurassic), 5: limestones with silex of Norian age, 6: volcano-sedimentary of the Triassic, 7: radiolarites and serpentines, 8: micro-breccia limestones of the Jurassic, 9: radiolarites and silicate shales, 10: pillow lavas and hyaloclastites, 11: dolerites and tuffs, 12: pre-Upper Cretaceous tectonic contact

Fig. 8.112 View of the Vatos unit in Central Crete directly above the Tripolis unit (from Papanikolaou and Vassilakis 2010). Due to the extensional detachment the intermediate nappes of the Pindos/Ethia, Arvi and Miamou are omitted

An impressive aspect is that the Vatos nappe reaches up to the southern edge of the Hellenides in Gavdos Island, where the meta-ophiolitic rocks of Upper Jurassic age of metamorphism are covered by Upper Cretaceous marls in the local **Kalypso** unit (Vicente 1970). This unit crops out in the form of tectonic klippen over the Pindos/Ethia unit. Therefore, although in Gavdos Island we are at the edge of the Hellenic margin towards the Hellenic trench, upper

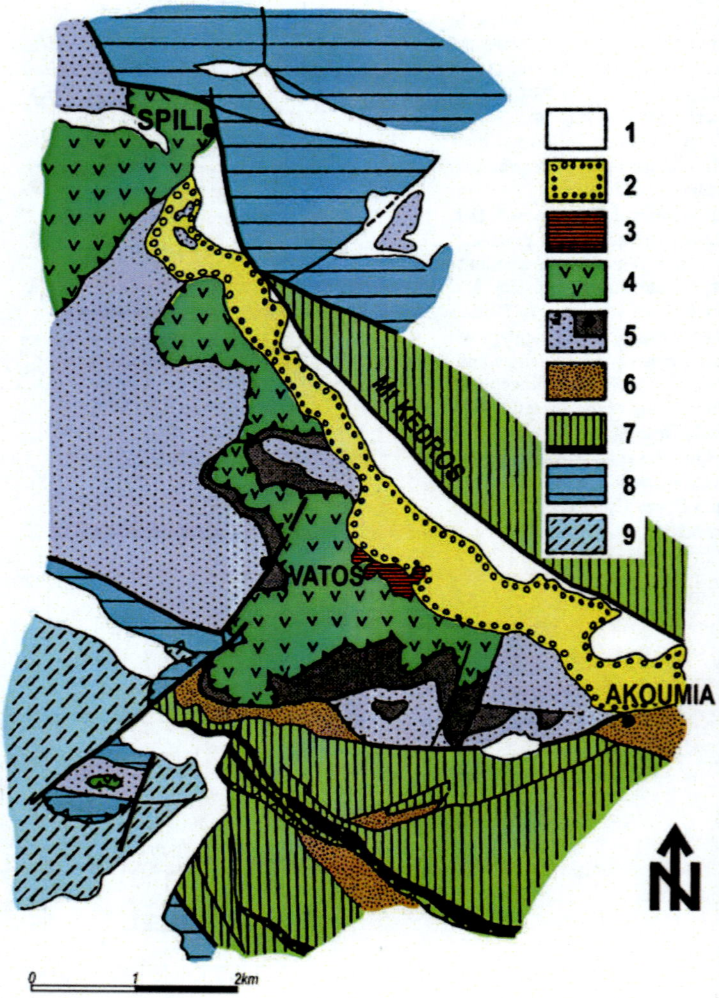

Fig. 8.113 Simplified geological map of the Spili area in Central Crete (from Bonneau et al. 1977, modified), showing the Vatos unit over the Tripolis and Pindos units. 1: Quaternary, 2: Miocene, 3: Upper Cretaceous with Globotruncanes, superjacent to the ophiolites, 4: ophiolites, 5: (a) schists of Vatos (partially of Permian, Jurassic), and (b) ophiolitic olisthostrome (Upper Jurassic), 6: Pindos flysch, 7: limestones and radiolarites of Pindos, 8: Tripolis limestones, 9: Permian-Triassic Tyros beds

tectonic nappes of internal origin are observed instead of the more external units of the Paxos or Ionian, like in the Ionian Islands, towards the northern part of the arc, or like in Kastellorizo, towards the eastern part of the arc.

8.4.3 The Ophiolite Nappes of Axios/Vardar

The oceanic crust of Axios H4 can be observed as an obducted ophiolite nappe in many areas of the Internal Hellenides (see also Fig. 8.57). It is usually observed above the chaotic structure of the Jurassic ophiolite mélanges, developed either over the continental margins of the H3 carbonate platform, such as the Sub-Pelagonian and Almopia units, or over the abyssal-pelagic sequences of H4, such as the Maliac unit. The Axios ophiolite outcrops are numerous both in Northern and Southern Greece, although they are usually of limited thickness. Characteristic Axios ophiolite outcrops occur at the thrust sheets of Aridea and Ano Garefi above the Almopia metamorphosed carbonate platform, with overlying Upper Cretaceous and Lower Cretaceous sediments respectively (Migiros and Galeos 1990; Robertson et al. 2013). Large ophiolite nappes, often 1–2 km thick, can be observed in the Vourinos mt over the Triassico-Jurassic marbles of the Almopia unit (Fig. 8.114). In Kallidromo mt and neighboring outcrops of Beotia the ophiolites are observed over the Sub-Pelagonian unit and in Eastern Othris and Central Evia over the Maliac unit.

In several Axios ophiolite occurrences the Upper Cretaceous transgression can be observed with laterites at the base of the transgressive conglomerates, as well as Fe-nickel-aluminum deposits (i.e. in Larymna). Upper Cretaceous neritic carbonate sediments with Rudists and Eocene flysch are overlying the laterites. The transition to the flysch is usually made through intermediate transitional pelagic carbonate sediments of small thickness with Globotruncanes of the Maastrichtian. Therefore, the final Alpine tectonism of the Axios ophiolites takes place at the end of the Late Cretaceous–Eocene, as observed all over the outcrops of the Eastern Greece unit. This Eocene tectonic event often includes detachment of ophiolite bodies from their initial tectonic basement and further sliding towards more external regions, i.e. the External Hellenides. This re-activation produces a "confusion" with the H2 ophiolites of the Pindos oceanic basin. Thus, ophiolite bodies are observed within the Eocene flysch above the Upper Cretaceous unconformity, besides the ophiolite outcrops observed in their primary tectonic emplacement above the Upper Jurassic formations but beneath the Upper Cretaceous unconformity, like for example in Central Eastern Evia (Fig. 8.115). Similar multiphase ophiolite mélanges resulting from the paleo-Alpine and Alpine tectonisms have been documented from the Axios/Vardar ophiolite region in the Dinarides in Yugoslavia (Dimitrievic and Dimitrievic 1976).

A separate issue concerns the existence and preservation of syngenetic sediments of the ophiolite complexes before their obduction upon the continental margins. These sediments are generally rare and occur in usually ambiguous outcrops of radiolarite sediments, which in the case of Vourinos have been dated as Middle Jurassic (Chiari et al. 2003). It is obvious that the age of these sediments changes along the paleogeographic position of the sea floor spreading oceanic area and therefore, each dating is indicative only of regional importance but not general for the entire ophiolite complex.

In some internal units of the Axios Ocean H4 an older unconformity of Late Jurassic can be observed beneath the Late Cretaceous unconformity. This older unconformity covers the ophiolitic rocks and dates an older tectonic emplacement of the ophiolites within the early stages of the paleo-Alpine orogenic phase with the beggining of the closure of H4. These Upper Jurassic sediments have been studied by Galeos et al. (1994) and they have been distinguished in the geological maps of the Almopia region (Mercier and Vergely 1984) (see also Fig. 8.81).

The Eocene tectonism of the Axios domain has been related by some researchers with the final disappearance and closure of H4, whose part is thought to have been preserved after the paleo-tectonism of the Late Jurassic–Early Cretaceous (Robertson et al. 2013). However, a direct post-Cenomanian dating–proof of ophiolites genesis in the Axios oceanic area does not exist. The totality of the H4 ophiolite outcrops are re-tectonised ophiolitic bodies of the initial paleo-tectonic phase, as shown in Aridaia and in the Kymi region of Central Evia (see also, Figs. 8.82 and 8.115). The Axios ophiolites have been transported during the late Eocene–Oligocene up to the external platform over the Tripolis Unit, as indicated by the Angelona outcrops overlying Permian and Triassic limestones near Monemvasia in SE Peloponnese (Gerolymatos et al. 1982). The occurrence of Permian limestones and the determination of Triassic radiolarites within the dismembered Angelona ophiolites (Danelian et al. 2000) point to their provenance from the Axios Ocean.

8.4.4 The Metamorphic Nappes with Ophiolites of the Northern Sporades

A distinct case of units in the H4 region are some metamorphic rocks associated with ophiolites and metamorphosed mafic volcanic rocks, underlain by the Upper Cretaceous unconformity. Their outcrops occur in the Northern Sporades islands and mainly in **Skyros** Island

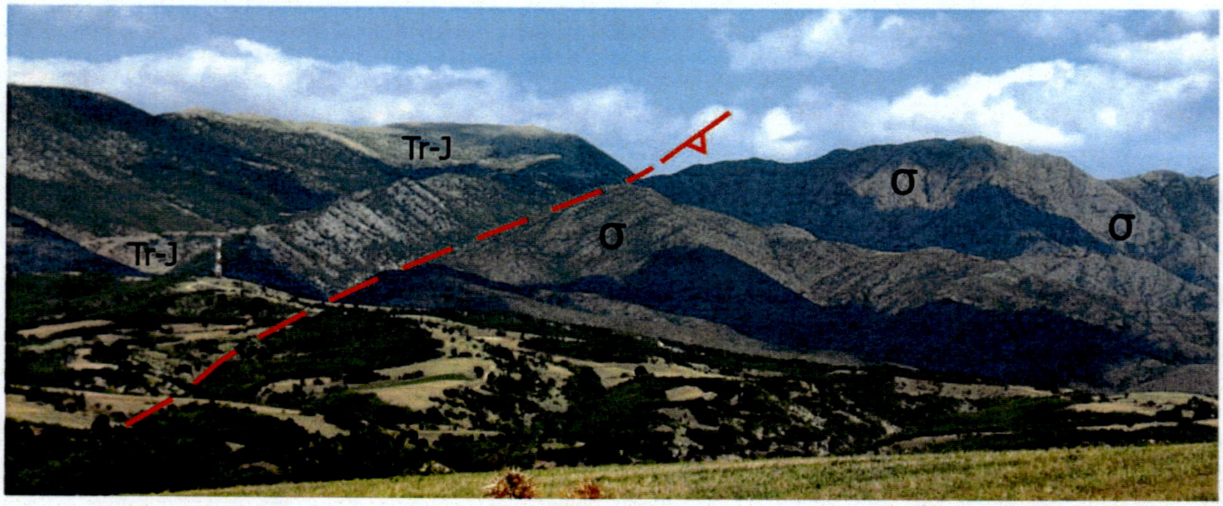

Fig. 8.114 View of the Vourinos ophiolites (σ) of H4, which have been tectonically emplaced upon the Almopia marbles (Tr-J) of H3 towards the north (left side of the photo) as seen from the Mesohellenic basin of the Grevena area

(Fig. 8.116) (Jacobshagen et al. 1983; Jacobshagen and Wallbrecher 1984; Jacobshagen 1986).

Greenschists and amphibolitic schists with some marble intercalations can be observed in the two nappes above the SubPelagonian. According to Jacobshagen and Wallbrecger (1984) they are the «Eo-Hellenic nappe» and the Skyros nappe. In fact, the Eo-Hellenic nappe comprises metamorphic rocks of a pelagic, volcano-sedimentary sequence, overlying tectonically the ophiolites, whereas the upper Skyros/Olympus nappe comprises medium grade rocks of mica schists and marbles above sheared ultramafic rocks with a chaotic structure. They crop out mainly at the northeastern part of Skyros and overlie the Sub-Pelagonian unit, observed all over the rest of the island. These metamorphic rocks are pre-Late Cretaceous tectonic nappes, which have been imbricated with the ophiolites of Axios and show a tectonic origin from the more internal segments of the arc. Unpublished recent geochronological data comprising 40Ar/39Ar analyses have shown a Late Cretaceous metamorphism for the upper tectonic nappe of the Skyros/Olympus tectonic unit of the local Olympus mt, around 96–70 Ma, with an early phase prior to 96 Ma, followed by a second phase at 84–88 Ma and by late alteration phases after 68–70 Ma (Boundi et al. 2019).

Thus, these metamorphic nappes probably represent segments of the H4 ocean, derived from its internal margin towards the more internal continental platform terrane H5. They were probably involved in the late stages of the paleo-Alpine orogenic phase, with subduction and subsequent rapid exhumation within the Late Cretaceous. However, the overlying non metamorphosed Upper Cretaceous limestones and Eocene flysch are involved in the final Alpine thrusting in between the paleotectonised units. Therefore, from the geodynamic point of view, they are associated to the paleo-Alpine Almopia subduction process, prior to the Upper Cretaceous unconformity. Nevertheless, the Skyros meta-sediments are pelagic with submarine basic volcanism, with Maliac affinities, possibly representing a late closure of the Axios oceanic basin. In fact, the Skyros relative autochthon certainly belongs to the internal carbonate platfrom H3 but its low crystallinity and intense deformation place it, from the geodynamic point of view, between the Almopia and the Sub-Pelagonian unit.

8.5 The Lesvos–Paikon Platform—H5

8.5.1 The Lesvos Unit

The northern part of the Lesvos Island comprises almost entirely volcanic rocks, together with some molasse sediments of Lower Miocene age. In the rest of the island two units can be observed, comprising metamorphic rocks of Lower Carboniferous to Triassic ages (Hecht 1970, 1972; Katsikatsos et al. 1986). Lithologically, they comprise low grade metamorphic clastic rocks with intercalations of tuffs and carbonate horizons of significant thickness. The metamorphism is Alpine, characterized as greenschist facies, with parageneses also of the pumpellyite–actinolite facies (Katagas and Panagos 1979). Moreover, there are large outcrops of ultramafic rocks and minor mafic bodies, dikes and tuffs, i.e. pieces of an ophiolite complex, which forms a tectonic nappe over the relatively autochthon metamorphic unit.

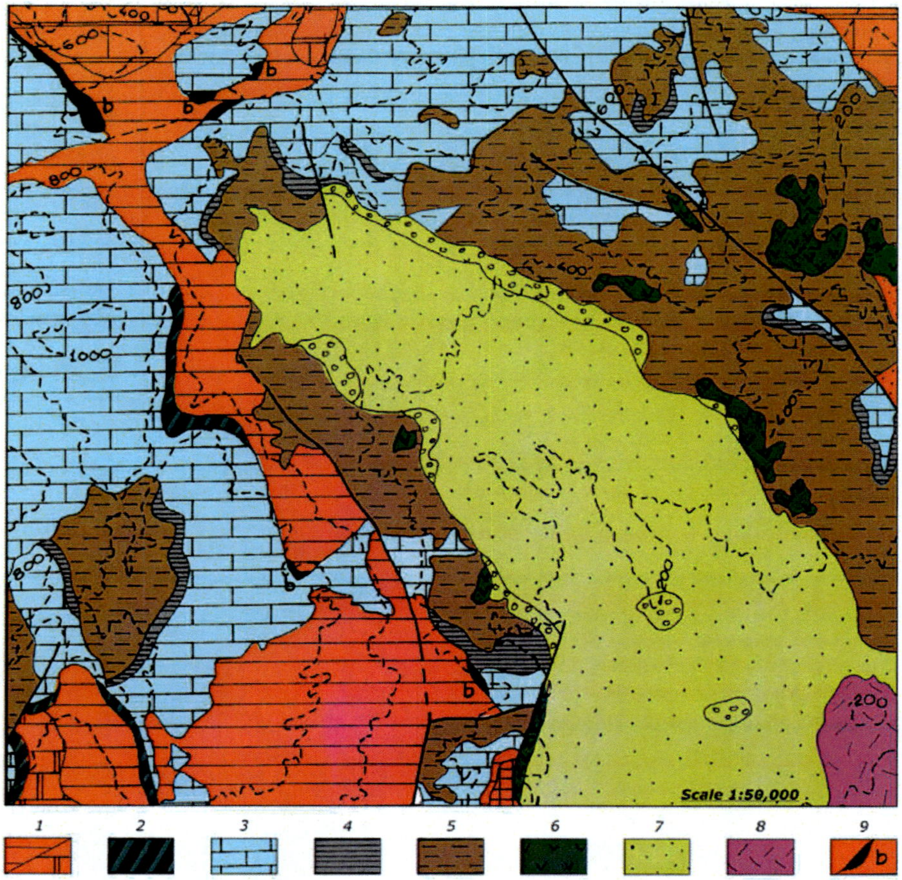

Fig. 8.115 Geological map of Central Eastern Evia from the Kymi region, showing ophiolite outcrops of H4 inside the Eocene flysch at the top of the Upper Cretaceous transgression on the Sub-Pelagonian unit. These outcrops contrast the ophiolites occurring inside the schist-chert formation of the Upper Jurassic (from Katsikatsos et al. 1981, modified). 1: Triassic-Jurassic limestones of the internal carbonate platform H3, with Megalodon at its base and Cladocoropsis at its upper layers, 2: schist-chert formation of the Upper Jurassic with small ophiolite bodies, 3: limestones, marly at the base and thick-bedded with rudists at the upper part, 4: thin-bedded marly limestones with Globotruncanes, 5: Paleocene–Eocene flysch, 6: serpentinised ophiolites, mainly harzburgites, 7: Neogene marls, marly limestones and sandstones with conglomerates at the base, of lacustrine facies, rich in lignites, 8: reddish lavas of dacitic and andesitic composition in the form of domes of Middle Miocene age. 9: bauxite

The two tectonic units of Lesvos Island correspond to two terranes, H5 and H6, which, together with the Chios Allochthon unit, belong to the Cimmerides. They are characterized by Permian–Triassic sequences, tectonized in the Late Triassic–Early Jurassic, with a characteristic unconformity in the Lias (Papanikolaou 1997, 1999, 2009). Therefore, this is an orogeny older than the paleo-Alpine events, characterized by the Liassic unconformity, instead of the Late Jurassic and Cenomanian unconformities, with a time difference of approximately 20 Myrs.

The stratigraphic column of the Lesvos autochthon includes: (i) a thick volcano-sedimentary complex of Upper Paleozoic age, (ii) a carbonate platform a few hundred meters thick, in which Permian horizons with Productus have been found near the base, while horizons with Megalodon of Upper Triassic–Liassic (?) age have been observed near the top, and (iii) an upper phyllitic schistose sequence together with meta-sandstones, probably representing a meta-flysch formation (Fig. 8.117).

The Lesvos autochthon is therefore an individual unique carbonate platform, located in a key position, as similar rocks occur at the opposite coastline of Asia Minor, within the Cimmerides region, bibliographically referred to the Pontides, characterized by tectonic activity in the Late Triassic–Liassic (Sengor 1984; Papanikolaou and Demirtasli 1987). The post-orogenic evolution of the Pontides is characterized of epicontinental sedimentation, interrupted by several unconformities of Lias, Malm and Late Cretaceous. The overall post-orogenic tectonism is simple, with successive episodes of arc volcanism in the Late Cretaceous, the

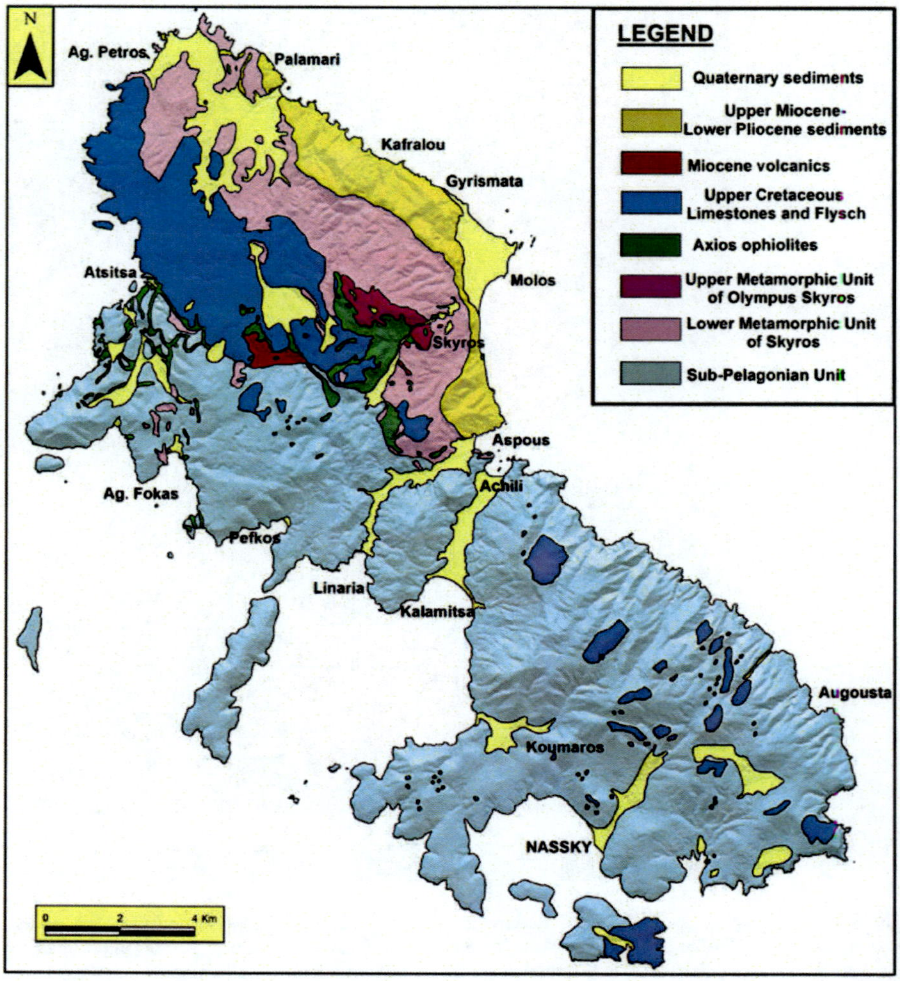

Fig. 8.116 Simplified geological map of Skyros Island, showing the two metamorphic nappes above the Sub-Pelagonian unit (based on Jacobshagen et al. 1983). These are tectonic nappes of the paleo-Alpine phase with deep geodynamic phenomena and ophiolite imbrications. The non metamorphosed Upper Cretaceous sediments are involved in the final tectonic structure of the Alpine tectonism

Eocene and the Oligocene (Fourquin 1975). Thus, this domain of the Pontides was throughout the Late Cretaceous–Oligocene period participating in successive volcanic arcs, as a result of the ongoing orogenic process which was migrating southwards in the Taurides.

It is noteworthy that in the village of Moria, north of Mytilini, an ancient quarry–marble laboratory has been preserved, inside the unique blue-gray Upper Triassic–Liassic marbles with Megalodon of the Lesvos autochthon. Columns and other rests of artifacts made of this marble quarry are observed both in Pergamos and Troy archaeological sites, at the opposite Asia Minor coastal areas (Fig. 8.118).

8.5.2 The Chios Allochthon Unit

In Chios Island, two units can be observed, mapped during the 1960s by a team of German geologists under the guidance of Professor V. Jacobshagen (Besenecker et al. 1968). The relatively autochthon unit includes a several km thick Permian clastic formation with olistholites of Silurian-Lower Carboniferous age overlain by a very thick Triassic-Jurassic carbonate platform (Papanikolaou and Sideris 1983), which belongs to the H3 platform, with a characteristic presence of Upper Jurassic limestones at the top of the column. The upper unit corresponds to a tectonic nappe belonging to H5 (Papanikolaou and Sideris 1992). In a restricted outcrop under the post-Alpine continental Miocene sediments in Southeastern Chios, ophiolites can be also observed, which probably belong to H4. The H5 outcrops are spread all along the island from the northeastern edge of Chios in Kardamyla till the southern part in Emborio and Mesta. The stratigraphic column comprises (Papanikolaou 2009): (i) non-metamorphosed Upper Paleozoic sediments with pelites, sandstones and thin intercalations of Carboniferous limestones at the base, (ii) a Permian carbonate platform, a few hundred meters thick (Baud et al. 1991),

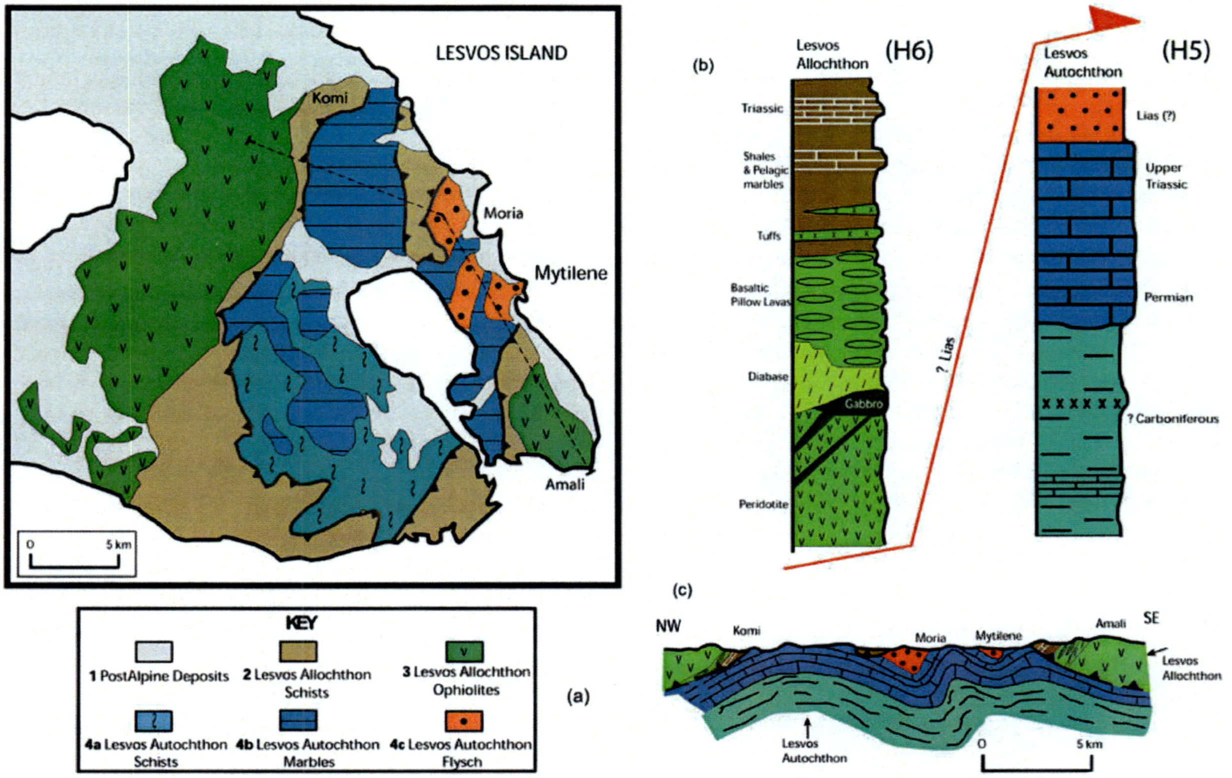

Fig. 8.117 a Simplified geological map of Southern Lesvos, showing the two tectonic units with the ophiolites in the Allochthon and the Permian–Triassic carbonate platform in the Autochthon. b Stratigraphic columns of the autochthon and the Allochthon units of Lesvos. c Transverse geological section of the two tectonic units (from Papanikolaou 2009)

Fig. 8.118 View of ancient columns made from the Triassic marbles with large Megalodons of the Lesvos autochthon in the archaeological site of Troy

(iii) a Liassic unconformity, marked by red clastics, approximatly 20 m thick, at its base and (iv) superjacent Upper Liassic platform carbonates of significant thickness (Fig. 8.119).

In the region east of Kardamyla, the Liassic unconformity assumes the features of a disconformity, with a stratigraphic hiatus comprising the entire Triassic. On the contrary, in southern Chios, the Liassic unconformity lies directly over the Carboniferous and the stratigraphic hiatus includes the entire Permian and Triassic periods, reaching almost 70 Myrs. In Southern Chios bauxite outcrops occur within the Liassic limestones, as well as overlying disconformable Upper Cretaceous limestones with Rudists (Papanikolaou and Soukis 2000). Flysch is not observed over the Liassic platform, but only younger unconformities in the Tertiary (Besenecker et al. 1968), similar to those described in the Pontides in Asia Minor (Rice et al. 2006). The sedimentation and orogeny in the southern central Pontides during the Early Cretaceous, described by Okay et al. (2013), shows the continuation of the paleo-Alpine tectonic phases of the Hellenides to the east in Minor Asia.

8.5.3 The Paikon Unit

Mercier (1966) had distinguished three sub-zones within the Axios zone, Almopia to the west, Paikon in the middle and Peonia to the east. Nevertheless, he also separated further individual subunits, due to the absence of a stable stratigraphy within each subunit. The Almopia subzone, observed to the west, has a distinctive stratigraphic column, composed of the Triassic–Jurassic carbonate platform in H3, overlain by the ophiolite nappe of the Axios H4 during the Late Jurassic–Early Cretaceous, exactly like its equivalent the non-metamorphosed Sub-Pelagonian platform to the south. The Paikon unit is located more internally than the Axios ophiolite suture of the H4 and shows characteristics of a volcanic arc during the Late Jurassic–Early Cretaceous. The Peonia unit is located even more internally and shows characteristics of a typical back-arc basin and marginal sea during the same period.

The Paikon unit is characterized by a stratigraphic column that according to Mercier (1968) includes from base to top (Fig. 8.120): (i) gneisses and mica schists, possibly of

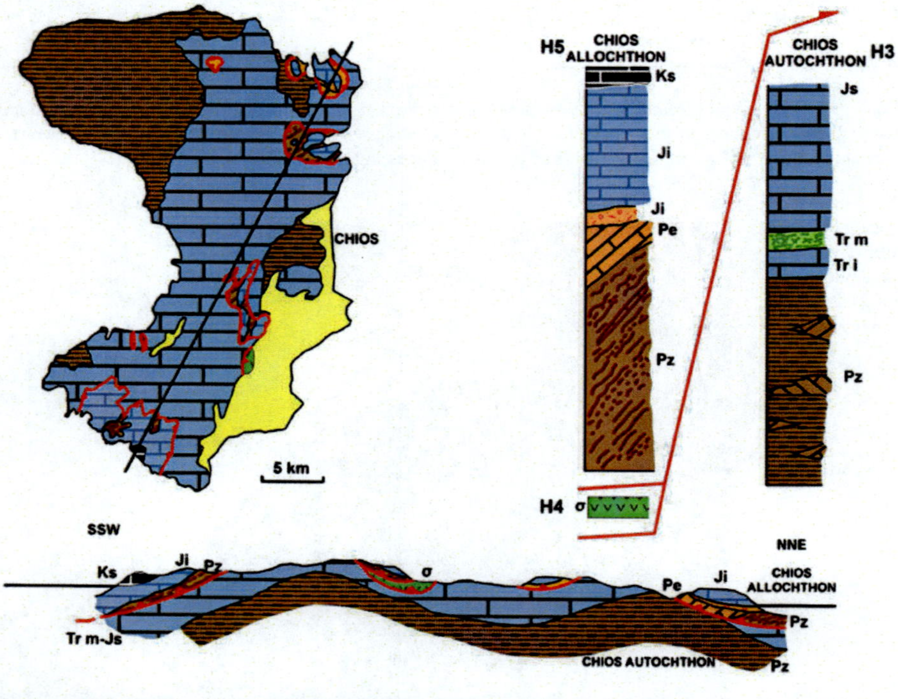

Fig. 8.119 **a** Schematic geological map of Chios Island with distinction of the two units, an autochthon internal platform of the Sub-Pelagonian of H3 and an allochthon unit with the Liassic unconformity of H5. **b** Schematic stratigraphic columns of the Chios units. **c** Schematic geological section of NNE-SSW orientation of Chios Island (from Papanikolaou 2009). In the northern section of the Allochthon outcrops only the Triassic is missing between the Carboniferous and the Upper Lias, whereas the Permian is also missing from the southern section, where the Liassic carbonates rest directly on top of the Carboniferous formations. At the southernmost outcrop of the Chios Allochthon Upper Cretaceous neritic limestones with rudists (Ks) have been reported (Papanikolaou and Soukis 2000)

Upper Paleozoic age. (ii) marbles of neritic facies, possibly of Upper Paleozoic–Triassic age. (iii) a metamorphosed pelagic sequence with cipollinic marbles and mica schists, possibly of Jurassic age. (iv) clastic rocks with acidic-basic volcanism in the Late Jurassic, including rhyolites, rhyolitic tuffs, and spilites-keratophyres, related to the volcanic arc of the paleo-Alpine orogeny. The neighboring Fanos granite of Upper Jurassic age (Borsi et al. 1966) represents a correspondent magmatic formation of the same geo-dynamic origin (Aubouin 1977). More recently, the Fanos granite has been considered as the result of an intra-oceanic subduction within the Axios/Vardar ocean (Michail et al. 2016). The Paikon stratigraphic column is concluded with non-metamorphosed Cretaceous formations, which include flyschoid clastics and volcanics of the Lower Cretaceous, and unconformably overlying neritic Upper Cretaceous limestones of great thickness.

Younger research has questioned the above description of Mercier (1968) and has provided various interpretations (Mercier and Vergely 1975; Godfriaux and Ricou 1991; Bebien et al. 1994; Ricou and Godfriaux 1995; Vergely and Mercier 2000; Brown and Robertson 2003; Tranos et al. 2007; Katrivanos et al. 2013; Brun et al. 2016). The general idea is the existence of tectonic window structure of the Paikon unit, below the paleo-Alpine and/or Alpine nappes. Thus, the outcrops of the Paikon area have been considered of Pelagonian or even more external origin, with tectonic transport to the east but also to the west. The above points do not abolish the general assumption of the existence of an Upper Jurassic acidic volcanism, which characterizes the

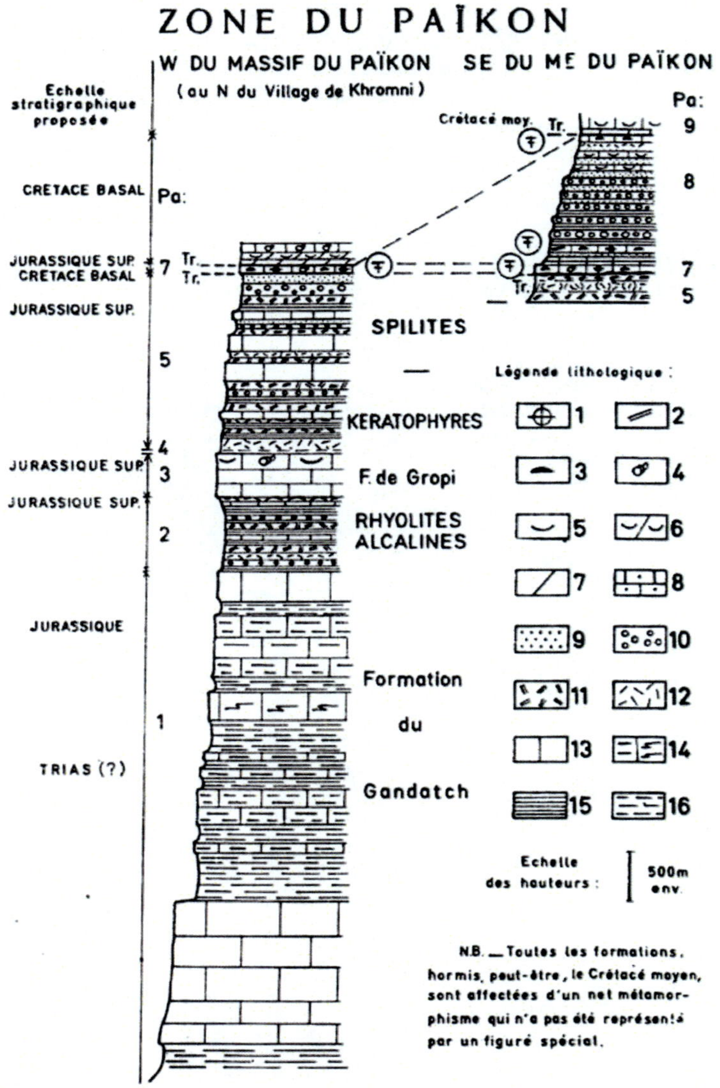

Fig. 8.120 Stratigraphic column of the Paikon unit (from Mercier 1968). 1: Cladocoropsis and corals, 2: algae, 3: foraminifera, 4: gastropods, 5: shell fragments, 6: dolomitic limestones with shells, 7: dolomites, 8: sandstone limestones, 9: sandstones, 10: conglomerates, 11: quartz keratophyres and sericite porphyries, 12: spilites and diabases, 13: marbles and crystalline limestones, 14: cipollines, 15: schists and pelites, 16: chloritic schists

Paikon unit for the paleo-Alpine orogeny. This volcanic arc signature may continue to corresponding rocks of the Peonia unit, where younger acidic intrusions penetrate into the basic rocks of the Peonia basin.

8.5.4 The Doubia Unit

Within the Circum-Rhodope Belt to the east of the Paikon unit there are some distict tectonic units with similar characteristics of neritic carbonate platforms, which should be also placed within the H5. One such characteristic outcrop occurs in the Doubia area in Chalkidiki, where Kauffmann et al. (1976) and Kockel et al. (1977) described a neritic carbonate sequence of Upper Triassic age with Megalodon bearing limestones and dolomites. These shallow water carbonates are overlying an Upper Permian–Lower Triassic volcano-sedimentary complex, with petrological–geochemical characteristics indicating a rift environment (Dimitriadis and Asvesta 1993). The presence of clastic meta-sediments of Upper Triassic–Liassic (?) age (Mercier 1968; Kockel et al. 1977), overlying the carbonate platform probably represent a flysch (Fig. 8.121).

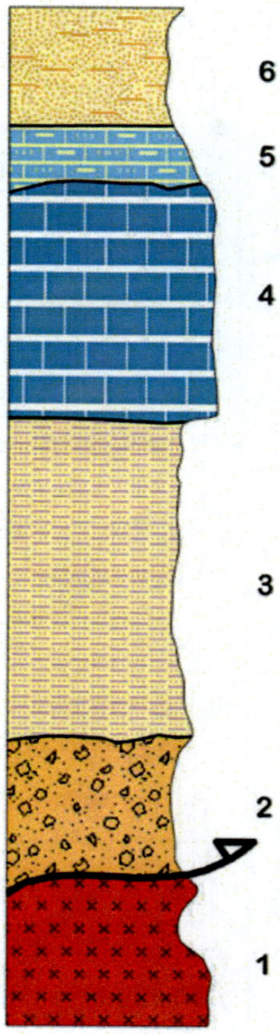

Fig. 8.121 Stratigraphic column of the Doubia unit, based on data by Kauffmann et al. (1976), and Kockel et al. (1977). 1: Vertiskos gneisses, 2: meta-clastic rocks of Permian age, Examili formation, 3: volcano-sedimentary rocks of Upper Permian–Middle Triassic, 4: carbonate platform of the Middle–Upper Triassic, 5: pelagic limestones with marly pelagic facies, 6: Liassic flysch, Melissochori–Svoula formation

Outcrops of similar units are also observed in other areas along the Circum-Rhodope belt, to the north of Thessaloniki, where Permian–Triassic sequences, including Triassic carbonate platforms are overlain by Uppermost Triassic–Lower Jurassic flysch formations of the Svoula/Melissochori type (Mercier 1968). In Oraiokastro and Nea Santa, there are descriptions of sequences with a volcano-sedimentary base in the Late Paleozoic, a carbonate platform in the Triassic, and flysch in the Early Jurassic (Fig. 8.122) (Stais and Ferriere 1991). In the case of Nea Santa, an early deepening of the carbonate platform during the Carnian is observed, with ammonitico rosso facies and pelagic carbonate sedimentation in the Norian–Rhaetian, prior to the arrival of the clastic material.

The Doubia unit shows no correlation with the other Circum Rhodope outcrops, which comprise pelagic formations with ophiolites of Triassic–Liassic age and flysch, corresponding to the oceanic terrane H6. Until today, there is no detailed research, to distinguish the formations of the two terranes of H5 and H6 inside the former Circum-Rhodope belt. Besides the western outcrops of the belt in Central Macedonia and Chalkidiki, there are also the eastern outcrops of the belt in Alexandroupolis. It is remarkable that in the western outcrops, the Circum-Rhodope formations underlie the metamorphic nappes of the Serbo-Macedonian sensu lato, whereas in the Athos peninsula and Alexandroupolis they are overlying the metamorphics of the Serbo-Macedonian and Eastern Rhodope respectively. The carbonate rocks forming the impressive cliff of the Athos mt belong to the Triassic carbonate platform of the Doubia unit. The overall structure of the Athos peninsula implies a complex tectonic relation with the underlying gneisses and schists, due to the Alpine deformation of the Eocene-Miocene period, involving also extensional low angle faults (Georgiadis et al. 2007).

A separate issue is the Upper Triassic Arnaia granite (De Wet et al. 1989), which could be part of the magmatic arc of the Cimmerides, distinct both from the Upper Paleozoic granites of the Variscan basement and from the Upper Jurassic granites of the paleo-Alpine orogeny.

8.6 Lesvos–Circum Rhodope Ocean—H6

8.6.1 The Lesvos Allochthon

The allochthonous unit of Lesvos includes large outcrops of an ophiolitic complex, consisting mainly of ultramafic rocks, and minor occurences of gabbroic dikes, diabases, basaltic pillow lavas and tuffs (Migiros et al. 2000). A metamorphic pelagic sequence of greenschists and pelagic marbles with Triassic conodonts are associated in syngenetic stratigraphic position with the ophiolite complex (Papanikolaou 1999,

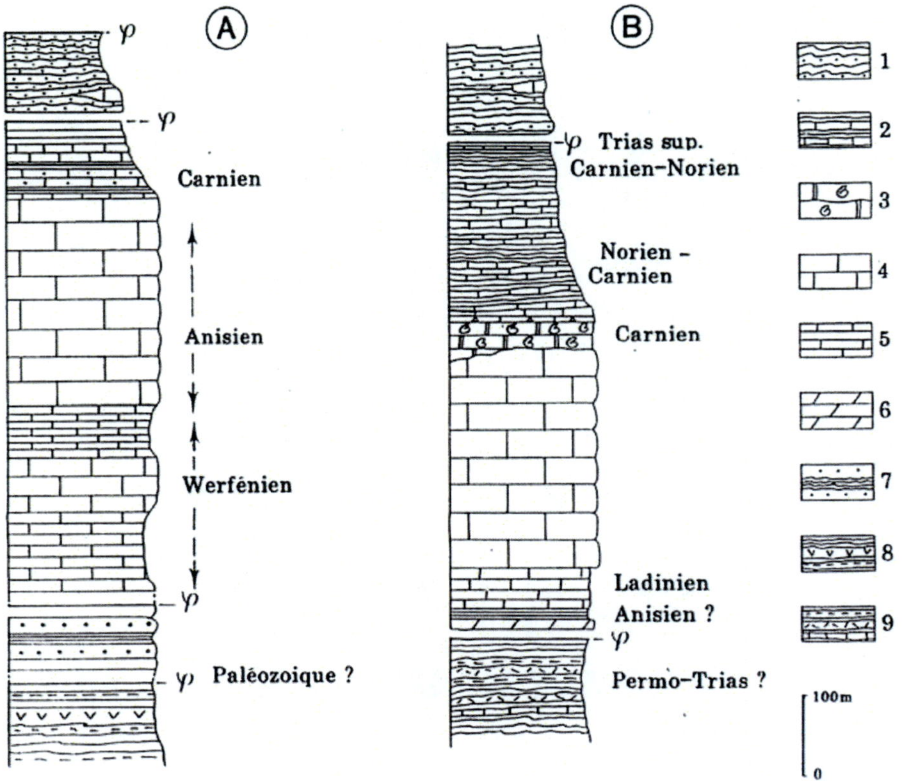

Fig. 8.122 Two stratigraphic columns with Triassic carbonate platforms from Oraiokastro (a) and Nea Santa (b) (from Stais and Ferriere 1991). 1: Melissochori flysch, 2: limestone-sandstone formations, 3: limestones of the ammonitico rosso facies, 4: neritic limestones, 5: platy limestones, 6: dolomitic limestones, 7: clastics of the Verrucano type (rifting), 8: volcano-sedimentary of Oraiokastro, 9: volcano-sedimentary of Nea Santa

2009) (see also Fig. 8.117). In the region of the Amali peninsula, in SE Lesvos, the pelagic sequence is observed above the ultramafic rocks, interfingering with the meta-lavas and the meta-tuffs in normal stratigraphic position, while northwards in the Komi area of central Lesvos, it is observed diping below the ultramafics in reverse position. Amphibolites occurring at the base of the ophiolitic nappe were dated at 153–158 Ma (Hatzipanagiotou and Pe-Piper 1995), an age that probably corresponds to the tectonic emplacement of the ophiolite nappe. From the metamorphic point of view, pressure values of 6–7 Kbars and temperatures of 1.100°–650 °C have been determined, with two rodingitization phases, while geochemical data show a marginal sea regime (Migiros et al. 2000; Hatzipanagiotou et al. 2003). The discovery of Triassic eclogites in northwest Turkey has indicated the Early Mesozoic subduction within the Cimmerides (Okay and Monie 1997).

8.6.2 The Peonia Unit

The Peonia unit was distinguished by Mercier (1968) as the eastern subzone of the Vardar oceanic area. It includes various mafic rocks, which following petrological and geochemical analyses have been proved to represent a marginal sea environment and not typical ophiolites. This is evident from the absence of ultramafic rocks and the presence only of the upper parts of the ophiolite complex, like gabbros, diabases, dike system, basaltic pillow lavas, etc. (Bebien 1982; Haenel-Remy and Bebien 1987; Saccani et al. 2008). Additionaly, it is the only outcrop of mafic rocks where transgressive sediments of the Upper Jurassic (Kimmeridgian–Tithonian) are overlying them, with thick basal conglomerates croping out in the Demir Kapija region, to the north of Geuvgeli (Papanikolaou and Stojanov 1983). Consequently, these mafic rocks are definitely of pre-Kimmeridgian age. In northern Greece Upper Jurassic limestones with Cladocoropsis mirabilis unconformably overlie the ophiolite nappe in the Oraiokastro area, underlain by the Doubia type neritic unit (Stais and Ferriere 1991). The unconformable sediments have remained non-metamorphosed, even though they have been tectonically transported together with their basement westwards, over the Paikon unit, during the Eocene. Essentially, the Peonia unit is mainly associated with Mercier's Pre-Peonian subunit, while the Eastern Peonia subunit belongs to the Circum-Rhodope belt.

In conclusion, the Peonia unit is an ephemeral paleogeographic region, chaqracterized by its geodynamic position within the paleo-Alpine orogenic arc. This is shown by

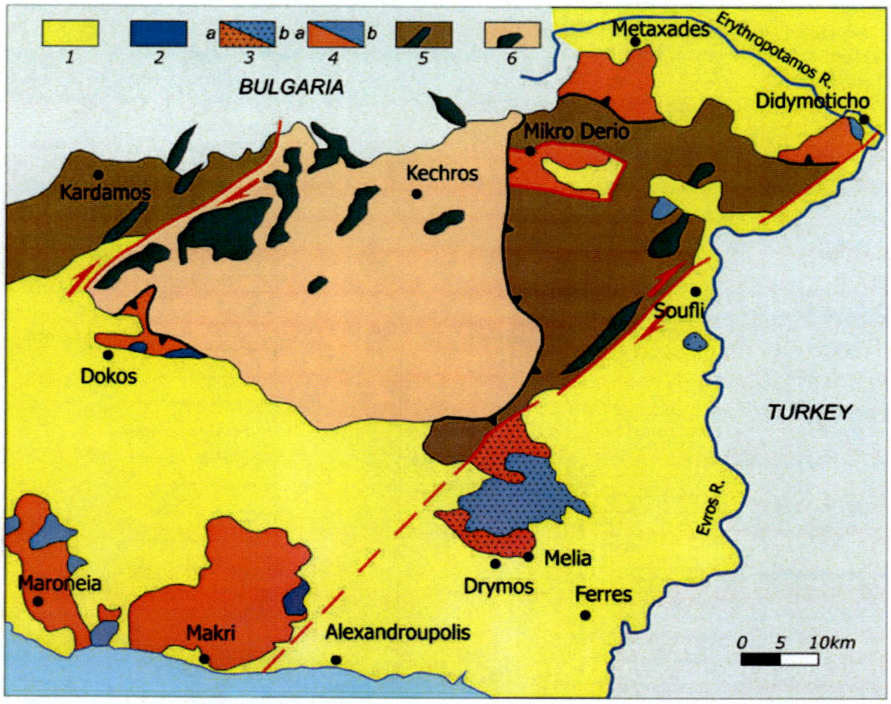

Fig. 8.123 Simplified geological map, showing the two Triassic–Jurassic units of the Circum-Rhodope belt in the Alexandroupolis region. Both units comprise mafic rocks of the H6. The underlying metamorphics of the East Rhodopean units, include also the ophiolites of Eastern Rhodope (H8). 1: Eocene–Oligocene molassic sediments of the Thrace Basin and post-Alpine sediments, 2: transgressive Lower Cretaceous limestones of Aliki, 3: Non-metamorphosed Melia unit, ophiolites (a) and flysch (b), 4: Makri unit, meta-sediments (a) and ophiolites (b), 5–6: metamorphic basement of Rhodope, upper unit Kardamos (5) and lower unit Kechros (6). The ophiolites of Eastern Rhodope H8 can be observed in both Rhodopean units

its marginal sea geotectonic setting, indicated by the Upper Jurassic mafic complex but also by the Upper Jurassic–Lower Cretaceous sediments of molasse type, deposited in a back-arc basin. This marginal basin opened during the Late Jurassic–Early Cretaceous, during the tectonic emplacement of the Axios ophiolites H4 on the Almopia carbonate platform H3, in the frame of the paleo-Alpine orogenic arc (Aubouin 1977). Since then, it was preserved in the area of the previously tectonised H5 and H6 terranes. The pre-Late Jurassic formations of the Peonia unit are considered by Ferriere et al. (2016) as belonging to the internal margin of the Maliac ocean (H4).

8.6.3 The Circum-Rhodope Unit

The Circum-Rhodope unit includes the most internal parts of Mercier's Peonia unit (1968), included in another tectonic unit, defined as «Circum-Rhodope belt» ten years later (Kauffmann et al. 1976; Kockel et al. 1977). In fact, it extends in a much wider region both in Chalkidiki to the south and to the Eastern Rhodope (VonBraun 1968).

The unit comprises metamorphic formations, of low-medium grade in the greenschist facies, with clastic sequences of mainly Triassic and Jurassic age, mixed with ophiolitic rocks belonging to the H6. Upper Permian–Scythian–Anisian horizons as well as Upper Triassic marbles and dolomites were initially also included in the Circum Rhodope belt, which were previously examined separately in the Doubia unit in the context of H5.

The Circum Rhodope outcrops are located in a western zone along its tectonic contact with the Serbo–Macedonian/Vertiskos unit, extending towards Chalkidiki in the south. Then they change direction towards the east in the Athos peninsula and then in the Thrace area, west of Alexandroupolis, but also in the northeast around Didimoticho and Orestias. Here, the low grade metamorphic phyllitic rocks ("semi-metamorphic formation of Rhodope") overlie the Eastern Rhodope metamorphic units (Maratos and Adronopoulos 1964, 1965; Davis 1963; Bonev and Stampfli 2011).

The term Circum-Rhodope belt was given initially to describe all the low metamorphic grade formations found in the periphery of the internal tectono-metamorphic belt of

Rhodope s.l. (including also the Serbo-Macedonian). The central idea was that they represent transgressive sequences over the old pre-Alpine massif. This idea was abandoned when it became evident that all the formations composing the Circum-Rhodope belt are separated by the Rhodopean metamorphic core complex by tectonic contacts. These contacts are often Tertiary extensional detachments or re-activated normal and strike-slip faults, involving also the Paleogene molasse sediments of the Thrace Basin (Papanikolaou 1988b; Papanikolaou and Triantaphyllou 2010; Kilias et al. 2013).

The general stratigraphic column of the Circum-Rhodope unit is not clearly determined, as there is some confusion due to the interference of the neritic formations of the Doubia unit. The most characteristic unit of the abyssal-pelagic Circum Rhodope unit H6 is known as the **Aspri Vrisi–Chortiatis** unit (Kockel et al. 1977). It is evident that there are Permian–Triassic and Jurassic pelagic sediments, in a syngenetic relation with mafic diabasic and volcanic rocks and tuffs. The various members of the ophiolitic complex can be observed in different locations, comprising both its ultra-mafic and mafic lithologies. The volcano-sedimentary character is present in successive periods, corresponding either to the rifting stage, mainly in the Permian–Triassic, or to the opening stage of the oceanic basin, mainly in the Late Triassic–Early Jurassic. The presence of clastic sequences of flysch type can be observed both in the Late Triassic as well as in the Early Jurassic. The ceiling of the column has not yet been determined, but it probably reaches the Late Lias–Dogger. The fact that transgressive sediments are not generally observed upon the Circum Rhodope unit does not permit the accurate age determination of tectonism. Two minor occurrences of sediments of Lower Cretaceous age have been reported in Chalkidiki (Kockel et al. 1977) and in the Aliki limestones of Western Alexandroupolis (Maratos and Adronopoulos 1965). More recent detailed biostratigraphic analysis has dated both these outcrops of the unconformable limestones as lowermost Cretaceous, clearly different from the Cenomanian unconformity of the Eastern Greece domain (Ivanova et al. 2015). It is characteristic that a middle (?)–Upper Jurassic age of the blueschist type metamorphism has been reported from the metabasic rocks of the Chortiatis unit (Michard et al. 1994, 1998). This age is younger than the Upper Triassic–Lower Jurassic volcano-sedimentary formations of the upper part of the unit and older than the postulated Early Cretaceous transgression. The Jurassic deformation structures of the Cimmeride and the paleo-Alpine orogenic phases have been greatly influenced by the Alpine deformation during Eocene–Oligocene. Then a reversal of the kinematics brought the Vertiskos unit of the Serbo-Macedonian over the Circum-Rhodope unit along the eastern margin of the Axios basin (Mercier 1968; Tranos et al. 1999).

In the Alexandroupolis region two tectonic units can be distinguished, with different stratigraphic and tectono-metamorphic characteristics (Fig. 8.123). Ophiolitic rocks of Lower (?) Jurassic age, exhibit geochemical characteristics of a marginal sea–volcanic arc (Maganas 2002) or supra-subduction zone ophiolites (Bonev and Stampfli 2009).

The **Makri** unit comprises low to medium metamorphic grade crystalline limestones, phyllites and greenschists, passing upwards to mafic igneous rocks of the Lower Jurassic (Maronia, Petrota). Triassic age has been dedtermined in the lower limestones (Maratos and Adronopoulos 1964). Their metamorphism occurred before the transgressive deposition of the Lower Cretaceous Aliki limestones. The second unit of the **Drymos–Melia** includes non-metamorphic rocks, with mafic volcanics at its base and clastic flyschoid sediments on top, where Middle Jurassic ammonites have been reported (Trikkalinos 1955; Maratos and Adronopoulos 1965).

A recent attempt to correlate the Circum-Rhodope outcrops of Chalkidiki with those of Makri and Drymos–Melia in Alexandroupolis showed three different stratigraphic columns, including mafic rocks in the Lias–Dogger and Jurassic clastic sediments of various origins (Meinhold and Kostopoulos 2013), metamorphosed in Chalkidiki and Makri and non-metamorphosed in the Drymos-Melia unit (Fig. 8.124).

In the Asia Minor region, the Circum-Rhodope is associated with the Karakaya type formations, of Permian–Triassic age, which are included in the Cimmerides (Sengor 1984; Papanikolaou and Demirtasli 1987; Pickett and Robertson 1996).

8.7 The Pangeon Platform—H7

8.7.1 The Pangeon Unit

In Rhodope, we can distinguish two large units, the tectonically underlying Pangeon unit, characterized by a carbonate platform, and the tectonically overlying Sidironero unit, which shows high grade metamorphic rocks with migmatites and anatectic granites, resulting in an upside down metamorphic zonation (Papanikolaou and Panagopoulos 1981; Papanikolaou 1984c, 1988b). Their boundary is a complex tectonic zone observed from Xanthi to the northeast of Nevrokopi, near the Greek-Bulgarian borders, which brings the Sidironero unit, i.e. the area north of the tectonic zone, over the Pangeon unit, i.e. the area south of the tectonic zone (Fig. 8.125). This tectonic zone was named later by some authors as the Nestos line, because for some part of it, it follows approximately the Nestos River, whereas the Pangeon unit was called Drama Unit (Ricou et al. 1998).

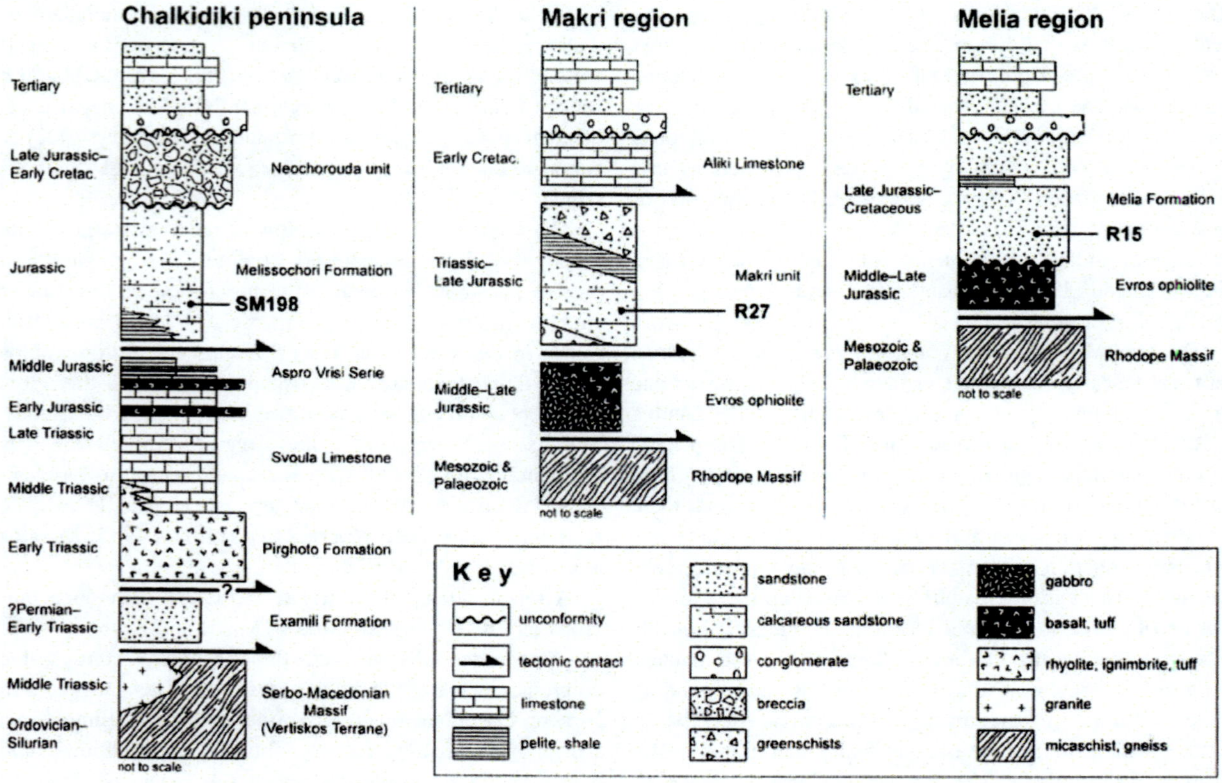

Fig. 8.124 Tectonostratigraphic columns of the Circum-Rhodope unit in the Chalkidiki, Makri, and Drymos-Melia areas (from Meinhold and Kostopoulos 2013). Their Jurassic clastics were geochronologically analyzed with U-Pb in zircons, showing a completely different origin of the clasts. The Jurassic ophiolitic rocks are included in all three units, though with a different correlation with their surrounding rock formations, while, the Evros type ophiolites correspond to H6

The Pangeon unit is characterized by a several hundred meterst hick formation of marbles. However, the marbles are intensively folded with multiple subhorizontal isoclinal folds with extreme synkinematic tectonic flow, producing thinning of the marbles down to only a few tens of meters thickness. This deformation is evident in the mountainous region between the Kavala and Palea Kavala area as shown in the detailed geological map (Papanikolaou 1984a) (Fig. 8.126). In several locations the marbles appear to wedge out laterally within the schists but this is due to the fold hinges of the isoclinal folding. The immediate occurrence of the Kavala granodiorite beneath the Pangeon sequence may explain the extreme tectonic flow observed in the marbles and mica schists. Above the marbles, another formation of mica schists with quartzites and thin marble horizons is observed, resembling a meta-flysch formation. Another formation consisting of amphibolites, gneisses and mica schists, occurs below the marbles, observed mainly at the cores of the large anticlines (e.g. around Kavala). Thus, the stratigraphic column of the Pangeon unit includes the regular trilogy observed in the Hellenides continental terranes, with: (i) a volcano-sedimentary complex at its base, (ii) a carbonate platform at the middle and (iii) an upper meta-flysch formation (Figs. 8.127 and 8.128).

At the base of the stratigraphic column large granite plutons can be observed, as uniquely shown at the core of the mega-anticline of Kavala, where the granodiorite has been dated with zircons as Carboniferous, at about 300 Ma (Kokkinakis 1978). However, the geochronological data from the micas of the granodiorite indicate a Miocene age, which is possibly due to the final deformation phase, when the granodiorite was uplifting near the contemporaneous magmatic-volcanic arc (Papanikolaou et al. 1982). The Late Carboniferous–Early Permian ages of the granites underlying the Pangeon platform have been confirmed in several other localities (Turpeaud and Reichmann 2010).

The age of the Pangeon formations is not known because fossils have not been found. In the Greek region, at the Falakron mt marbles, some poorly-preserved corals have been reported, which may belong to the Paleozoic or even the Mesozoic (Meyer and Pilger 1963). However, these corals were later considered as probably Silurian–Carboniferous (Jordan 1969). In the Bulgarian Rhodope some lamellibranches have been found of possible Ordovician–

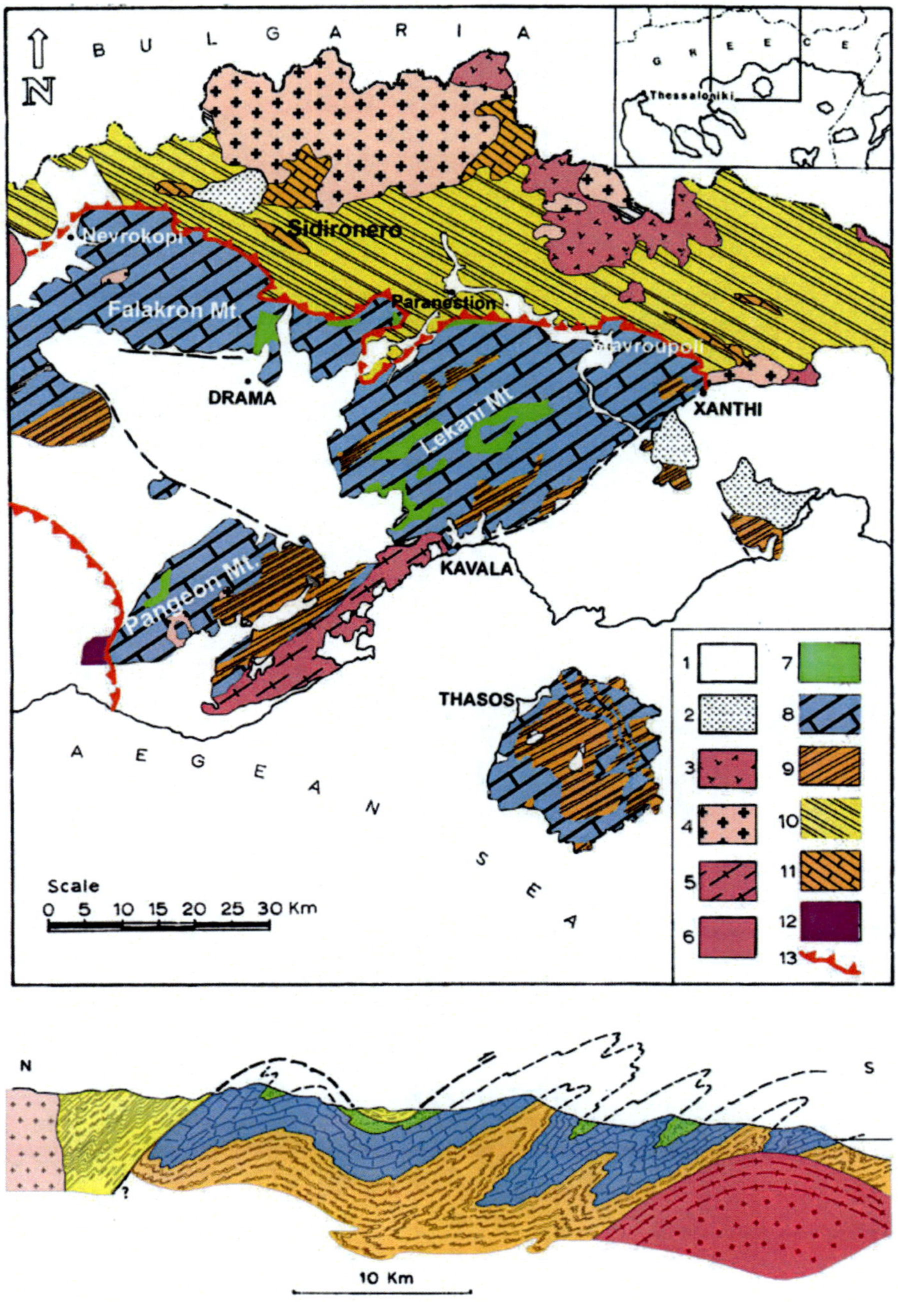

Fig. 8.125 **a** Simplified geological map of the main part of the Rhodope massif in Greece (from Papanikolaou and Panagopoulos 1981). 1: Neogene and Quaternary, 2: Paleogene molasse, 3: Paleogene acid volcanics, 4: Post-tectonic granites, 5: Foliated Kavala granodiorite, 6: foliated granites. 7: Chlorite mica schists, 8: marbles, 9: Gneisses, amphibolites, mica schists, 10: Augen-gneisses, amphibolites, migmatites, marbles within 10, 12: Gneisses of the Serbo-Macedonian belt, 13: complex tectonic zone (nos 5, 7, 8, 9 belong to the Pangeon unit, whereas nos 6, 10, 11 belong to the Sidironero unit). **b** Schematic cross section of Rhodope in an almost N-S orientation, through Sidironero–Kavala, showing the thrusting of the Sidironero unit over the Pangeon platform, and the general structure of the km scale ENE-WSW isoclinal folds, with an asymmetry towards the south (from Papanikolaou and Panagopoulos 1981)

Lower Carboniferous age (Ancirev et al. 1980). However, these fossiliferous marbles do not belong to the Pangeon unit, but are possibly associated with the marbles of the overlying Sidironero unit. Pre-Cambrian ages have been reported for the Bulgarian part of Rhodope, based on phytoplankton found in all the marbles bearing formations, (Kozhukharov and Timoveev 1980). These age determinations are dubious, because the method used is not considered reliable, as the organic matter is not possible to be preserved in great pressures and especially in high temperatures, such as those of the medium–high grade metamorphic units of the Rhodope (Papanikolaou 1981b). Most recently, U-Pb dating of detrital zircons and Sr isotopic constraints from the marbles of Pirin mt in SW Bulgaria

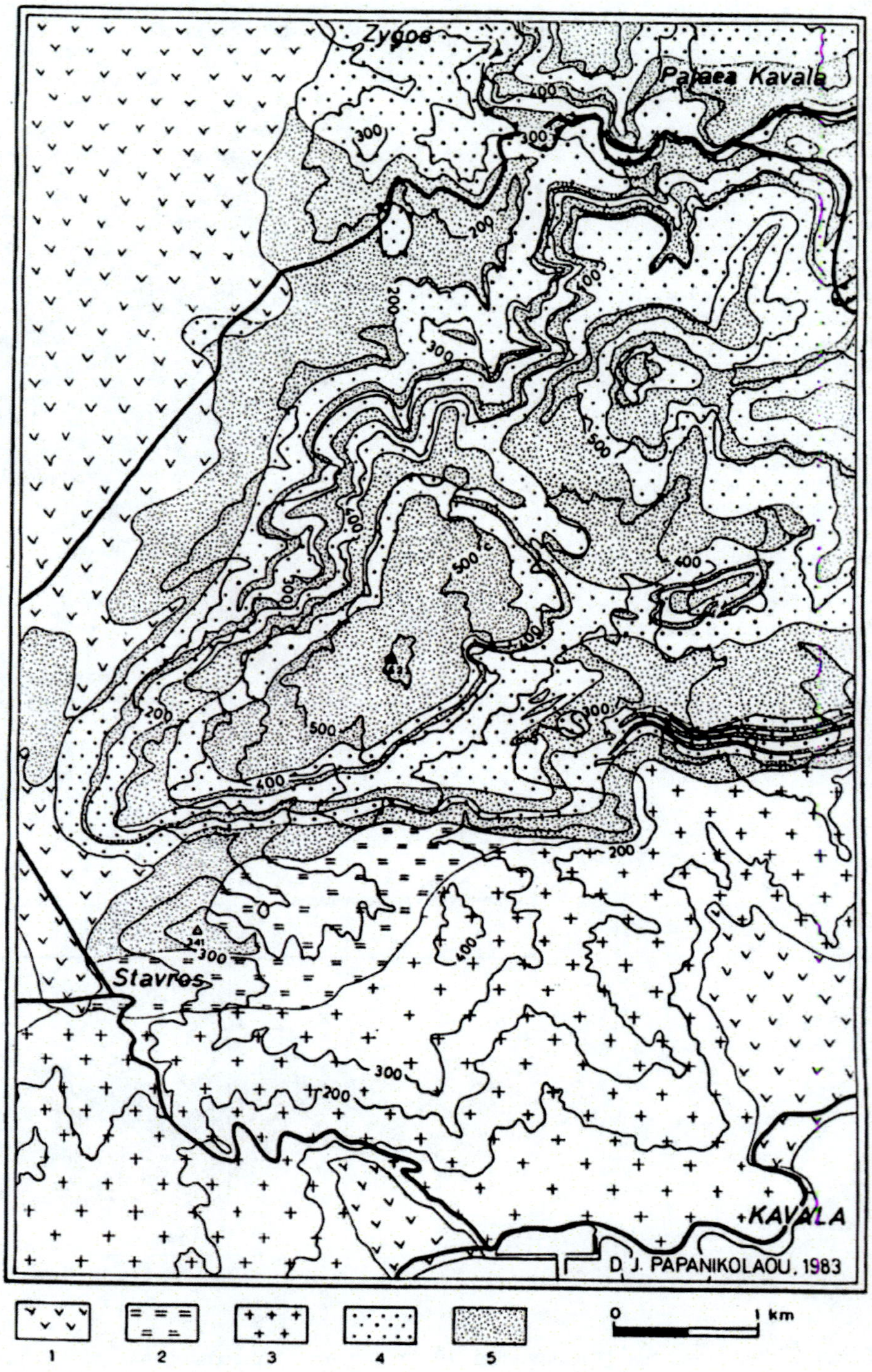

Fig. 8.126 Geological map of the Kavala–Palea Kavala region, where the Pangeon unit crops out. The main deformation characteristic is the repetition of the same marble and schist formations due to isoclical folding and thrusting (from Papanikolaou 1984a). 1: Quaternary, 2: gneisses, 3: granodiorite, 4: mica schists and quartzites, 5: marbles

proposed a Middle Permian age for part of the marbles (Bonev et al. 2019).

The deformation of the Pangeon unit includes at least two synmetamorphic phases of isoclinal ductile flow folds in NE-SW orientation (Papanikolaou and Panagopoulos 1981), while the metamorphism of the upper group of the marbles-schists, corresponds to the greenschist facies (Papanikolaou et al. 1982). The age of metamorphism is considered Cretaceous, but it concerns only the younger amphibolitic facies, which is associated with the

Fig. 8.127 View of the marbles of the carbonate platform (2) of the Pangeon unit in Thasos Island, over the gneisses-schists-amphibolites (1) of the underlying volcano-sedimentary complex (from Papanikolaou 2013)

migmatization and the extrusion of anatectic granites, such as the Elatia granite in the Kara-Dere forest of Northern Drama area (Soldatos and Christofides 1986). In general detailed studies on the P/T conditions of metamorphism and radiochronological data are not available from the Pangeon Unit, contrary to the overlying Sidironero nappe, where there are several studies especially from the eastern part of Rhodope around Kimi area (see relative Sect. 8.9.2).

In conclusion, the Pangeon unit comprises a metamorphosed carbonate platform within the frame of a continental terrane, the age of which possibly includes the Upper Paleozoic–Lower Mesozoic, characterized by multiple tectono-metamorphic phases, which last from the Middle Mesozoic to the Miocene.

8.7.2 The Kerdylia Unit

The Serbo-Macedonian massif includes, according to Kockel et al. (1977), two large units; the lower unit of Kerdylia and the upper unit of Vertiskos. The two units were separated by a chartographic boundary–contact. Later studies in the Northern Chalkidiki region have distinguished three units in the former Serbo-Macedonian, by the additional distinction of the Volvi ophiolites, in between the Kerdylia and Vertiskos units (Dixon and Dimitriadis 1984). These three units are associated with the three terranes, H7, H8 and H9 (Papanikolaou 2009) (Fig. 8.129).

The deeper unit of Kerdylia comprises rocks of pre-Alpine continental crust and overlying marbles. These rocks are overlain by the Volvi ophiolite unit, separated by a complex deeply folded tectonic zone. The Vertiskos unit is overlying the previous tectonic units, comprising pre-Alpine gneissic basement rocks.

The tectonic contacts separating the three tectonic units in the Chalkidiki area have been developed in the ductile lower crust, characterized by mylonites, phyllonites, super-mylonites and blastomylonites. This is contrasting the tectonic macro- and micro- fault gouge and breccias occurring in other tectonic contacts at the brittle shallow crust.

Stratigraphically, the Kerdylia unit consists of a lower group of several km thickness, with gneisses, amphibolites, leptynites, and mica schists, intruded by granitic veins and pegmatites, and an upper group of neritic marbles of several hundred meters thickness. A Late Carboniferous–Early Permian age has been proposed for the Kerdylia granites (Himmerkus et al. 2012). The overlying marbles should be post-Early Permian in age. They are especially observed at the periphery of the gneissic core, with multiple isoclinal folding at the km scale. Interestingly, no mica schists and related metasediments have been found on top of the marbles or in the synclines of their isoclinal folds that might represent a metaflysch formation. Therefore, the attribution of the Kerdylia unit in the same terrane as the Pangeon unit is based mainly on the existence of the carbonate platform over the pre-Alpine crystalline basement and also on its tectonic position in the form of a tectonic window under the Volvi and Vertiskos units. It should be noted that the contact of the former Serbo-Macedonian units with the Rhodope units is located along the Strymon River valley (Kockel and Walther 1965), where a normal extensional detachment fault, has been described (Dinter and Royden 1993). This

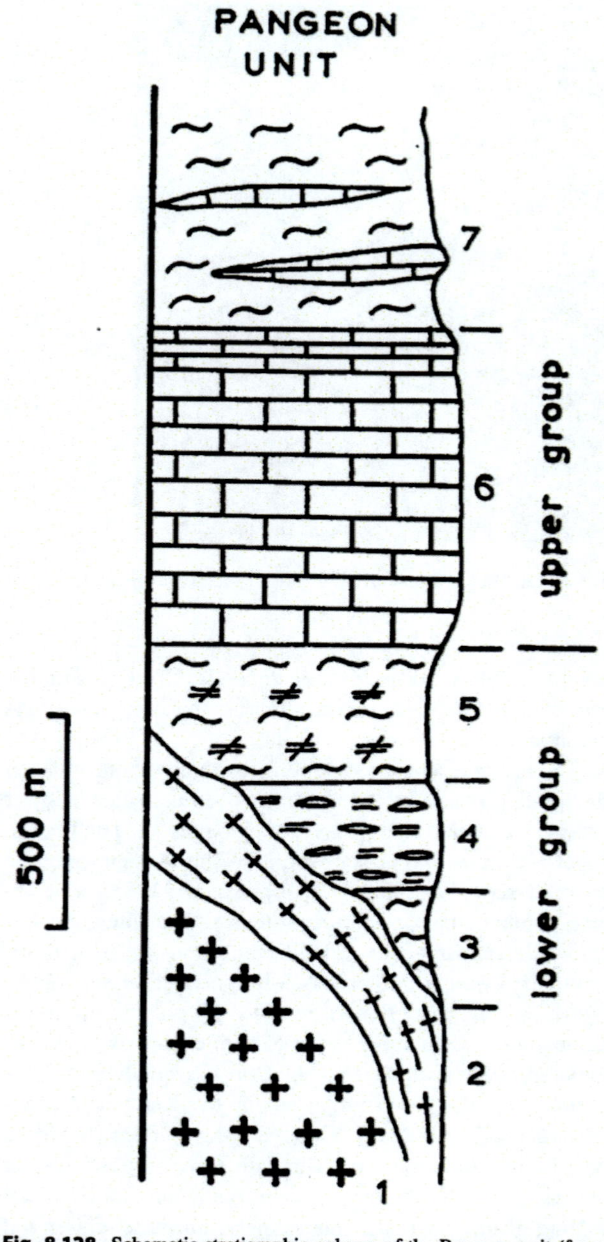

Fig. 8.128 Schematic stratigraphic column of the Pangeon unit (from Papanikolaou 1988b). 1: granite, 2: orthogneiss, 3: mica schist, 4: augengneiss, 5: amphibolites and mica schists, 6: marbles, 7: mica schists and quartzites with thin layers of marbles (meta-flysch)

detachment system denudates the Rhodopean core to the east and subsides the Serbo-Macedonian units to the west. The tectonic contact between the Kerdylia and the Pangeon units can be observed at the western slopes of the Pangeon mt in the area of Amphipolis (see also the cross section of Fig. 8.130).

8.8 The Volvi–Eastern Rhodope Ocean—H8

The presence of metamorphic rocks belonging to the ophiolite complex in Rhodope sensu lato was historically known since earlier periods (Kossmat 1924). However, the «Volvi ophiolites» were distinguished after the work of Dimitriadis (1980) and Dixon and Dimitriadis (1984), who studied the large outcrops around the **Volvi** lake. The meta-ophiolite rocks preserve their petrographic character relatively well in some areas, while in others, they are strongly schistosed and nowadays correspond to amphibolites and amphibolitic schists. This is mainly true for the diabasic and basaltic rocks, as well as for tuffs, while the peridotites show intense serpentinisation. The gabbros on the other hand show transitional shear-schistosity zones and they are transformed to amphibolites, although preservation of protoliths in between the shearing zones is also observed. The Volvi ophiolite belt is observed over significant area, with outcrops of several square kilometers (see also the map on Fig. 8.129), with a thickness reaching 1–2 km. The main outcrops create a tectonic envelope zone above and around the underlying Kerdylia unit. From a chronological point of view, their tectonism precedes the extrusion of the Late Jurassic granite rocks (Fig. 8.130).

The overall tectonic position of the Volvi ophiolites within the Alpine orogenic cycle has been interpreted as the last members of the Paleo-Tethys, tectonised in the Triassic within the Cimmeride orogeny (Sengor et al. 1984). The P/T conditions of the blueschist metamorphism of the Volvi ophiolite rocks were determined at 12 kbars and 530 °C in eclogites, followed by retrograde greenschist facies (Dimitriadis and Godelitsas 1991).

The other significant outcrop of H8 ophiolites in the region of the internal tectono-metamorphic belt is located in **Eastern Rhodope**, mainly in the Soufli area (Maratos 1960) (see also Fig. 8.123). These outcrops of metamorphosed ophiolites occur inside the metamorphic rocks of Rhodope; this is the reason why they are separated from the ophiolite outcrops that belong to the Circum-Rhodope unit, which overlie the metamorphic rocks of Rhodope. The Eastern Rhodope ophiolites comprise serpentinised peridotites, gabbros, gabbro-pegmatites, amphibolitic gneisses and amphibolitic schists (Maratos 1960).

In the intermediate central area of Rhodope, between the western outcrops of the Volvi ophiolites and the eastern outcrops of the Eastern Rhodope ophiolites, there are scattered ophiolite, eclogite, or amphibolite outcrops. These rocks originate from the metamorphism in blueschist metamorphic conditions, and ususally occur in the form of tectonic lenses due to boudinage, both in Southern Bulgarian Rhodope (Kozhukharova 1980, 1984; Kozhukharov et al. 1988) and in the Northern Greek Rhodope (Liati and

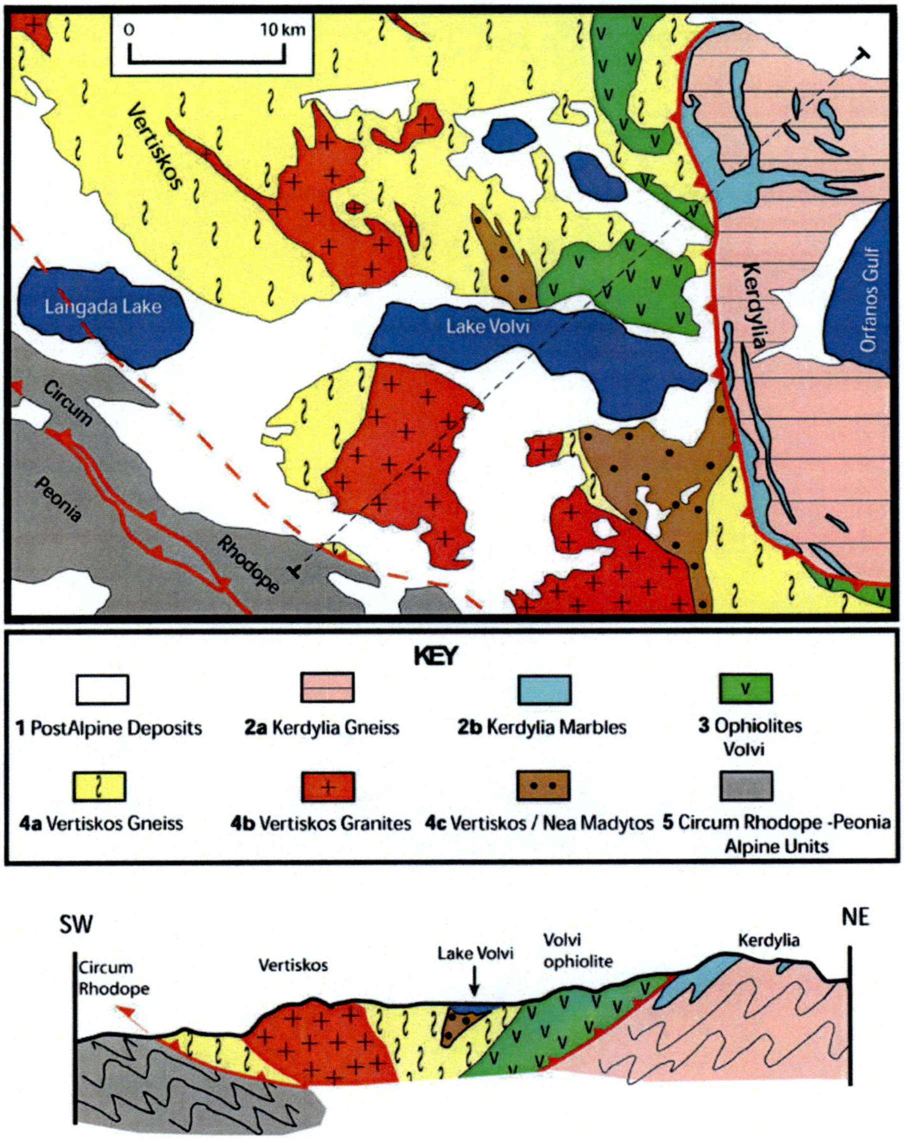

Fig. 8.129 Simplified geological map of the Chalkidiki area and schematic geological section, showing the position of the Volvi ophiolites (H8) above the relatively autochthon Kerdylia unit (H7) and below the base of the Vertiskos Allochthon (H9) (from Papanikolaou 2009)

Mposkos 1990; Liati and Seidel 1996). Their overall tectono-metamorphic evolution shows extremely high pressure values (UHP) in the Jurassic, with subduction at great depths and subsequent exhumation of the rocks into the upper crust with temperature increase, untill the anatectic conditions during the Cretaceous (Liati et al. 2011).

Besides the ophiolite rocks, it would be important to determine some metamorphic pelagic sequence belonging to H8. A strong candidate is the sequence described in the area of **Nea Madytos**, where low grade metamorphic meta-sediments, cipollinic marbles, and mica schists have been mapped, representing possible Upper Paleozoic–Mesozoic sediments (Sakellariou 1989) (see also Fig. 8.130). This particular sequence has pelagic features and is located above the meta-ophiolites of Volvi at the base of the Vertiskos unit. Therefore, it might be included in the oceanic terrane H8.

8.9 The Allochthonous Pre-alpine Basement of Rhodope—H9

8.9.1 The Vertiskos Unit

The Vertiskos unit comprises gneisses, migmatites and granites, while it does not contain any marbles. At its base it is delimited by large metamorphosed and schistosed mafic and ultramafic bodies, members of the Volvi ophiolite complex.

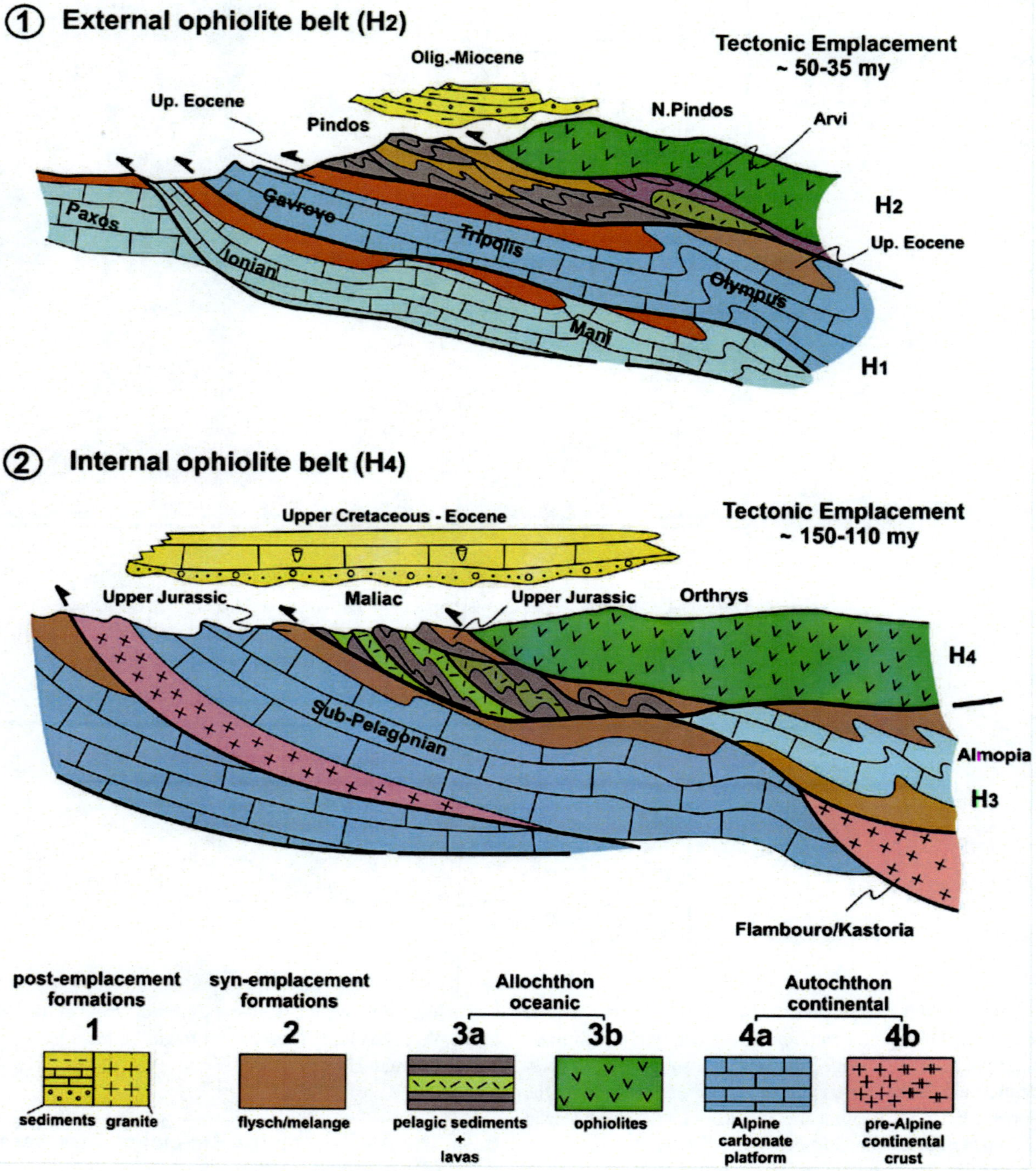

Fig. 8.130 Schematic geological cross sections showing the tectonic emplacement of the ophiolites H2, H4, H6 and H8 over the adjacent platforms to the south, corresponding to the terranes H1, H3, H5 and H7 respectively (from Papanikolaou 2009). The dating is based on the unconformable transgressive sediments, which cover the tectonic contacts. An exceptional case is the contact between the H8 over the H7, which is post-dated by the Late Jurassic granite intrusions

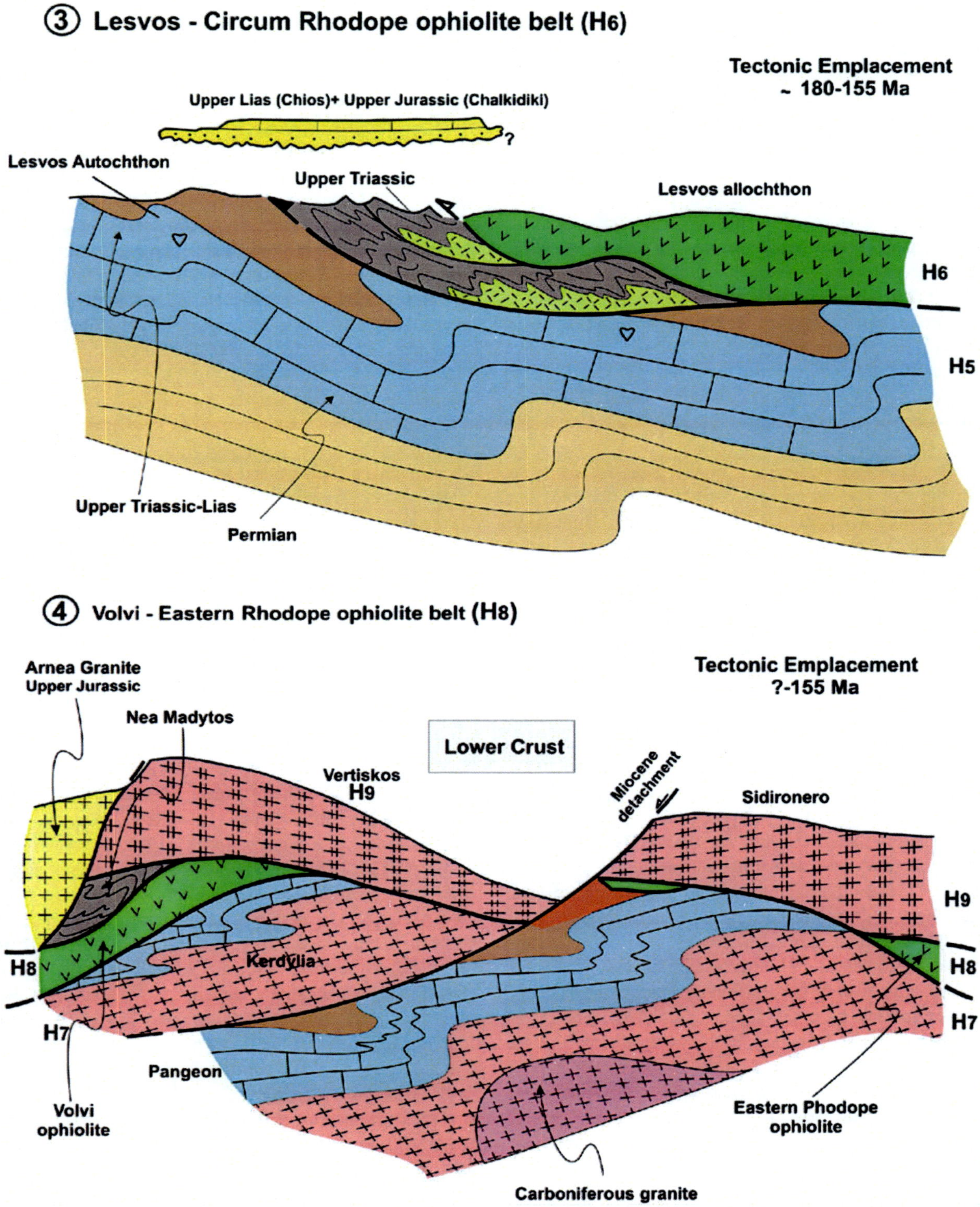

Fig. 8.130 (continued)

The gneissic formations of the Vertiskos unit show a Variscan metamorphism of about 300 Ma, followed by intense paleo-Alpine and Alpine metamorphic phases (Mercier 1968; Borsi et al. 1964; Harre et al. 1968; Zervas 1980). Recent geochronological data showed the presence of Lower Paleozoic granitic formations in the Vertiskos unit, around the Ordovician–Silurian (Himmerkus et al. 2006, 2009a).

The Paleozoic granites have been generally gneissified, but there are still granitic vein intrusions and pegmatites, determined as Jurassic in age (Zervas 1980). These are acidic magmatic phenomena, chronologically and geochemically associated with the events already mentioned in the Paikon unit, within the context of the Late Jurassic–Early Cretaceous orogenic arc. Therefore, while in the Paikon unit we have a surficial volcanic edifice, in the Vertiskos unit there is a deep magmatic province of the same volcanic/magmatic arc.

Besides the lower Paleozoic granites of the Caledonian cycle, the Upper Paleozoic granites of the Variscan cycle and the Late Jurassic granites of the paleo-Alpine cycle, there are also Triassic granites, especially in the Arnea area of Northern Chalkidiki (De Wet et al. 1989; Himmerkus et al. 2009b). These Triassic granites might be correlated to the magmatic/volcanic arc of the Cimmerides or to a Triassic rifting. However, Mesozoic eclogite facies preceding Barrovian metamorphism has been reported more recently from the Chalkidiki peninsula (Kydonakis et al. 2015), although the final exhumation of the metamorphic rocks occurred only in the Late Cretaceous (Kydonakis et al. 2014a).

The boundary between the Serbo-Macedonian and the Rhodope had been placed along the Strymon River, which is following a neotectonic graben (Kockel and Wallther 1965). However, the contact is observed only in the western slopes of the Pangeon mt at the Amphipolis area, where the Kerdylia gneisses are overlying the Pangeon marbles and schists. More recent research has shown that the contact of the two metamorphic groups is a Miocene extensional detachment fault, running along the eastern margin of the Miocene Strymon supra-detachment basin (Dinter and Royden 1993). Thus, the Vertiskos unit has been preserved at the hanging wall of the detachment whereas the Rhodope units are making the high Vrondou and the Pangeon mts at the footwall. On the contrary, at the western contact of the Vertiskos unit towards the Circum Rhodope unit a steep thrust diping 50°–60° to the east is observed along the Doirani–Kilkis–Langada tectonic zone. This tectonic movement is of Late Eocene–Oligocene age as there are molassic sediments of Eocene age involved in these structures (Mercier 1968).

It should also be noted that there are outcrops of pre-Alpine basement even further to the west of the tectonic contact between the Vertiskos and Circum Rhodope. These outcrops may represent the basement of the Paikon/ Doubia units, such as the gneisses located in the **Karathodoro** area (Mercier 1968; Mercier and Bebien 1977). Unfortunately, these outcrops are generally very poor due to the cover of the post-Alpine sediments and the recent deposits of the Axios River.

The eventual Mesozoic sedimentary cover of the Vertiskos unit is still questionable and a possible stratigraphic relation with some Mesozoic sediments has not yet been described. Further north, in Southwestern Bulgaria, the orogenic group of the **Kraistides** is observed as a probable extension of the Vertiskos unit. This group has similar pre-Alpine basement, but it also preserves its Mesozoic cover sequence, which is of Germanic facies, similar to the Balkanides units to the north (Boncev 1958). Therefore, these sedimentary formations are observed all over the European continent during the end of the Variscan cycle, with epicontinental sediments from the erosion of the Meso-Europa. Unique stratigraphic facies are well established, such as the Bundsandstein, Muschenkalk, etc. of Lower–Middle Triassic. Thus, the Vertiskos unit possibly represents the only unit in the Hellenides (H9 terrane) that, during the initiation of the Alpine cycle in the Early Triassic, was not part of the Tethys ocean or the Gondwana margin, as all the other Hellenides to the south, but it had already been integrated in Meso-Europa during the Variscan orogeny.

8.9.2 The Sidironero Unit

The Sidironero unit is the upper tectonic unit of Rhodope, comprising medium–high grade metamorphic rocks, observed above the relative autochthon Pangeon unit (Papanikolaou and Panagopoulos 1981; Papanikolaou 1984c, 1988b). This unit was previously considered as the upper horizons of the Rhodope massif by Kronberg et al. (1970) (Fig. 8.131), despite the fact that migmatites were overlying greenschists. This was a classical case of an oversimplified tectonic structure of metamorphic rocks, forming a monoclinal sequence, without isoclinal mega-folds and internal imbrications. Thus, the entire stratigraphic column of the Rhodope massif was considered to include more than 12 km thick metamorphic rocks with the thickness of the marbles reaching 6 km! This was due: (i) to the consideration of only one continuous metamorphic stratigraphic sequence, without the distinction of the Sidironero and Pangeon units and (ii) the consideration of a simple monoclinal stratigraphic sequence without repetitions of the marbles and schists due to isoclinal folding. On the contrary, the Pangeon marbles together with their adjoining gneisses and mica schists are isoclinally folded at the km

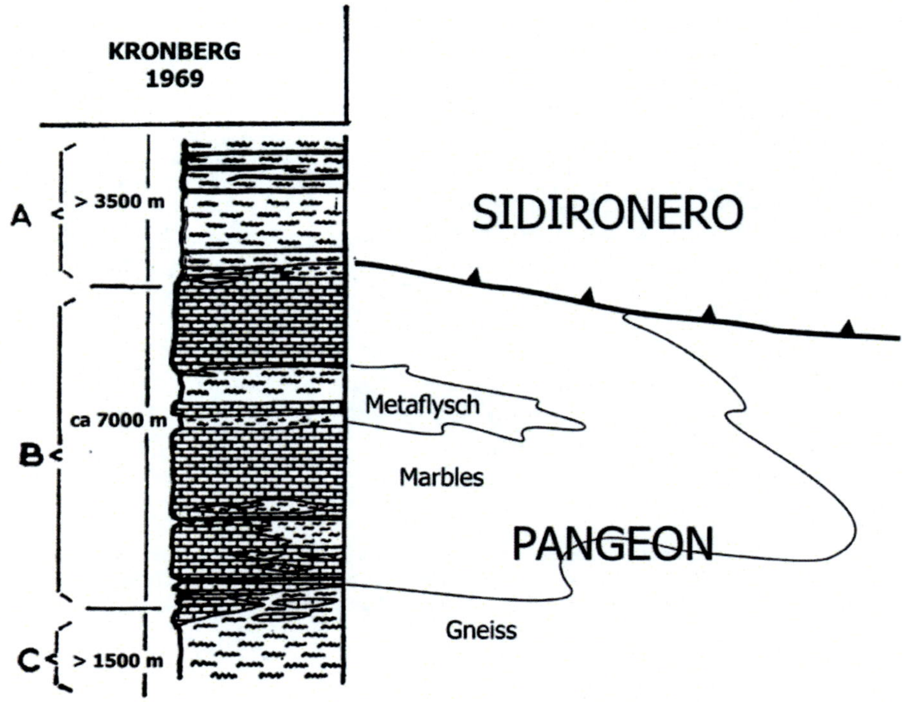

Fig. 8.131 Schematic stratigraphic column of the Rhodope massif by Kronberg (1969), where the overall thickness exceeds 12 km. In fact, after redefining the tectonic structure, following Papanikolaou and Panagopoulos (1981), the upper gneiss formation belongs to the separate Sidironero upper tectonic unit, while the two marble horizons are repetitions due to isoclinal folding of the same marble horizons of the lower Pangeon unit. The intermediate schists between the two marble horizons are considered as the uppermost stratigraphic formation of the Pangeon meta-flysch (modified from Papanikolaou 1986c)

scale, as shown in the N-S geological cross section of the Southern Rhodope from Sidironero to Kavala (Papanikolaou and Panagopoulos 1981) (see also Fig. 8.125).

The stratigraphic succession of the Sidironero unit is very hard to be defined. In the area of Potamoi–Sidironero–Paranesti a reversed sequence can be observed, which includes (Papanikolaou 1988b): (i) muscovite gneisses, (ii) biotite gneisses, (iii) augengneisses, (iv) biotite augengneisses, (v) amphibolites with marble intercalations and muscovite gneisses (vi) migmatites, and (vii) anatectic granites. The anatectic granites of Elatia (Kara Dere), occurring to the north of the Sidironero village up to the Greek-Bulgarian borders, are located in the center of a large anticlinal structure of the metamorphic core complex, where a gradual transition within 1–2 km is observed from the migmatites to the granites (Papanikolaou and Panagopoulos 1981; Papanikolaou 1984a, 1988b). Their age was determined as Upper Cretaceous (Soldatos and Christofides 1986). In contrast, the gneissic granites of the pre-Alpine basement of the Sidironero unit and its equivalent outcrops in SW Bulgaria have been dated as Ordovician–Silurian (Peytcheva et al. 2009). These ages are similar to those of the Vertiskos unit in the Serbo-Macedonian belt, implying a probable common origin of the two basement units occurring above the Pangeon-Kerdylia tectonic windows.

The Sidironero unit extends to the north into Bulgaria, where it is considered as the lower tectonic unit of Rhodope, known with the name of Median Rhodope unit (Unite des Rhodopes Moyennes), under the nappes of Madan, Asenica and Marica (Ivanov 1985, 1988) (Fig. 8.132).

In SE Bulgaria, the Thrace unit can be observed under the Madan upper nappe, but not also under the Median Rhodope unit (in the figure, the contact was incorrectly extended westwards with question marks). Therefore, the relatively autochthon of the entire Rhodope structure is represented by the Pangeon unit, which is not present in Bulgaria, except for the Pirin mt in the SW. An important aspect is the observation of slightly schistosed sediments of Upper Cretaceous and Paleocene age between some tectonic contacts. This indicates that in the Rhodope area some metamorphic nappes form a surficial thin skin tectonics structure. The above scheme of the Bulgarian Rhodope of Ivanov was considered as revolutionary in 1985, since previously Rhodope was considered Pre-Cambrian or even Archaean, with minimum Alpine influence (e.g. Boncev 1966, 1986). Some of the tectonic contacts between the Rhodopean units and the intermediate schistosed sediments were later interpreted as low angle normal faults and not as overthrusts (Burchfiel et al. 2003, 2008). However, this late Tertiary extension does not change the previous basic succession of the tectonic nappe pile. More recently, structural and geochronological evidence for Paleogene thrusting has been also presented by Jahn-Awe et al. (2010) in the western Bulgarian Rhodope.

The metamorphic features of the Sidironero unit show high grade metamorphism, as characteristically shown by the migmatites and the anatectic granites, which are the

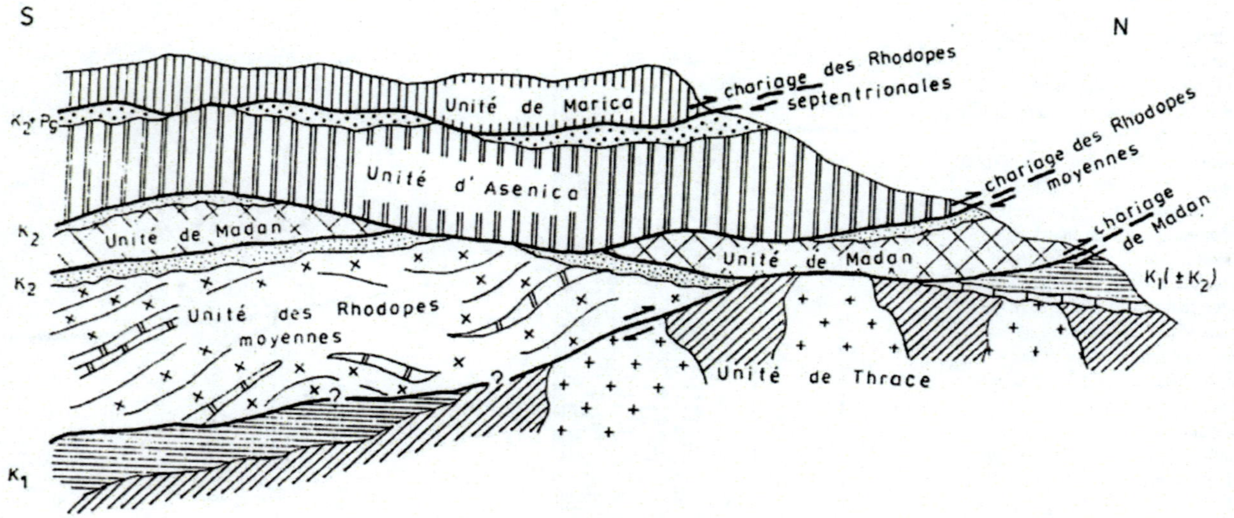

Fig. 8.132 Synthetic schematic representation of the general tectonic structure of the Bulgarian Rhodope, which, based on data by Ivanov (1985), comprises five Alpine tectonic units. The final deformation phase occurred in the Late Cretaceous–Eocene, as indicated by the involvement of sediments (K1, K2, Pc) in the tectonic contacts

result of the Cretaceous tectono-metamorphic-magmatic evolution. However, remnants of an older blueschist metamorphism of Jurassic (?) age are preserved, with especially very high pressure values and low temperatures of the eclogitic type (ultra high pressure/ low temperature eclogites) in addition to the presence of micro-diamonds (Mposkos and Kostopoulos 2001). The above data indicate that the Sidironero unit was subjected to an initial extreme subduction at more than 100 km depth and gradually was exhumed to the upper crust during the Late Cretaceous—Early Tertiary. During its late exhumation process it suffered migmatisation and anatectic phenomena and was tectonically incorporated with the underlying less metamorphosed Pangeon unit (Barr et al. 1999). The final magmatic–volcanic activity occurred during the Late Eocene–Oligocene at the early stages of the modern Aegean volcanic arc, when the tectono-magmatic cycle of Rhodope was closed (Papanikolaou 1993). The intermediate stages of the tectono-metamorphic path during the exhumation process may be represented by the Jurassic–Early Cretaceous successive metamorphic events, proposed by other researchers (Wawrzenitz and Mposkos 1997; Liati et al. 2011). These stages comprise metamorphic events of ultra high pressure/low temperature, high pressure/high temperature, high temperature/low pressure. More recently, three distinct UHP and HP metamorphic events have been recognised in the Sidironero Unit and its equivalent units in the Eastern Rhodope, based on zircons dating, that showed an Early-Middle Jurassic UHP event at 158 Ma on a protolith of Permian age, followed by two more HP events in the Late Cretaceous at 74 Ma and in the Late Eocene at 42 Ma (Liati et al. 2016). Thus, the allochthon basement nappe of Rhodope has undergone repeated subduction processes for a long time period and not one single subduction. A major concern, however, is whether the above P/T estimates of the metamorphic events as well as the obtained metamorphic ages are representative of the Sidironero unit or they may be mixed with the equivalent metamorphic formations of the Volvi oceanic basin. This is plausible because there is no chartographic distinction in the geological maps of the Volvi lithologies from those of the Sidironero unit in the intermediate area between the Chalkidiki and Eastern Rhodope.

A different over-simplified approach was presented by Ricou et al. (1998) who suggested that the allochthon Vertiskos and Sidironero units are linked to the Vardar allochthon complex and that the underlying Pangeon and Kerdylia units are tectonic windows of Pelagonian affinity. Thus, they correlated: (i) the Paleozoic Kataphygion granite and associated metamorphic rocks of the Pelagonian basement nappe in Pieria mts with the Vertiskos ortho-gneisses and (ii) the Vardar Mesozoic ophiolite olisthostrome complex with the ophiolitic rocks of Volvi and the metabasic rocks occuring within the Sidironero unit, especially those croping out in Southern Bulgaria, earlier described by Kozhukharova (1980, 1984). Another more revolutionary approach was proposed by Froitzheim et al. (2014) who considered the relative autochthon of Rhodope as as tectonic window of the Apulian plate, equivalent to the external Hellenides, whereas the Volvi and Sidironero units should be units of the internal Hellenides, equivalent to the Axios/Vardar ocean. This interpretation was mainly based on a Jurassic age determination of a meta-ophiolite from the western Bulgarian Rhodope, considered as part of the Axios/Vardar oceanic domain.

References

Altherr, R., Kreuzer, H., Wendt, I., Lenz, H., Wagner, G., Keller, J., Harre, W. & Hohndorf, A. 1982. A late Oligocene / Early Miocene High Temperature Belt in the Attic - Cycladic Crystalline Complex (SE Pelagonian, Greece). Geol Jb., E 23, 97–164.

Ancirev, A. Gorozanin, O., Velickov, D. & Bogoyavleskaya, O. 1980. About a find of Faunistic remains in the metamorphic rocks of the western Rhodopes. Geol. Balc., 10, 1, 29–32.

Andriessen, P. A. M., Banga, G. & Helbeda, E. H. 1987. Isotopic age study of pre-Alpine rocks in the basal units on Naxos, Sikinos and Ios, Greek Cyclades. Geol. En Mijn., 66, 3–14.

Andriessen, P., Boelrijk, N., Hebeba, E., Priem, N., Verdurmen, E. & Vershure, R. 1979. Dating the events of Metamorphism and Granitic Magmatism in the Alpine Orogen of Naxos (Cyclades, Greece). Contrib. Mineral. Petrol., 69, 215–225.

Angelier, J., Glaçon, G. & Müller, C. 1978. Sur I' existence et la position tectonique du Miocène inferieur marin dans I' archipel de Naxos (Cyclades, Grèce). C. R. Acad. Sc. Paris, 286, 21–24.

Aranitis, S. 1963. Die Entstehung der Erupticgesteine von Hermioni-Gebiet und die mit ihnen verburdene Vererzung. Ann. Geol. Pays Hellen., 14, 213–304.

Ardaens, R., Colin, J.P. & Kozur, H. 1979. Sur la présence du Scythien supérieur fossilifère dans la chaine du Vardoussia (Grèce continentale). Conséquences paléogéographiques. C. R. somm. Soc. Géol. France, 132–135.

Argyriadis, I. 1967. Sur le problème des relations structurales entre formations métamorphiques en Attique et Eubée. C. R. Acad. Sc. Paris, 264, 438–441.

Arikas, K., Pape, M. Serelis, K. & Tsagarlidis, A. 2001. Petrological – Mineralogical study of metabasites (prasinites) of Laurium and their genetic geotectonic environment. Bull. Geol. Soc. Greece, 34/3, 901–910 (in greek).

Aronis, G., Panagiotidis, G. in collaboration with Monopolis, D. & Morikis, A. 1964. Delphi sheet. Geol. Map of Greece, at scale 1/50.000, IGME.

Arsovski, M., Dumurdzanov, N., Hristov, S., Ivanov, T., Ivanova, V., Petkovski, P. & Stojanov, R. 1979. Correlation of the Pre-Cambrian complexes of the Pelagonian massif, Vardar zone and Serbo-Macedonian massif. VI Coll. Geol. Aegean Region, Athens 1977, Proc. II, 549–557.

Aubouin, J. 1955. Les couches de passage au flysch dans l' E du Pinde meridionale (synclinal de Tirna-Perliango, Thessalie, Grece). C. R. somm. Soc. Geol. France, 137–141.

Aubouin, J. 1959. Contribution à I' étude géologique de la Grèce septentrionale: Les confins de I'Epire et de la Thessalie. Ann. Géol. Pays Hellén., 10, 1–483.

Aubouin, J. 1977. Alpine Tectonics and Plate Tectonics: Thoughts about the Eastern Mediterranean. In: Europe from crust to core, 143–158, J. Wiley.

Aubouin, J. & Dercourt, J. 1975. Les transversales dinariques derivent elles de paleofailles transformantes? C. R. Acad. Sci. Paris, 281, 347–350.

Aubouin, J., Blanchet, R., Cadet, J.P., Celet, P., Charvet, J., Chorowitz J., Cousin, M. & Rampnoux, J.P. 1970. Essai sur la geologie des Dinarides. Bull.Soc. Geol. France, 12, 1060–1095.

Aubouin, J., Bonneau, M., Davidson, J., Leboulenger, P., Matesco, S.& Zambetakis, A. 1976. Esquisse structurale de l arc égéen externe: des Dinarides aux Taurides. Bull. Soc. Géol. France, XVIII, 2, 327–336.

Aubouin, J., Le Pichon, X., Winterer, E. & Bonneau, M. 1979. Les Hellenides dans I' optique de la tectonique des plaques. VI Coll. Geol. Aegean Region, Athens 1977, Proc., 3, 1333–1354.

Aubouin, J., Desprairies, A. & Terry, J. 1977. Le synclinal d' Epire – Akarnanie, la nappe du Pinde-Olonos & la nappe ophiolitique. Bull. Soc. Geol. France, 19, 20–27.

Augier, R., Jolivet, L., Gadenne, L., Lahfid, A., Driussi, O., 2015. Exhumation kinematics of the Cycladic Blueschists unit and back-arc extension, insight from the Southern Cyclades (Sikinos and Folegandros Islands, Greece). Tectonics, 34(1), 152–185.

Avigad, D., 1998. High-pressure metamorphism and cooling on SE Naxos (Cyclades, Greece). Europ. J. Mineral., 10(6), 1309–1319.

Avigad, D. & Garfunkel, Z. 1989. Low-angle faults above and below a blueschist belt: Tinos Island, Cyclades, Greece. Terra Nova, 1, 182–187.

Baltatzis, E. 1996. Blueschist to greenschist transition and the P-T path of Prasinites from the Lavrion area, Greece. Min. Mag., 60, 551–561.

Barr, S.R., Temperley, S. & Tarney, J., 1999. Lateral growth of the continental crust through deep level subduction–accretion: a re-evaluation of central Greek Rhodope. Lithos, 46, 69–94.

Baud, A. & Papanikolaou, D., 1981. Olisthotites and flysch facies at the Permo-Triassic transition from Attica, Eastern Greece. IGCP Project 106, Sarajevo, 33–34.

Baud, A., Jenny, C., Papanikolaou, D., Sideris, C., & Stampfli, G., 1991. New observations on Permian stratigraphy in Greece and geodynamic interpretation. Bull. Geol. Soc. Greece, XXV, 187–206.

Baumgartner, P.O., 1985. Jurassic sedimentary evolution and nappe emplacement in the Argolis Peninsula (Peloponnesus, Greece), 99. Birkhauser, Basel.

Baziotis, I., Proyer, A. & Mposkos, E. 2009. High-pressure/low-temperature metamorphism from the basalts in Lavrion (Greece): implications for the presentation of peak metamorphic assemblages in blueschists and greenschists. European J. Mineralogy, 21, 1, 133–148.

Bebien, J. 1982. L'association ignée de Geuvgeli (Macédoine, Grèce). Expression d'un magmatisme ophiolithique dans une déchirure continentale. Thèse, Univ. Nancy, 467p.

Bebien, J., Platevoet, B. & Mercier, J.L. 1994. Geodynamic significance of the Paikon Massif in the Hellenides. Contribution of the volcanic rock studies. Bull. Geol. Soc. Greece, 30, 63–67.

Bernoulli, D., De Graciansky, P.C. & Monod, O. 1974. The Extension of the Lycian Nappes (SW Turkey) into the Southeastern Aegean Islands. Eclogae geol. Helv., 67,1, 39–90.

Bernoulli, D. & H. Laubscher 1972. The palinspastic problem of the Hellenides. Eclogae geol. Helv., 24, 347–372.

Besenecker, H., Dürr, S., Herget, G., Jacobshagen, V., Kauffmann, G., Lüdtke, G., Roth, W. & Tietze, K.W. 1968. Geologie von Chios (Ägäis). Geologica et Paleontologica, 2, 121–150.

Blake, M.C., Bonneau, M., Geyssant, J., Kienast, J.R., Lepvrier, C., Maluski, H. & Papanikolaou, D., 1981. A geologic reconnaissance of the Cycladic Blueschist Belt, Greece. Bull. Geol. Soc. Amer., 92, 247–254.

Blumor, T., Dollinger, J., Knobel, M., Mutter, A., Zarda, S. & Kowalczyk, G. 1994. Plattenkalk series and Kastania phyllites of the Taygetos mts., new results on structure and succession. Bull. Geol. Soc. Greece, 30 (2), 83–92.

Boncev, E. 1958. Uber die tektonische Ausbildung der Kraistiden. Geologie, Berlin.

Boncev, E. 1966. Revue general de la structure geologique de la Bulgarie. Bull. Inst. Geol. Sofia, 15, 5–24.

Boncev, E. 1986. Apercu general sur la tectonique des Balkans. Geol. Balc., 16.2, 3–32.

Bonev, N. & Stampfli, G., 2009. Gabbro, plagiogranite and associated dykes in the supra-subduction zone Evros Ophiolites, NE Greece. Geol. Mag., 146, 72–91.

Bonev, N. & Stampfli, G., 2011. Alpine tectonic evolution of a Jurassic subduction accretionary complex: deformation, kinematics and 40Ar/39Ar age constraints on the Mesozoic low-grade schists of the Circum-Rhodope Belt in the eastern Rhodope–Thrace region, Bulgaria–Greece. J. Geod., 52, 143–167.

Bonev, N., Filipov, P., Raicheva, R., Chiaradia, M. & Moritz, R. 2019. Detrital zircon age and Sr isotopic constraints for a Late Paleozoic carbonate platform of the lower Rhodope thrust system, Pirin, SW Bulgaria. Geological Magazine, in print.

Bonneau, M. 1972. La nappe métamorphique de I' Asteroussia, lambeau d' affinités pélagoniennes charrié jusque sur la zone de Tripolitza de la Crète moyenne (Grèce). C.R. Acad. Sc. Paris, 275, 2303–2306.

Bonneau, M. 1973a. Sur les affinités ioniennes des «Calcaires en plaquettes» épimetamorphiques de la Crète, le charriage de la série de Gavrovo-Tripolitsa et la structure de I' arc égéen. C.R. Acad. Sc. Paris, 277, 2453–2456.

Bonneau, M. 1973b. Les differents "series ophiolitiques" de la Crete: une mise au point. C. R. Ac. Sci., Paris, 276, 1249–1252.

Bonneau, M. 1976. Esquisse structural de la Crète alpine. Bull Soc. Géol. France, 8, 2, 351–353.

Bonneau, M. 1984. Correlation of the Hellenide nappes in the south-east Aegean and their tectonic reconstruction. Geol. Soc. London, Sp. Publ.,17, 517–527.

Bonneau, M., Geyssant, J.& Lepvrier, C. 1978. Tectonique alpine dans le massif d'Attique-Cyclades. Plis couchés kilométriques dans I' île de Naxos, conséquences. Rev. Géogr. Phys. Géol. Dyn., 20, 1, 109–122.

Bonneau, M. & Karakitsios, V. 1979. Les niveaux inférieurs (Trias supérieur) de la nappe de Tripolitza en Crète moyenne (Grèce) et leurs relations avec la nappe des phyllades. Problèmes stratigraphiques, tectoniques et de métamorphisme. C. R. Acad. Sci., Paris, D 288, 15–18.

Bonneau, M. & Kienast, J.R. 1982. Subduction, collision et schistes bleus: l' exemple de I' Egée (Grèce). Bull. Soc. Géol. France, XXIV, 4, 785–791.

Bonneau, M. & Zambetakis, A. 1975. La serie de Mangassa de la Crete orientale (Grece): une klippe d'origine pindique externe. C. R. Acad. Sci., Paris, 281, 17–19.

Bonneau, M., Angelier, J. & Epting, M. 1977. Réunion extraordinaire de la société géologique de France en Crète. Bull. Soc. Géol. France, 19, 87–102.

Bonneau, M., Blake, M.C.Jr., Geyssant, J., Kienast, J.R., Lepvrier, C., Maluski, H. & Papanikolaou, D. 1980. Sur la signification des séries métamorphiques (schistes bleus) des Cyclades (Hellénides, Grèce). L'exemple de l' île de Syros. C. R. Acad. Sci., Paris, 290, 1463–1466.

Bornovas, I. 1964. The geology of Lefkas Island. Geol. Geophys. Res., IGME, 10/1, 143 p (in greek).

Borsi, S., Ferrara, G., Mercier, J. 1964. Détermination de I' âge des séries métamorphiques du Massif Serbo-Macédonien au Nord-Est de Thessalonique (Grèce) par les méthodes Rb/Sr et K/Ar. Ann. Soc. Géol. Nord, LXXXIV, 223–225.

Borsi, S., Ferrara, G. & Merger, J. 1966. Mesures d' âge par la méthode Rb/Sr des granites filoniens de Karathodoro (zones internes des Hellénides, Macédoine centrale, Grèce). Savez Geol. Drus. Szr., Jugoslavije, 5–10.

Bortolotti, V., Carras, N., Chiari, M., Fazzuoli, M., Marcucci, M., Nirta, G., Principi, G. & Sacani, E. 2009. The ophiolite bearing melange in the early Tertiary Pindos flysch of Etolia (Central Greece). Ofioliti, 34, 83–94.

Boundi, D., Papanikolaou, D., Bosio, G., Villa, I. M. & Montemagni, S. 2019. Upper Cretaceous metamorphism in the upper tectonic unit of Skyros: implications for the geotectonic evolution of the internal Hellenides. manuscript.

British Petroleum Co, 1971. The geological results of petroleum exploration in Western Greece. Geology of Greece, I.G.R.S., No 10, 72 p.

Brocker, M. & Franz, L. 1998. Rb-Sr isotope studies on Tinos island (Cyclades, Greece): additional time constraints for metamorphism, extent of infiltration-controlled overprinting and deformational activity. Geological magazine, 135, 369–382.

Brocker, M. & Franz, L. 2006. Dating metamorphism and tectonic juxtaposition on Andros island (Cyclades, Greece): results of a Rb-Sr study. Geological Magazine, 143, 609–620.

Brocker, M. & Pidgeon, R.T., 2007. Protolith ages of meta-igneous and metatuffaceous rocks from the Cycladic blueschist unit, Greece: results of a reconnaissance U–Pb zircon study. J. Geol., 115, 83–98.

Brown, S. A. M. & Robertson, A.H.F., 2003. Sedimentary geology as a key to understanding the tectonic evolution of the Mesozoic – Early Tertiary Paikon Massif, Vardar suture zone, N. Greece. Sedimentary Geology, 160, 179–212.

Brunn, J. 1956. Contribution à I'étude Géologique du Pinde Septentrional et d'une partie de la Macédoine Occidental. Ann. Géol. Pays Hellén., 7, 1–358.

Brunn, J.H., Faugeres, L. & Robert, P. 1970. Existence dans la région de Kozani du prolongement occidental de la nappe du Haut Vermion (Macédoine, Grèce). C. R. somm. Soc. Géol. France, 253–255.

Brunn, J. H., Argyriadis, I. Ricou, L., Poisson, A., Marcoux, J. & Degraciansky, P.C. 1976. Elements majeurs de liaison entre Taurides et Hellenides. Bull. Soc. Geol. France, 18, 481–497.

Brun, J.P., Faccenna, C., Gueydan, F., Sokoutis, D., Philippon, M., Kydonakis, K. & Gorini, C. 2016. The two - stage Aegean extension, from localised to distributed, a result of slab rollback acceleration. Canadian Journal of Earth Sciences, 53, 11, 1142–1157.,

Buick, I. S. & Holland, T. J. B. 1989. The PTt path associated with crustal extension, Naxos, Cyclades, Greece. Geol. Soc., London, Sp. Publ., 43(1), 365–369.

Bulle, F., Brocker, M., Gartner, C. & Keasling, A., 2010. Geochemistry and geochronology of HP mélanges from Tinos and Andros, cycladic blueschist belt, Greece. Lithos, 117, 61–81.

Burchfiel, B.C., Nakov, R., Dumurtzanov, D., Papanikolaou, D., Tzankov, T., Serafimovski, T., King, R.W., Kotzev, V., Todosov, A. & Nurce, B. 2008. Evolution and Dynamics of the Cenozoic tectonics of the South Balkan extensional system. Geosphere, 4, 6, 917–938.

Burchfiel, B. C., Nakov, R. & Tzankov, T. 2003. Evidence from the Mesta half-graben, SW Bulgaria for the Late Eocene beggining of Aegean extension in the central Balkan Peninsula. Tectonophysics, 375, 61–76.

Burchfiel, B.C., Royden, L.H., Papanikolaou, D. & Pearce, F.D. 2018. Crustal development within a retreating subduction system: the Hellenides. Geosphere, 14, 2.

Capedri, S., Lekkas, E., Papanikolaou, D., Skarpelis, N., Venturelli, G. & Gallo, J. 1985. The ophiolite of the Koziakas range, Western Thessaly (Greece). N. Jb. Miner. Abh., 152, 1, 45–64.

Cayeux, L 1903. Phénomènes de charriage dans la Méditerranée orientale. C.R. Acad. Sci. Paris, 136, 474–476.

Cayeux, L. 1911. Existence des calcaires a gyroporelles dans les Cyclades. C. R. Acad. Sci., Paris, 152, 292–293.

Celet, P. 1962. Contribution à l'étude géologique du Parnasse-Kiona et d'une partie des régions méridionales de la Grèce continentale. Ann. Géol. Pays Hellén. 7, 1–358.

Celet, P. 1979. Les bordures de la zone du Parnasse. Evolution paléogéographique au Mésozoique et caractères structuraux. VI Coll. Geol. Aegean Region, Athens 1977, Proc. 2, 725–740.

Celet, P. & Clement, B. 1971. Sur la présence d'une nouvelle unité paléogéographique et structurale en Grèce continentale du Sud: l' unite du flysch béotien. C. R. somm. Soc. Géol. France, 43–47.

Celet, P. & Ferriere, J. 1978. Les Hellénides internes: Le Pélagonien. Eclogae geol. Helv.,71/3, 467–495.

Celet, P., Clement, B. & Ferriere, J. 1976. La zone béotienne en Grèce: Implications paléogéographiques et structurales. Eclogae géol. Helv., 69, 577–599.

Channel J.E.T., D' Argenio, B. & Horvath, F. 1979. Adria, the African Promontory in Mesozoic Mediterranean Paleogeography. Earth Science Reviews, 15, 213–292.

Chatzaras, V., Xypolias, P., Kokkalas, S., & Koukouvelas, I. 2011. Oligocene–Miocene thrusting in central Aegean: insights from the Cycladic island of Amorgos. Geological Journal, 46(6), 619–636.

Chatzaras, V., Dorr, W., Finger, F., Xypolias, P. & Zulauf, G. 2013. U-Pb single zircon ages and geochemistry of metagranitoid rocks in the Cycladic Blueschists (Evia Island): implications for the Triassic tectonic setting of Greece. Tectonophysics 595–596, 125–139.

Chen, G. 1995. Evolution of the high and medium pressure metamorphic rocks on the island of Samos, Greece. Ann. Geol. Pays Hellen., 36, 799–915.

Chiari, M., Bortolotti, V., Marcucci, M., Photiades, A., Principi, G. & Saccani, E. 2012. Radiolarian biostratigraphy and geochemistry of the Koziakas massif ophiolites (Greece). Bull. Geol. Soc. France, 183, 4, 287–306.

Chiari, M., Bortolotti, V., Marcucci, M., Photiades, A. & Principi, G., 2003. The middle Jurassic siliceous sedimentary cover at the top of the Vourinos ophiolite (Greece). Ofioliti, 28, 95–103.

Christodoulou, G. 1972. Observations on the geological structure of Kastellorizo Island (Dodekanese) and the age of its limestones. Bull. Geol. Soc. Greece, 9/2, 387–412 (in greek).

Clement, B. 1971. Découverte d'un flysch éocrétacé en Béotie (Grèce continentale).C.R. Acad. Sci. Paris, 272, 791–792.

Clift, P. & Robertson, A., 1989. Evidence of a late Mesozoic ocean basin and subduction–accretion in the southern Greek Neo-Tethys. Geology, 17, 559–563.

Creutzburg, N. 1958. Probleme des Gebirgbaus und der Moerphogenese auf der Insel Kreta. Freiburger Univeristatsreden, 26, 46p.

Creutzburg, N., Papastamatiou, J., 1969. Die Ethia-serie des sudlichen Mittelkretas und ihre Ophiolithvorkommen. Heidelberger Akademie der Wissenschaften, Mathematisch-Naturwissenschaftliche Klasse, 1, 1–63.

Creutzburg, N. & Seidel, E. 1975. Zum Stand der Geologie des Präneogens auf Kreta.N. Jb. Geol. Paläont. Abh., 149, 363–383.

Danelian, T. & Robertson, A. H. F., 1995. Palaeogeographic implications of the age of radiolarian-rich sediments in Beotia (Greece). Sp. Publ. Geol. Soc. Greece, 4, 175–179.

Danelian, T., Lekkas, S. & Alexopoulos, A. 2000. Decouverte de radiolarites triasiques dans un complexe ophiolitique a l' extreme-Sud du Peloponnese (Angelona, Laconie, Grece). C. R. Acad. Sc., Paris, 330, 639–644.

Davis, E. 1963. Geological structure of Samothraki Island. Ann. Geol. Pays Hellen., 14, 133–212 (in greek).

Davis, E. 1967. Occurence of granitic rocks in the metamorphic system of Asteroussia Mts of Southern Crete. Proc. Acad. Athens, 42, 253–270 (in greek).

De Bono, A., Martini, R., Zaninetti, L., Hirsch, F., Stamfpli, G. & Vavassis, I., 2001. Permo-Triassic stratigraphy of the Pelagonian zone in central Evia island (Greece). Eclogae Geol. Helv., 94, 289–311.

Delaloye, M., Economou, C. & Skounakis, S. 1977. Ages radiometriques de quelques roches ophiolitiques de l ' ile de Crete. 6th Coll. Aegean Region, Proc., 1, 129–135.

Del Ben, A., Mocnik, A., Volpi, V. & Karvelis, P. 2015. Old domains in the south Adria plate and their relationship with the west Hellenic front. J. Geodynamics, 89, 15–28.

Dercourt, J. 1964. Contribution à l' étude géologique d'un secteur du Péloponnèse septentrional. Ann. Géol. Pays Hellén.,15, 1–418.

Dercourt, J., Makris, J. & Melidonis, N., 1980. Grece. Introduction a la geologie generale. 26th International Geological Congress, Livret-Guide, G12, 159p.

Dermitzakis, M. 1978. Stratigraphy and Sedimentary History of the Miocene of Zakynthos (Ionian Islands, Greece). Ann. Géol. Pays Hellén., 29/1, 47–186.

Dermitzakis, M. & Papanikolaou, D. with contr. of S. Theodoridis and R. Mirkou 1980. The Molasse of Paros Island, Aegean Sea. Ann. Naturhist. Mus. Wien, 83, 59–71.

Derycke, F. & Godfriaux, I. 1978. Decouverte de microfaunes paleogenes dans le flysch metamorphique de Spilia (Ossa, Grece). C. R. Acad. Sci. Paris, 286, 555–558.

Desio, A. 1931. Le isole italiane dell' Egeo. Mem. Carta geol. Ital., 24, 534 p.

De Wet, A. P., Miller, J. A., Bickle, M. J. & Chapman, H. J., 1989. Geology and geochronology of the Arnea, Sithonia and Ouranopolis intrusions, Chalkidiki peninsula, northern Greece. Tectonophysics, 161, 65–79.

DeWewer, P. 1976. La "formation a blocs" olistostrome chevauche par la nappe du Pinde-Olonos (Peloponnese, Grece). C. R. Ac. Sci. Paris, 282, 21–24.

Dimitriadis, S. 1980. A possible paleomargin evolution of the southernmost part of Serbo-Macedonian massif. Proc. 26th Int. Geol. Congress, Paris 1980, Abstracts, I, 335.

Dimitriadis, S. & Asvesta, A., 1993. Sedimentation and magmatism related to the Triassic rifting and later events in the Vardar–Axios Zone. Bul. Geol. Soc. Greece, 28, 149–168.

Dimitriadis, S. & Godelitsas, A., 1991. Evidence of high pressure metamorphism in the Vertiskos Group of the Serbomacedonian massif: the eclogite of Nea Roda, Chalkidiki. Bull. Geol. Soc. Greece, 25, 67–80.

Dimitrievic, M.D. & Dimitrievic, M. N. 1976. The polyphase melange of the Vardar zone. Bull. Soc. Geol. France, 18, 205–208.

Dinter, D.A. & Royden, L.H., 1993. Late Cenozoic extension in NE Greece: Strymon Valley detachment and Rhodope Core Complex. Geology, 21, 45–48.

Dixon, J.E. & Dimitriadis, S. 1984. Metamorphosed ophiolitic rocks from the Serbo-Macedonian Massif, near Lake Volvi, North-east Greece. Geol. Soc. London, Sp. Publ. 17, 603–618.

Dubois, R. & Bignot, G. 1979. Presence d'un «hard-ground» nummulitique au sommet de la serie cretace d' Almyropotamos (Eubee meridionale, Grece). Consequences. C. R. Acad. Sci., Paris, 289, 993–995.

Duchene, S., Aissa, R. & Vanderhaeghe, O. 2006. Presssure-temperature-time evolution of metamorphic rocks from Naxos (Cyclades, Greece): constraints from Thermobarometry and Rb-Sr dating. Geodynamica Acta, 19, 299–319.

Dürr, St. 1975. Über Alter und geotektonische stellung des Menderes-Kristallins / SW-Anatolien und seine aequivalente in der mittleren Aegais. Habil. schrift., Marburg / Lahn, 107 p.

Dürr, St. & Altherr, R. 1979. Existence de klippes d' une nappe composite néogène dans I' île de Myconos / Cyclades (Grèce). Rapp. Comm. int. Mer Médit., 25/26, 2a, 33–34.

Dürr, S. & Flugel, E. 1979. Contribution à la stratigraphie du cristallin des Cyclades: Mise en évidence du Trias Supérieur dans les marbres de Naxos (Grèce). Rapp.Comm. int. Mer Médit, 25/26,2a, 31–32.

Dürr, St., Altherr, R., Keller, J., Okrusch, M.& Seidel, E. 1978. The median Aegean crystalline belt: Stratigraphy, Structure, Metamorphism, Magmatism. In: Alps, Apennines, Hellenides, 455–477.

Engel, M. & Reischmann, T. 1998. Single zircon geochronology of orthogneisses from Paros, Greece. Bull. Geol. Soc. Greece, 32, 91–99.

Epting, M., Kudrass, R.& Schafer, A. 1972. Stratigraphie et position des séries métamorphiques aux Talea Ori/Crète. Z. Deutsch. Geol. Ges., 123, 365–370.

Ferriere, J. 1976. Sur la signification des séries du massif de I'Othris (Grèce continentale centrale): la zone isopique maliaque. Ann. Soc. Géol. Nord, 96/2, 121–134.

Ferriere, J. 1977. L' Othrys et le Pelion. Reunion Extraordinaire en Grece, Bull.Soc. Geol. France, XIX, 1, 49–53.

Ferriere, J. 1979. Le secteur meridional du «massif metamorphique de Thessalie» . Le massif du Pelion et ses environs. VI Coll. Geol, Aegean Region, Athens 1977, Proc. I, 291–309.

Ferriere, J. 1982. Paléogéographies et tectoniques superposées dans les Hellénides internes: les massifs de I'Othrys et de Pélion. Soc. Géol. Nord, Publ. 8, 970 p.

Ferriere, J., Baumgartner, P. & Chanier, 2016. The Maliac Ocean: the origin of the Tethyan Hellenic ophiolites. Int. J. Earth Sci., 105, 1941–1963. https://doi.org/10.1007/s00531-016-1303-6.

Flansburg, M., Stockli, D., Poulaki, E. & Soukis, K. 2019. Tectonomagmatic and stratigraphic evolution of the Cycladic basement, Ios Island, Greece. Tectonics, 38, https://doi.org/10.1029/2018tc005436.

Fleury, J.J. 1977. De Lamia à Messolonghi. La nappe de Pinde-Olonos et l'unité Meghdovas. Bull. Soc. Géol. France, 19,1, 53–60.

Fleury, J.J. 1980. Les zones de Gavrovo-Tripolitza et du Pinde-Olonos (Grèce continentale et Péloponnèse du Nord). Evolution d'une plateforme et d' un bassin dans le cadre alpin. Publ. Soc. Géol. Nord, 4.

Forster, M. & Lister, G.S., 1999. Detachment faults in the Aegean core complex of Ios, Cyclades, Greece. In: Ring, U., Brandon, M.T., Lister, G.S., Willett, S.D. (Eds.), Exhumation Processes: Normal Faulting, Ductile Flow and Erosion. Sp. Publ., Geol. Soc., London, 305–324.

Fourquin, C. 1975. L'Anatolie du NW, marge méridionale du continent européen, histoire paléogéographique, tectonique et magmatique durant le Secondaire et le Tertiaire. Bull. Soc. Géol. France, (7), 17, 1058–1070.

Franz, L., Okrusch, M., Seidel, E. & Kreuzer, H. 2005. Polymetamorphic evolution of pre-Alpidic basement relics in the external Hellenides, Greece. Neues Jahrbuch für Mineralogie - Abhandlungen, 181/2, 147–172.

Froitzheim, N., Jahn-Awe, S., Frei, D., Wainright, A.N., Maas, R., Georgiev, N., Nagel, T.Z. & Pleuger, J. 2014. Age and composition of meta-ophiolite from the Rhodope Middle Allochthon (Satovcha, Bulgaria). A test for the maximum-allochthony hypothesis of the Hellenides. Tectonics, 33, 1477–1500.

Fytrolakis, N. 1972. The effect of some orogenetic mouvements and the formation of gypsum in Eastern Crete (Sitia district). Bull. Geol. Soc. Greece, 9, 81–100 (in greek).

Fytrolakis, N. 1980. The geological structure of Crete. Problems, observations and conclusions. Habilitation Thesis, National Polytechnical University, 147 p (in greek).

Fytrolakis, N. & Papanikolaou, D. in collabor. with Panagopoulos, A. 1981. Stratigraphy and structure of Amorgos Island, Aegean Sea. Ann. Géol. Pays Hellén., 30/2, 455–472.

Galeos, A. 1986. Kastellorizo sheet. Geological Map of Greece, at scale 1/50,000. IGME.

Galeos, A., Pomoni-Papaioannou, F., Tsaila-Monopolis, S., Turnsek, D. & Ioakim, C. 1994. Upper Jurassic–Lower Cretaceous "molassic-type" sedimentation in the western part of the Almopia subzone, Aridhea Loutra Unit (northern Greece). Bull. Geol. Soc. Greece, 30, 171–184.

Georgiadis, G. A., Tranos, M. D. & Mountrakis, D. 2007. Late- and Post-Alpine tectonic evolution of the southern part of the Athos Peninsula, northern Greece. Bull. Geol. Soc. Greece, 40, 309–320.

Gerolymatos, I., Dornsiepen, U. & Trifonova, E. 1982. Vorkommen von Hoheren Zentralhellenischen Deckeneinheiten in Lakonien (Peloponnes, Griechenland). Proc. Acad. Athens, 57, 247–301.

Godfriaux, I. 1962. Sur la presence de l'Eocene dans les calcaires superieurs de l ' Olympe (Thessalie septentrionale, Grece). Bull. Soc. Geol. France, 4, 49–54.

Godfriaux, I. 1968. Etude géologique de la région de I'Olympe (Grèce). Ann. Géol. Pays Hellén., 19, 1–281.

Godfriaux, I. 1977. L' Olympe. Bull, Soc. Géol France, XIX, 1, 45–49.

Godfriaux, I. & Ricou, L.E., 1991. The Paikon, a tectonic window within the Internal Hellenides, Macedonia, Greece. C. R. Acad. Sci., 313(12), 1479–1484.

Gouliotis, L. 2014. The tectonic structure of Giona Mt and its hydrogeological applications. PhD Thesis, University of Athens, 433 p (in greek).

Gutnic, M., Monod, O., Poisson, A. & Dumont, J.F. 1979. Géologie des Taurides Occidentales (Turquie). Mém. Soc. Géol. France, 137, 1–109.

Haenel-Remy, S. & Bebien, J., 1987. Basaltes et dolerites riches en magnesium dans l' association ignee de Guevgueli (Macedoine, Grece): les temoins d' une evolution depuis des tholeites abyssales jusqu' aux basaltes continentaux. Ofioliti, 12, 91–106.

Hall, R., Audley-Charles, M G. & Carter, D.J. 1984. The significance of Crete for the evolution of the Eastern Mediterranean. Geol. Soc. London, Sp. Publ. 17, 499–516.

Harre, W., Kockel, F., Kreuzer, H., Lenz, H., Müller, P. & Walther, H. W. 1968. Über Rejuventationen im Serbo-Mazedonischen Massiv (Deutung radiometrischerAlters-Bestimmungen). (XXIII Int Geol. Congress 6).Geologica et Palaeontologica, 2, 193–194.

Hatzipanagiotou, K., 1988. Einbindung der obersten Einheit von Rhodos und Karpathos (Griechenland) in der alpinischen Ophiolith-Gürtel. N. Jahr. Geol. Paläont., Abh., 176, 395–442.

Hatzipanagiotou, K. & Pe-Piper, G., 1995. Ophiolitic and sub-ophiolitic metamorphic rocks of the Vatera area, southern Lesbos (Greece): geochemistry and geochronology. Ofioliti, 20, 17–29.

Hatzipanagiotou, K., Tsikouras, B., Migiros, G., Gartzos, E. & Serelis, K., 2003. Origin of rodingites in ultramafic rocks from Lesvos Island (NE Aegean, Greece). Ofioliti, 28(1), 13–23.

Hausmann, J.F.L. 1845. Beitrage zur Oryktographie von Syra und ein neues Mineral, der glaucophan. Göttingen Geol. Anz., 193–198.

Hecht, J. 1970. Zur Geologie von südost-Lesbos (Griechenland). Dissert., Univ. München, 64 p.

Hecht, J. 1972. Lesbos Island. Geol. map at scale 1 /50,000, IGME.

Henzes-Kunst, F. & Kreuzer, H., 1982. Isotopic dating of pre-Alpidic rocks from the island of Ios (Cyclades, Greece). Contr. Mineral. Petr., 80, 245–253.

Henzes-Kunst, F. & Okrusch, M. 1978. Polymetamorphose auf los, Kykladen – Kristallin (Griechenland). Fortschr. Miner., 56, 1, 38–39.

Himmerkus, F., Reischmann, T. & Kostopoulos, D. 2006. Late Proterozoic and Silurian basement units within the Serbo-Macedonian Massif, northern Greece: the significance of terrane accretion in the Hellenides. Geol. Soc., London, Sp. Publ., 260, 35–50.

Himmerkus, F., Reischmann, T. & Kostopoulos, D. 2009a. Serbo-Macedonian revisited: a Silurian basement terrane from northern Gondwana in the Internal Hellenides, Greece. Tectonophysics, 473, 20–35.

Himmerkus, F., Reischmann, T. & Kostopoulos, D. 2009b. Triassic rift-related meta-granites in the Internal Hellenides, Greece. Geol. Mag., 146, 252–265.

Himmerkus, F., Zachariadis, P., Reischmann, T. & Kostopoulos, D. 2012. The basement of the Mount Athos peninsula, northern Greece: insights from geochemistry and zircon ages. Intern. J. Earth Sciences, 101, 1467–1485.

Huet, B., Labrousse, C. & Jolivet, L. 2009. Thrust or detachment? Exhumation processes in the Aegean. Insight from a field study of Ios (Cyclades, Greece). Tectonics, 28, TC3007.

Huet, B., Labrousse, L., Monie, P., Malvoisin, B. & Jolivet, L., 2014. Coupled phengite 40Ar–39Ar geochronology and thermobarometry: P-T-t evolution of Andros Island (Cyclades, Greece). Geological Magazine.

Hynes. A. J., Nisbet, E.G., Smith, A.G., Welland. M. J. P. & Rex, D. C. 1972. Spreading and emplacement ages of some ophiolites in the Othris region (eastern central Greece). Zeit. deuts. Geol. Ges. 123, 455–468.

I.G.S.R. & I.F.P. 1966. Etude géologique de I'Epire. Technip., 306 p.

Ivanov, Z. 1985. Position tectonique, structure géologique et évolution alpidique du massif des Rhodopes. Reun. extr. Soc. Géol. France en Bulgarie, Guide, 1–31.

Ivanov, Z., 1988. General framework of the Rhodope massif geological and structural development, in the Balkanides setting. Bull. Soc. Geol. France, 4(2), 227–240.

Ivanova, D., Bonev, N. & Chatalov, A 2015. Biostratigraphy and tectonic significance of lowermost Cretaceous carbonate rocks of the Circum-Rhodope belt (Chalkidijki peninsula and Thrace region, NE Greece). Cretaceous Research, 52, 25–63.

Jacobshagen, V. 1979. Structure and geotectonic evolution of the Hellenides. VI Coll. Geol. Aegean Region, Athens 1977, Proc. 3, 1355–1367.

Jacobshagen, V. 1980. Die eozäne Orogenese in the Ägäis. Berl. Geow. Abh., 21–33.

Jacobshagen, V. 1986. Geologie von Griechenland. Gebruder Borntraeger, 363 pp.

Jacobshagen, V. & Wallbrecher, E. 1984. Pre-Neogene nappe structure and metamorphism of the North Sporades and the southern Pelion peninsula. Geol. Soc. London, Sp. Publ., 17, 591–602.

Jacobshagen, V., Matarangas, D. and 11 co-workers 1983. Skyros Island. Geological Map of Greece at scale 1/50,000. IGME.

Jahn-Awe, S., Froitzheim, N., Nagel, T.Z., Frei, D., Georgiev, N. & Pleuger, J. 2010. Structural and geochronological evidence for Paleogene thrusting in the western Rhodopes, SW Bulgaria: Elements for a new tectonic model of the Rhodope Metamorphic Province. Tectonics, 29, 1–30.

Jansen, J.B.H. 1977. The Geology of Naxos. Geol. Geoph. Res. I.G.M. R., XIX, 1, 1–100.

Jansen, J.H.B. & Schuiling, R.D. 1976. Metamorphism on Naxos: Petrology and geothermal gradients. Amer. J. Sci., 276, 1225–1253.

Jolivet, L., Goffe, B., Monie, P., Truffert, C., Patriat, M. & Bonneau, M., 1996. Miocene detachment in Crete and exhumation P–T–t paths of high pressure metamorphic rocks. Tectonics 15, 1129–1153.

Jolivet, L., Faccenna, C., Huet, B., Labrousse, L., Le Pourhiet, L., Lacombe, O., Lecomte, E., Burov, E., Denele, Y., Brun, J.P., Philippon, M., Paul, A., Salaun, G., Karabulut, H., Piromallo, C., Monie, P., Gueydan, F., Okay, A.I., Oberhansli, R., Pourteau, A., Augier, R., Gadenne, L. & Driussi, O., 2013. Aegean tectonics: Strain localisation, slab tearing and trench retreat. Tectonophysics, 597–598, 1–33.

Jolivet, L., Menant, A., Sternai, P., Rabillard, A., Arbaret, L., Augier, R., Laurent, V., Beaudouin, A., Grasement, A., Huet, B., Labrousse, L. & Le Pourhiet, L. 2015. The geological signature of a slab tear below the Aegean. Tectonophysics, 659, 166–182.

Jones, G. & Robertson, A.H.F., 1991. Tectono-stratigraphy and evolution of the Mesozoic Pindos ophiolite and related units, northwestern Greece. J. Geol. Soc. 148, 267–288.

Jordan, H. 1969. Geologie und Petrographie im Zentralteil des Bos Dag (Drama, Griechisch Makedonien). Geotektonische Forschungen 31, 50–85.

Kalaitzidis, S., Siavalas, G., Skarpelis, N., Araujo, C. V. & Christanis, K. 2010. Late Cretaceous coal overlying karstic bauxite deposits in the Parnassus-Giona unit, central Greece: Coal characteristics and depositional environment. Int. J. Coal Geol., 81, 211–226.

Kalpakis, G. 1979. Contribution to the sedimentological study of biomicrites, stromatolites, hard-grounds, of the Parnassos-Giona zone in the Late Cretaceous-Paleogene boundary. PhD Thesis, University of Athens, 129 p (in greek).

Kaplanis, A. Koukouvelas, I., Xypolias, P. & Kokkalas, S. 2013. Kinematics and Ophiolite obduction in the Gerania and Helikon Mountains, central Greece. Tectonophysics, 595–596, 215-234.

Karakitsios, V. 1979. Contribution à l'étude géologique des Hellénides. Etude de la région de Sellia (Crète moyenne-occidentale, Grèce). Thèse 3e cycle, Paris.

Karakitsios, V. 1995. The influence of preexisting structure and halokinesis on organic matter preservation and thrust system evolution in the Ionian Basin, northwest Greece. Bull. Am. Ass. Petr. Geol., 79, 960–980.

Karotsieris, Z. 1981. Geological research in Vytina area (Central Peloponnesus). PhD Thesis, University of Athens, 202 p (in greek).

Katagas, C. 1984. High pressure metamorphism in Ghiaros Island, Cyclades, Greece. Geol. Soc. London, Sp. Publ. 17, 537–544.

Katagas, C. & Panagos, A. 1979. Pumpellyite - actinolite and greenschist facies metamorphism in Lesvos Island (Greece). Tschem. Min. Pet. Mitt., 26, 235–254.

Katerinopoulos, A. 1983. Contribution to the study of ploutonic rocks of West Varnous. PhD Thesis, University of Athens, 182 p (in greek).

Katerinopoulos, A. 2008. Variscan basic dykes in the Pelagonian (Northern Greece and South F.Y.R.O.M.): Geodynamic significance based on petrological, geochemical and geochronogical studies. Chemie der Erde, 68, 93–103.

Katrivanos, E., Kilias, A. & Mountrakis, D. 2013. Kinematics of deformation and structural evolution of the Paikon Massif (Central Macedonia, Greece): A Pelagonian tectonic window? N. Jahrb. Geol. Paleontol. Abh., 269, 149–171.

Katsikatsos, G. 1969. The age of the metamorphic system of Southern Evia and its stratigraphic succession. Proc. Acad. Athens, 44, 223–238 (in greek).

Katsikatsos, G. 1979. La structure tectonique de l' Attique et de l' île d'Eubée. VI Coll. Geol. Aegean Region, Athens 1977, Proc. I, 211–228.

Katsikatsos, G., Mercier, J.L. & Vergely, P. 1976a. L' Eubée méridionale: une double fenêtre polyphasée dans les Hellénides internes. C. R. Acad. Sci. Paris, 283, 459–462.

Katsikatsos, G., Mercier, J. L & Vergely, P. 1976b. La fenêtre d' Attique - Cyclades et les fenêtres métamorphiques des Hellénides internes. C. R. Acad. Sci. Paris, 283, 1613–1616.

Katsikatsos. G., Fytikas, M. & Koukis, G. 1981. Kymi sheet. Geological Map of Greece at scale 1/50,000. IGME.

Katsikatsos, G., Migiros, G. & Vidakis, M. 1982. Structure géologique de la région de Thessalie orientale (Grèce). Ann. Soc. Géol. Nord, C1, 177–188.

Katsikatsos, G., Migiros, G., Triantaphyllis, M. & Mettos, A. 1986. Geological structure of internal Hellenides (E.Thessaly–SW. Macedonia, Euboea–Attica–Northern Cyclades islands and Lesvos). IGME Geological and Geophysical Research, Prof. Papastamatiou Special Issue, 191–212.

Katzir, Y., Garfunkel, Z., Avigad, D. & Matthews, A. 2007. The geodynamic evolution of the Alpine orogen in the Cyclades (Aegean Sea, Greece): insights from diverse origins and modes of emplacement of ultramafic rocks. Geol. Soc., London, Sp. Publ., 291, 17–40.

Kauffmann, G., Kockel, F., Mollat, H. 1976. Notes on the stratigraphic and paleogeographic position of the Svoula Formation in the Innermost zone of the Hellenides (Northern Greece). Bull. Soc. Géol. France, 18, 2, 225–230.

Kilias, A. 1980. Geological and tectonicv study of the Eastern Varnous area (NW Macedonia). PhD Thesis, University of Thessaloniki, 271 p (in greek).

Kilias, A. 1996. Tectonic evolution of the Olympus – Ossa Mountains: emplacement of the blueschists Unit in Eastern Thessaly and exhumation of Olympus-Ossa carbonate dome as a result of Tertiary extension (central Greece). Mineral Wealth, 96, 7–22.

Kilias, A. & Mountrakis, D. 1985. Das Rizomata-Fenster im nordoestlichen Pieria-Gebirge. Neue Daten zur geologischen Grenze der Pelagonischen und der Axios-Zone in Griechenland. N. Jahrb. Geol. Palaeont. Monatshefte, 249–256.

Kilias, A., Frisch, W., Ratschbacher, L., & Sfeikos, A. 1991a. Structural evolution and P/T conditions of metamorphism of blueschists of Ambelakia nappe, Eastern Thessaly. Bull. Geol. Soc. Greece, 25, 81–99.

Kilias, A., Fasoulas, CH., Priniotakis, M., Sfeikos, A. & Frisch, W. 1991b. Deformation and HP/LT metamorphic conditions at the Krania tectonic window, Western Thessaly (Northern/Central Greece). Z. Deut. Geol. Ges., 87–96 and Bull. Geol. Soc. Greece, 25, 275–292 (in greek).

Kilias, A., Frisch, W., Avgerinas, A., Dunkl, I., Falalakis, G. & Mountrakis, D., 2010. The Pelagonian nappe pile in the Northern Greece and FYROM. Structural evolution during the Alpine orogeny: a new approach. Bull. Geol. Soc. Greece, 43, 276–288.

Kilias, A., Falalakis, G., Sfeikos, A., Papadimitriou, E., Vambaka, A. & Gkarlaouni, C. 2013. The Thrace basin in the Rhodope province of NE Greece – A tertiary supradetachment basin and its geodynamic implications. Tectonophysics, 595–596, 90–105.

Kiskyras, D. 1962. Observations on the geology of the bauxite deposits of Western Greece. Proc. Acad. Athens, 37, 370–378 (in greek).

Kiskyras, D. 1972. The geotectonic zones of the alpine orogeny in Greece. Bull. Geol. Soc. Greece, 9/2, 93–110 (in greek).

Kockel, F.& Walther, H.W. 1965. Die Strimon-linie als Grenze zwischen Serbo-Mazedonischen und Rila-Rhodope Massiv in Ost-Mazedonien. Geol. Jb., 83, 576–602.

Kockel, F., Mollat, H. & Walther, H.W. 1977. Erläuterungen zur geologischen karteder Chalkidiki und angrenzender Gebiete 1/100,000 (Nord Griechenland). Bund. Für Geowiss. v. Rohstoffe, 110 p. Hannover.

Koepke, J., & Seidel, E. 1984. Oberkretazisches Kristallin an der Basis von Ophiolithen der Südägäis: Charakterisierung der Metamorphose-Fazies. Tschermaks mineralogische und petrographische Mitteilungen, 33(4), 263–286.

Koepke, J., Seidel, E. & Kreuzer, H. 2002. Ophiolites on the Southern Aegean islands Crete, Karpathos and Rhodes: composition, geochronology and position within the ophiolite belts of the Eastern Mediterranean. Lithos, 65, 183–203.

Kokinou, E., Kamberis, E., Vafidis, A., Monopolis, D., Ananiadis, G. & Zelilidis, A. 2005. Deep seismic reflection data from offshore western Greece: A new crustal model for the Ionian Sea. Journal Petroleum Geology, 28 (2), 185–202.

Kokkinakis, A. 1978. Das Inrtusivgebiet des Symvolon-Gebirges und von Kavalla in Ostmakedonien, Griechenland. Dr. Thesis, Univ. München.

Komprobst, J., Kienast, J. R., & Vilminot, J. C. 1979. The high-pressure assemblages at Milos, Greece. Contributions to Mineralogy and Petrology, 69(1), 49–63.

Kopp, K.O. & Ott, E. 1977. Spezialkartierungen im Umkreis neuer Fossilfunde in Trypali- und Tripolitzakalken Westkretas. N. Jb. Geol. Pal. Mh., 217–238.

Kopp, K.O. & Wernado, G. 1983. Uber eine intra-triadische Deckenbewegung auf Kreta. Geol. Rundschau, 72, 3, 895–910.

Koroneos, A., Kilias, A. & Avgerinas, A. 2013. Hercynian plutonic rocks of Voras Mountain, Macedonia, Northern Greece: their structure, petrogenesis and tectonic significance. Intern. Geol. Rev., 55, 994–1016.

Kossmat, F. 1924. Geologie der Zentral Balkanhalbinsel. In: Die Kriegsschanplätz 1914–1918 geologisch dargestelt, 13, 198 p., Berlin.

Kozhoukharov, D. & Timofeev, B. 1980. First finds of microphytofossils in the Precambrian of the Rhodope massif. In: I.G.C. P. Project 22, The Precambrian in South Bulgaria, Sofia Meeting 1980, 27–32.

Kozhoukharov, D., Kozhoukharov, E. & Papanikolaou, D. 1988. Precambrian in the Rhodope massif. In: Precambrian in younger fold belts, Joubek et al editors, 723–778, Wiley.

Kozhukharova, E. 1980. Eclogites in the Precambrian from the eastern Rhodope block. Comptes rendus de l'Académie bulgare des sciences, 33, 375–378.

Kozhukharova, E. 1984. Origin and structural position of the serpentinized ultrabasic rocks of the Precambrian ophiolitic association in the Rhodope Massif: I. Geologic position and composition of ophiolite association (in Russian). Geol. Balc.,14, 9–36.

Krahl, J., Kauffmann, G., Kozur, H., Richter, D., Forster, O. & Heinritzi, F. 1983. Neue Daten zur Biostratigraphie und zur tektonischen Lagerung der Phyllit-Gruppe und der Trypali-Gruppe auf der Insel Kreta (Griechenland). Geol. Rund., 72,1147–1166.

Krahl, J. Kauffman, G., Richter, D., Kozur, H., Moeller, I., Foerster, O., Heinritzi, F. & Dornsiepen, U. 1986. Neue Fossilfunde in der Phyllit-Gruppe Ostkretas (Griechenland). Z. Dt. Geol. Ges., 137, 523–536.

Kranis, H. & Papanikolaou, D. 2001. Evidence for detachment faulting on the Parnassos Mountain front (Central Greece). Bull. Geol. Soc. Greece, 34/1, 281–287.

Kreüzer, H., Harre, W., Lenz, H., Wendt, L, Henzes-Kunst, F. & Okrusch, M.1978. K /Ar und Rb/Sr Daten von Mineralen aus dem polymetamorphen kristallin der Kycladen-lnsel los (Griechenland). Fortscchr. Miner., 56, 1, 69–70.

Kronberg, P 1969. Gliederung, Petrographie & Tektogenese des Rhodopen – Kristallinsim Tsal-Dag, Simvolon und Ost-Pangäon in Griechisch-Mazedonien. Geotekt.Forsch., 31, 1–49.

Kronberg, P., Meyer, W.& Pilger, A. 1970. Geologie der Rila-Rhodope-Masse zwischen Strimon und Nestos (Nordgriechenland). Beih. geol. Jb., 88, 133–180.

Ktenas, K. 1906. Overthrusting in the greek mountains. Bulletin of Physiod. Soc., 6 (in greek).

Ktenas, K. 1907a. Die Einlagerungen im Krystallinen Gebirge der Kykladen auf Syra und Siphnos. Tschermaks miner, petrog. Mitt., 25, 257–320.

Ktenas, C. 1907b. La formation de la jadeite et des provinces mineralogiques sodiques dans les schistes cristallins. C. R. Acad. Sci. Paris, 147, p. 254.

Ktenas, C. 1907c. Sur l'age des terrains calcaires des environs d'Athenes. C. R. Acad. Sci., Paris, 144, p. 1235.

Ktenas, K. 1909. Uber die eruptiven Bildungen des Parnes gebirges in Attika. Centralblatt Mineralogie etc., p. 557.

Ktenas, C. 1924a. Formations primaires semimétamorphiques au Péloponnèse central. C.R. somm. Soc. Géol. France, 61–63.

Ktenas, C. 1924b. L'age des formations volcaniques du massif de Parnes (Attique). C. R. somm. Soc. Geol. France, p. 60.

Kumerics, C., Ring, U., Brichau, S., Glodny, J., & Monie, P. 2005. The extensional Messaria shear zone and associated brittle detachment faults, Aegean Sea, Greece. Journal of the Geological Society, 162 (4), 701–721.

Kydonakis, K., Gallagher, K., Brun, J.P., Jolivet, L., Gueydan, F. & Kostopoulos, D. 2014a. Upper Cretaceous exhumation of the western Rhodope metamorphic provence (Chalkidiki Peninsula, northern Greece).Tectonics, 33, 1113–1132, https://doi.org/10.1002/2014tc003572.

Kydonakis, K., D. Kostopoulos, M. Poujol, J. P. Brun, D. Papanikolaou & J. L. Paquette, 2014b. The dispersal of the Gondwana Super-fan System in the eastern Mediterranean: New insights from detrital zircon geochronology. Gondwana Res., 25, 1230–1241.

Kydonakis, K. Moulas, E., Chatzitheodoridis, E., Brun, J.P. & Kostopoulos, D. 2015. First report on Mesozoic eclogite-facies metamorphism preceding Barrovian overprint from the western Rhodope (Chalkidiki, northern Greece). Lithos, 220–223, 147–163.

Lalechos, N. in collaboration Christodoulou, G. 1973. Kato Figalia sheet. Geol. Map of Greece at scale 1/50,000, IGME.

Lekkas, E. 1986. The occurrence of clastic Triassic at the base of the Koziakas sequence, Western Thessaly. Geol. Geophys. Res., IGME, special vol. Prof. Papastamatiou, 235–242 (in greek).

Lekkas, E. 1987. Geological structure and geodynamic development of the Koziakas mountain range (Western Thessaly). PhD Thesis, University of Athens, 274 p. (in greek).

Lekkas, E., Danamos, G., Skourtsos, E. & Sakellariou, D. 2001. The occurrence and geodynamic significance of the volcano-sedimentary Tyros Beds (base of the Gavrovo-Tripolis Unit) on Rhodes Island, Dodekanese, SE Aegean. Bull. Geol. Soc. Greece, 34/1, 83–89.

Lekkas, S. 1978. Contribution to the geological structure of the area SE of Tripolis. PhD Thesis, University of Athens, 192 p. (in greek).

Lekkas, S. 1979. Observations sur le mecanisme du chiarriage de la nappe d' Arcadie sur la zone de Tripolitza. VI Coll. Aegean Region, Athens 1977, Proc. II, 675–680.

Lekkas, S. & Lozios, S. 2000. Tectonic structure of Mt Hymittos. Ann. Geol. Pays Hellen., 38, 47–62.

Lekkas, S. & Papanikolaou, D. 1978. On the phyilite problem in Peloponnesus. Ann. Géol. Pays Hellén., 29/1, 395–410.

Leleu, M. & Neumman, M. 1969. L' âge des formations d' Attique: du paléozoique au mésozoique. C.R. Acad. Sci. Paris, 268, 1361–1363.

Lepsius, R. 1893. Geologie von Attika. 196 S, Berlin.

Liati, A. & Mposkos, E., 1990. Evolution of the eclogites in the Rhodope Zone of northern Greece. Lithos, 25, 89–99.

Liati, A. & Seidel, E. 1996. Metamorphic evolution and geochemistry of kyanite eclogites in central Rhodope, northern Greece. Contr. Mineral. Petr.,123, 293–307.

Liati, A., Gebauer, D. & Fanning, C.M. 2011. Geochronology of the Alpine UHP Rhodope zone: a review of isotopic ages and constraints on the geodynamic evolution. In: Dobrzhinetskaya, L. F., Faryad, S.W., Wallis, S., Cuthbert, S. (Eds.), Ultrahigh-pressure Metamorphism 25 Years after the Discovery of Coesite and Diamond. Elsevier, Amsterdam, 295–324.

Liati, A., Skarpelis, N. & Fanning, C.M. 2013. Late Permian-Early Triassic igneous activity in the Attic Cycladic Belt (Attica): new geochronological data and geodynamic implications. Tectonophysics, 595–596, 140–147.

Liati, A., Theye, T., Fanning, M., Gebauer, D. & Rayner, D. 2016. Multiple subduction cycles in the Alpine orogeny as recorded in single zircon crystals (Rhodope zone, Greece). Gondwana Res., 29 (1), 199–207.

Lippolt, H.J. & Baranyi, 1. 1976. Oberkretazische Biotit und Gesteinsalter aus Kreta. N. Jb. Geol. Pal. Mh. 404–414.

Lozios, S. 1993. Tectonic analysis of the metamorphic formations of Northeastern Attica. PhD Thesis, University of Athens, 329 p. (in greek).

Lugeon, M. 1902. Les grandes nappes de recouvrement des Alpes du Chablais et de la Suisse. Bull. Soc. Géol. France (4), 1, 723–825.

Maganas, A. 2002. Constraints on the petrogenesis of Evros ophiolite extrusives, NE Greece. Lithos, 65, 171–182.

Maillot, H. 1979. Etude sedimentologique du «premier flysch» en Peloponnese occidentale (Grece): La Formation d' Andritsena. Bull. Geol. Soc. Greece, XIV, 1, 94–115.

Marakis, G. 1970. Geochronological dating of Macedonia granites. Ann. Géol. Pays Héllen., 21, 121–152 (in greek).

Marakis, G. 1972. Datation des roches des zones internes de la Grèce. C.R. Soc. Phy., Hist. nat. Genève, 7, 2–3, 52–58.

Maratos, G. 1960. The ophiolites of the Soufli area. Geol. Geoph. Res., IGME, 6/2, 83–178 (in greek).

Maratos, G. & Andronopoulos, B. 1964. Nouvelles donnees sur l'age des phyllites du Rhodope. Bull. Geol. Soc. Greece, 6/1, 113–132 (in greek).

Maratos, G. & Andronopoulos, B. 1965. The fauna discovered in the limestones of Aliki – Alexandroupolis (Rhodope phyllites). Bull. Geol. Soc. Greece, 6/2, 348–352 (in greek).

Marinos, G. 1955. Granitic occurrences in Argolis (Ermioni) and the age of the schist-psammitic complex. Bull. Geol. Soc. Greece, 2, 121–123 (in greek).

Marinos, G., & Petrascheck, W.E. 1956. Laurium. Geol. Gephys. Res., IGME, IV, 1–247 (in greek).

Marinos, G., Katsikatsos, G., Georgiadou-Dikeoulia, E. & Mirkou, R. 1971. The system of the Athens schists I: Stratigraphy and Tectonics. Ann. Géol.Pays Hellen., 23, 183–216 (in greek).

Marinos, G., Katsikatsos, G., & Mirkou-Peripopoulou, R. 1974. The system of the Athens schists II: Stratigraphy and Tectonics. Ann. Géol. Pays Héllen., 25, 439–444 (in greek).

Mariolakos, I. 1971. Tectonic analysis of the joint systems of North Hymetos. Ann. Géol. Pays Hellén., 23, 323–379 (in greek).

Mariolakos, I. & Papanikolaou, D. 1973. Observations on the structural geology of Western Pentelikon – Attica. Bull. Geol. Soc. Greece, X, 2, 134–179 (in greek).

Mariolakos, I. & Papanikolaou, D. 1982. Cretaceous Emplacement of the Ophiolites in Northern Greece and Related Tectonosedimentary Environments. Geol. Evol. East. Medit., Edinburgh 1982, Abstracts, p. 71.

Marks, P. & Schuiling, R.D. 1965. Sur la présence du Permien superieur non-métamorphique à Naxos. Proc. Acad. Athens, 40, 96–99.

Marnelis, P. & Bonneau, M. 1979. Stratigraphie et structure de I' île d' Astypalea. VI Coll. Geol. Aegean Region, Athens 1977, Proc. I, 323–332.

Martha, S. O., Doerr, W., Gerdes, A., Petschick, R., Schastok, J., Xypolias, P & Zulauf, G. 2016. New structural and U-Pb zircon data from Anafi crystalline basement (Cyclades, Greece): constraints on the evolution of a Late Cretaceous magmatic arc in the Internal Hellenidres. Intern. J. Earth Science, 105, 2031–2060.

Masson, H. 1972. Sur l'origine de la corngeule par fracturation hydraulique. Eclogae geol. Helv., 65/1, 27–41.

Masson, H. 1976. Un siecle de geologie des Prealpes: de la decouverte des nappes a la recherche de leur dynamique. Eclogae geol. Helv. 69/2, 527–575.

Matthews, A. & Schliestedt, M. 1984. Evolution of the blueschist and greenschist facies rocks of Sifnos, Cyclades, Greece. Contr. Miner. Petr., 88, 150–163.

Mavridis, A. & Matarangas, B. 1979. Aghiophyllo sheet. Geological Map of Greece at scale 1/50,000, IGME.

Mavridis A., Skourtsis-Koroneou, V. & Tsaila-Monopolis, S. 1979. Contribution to the geology of the Subpelagonian zone (Vourinos area, western Macedonia).VI. Coll. Geol. Aegean Region. Athens 1977, Proc. I, 175–195.

Meinhold, G. & Kostopoulos, D. 2013. The Circum-Rhodope Belt, northern Greece: Age, provenance and tectonic setting. Tectonophysics, 595–596, 55-68.

Melidonis, N. 1963. The geology of Anafi Island. Geol. Geophys. Res., IGME, 8, 61–308 (in greek).

Melidonis, N. 1980. The geological structure and the metalliferous occurrences of Tinos Island. Special Studies Geol. Greece, IGME, 13, 1–80 (in greek).

Mercier, J. 1966. Paléogéographie, orogenèse, métamorphisme et magmatisme des zones internes des Hellénides en Macédoine (Grece): vue d'ensemble. Bull. Soc.Géol. France, 8, 1020–1049.

Mercier, J. 1968. Etude géologique des zones internes des Hellénides en Macédoine centrale (Grèce). Ann. Géol. Pays Hellén., 20, 1–792.

Mercier, J. L. & Bebien, J. 1977. Zones du Paikon, de Peonias et Serbo-Macedonien. Bull. Soc. Geol. France, 19/1, 41–45.

Mercier, J. & Vergely, P. 1972. Les mélanges ophiolitiques de Macédoine (Grèce). Décrochements d' âge anté-crétacé supérieur. Zeit. deuts. Geol. Ges., 128, 469–489.

Mercier, J. & Vergely, P. 1975. Le massif du Paikon est-il une fenetre tectonique dans la zone d' Axios-Vardar (Macedoine, Grece)? C. R. Acad. Sci. Paris, 320, 765–772.

Mercier, J. & Vergely, P. 1977. La zone d'Almopias. Bull. Soc. Geol. France, 19/1, 34–40.

Mercier, J. & Vergely, P. 1984. Edessa sheet. Geological Map of Greece at scale 1/50,000. IGME.

Meyer, W., Pilger, A. 1963. Zur Geologie de Gebietes zwischen Strymon und Nestos (Rhodopen Massiv) in Griechisch Mazedonien. N. Jb. Geol. Paläont. Abh., 118, 272–280.

Michail, M., Pipera, K., Koroneos, A., Kilias, A. & Ntaflos, T. 2016. New perspectives on the origin and emplacement of the late Jurassic Fanos granite, associated with an intra-oceanic subduction within the Neo-Tethyan Axios-Vardar ocean. Int. J. Earth Sci., 105, 1965–1983.

Michard, A., Goffe, B., Liati, A. & Mountrakis, D. 1994. Blueschist-facies assemblages in the peri-Rhodopian zone and hints for an Eohellenic HP/LT belt in Northern Greece. Bull. Geol. Soc. Greece, 30, 185–192.

Michard, A., Feinberg, H. & Montigny, R.1998. Supra-ophiolitic formations from the Thessaloniki nappe (Greece) and associated magmatism: an intra-oceanic subduction predates the Vardar obduction. C. R. Acad. Sci., 327, 493–499.

Migiros, G. 1983. Geological study of the Kato Olympus area, Thessaly. PhD Thesis, University of Patras, 204 p (in greek).

Migiros, G. & Galeos, A. 1990. Tectonic and stratigraphic significance of the Ano Garefi ophiolitic rocks (northern Greece). In: Malpas, J., Moores, E.M., Panayiotou, A., Xenophontos, C. (Eds.), Ophiolites: Oceanic Crustal Analogues. Geological Survey Department, Nicosia, 279–284.

Migiros, G. & Tselepidis, V. 1990. Der erste Nachweis von Hallstater kalken in der Nord-Pindos Decke (NW Griechenland). N. Jahr. Geol. Paläont., H4, 248–256.

Migiros, G., Hatzipanagiotou, K., Gartzos, E., Serelis, K. & Tsikouras, B. 2000. Petrogenetic evolution of ultramafic rocks from Lesvos Island (NE Aegean, Greece). Chemie der Erde - Geochemistry, 60 (1), 27–46.

Minoux, L. 1981. Contribution à l' étude géologique des Hellénides: Stratigraphie, structure et métamorphisme de l' île d'Amorgos (Cyclades, Grèce). Thèse 3eme cycle, Université P. et M. Curie, Paris, 181 p.

Monopolis, D. & Bruneton, A. 1982. Ionian Sea (Western Greece): its structural outline deduced from drilling and geophysical data. Tectonophysics, 83, 227–242.

Mountrakis, D. 1983. The geological structure of the Northern Pelagonian zone and the geotectonic evolution of the Internal Hellenides. Habilitation Thesis, University of Thessaloniki, 289 p (in greek).

Mountrakis, D. 1984. Structural evolution of the Pelagonian zone in Northwestern Macedonia, Greece. Geol. Soc. London, Sp. Publ. 17, 581–590.

Mountrakis, D. 1986. The Pelagonian Zone in Greece: a polyphase-deformed fragment of the Cimmerian continent and its role in the geotectonic evolution of the eastern Mediterranean. J. Geol., 94, 335–347.

Mposkos, E. 1978. Diasporit und Smirgelforkommen der Insel Samos (Griechenland). In: 4th Intern. Congress for the Study of Bauxites, Proc. 2, 614–631.

Mposkos, E.D. & Kostopoulos, D.K., 2001. Diamond, former coesite and supersilicic garnet in metasedimentary rocks from the Greek Rhodope: a new ultrahigh-pressure metamorphic province established. Earth & Plan. Sci. Let., 192,497–506.

Mposkos, E. & Perraki, M. 2001. High pressure Alpine metamorphism of the Pelagonian allochthon in the Kastania area (Southern Vermion), Greece. Bull. Geol. Soc. Greece, 939–947.

Mutti, E., Orobelli, G. & Pozzi, P., 1970. Geological studies on the Dodekanese Islands (Aegean Sea). IX. Geological map of the Island of Rhodes (Greece). Explanatory notes. Ann. Géol. Pays Hellén., 22, 77–226.

Negris, Ph. 1906a. Sur la nappe charriée de Péloponnèse C.R. Acad. Sci. Paris, 142,182–184.

Negris, Ph. 1906b. Sur les racines de la nappe de charriage du Péloponnèse. C.R. Acad.Sci., Paris, 142, 308–310.

Negris, Ph. 1909. Sur les brèches de friction dans les surfaces de charriage du Péloponnèse. C. R. Acad. Sci. Paris, 148, 1700–1702.

Negris, Ph. 1915. Roches cristallophylliennes et tectonique de la Grèce. 123 p. Athènes.

Negris, Ph. & Boussac, J. 1914, Découverte du Miocène a Naxos (Cyclades). C.R.somm. Soc. Géol. France, (4), 14, 143–1.

Okay, A.I. & Monie, P. 1997. Early Mesozoic subduction in the Eastern Mediterranean: evidence from Triassic eclogite in northwest Turkey. Geology, 25, 595–598.

Okay, A.I., Sunal, G., Sherlock, S., Altiner, D., Tuysuz, O., Kylander-Clarkl, A.R.C., & Aygul, M. 2013. Early Cretaceous sedimentation and orogeny on the active margin of Eurasia: Southern Central Pontides, Turkey. Tectonics 32, 1247–1271.

Okrusch, M., Seidel, E. & Davis, E. 1978. The assemblage jadeite-quartz in the glaucophane rocks of Sifnos (Cyclades, Greece). N. Jb. Miner. Abh., 132, 284–308.

Palamakumbura, R. N., Robertson, A. H. F. & Dixon, J. 2013. Geochemical, sedimentary and micropaleontological evidence for a

Late Maastrichtian oceanic seamount within the Pindos ocean (Arvi Unit, S. Crete, Greece). Tectonophysics, 595–596, 250–262.

Papageorgakis, I. 1968a. About a Cretaceous outcrop on Paros Island. Proc. Acad. Athens, 43, 163–174 (in greek).

Papageorgakis, I. 1968b. Occurence of marine Miocene on Paros Island. Proc. Acad. Athens, 43, 368–376 (in greek).

Papanikolaou, D. 1976. The age of the metamorphics in Andros Island. Proc. Acad. Athens, 51, 292–301 (in greek).

Papanikolaou, D. 1977. Structural geology and tectonics of Paros Island (Aegean Sea). Ann. Geol. Pays Hellen., 28, 450–464.

Papanikolaou, D. 1978a. Andros Island (Gavrion, Andros and Piso Meria sheets). Geological Map of Greece, at scale 1/50,000, I. G.M.E.

Papanikolaou, D. 1978b. Contribution to the geology of Ikaria Island, Aegean Sea. Ann. Géol. Pays Hellén., 29/1, 1–28 (in greek).

Papanikolaou, D. 1978c. Geological Research in Andros Island. PhD Thesis, University of Athens, 231 p (in greek).

Papanikolaou, D. 1978d. Contribution to the Geology of Aegean Sea. The island of Andros. Ann. Géol. Pays Hellén., 29/2, 477–553.

Papanikolaou, D. 1979a. Unités tectoniques et phases de déformation dans l' île de Samos, Mer Egée, Grèce. Bull. Soc. Géol. France, (7), XXI, 6, 745–752.

Papanikolaou, D. 1979b. Stratigraphy and structure of the Paleozoic rocks in Greece: an introduction. In: Sassi, F.P. (Ed.), IGCP Project No. 5 "Correlation of Prevariscan and Variscan Events of the Alpine–Mediterranean Mountain Belt, 93–102.

Papanikolaou, D. 1980a. Contribution to the geology of the Aegean Sea. The island of Paros. Ann. Géol. Pays Hellén. 30/1, 65–96.

Papanikolaou, D. 1980b. Les écailles de Thymaena: témoins d' un mouvement tectonique miocène vers l'intérieur de l'arc égéen. C.R. Acad. Sc. Paris, 290, 307–310.

Papanikolaou, D. 1980c. The Metamorphic Hellenides. 26th Int. Geol. Congress., Paris, 1980, Abstracts, I, 371.

Papanikolaou, D. 1981a. Remarks on the kinematic interpretation of folds from some cases of the Western Swiss Alps and of the Hellenides. Ann. Géol. Pays Hellén., 30/2, 741–762.

Papanikolaou, D. 1981b. Some problems concerning correlations within the metamorphic belts of the Pelagonian and the Rhodope. In: Sassi F.P. (editor), IGCP Project No 5, Newsletter; 3, 119–123.

Papanikolaou, D. 1984a. Introduction to the Geology of Greece: The pre-Alpine Units. In: IGCP No 5, 1984 Field meeting in Greece, Fieldguide, Part, I, 3–35.

Papanikolaou, D. 1984b. Tectonic Evolution of the Hellenides. 27th Intern. Geological Congress, Moscow 1984, Abstracts III, 351–352.

Papanikolaou, D. 1984c. The three metamorphic belts of the Hellenides: a review and a kinematic interpretation. Geol. Soc. London, Spec. Publ., 17, 551–561.

Papanikolaou, D. 1986a. Late Cretaceous Paleogeography of the Metamorphic Hellenides. Geol. Geoph. Res., IGME, Special volume in honor of Prof. Papastamatiou, 315–328.

Papanikolaou, D. 1986b. The medial tectonometamorphic belt of the Hellenides. Bull. Geol. Soc. Greece, 20, 101–120 (in greek).

Papanikolaou, D. 1986c. Geology of Greece. Eptalofos Publications, 240 p. Athens (in greek).

Papanikolaou, D., 1987. Tectonic evolution of the Cycladic blueschist belt (Aegean Sea, Greece). In: Helgeson, H.C. (Ed.), Chemical Transport in Metasomatic Processes: NATO ASI Series, 429–450.

Papanikolaou, D. 1988a. Precambrian in the Hellenides (Pelagonian, Cyclades, Peloponnesus–Crete). In: Zoubek, V., Cogne, J., Kozhoukharov, D., Krautner, H. (Eds.), IGCP No 22, «Precambrian in Younger Fold Belts» , 821–840, Wiley and Sons, New York.

Papanikolaou, D. 1988b. The Rhodope Massif in Greece. In: Zoubek, V., Cogne, J., Kozhoukharov, D., Krautner, H. (Eds.), IGCP No 22, "Precambrian in younger fold belts", 765–778, Willey & sons, New York.

Papanikolaou, D. 1988c. Introduction to the Geology of Crete and Field guide. IGCP No 276, October 1988 FieldMeeting in Crete, 3–36.

Papanikolaou, D. 1989a. Occurrence of Arvi, Western Thessaly and Orliakas type formations in Argolis. Bull. Geol. Soc. Greece, 24, 71–84 (in greek).

Papanikolaou, D., 1989b. Are the medial crystalline massifs of the eastern Mediterranean drifted Gondwanian fragments? In: Papanikolaou, D., Sassi, F.P. (Eds.), Sp. Publ. Geol. Soc. Greece, 1,63–90 and IGCP 276, Newsletter, No 1, Athens.

Papanikolaou, D. 1990. Probable geodynamic interpretation of the schist-chert formations of the Hellenides. Bull. Geol. Soc. Greece, 24, 135–148 (in greek).

Papanikolaou, D. 1993. Geotectonic evolution of the Aegean. Bull. Geol. Soc. Greece, 28, 33–48.

Papanikolaou, D. 1996. Paros Island. Geological Map of Greece at scale 1/50,000. IGME.

Papanikolaou, D. 1997. The tectonostratigraphic terranes of the Hellenides. Ann. Géol. Pays Hellén., 37, 495–514.

Papanikolaou, D. 1999. The Triassic ophiolites of Lesvos Island within the Cimmeride orgenic event. E.U.G., 10, Strasbourg, Abs, 315.

Papanikolaou, D. 2009. Timing of tectonic emplacement of the ophiolites and terrane paleogeography in the Hellenides. Lithos, 108, 262–280.

Papanikolaou, D. 2013. Tectonostratigraphic models of the Alpine terranes and subduction history of the Hellenides. Tectonophysics, 595–596, 1-24.

Papanikolaou, D. & Baud, A. 1982. Complexes à blocs et séries à caractère flysch au passage Permien - Trias en Attique (Grèce orientale). 9e R.A.S.T., Paris 1982, p. 492.

Papanikolaou, D., Bargathi, H., Dabovski, C., Dimitriu, R., EL-Hawat, A., Ioane, D., Kranis, H., Obeidi, A., Oaie, C., Seghedi, A. & Zagorchev, I. 2004a. TRANSMED Transect VII: East European Craton–Scythian Platform–Dobrogea–Balkanides–Rhodope Massif–Hellenides–East Mediterranean–Cyrenaica. In: Cavazza, W., Roure, F., Spakman, W., Stampfli, G., Ziegler, P. (Eds.), The TRANSMED Atlas: the Mediterranean Region from Crust to Mantle. Springer-Verlag, Heidelberg.

Papanikolaou, D. & Demirtasli, E. 1987. Geological correlations between the Alpine segments of the Hellenides–Balkanides and Taurides–Pontides. In: Flügel, H.W., Sassi, F.P., Grecula, P. (Eds.), IGCP No 5, «Pre-Variscan and Variscan Events in the Alpine–Mediterranean Mountain Belts.» Alfa Publishers, Bratislava, 387–396.

Papanikolaou D. & Ebner, F. 1997.Tectono-stratigraphic maps and terrane descriptions. Ann. Géol. Pays Hellén., 37, 195–197.

Papanikolaou, D., Gouliotis, L. & Triantaphyllou, M. 2009. The Itea–Amfissa detachment: a pre-Corinth rift Miocene extensional structure in central Greece. Geol. Soc.,London, Sp. Publ., 311, 293–310.

Papanikolaou, D. & Lekkas, E., 1979. Lateral transition between the Pindos zone and the Western Thessaly unit in the area of Tavropos. Bull. Geol.,Soc. Greece, XIV, 1, 70–84 (in greek).

Papanikolaou, D., Lekkas, E. & Sakellariou, D. 1995. Tectonic units and terrane analysis in Rhodes and adjacent Dodekanese islands, Greece. XV Congress Carpatho-Balkan Geol. Assoc., Symposium on Tectono-Stratigraphic Terranes, Athens September 1995, Abstracts, p. 20.

Papanikolaou, D., Lozios, S., Souκis, K & Skourtsos, E. 2004b. The geological structure of the allochthon system of the «Schists of Athens» . Bull. Geol. Soc. Greece, 36, 1550–1559 (in greek).

Papanikolaou, D. & Nomikou, P. 1998. The Paleozoic of Kos: a low grade metamorphic unit of the basement of the external Hellenides terrane. IGCP 276, Newsletter 6, Sp. Publ. Geol. Soc. Greece, 3, 155–166.

Papanikolaou, D. & Panagopoulos, A. 1981. On the structural style of Southern Rhodope. Geol. Balc., 11.3., 13–22.

Papanikolaou, D. & Royden, L. 2007. Disruption of the Hellenic arc: Late Miocene extensional detachment faults and steep Pliocene–Quaternary normal faults — or what happened at Corinth? Tectonics, 26.

Papanikolaou, D., Sassi, F.P. & Skarpelis, N. 1982. Outlines of the Pre-Alpine Metamorphisms in Greece. In: Sassi & Varga (editors), I.G.C.P. No 5, Newsletter, 4, 56–62 and Ann. Géol. Pays Hellén., 31/1, 16-31.

Papanikolaou, D. & Sideris, C. 1979. Sur la signification des zones «ultrapindique» et «béotienne» d'après la géologie de la région de Karditsa: l'unité de Thessalie Occidentale. Eclogae Geol. Helv., 72, 251–261.

Papanikolaou, D. & Sideris, C. 1992. Introduction to the geology of Chios island—fieldguide. 6th Congress of the Geological Society of Greece on «The Geology of the Aegean» and IGCP 276, 1992 FieldMeeting, Athens, 15 p.

Papanikolaou, D. & Sideris, Ch. 1983. Le Paléozoique de l' autochtone de Chios: Une formation à blocs de type wildflysch d' âge Permien (pro parte). C. R. Acad. Sc.Paris, 297, 603–606.

Papanikolaou, D., Sideris. Ch., Marcopoulos, Th., Katerinopoulos, A. & DImitriadis, S. 1984. IGCP No 5, Field guide, Part, II, Description of the Itinerary (Stops 1-26) 36–62.

Papanikolaou, D. & Skarpelis, N. 1987. The blueschists in the external metamorphic belt of the Hellenides: composition, structure and geotectonic significance of the Arna Unit. Ann. Géol. Pays Hellén., 33, 47–68.

Papanikolaou, D. & Soukis, K. 2000. Two Mesozoic unconformities in the Chios Allochthon. IESCA, Izmir, Abstracts, p 24.

Papanikolaou, D. & Stojanov, R. 1983. Geological correlations between the Greek and the Yugoslave part of the Pelagonian Metamorphic Belt. In: Sassi F.P. (ed), IGCP No 5, Newsletter, 5, 146–152.

Papanikolaou, D. & Triantaphyllou, M. 2010. Tectonostratigraphic observations in the western Thrace Basin in Greece and correlations with the eastern part in Turkey. Geol. Balc., 39/1–2, 293-294.

Papanikolaou, D. & Triantaphyllou, M. 2013. Growth folding and uplift of Lower and Middle Pleistocene marine terraces in Kephalonia: implications to active tectonics. 4th Intern. INQUA meeting on Paleoseismology, Active Tectonics and Archaeoseismology, Aachen, Proc., 185–188.

Papanikolaou, D. & Vassilakis, E. 2010. Thrust faults and extensional detachment faults in Cretan tectono-stratigraphy: implications for Middle Miocene extension. Tectonophysics, 488, 233–247.

Papanikolaou, D. & Zambetakis-Lekkas, A. 1980. Nouvelles observations et datation de la base de la série pélagonienne (s.s.) dans la région de Kastoria, Grèce. C.R. Acad. Sc. Paris, 291, 155–158.

Papanikolaou, I. & Lekkas, E. 2001. Synsedimentary tectonism during the transition from carbonate to clastic sedimentation in the Ionian unit (Messolongi area). Bull. Geol. Soc. Greece, 34, 191–198 (in greek).

Papanikolaou, M., Papanikolaou, D. & Triantaphyllou, M. 2010. Post-Alpine Late Pliocene – Middle Pleistocene uplifted marine sequences in Zakynthos Island. Bull. Geol. Soc. Greece, 43/1, 475–485.

Papastamatiou, J. 1951. The emery of Naxos. Geol. Geophys. Res., IGME, 1, 37–69 (in greek).

Papastamatiou, J. 1958. On the age of the crystalline limestones of Thera Island. Bull. Geol. Soc. Greece, Ill, 1, 104–113 (in greek).

Papastamatiou, J. 1960. La Géologie de la région montagneuse du Parnasse – Kiona - Oeta. Bull. Soc. Géol. France, 2, 398–409.

Papastamatiou, J. 1963. Sur la présence de roches sédimentaires d' âge pré-triasique à Myconos (Archipel des Cyclades, Grèce). C. R. Acad. Sc. Paris, 256, 5167–5169.

Papastamatiou, J. & Tataris, A. 1963. Tectonics of the transitional sediments from the Parnassos-Giona zone to the Olonos-Pindos zone. Bull. Geol. Soc. Greece, 5, 183–188 (in greek).

Papastamatiou, I., Betoulis, D. & Tataris, A. 1962. Kallidromon mt.: its geology and its relation with Parnassos mt. Bull Geol. Soc. Greece, 5/1, 43–52 (in greek).

Papastamatiou, I., Bornovas, I., Betoulis, D. & Christodoulou, G. 1959. Ziros sheet, geological map at scale 1/50,000. IGME

Papastamatiou, J. & Reichel, M. 1956. Sur l'age des phyllades de l'ile de Crete. Eclog, Geol. Helv., 49, 147–149.

Papavasiliou, S. 1914. Die Smirgellagerstätten von Naxos, nebst denjenigen von Iraklia und Sikinos. Z. Deutsch. geol. Ges., 65, 1–123.

Paraskevaidis, I. & Chorianopoulou, P. 1978. A section from Aegaleo Mt, the Schist of Athens and the Athenian hills. Bull. Geol. Soc. Greece, 13/2, 116–141 (in greek).

Para, T., Vidal, O. & Jolivet, L. 2002. Relation between the intensity of deformation and retrogression in blueschists metapelites of Tinos Island (Greece) evidenced by chlorite-mica local equilibria. Lithos, 63, 41–66.

Patzak, M., Okrusch, M., & Kreuzer, H. 1994. The Akrotiri Unit on the island of Tinos, Cyclades, Greece: Witness to a lost terrane of Late Cretaceous age. Neues Jahrbuch fur Geologie und Palaontologie-Abhandlungen, 194(2), 211–252.

Pe-Piper, G. & Piper, D. 1984. Tectonic setting of the Mesozoic Pindos basin of the Peloponnese, Greece. Geol. Soc. London, Sp. Publ. 17, 563–567.

Pe-Piper, G. & Piper, D. 2002. The Igneous Rocks of Greece. The Anatomy of an Orogen. Gebrueder Borntraeger, Berlin/Stuttgart.

Perrier, R. & Koukouzas K. 1967.Philiates sheet. Geol. Map of Greece at scale 1/50,000, IGME.

Peytcheva, I., Von Quadt, A., Sarov, S., Voinova, E. & Kolcheva, K. 2009. Ordovician protoliths of metamorphic rocks in eastern Pirin-western Rhodopes: Are they part of the Ograzhden unit? Proc. An. Conf. Bulg. Geol. Soc., 17–18.

Philippon, M., Brun, J.P. & Gueydan, F. 2012. Deciphering subduction from exhumation in the segmented Cycladic Blueschist Unit (Central Aegean, Greece). Tectonophysics, 524–525, 116–134.

Philippson, A. 1892. Der Peloponnes. Verlag Friedländer, 642, S., Berlin.

Philippson, A. 1901. Beiträge zur Kenntnis der griechischen Inselwelt. Peterm. Milt. Erganzunheft,134, 1–172, Gotha.

Philippson, A. 1959. Die Griechischen Landschaften. Volumes I-V, V. Klostermann, Frankfurt.

Pichon, J.F. 1976. Condition de gisement des ophiolites sur la bordure occidentale du Vermion (zone pélagonienne, Grèce). Thèse Univ. Paris-Sud, 201 p.

Pickett, E. & Robertson, A. 1996. Formation of the Late Palaeozoic–Early Mesozoic Karakaya Complex and related ophiolites in NW Turkey by Palaeotethyan subduction–accretion. J. Geol. Soc., 153, 995–1009.

Pomonis, P., Tsitouras, B. & Hatjipanagiotou, K. 2002. Origin, evolution and radiometric dating of sub-ophiolitic rocks from the Koziakas ophiolite. N. J. Min. Abh., 177, 255–276.

Poulaki, E., Stockli, D., Flansburg, M. & Soukis, K. 2019. Zircon U-Pb chronostratigraphy and provenance of the Cycladic Blueschist Unit and the nature of the contact with the Cycladic basement on Sikinos and Ios Islands, Greece. Tectonics, https://doi.org/10.1024/2018.

Psonis, K. & Latsoudas, Ch. 1982. Kardamyli sheet. Geological Map of Greece at scale 1/50,000, IGME.

Reinecke, TH., Altherr, R., Hartung, B., Hatzipanagiotou, K., Kreuzer, H., Harre, W., Klein, H., Keller, J., Geenen, E. & Böger, H. 1982. Remnants of a Late Cretaceous High Temperature Belt on the Island of Anafi (Cyclades, Greece). N.Jb. Miner. Abh., 145, 2, 157–182.

Reischmann, T., Kostopoulos, D.K., Loos, S., Anders, B., Avgerinas, A. & Sklavounos, S.A. 2001. Late Palaeozoic magmatism in the basement rocks southwest of Mt Olympos, central Pelagonian zone, Greece: Remnants of a Permo-Carboniferous magmatic arc. Bull. Geol. Soc. Greece, 43/3, 985–993.

Renz, C. 1908. Sur les preuves de l'existence du Carbonifère et du Trias dans l' Attique. Bull. Soc. Géol. France. 8, 519–523.

Renz, C. 1940. Die Tektonik der griechischen Gebirge. Pragm. Akad. Athinon, 8.

Renz, C. 1955. Die vorneogene Stratigraphie der normal sedimentären Formationen Griechenlands. I.G.S.R., 637 p., Athens.

Rice, S.P., Robertson, A.H.F. & Ustaomer, T. 2006. Late Cretaceous–Early Cenozoic tectonic evolution of the Eurasian active margin in the Central and Eastern Pontides, northernTurkey. Geol. Soc., London, Sp. Publ., 260, 413–445.

Richter, D. & Mariolakos, I. 1972.Paläomorphologie und eozäne Verkastung der Gavrovo-Tripolis Zone auf dem Peloponnes (Griechenland). Bull. Geol. Soc. Greece, 9/2, 206–228.

Richter, D. & Mariolakos, I. 1974. Die Beziehung zwischen Tripolitza - kalk und Flysch in der Gavrovo — Tripolis zone im Gebiet Nördlich Argos (Peloponnes). Ann. Géol Pays Hellén., 25, 1–12.

Richter, D. & Mariolakos, I. 1975. Stratigraphische Untersuchungen an Kreide/Tertiär - Wende Im Gebiet von Delfi - Amfissa - Amfiklia (Parnass – Giona zone Griechenland). Ann. Géol Pays Hellén., 26, 417–434.

Ricou, L.E., 1980. La tectonique de coin et la genèse de l' arc égéen. Rev Géol. dyn. Géogr. Phys. 22, 147–155.

Ricou, L.E. & Godfriaux, I. 1995. Review on the Paikon multiple window and the structure of the Vardar in Greece. C. R. Acad. Sci. Paris, 321(7), 601–608.

Ricou, L. E., Burg, J. P., Godfriaux, I., & Ivanov, Z. 1998. Rhodope and Vardar: the metamorphic and the olistostromic paired belts related to the Cretaceous subduction under Europe. Geodinamica Acta, 11(6), 285–309.

Rimmelé, G., Oberhänsli, R., Candan, O., Goffé, B., & Jolivet, L. 2006. The wide distribution of HP-LT rocks in the Lycian Belt (Western Turkey): implications for accretionary wedge geometry. Geological Society, London, Special Publications, 260(1), 447–466.

Ring, U., Laws, S. & Bernet, M. 1999. Structural analysis of a complex nappe sequence and late orogenic basins from the Aegean island of Samos, Greece. Journal of Structural Geology, 21, 1575–601.

Robert, U. & Bonneau, M. 1982. Les basaltes des nappes du Pinde et d'Arvi (Crete) et leur signification dans l'evolution geodynamique de la Mediterranee orientale. Ann. Geol. Pays Hellen., 31, 373–408.

Robertson, A.H.F., Trivic, B., Deric, N. & Bucur, I. I. 2013. Tectonic development of the Vardar ocean and its margins: Evidence from the Republic of Macedonia and Greek Macedonia. Tectonophysics, 595–596, 25–54.

Roche, V., Laurent, V., Cardello, G. L., Jolivet, L., & Scaillet, S. 2016. Anatomy of the Cycladic Blueschist Unit on Sifnos Island (Cyclades, Greece). Journal of Geodynamics, 97, 62–87.

Roche, V., Jolivet, L., Papanikolaou, D., Bozkurt, E., Menant, A. & Rimmele, G. 2019. Slab fragmentation beneath the Aegean/Anatolia transition zone: insights from the tectonic and metamorphic evolution of the Eastern Aegean region. Tectonophysics, 754, 101–129.

Romano, S., Brix, M., Dorr, W., Fiala, J., Krenn, E. & Zulauf, G. 2006. The Carboniferous to Jurassic evolution of the pre-Alpine basement of Crete: constraints from U–Pb and U–(Th)–Pb dating of orthogneiss, fission-track dating of zircon, structural and petrological data. Geol. Soc., London, Sp. Publ., 260, 69–90.

Romano, S.S., Dorr, W. & Zulauf,G. 2004. Cambrian granitoids in the pre-Alpine basement of Crete (Greece): evidence from U-Pb dating of zircon. Intern. J. Earth Sciences, 93, 844–859.

Rosenbaum, G., Ring, U., & Kühn, A. 2007. Tectono-metamorphic evolution of high-pressure rocks from the island of Amorgos (Central Aegean, Greece). Journal of the Geological Society, 164 (2), 425–438.

Saccani, E., Bortolotti, V., Marroni, M., Pandofli, L., Photiades, A. & Principi, G. 2008. The jurassic association of back arc basin ophiolites and calc-alkaline volcanics in the Guevgeuli Complex (Northern Greece): implication for the evolution of the Vardar zone. Ofioliti, 33, 2, 209–227.

Sakellariou, D. 1989. Geologie des SerboMazedonischen Massivs in der Nordostlichen Chalkidiki, N-Griechenland. Deformation und Metamorphose. Dissertation Universitat Mainz, 177p.

Sannemann, W. & Seidel, E. 1976. Die Trias – Schichten von Rawdoucha / NW Kreta. Ihre Stellung im kretischen Deckenbau. N. Jb. Min. Abh., 126, 221–228.

Savoyat, E. et al. 1970. Raftopoulo Sheet. Geological Map of Greece, at scale 1/50,000, IGME, Athens.

Schardt, H. 1898. Les regions exotiques du versant N des Alpes suisses. Bull. Soc. Vaud. Sci. Nat., 34, 114–219.

Schermer, E.R. 1990. Mechanisms of blueschist creation and preservation in an A-type subduction zone, Mount Olympus region, Greece. Geology 18, 1130–1133.

Schermer, E.R., Lux, D.R. & Burchfiel, B.C. 1993. Temperature – time history of subducted continental crust, Mt Olympos region, Greece. Tectonics, 9, 1165–1195.

Schliestedt, M., Altherr, R. & Matthews, A. 1987. Evolution of the Cycladic crystalline complex: petrology, isotope geochemistry and geochronology. In: Helgeson, H.G. (Ed.), Chemical Transport in Metasomatic Processes, Reidel Publishers, Dordrecht, 389–428.

Schmitt, A. 1983. Nouvelles contributions à l' étude géologique des Pieria, de l'Olympe et le l' Ossa (Grèce du Nord). Thèse, Fac. Polyt. de Mons.

Schneider, D., Grasemann, B., Lion, A., Soukis, K. & Draganits, E. 2018. Geodynamic significance of the Santorini Detachment System (Cyclades, Greece). Terra Nova, 2018; 1–9, DOI: 10.1111/ter.12357.

Seidel, E., Okrusch, M., Kreuzer, H., Raschka, H.& Harre, W. 1976. Eo-Alpine Metamorphism in the Uppermost Unit of the Cretan Nappe System. Petrology and Geochronology. Contrib. Mineral. Petrol., 57, 259–275.

Seidel, E., Kreuzer, H. & Harre, W. 1977. K/Ar Dates of the Cretan Phyllite-Quartzite Series. VI Coll. Geol. Aegean Region, Athens, 1977, Proc. I, 121–127.

Seidel, E., Okrusch, M., Kreuzer, H., Raschka, H. & Harre, W. 1981. Eo-alpine metamorphism in the uppermost unit of the Cretan nappe system, petrology and geochronology: part 2. Synopsis of high temperature metamorphics and associated ophiolites. Contrib. Mineral. Petrol., 76, 351–361.

Seidel, E., Kreuzer, H. & Harre, W. 1982. A late Oligocene / Early Miocene High Pressure Belt in the External Hellenides. Geol. Jb. E 23, 165–206.

Sengor, A.M.C. 1984. The Cimmeride orogenic system and the tectonics of Eurasia. Geol. Soc. Am. Sp. Papers, 195, 74p.

Sengor, A.M.C., Yilmaz, Y., Sungurlu, O. 1984. Tectonics of the Mediterranean Cimmerides: nature and evolution of the western termination of Palaeo-Tethys. Geol. Soc. London, Sp. Publ. 17, 77–112.

Shaked, Y., Avigad, D. & Garfunkel, Z. 2000. Alpine high-pressure metamorphism at the Almyropotamos window (southern Evia, Greece). Geol. Mag., 137, 367–380.

Sharp, I.R. & Robertson, A.H.F. 2006. Tectonic-sedimentary evolution of the western margin of the Mesozoic Vardar Ocean: evidence from the Pelagonian and Almopias zones, northern Greece. Geol. Soc., London, Sp. Publ., 260, 373–412.

Skarpelis, N. 1982. Metallogeny of masive sulfide deposits and petrology of the external metamorphic belt of the Hellenides (SE Peloponnesus). PhD Thesis, University of Athens, 149 p (in greek).

Skourtsos, E., Alexopoulos, A., Zambetaki-Lekka, A. & Lekkas, S. 2001. The occurrence of the Internal Hellenides on Parnon mountain range, central eastern Peloponnesus. Bull. Geol. Soc. Greece, 34/1, 47–54 (in greek).

Skourtsos, E. & Lekkas, S. 2004. Kosmas-Gythio Unit: a metamorphic carbonate sequence overlying the phyllites-quartzites unit in southern Peloponnesus, Greece. Bull. Geol. Soc. Greece, 36, 1679–1687 (in greek).

Smith, A.G. & Woodcock, N.H., 1976. The earliest Mesozoic structures in the Othris region, east central Greece. Bull. Soc. Geol. France, 7, 245–251.

Smith, A.G., Hynes, A.J., Menzies, M., Nisbet, E.G., Price, I., Welland, M.J. & Ferriere, J. 1975. Stratigraphy of the Othris Mountains, eastern central Greece: a deformed mesozoic continental margin sequence. Eclogae Geol Helv., 68, 463–482.

Soldatos, I. & Christofides, G. 1986. Rb–Sr geochronology and origin of the Elatia pluton, Central Rhodope, North Greece. Geol. Balc., 16, 15–23.

Soukis, K. 2011. Deformation of granitic rocks in the Aegean area. PhD Thesis, University of Athens, 439 p (in greek).

Soukis, K. & Papanikolaou, D. 2004. Contrasting geometry between Alpine and Late- to Post-Alpine tectonic structures in Anafi Island (Cyclades). Bull. Geol. Soc. Greece, 36, 1688–1696.

Soukis, K. & Stockli, D.F. 2013. Structural and thermochronometric evidence for multi-stage exhumation of southern Syros, Cycladic Islands, Greece. Tectonophysics, 595–596, 148–164.

Spray, J.G. & Roddick, J.C., 1980. Petrology and $^{40}Ar/^{39}Ar$ Geochronology of some Hellenic sub-Ophiolite metamorphic Rocks. Contrib. Mineral. Petrol., 72, 43–55.

Stais, A. & Ferriere, J. 1991. Nouvelles donnees sur la paleogeographie Mesozoique du domain Vardarien: les bassins d'Almopias et de Peonias (Macedoine, Hellenides internes septentrionales). Bull. Soc. Geol. Grece, 26/1, 491–507.

Stampfli, G., Vavassis, I., De Bono, A., Rosselet, F., Matti, B. & Belllini, M. 2003. Remnants of the Paleotethys oceanic suture-zone in the western Tethyan area. Boll. Soc. Geol. Ital., sp. Vol. 2, 1–24.

Tataris, A. 1964. About the presence of the Olonos-Pindos zone in the Symi-Viannos area (Eastern Crete) and the age of its spilites. Proc. Acad. Athens, 39 (in greek).

Tataris, A. 1965. About the presence of Eocene in the semimetamorphic basement of Thera Island. Bull. Geol. Soc. Greece, 6, 232–238 (in greek).

Tataris, A. 1967a. New research on the structure of Salamis Island and the opposite area of Perama (Attica). Bull. Geol. Soc. Greece, 7/1, 36–51 (in greek).

Tataris, A. 1967b. Observations on the structure of the area Skaramaga-Aegaleo Mt-Pireas-Athens (Attica). Bull. Geol. Soc. Greece, 7/1, 52–88 (in greek).

Tataris, A. 1972. New results on the geology of Salamis Island and the area of Attica. Bull. Geol. Soc. Greece, 9/2, 482–514 (in greek).

Tataris, A. 1975. Some questions on the «itinerary» of the younger (sh2) schist-psammite-chert complex and the relation Pelion – Olympus. Bull. Geol. Soc. Greece, 12, 95–112 (in greek).

Tataris, A. & Christodoulou, G. 1965. On the geological structure of the Lefka Ori (Western Crete). Bull. Geol. Soc. Greece, VI/2, 319–347 (in greek).

Tataris, A. & Kallergis, G. 1965. The geological structure of Trapezona – Arachnaion massif and the Nafplion – Lygourio areas. IGME, Geol. Studies, 9/6, 195–220 (in greek).

Terry, J. 1971. Sur l'age triasique de laves associees á la nappe ophiolitique du Pinde septentrionale (Epire & Macedoine, Grece). C. R. som. Soc. Geol. France, 384–385.

Theodoropoulos, D. 1979. Limin Vatheos sheet. Geological Map of Greece at scale 1/50,000, IGME.

Thiebault, F. 1977. Etablissement du caractère ionien de la série des calcschistes et marbres (Plattenkalk) en fenêtre dans le massif du Taygète (Péloponnèse, Grèce). C. R. somm. Soc. Géol. France, 3, 159–161.

Thiebault, F. 1982. L' évolution géodynamique des Hellenides externes en Péloponnèse méridional. Publ. Soc. Géol. Nord., 6, 574 p.

Thiebault, F. & Kozur, H. 1979. Precisions sur l'age de la formation de Tyros (Paleozoique superieur – Carnien) & la base de la serie de Gavrovo-Tripolitza (Carnien) (Peloponnese meridional, Grece). C. R. Acad. Sc. Paris, 288, 23–26.

Tranos, M. D., Plougarlis, A. P. & Mountrakis, D. M. 2007. A new consideration about the Almopia – Paikon boundary based on the geological mapping of the area of Nerostoma-Lakka (Central Macedonia, Greece). Bull. Geol. Soc. Greece, 40/1, 488–499.

Tranos, M. D., Kilias, A. A. & Mountrakis, D. M. 1999. Geometry and kinematics of the Tertiary post-metamorphic Circum Rhodope Belt Thrust System (CRBTS), Northern Greece. Bull. Geol. Soc. Greece, 33, 5–16.

Triantaphyllou, M.V. 2013. Calcareous nannofossil dating of Ionian and Gavrovo flysch deposits in the External Hellenides Carbonate Platform (Greece): overview and implications. Tectonophysics, 595–596, 235–249.

Trikkalinos, J. 1942. Über die ob. Kreidetransgression auf den Kristallinen Schichten der insel Paros. Ann. Géol. Pays HeIIén.,1, 1–6.

Trikkalinos, J. K. 1955. Beiträge zur Erforschung des tektonischen Baues Griechenlands. Über das Alter der vortertiaren Schichten des Gebietes von Alexandroupolis–Didymotichon–Westthrazien. Ann. Géol. Pays Hellén., 6, 81–82.

Trotet, F., Jolivet, L. & Vidal, O., 2001. Tectono-metamorphic evolution of Syros and Sifnos islands (Cyclades, Greece). Tectonophysics, 338, 179–206.

Trotet, F., Goffé, B., Vidal, O. & Jolivet, L., 2006. Evidence of retrograde Mg-carpholite in the Phyllite–Quartzite nappe of Peloponnese from thermobarometric modelisation — geodynamic implications. Geodinamica Acta, 19 (5), 323–343.

Tsaila-Monopolis, S. & Galeos, A. 1984. Etude geologique et stratigraphique des ilots de Dodekanese, Ro et Strongyli. Correlation a l' ile de Kastellorizo. Marine Geology, 55/3–4, 479-486.

Turpaud, P. & Reischmann, T. 2010. Characterization of igneous terranes by zircon dating: implications for UHP occurences and suture identification in the central Rhodope, northern Greece. Intern. J. Earth Sciences, 99, 567–591.

Underhill, J. 1989. Late Cenozoic defromation of the Hellenide foreland, western Greece. Bull. Geol. Soc. Am., 101, 613–634.

Ustaszewski, K., Schmid, S.M., Lugovic, B., Schuster, R., Schategger, U., Bernoulli, D., Hottinger, L., Kounov, A., Fugenschuh, B. & Schefer, S. 2009. Late Cretaceous intra-oceanic magmatism in the internal Dinarides (northern Bosnia and Herzegovina): implications for the collision of the Adriatic and European plates. Lithos, 108, 106–125.

Valkaniotis, S., Ganas, A., Papathanassiou, G. & Papanikolaou, M. 2014. Field observations of geological effects triggered by the January-February 2014 Cephalonia earthquakes (Ionian Sea, Greece). Tectonophysics.

Van Der Maar, P.A. & Jansen, J.B. 1983. The geology of the polymetamorphic complex of Ios, Cyclades, Greece and its significance for the Cycladic Massif. Geol. Rundschau, 72, 1, 283–299.

Vergely, P. 1976. Chevauchement vers I' Ouest et retrocharriage vers I' Est des ophiolites: deux phases tectoniques au cours du Jurassique supérieur - Eocrétacé dans les Hellenides internes. Bull. Soc. Géol. France, 18, 233–246.

Vergely, P. 1984. Tectonique des ophiolites dans les Hellenides internes. Conséquences sur l' évolution des régions téthysiennes occidentales. Thèse, Univ. Paris - Sud, vol. 1, 2.

Vergely, P. & Mercier, J.L. 2000. Donnees nouvelles sur les chevauchements d' age post-Cretace superieur dans le massif du Paikon (zone de l' Axios-Vardar, Macedoine, Grece): un nouveau modele structural. C. R. Acad. Sci., Paris, 330, 555–561.

Vicente, J.C. 1970. Etude geologique de l' ile de Gavdos (Grece), la plus meridionale de l' Europe. Bull. Soc. Geol. France, 7, 12, 481–495.

Von Braun, E. 1968. Die mesozoischen Hüllgesteine der SE-Rhodopen in Westthrazien, (Griechenland). Geol. Jb., 85.

Voreadis, G. 1932. The schist-chert complex of Salamis and its mafic and ultramafic erruptions. Greek Geological Service, Publ. 19 (in greek).

Voreadis. G. 1938. Neocimmerian folding in Eastern Greece and the phases of the Alpine orogeny in Greece. Habilitation Thesis, Athens (in greek).

Vrielynck, B. 1982. Evolution paleogeographique et structurale de la presque ile d'Argolide (Grece). Rev. Geol. Dyn. Geogr. Phys., 23/4, 277–288.

Wachendorf, H., Best, G. & Gwosdz, W. 1975. Geodynamische Interpretation Ost-Kretas. Geol. Rundschau, 64, 728–750.

Wawrzenitz, N. & Mposkos, E. 1997. First evidence for Lower Cretaceous HP/HT metamorphism in the eastern Rhodope, North Aegean region, North-east Greece. Europ. J. Mineral., 9, 659–664.

Wijbrans, J. R. & McDougall, I. 1988. Metamorphic evolution of the Attic-Cycladic Mretamorphic Belt on Naxos (Cyclades, Greece) utilizing 40Ar/39Ar age spectrum measurements. J. Metam. Geol., 6, 571–594.

Wunderlich, H.G. 1971. Dinariden, Helleniden, Minoiden - Ammer Kungen zur vergleichenden Geodynamik des ostmeriterranen Raumes. N. Jb. Geol. Paläont. Mh, 566–579.

Xypolias, P., Dorr, W. & Zulauf, G. 2006. Late Carbgoniferous plutonism within the pre-Alpine basement of the External Hellenides (Kithira, Greece); evidence from U-Pb zircon dating. J. Geol. Soc., London, 163, 539–547.

Yarwood, G.A. & Aftalion, M. 1976. Field relation and U-Pb geochronology of a granite from the Pelagonian zone of the Hellenides (High Pierria, Greece). Bull. Soc. Géol. France, (7), 18, 2, 259–264.

Yarwood, G.A. & Dixon, J.E., 1977. Lower Cretaceous and younger thrusting, in the Pelagonian Rocks of the High Pieria, Greece. VI Coll. Geol. Aegean Region, Athens 1977, Proc. I, 269–280.

Zeffren, S., Avigad, D., Heimann, A. & Gvirtzman, Z., 2005. Age resetting of hanging wall rocks above a low-angle detachment fault: Tinos Island (Aegean Sea). Tectonophysics, 400(1–4), 1–25.

Zervas, S. 1980. Age determination by the ^{87}Rb-^{87}Sr method of some pegmatites in the area of Lagada (Macedonia, Greece). Ann. Géol. Pays Hellén, 30/1, 143–153.

9 The Pre-orogenic Evolution of the Hellenides—Paleogeographic Reconstruction

9.1 Incorporating the Tectonostratigraphic Terranes in the Tethys Region

The reconstruction of the paleogeographic organization of the Hellenides can be accomplished with the restoration of the large geotectonic units in their initial location and the gradual description of their evolution, from the primary early stages of Tethys during the Early Triassic until today. The contemporary location of the Hellenides in the European margin had created an illusion in the early views of the geoscientists, that the paleogeographic region of the Hellenides was related to southern Europe, even though they belong to the southern orogenic chain of the Alpine system, where similarities with the units of the Taurides to the east had already been identified, showing clearly Gondwanian features. This was true for both the Mesozoic—Lower Cenozoic sequences, but also for the Paleozoic ones, which were unsuccessfully correlated with Variscan Europe. The application of the new theory of the tectono-stratigraphic terranes since the 1980s in the Mediterranean area had a dramatic impact and changed this outdated point of view. Thus, the Hellenides are now considered to be paleogeographic regions of the Tethyan Ocean, but they also comprise fragments of pre-Alpine African crust, which were rifted and detached from the northern margin of Africa and then gradually drifted through Tethys until they were finally integrated in the European margin (Papanikolaou 1989, 2013). This new notion, combined with the three stages of their paleogeographic and paleo-geodynamic evolution previously described, provides the possibility of their relative incorporation in the Tethyan paleogeographic region. Additionally, it offers the possibility to propose the chronological succession of the events, based on certain rational models and not just a random succession of geotectonic movements and processes.

Therefore, the beginning of the stratigraphic column beneath the carbonate platform of every continental terrane defines its position in Africa during the initiation of the rifting stage, while the beginning of the column of each oceanic basin defines the initiation of the embryonic stage of rifting, prior to the complete opening of the basin and its oceanization. The diagrams of the respective tectonostratigraphic models of the continental and oceanic terranes of the Hellenides (Fig. 8.1) provide data for their placement in a synthetic map (Fig. 7.11). The final stage of the integration of each terrane in the European margin can be documented even better, both by the flysch-type sedimentation as well as by the other orogenic tectonic events, that is, the molassic sedimentation, the metamorphism, and the magmatism—volcanism within the orogenic arc.

Naturally, during the entire procedure of the palinspastic paleogeographic reconstructions, the syn-orogenic movements within the arc are taken into account, both at shallow and deep level. Thus, outcrops of metamorphic units in tectonic windows and their equivalent non-metamorphic units, belonging, however, to the same pre-orogenic paleogeographic region can now be correlated (e.g. the Mani and Ionian units) offering a complete view of the geotectonic evolution.

9.2 Incorporating the Tectonic Units in Their Pre-orogenic Paleogeographic Location

During the orogeny, the pre-existing paleogeographic organization is terminated and highly deformed through the creation of tectonic nappes, disrupting various segments of the pre-orogenic area, each of which follows its own, unique orogenic evolution. Therefore, adjacent segments of the pre-orogenic region may end up in different positions inside the island arc or the rest synorogenic and post-orogenic areas behind it. At the same time, a tectonic unit may remain in almost surface conditions, experiencing only the shallow geodynamic phenomena of the orogenic arc with a simple deformation, while another unit may be subducted several tens of kilometers deep under the island arc and then undergo intense metamorphism and deep ductile deformation, followed by more shallow deformation structures during its exhumation to the surface.

D. I. Papanikolaou, *The Geology of Greece*, Regional Geology Reviews,
https://doi.org/10.1007/978-3-030-60731-9_9

A vital factor regarding the creation of large or small tectonic units from respective segments of the pre-orogenic paleogeography and the resulting simple or complex, shallow or deep level evolution of each unit inside the orogenic arc, is the pre-existing discontinuities—lateral transitions. For example, a boundary separating the shallow water carbonate platform from the adjacent oceanic basin, with pelagic-abyssal sedimentation forms a common discontinuity. The pre-existing heterogeneity in the lithology of the upper crust defines where the stresses will be relieved in the orogenic arc, usually resulting in certain stratigraphic columns, of the tectonic nappes, reflecting the sedimentary facies corresponding to the pre-orogenic paleogeography. Thus, the stratigraphic correlation between the corresponding formations of the adjacent tectonic units is a major source of information. The main point though is to define—mainly chronologically—the destruction of the pre-existing paleogeography during its insertion in the orogenic arc. Therefore, the pre-orogenic paleogeographic organization is usually preserved until the flysch sedimentation stage in the paleo-trench area. The end of the flysch sedimentation usually coincides with the detachment of the unit from its basement and the creation of independent tectonic units. Later on, further disruption and creation of new, progressively smaller, tectonic units during the subsequent orogenic stages until the extensional detachments at the inner arc region, create the final present tectonic units. Thus, the tectonic boundaries of the tectonic units are formed either through the early compressional thrusting of through the late extensional normal faulting. The transition from the pre-orogenic sedimentation to the synorogenic sedimentation of flysch is the first and main criterion for the paleogeographic organization. This transition shows the input sequence of the various isotopic zones inside the orogenic wedge of the arc, from the first paleogeographic areas that were incorporated in the orogenic arc to the more recent ones. This phenomenon is known as **orogenic migration** from the interior to the exterior of the arc. A good example is the case of the Pindos and Parnassos units, which enter the paleotrench in the Maastrichtian–Danian, while the Gavrovo and Ionian units enter much later in the Lutetian, as described in a chronological table by Aubouin (1959) (Fig. 9.1).

If the age of the flysch is similar or when there are no accurate chronological data, due to metamorphism for example, then the effort is focused in defining the age of

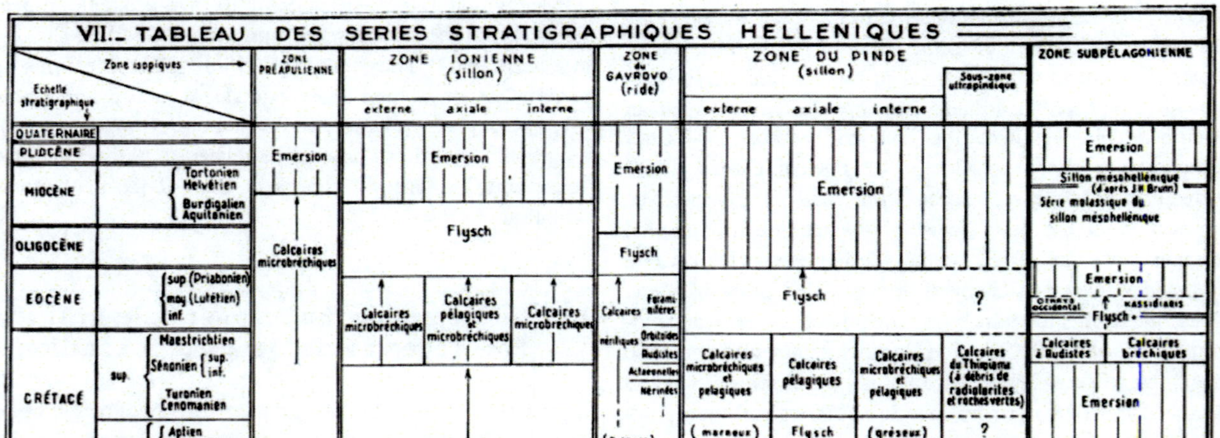

Fig. 9.1 Table showing the orogenic migration in the Hellenides, based on the flysch ages and on the subsequent emersion of the Hellenides (from Aubouin 1959)

metamorphism, deformation and magmatism, within the context of the tectono-metamorphic evolution of each unit. This will result into a logical sequence of geodynamic events in the orogenic arc, in agreement with a kinematic model, thus reconstructing the initial paleo-geographic location with higher or lower certainty. It is evident that the lesser data we dispose, the more alternative hypotheses can be formulated.

The general assumption is that the definition of the pre-orogenic paleo-geography is based almost exclusively on the tectonic analysis and the resulting tectonic evolution. It is characteristic that prior to the discovery of the metamorphic Hellenides, the tectonic basis of the methodology about the paleogeographic reconstruction was the simple retro-motion of each tectonic nappe next to the underlying tectonic unit. Thus, the paleogeographic organization was absolutely the same with the contemporary geotectonic zonation, as a result of the false interpretation globally known as "cylindrism" (Fig. 9.2).

Today, it is known that many units have been buried under tectonic nappes and that the tectonic transport along the thrusts can differ significantly up to tenfold (from 10–20 km up to 150 km or more). Therefore, today's adjacent position of two tectonic units does not pre-conceive the paleogeographic neighborhood of them. The criteria for the tracking of large- or small- scale thrusts are focused on the differences between the geodynamic deep and shallow level phenomena of the two units, separated by the thrust and the characteristics of their shear zones (Papanikolaou 1984c). Consequently, the following cases can be interpreted as large-scale thrusts: (1) The observation of the Pindos nappe over the Ionian unit, showing one unit, with the beginning of flysch sedimentation in the Maastrichtian–Danian, over another unit, with the beginning of flysch sedimentation in the Late Eocene–Oligocene, i.e. with a time difference of more than 25 million years. (2) The observation of the non-metamorphic Eastern Greece unit in the Cyclades–Cycladic unit—over the metamorphic units of Northern and Southern Cyclades. This superposition shows units that experienced only shallow conditions and units found at depths of at least 40–60 km respectively during the Eocene, when their blueschist assemblages were formed. (3) The observation of the Axios ophiolite nappe over the metamorphic carbonate platform of the Almopia unit or over the non-metamorphic Sub-Pelagonian unit, showing an oceanic region over a shallow carbonate platform of a (micro-)continental margin. On the other hand, the following cases can be considered as small-scale thrusts: (1) The observation of the Parnassos unit on the Pindos unit, with the transitional Vardousia unit imbrications in between, which share the same age for their flysch sedimentation and diachronically transitional slope facies between the two paleogeographic areas (Celet 1979). (2) The observation of the Western Thessaly unit over the Pindos unit, along the Koziakas mountain range, where the lateral transition has been described in the Late Cretaceous–Eocene, between the Upper Cretaceous pelagic limestones and the Paleocene-Eocene flysch of both units (Papanikolaou and Lekkas 1979).

The complexity of the palinspastic procedure, aiming to reconstruct the pre-orogenic paleogeography and the evolution of the orogenic events, becomes much greater if models including two or more opposed convergent plate boundaries are used, as this was proposed for the paleo-Alpine phase of the Late Jurassic (Hynes et al. 1972; Smith 1979, 1993; Vergely 1976; Robertson 2004; Rassios and Smith 2000; Rassios and Moores 2006; Kaplanis et al. 2013). Such an opposite direction of tectonic emplacement was proposed for certain cases of ophiolite nappes for the Late Jurassic events. However, even if these interpretations are valid, they represent an intra-oceanic compression event of local significance, involving minor displacement of the ophiolite nappes up to the adjacent (micro-) continental margin, without the creation of an orogenic arc and transition of the passive margin into an active one. This argument is validated by the absence of other geodynamic phenomena, such as arc volcanism, metamorphism, morphogenesis, etc. in the overriding plate.

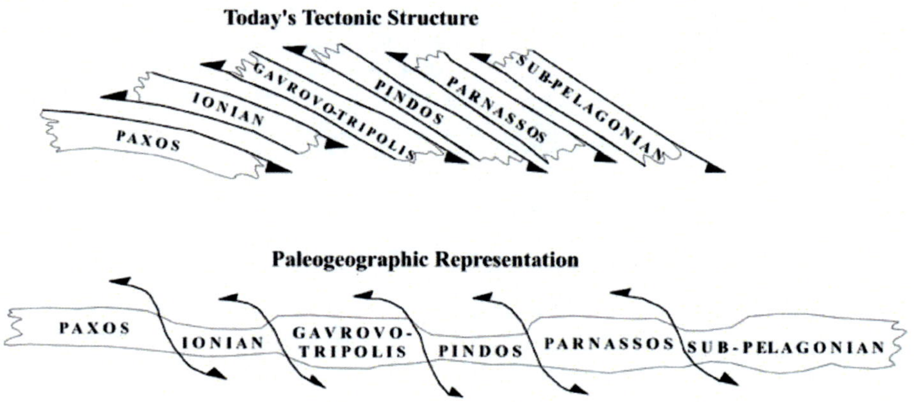

Fig. 9.2 Schematic representation of cylindrism in the Hellenides, where each tectonic nappe was considered as paleogeographically originating from the area directly adjacent to its relatively autochthonous unit (from Papanikolaou 1986c)

9.3 Pre-orogenic Paleo-Geographic Organization of the Hellenides

9.3.1 The Geosynclinal Period

During the end of the 1950s, Aubouin (1959) created a model based on the existing views regarding the paleogeography of the Hellenides, by using his own observations in addition to previous ones and especially those made by Philippson (1898, 1959) and Renz (1940, 1955). The paleogeographic organization of the Hellenides was also proposed by Aubouin (1965) as an international model that can be used as a pattern for the organization of the geosynclines (Fig. 9.3). This model distinguished the following domains:

(i) the foreland, represented by the pre-Apulian zone (Paxos),
(ii) the miogeosynclinal furrow, represented by the Ionian zone,
(iii) the miogeosynclinal ridge, represented by the Gavrovo zone,
(iv) the eugeosynclinal furrow, represented by the Pindos and Sub-Pelagonian zones with the ophiolite extrusions,
(v) the eugeanticlinal ridge, represented by the Pelagonian zone,
(vi) the oceanic furrow, represented by the Axios/Vardar zone.

The foreland (Apulia) is considered as a continental region (Hochkraton), while an oceanic region (Tiefkraton) is developed after the eugeanticlinal ridge, a region representing the Axios zone. Finally, Rhodope represents the "intermediate mountains" (zwischengebirge), where we have the axial divergence zone of the southern mountain chain of Tethys (Hellenides, Dinarides) from the northern chain (Balkanides).

In a series of thirteen transverse paleo-geographic cross sections of the Hellenides, which represent various periods, from the Late Triassic until the Middle Miocene, Aubouin (1959) records the successive phases of the—then—known geotectonic zones (from which the Parnassos-Giona zone had been omitted) (Fig. 9.4). His methodology is based on the major paleogeographic character of each geotectonic unit/zone, as reflected by the sedimentary facies and (micro-) paleontological findings of its stratigraphic column.

The major evolutional characteristic of the Hellenides is their configuration mainly during the Late Lias, when the miogeosynclinal furrow of the Ionian zone was created. The development of the isopic zones remains the same later on, with almost the same stratigraphic/sedimentary facies until their gradual participation in the orogenic wave. The orogeny is shown to begin during the Barremian in the eugeanticlinal ridge of the Pelagonian, and gradually migrates until the foreland during the Middle Miocene. The main difference between the mio- and the eu-geosynclinal furrow of Aubouin (1959, 1965) is the lack of ophiolites in the Ionian zone, while a secondary difference is the younger flysch sedimentation and overall tectogenesis of the Ionian in respect to the Pindos zone.

The information presented in this sub-chapter clearly does not reflect the present day realities and views. However, they may be useful for those studying the evolution of geology through the years and also for all scientists retreating the old publications, in order to be in line with the existing theories of the time and obtain a better understanding of the data and views presented (see also sub-chapters 4.1 and 7.1).

9.3.2 The Plate Tectonics Period

The previous models regarding the paleogeographic organization of the Hellenides have been adapted to the plate tectonics theory in the 1970s (Dercourt 1970, 1972; Aubouin

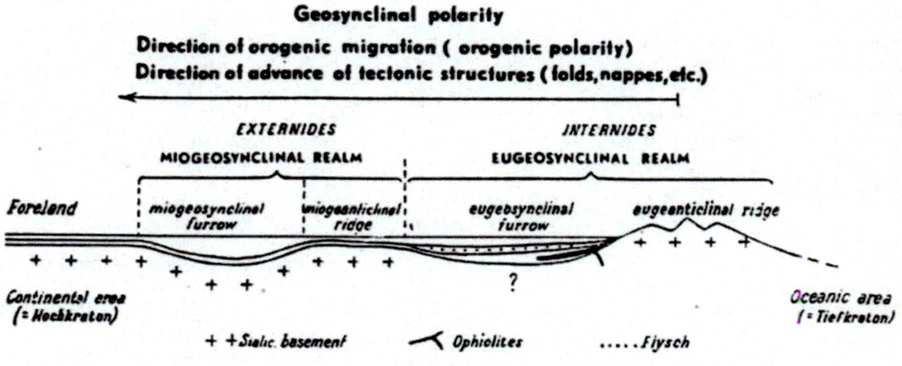

Fig. 9.3 Diagram of a transverse section of the Hellenides as a model of the geosyncline organization, according to Aubouin (1965)

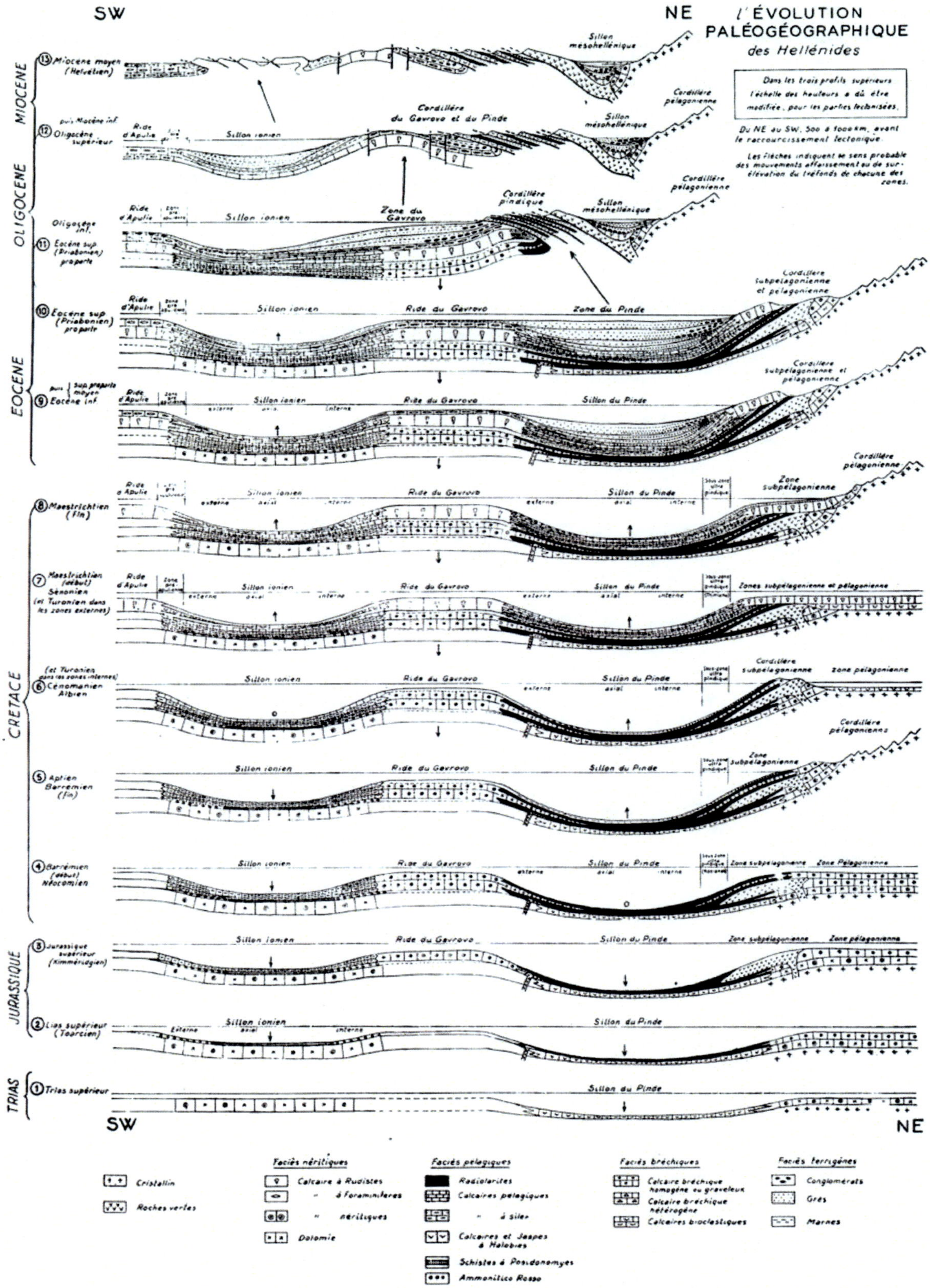

Fig. 9.4 The paleogeographic evolution of the Hellenides, from the Triassic to the Miocene (from Aubouin 1959)

1974, 1977; Aubouin et al. 1979; Jacobshagen et al. 1976; 1978; Jacobshagen 1979). Thus, Jacobshagen (1979, 1986) comes to a conclusion with a similar succession of paleogeographic regions, but with the addition of some units that were distinguished in the meantime, as well as a similar succession of the various stages of tectogenesis, in a series of four schematic cross sections of the Hellenides (Fig. 9.5).

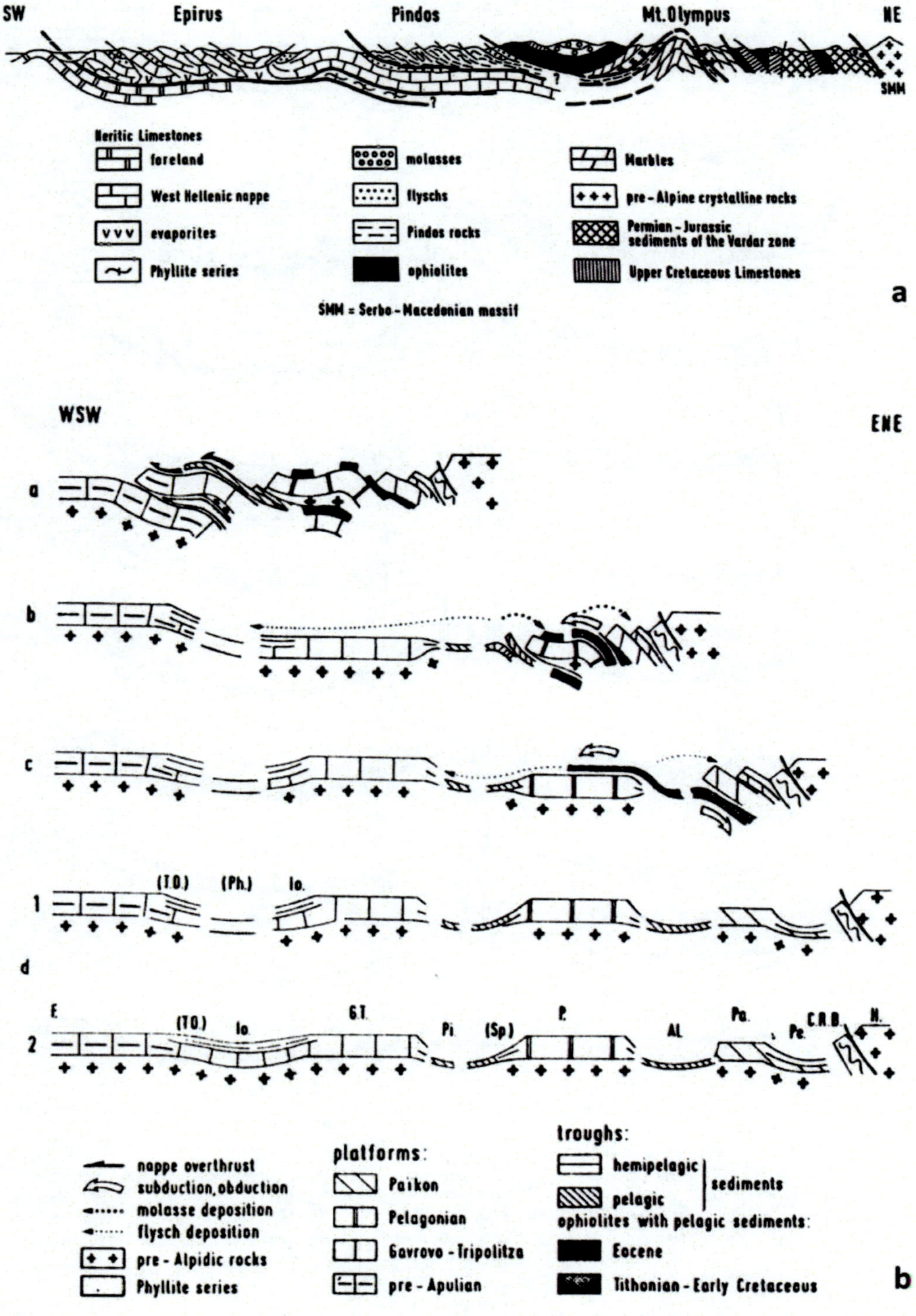

Fig. 9.5 **a** Cross—section of the Hellenides through northern Greece (based on Aubouin 1974, re-interpreted from Jacobshagen 1979). **b** Stages of the paleogeographic—orogenic evolution of the Hellenides, according to Jacobshagen (1979). **a**: Middle Miocene, **b**: Eocene, **c**: Tithonian—Early Cretaceous, **d**: Early Malm. The phyllitic unit (Arna) is considered as a separate furrow (1) or as the basement of the carbonate sediments of the Ionian—Gavrovo (2)

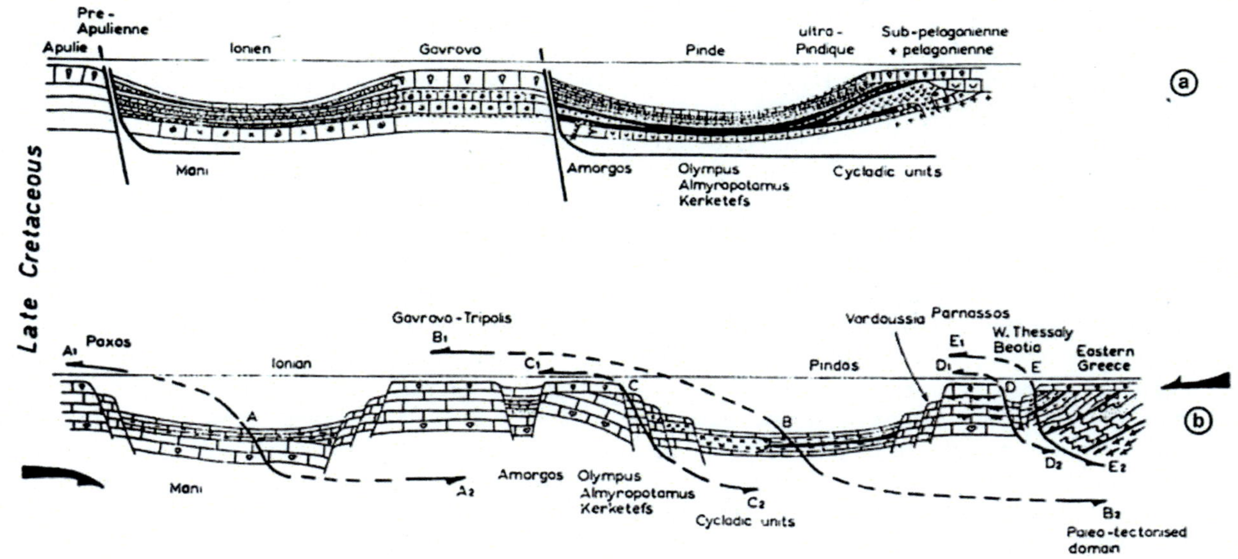

Fig. 9.6 Schematic paleo-geographic representation of the Hellenides in a transverse section during the Late Cretaceous. **a**: According to Aubouin (1959), modified so that the two main break zones could be visible, where the metamorphic Hellenides should be placed. **b**: According to Papanikolaou (1984b, 1986a), indicating the paleogeographic organization involving also the metamorphic Hellenides and the overall relative tectonic transport along the overthrusts that created the tectonic nappes of the External Hellenides during the Eocene–Miocene

Therefore, the data from the Axios zone towards the interior of the arc have been completed, as proposed by Mercier (1966) and from the Circum-Rhodope and Serbo-Macedonian zones as proposed by Kockel and Walther (1965) and Kauffmann et al. (1976). Additionally, the external section of the arc has been completed with the Talea Ori (Mani unit), as proposed by Epting et al. (1972) and Thiebault (1977), which was placed between the pre-Apulian and the Ionian zones. Lastly, two possible locations for the phyllitic zone (Arna unit) are noted; the first as a separate basin between the Ionian and the pre-Apulian zones and the other as the basement of the Ionian zone. The main difference of the paleo-geographic sections of the Hellenides by Jacobshagen in between the relation to the previous ones made by Aubouin is the interrupted continuity between several adjacent units. This is due to the realization that there are many units, mainly metamorphosed, which should be placed in between the until then known non-metamorphic units, as this was already the case in the example of the metamorphic Ionian–Mani unit.

Since the end of the 1970s for each new tectonic unit that is introduced, especially in the metamorphic domain, a paleo-geographic position is proposed. This is often different according to various researchers of even by the same researcher in different times, depending on the data availability.

In a first synthesis of the metamorphic Hellenides by Papanikolaou (1986a), there is a systematic analysis—investigation regarding their paleogeographic position inside the classical figure of the non-metamorphic Hellenides, established by Aubouin (1959) (Fig. 9.6).

The Late Cretaceous was chosen as the period of the paleo-geographic representation because: (1) After the Late Cretaceous the paleo-geography was disrupted by the main Alpine orogenic phase, and (2) Prior to the Late Cretaceous, the paleo-Alpine phase was taking place, during which many other units are involved and more specifically, the issue of the ophiolite sutures arises, that produces a large number of alternative hypotheses. Thus, the chosen period of the Late Cretaceous encounters the entire area of the Internal Hellenides as an already tectonised domain, whose most external region is the shallow water carbonate platform of the Eastern Greece composite unit.

From the four lateral transitions between the ridges and the furrows of the Hellenides shown by Aubouin (1959) in his figure, two regarding the Ionian/Gavrovo and Pindos/(Parnassos) UltraPindic are retained as possible, while the other two, regarding the pre-Apulian/Ionian and Gavrovo/Pindos ones, are considered as large tectonic breaks of the paleo-geographic organization, in which several units of the metamorphic Hellenides should be placed. More specifically:

(1) In the **Paxos (pre-Apulian) / Ionian** boundary, the Mani unit should be inserted, which litho-stratigraphically comprises an intermediate space

between the two units. Its present tectonic position in the form of tectonic windows under the Ionian, Gavrovo–Tripolis, and Pindos nappes in Peloponnese–Crete–Dodecanese is explained by its subduction during the Oligocene—Early Miocene and its consequent exhumation in the Middle/Late Miocene. Thus, the paleo-geographic boundary between the Mani and the Ionian units, shown in the section as A, corresponds to the two opposite displaced tectonic margins, A1 of the Ionian unit over the Paxos (e.g. Zakynthos, Cephalonia, Lefkada), which moved at shallow level, and A2 of the Mani unit under Tripolis (e.g. Taygetus and Eastern Parnon mountains), which moved at a depth of more than 15–20 km and acquired its crystallinity and ductile/flow deformation (Fig. 9.6). Today, in this particular break of the Hellenides paleogeography, the Western Crete unit has been also inserted (Papanikolaou 1988c; Papanikolaou and Vassilakis 2010) between the Mani and Ionian units.

(2) The **Ionian/Gavrovo** boundary is probably a lateral paleo-geographic transition, initially created during the rifting stage of the external carbonate platform that created the Ionian and Mani units in the Late Lias. This seems probable due to the existence of a continuous flysch formation over the two units in Epirus–Western Sterea, without any observation of a significant thrust of the Gavrovo unit over the Ionian one. Especially in the area between Messolonghi and Klokova-Varasova, the basal formation of the flysch (a), which can be observed over the pelagic limestones of the Ionian unit, is absent over the neritic limestones of Gavrovo (BP 1971) (see also Fig. 8.17). Therefore, it is possible that a buried thrust under the formation b of the flysch exists. This blind thrust may constitute syn-sedimentary lateral tectonism during the timeframe of the sedimentation of the formation (a) of the flysch and it is possibly related to the micro-emergence events observed in Klokova during the Late Eocene, with bauxites and Microcodium horizons (Fleury 1980).

(3) In the **Gavrovo / Pindos** boundary, the Amorgos, Olympus- Almyropotamos–Kerketeas, and the Cycladic blueschist units (mainly the Northern Cyclades) should be inserted. The Amorgos and Olympus–Almyropotamos–Kerketeas units were a region of neritic carbonate sedimentation with Nummulites during the Middle-Late Eocene, when the flysch sedimentation had already ended in the Pindos unit and its tectonism had started. Additionally, the stratigraphy of the Cycladic units seems to reach up to the Upper Cretaceous, and the fact that they have continuous stratigraphic columns, places them among the External Hellenides. This is supported by the contemporaneous transition of the Pindos flysch with that of Parnassos (Celet 1962, 1979) as well as that of Western Thessaly (Papanikolaou and Lekkas 1979). The lateral transitions documented between the above three tectonic units as well as with the Eastern Greece area endorce the same conclusion. Moreover, the geochronological dating of the blueschists in the Cyclades at 70–40 Ma, with an average age in the Late Eocene (Altherr et al. 1979, 1982), suggest that their stratigraphic columns enter the Early Cenozoic with flysch formations. Finally, the aforementioned large tectonic transport of the Pindos nappe, which lasted at least 25 Myrs, as well as the completely different paleo-geographic facies of the pelagic Pindos ("eu-geosyncline") from the external ridge of the Gavrovo carbonate platform end up to the same conclusion that the Gavrovo (Tripolis)/Pindos boundary constitutes a large paleogeographic break/gap. This break was created during the Eocene subduction of the external carbonate platform of Gavrovo–Tripolis–Amorgos–Olympus (—Almyropotamos–Kerketeas) and the transitional—pelagic region of the Cycladic units under the Pindos and the more internal units. Thus, the paleo-geographic boundary B was tectonically shifted towards the external area on B1, over Gavrovo and up to the internal Ionian thrusts, where the front of the Pindos nappe is located today (Fig. 9.6). Its shallow journey through the platform units created formations of wild flysch. On B2, it was shifted towards the internal region, with subduction and metamorphic events, where the Cycladic blueschists can be observed today under the Eastern Greece unit, together with its pre-Alpine basement. Various smaller thrusts segmented the paleo-geographic region that was subducted, with a characteristic case that of the Cycladic blueschists thrust over the Olympus (—Almyropotamos–Kerketeas) unit, as depicted by C1, C, C2 (Fig. 9.6).

(4) The **Pindos/Sub-Pelagonian** boundary has practically no value today, after been aware of the allochthonous nature of the Sub-Pelagonian ophiolites. However, the lateral transition between the Pindos (-Parnassos)–Western Thessaly (-Beotia) —Eastern Greece units during Late Cretaceous is undisputable, based on the available data. Only thrusts with relatively small displacement have been created, such as those of the Western Thessaly unit over the Pindos unit, of the Beotia unit over the Parnassos unit (D1, D, D2) and of the Eastern Greece unit over all the others (E1, E, E2) (Fig. 9.6).

Based on the data mentioned above during the beginning of the 1980s and choosing the simplest model of the Hellenides, according to which there is one major ophiolite suture in the Axios region and a secondary oceanic basin in the Pindos region, Papanikolaou (1984a, 1986a) proposed the following paleo-geographic organization and evolution of the Hellenides (Fig. 9.7):

- During the **Lias**, two neritic carbonate platforms are developed, separated by the Pindos trough into: (i) an external part, which includes the subsequent Paxos, Mani, Ionian, Gavrovo, Tripolis, Amorgos, Olympus (Almyropotamos, Kerketeas) and part of the Cycladic units and (ii) an internal part, which includes the subsequent Parnassos, Beotia, Sub-Pelagonian, and Almopia units. In the internal margin of the internal carbonate platform, the Maliac unit is developed as an independent unit, separating the more internal Paikon unit, while the Peonia and the Circum-Rhodope units comprise sediments deposited adjacent to the Tethyan ophiolites.
- During the **Dogger**, intense rifting phenomena with taphrogenesis can be observed, especially in the external carbonate platform, with the creation of: (i) The Mani–Ionian graben,

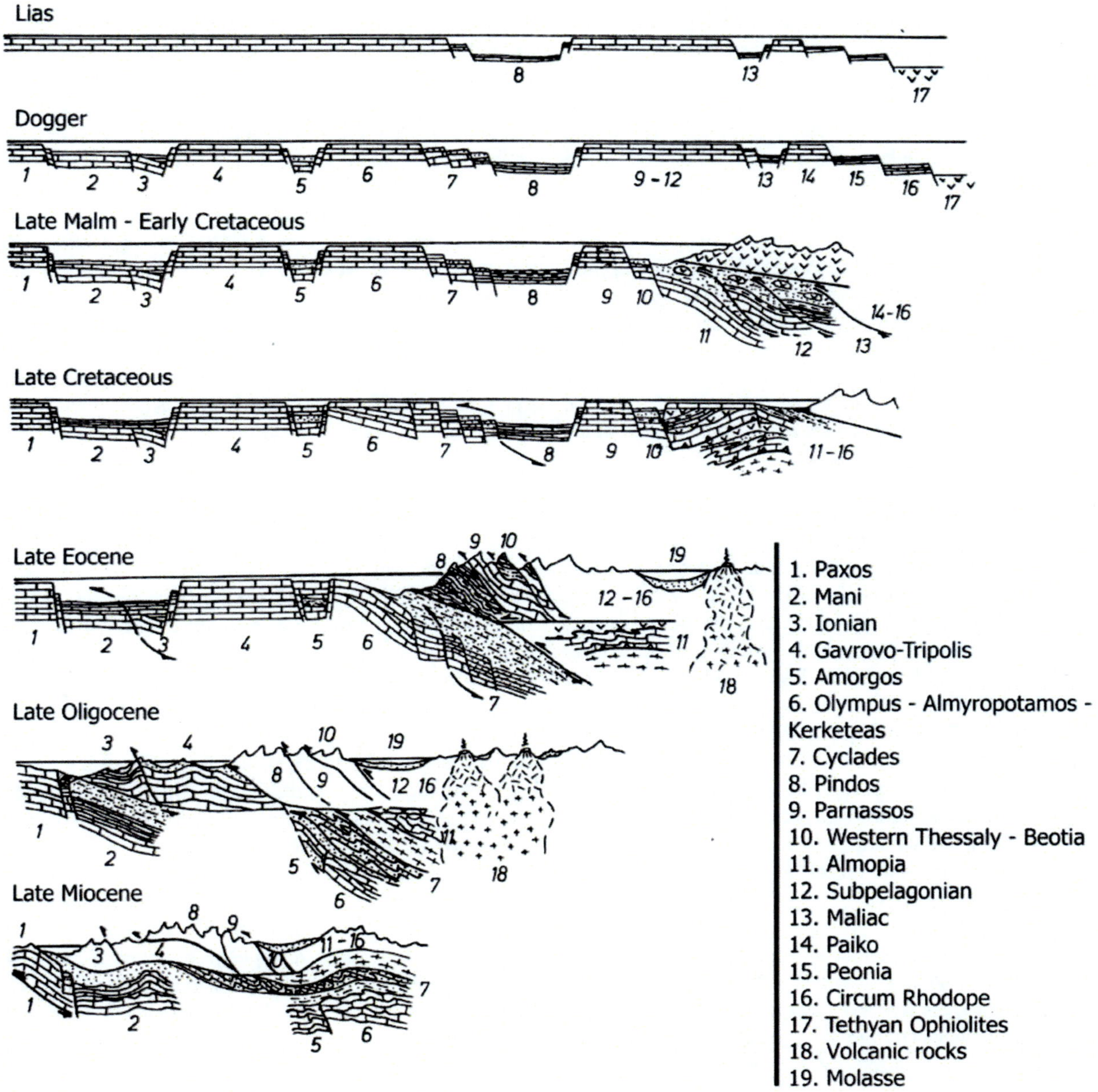

Fig. 9.7 Paleogeographic organization and tectonic evolution of the Hellenides, during the Lias—Late Miocene (from Papanikolaou 1986c)

which distinguishes the tectonic horsts with the Paxos unit externally and the Gavrovo and Tripolis units internally. (ii) The Amorgos graben, which is developed between the horsts of the Gavrovo–Tripolis externally, and the Olympus (Almyropotamos–Kerketeas) internally. (iii) The submergence of the external platform margin towards the Pindos trough, where the Cycladic units are developed (mainly the Northern Cyclades). (iv) The submergence of the Sub-Pelagonian margin towards the Maliac unit.

- During the **Late Malm—Early Cretaceous**, the previous paleo-geographic structure seems to be preserved in the external area, while the paleo-Alpine orogeny initiated in the internal area, with the tectonic emplacement of the ophiolite nappes over the Maliac, Sub-Pelagonian, Almopia etc. units, and the creation of the orogenic arc system with flysch sedimentation in the area of the Western Thessaly—Beotia units. Thus, only the Parnassos unit had been left non-tectonised from the area of the internal carbonate platform (apart from its vertical movements, which are related to the bauxite genesis).
- During the **Late Cretaceous**, the paleogeographic organization remains almost the same, with the addition of a new huge neritic platform in the internal area, which covers the largest part of the paleo-tectonised Hellenides since the Cenomanian and creates the Eastern Greece domain. At the end of the Cretaceous, we have the initiation of the subduction of segments of the Cycladic units under the Pindos and the more internal units, which continue their orogenic evolution at shallow level.
- During the **Late Eocene**, the paleogeographic pre-orogenic region is visibly limited with the gradual subduction of the Amorgos, Olympus (Almyropotamos–Kerketeas), and the Cycladic units, with a simultaneous emergence of the Pindos and all the more internal units. At the same time, molasse deposition can be observed in the back arc area.
- During the **Late Oligocene**, the pre-orogenic paleogeographic region of the Hellenides is limited only in the Paxos unit, while the Mani unit has been subducted under the Ionian, Gavrovo-Tripolis, Pindos, and all the more internal units, which are uplifted and emerged.
- During the **Late Miocene**, the pre-orogenic paleogeographic evolution of the Hellenides comes to an end, with the uplifting and participation in the orogenic arc, of the last, most external unit - foreland, the Paxos, beneath the Ionian nappe.

The above paleogeographic evolution is the last synthesis (Papanikolaou 1984a, 1986a) before the application of the tectono-stratigraphic terranes concept in the Hellenides, and it is therefore characterized by a certain degree of static view, certainly influenced by the previous models and especially of those proposed by Aubouin (1959, 1965).

During the same period the Hellenides section of the Alpine system was also included in a synthesis of the paleogeographic evolution of the Tethyan segment between the Atlantic and the Pamir, from Lias to present, by the French-Soviet scientific cooperation, published in paleogeographic maps at 1/20,000,000 scale (Dercourt et al. 1985; Ricou et al. 1985; Savostin et al. 1986). Irrespectively to various disagreements and interpretations for various sections of that region, and for the various periods between the rest of the researchers (but also between the authors), and especially regarding the number of oceanic basins, these maps were a valuable basis, at the very least for discussion. The main elements of the paleo-geographic maps concerning the Hellenides have been included in seven Paleo-geographic sketches (Fig. 9.8). Therefore, the paleogeographic organization and evolution of the Hellenides within the Tethyan frame by Dercourt et al. (1985) and Ricou et al. (1985) can be summarized in the following main events:

During the **Lias** (190 Ma, Fig. 9.8), the paleogeographic region of the Hellenides was depicted as calm, with neritic carbonate sedimentation over a continental crust, except for two–three grabens, where the crust was thinner and pelagic sedimentation could be observed, like in the Ionian and Pindos. In the Axios/Vardar oceanic area, Tethys could be observed with a width a little shorter than that of the Hellenides. The southern margin of Tethys was a passive one, while the northern margin seems to be active, with a subduction zone under the Rhodope area.

During the **Jurassic/Cretaceous** boundary (130 Ma) a subduction of the internal part of the Hellenides can be observed under the Axios oceanic crust, with a local emergence due to the orogeny, although it only occurs as local event, without any continuity eastwards. At the same time, the Ionian and Pindos basins have been developed, as well as the basin to the south of Paxos, a precursor of the contemporary Eastern Mediterranean basin. In the Rhodope region, the northern margin is still considered as active, with a presence of a volcanic arc.

During the **Aptian** (110 Ma), the opening of two oceans can be observed, one in the eastern continuity of the Pindos basin, between the Menderes and Kirsehir massifs and another to the south of Paxos, in the region of the present Eastern Mediterranean, between Africa and the Hellenides, which terminates in Arabia to the east. In the meantime, the northern margin is still active in the Rhodope region.

During the **Cretaceous/Paleocene** boundary (65 Ma) there are no large-scale disruptions observed in the Hellenides area, apart from the approaching Rhodope and the fossilisation of the Ocean of the Eastern Mediterranean. On the other hand, towards the east, the major tectonic emplacement of the ophiolites can be observed with the closure of the eastern prolongation of the Pindos Ocean between the Menderes and Kirsehir massifs. Moreover, the

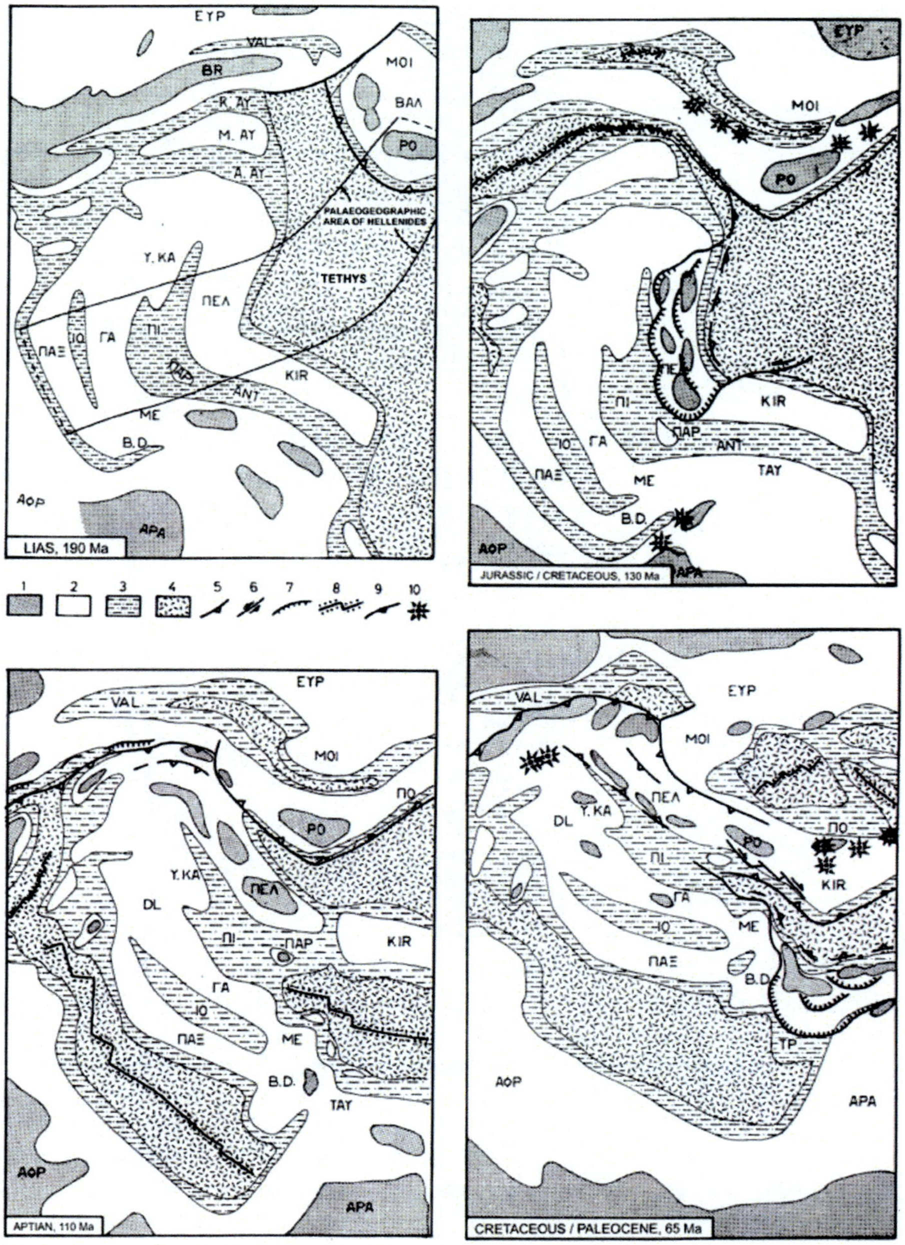

Fig. 9.8 Paleogeographic maps of a Tethyan segment, including the Hellenides and the adjacent areas for the timeframe from Lias to present (from Dercourt et al. 1985, simplified). EΥP: Europe, AΦP: Africa, APA: Arabia, VAL: Valais, BR: Brianconnais, K.AY: Lower Austro-Alpides, M.AY: Middle Austro-Alpides, Y.KA: Upper Karst, MOI: Moesia platform, BAΛ: Balkanides, PO: Rhodope, ΠΕΛ: Pelagonian, ΠΙ: Pindos, ΓΑ: Gavrovo, IO: Ionian, ΠΑΞ: Paxos, ΠΑΡ: Parnassos, KIR: Kirsehir, ANT: Antalya, ME: Menderes, B.D.: Bey Daglari, ΠΟ: Pontides, TAY: Taurides, DL: Dalmatia, TP: Troodos, AΛ: Alps, ΔΕΙ: Dinarides, ΕΛΛ: Hellenides, KAY: Caucasus, KAP: Carpathians, AΠ: Apennines, KAΛ: Calabria 1: land, regardless of crustal type, 2: thick continental crust, 3: thin continental crust, 4: oceanic crust, 5: subduction zones, 6: transform zones with horizontal slip and large shears of the lithosphere, 7: obduction of oceanic crust, 8: mid-ocean ridge, 9: overthrusts, 10: volcanoes (of orogenic arc).

opening along the southern European margin of the Black Sea Ocean is noteworthy.

During the **Priabonian / Oligocene** (35 Ma), no opening of oceans can be observed. Both the East Mediterranean ocean and the Black Sea ocean are inactive, while the orogeny is dominant in between them. The ongoing orogeny has caused the uplift and erosion of the Hellenides, except for the Gavrovo–Ionian –Paxos area. Volcanic arcs can also be observed in the regions of the back-arc basins.

During the **Tortonian** (10 Ma) the situation remains the same, apart from the migration of the orogeny towards the entire Hellenides region, up to the Paxos in the south and to the boundary with the East Mediterranean Ocean. The collision of Arabia with the folded Eurasian margin can be observed to the east.

During the **present** period the orogeny continues migrating southwards, with the subduction and closure of the Eastern Mediterranean Ocean, while the Black Sea remains

Fig. 9.8 (continued)

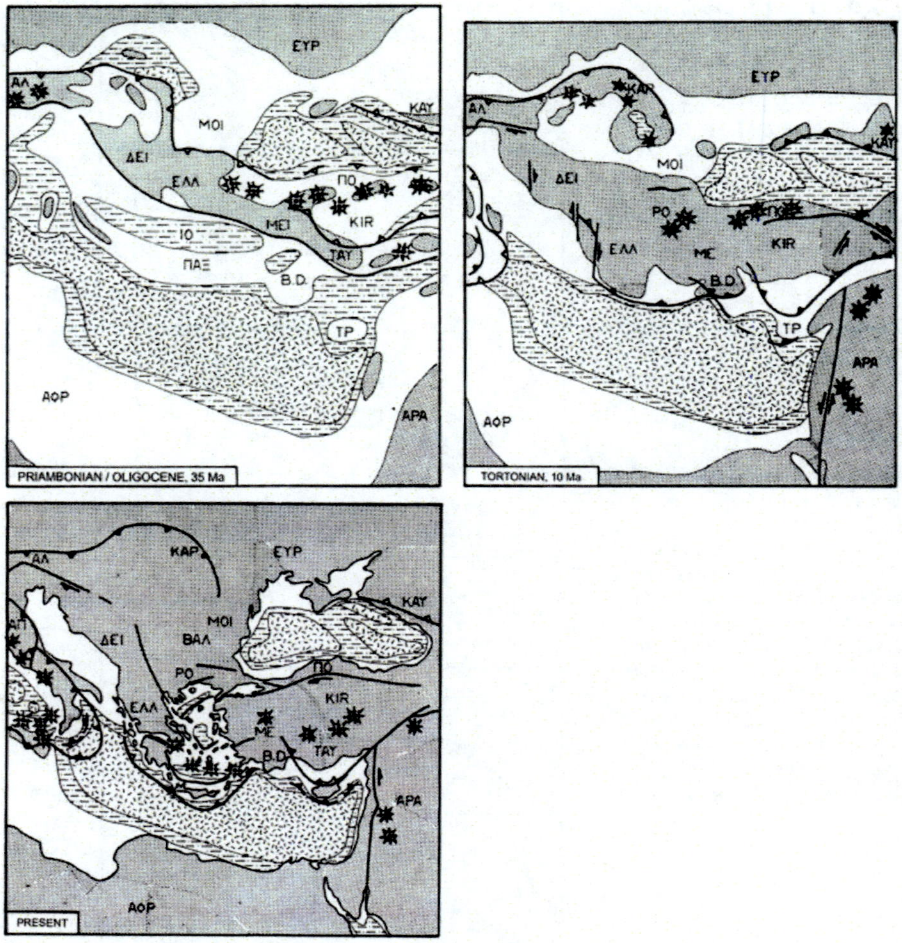

fossilified and inactive. Orogenic arcs can now be observed exclusively in the southern section of Tethys, while the northern section along Europe tends to become inactive. Finally, the effect of the collision of Arabia towards the north is beginning to be felt with major strike slip zones, along with its simultaneous divergence from Africa, with the opening of the Red Sea, which extends towards the Eastern Mediterranean.

9.3.3 The Tectono-Stratigraphic Terranes Period

The application of the tectono-stratigraphic terrane theory in the Hellenides changed the overall model of the static perception to a dynamic one, with constant terrane movement from the south to the north. Thus, on the one hand, the continental blocks were moving northwards, while on the other hand, the oceanic basins opened in between the continental blocks and closed by reaching the European margin (from the early Jurassic to present). The new point of view shows that the Hellenides as a mountain range in the active European margin are a result of long subduction periods of the oceanic crust of the H0, H2, H4, H6, and H8 basins, and of shorter periods of micro-collisions, during the arrival and accretion of the continental blocks of H1, H3, H5, and H7 to the European margin. The most recent analysis of the paleogeographic evolution of the Hellenides in a series of maps according to Papanikolaou (1989, 2013) (Fig. 9.9) shows that the maximum extent of their paleogeographic area is observed in the Lias–Dogger, while the present day dimensions along a transverse section are limited only to one fourth.

Therefore, in the **Early Triassic**, the oceanic basins H4, H6, and H8 can be simultaneously observed, from which the H8 is in a subduction stage, while the H4 and H6 are in an opening stage. The subsequent basins H0 and H2 were still in the rift valley stage. The H5 and H7 platforms are in a drifting stage, while the subsequent H1 and H3 platforms are in a rifted margin stage. It is noteworthy that the Permian–Triassic age of the initiation of the rift valley stage to the north of Africa was confirmed by geophysical data extracted

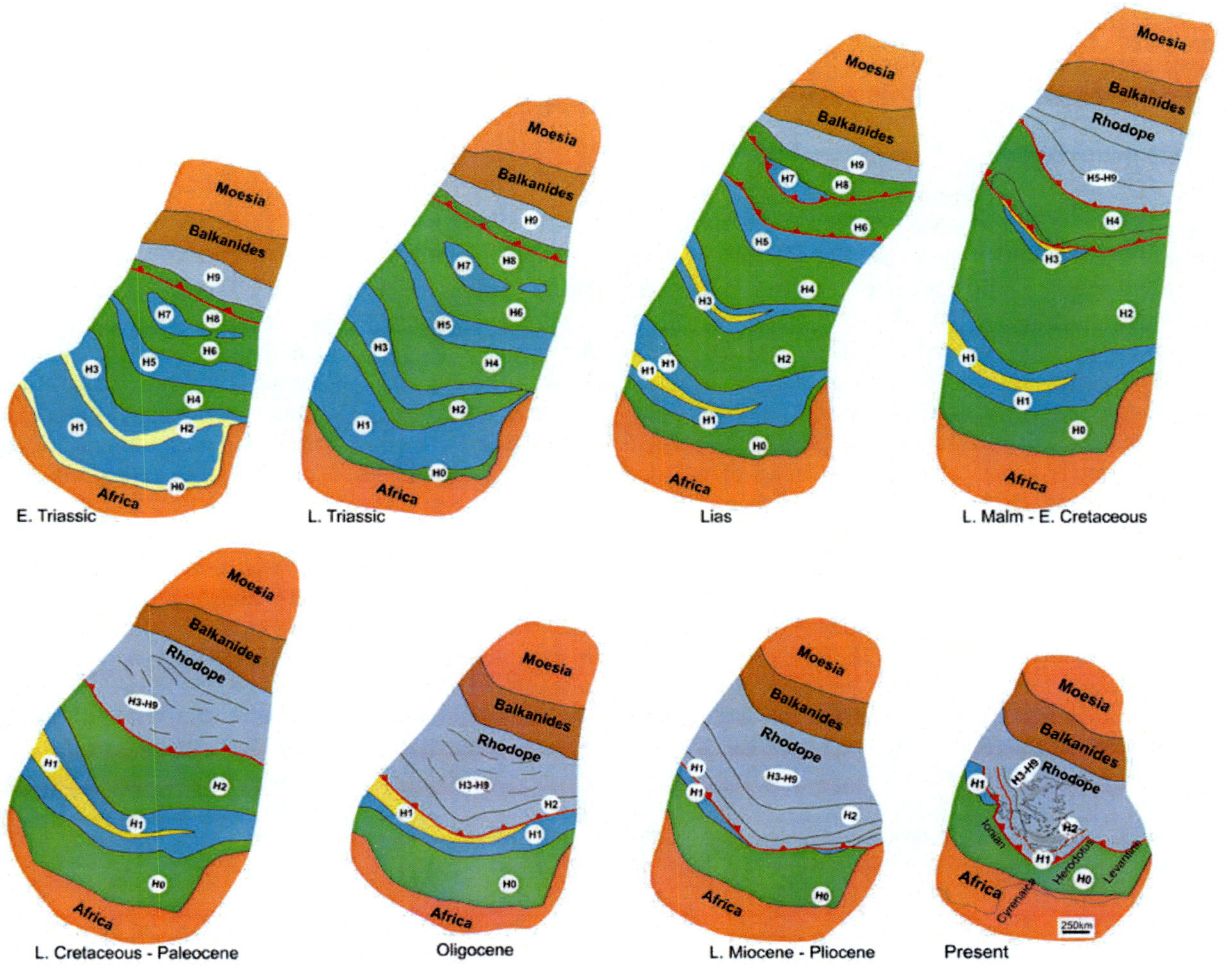

Fig. 9.9 Schematic paleogeographic sketches of the evolution of the Hellenides in the Tethys region, with the drifting motions of the continental terranes and the successive opening and closure stages of the oceanic basins (from Papanikolaou 2013)

from the submarine investigation of the Levantine basin (e.g. Gardosh and Drunckman 2006).

In the **Late Triassic**, a general expansion of the Hellenides region can be observed, mainly due to the further opening of the H6 and H4 basins, but also due to the beginning of the opening of the H0 and H2 ones. The shallow carbonate sedimentation has been expanded all over the H1, H3, H5, and H7 platforms.

In the **Lias**, subduction and orogenic arcs can be initially observed in H8 and H7 respectively, and later on also in H6 and H5 as part of the Cimmerides orogeny. The H0, H2, and H4 basins have been significantly expanded, while a rifting and subsidence in parts of both the external platform (Mani–Ionian, Amorgos) and the internal platform (Sub-Pelagonian B) can be observed at the end of the Lias.

At the **Malm/Early Cretaceous** boundary, the H4 starts to subduct, and the paleo-Alpine orogeny comprises ophiolite obduction of segments of the H4 over the internal part of the H3 platform. At the same time, the more internal terranes of H5, H6, H7, H8, and H9 have been integrated in the active European margin with arc volcanism and back arc basins. The Pindos–Cyclades Ocean (H2) has almost reached its maximum extent, while in some locations it links with some remnants of the Axios (H4). At the external margin of the H3 platform the sedimentation of the Parnassos platform is still undergoing with minor emersions and bauxite formations, while more internally, the Beotian flysch is deposited along a paleotrench.

At the **Late Cretaceous / Paleocene** boundary, the orogeny involved the entire H3 platform and a large part of the H2 oceanic basin with initiation of subduction of the Cycladic blueschists. All the more internal units of Pindos, from H3 to H9, have been uplifted and incorporated to the arc structures.

At the **Oligocene**, the Pindos- Cyclades Ocean H2 closed and parts of the external platform H1 were subducted (Olympus, Amorgos, Western Crete, Mani). The structure of Greece includes the terranes of H2 to H9 and was significantly reduced, due to the ongoing shortening.

At the **Late Miocene / Pliocene** boundary, the entire Hellenides have been subducted and participated in the orogenic evolution. The initiation of the subduction of the Ionian oceanic basin H0 under the external margin of the external platform H1 in Paxos, configurates the present geotectonic process.

At present, in the **Holocene**, the largest part of H0 has been subducted under the Hellenides, which consist of the amalgamated H1 through H9 terranes. The front of the Hellenic arc including the accretionary prism of the Eastern Mediterranean is approaching the beginning of collision south of Crete with Cyrenaica (Papanikolaou et al. 2004). At the same time, a differentiation can be observed along the Hellenic arc, with the Ionian subduction zone of the H0 in the region from Preveza to Rhodes, and a continental subduction—microcollision of the Apulian platform to the north, with Epirus and Albania. The Cyrpus arc east of Rhodes Island exhibits low activity, despite the fact that the Eastern Mediterranean Ocean extends to the east with the Levantine basin. This is possibly due to the lateral movement of Cyprus, with a western vector, the same as the Anatolian micro-plate, as shown by the geodetic GPS calculations (Reilinger et al. 2006) and thus, there is no significant convergence between the Cyprus arc and the Levantine basin. This also explains the absence of a modern volcanic arc in the Cappadocia region, behind the Taurides, although the volcanic arc used to be active throughout the Miocene—Pliocene. It is characteristic that the impressive Kayseri volcano is today dormant (Fig. 9.10). At the same time, in the intermediate space between the Hellenic and the Cyprus arc, the large Bey Daglari platform intervenes, forming an indentor, through sinistral strike slip motions, that rendered the contact of Southwestern Anatolia with the African plate inactive.

The shortening of the Hellenides region was conducted through the almost complete disappearance of the paleogeographic region of the oceanic basins, the rocks of which (ophiolites and sediments) are found imbricated inside the continental fragments of the terranes. These deep pelagic oceanic sediments and lavas comprise the geological bedrock of several high mountains in Greece. The continental blocks themselves show only small-scale internal imbrications and a relatively small tectonic repetition of their pre-Alpine crusts, produced during continental subduction. This can be depicted in the generalised N-S section, from Rhodope to the Hellenic trenches, where the Hellenic/Aegean crust is basically the sum of the continental crusts of the H1, H3, H5, H7 and H9 terranes (Papanikolaou et al. 2004) (Fig. 9.11).

A similar structure for the Hellenic crust was proposed by Van Hinsbergen et al. (2005) as a result of the successive subduction of the oceanic and continental crust of the Hellenides under the European margin, from Jurassic to present. A more detailed model of the successive stages of crustal development within the Hellenic retreating subduction system has been presented by Burchfiel et al. (2018), where the detachment surfaces are initiated within the subducting lithosphere at distinct levels, depending on the nature of the crust (See also Fig. 4.2, referring to the creation of tectonic units).

9.4 Characteristic Stratigraphic/Sedimentological Facies of the Hellenides

Since the beginning of the Alpine sedimentation cycle of the Hellenides during the Permian- Triassic until its termination during the Miocene, certain characteristic sedimentation facies were developed, some neritic, some pelagic, and some

Fig. 9.10 The dormant volcano of Kayseri in Cappadocia, a volcano that was active until the Early Pleistocene

transitional slope facies. Additional special facies are added depending on specific cases of syn-sedimentary tectonism, when we have the presence of clastic formations, or of volcanism, when volcano-sedimentary sequences are involved.

Clastic formations in the Hellenides during the pre-orogenic stage are traced: (i) in almost all of the known stratigraphic columns during the Permian–Triassic, with their uppermost layers observed under the base of carbonate platform sediments, usually in the Middle Triassic (Papanikolaou 1979), (ii) in the Amorgos unit, as well as in some Cycladic units during the Middle-Late Jurassic (Fytrolakis and Papanikolaou 1981; Papanikolaou 1986a), (iii) in the Western Thessaly—Beotia units in the Late Jurassic—Early Cretaceous (Celet et al. 1976; Papanikolaou and Sideris 1979), and later on in the Cenomanian in the Pindos unit (Aubouin 1959; Maillot 1979), as early events of flysch sedimentation ("beotian flysch" and "first flysch" respectively), (iv) in the Sub-Pelagonian region during the Late Jurassic—Early Cretaceous (Papanikolaou 1990), and in their associated metamorphosed formations in Almopia, where the schist-sandstone-chert formations are developed, although these are mélanges of the paleo-Alpine orogeny, (v) in the Eastern Greece region during the Cenomanian—Turonian, when flyschoid formations are created within syn-sedimentary grabens (Mercier and Vergely 1977; Mariolakos and Papanikolaou 1982), between the neritic carbonate platform sedimentation, especially in the area above the Almopia unit (e.g. Vermio, Vourinos).

Volcano-sedimentary sequences are traced: (i) in almost all of the known stratigraphic columns, except of the Ionian unit and those of the more external units, in the Permian–Triassic, but more especially in the Middle Triassic (Papanikolaou 1979). The Tyros beds in the Peloponnese (Ktenas 1924a) are probably the most prominent example, which include the well-known andesitic formation "Lapis Lacedaemonius" (Krokeischer Stein), as well as some unique outcrops of Parnitha mt in Attica (Ktenas 1909, 1924b). These outcrops clearly record the rifting stage. (ii) in the Triassic—Jurassic formations of the Maliac unit, which indicate intra-oceanic environment (Ferriere 1982). (iii) in the metamorphosed Jurassic–Cretaceous (?) formations of the Northern Cyclades (Papanikolaou 1978, 1986a), as well as the Jurassic formations of the Circum-Rhodope (Kockel et al. 1977, Bonev and Stampfli 2009) which are related to abyssal-pelagic environments. (iv) the Late Jurassic formations of Paikon (Mercier 1968), which are also unique since they are the only ones that relate to acidic type volcanism (while all the others are mainly of basic type) and characterize a volcanic arc. (v) the Upper Cretaceous formations of the Arvi unit in Crete (Tataris 1964; Bonneau 1976), which are interbeded with basalts in an oceanic environment.

A characteristic facies of condensed stratigraphic horizons is that of the red-violet nodular marly limestones, also known as the ammonitico rosso facies as it often includes ammonites (Frech 1907) (e.g. in Epidavros). This is a pelagic facies near slopes and is usually observed in tectonically restless areas with syn-sedimentary tectonism and/or volcanism. In the Hellenides, it can often be observed during the Early–Middle Triassic in the Sub-Pelagonian, Maliac, Almopia, Vardousia, Circum-Rhodope etc. units, and during the Late Lias–Dogger in the Ionian and Sub-Pelagonian B units (Renz 1955). The

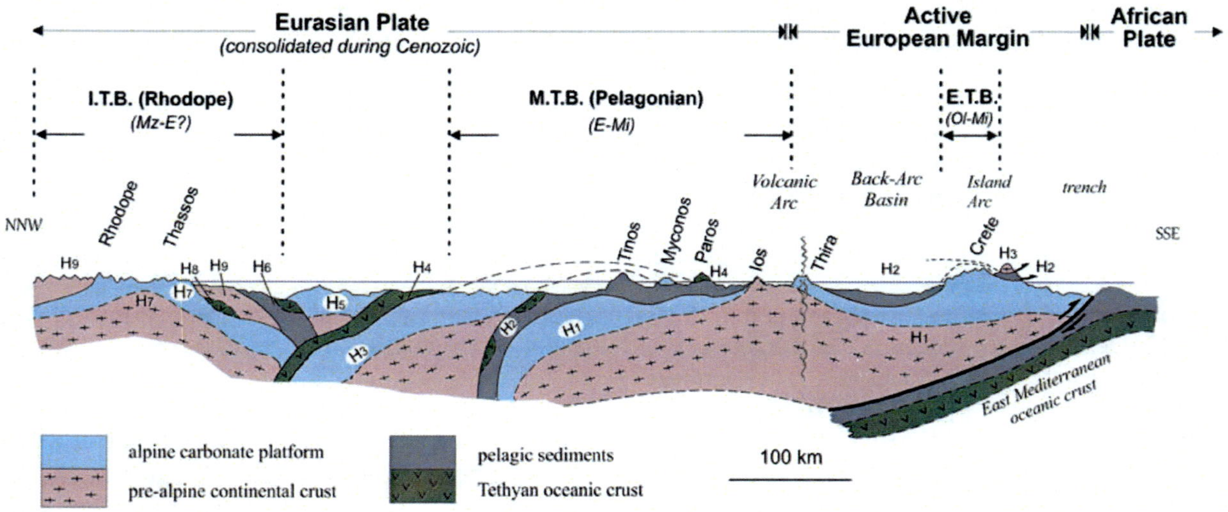

Fig. 9.11 Schematic tectonic N-S section through the Hellenic crust, showing its composition basically from the accreted crustal fragments of the continental terranes with only thin intermediate layers of the oceanic terranes (from Papanikolaou et al. 2004)

Middle Triassic facies is also known as the Hallstatt facies, name derived from the Alps, where the same formation, of the same age can be observed. The Middle Triassic facies corresponds to a rifting geodynamic environment, with associated phenomena of volcanism, while the Middle Jurassic facies is associated with a collapse during a rifting stage of pre-existing carbonate platforms without any volcanism (Papanikolaou 2013).

Characteristic **transitional slope facies** with micro-breccia carbonate formations are traced in various times and locations. The Vardousia transitional unit is characterized by such transition slope facies in its entirety, throughout the sedimentation process from the Late Triassic to the Late Cretaceous (Celet 1979). Therefore, during the pelagic sedimentation breccia horizons of neritic fragments are supplied from the Parnassos platform. At the same time, the thickness is in transitional intermediate level, between the small thickness of a few hundreds of meters of the Pindos pelagic sequence and the large thickness of 2–3 km of the Parnassos carbonate platform (Gouliotis 2014). Apart from the Vardousia unit, the other well-known slope facies can be observed in the Ionian, during the Senonian, with micro-breccia horizons with rudist fragments (Renz 1955), supplied from the Gavrovo carbonate platform, embedded in pelagic micritic limestones with Globotruncanes. The same is observed in the internal Pindos, the Western Thessaly and the Beotia units, where there is a dominant presence of breccias derived from ophiolites and cherts, such as the Thymiama facies (Aubouin 1959; Papanikolaou and Sideris 1979). In Western Thessaly (former UltraPindic zone) and in Beotia, the micro-breccia limestone facies with ophiolite and chert fragments began in the Malm and continued throughout the Cretaceous up to the Maastrichtian. Similar facies can be observed in the Maliac unit, although only during the Jurassic (Ferriere 1982). In the breccias facies mentioned above, we should distinguish the relatively autochthonous or parautochthonous cases, where the clast material originates from the same paleo-geographic region or the directly adjacent ridge and is caused by syn-sedimentary tectonism. In other cases the clastic material is allochthonous in origin, transported from further away, therefore the tectonic event influences a different area, than the sedimentation area, which is passively accepting the transported sediment. Another distinction can be made between the clasts that are created inside the sedimentation basin from local phenomena of instability and the clasts that originate from the erosion of terrigenous regions, transported in the sedimentation area usually from the evolving island arc.

The **pelagic facies** in the Hellenides include pelagic limestones, often with siliceous layers or nodules, or chert and especially radiolarites, often with interbeddings of red pelites and pelagic limestone layers. The radiolarites can be observed mainly in the Jurassic of the Pindos, Maliac, Sub-Pelagonian B, and Western Thessaly units, as well as in the Triassic of the Maliac (Fleury 1980; Ferriere 1982). In the regions of the Ionian, Mani, Amorgos, and parts of the Northern Cyclades, the well-known facies of the Vigla limestones is being developed, a formation consisting of alternations of pelagic limestones with Calpionelles or Globotruncanes and white cherts (Renz 1955; Papanikolaou 1986a). Apart from the Northern Cyclades, where no stratigraphic data are available (due to the metamorphism), but only lithologic similarities, in all the other cases, the age of the facies begins in the Malm and ends at the beggining of the Late Cretaceous. The pelagic facies of the Late Cretaceous limestones with Globotruncanes is another characteristic formation. It is traced in the Ionian, and Pindos units, as well as in other more internal units, in the transitional beds towards the flysch (Philippson 1892, 1956–1959; Renz 1955) (i.e. Eastern Greece, Vermio, Parnassos, Western Thessaly). Finally, another characteristic pelagic facies is that of the Upper Triassic limestones with Halobia in the Pindos and Western Thessaly units (Renz 1955; Aubouin 1959; Lekkas 1986).

In the **neritic facies** of the Hellenides, no significant differences exist, despite the usual alternation of environments based on local paleogeographic conditions (intra-tidal, supra tidal, etc.) (Kalpakis and Sideris 1981; Kalpakis and Lekkas 1982). One could distinguish: (i) the stromatolitic limestones in the Upper Triassic–Lias of Tripolis, Sub-Pelagonian, etc., (ii) the dolomites and the thick-bedded limestones with algae, corals, gastropods (Diplopora, Gyroporelia, Paleodasycladus, Megalodon, etc.), that co-create the Pantokrator facies of Renz (1955) described in the Ionian, though it can be seen all over the neritic Hellenides during the Late Triassic–Lias, (iii) the Upper Jurassic limestones with Cladocoropsis Mirabilis, which are traced in the Tripolis, Parnassos and the Sub-Pelagonian and are valuable as bauxite-bearing indicators (Papastamastiou 1960; Tataris 1975) (ceiling of b1), (iv) the Upper Cretaceous bituminous, black, rudist-bearing limestones, which can be observed in the Gavrovo, Tripolis, Amorgos, Parnassos, Olympus, Almyropotamos, Eastern Greece, etc., (Philippson 1892, 1956–1959; Renz 1955; Fleury 1980). (v) the Nummulite-bearing limestones, which are commonly traced in the Paxos, Gavrovo, Tripolis, Olympus, Almyropotamos and Amorgos units (Philippson 1892, 1956–1959; Renz 1955; Fleury 1980; Godfriaux 1968; Fytrolakis and Papanikolaou 1981).

Lastly, another distinctive formation relates to syngenetic sediments associated to basic volcanism, linked with the ophiolite complexes of the Hellenides, such as: (i) the Jurassic sediments of the Circum-Rhodope (Kauffman et al. 1976), (ii) the Jurassic sediments of the Northern Cyclades Papanikolaou 1986a), (iii) the Lower Triassic–Jurassic sediments of the Maliac (Ferriere 1982), (iv) the Upper Jurassic sediments of Western Thessaly (Papanikolaou and Sideris 1979), (v) the Upper Jurassic sediments of Peonia (Mercier 1968), (vi) the Upper Cretaceous sediments of Arvi in Crete (Bonneau 1976).

References

Altherr, R., Schliestedt, M., Okrusch, M., Seidel E., Kreuzer, H., Harre, W., Lenz, H., Wendt, H & Wagner, G. 1979. Geochronology of High - Pressure Rocks on Sifnos (Cyclades, Greece). Contrib. Mineral. Petrol., 70, 245–255.

Altherr, R., Kreuzer, H.. Wendt, I., Lenz, H., Wagner, G., Keller, J., Harre, W. & Hohndorf, A. 1982. A late Oligocene / Early Miocene High Temperature Belt in the Attic - Cycladic Crystalline Complex (SE Pelagonian, Greece). Geol Jb., E 23, 97–164.

Aubouin, J. 1959. Contribution à l' étude géologique de la Grèce septentrionale: Les confins de l'Epire et de la Thessalie. Ann. Géol. Pays Hellén.,10, 1–483.

Aubouin, J. 1965. Géosynclines. Devel. Geotectonics, 1, 335 p., Elsevier.

Aubouin, J. 1974. Des tectoniques superposées et leur signification par rapport aux modèles géophysiques. L'exemple des Dinarides, paléotectonique, tectonique, tarditectonique, néotectonique. Bull. Soc. géol. France, XV, 426–460.

Aubouin, J. 1977. Alpine Tectonics and Plate Tectonics: Thoughts about the Eastern Mediterranean. In: Europe from crust to core, 143–158, J. Wiley.

Aubouin, J., Le Pichon, X., Winterer, E. & Bonneau, M. 1979. Les Hellenides dans l' optique de la tectonique des plaques. VI Coll. Geol. Aegean Region, Athens 1977, Proc., 3, 1333–1354.

Bonev, N. & Stampfli, G., 2009. Gabbro, plagiogranite and associated dykes in the supra-subduction zone Evros Ophiolites, NE Greece. Geol. Mag., 146, 72–91.

Bonneau, M. 1976. Esquisse structural de la Crète alpine. Bull Soc. Géol. France, 8, 2, 351–353.

British Petroleum Co, 1971. The geological results of petroleum exploration in Western Greece. Geology of Greece, I.G.R.S., No 10, 72 p.

Burchfiel, B.C., Royden, L.H., Papanikolaou, D. & Pearce, F.D. 2018. Crustal development within a retreating subduction system: the Hellenides. Geosphere, 14, 2.

Celet, P. 1962. Contribution à l'étude géologique du Parnasse-Kiona et d'une partie des régions méridionales de la Grèce continentale. Ann. Géol. Pays Hellén. 7, 1–358.

Celet, P. 1979. Les bordures de la zone du Parnasse. Evolution paléogéographique au Mésozoique et caractères structuraux. VI Coll. Geol. Aegean Region, Athens 1977, Proc. 2, 725–740.

Celet, P., Clement, B. & Ferriere, J. 1976. La zone béotienne en Grèce: Implications paléogéographiques et structurales. Eclogae géol. Helv., 69, 577–599.

Dercourt, J., 1970. L'expansion océanique actuelle et fossile; ses implications géotectoniques. Bull. Soc. Geol. France, 7, 261–317.

Dercourt, J. 1972. The Canadian Cordilliera, the Hellenides and the sea-floor spreading Theory. Canad. Jour. Earth Sci, 9, 709–743.

Dercourt, J., Zonenshain, LP., Ricou, LE., Kazmin, V.G., Le Pichon, X., Kνιpper, A.L, Grandjacquet, C., Sborshchikov, I.M., Boulin, J., Sorokhtin, O.,Geyssant, J., Lepvrier, C., Bizu-Duval, B., Sibuet, J. C., Savostin, LA., Westphal,M. & Lauer, J.P. 1985. Présentation de 9 cartes paléogéographiques au1/20,000,000 étendant de l' Atlantique au Pamir pour la période du Lias à l' actuel. Bull. Soc. Géol. France, I, 5, 637–652.

Epting, M., Kudrass, R. & Schafer, A. 1972. Stratigraphie et position des séries métamorphiques aux Talea Ori/Crète. Z. Deutsch. Geol. Ges., 123, 365–370.

Ferriere, J. 1982. Paléogéographies et tectoniques superposées dans les Hellénides internes: les massifs de l'Othrys et de Pélion. Soc. Géol. Nord, Publ. 8, 970 p

Fleury, J.J. 1980. Les zones de Gavrovo-Tripolitza et du Pinde-Olonos (Grèce continentale et Péloponnèse du Nord). Evolution d'une plateforme et d' un bassin dans le cadre alpin. Publ. Soc. Géol. Nord, 4.

Frech, F. 1907. Die hallstatter Kalke bei Epidauros (Argolis) und ihre Cephalopoden. N. J. B. Min., Festband, 1–32.

Fytrolakis, N. & Papanikolaou, D. in collabor. with Panagopoulos, A. 1981. Stratigraphy and structure of Amorgos Island, Aegean Sea. Ann. Géol. Pays Hellέη., 30/2, 455–472.

Gardosh, M. A. & Drunckman, Y., 2006. Seismic stratigraphy, structure and tectonic evolution of the Levantine Basin, offshore Israel. Geol. Soc, London, Sp. Publ., 260, 201–227.

Godfriaux, I. 1968. Etude géologique de la région de l'Olympe (Grèce). Ann. Géol. Pays Hellέη., 19, 1–281.

Gouliotis, L. 2014. The tectonic structure of Giona Mt and its hydrogeological applications. PhD Thesis, University of Athens, 433 p (in greek).

Hynes. A. J., Nisbet, E.G., Smith, A.G., Welland. M. J. P. & Rex, D. C. 1972. Spreading and emplacement ages of some ophiolites in the Othris region (eastern central Greece). Zeit. deuts. Geol. Ges. 123, 455–468.

Jacobshagen, V. 1979. Structure and geotectonic evolution of the Hellenides. VI Coll. Geol. Aegean Region, Athens 1977, Proc. 3, 1355–1367.

Jacobshagen, V. 1986. Geologie von Griechenland. Gebruder Borntraeger, 363 pp.

Jacobshagen, V., Dürr, ST., Kockel F., Kopp, K.O., Kowalcyk, G. with contr. Berckhemer H., Büttner, D. 1978. Structure and Geodynamic Evolution of the Aegean Region. In: Alps, Apennines, Hellenides, 537–564, Stuttgart.

Jacobshagen, V., Risch, H. & Roeder, D. 1976. Die eohellenische Phase. Definition & Interpretation. Z. Deut. Geol. Ges., 127, 133–145.

Kalpakis, G. & Lekkas, S. 1982. Loferites of the Pigadakia area of Peloponnesus. Ann. Géol. Pays Hellén., 31, 73–88 (in greek).

Kalpakis, G. & Sideris, Ch. 1981. Diagenetic phenomena in Upper Juraqssic- Lower Cretaceous limestones of Parnassos-Giona zone and their importance. Ann. Géol. Pays Hellén., 30/2, 778–792 (in greek).

Kaplanis, A. Koukouvelas, I., Xypolias, P. & Kokkalas, S. 2013. Kinematics and Ophiolite obduction in the Gerania and Helikon Mountains, central Greece. Tectonophysics, 595–596, 215–234.

Kauffmann, G. 1976. Perm und Trias im östlichen Mittelgriechenland und auf einigen ägäischen Inseln. Z. Deutsch. Geol. Ges., 127, 387–398.

Kauffmann, G., Kockel, F., Mollat, H. 1976. Notes on the stratigraphicand paleogeographic position of the Svoula Formation in the Innermost zone of the Hellenides (Northern Greece). Bull. Soc. Géol. France, 18, 2, 225–230.

Kockel, F.& Walther, H.W. 1965. Die Strimon-linie als Grenze zwischen Serbo-Mazedonischen und Rila-Rhodope Massiv in Ost-Mazedonien. Geol. Jb., 83, 576–602.

Kockel, F., Mollat, H. & Walther, H.W. 1977. Erläuterungen zur geologischen karteder Chalkidiki und angrenzender Gebiete 1/100,000 (Nord Griechenland). Bund. Für Geowiss. v. Rohstoffe, 110 p. Hannover.

Ktenas, K. 1909. Uber die eruptiven Bildungen des Parnes gebirges in Attika. Centralblatt Mineralogie etc., p. 557.

Ktenas, C. 1924a. Formations primaires semimétamorphiques au Péloponnèse central. C.R. somm. Soc. Géol. France, 61–63.

Ktenas, C. 1924b. L'age des formations volcaniques du massif de Parnes (Attique). C. R. somm. Soc. Geol. France, p. 60.

Lekkas, E. 1986. The occurrence of clastic Triassic at the base of theKoziakas sequence, Western Thessaly. Geol. Geophys. Res., IGME, special vol. Prof. Papastamatiou, 235–242 (in greek).

Maillot, H. 1979. Etude sedimentologique du «premier flysch» en Peloponnese occidentale (Grece) : La Formation d' Andritsena. Bull. Geol. Soc. Greece, XIV, 1, 94–115.

Mariolakos, I. & Papanikolaou, D. 1982. Cretaceous Emplacement of the Ophiolites in Northern Greece and Related Tectonosedimentary Environments. Geol. Evol. East. Medit., Edinburgh 1982, Abstracts, p. 71.

Mercier, J. 1966. Paléogéographie, orogenèse, métamorphisme et magmatisme des zones internes des Hellénides en Macédoine (Grece): vue d'ensemble. Bull. Soc.Géol. France, 8, 1020–1049.

Mercier, J. 1968. Etude géologique des zones internes des Hellénides en Macédoine centrale (Grèce). Ann. Géol. Pays Hellén., 20, 1–792.

Mercier, J. & Vergely, P. 1977. La zone d'Almopias. Bull. Soc. Geol. France, 19/1, 34–40.

Papanikolaou, D. 1978. Contribution to the Geology of Aegean Sea. The island of Andros. Ann. Géol. Pays Hellén., 29/2, 477–553.

Papanikolaou, D. 1979. Stratigraphy and structure of the Paleozoic rocks in Greece: an introduction. In: Sassi, F.P. (Ed.), IGCP Project No. 5 "Correlation of Prevariscan andVariscan Events of the Alpine–Mediterranean Mountain Belt, 93–102.

Papanikolaou, D. 1984a. Tectonic Evolution of the Hellenides. 27th Intern. Geological Congress, Moscow 1984, Abstracts III, 351–352.

Papanikolaou, D. 1984b. The three metamorphic belts of the Hellenides: a review and a kinematic interpretation. Geol. Soc. London, Spec. Publ., 17, 551–561

Papanikolaou, D. 1986a. Late Cretaceous Paleogeography of the Metamorphic Hellenides. Geol. Geoph. Res., IGME, Special volume in honor of Prof. Papastamatiou, 315–328.

Papanikolaou, D. 1986b. Geology of Greece. Eptalofos Publications, 240 p. Athens (in greek).

Papanikolaou, D. 1986c. The medial tectonometamorphic belt of the Hellenides. Bull. Geol. Soc. Greece, 20, 101–120 (in greek).

Papanikolaou, D. 1988. Introduction to the Geology of Crete and Field guide. IGCP No 276, October 1988 FieldMeeting in Crete, 3–36.

Papanikolaou, D., 1989. Are the medial crystalline massifs of the eastern Mediterranean drifted Gondwanian fragments? In: Papanikolaou, D., Sassi, F.P. (Eds.), Sp. Publ. Geol. Soc. Greece, 1,63–90 and IGCP 276, Newsletter, No 1, Athens.

Papanikolaou, D. 1990. Probable geodynamic interpretation of the schist-chert formations of the Hellenides. Bull. Geol. Soc. Greece, 24, 135–148 (in greek).

Papanikolaou, D. 2013. Tectonostratigraphic models of the Alpine terranes and subduction history of the Hellenides. Tectonophysics, 595–596, 1–24.

Papanikolaou, D.& Sideris, C. 1979. Sur la signification des zones «ultrapindique» et «béotienne» d'après la géologie de la région de Karditsa: l'unité de Thessalie Occidentale. Eclogae Geol. Helv., 72, 251–261.

Papanikolaou, D. & Lekkas, E., 1979. Lateral transition between the Pindos zone and the Western Thessaly unit in the area of Tavropos. Bull. Geol.,Soc. Greece, XIV, 1, 70–84 (in greek).

Papanikolaou, D. & Vassilakis, E. 2010. Thrust faults and extensional detachment faults in Cretan tectono-stratigraphy: implications for Middle Miocene extension. Tectonophysics, 488, 233–247.

Papanikolaou, D., Bargathi, H., Dabovski, C., Dimitriu, R., El-Hawat, A., Ioane, D., Kranis, H., Obeidi, A., Oaie, C., Seghedi, A. & Zagorchev, I. 2004. Transmed Transect VII: East European Craton–Scythian Platform–Dobrogea–Balkanides–Rhodope Massif–Hellenides–East Mediterranean–Cyrenaica. In: Cavazza, W., Roure, F., Spakman, W., Stampfli, G., Ziegler, P. (Eds.), The TRANSMED Atlas: the Mediterranean Region from Crust to Mantle. Springer-Verlag, Heidelberg.

Papastamatiou, J. 1960. La Géologie de la région montagneuse du Parnasse–Kiona–Oeta. Bull. Soc. Géol. France, 2, 398–409.

Papastamatiou and Reichel, 1956. Sur l'age des phyllades de l'ile de Crete. Eclog, Geol. Helv., 49, 147–149.

Philippson, A. 1892. Der Peloponnes. Verlag Friedländer, 642, S., Berlin.

Philippson, A. 1898. La tectonique de l' Egéide. Ann. de Géographie, 112–141.

Philippson, A. 1956–1959. Die Griechischen Landschaften. Volumes I-V, V. Klostermann, Frankfurt.

Rassios, A.H.E. & Moores, E.M., 2006. Heterogeneous mantle complex, crustal processes, and obduction kinematics in a unified Pindos - Vourinos ophiolitic slab (northern Greece). Geol. Soc. Sp. Publ. London, 237–266.

Rassios, A. & Smith, A.G., 2000. Constraints on the formation and emplacement age of western Greek ophiolites (Vourinos, Pindos, and Othris) inferred from deformation structures in peridotites. Geol. Soc. Am., Special Paper, 473–483.

Reilinger et al. 2006. GPS constraints on continental deformation in the Africa–Arabia–Eurasia continental collision zone and implications for the dynamics of plate interactions. J. Geoph. Res. 111, B05411.

Renz, C. 1940. Die Tektonik der griechischen Gebirge. Pragm. Akad. Athinon, 8.

Renz, C. 1955. Die vorneogene Stratigraphie der normal sedimentären Formationen Griechenlands. I.G.S.R., 637 p., Athens.

Ricou, L.E., Zonenshain, L.P., Dercourt, J., Kazmin, V.G., Le Pichon, X., Knipper, A.L., Grandjacquet, C., Sborshchikov, I.M., Geyssant, J., Lepvrier, C., Perchersky, D.M., Boulin, J., Sibuet, J.C., Savostin, LA., Sorokhtin, O., Westphal, M., Bazhenov, M.L., Lauer, J.P. & Bizu-Duval, B. 1985. Methodes pour l' établissement de neuf cartes paléogéographiques de l'Atlantique au Pamir depuis le Lias. Bull. Soc. Géol. France, I, 5, 625–635.

Robertson, A., 2004. Development of concepts concerning the genesis and emplacement of Tethyan ophiolites in the Eastern Mediterranean and Oman regions. Earth Science Reviews, 66, 331–387.

Savostin, L.A., Sibuet, J.C., Zonenshain, L.P., Le Pichon, X. & Roulet, M.J. 1986. Kinematic evolution of the Tethys belt from the Atlanticocean to the Pamirs since the Triassic. Tectonophysics, 123, 1–35.

Smith, A.G. 1979. Othris, Pindos and Vourinos ophiolites and the Pelagonian Zone. VI.Coll. Geol. Aegean Region, Athens 1977, Proc. Ill, 1369–1374.

Smith, A.G., 1993. Tectonic significance of the Hellenic–Dinaric ophiolites. Geol. Soc., London, Sp. Publ., 76, 213–243.

Tataris, A. 1964. About the presence of the Olonos-Pindos zone in the Symi-Viannos area (Eastern Crete) and the age of its spilites. Proc. Acad. Athens, 39 (in greek).

Tataris, A. 1975. Some questions on the «itinerary» of the younger (sh2) schist-psammite-chert complex and the relation Pelion – Olympus. Bull. Geol. Soc. Greece, 12, 95–112 (in greek).

Thiebault, F. 1977. Etablissement du caractère ionien de la série des calcschistes et marbres (Plattenkalk) en fenêtre dans le massif du Taygète (Péloponnèse, Grèce). C. R. somm. Soc. Géol. France, 3, 159–161.

Vergely, P. 1976. Chevauchement vers I' Ouest et retrocharriage vers I' Est des ophiolites: deux phases tectoniques au cours du Jurassique supérieur - Eocrétacé dans les Hellenides internes. Bull. Soc. Géol. France, 18, 233–246.

Orogenic Evolution of the Hellenides

10

10.1 The Integration of the Hellenic Tectonostratigraphic Terranes in the European Margin

The orogenic evolution of the Hellenides took place exclusively in the European margin, which was active throughout the Mesozoic—Cenozoic, in contrast to the African margin, which remained passive, without any orogenic phenomena (Papanikolaou 1989b, 1997; Stampfli et al. 1991; Papanikolaou et al. 2004). This can be visualized in a series of geotectonic profiles, including the distinction of the nine Hellenic terranes, showing the drifting of the continental terranes from Africa to Europe, as well as the opening of the oceanic basins and their subsequent subduction and closure before their accretion to the European Margin (Papanikolaou 2013) (Fig. 10.1).

Already since the Paleozoic, tectono-stratigraphic terranes of Gondwanian origin had been integrated in the Eurasian margin, both in the Balkanides and the Pontides, within the frame of the evolution of paleo-Tethys, the closure of which is considered to have occurred in the Triassic (Sengor et al. 1988; Yanev 1993; Haydoutov et al. 1997; Goncuoglu et al. 1997; Karamata et al. 1997; Haydoutov 2002). The evolution of the Cimmerides during the Late Triassic—Lias may be linked with the closure of paleo-Tethys, although it might also be related with the closure of another, smaller oceanic basin located southwards. In the orogenic evolution of the Hellenides, this initial stage of the Alpine cycle is related with the evolution of the H6 and H8 basins, regarding the large scale geotectonic structure. Based on the present geotectonic position of the oceanic sutures of H6 and H8, it would be logical to assume that H8 was the last remnant of paleo-Tethys, while H6 would be a basin developed directly to the south, closed during the Cimmeride orogeny, that followed the general northward drift of the H5 terrane. This northward drift is not a process occurring only in the Mesozoic—Cenozoic history, but also extends back into the Paleozoic, both in the region of Europe as well as in Asia (Sengor 1989).

However, the available chronologies for the H6 and H8 tectonic events are similar, constrained between the Early and the Late Jurassic. The differentiation between H6 and H8 is based on the intermediate continental terrane of the Rhodope autochthon H7, which surely had finite dimensions and as such, in its lateral ends, the two basins of H6 and H8 are expected to merge into a single basin. A similar process is expected for the oceanic terranes of the Axios H4 and Lesvos/Circum Rhodope H6 on either side of the Paikon/Lesvos autochthon continental terrane H5, as well as for the Pindos-Cyclades H2 and the Axios H4 oceanic basins, which are separated by the Pelagonian terrane H3, which are clearly linked into a single basin to the north of H3 towards Serbia (Papanikolaou 2009). The Rhodope area has undergone thermal effects of extreme intensity, with amphibolitic facies metamorphism, migmatization, and extrusion of anatectic granites in the Cretaceous, which have overprinted and erased the previous processes of HP/LT metamorphic conditions. Thus, a detailed mapping and tectono-stratigraphic, as well as tectono-metamorphic analysis, combined with new geochronological datings, will allow the understanding of the complex processes that occurred during the Triassic—Jurassic, throughout the area of the terranes H5, H6, H7, H8, and H9, which are involved in the internal tectono-metamorphic belt of the Hellenides.

During the Late Jurassic—Early Cretaceous, it is certain that all the terranes H5-H9 were participating in the inner part of the orogenic arc of the paleo-Alpine orogenic phase, and have experienced the processes and effects of the volcanic arc. Here, it should be noted that the Paikon H5 and the Peonia H6 domains seem to have undergone surficial processes dominated by volcanism, while the areas of the metamorphics in the regions of H7, H8, and H9 have undergone deep-level magmatic processes, which can be documented on the surface today due to the subsequent considerable uplifting and erosion of the overlying formations. The main geotectonic event of the Palao-Alpine orogeny is the obduction of the Axios H4 oceanic basin onto the H3 Internal platform, with the ophiolite nappes emplaced

D. I. Papanikolaou, *The Geology of Greece*, Regional Geology Reviews,
https://doi.org/10.1007/978-3-030-60731-9_10

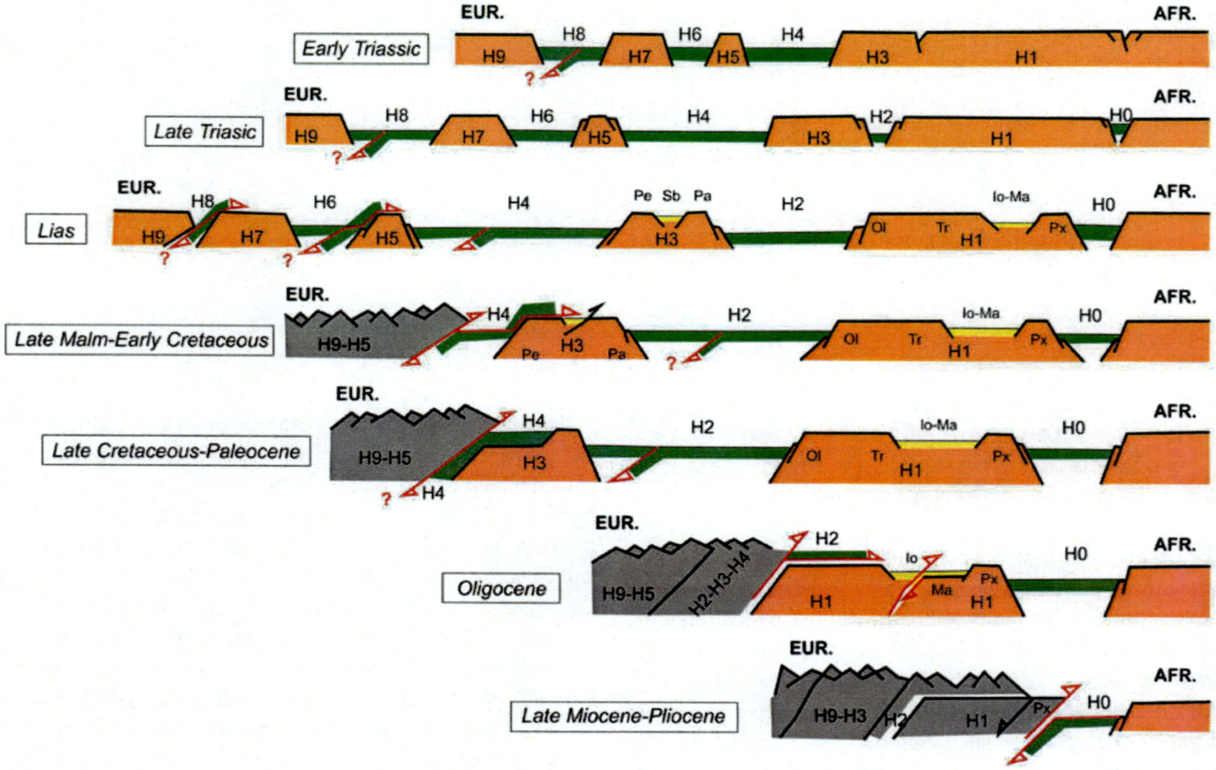

Fig. 10.1 Schematic tectonic sections, depicting the geodynamic evolution of the terranes H1—H9, through the paleogeographic region of the Hellenides in Tethys, from the Early Triassic to present (from Papanikolaou 2013). Pe: Pelagonian, Sb: Sub-Pelagonian, Pa: Parnassus, Ol: Olympus, Tr: Tripolis, Io: Ionian, Ma: Mani, Px: Paxos

over the Pelagonian (Pe) and SubPelagonian (Sb) domains, but not over the more external Parnassos (Pa) platform (Fig. 10.1). However, it is not certain whether the H4 ocean definetely closed or some part of it remained open also in the Late Cretaceous. It is noteworthy that since Malm, the paleogeographic region of the Hellenides is shrinking, with the subduction and integration of terranes in the European margin, while the region has been extended into a maximum in the Lias.

At the end of the Cretaceous—Paleocene, the Axios Ocean H4 definetely closed and the subduction of the Pindos Ocean H2 has already begun. However, there are certain geochronological data derived from the amphibolites occuring at the base of the ophiolite nappes, showing tectonic processes in the H4 and H2 basins already since the Lias, possibly due to intra-oceanic subduction.

In the Oligocene, the Pindos H2 area has already been closed and tectonically emplaced over the internal section of Olympus (Ol) and Tripolis (Tr) of the external platform H1 (Fig. 10.1). At the same time, the internal imbrication of the platform H1 has already begun with the Mani (Ma) and western Crete units been subducted under the internal section of Ionian (Io) and Tripolis (Tr) of H1. At the end of the Miocene—Pliocene, the orogenic evolution of the Hellenides has been completed and the contemporary subduction process of the Ionian oceanic basin H0 has begun, beneath the external margin of the Paxos (Px) in the H1. The subduction of H0 is occurring beneath the amalgamated region of H1-H9 terranes of the newly formed Aegean microplate, with an almost total disappearance of the Tethyan realm, created during the Early Mesozoic.

10.2 History of the Hellenic Subduction Zone

The dominant tectonic process of the orogeny of the Hellenides is the convergence between the African and European plates and the Hellenic subduction zone, whose history begins in the Lias and continues up to the present day (Papanikolaou 2013). The paleogeographical reconstruction implies that the Hellenides experienced their maximum geographical extent during the Lias (Dercourt et al. 1985; Papanikolaou 2013). Nevertheless, additional oceanic segments have been integraded during the Middle Jurassic—Late Cretaceous opening in the H2 and H0 terranes, contemporaneously with the subduction and closure of the previous H6 and H4 oceanic basins. The overall size of the Hellenides by considering the sum of the different chronological periods is estimated about 4.000 km (Fig. 10.2b), but it has never overcome approximately the half of it for each period. The relative motions between Africa and Europe obtained from paleomagnetic studies (Bullard et al. 1965; Smith 1971, 2006; Channel et al. 1979; Jolivet and Faccenna 2000; Rosenbaum et al. 2002; LePichon et al. 2019) have shown that during the early period of the Tethyan history opening and sinistral strike slip kinematics along an E-W megashear were prevailing, whereas from the beginning of Late Cretaceous N-S convergence dominated with rapid reduction of the Tethyan realm until the Early Miocene, when convergence slowed down to the actual rates extracted by the GPS measurements.

The size estimation process of the nine Hellenic terranes was conducted with a good approximation for the continental terranes H1, H3, H5, H7 and H9, which are generally non-compressed and they are preserved, even when internal imbrications are involved. On the contrary, it was conducted with a fairly generalised approximation for the oceanic basins of the terranes H0, H2, H4, H6, and H8, which have disappeared within the subduction process and only small fragments have been preserved in their final tectonic emplacement/obduction during the closure of the basins. At the same time, the pelagic sediments of the basins are intensively folded and imbricated, due to their plastic/easily deformed nature, so estimates of the initial size are characterized by significant uncertainty. The schematic representation of a characteristic seismic tomography of the Hellenic arc allow the correlation of the subducted sections deep within the asthenosphere and the mesosphere with the reproduced section of the Hellenides prior to tectonism (Papanikolaou 2013) (Fig. 10.2a). In the meantime, the comparison of the present cross Sect. (1a) with the palinspastic Sect. (1b), illustrates the magnitude of the shrinking of the paleogeographic region, and reveals the regions that have essentially disappeared. It should be reminded that the main tectonic process during the subduction of the Hellenides is the detachment of the upper sections of the crust from the subducted African plate (of a few kilometers thickness in the basins and of a few tens of kilometers thickness in the carbonate platforms), with an uninterrupted continuity of the subduction of the remaining lithospheric slab towards the inner part (see also Figs. 2.8. and 4.2).

The history of the Hellenic subduction zone may be schematically reproduced in chronological order by distinguishing the rifting stage period for each terrane, the period of the continental terrane drifting, as well as the simultaneous ocean basin opening, and finally, the subduction period for each terrane (Papanikolaou 2013) (Fig. 10.2c). Therefore, combined with the tectono-stratigraphic diagrams of the continental terranes and the oceanic basins (see Fig. 8.1), the individual stages for each terrane have been chronologically constrained. The rifting stage in the Hellenides started in the Carboniferous for all the terranes, and terminated in the Early Permian for the terranes H5-H8, in the Middle Triassic for the terranes H3 and H4, and in the Carnian for the terranes H1 and H2. The termination of the rifting stage is associated with the beginning of the carbonate platform sedimentation in the continental terranes, and of the pelagic sedimentation in the oceanic basins respectively. The initiation of the flysch sedimentation signals the arrival of the terranes in the Hellenic orogenic trench and the beginning of the subduction, while the final uplifting of each unit signals either the accretion of the continental terranes in the European margin, or the disappearance of the ocean and its eventual tectonic emplacement on the adjacent margin towards the south (obduction).

The subduction process characteristically includes periods corresponding to distinct orogenic phases (Papanikolaou 2013) such as: (1) the paleo-Alpine orogenic phase, with a subduction event at 140 Ma and a completion at 110 Ma corresponding to the Cenomanian unconformity, separating H3 in two sections, one paleo-tectonized and one of a main Alpine phase, (2) the beginning of the subduction of Pindos in the Maastrichtian, at 65 Ma, and the closure of the basin in the Late Eocene at 40 Ma, (3) the subduction of Mani under the Ionian and other units of H1 during the end of the Oligocene at 25 Ma, (4) the accretion of H1 in the Hellenic arc and the beginning of the subduction of H0 in the Miocene/ Pliocene boundary at 5 Ma.

By examining each oceanic basin separately, we can differentiate the age of the tectonic emplacement of its ophiolites on the adjacent platforms to the south (Papanikolaou 2009), as documented in the four obduction phases, depicting the emplacement of H8 over H7 during the Jurassic, of H6 over H5 in the Lias, of H4 over H3 in the Late Jurassic—Early Cretaceous, and of H2 over H1 in the Late Eocene (See also Fig. 8.130). The above events of subduction/obduction of the ophiolites do not contradict the continuity of the subduction process. However, it is important to note that the subduction process is different, based on whether we have oceanic or continental crust that is being subducted. In particular, the subduction rate of the oceanic crust is usually 6–8 cm/year, while the rate of the continental crust is reduced to 1 cm/year (Royden and Papanikolaou 2011). These subduction rate alternations impact significantly on the arc and related phenomena in both plates (Lallemand et al. 2005; Royden and Husson 2006, 2009).

The examination of the geotectonic evolution east of the Aegean area shows that the Neotethyan closure history of Western Anatolia comprises two successive phases of subduction of a single lithospheric slab and episodic accretion of

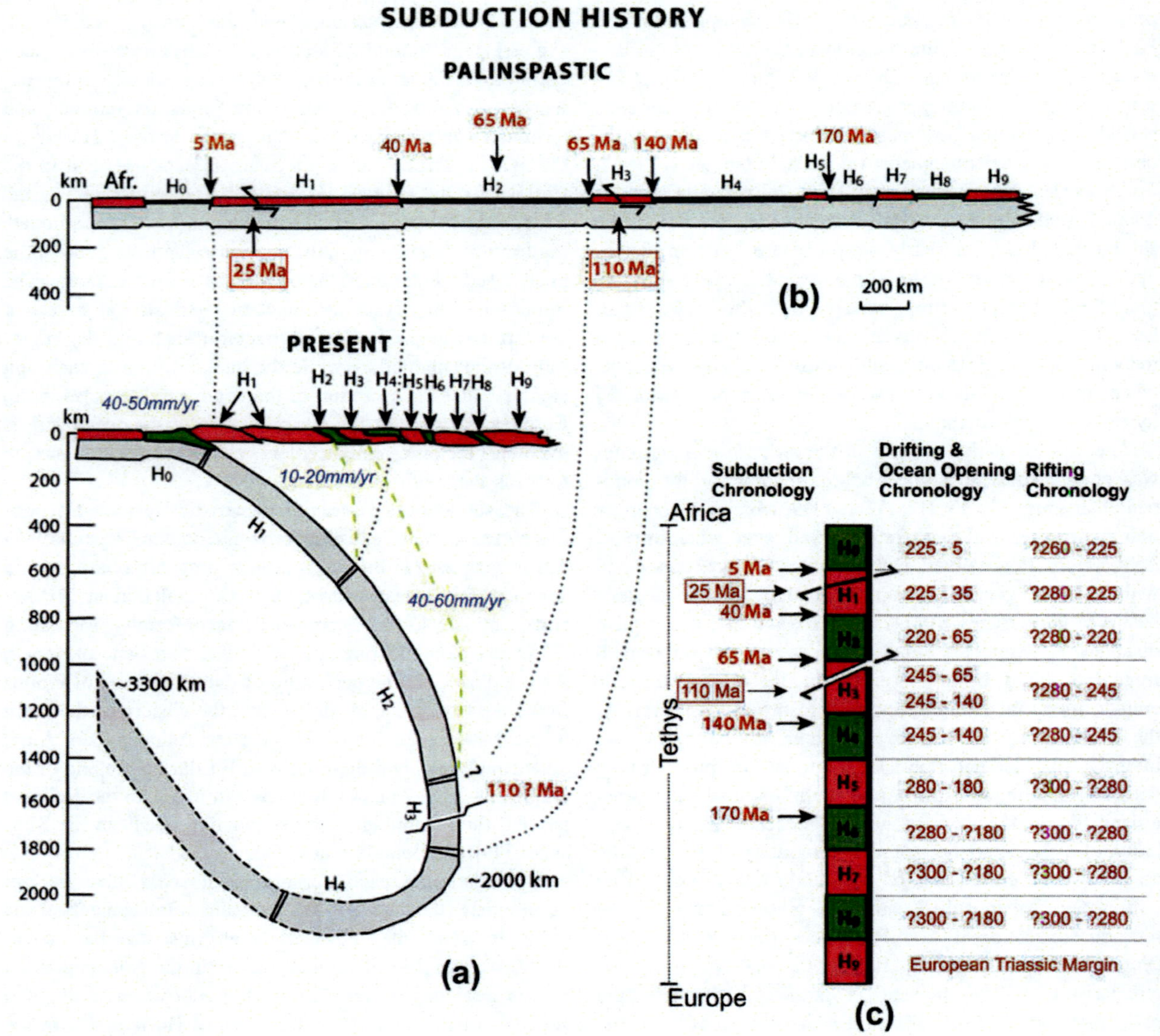

Fig. 10.2 The structure and history of the Hellenic subduction zone according to Papanikolaou (2013). **a** Schematic diagram of the seismic tomography of the Hellenic subducted lithosphere and the Hellenic terranes along the present section of the Hellenides, over the seismic tomography that was granted by Spakman and interpreted by Papanikolaou (2004). **b** Palinspastic representation of the subducted lithosphere and placement of the Hellenic terranes on it. The correlation of the subducted sections with the palinspastic section is depicted through the use of thin doted lines. **c** Schematic representation of the simplified chronology of the three stages of rifting, drifting and subduction/accretion of the terranes, with the main tectonic and geodynamic events in the Hellenides highlighted

two continental domains, separated by a continental trough, representing the eastern end of the Cycladic ocean of the Aegean (H2) (Pourteau et al. 2016). Their tectonic scenario suggests a continuous subduction since $\sim$110 Ma, marked by roll-back episodes in the Paleocene and the Oligo-Miocene and slab tearing below Western Anatolia during the Miocene.

10.3 Documentation of the Orogenic Arcs in the Hellenides

The older Alpine orogeny that can be documented in the Hellenides is the Cimmeride orogeny, related to the emplacement of H6 over H5 during the **Late Triassic/Lias** boundary. The completion of the orogeny is sealed by the Lias unconformity, which in Greece can be observed mainly in the Chios Allochthon. The orogenic arc of the Cimmerides has been documented in the NW Asia Minor, with characteristic outcrops in the Kutahya area, where granitic rocks of the Late Triassic are observed below the Lias unconformity. At the same time, the Permian–Triassic formations known as Karakaya have undergone metamorphism, and emplacement of ophiolites of Permian age, probably associated with those of the Lesvos Allochthon (Papanikolaou and Demirtasli 1987; Papanikolaou 2009; Robertson and Ustaomer 2012). Moreover, flysch formations of Late Triassic age can be observed, metamorphosed or not, while Triassic blueschists have also been reported (Okay et al. 2002). Overall, there is plethora of data documenting the Cimmeride orogeny, although its reconstruction is not possible in the Greek region, but mainly in Asia Minor.

The most important early orogeny in the Hellenides is the so-called paleo-Alpine, which occurred during the **Malm-Early Cretaceous**, i.e. the Neo-Cimmerian and the Austrian foldings of Stille (1936). During this orogeny, which Aubouin had already analysed since the 1970s, the tectonic emplacement of the Axios ophiolite nappes took place over the internal units, forming the dominant geotectonic event. Available data strongly support the formation of an orogenic arc during this period with (Fig. 10.3):

- trenches with flysch sedimentation traced in all the internal units: Maliac, Sub-Pelagonian, Almopia, etc., during the Malm, with the areas of the Western Thessaly —Beotia as external boundary.
- island arcs, mainly made of outcrops of ophiolites and radiolarites that supply the trenches with clastic material.
- back-arc basins in the Peonia region, with the creation of a marginal sea with a crust made of basic rocks followed by molassic sedimentation.
- Volcanic/magmatic arcs with outcrops mainly in the Paikon unit, as well as in the Vertiskos and Kerdylia units of the Serbo-Macedonian.

Additionally, during this orogenic event, deep-level geodynamic phenomena with metamorphism and ductile deformation take place in the Almopia unit, as well as in other internal units, such as the allochthonous metamorphic units of Skyros Island (Vergely 1984; Jacobshagen and Wallbrecher 1984). The same deep geodynamic phenomena are also observed in the pre-Alpine basement rocks of the Kastoria, Flambouro, Vertiskos, and Kerdylia units, overprinting older Variscan events (Mercier 1968; Schermer et al. 1993; Mposkos and Perraki 2001).

Nevertheless, the paleo-Alpine orogeny lasted for approximately 40 Myrs from Malm and Early Cretaceous up to Cenomanian. At the same time, at least two distinct well defined stratigraphic unconformities have been described in the H4 region, with the most dominant one corresponding to the Cenomanian unconformity and the other one during the Late Jurassic (see Sects. 8.3.5 and 8.4.3). Therefore, the so-called paleo-Alpine orogenic phase could be distinguished in two sub-phases, both of them referring to the Axios region. The older unconformity of the Late Jurassic affected the more internal sections of H4, which started the subduction and the creation of a first orogenic arc and were completed in the Early Cretaceous by the second phase. The entire paleotectonised domain was covered by the Cenomanian unconformity after the closure of H4. Furthermore, the Late Jurassic phase probably affected sections of the

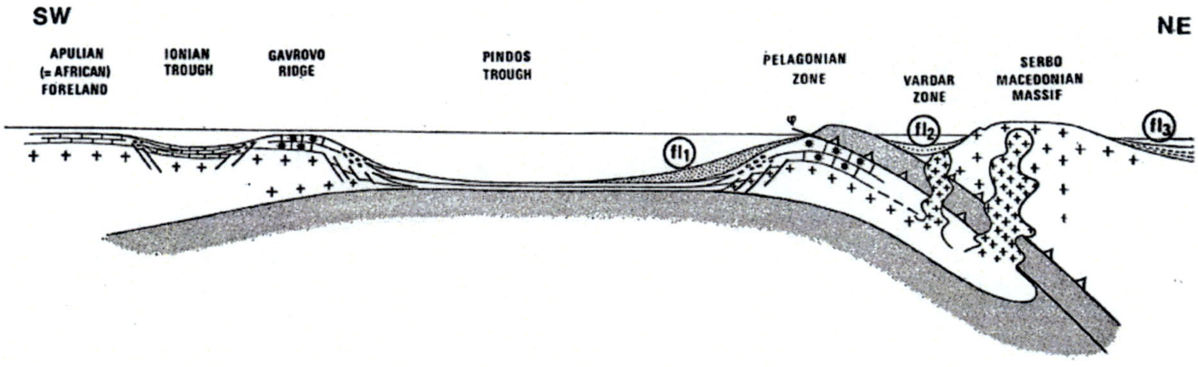

Fig. 10.3 Paleo-geodynamic scheme during the Jurassic /Cretaceous boundary in the Hellenides, according to Aubouin (1977), with documentation of the orogenic arc in the Axios and Pelagonian regions, where the ophiolites are tectonically emplaced, providing clastic flysch type material in the troughs (fl1, fl2, fl3), while the volcanic arc can be seen in the region of the internal Axios and on the Serbo-Macedonian. The Pindos basin is considered as a marginal sea, while the rest of the External Hellenides are the passive margins of the Apulian (Africa) of Atlantic type, in contrast to the European margin, which is an active margin of the Pacific type—the Andes

more internal part of H6, which may have not disappeared completely during the Late Triassic-Liassic phase of the Cimmerides.

The next orogenic arc that can be documented is in the **Eocene** and includes (Jacobshagen 1979; Papanikolaou 1984) (Fig. 10.4):

- the trenches with successive flysch sedimentation in the Eastern Greece, Western Thessaly—Beotia, Parnassos, Pindos and Cyclades units,
- the island arc, in the area of the internal units, with the progressive addition of the Western Thessaly—Beotia, Parnassos, and Pindos units,
- the back-arc basin, with molasse deposition in the Rhodope—North Aegean area,
- the volcanic arc, in the Rhodope—North Aegean area.

In the pre-orogenic area of the subducting plate, the units of the external platform enter into the trench during the Late Eocene in a successive order, from Olympus, Tripolis, Gavrovo, and then the Ionian. Deep-level phenomena with blueschist metamorphism can also be observed in the Cyclades, in the subduction zone, and greenschist metamorphism in Rhodope, with multiple phases of ductile deformation. This phase may have already started in the Late Cretaceous, with the subduction of the units that can be seen today overlying the Northern Cyclades, like the Makrotantalon/Ochi unit, which are later integrated at depth with the Cycladic blueschists during the Late Eocene.

Subsequently, the orogenic arc migrated during the **Oligocene—Early Miocene** in a more external position (Papanikolaou and Dermitzakis 1981; Papanikolaou 1986) with (Fig. 10.5):

- a trench, with flysch sedimentation in the Ionian, Mani, Gavrovo and Tripolis units,
- an island arc, with the Pindos, Parnassos, Western Thessaly—Beotia and sections of Eastern Greece units,
- a back-arc basin, with molasse deposition in the Meso Hellenic Trough and the Cyclades,
- a volcanic arc, in the North Aegean with massive volcanic extrusions in the Limnos, Lesvos, and Aghios Efstratios islands.

At the same time, the blueschist metamorphism in the Arna unit occurs, in addition to a low grade metamorphism in Mani and Western Crete units, as well as the greenschist metamorphism and migmatization in the Cyclades.

Immediately afterwards, during the Late Miocene, in the **Messinian**, the migration of the arc to a more external position occurs, as well as the initiation of the subduction of the last remnant of the Tethys Ocean of H0, resulting in the reorganization of the arc, with the restriction of the contemporary arc between Lefkada and Rhodes (see also

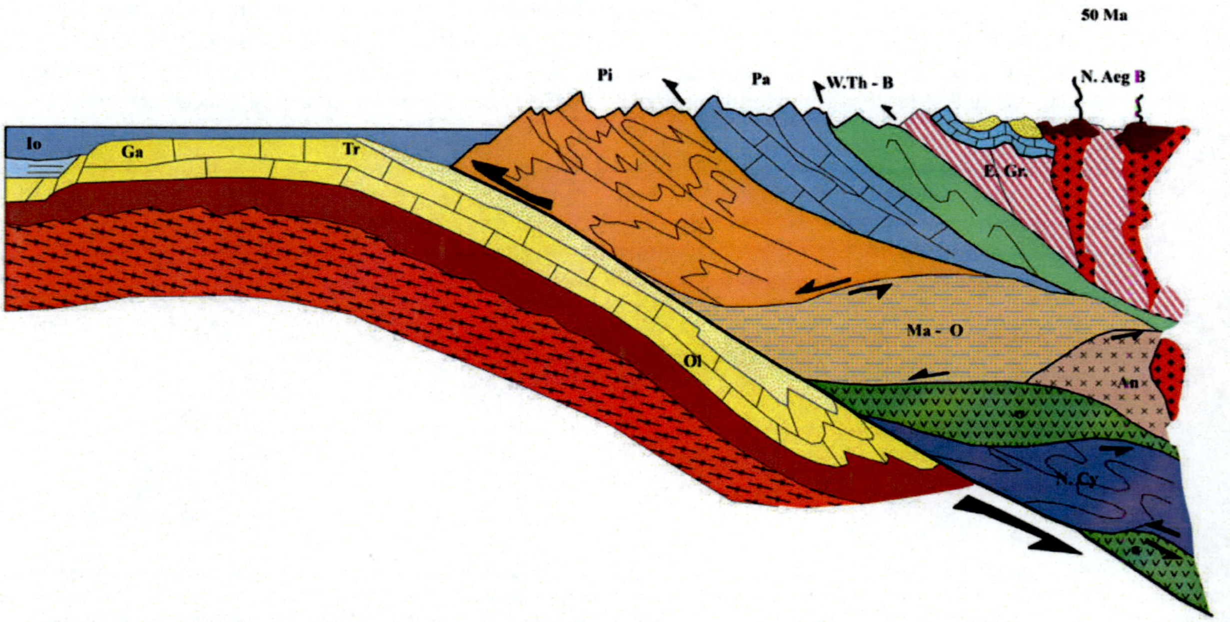

Fig. 10.4 Schematic representation of the Hellenic orogenic arc during the Eocene. Io: Ionian, Ga: Gavrovo, Tr: Tripolis, Ol: Olympus, Pi: Pindos, Pa: Parnassos, W.Th.-B: Western Thessaly—Beotia, E.Gr: Eastern Greece, Ma-O: Makrotantalon—Ochi, An: Anafi, N.Cy: Northern Cyclades, N.Aeg.B: Northern Aegean Basin, σ: ophiolites H2

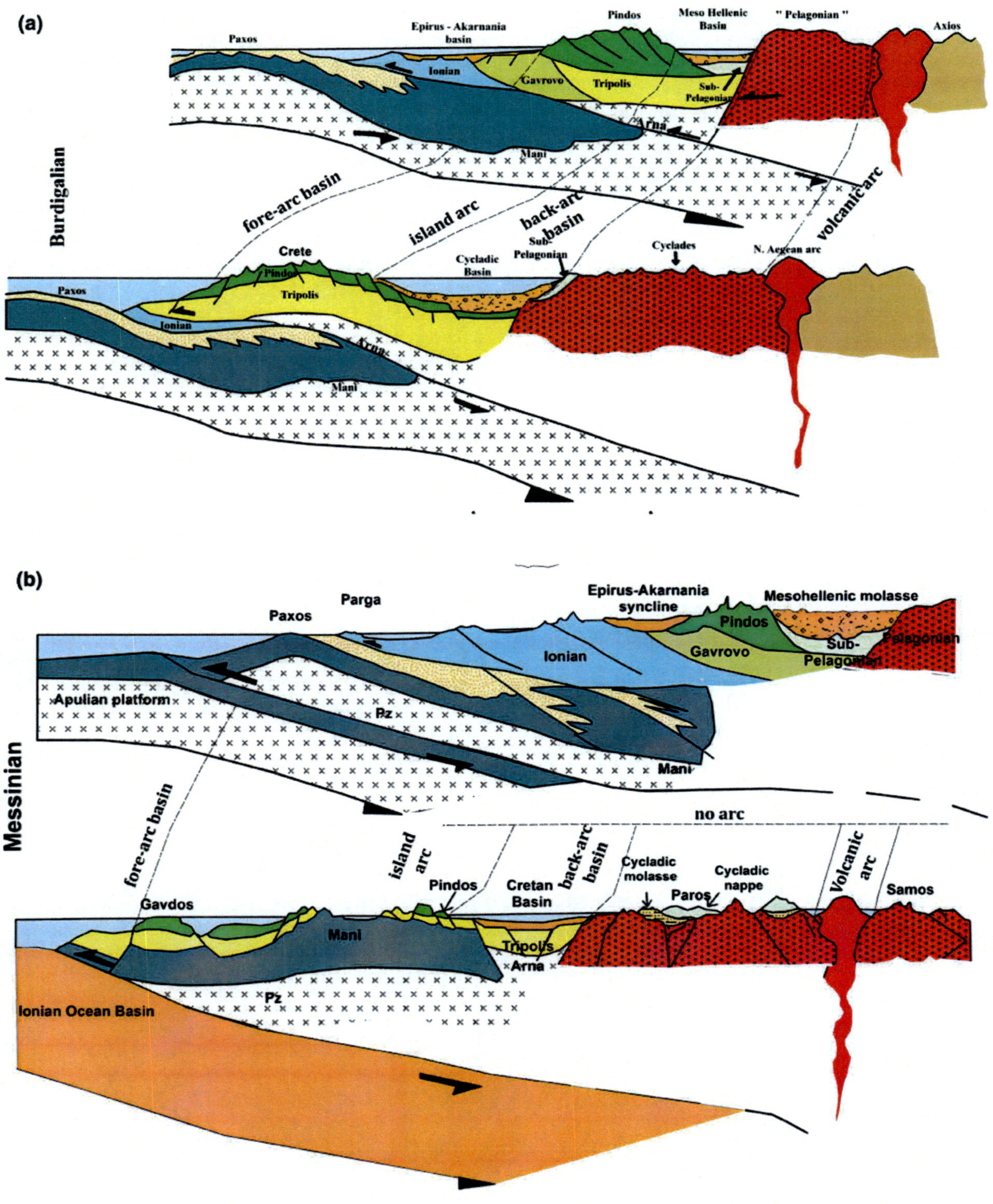

Fig. 10.5 The Hellenic orogenic arc, as shown in two transverse sections of the External Hellenides area during the Oligocene—Early Miocene (mainly in the Burdigalian) and during the Late Miocene (mainly in the Messinian) (based on Papanikolaou and Dermitzakis 1981). The sections of the northern sector show that the arc was rendered inactive during the Tortonian, while the southern sector of the arc migrated to a new, more external position, above the newly established oceanic subduction of the Ionian basin and to the creation of the Cretan back arc basin

Figs. 11.2 and 11.3). Thus, the areas of the arc north of the CHSZ, become inactive and there is no longer trench, island arc, back-arc basin and volcanic arc. The two sections of the arc, one through the Northern Hellenides and the other through the Southern Hellenides, are radically different, as shown in Fig. 10.5.

Finally, we reach the **present** geometry of the Hellenic orogenic arc, as documented by Angelier (1979), with (Fig. 10.6):

- a trench, in the area of the Ionian Sea and south of Crete in the Pliny and Strabo trenches,
- an island arc, that includes the Peloponnese, Crete, and the Dodecanese,
- a back-arc basin, in the area of the Cretan basin,
- a volcanic arc, in the area of the southern slopes of the Cycladic plateau along the northern margin of the Cretan basin.

The above descriptions demonstrate that the orogeny of the Hellenides is essentially continuous from Lias to present, despite the fact that characteristic orogenic phenomena can be observed in different sections of their paleogeographic organization. That is to say, since the Lias, we have a continuous plate convergence, with the southern margin of the Eurasian plate being permanently active. It is under this margin that the Tethyan intermediate micro-plates of the tectono-stratigraphic terranes H8-H1 have been subducted and accreted, until the present subduction of the African plate under the Aegean micro-plate, which was differentiated approximately 10 Ma ago in the southern margin of the Eurasian plate.

The northern boundary of the Anatolian and Aegean microplates is expressed by the North Anatolian Fault and its western prolongation into the Aegean (McKenzie 1970, 1972; Armijo et al. 1999; Sengor et al. 2005). Thus, the two basins of the North Aegean Sea, the North Aegean Basin (NAB) and the Skyros Basin (SKB), form the actual plate boundary of the Aegean microplate within the hinterland of the Hellenic arc and trench system. Based on detailed swath mapping and air gun lithoseismic profiles the overall structure of the two basins has been described (Papanikolaou et al. 2002, 2006, 2019b) with distinction of three segments: (i) an eastern segment, where the dominant tectonic trend is ENE-WSW, with relatively shallow depths, forming narrow basins with dextral strike slip deformation. (ii) a central segment, where the tectonic trend becomes NE-SW, the depth increases and the dominant deformation becomes oblique faulting and (iii) a western segment, where the tectonic trend remains NE-SW, but the width becomes larger, the depth becomes maximum and the deformation acquires prevailing characteristics of normal faulting with NW–SE orientation (Fig. 10.7). The overall process indicates a gradual change from strike slip to normal faulting as approaching to the Greek mainland. This change observed from NE to SW, has produced gradual subsidence and gradual widening of the basins and gradual growth of fault throws.

It is noteworthy that both North Aegean basins are gradually expanding and deepening from east to west) (Papanikolaou et al. 2019a). This is reflected on: (1) Their seafloor morphology, with the maximum depths ranging from 700 to 1650 m (NAB) and from 500 to 1200 m (SKB) respectively; (2) Their length (350 km (NAB) versus

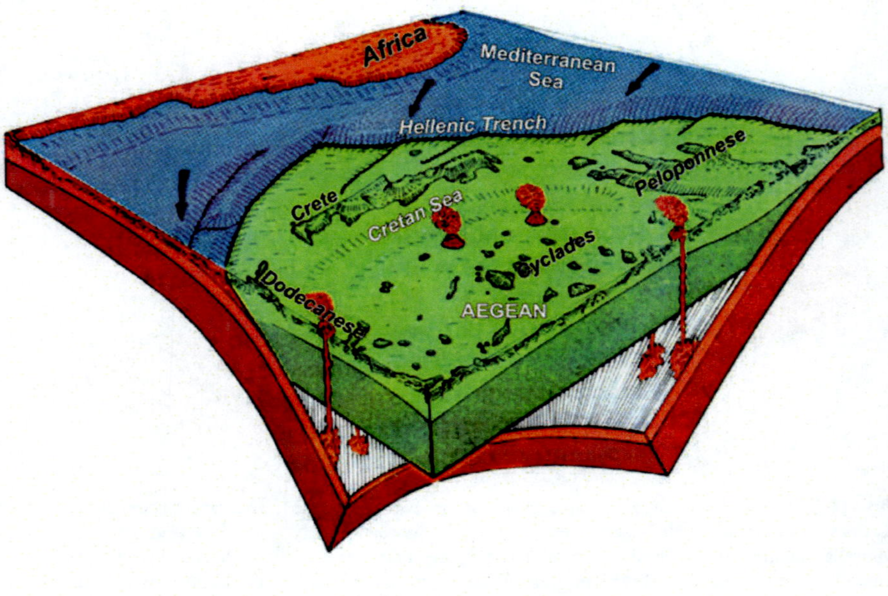

Fig. 10.6 The actualistic model of the contemporary orogenic arc of the Hellenides, in a three-dimensional representation, according to Angelier (1979)

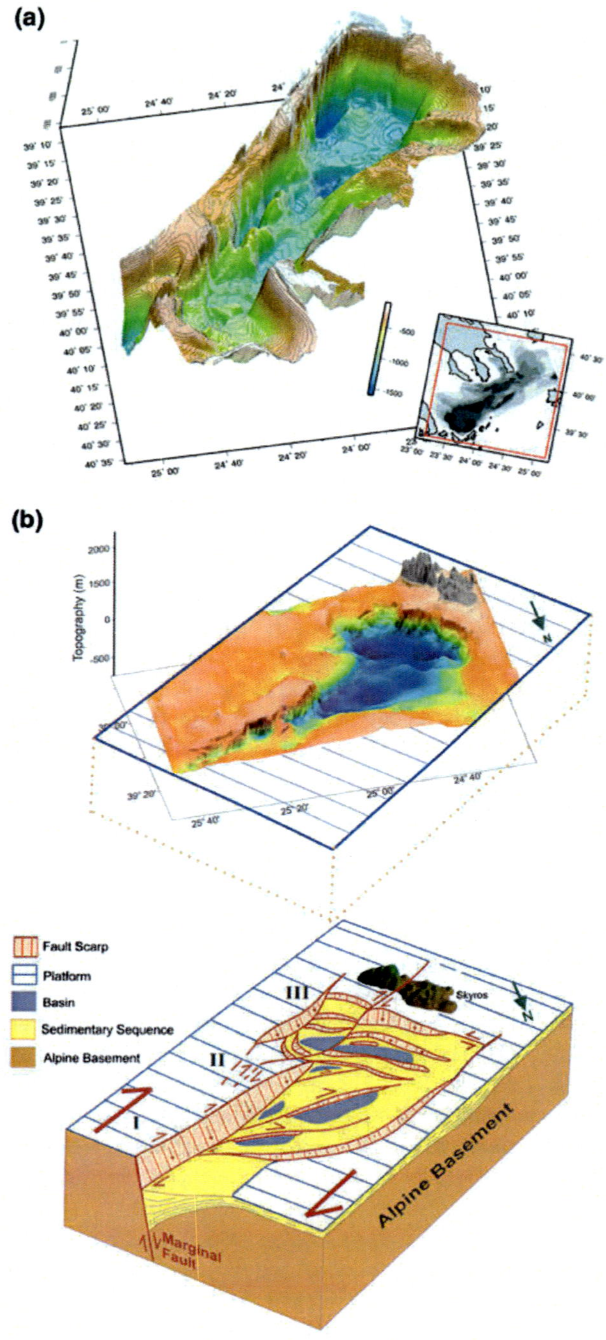

Fig. 10.7 Stereographic diagrams of the North Aegean basin bathymetry (a) and Skyros Basin bathymetry and tectonic structure (b), showing the increase of their depth towards the W-SW and their deformation from ENE-WSW strike slip faulting to NW–SE normal faulting (based on data from Papanikolaou et al. 2002, 2006, 2019b)

200 km (SKB)) and width (from 25 to 50 km and from 10 to 40 km); (3) Their maximum throws (from >800 m to >1700 m versus >1100 m to >1200 m). It is remarkable that the two basins are similar, both regarding their geometry and their geodynamic characteristics with an approximate 2/1 (NAB/SKB) ratio. This is shown by: (a) Their maximum submarine escarpments (1400 vs. 800 m); (b) Their subsidence, as indicated by the depth of the Alpine basement beneath the syntectonic sedimentary sequences (>2500 m vs. >1900 m); (c) Their seismicity, with much higher activity in the northern basin; (d) Their GPS ratios, showing oblique extension to the SSW, with ~16 mm/year versus ~8 mm/year (Mcklusky et al. 2000; Muller et al. 2013).

The western termination of the tectonic boundary of the Aegean plate is represented by the CHSZ through continental Greece from the Maliac Gulf to the Amvrakikos Gulf and the Cephalonia transform fault (Papanikolaou and Royden 2007).

10.4 Migration of the Orogenic Arc

Since the creation of the first orogenic arc in the Hellenides and up to the present, a gradual migration of the arc has been observed from the hinderland in the area of the Balkanides, north of Rhodope, up to the contemporary foreland, south of the present Hellenic trench system. Therefore, all the individual geodynamic deep-level and surficial phenomena associated with the orogenic arc, show a general gradation from the internal part of the arc, where the oldest formations are located, towards the circumference, where the younger ones are located. This is generally true for the flysch formations, the molasse formations, the volcanics and their associated plutons of the calc-alkaline type, which point to magmatic arcs, the folding of the sediments, the metamorphic phenomena, and the deep ductile synmetamorphic deformations (Papanikolaou 1986c, 1993). Even though there are certain periods during which the geometry of the orogenic arc is stabilized in certain paleogeographic settings, the overall continuous movement and transformation of the arc is evident with the entry of new domains, which is the result of the continuous convergence of the plates in the Tethyan region. This means that while the geometry of the arc remains approximately stable, the terranes and the units that take part in each section of the arc are constantly changing. Thus, since the Lias, the orogenic arc has absorbed the entire paleogeographic region of the Hellenides, i.e. with conservative calculations, an area of about 4.000 km in a timeframe of 180 million years, which corresponds to an average integration of 2.2 cm/year of the crust between Eurasia and Africa. This value is compatible with the contemporary convergence rate between Europe and Africa, which is 1 cm/year, and with the (oceanic) subduction rate of the Ionian basin, which is 4–5 cm/year. Additionally, it can also be compared with the convergence rates between

Europe and Africa, resulted from paleo-magnetic data of previous geologic periods (e.g. Smith 2006).

The migration of the orogenic arc can be recognized in paleogeographic maps either based on the contemporary geography or on palinspastic models, and it is facilitated from the migration of the volcanic arc. The volcanic arc is the most important feature of the arc because it is the latest and seals the entire geometry for each period. Additionally, the volcanic arc is preserved, regardless the time that has elapsed since it builds the rocks from the surface down to great depths. Therefore, even if significant erosion of several kilometers has occurred, eroding all the surficial volcanic rocks (e.g. tufs, pyroclastics, lavas, pumice etc.) the sub-volcanics and plutons (e.g. granites) that lie below, testify the presence of the volcanic arc. In a few words, the andesitic type volcanic rocks may be absent, though the underlying granitic ones may be preserved. Following the above remarks, maps showing the migration of the volcanic arc in the Aegean region have been created, which show its gradual migration from the Late Eocene, from Northern Greece, to the Quaternary, at its present location in the northern margin of the Cretan basin (Papanikolaou 1993) (Fig. 10.8). The above migration of the volcanic arc of about 400 km in 40 Ma, reveals a rate of about 10 km per one million years. Therefore, in a smaller scale, for example in the area of Attica and the Saronic Gulf, we have the granites of Lavrion at 10–8 Ma, the volcanics of Aigina at 3.0–1.5 Ma and the volcanics in Methana at the last interval of 0.7 Ma, with active volcanism even until today (Pe-Piper and Piper 2002). These data show a local westward migration of the volcanic arc within the Saronic Gulf of about 80 km in 10 Ma.

It is interesting to note that, in the Late Cretaceous the large scale volcanic arc of the Balkanides is located mainly along the Sredno Gorie belt, with characteristic granodiorites preserved in Philippoupolis (Plovdiv), which continue to the east of Constantinople (Istanbul) through the entire region of the Northern Pontides (Bocaletti et al. 1978). These are volcanic arcs of the Late Cretaceous created by the subduction of parts of the Hellenides and the Taurides. These volcanics are spatiotemporally related with the anatectic granites in the Rhodope region, such as in the Sidironero unit (Papanikolaou and Panagopoulos 1981; Soldatos and Christofidis 1986). Therefore, the Late Cretaceous subduction is not easily traced in the Hellenides, since it appears as a hiatus between the paleo-Alpine and the Alpine orogeny. However, the Late Cretaceous volcanic arcs, participating in the Late Cretaceous orogeny are dominant in the Balkanides as well as the neighboring Pontides (Fourquin 1975). A description of the magmatic evolution since the Late Cretaceous in relation to crustal and mantle dynamics in the Eastern Mediterranean region was recently presented by Menant et al. (2016).

Even more impressive is the fact that the volcanic arc of the Late Jurassic—Early Cretaceous is not located to the north of the arc of the Late Cretaceous in the Balkanides, but in the Greek region, between the terranes H5–H8, and especially in the Paikon unit. This shows another type of allochthony of the volcanic arc, that is a large-scale transportation to a more external position after its creation. Its current position corresponds to the place where the Late Eocene arc was developed (Papanikolaou 1993). This indicates that the tectonic transport of the arc occurred prior to that, most probably in the Late Cretaceous, maybe in relation to the large strike-slip movements in the Black Sea area (Papanikolaou et al. 2004). Another more revolutionary idea is that the entire Axios/Vardar domain is allochthonous, being transported during Cretaceous above the Rhodope autochthon units, which form a tectonic window, probably equivalent to the Pelagonian (Ricou et al. 1998). In this case, the Late Jurassic—Early Cretaceous magmatic/volcanic arc of Paikon-Peonia-Vertiskos is also allochthonous and its roots should be recovered within the Rhodope complex, north of the Pangeon carbonate platform (Papanikolaou 1985,1993; Ricou et al. 1998).

The geochemical composition of the volcanic rocks of the arc also depends, among other factors, from the type of crust that is being subducted in each period. The External Hellenides, for example, experienced the transition from an oceanic subduction of the H2 to a continental subduction of the H1 during the Late Eocene—Oligocene, and the opposite one, from the continental subduction of H1 to the oceanic subduction of the H0 during the Late Miocene—Pliocene. The time lag of the emergence of the volcanic formations at the surface since the beginning of the subduction of an area, is estimated at 10–15 Ma. This duration results from the subduction rate and the depth of melting of the anatectic magmas. Thus, today, the subduction rate of 4–5 cm/year, the anatexis depth maximum about 180–200 km and the subduction dip of 35–40° provide some major constraints for such calculations. Therefore, it would take about 7–8 Myrs to descent and at least 5–6 Myrs for their uplift and extrusion at the surface. This means that the contemporary Aegean volcanism is associated to sections of the African plate that entered the subduction zone about 12–15 Ma ago. The ascending curve of the volcanic/magmatic rocks can be understood by the observation of granitic and volcanic rocks in the same area, such as the case of Santorini, where Upper Miocene granitic bodies have been described within the metamorphic basement of Athinios at the present Santorini volcano, dated at 7–9 Ma (Skarpelis et al. 1992). This implies that an uplift of about 5–8 km has occurred during the last 6–7 Ma, judging from the crystallisation depth of the granites. The same conclusion can be obtained also from the Upper Miocene Kos monzonite, below the local volcanic field (Altherr et al. 1982). Primitive Santorini basaltic lavas

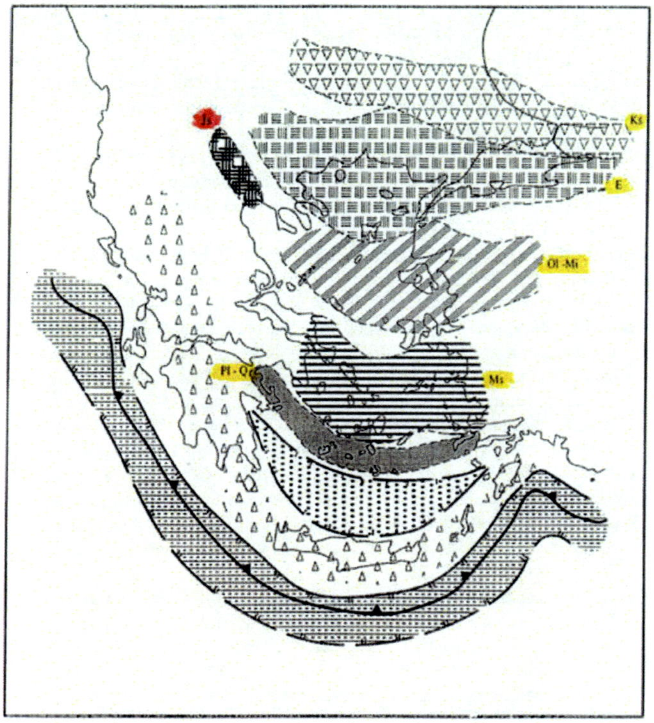

Fig. 10.8 The migration of the volcanic arc as part of the migration of the Hellenic orogenic arc, since the Cretaceous (from Papanikolaou 1993). The location of the arc during the Late Jurassic—Early Cretaceous (Js in red) is "irregular" and approximately coincides with the location of the Eocene volcanic arc, instead of its "regular" position in Northern Bulgaria, to the north of the Late Cretaceous arc. Ks: Late Cretaceous, E: Eocene, Ol-Mi: Oligocene-Early Miocene, Ms: Late Miocene, Pl-Q: Pliocene–Quaternary

formed by 18% fluxed partial melting at 1323 °C and 1.7 GPa show that slab flux to Santorini source derives from 145–150 km depth, that is in agreement with geophysical data from the Wadati-Benioff zone (Baziotis et al. 2018).

A different approach has been developed more recently regarding the magmatic acid igneous bodies, occurring in the central Cyclades in the form of core complexes during the Middle-Late Miocene (15–8 Ma), which show some different geochemical signature from high K calc-alkaline to alkaline (Jolivet et al. 2013, 2015; Menant et al. 2016; Brun et al. 2016). This process formed the **thermal Cycladic zone**, where a migration of the magmatic activity is observed from the Menderes massif to the Cyclades, probably controlled by the tearing of the slab in the Aegean—Anatolian transition zone and the westward directed asthenospheric flow (Menant et al. 2016; Brun et al. 2016). However, the location of this Middle-Late Miocene thermal zone fits well with the overall southward migration of the Aegean volcanic arc since the Early Tertiary and does not disrupt the whole process. Nevertheless, the impressive fast uplift and erosion of the overlying rocks along the Late Miocene granitic core complexes in the Cyclades islands would be justified (See also Fig. 8.99 in Paros Island).

The migration of the volcanic arc has an immediate effect on the **geothermy** of Greece, as it controls the areas of high, intermediate, and low enthalpy of the geothermal fields (Papanikolaou 1989a). Therefore, the geothermal fields of high enthalpy are essentially developed around the contemporary active volcanoes along the Aegean volcanic arc, while the intermediate enthalpy fields are located in volcanic centers of the previous volcanic arcs of the Neogene; the low enthalpy fields are developed in the older volcanic centers of the Paleogene. Naturally, apart from the presence of volcanic formations, the presence of large faults in the upper crust plays an important role in the development of the geothermal fields, since they favour the deep circulation of the surface water and subsequently act as pathways for several **thermo-metallic springs**. The heat flow of the crust, is characterized by the geothermal gradient. Unfortunately, in Greece, there is no detailed map of thermal flow, except from the relatively general data given by Jongsma (1974) for the Aegean Sea and by Cermak (1979) for the Eastern Mediterranean.

References

Altherr, R., Kreuzer, H.. Wendt, I., Lenz, H., Wagner, G., Keller, J., Harre, W. & Hohndorf, A. 1982. A late Oligocene / Early Miocene High Temperature Belt in the Attic - Cycladic Crystalline Complex (SE Pelagonian, Greece). Geol Jb., E 23, 97–164.

Angelier, J. 1979. Néotectonique de l' arc Egéen. Soc. Geol. Nord, Publ., 3, 1–417.

Armijo, R., Meyer, B., Hubert, A. & Barka, A. 1999. Westward propagation of the North Anatolian fault into the northern Aegean: timing and kinematics. Geology, 27, 267–270.

Aubouin, J. 1977. Alpine Tectonics and Plate Tectonics: Thoughts about the Eastern Mediterranean. In: Europe from crust to core, 143–158, J. Wiley.

Baziotis, I., Kimura, J., Pantazidis, A., Klemme, S., Berndt, J. & Asimow, P. D. 2018. Geophysical source conditions for basaltic lava from Santorini volcano based on geochemical modelling. Lithos, 316-317, 295–303.

Boccaletti, M., Manetti, P., Peccerillo, A. & Stanicheva-Vassileva, G. 1978. Late Cretaceous high-potassium volcanism in eastern Srednogorie, Bulgaria. Geol. Soc. Am. Bull., 89, 439–447.

Brun, J.P., Faccenna, C., Gueydan, F., Sokoutis, D., Philippon, M., Kydonakis, K. & Gorini, C. 2016. The two—stage Aegean extension, from localised to distributed, a result of slab rollback acceleration. Canadian Journal of Earth Sciences, 53, 11, 1142–1157.

Bullard, E., Everett, J. & Smith, A. 1965. The Fit of the Continents around the Atlantic. Philos. Trans., R. S. London, A258, 41–51.

Cermak, V. 1979. Heat flow map of Europe. In: Cermak & Rybach editors, Terrestial Heat Flow Map of Europe., Springer Verlag, 3–40.

Channel J.E.T., D' Argenio, B. & Horvath, F. 1979. Adria, the African Promontory in Mesozoic Mediterranean Paleogeography. Earth Science Reviews, 15, 213–292.

Dercourt, J., Zonenshain, L.P., Ricou, L.E., Kazmin, V.G., Le Pichon, X., Knipper, A.L, Grandjacquet, C., Sborshchikov, I.M., Boulin, J., Sorokhtin, O.,Geyssant, J., Lepvrier, C., Bizu-Duval, B., Sibuet, J. C., Savostin, LA., Westphal, M. & Lauer, J.P. 1985. Présentation de 9 cartes paléogéographiques au1/20,000,000 étendant de I' Atlantique au Pamir pour la période du Lias à I' actuel. Bull. Soc. Géol. France, I, 5, 637–652.

Fourquin, C. 1975. L'Anatolie du NW, marge méridionale du continent européen, histoire paléogéographique, tectonique et magmatique durant le Secondaire et le Tertiaire. Bull. Soc. Géol. France, (7), 17, 1058–1070.

Goncuoglu, C., Dirik, K. & Kozlu, H., 1997. Pre-Alpine and Alpine terranes in Turkey : explanatory notes to the terrane map of Turkey. Ann. Géol. Pays Hellén., 37, 515–535.

Haydoutov, I., 2002. Peri-Gondwanan terranes in the pre-Palaeozoicbasement of Bulgaria. Geol. Balc., 32, 2–4.

Haydoutov, I., Gochev, P., Kozhukharov, D. & Yanev, S. 1997. Terranes in the Balkan area. In: Papanikolaou, D. (Ed.), IGCP Project 276. Terrane Map and Terrane Descriptions, Ann. Geol. Pays Hellen., 37, 479–494.

Jacobshagen, V. 1979. Structure and geotectonic evolution of the Hellenides. VI Coll. Geol. Aegean Region, Athens 1977, Proc. 3, 1355–1367.

Jacobshagen, V. & Wallbrecher, E. 1984. Pre-Neogene nappe structure and metamorphism of the North Sporades and the southern Pelion peninsula. Geol. Soc. London, Sp. Publ., 17, 591–602.

Jolivet, L. & Faccenna, C. 2000. Mediterranean extension and the Africa—Eurasia collision. Tectonics, 19, 6, 1095–1106.

Jolivet, L., Faccenna, C., Huet, B., Labrousse, L., Le Pourhiet, L., Lacombe, O., Lecomte, E., Burov, E., Denele, Y., Brun, J.P., Philippon, M., Paul, A., Salaun, G., Karabulut, H., Piromallo, C., Monie, P., Gueydan, F., Okay, A.I., Oberhansli, R., Pourteau, A., Augier, R., Gadenne, L. & Driussi, O., 2013. Aegean tectonics: Strain localisation, slab tearing and trench retreat. Tectonophysics, 597-598, 1–33.

Jolivet, L., Menant, A., Sternai, P., Rabillard, A., Arbaret, L., Augier, R., Laurent, V., Beaudouin, A., Grasement, A., Huet, B., Labrousse, L. & Le Pourhiet, L. 2015. The geological signature of a slab tear below the Aegean. Tectonophysics, 659, 166–182.

Jongsma, D. 1974. Heat flow in the Aegean Sea. Geophys. J. Roy. Astron. Soc., 37, 337–346.

Karamata, S., Krstic, B., Dimitrievic, D.M., Dimitrievic, M.N., Kneevic, V., Stojanov, R. & Filipovic, I. 1997. Terranes between the Moesian plate and the Adriatic Sea. In: Papanikolaou, D. (Ed.), IGCP Project 276. Terrane Map and Terrane Descriptions, Ann. Geol. Pays Hellen., 37, 429–477.

Lallemand, S., Heuret, A. & Boutelier, D., 2005. On the relationships between slab dip, back-arc stress, upper plate absolute motion, and crustal nature in subduction zones. Geochemistry, Geophysics, Geosystems, 6.

Le Pichon, X., Sengor, A. M. C. & Imren, C. 2019. A new approach to the opening of the Eastern Mediterranean Sea and the origin of the Hellenic subduction zone: Part 2: The Hellenic subduction zone. Canadian J. Earth Sciences, in print.

Mckenzie, D.P. 1970. Plate tectonics in the Mediterranean Region. Nature, 226, 239–243.

Mckenzie, D.P. 1972. Active tectonics of the Mediterranean Region. Geoph. J.R. Astron. Soc., 30, 109–185.

McKlusky, S. C. et al, 2000. Global Positioning System constraints onplate kinematics and dynamics in the Eastern Mediterranean andCaucasus. J. Geoph. Res., 105, 5695–5719.

Menant, A., Jolivet, L., & Vrielynck, B. 2016. Kinematic reconstructions and magmatic evolution illuminating crustal and mantle dynamics of the eastern Mediterranean region since the late Cretaceous. Tectonophysics, 675, 103–140.

Mercier, J. 1968. Etude géologique des zones internes des Hellénides en Macédoine centrale (Grèce). Ann. Géol. Pays Hellén., 20, 1–792.

Mposkos, E. & Perraki, M. 2001. High pressure Alpine metamorphism of the Pelagonian allochthon in the Kastania area (Southern Vermion), Greece. Bull. Geol. Soc. Greece, 939–947.

Muller, M. D., Geiger, A., Kahle, H. G., Veis, G., Billiris, H., Paradissis, D. & Felekis, S. 2013. Velocity and deformation fields inthe North Aegean domain, Greece and implications for faultkinematics, derived from GPS data 1993–2009. Tectonophysics, 597-598, 34-49.

Okay, A.I., Monod, O. & Monié, P. 2002. Triassic blueschists and eclogites from northwest Turkey: Vestiges of the Paleo-Tethyan subduction. Lithos, 64(3-4), 155-178.

Papanikolaou, D. 1984. The three metamorphic belts of the Hellenides: a review and a kinematic interpretation. Geol. Soc. London, Spec. Publ., 17, 551–561.

Papanikolaou, D. 1985. A new approach on the geotectonic significance of Rhodope and on the ophiolite suture zones of the Hellenides. In: NATO A.S.I., Ketin Symposium, Istanbul, 1985, Abstracts, 28.

Papanikolaou, D. 1986. Geology of Greece. Eptalofos Publications, 240 p. Athens (in greek).

Papanikolaou, D. 1989a. The genesis of geothermal fields within the geotectonic evolution of the Hellenic arc. TEE, 10-11/4/1989, Technical Chamber of Greece, Congress on: «The exploitation of the Greek Geothermal Potential: Present status and perspectives».

Papanikolaou, D., 1989b. Are the medial crystalline massifs of the eastern Mediterranean drifted Gondwanian fragments? In: Papanikolaou, D., Sassi, F.P. (Eds.), Sp. Publ. Geol. Soc. Greece, 1,63–90 and IGCP 276, Newsletter, No 1, Athens.

Papanikolaou, D. 1993. Geotectonic evolution of the Aegean. Bull. Geol. Soc. Greece, 28, 33–48.

Papanikolaou, D. 1997. The tectonostratigraphic terranes of the Hellenides. Ann. Géol. Pays Hellén., 37, 495–514.

Papanikolaou, D. 2009. Timing of tectonic emplacement of the ophiolites and terrane paleogeography in the Hellenides. Lithos, 108, 262–280.

Papanikolaou, D. 2013. Tectonostratigraphic models of the Alpine terranes and subduction history of the Hellenides. Tectonophysics, 595–596, 1–24.

Papanikolaou, D. & Dermitzakis, M. 1981. Major changes from the last stage of the Hellenides to the actual Hellenic Arc and Trench system. Intern.Symp. on the Hellenic Arc and Trench (H.E.A.T.), Athens 1981, Proceedings, II, 57–73.

Papanikolaou, D. & Panagopoulos, A. 1981. On the structural style of Southern Rhodope. Geol. Balc., 11.3., 13–22.

Papanikolaou, D. & Demirtasli, E. 1987. Geological correlations between the Alpine segments of the Hellenides–Balkanides and Taurides–Pontides. In: Flügel, H.W., Sassi, F.P., Grecula, P. (Eds.), IGCP No 5, «Pre-Variscan and Variscan Events in the Alpine–Mediterranean Mountain Belts.» Alfa Publishers, Bratislava, 387–396.

Papanikolaou, D. & Royden, L. 2007. Disruption of the Hellenic arc: Late Miocene extensional detachment faults and steep Pliocene–Quaternary normal faults—or what happened at Corinth? Tectonics, 26.

Papanikolaou, D., Alexandri, M., Nomikou, P. & Ballas, D. 2002. Morphotectonic structure of the Western Part of the North Aegean Basin based on swath bathymetry. Marine Geology, 190, 465–492.

Papanikolaou, D., Alexandri, M. & Nomikou, P. 2006. Active Faulting in the North Aegean Basin. Geol. Soc. Amer. Sp. Paper, 409, 189–209.

Papanikolaou, D., Bargathi, H., Dabovski, C., Dimitriu, R., El-Hawat, A., Ioane, D., Kranis, H., Obeidi, A., Oaie, C., Seghedi, A. & Zagorchev, I. 2004. Transmed Transect VII: East European Craton–Scythian Platform–Dobrogea–Balkanides–Rhodope Massif–Hellenides–East Mediterranean–Cyrenaica. In: Cavazza, W., Roure, F., Spakman, W., Stampfli, G., Ziegler, P. (Eds.), The TRANSMED Atlas: the Mediterranean Region from Crust to Mantle. Springer-Verlag, Heidelberg.

Papanikolaou, D., Nomikou, E. & Papanikolaou, I. 2019a. The western transformation of the North Anatolian Fault in the North Aegean Basins. 23 ATAG, 20th Anniversary of 1999 Earthquakes, Istanbul.

Papanikolaou, D., Nomikou, P., Papanikolaou, I., Lampridou, D., Rousakis, G. & Alexandri, S. 2019b. Active tectonics and sesmic hazard in Skyros Basin, North Aegean Sea, Greece. Marine Geology, 407, 94–110.

Pe-Piper, G. & Piper, D. 2002. The Igneous Rocks of Greece. The Anatomy of an Orogen. Gebrueder Borntraeger, Berlin/Stuttgart.

Pourteau, A., Oberhänsli, R., Candan, O., Barrier, E., & Vrielynck, B. 2016. Neotethyan closure history of western Anatolia: a geodynamic discussion. International Journal of Earth Sciences, 105(1), 203–224.

Ricou, L. E., Burg, J. P., Godfriaux, I., & Ivanov, Z. 1998. Rhodope and Vardar: the metamorphic and the olistostromic paired belts related to the Cretaceous subduction under Europe. Geodinamica Acta, 11(6), 285–309.

Robertson, A.H.F. & Ustaömer, T., 2012. Testing alternative tectono-stratigraphic interpretations of the late palaeozoic? Early mesozoic karakaya complex in NW turkey: Support for an accretionary origin related to northward subduction of palaeotethys. Turkish Journal of Earth Sciences, 21(6), 961–1007.

Rosenbaum, G., Lister, G. & Duboz, C. 2002. Relative motions of Africa, Iberia and Europe during Alpine orogeny. Tectonophysics, 359, 117–129.

Royden, L.H. & Papanikolaou, D.J., 2011. Slab segmentation and late Cenozoic disruption of the Hellenic arc. Geochemistry, Geophysics, Geosystems, 12, Q03010.

Royden, L. & Husson, L. 2006. Trench motion, slab geometry and viscous stresses in subduction systems. Geoph. J. Intern., 167, 881–905.

Royden, L.H. & Husson, L. 2009. Subduction with variations in slab buoyancy: models and application to the Banda and Apennine systems. In: Lallemand, S.E., Funiciello, F. (Eds.), Subduction Zone Geodynamics, Springer, Berlin Heidelberg, 35–45.

Schermer, E.R., Lux, D.R. & Burchfiel, B.C. 1993. Temperature—time history of subducted continental crust, Mt Olympos region, Greece. Tectonics, 9, 1165–1195.

Sengor, A.M.C. 1989. The Tethyside orogenic system: an introduction. In: NATO (Ed.), Tectonic Evolution of the Tethyan Region, 1–22.

Sengor, A.M.C., Tuysuz, O., Imren, C., Sakinc, M., Eyidogan, H., Gorur, N., Le Pichon, X. & Rangin, C. 2005. The North Anatolian Fault: A new look. Ann. Rev. Earth Planet. Sci., 33, 37–112.

Sengor, A.M.C., Altiner, D., Cin, A., Ustaomer, T. & Hsu, K.J. 1988. Origin and assembly of the Tethyside orogenic collage at the expense of Gondwana Land. Geol. Soc.,London, Sp. Publ., 37, 119–181.

Skarpelis, N., Kyriakopoulos, K. & Villa, I. 1992. Occurrence and 40Ar/39Ar dating of a granite in Thera (Santorini, Greece). Geol. Rundschau, 81, 729–735.

Smith, A.G. 1971. Alpine deformation and the oceanic areas of the Tethys, Mediterranean and Atlantic. Geol. Soc. Am. Bull., 82, 2039–2070.

Smith, A.G. 2006. Tethyan ophiolite emplacement, Africa to Europe motions, and Atlantic spreading. Geol. Soc., London, Sp. Publ., 260, 11–34.

Soldatos, I. & Christofides, G. 1986. Rb–Sr geochronology and origin of the Elatia pluton, Central Rhodope, North Greece. Geol. Balc., 16, 15–23.

Stampfli, G., Marcoux, J. & Baud, A. 1991. Tethyan margins in space and time. Palaeogeography, Palaeoclimatology, Palaeoecology, 87, 373–409.

Stille, H., 1936. Wege und Ergebnisse der geologisch-tektonischen Forchungen. 25Jahre Kaiser Wilhelm—Gesellschaft zur Forderung der Wiss., 11, Die Naturw.

Vergely, P. 1984. Tectonique des ophiolites dans les Hellenides internes. Conséquences sur l' évolution des régions téthysiennes occidentales. Thèse, Univ. Paris—Sud, vol. 1,2.

Yanev, S. 1993. Gondwana Palaeozoic terranes in the Alpine Collage System on the Balkans. J. Himal. Geol., 4, 257–270.

11 Neotectonics and Recent Paleogeography

11.1 From the Hellenides to the Present Hellenic Arc and Trench System

11.1.1 The Aegean (Micro-)Plate and the Distinction of the Northern and Southern Hellenides

The actual Hellenic arc and trench system is at present day constrained to the southern part of the Hellenides (see also Fig. 1.7). This marks a significant difference with the previous orogenic arcs, that extended from Lias to Miocene, along the entire length of the Hellenides and even further away, on both sides of Tethys, i.e. the Dinarides and the Alps to the north and the Taurides and the Iranides to the east. The disruption of the present active Hellenic arc from the former orogenic chain of Tethys and its subsequent unique and exclusive evolution occurred during the Middle/Late Miocene. During this period, two major events took place, modifying the entire setting:

(1) The Arabian plate, moved northwards with a higher velocity than that of the African plate, through the Dead Sea—Lebanon strike slip fault, which has a cumulative slip of several tens of km, and collided with Eurasia at the area south of the Caucasus. This collision resulted into lateral stress transmission and movement of Anatolia to the west (Brunn 1976; Molnar and Tapponier 1977; Mercier 1979; LePichon and Angelier 1979) (see also Chap. 3.8). This movement was effected through the North Anatolian fault, which has a right-lateral strike-slip component of several tens of km (e.g. Sengor et al 2005). The extension of the North Anatolian fault in the North Aegean (e.g. Armijo et al 1999) coincides with the northern boundary of the active section of the Hellenic arc, represented by the Aegean micro-plate (McKenzie 1970, 1972, 1978; Galanopoulos 1972). Simultaneously, the collision of the Adria micro-plate (known as the mole of Africa) occurred with Eurasia along the Alps, the Carpathian Mountains and the Dinarides (e.g. Channel et al. 1979).

(2) The Hellenic Subduction Zone changes, from the slow-rate continental type subduction of the external carbonate platform H1 under the Hellenic margin, to the accelerated subduction of the Ionian oceanic crust of H0 under the Hellenic arc, including also H1 (Royden and Papanikolaou 2011) (Fig. 11.1). This movement is bounded by the H0 paleogeographic area, which is terminated to the north by the southern margin of the Apulian platform, which extends up to the area west of Corfu and Paxos islands (Finetti et al. 1991; Finetti and Del Ben 2005). The 4–5 cm/year subduction rate of the Ionian oceanic crust H0, compared with the only 1 cm/year convergence rate of Europe—Africa, implies that tensile stresses are produced in the Aegean region. The tensile stresses that have been developed resulted in its subsidence of the Aegean, supported also by the plate boundary retreat and the roll back of the subduction zone generating an increase of the subduction angle (Royden and Papanikolaou 2011; Jolivet et al. 2013).

Therefore, these two independent mechanisms are interfingering in the intermediate area between the northern boundary of the Hellenic Subduction Zone (and of the Hellenic arc) in the Preveza area, and the westward extension of the North Anatolian fault in the North Aegean (Flerit et al. 2004; Papanikolaou et al. 2006, 2019a; Philippon et al. 2014; Beniest et al. 2016; LePichon et al. 2016). The intermediate Central Hellenic Shear Zone links the two areas, transforms the two mechanisms and accomodates most of the released strain (Papanikolaou and Royden 2007).

A disorder in the geometry of the Hellenic arc has been clearly traced during the Middle Miocene. This is revealed by the destruction of the previous geometry during the Oligocene—Early Miocene, all along the arc from the Scutari-Pec zone up to the Antalya Gulf. Approximately, the same geometry is re-established since the Late Miocene, in a

D. I. Papanikolaou, *The Geology of Greece*, Regional Geology Reviews,
https://doi.org/10.1007/978-3-030-60731-9_11

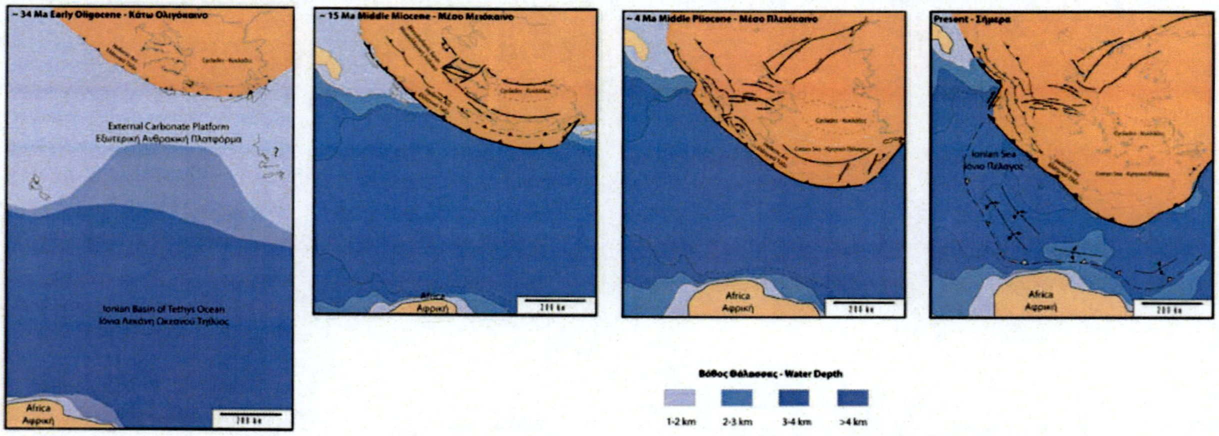

Fig. 11.1 Paleogeographic sketches of the evolution of the Hellenic arc since the Oligocene (according to Royden and Papanikolaou 2011)

new restricted scheme, at about the present arc dimensions (Papanikolaou and Dermitzakis 1981) (Fig. 11.2).

Thus, the uplift and termination of the molassic sedimentation in the Meso-Hellenic trough is constrained during Late Miocene, as well as in the analogous Tavas–Kale trough, south of Menderes in Asia Minor. At the same time the Cycladic molasse is tectonized and along with its non-metamorphosed Alpine basement becomes a surficial gravitational nappe (Dermitzakis and Papanikolaou 1979). Since the Late Miocene and especially in the Pliocene, when the orogenic geometry is restored and reaches approximately its present shape and location, both the back-arc basin of the Cretan Sea and the Aegean volcanic arc are restricted within the boundaries of the Aegean micro-plate; south of the Sperchios valley and west of the Asia Minor coasts.

The distinction of the Aegean microplate in the southern margin of the Eurasian plate, proposed in the early 70s (McKenzie 1970, 1972; Galanopoulos 1972), is due to the onset of the subduction of the Eastern Mediterranean oceanic crust, and especially of the Ionian Sea basin during the Late Miocene. The entry of the oceanic crust of the Ionian Basin in the Hellenic trench, produced an increase of the dip angle of the Hellenic subduction zone and its roll back beneath the newly formed Aegean micro-plate, as well as the development of retreating plate boundaries, which caused the extension and collapse of the Aegean domain (Royden 1993; Jolivet and Faccenna 2000; Faccenna et al. 2003; Royden and Papanikolaou 2011; Jolivet et al. 2013). The previously, continuous Hellenic orogenic arc was then differentiated in its northern section of the **Northern Hellenides**, where the slow convergence between the Apulian platform and the European margin is still undergoing, with a rate of 8 mm/year, and in the southern section of the **Southern Hellenides**, where the present Hellenic orogenic arc has been developed, with a rapid subduction rate of 40–50 mm/year (Royden and Papanikolaou 2011). The differentiation of the convergence rate, due to the different nature of the subducted crust to the north and south of Preveza, created a fracture in the form of a tear of the crust, and superficially created a fault of right-lateral horizontal sliding along the formatted tectonic zone, reaching about 100 km of horizontal slip in the Cephalonia fault (Fig. 11.3).

The geological differences between the Northern and the Southern Hellenides have been analysed by Papanikolaou (2010) and can be summarised in a dozen of key notes as follows (Fig. 11.4):

1. The bathymetry of the fore deep basins is dramatically different. The trench is actually missing, having depths that do not exceed 1 km west of Corfu in the Northern Hellenides, in contrast to the 4–5 km deep Hellenic trench in the Southern Hellenides.
2. The tectonic activity and related seismicity is very intense in the south, but weak in the north, where the only important active structure corresponds to the overthrust of the Ionian unit on the eastern margin of the Apulian platform in the area west of Corfu (Monopolis and Bruneton 1982; DelBen et al. 2015). The thrusting and strike slip focal mechanisms of numerous great earthquakes in the Ionian islands of Lefkada, Cephalonia and Zakynthos are very well known also due to their severe damages and fatalities (Papazachos et al. 2000a, b).
3. The Miocene accretionary wedge and the backstop of the East Mediterranean ridge do not continue northwards.
4. The front of the Hellenic nappes has advanced for several tens of km in the south with respect to their position in the north. This is illustrated by the offset of the Late Miocene thrusts of the Ionian unit, involving the Messinian evaporites (Ms in Fig. 11.4).

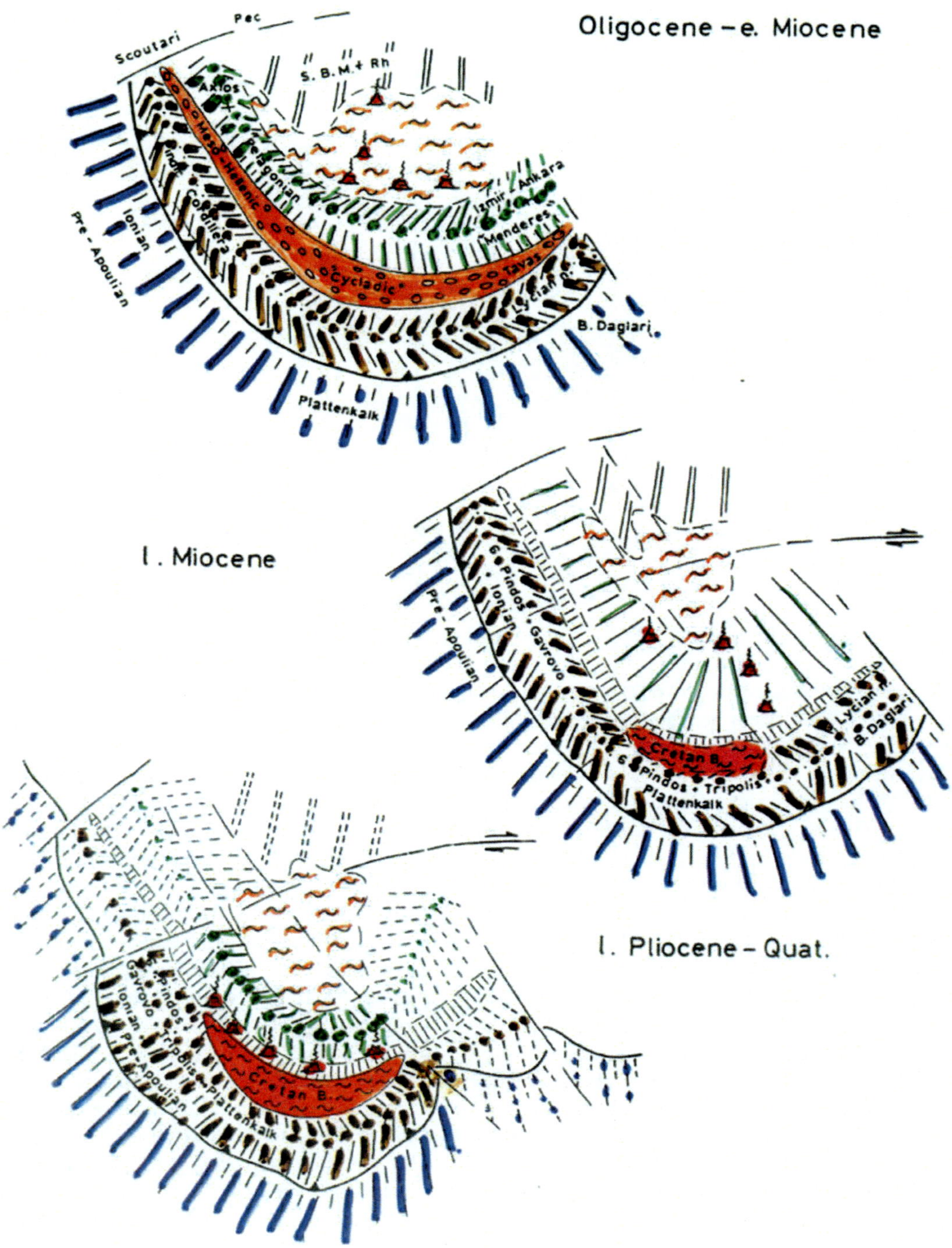

Fig. 11.2 Schematic representation of the orogenic arc of the Hellenides during: (a) the Oligocene—Early Miocene, (b) Late Miocene, (c) Late Pliocene—Quaternary. The tectonic units that took part in each period in the various arc segments are also indicated. The overall evolution shows the restriction of the arc into its present position, which happened after the Late Miocene (from Papanikolaou and Dermitzakis 1981)

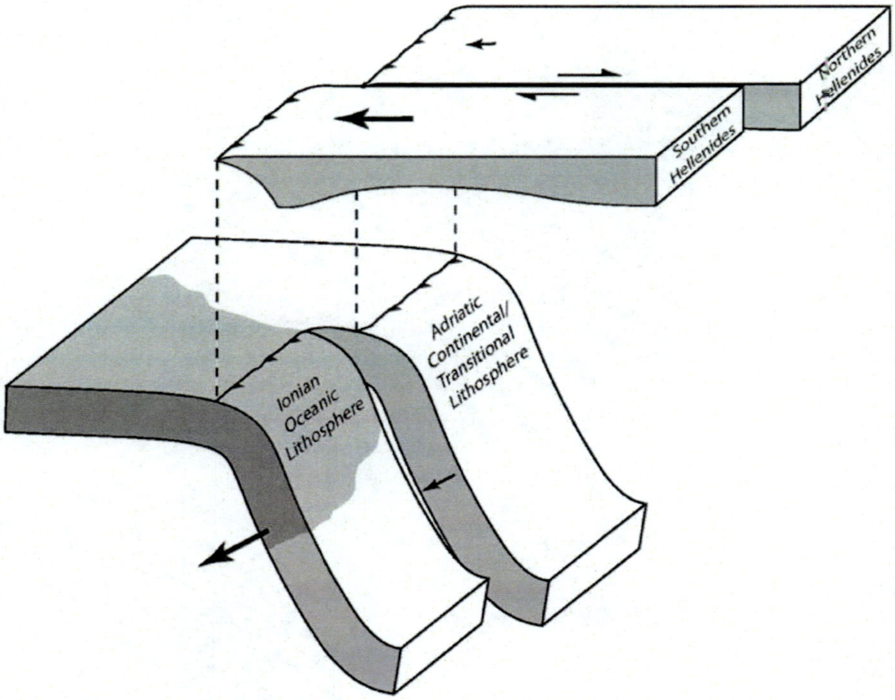

Fig. 11.3 Schematic representation of the differentiation of the Northern Hellenides from the Southern Hellenides, on either side of the Preveza—Lefkada zone, where the nature of the plate convergence changed, with slow continental subduction to the north and rapid oceanic subduction to the south. The result was the strike-slip fault of Cephalonia, the increase of the dip angle of the subduction zone in the south, where the Aegean micro-plate was created, as well as the present arc and trench system, in contrast to the northern Hellenides (from Royden and Papanikolaou 2011)

5. The front of the Pindos nappe and of the more internal units in Epirus and western Sterea runs parallel to the front of the Northern Hellenides at a distance of 100 km, whereas to the south it reaches the trench at a distance of 25–40 km and in particular towards the south in the Gavdos Island block it lies directly on the front of the arc.
6. Across the external Hellenides in the north there are no Plio-Quaternary basins, but only the Upper Miocene lignite basins (Ms in white, in Fig. 11.4) east of the Mesohellenic molassic basin. On the contrary, in the south there are several Plio-Quaternary basins from western Peloponnese to the central Aegean in Attica and southern Evia, both continental and marine.
7. The 3–5 km thick Oligo-Miocene Mesohellenic molassic sequence is uplifted. In the north, the top stratigraphic formation of Ontria lies at an altitude of 800 m, indicating an uplift of more than 1 km (see also Fig. 6.7). During Plio-Quaternaty it has been covered by fluvial/continental deposits on both sides of the primary Aliakmon River, draining the former basin. On the contrary, south of the Sperchios valley the whole molassic sequence is eroded and only some minor relics have been preserved in Othris and Oeti mts.
8. The Cretan molassic basin opened since Middle—Late Miocene and the Alpine basement has collapsed to more than 2–3 km depth.
9. The Miocene volcanic arc became inactive during Plio-Quaternary in the north, whereas it continued at more external position in the south with very important volcanic centres active throughout the Quaternary (Methana, Milos, Santorini in Fig. 11.4).
10. A system of arc parallel extensional detachments has been activated in the south with the formation of tectonic windows such as in the Taygetus and Parnon mts in Peloponnese, where the metamorphic units have been exhumed, beneath the broad tectonic window of the Tripolis carbonate platform (Mani, Kythera and Crete within the Tripolis shown in blue, in Fig. 11.4). On the contrary, in the north the Miocene fold and thrust belt remains intact without major extensional structures.
11. In the south, E–W trending normal faults have been formed during Late Pliocene—Pleistocene, disrupting the previous NNW–SSE trending structures, forming the active tectonic grabens of Corinth, Evoikos etc.
12. These transverse E–W faults are the seismically active faults that have been very catastrophic and caused many casualties (Papazachos et al. 2000a, b). Their absence from the north, implies a much lower seismic activity in the area.

Besides the above geological differences the crustal structure between the two segments of the Hellenides is different, as it can be compared along two parallel transverse

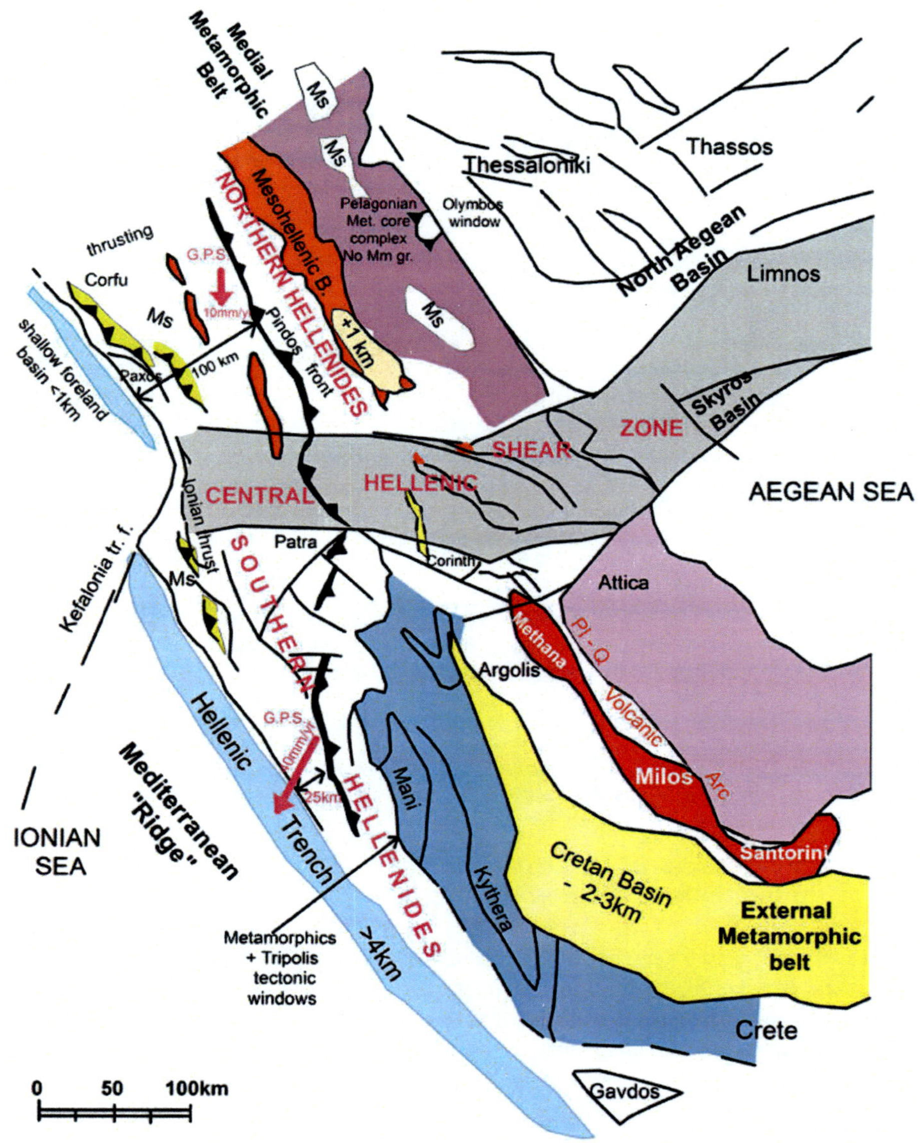

Fig. 11.4 The differences between the Northern and the Southern Hellenides in continental Greece (based on Papanikolaou 2010) (Explanation in the text)

sections, corresponding to the two segments, from the Ionian Sea to the Aegean Sea (Fig. 11.5).

The first major difference concerns the crustal length along the two subparallel profiles, which is significantly larger in the Southern Hellenides, due to the neotectonic extension of the Aegean micro-plate. Additionally, the average elevation is higher in the Northern than in the Southern Hellenides, because there are no marine basins with negative altitudes along the profile, but only mountain ranges like the Pindos. On the contrary, in the Southern Hellenides there are marine basins as well as other low elevation continental basins, representing tectonic grabens, filled with Plio-Pleistocene sediments. The above differences are accompanied by the crustal thickness, which is around 40 km in the Northern, but only 30 km in the Southern Hellenides. Another major difference resulting from the intense extension of the crust both with extensional detachments and related normal faulting in the Southern Hellenides is the formation of tectonic windows and of the recent rift structures of the gulfs of Corinth and Evia. In contrast, across the Northern Hellenides there is only a Miocene thin skin tectonic structure with shallow thrusting and folding from Corfu to West Thessaly.

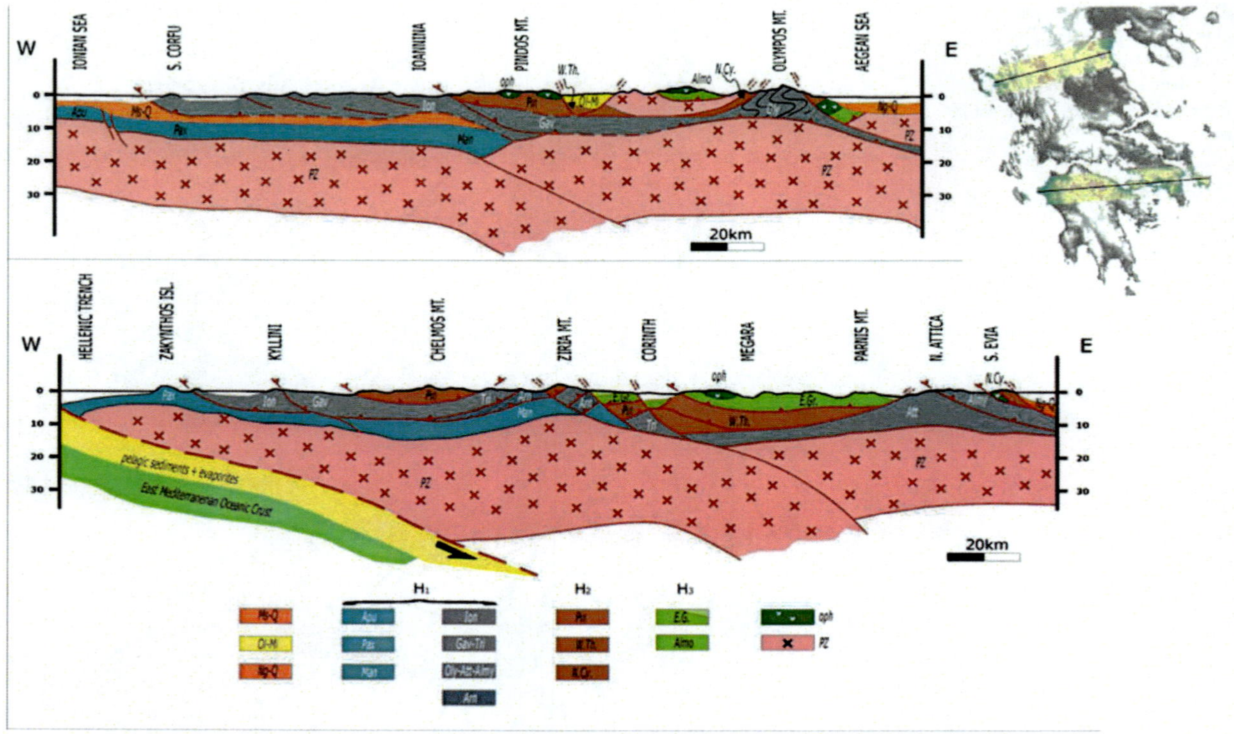

Fig. 11.5 Transverse tectonic profiles of continental Greece, showing the different crustal structures of the Northern and the Southern Hellenides with continental subduction and no arc structure in the north, but oceanic subduction and arc and trench structure in the south

11.1.2 The Extensional Detachments in the Hellenic Arc and the Aegean

The extensional detachments have been distinguished from the general group of the tectonic contacts in the early 1980s in the «Basin and Range» of the western U.S. (e.g. Wernicke 1981; Wernicke and Burchfiel 1982) as major low angle normal faults with displacement of several km or even tens of km, in the form of nappes within an extensional tectonic regime (see also Sect. 4.9). Their presence has been rapidly recognised worldwide especially in areas of extension, usually in the back arc area of fold and thrust belts along convergent plate boundaries. Their major characteristic is the structural omission of tectonic units or stratigraphic horizons between the footwall and the hanging wall, contrary to the thrust planes, where there is a duplication of geological formations, due to compression. Their presence in the Hellenic arc and the Aegean usually follows the Tertiary outcrops of the tectonic windows, especially in the metamorphic core complexes, where they separate the overlying non metamorphic units and syntectonic sediments from the underlying metamorphic and magmatic rocks.

The Strymon extensional detachment separating the Rhodope metamorphic rocks in the footwall from the Serbo-Macedonian metamorphic rocks and overlying Miocene sediments in the hangingwall was the first detachment to be described in Greece (Dinter and Royden 1993). The description of several other detachments followed in the Cyclades, Crete and Olympus (Lister et al. 1984; Jolivet et al. 1996, 2010; Kilias 1996; Kilias et al. 2002; Papanikolaou and Vassilakis 2010; Grasemann et al. 2012, VanHinsbergen and Meulenkamp 2006).

In the Peloponnese and central Sterea Papanikolaou and Royden (2007) described two major detachment systems; the East Peloponnese detachment system and the East Sterea detachment system (Fig. 11.6). The East Peloponnese detachment system has disrupted the tectonic nappe pile, comprising five tectonic units (Fig. 11.7a): the relative autochthon Mani unit, the metamorphic Arna unit, the Tripolis unit (including the Permo-Triassic Tyros Beds at its base), the Pindos unit and the higher nappes of internal Hellenides origin. The base of the detachment system follows the basal thrust of the Arna unit or of the Tripolis unit (Fig. 11.7b). Thus, the structural omission produced by the detachment comprises either the approximately 6–8 km thick nappe stack of the Arna-Tripolis-Pindos units or the approximately 4–6 km thick nappe stack of the Tripolis-Pindos units.

The structural omission is not constant along the East Peloponnese Detachment System, but it systematically

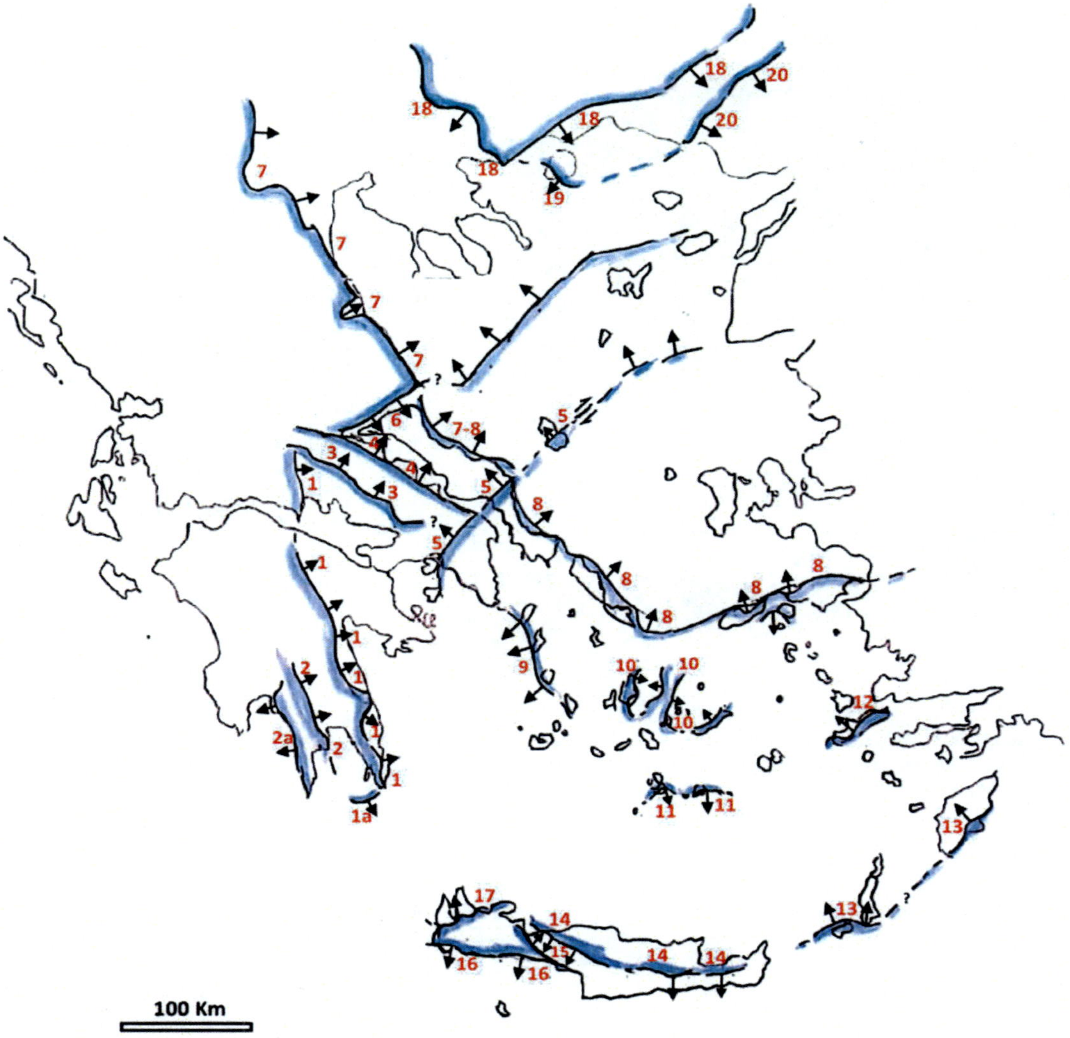

Fig. 11.6 Simplified map of the extensional detachment faults of the Hellenic system. The footwall along the detachment faults is marked in blue and the arrows point to the hangingwall (based on Papanikolaou and Royden 2007, for continental Greece and Papanikolaou and Vassilakis 2010 for Crete). 1: East Peloponnese (Parnon) Detachment, 2: Taygetus Detachment, 3: East Sterea/Parnassos Detachment, 4: East Sterea/Kallidromon Detachment, 5: Northern Attica—Southern Evia—Skyros Detachment, 6: Maliac Detachment, 7: Olympus Detachment, 8: Northern Cycladic Detachment, 9: Western Cycladic Detachment, 10: Central Cycladic Detachment, 11: Santorini-Anafi Detachment, 12: Kos Detachment, 13: Kasos—Lindos Detachment, 14: Psiloritis—Dikti Detachment, 15: Vatos Detachment, 16: Southwest Cretan Detachment, 17: Northwest Cretan Detachment, 18: Strymon—Xanthi Detachment, 19: Western Thassos Detachment, 20: East Rhodope Detachment

decreases from south to north. This is illustrated in the detachment parallel diagram from the Kyparissi segment in SE Peloponnese, through the segments of Central Parnon mt, Dolliana, Mercouri, Feneos, Galaxidi up to the northern segment of Prosilio in northern Giona mt (Fig. 11.8).

The thickness of the structurally omitted nappe pile along the East Peloponnese Detachment decreases from 8–9 km in the south to 1–2 km in the north (Fig. 11.8). The change is more abrupt on both sides of the Corinth Gulf, where from 4–5 km thickness omitted in the Feneos segment it is reduced to 2 km in the Galaxidi segment. However, the hangingwall of the detachment comprises on top of the Alpine units several hundred meters of Middle-Upper Miocene molassic sediments of the Itea-Amfissa basin (Papanikolaou et al. 2009) (see also Figs. 6.9 and 6.10). It should be noted that the total displacement along the detachment fault is a few tens of km, depending on the dip of the detachment surface. Usually, the detachment faults are low-angle normal faults with dips less than 30°, as this can be observed in the pictures showing the detachments of Parnon mt (Fig. 4.29), Kallithea in western Samos Island (Fig. 8.85), Psiloritis mt (Fig. 8.101), Vatos in central Crete

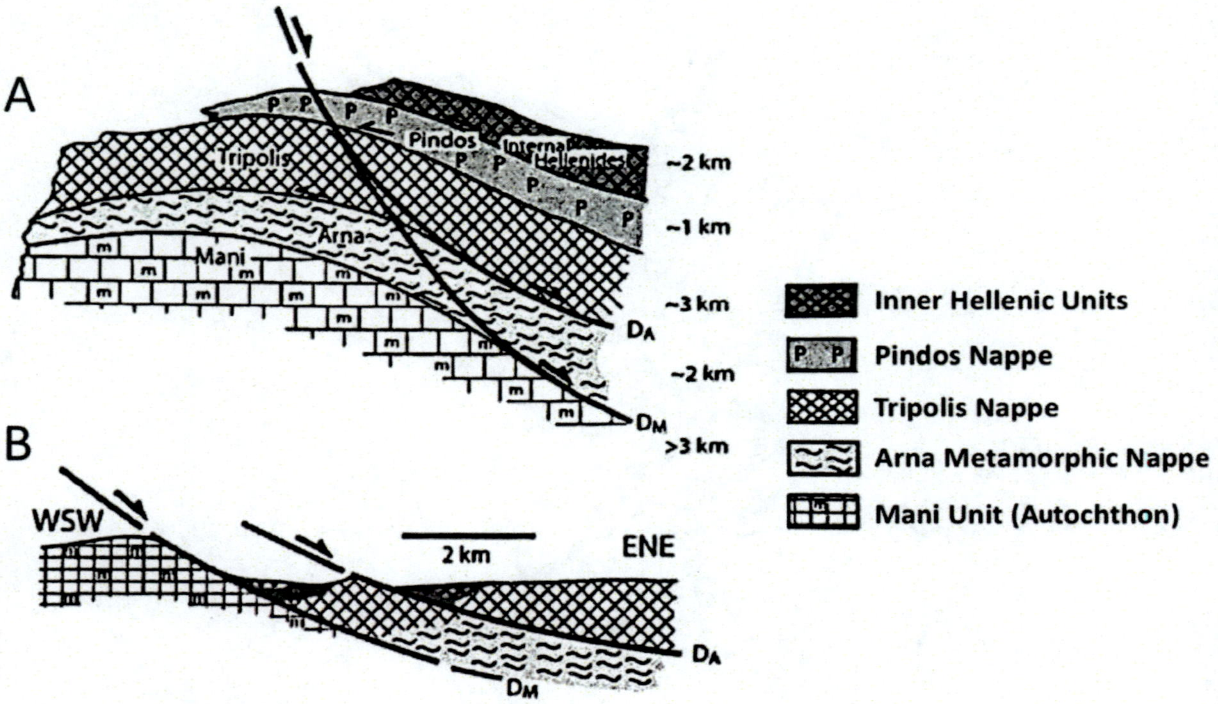

Fig. 11.7 a Generalized cross section of unextended portions of the Hellenic thrust belt, with approximate thicknesses and typical position of superposed extensional detachment faults, commonly detaching within the Arna (D_A) or Mani (D_M) units. **b** Schematic cross section of the East Peloponnese detachment system in the Parnon mt (from Papanikolaou and Royden 2007)

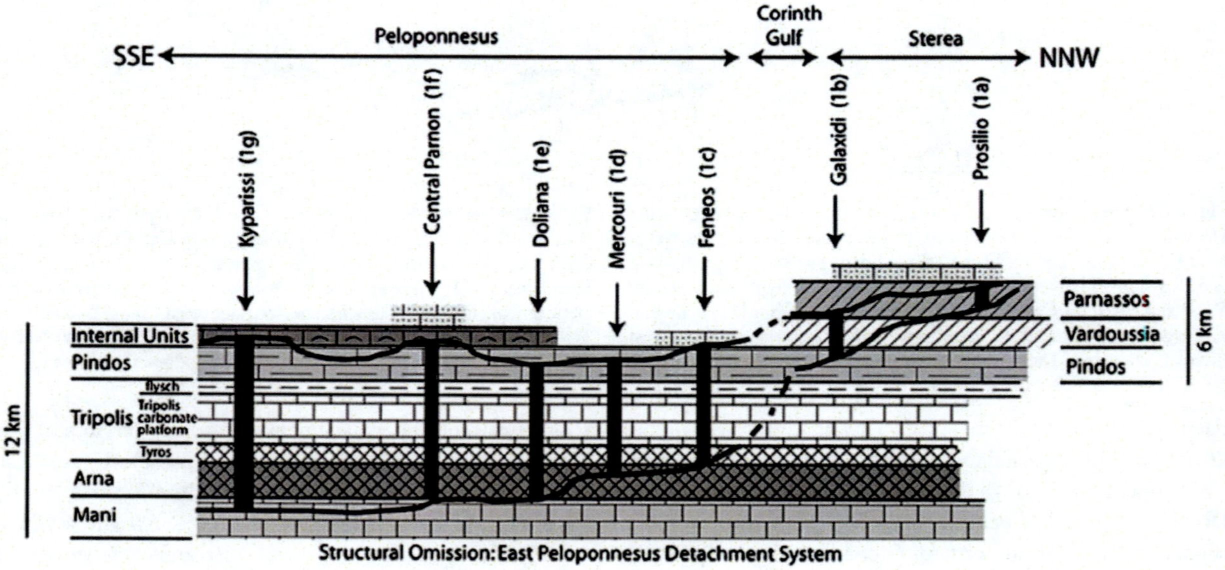

Fig. 11.8 Stacking sequence and approximate thicknesses of tectonic units of the Hellenides along the trend of the East Peloponnese Detachment System. Crustal omission caused by movement on the detachment is shown in seven localities along the detachment. Units between the upper and lower black lines are missing across the detachment surface; Missing sequences at the specific localities are indicated by vertical black bars (from Papanikolaou and Royden 2007)

(Fig. 8.112) and Anafi Island (Fig. 8.105). Thus, a structural omission of 6–8 km thick nappe stack may correspond to total displacement of 15–25 km. A minor detachment sub-parallel to the East Peloponnese detachment occurs in the southern Peloponnese along the eastern slopes of the Taygetus mt (Fig. 11.6). It disrupts the same nappe pile as that of the Parnon mt but its length is about half, limited between the Megalopolis basin in the north and the Taenaron Cape in the southern Mani peninsula.

The East Sterea detachment system has a NW–SE general tectonic trend comprising two sub-parallel faults bounding the Parnassos and the Kallidromon mts respectively (Fig. 11.6). The two arc-parallel detachments are bounded by two transverse NE–SW major faults, defining the CHSZ in eastern continental Greece. The Northern Attica–Southern Evia–Skyros Island fault zone in the south and the Maliac–Northern Sporades islands fault zone in the north. Indeed, it is interesting to note that the Attica detachment (southern boundary of the CHSZ) despite being inactive today, it influences the geometry, style and intensity of deformation, as expressed from the fault geometry, the fault slip-rates, the seismicity and geodetic rates on either side of the detachment (Papanikoloau and Papanikolaou 2007; Foumelis et al. 2013). Seismic tomography indicates that the detachment extends towards the southwest at the Saronikos Gulf (Drakatos et al. 2005). These transverse tectonic fault zones are characterized also by important dextral strike slip motion. Nevertheless, their normal displacement is remarkable, as indicated by the occurrence of non-metamorphic rocks of the internal Hellenides within the graben-like structure in between the two transverse fault zones. On the contrary, metamorphic rocks of the blueschits units and overlying «Pelagonian» marbles occur north of the Maliac zone and the Almyropotamos tectonic window with overlying blueschists south of the Northern Attica–Southern Evia zone. Thus, the two transverse detachment systems bring in contact non-metamorphic rocks of the H3 and H4 terranes with metamorphic units of the H1 and H2 terranes. These metamorphic units were subducted during Eocene to more than 40 km depth and came to surface during their exhumation in Miocene.

In the central Aegean area the emplacement of the superficial late gravitational nappe of the non metamorphic Cycladic unit over the underlying metamorphics during Late Burdigalian–Langhian was described by Dermitzakis & Papanikolaou (1979) and Papanikolaou (1980b, c). The central Aegean outcrops of the non metamorphic Cycladic nappe were considered as upper plate tectonic relics above the Cycladic blueschists mega-shear zone (Papanikolaou 1986b, 1987) (see also Figs. 4.28 and 8.54). Both in the Olympus area and the Cyclades the dome structure of the core complexes is characterized by a major detachment system with displacement towards the N–NE and a minor system towards the W–SW (Lister et al. 1984; Kilias 1996; Kilias et al. 2002; Jolivet et al. 2010; Grasemann et al. 2012). The same dome structure but more elongated in the E–W direction is observed in Crete, where the detachments dip both to the north and to the south (Fassoulas et al. 1994; Jolivet et al. 1996; Papanikolaou and Vassilakis 2010). Nevertheless, in several cases extensional detachment faults have been proposed without any structural omission, representing usual decollement and disharmonic structures within the same tectono-stratigraphic sequences (Papanikolaou and Vassilakis 2010).

11.1.3 Mantle Flow Dynamics in the Aegean

The deep tectonic processes of the Aegean microplate, observed at the lithospheric scale involving mantle flow dynamics are in the spotlight over the last years. Several methodologies and models were applied regarding the general flow direction, the rate of the subducting East Mediterranean slab and the kinematics of torroidal flow along the plate boundaries, which coincide with slab tears (e.g. Govers and Wortel 2005; Guillaume et al. 2013; Jolivet et al. 2009, 2013). Pearce et al. (2012) have compared the subduction process in the two segments of the Hellenides on both sides of the CHSZ, through high-resolution seismic images down to 100 km depth and determined a ~8 km thick layer beneath the Southern Hellenides, interpreted as subducted oceanic crust and a ~20 km thick layer beneath the Northern Hellenides, interpreted as subducted continental crust. The relative position of the two subducted crusts implies a ~70–85 km additional slab retreat in the south relative to the north. More recently, Evangelidis (2017) studied both tectonic boundaries of the Aegean microplate using the source-side splitting method and SKS shear-wave splitting measurements and concluded that: (1) The passage from continental to oceanic subduction in the western Hellenic arc is illustrated by a fore arc transitional anisotropy pattern, with NE–SW sub-slab mantle flow parallel to a smooth ramp, linking the two subducted slabs. However, the dominant feature is the trench-parallel mantle flow all along the fore arc zone, whereas trench-normal flow is observed at the sub-slab zone of the trench and in the mantle wedge beneath the back arc area. (2) In the eastern subduction zone of the Hellenic arc a general trench-normal anisotropy pattern is observed at depth, associated with the ongoing slab tear of the oceanic lithosphere, whereas trench-parallel directions are observed at shallower depths. Generally, the eastern slab tear has not been precisely located (Jolivet et al. 2015), although there have been several reports from various geophysical sources (Piromallo and Morelli 2003; Biryol et al. 2011; Salaun et al. 2012). The eastern slab tear running approximately along the transition zone of the Aegean /Anatolian microplates has been regarded as the cause for the acceleration of the Hellenic trench retreat during the last 10 Myrs (Brun and Sokoutis 2010; VanHinsbergen et al. 2010; Brun et al. 2016) as well as for the rotation of the Aegean microplate in relation to the Hellenic slab roll back in the west (LePichon et al. 2016). The geological signature of the eastern slab tear approximately along the Aegean/Anatolian transition zone has been investigated as far as the asthenospheric flow, the extension gradients and the ascending magmatic bodies in the Aegean and the Menderes massif (Jolivet et al. 2015; Brun et al. 2016). Its initiation has been proposed to occur in Early/Middle Miocene, affecting also the Thermal Cycladic event. More recently, Roche et al. (2019) concluded that the extensional flow above the Aegean/Anatolian tear has been constant in the NNE–SSW direction since Middle Miocene, without a localised zone of strike slip motion.

The overall tectonic processes of the Mediterranean are controlled by the lithospheric dynamics with their driving forces initiated through mantle convection at depth, affecting surface geology (Faccenna et al. 2014). Thus, two almost symmetric upper mantle convection cells dominate in the whole Mediterranean Basin. The downwellings are found in the center of the Mediterranean with opposite direction and are associated with the descent of the Tyrrhenian and Hellenic slabs. A return flow of the asthenosphere from the back arc regions is created toward the subduction zones, expressed in two upwellings beneath Anatolia and Eastern Iberia.

In conclusion, the two Aegean microplate slab tears are different regarding:

(1) The nature of the subducting crust, which is continental/oceanic in the west, but oceanic/oceanic in the east.
(2) The active kinematics, as expressed by the GPS rates, which are subparallel to the tear, but highly differentiated in the two segments in the west, contrary to the east, where they are normal to the tear with increasing rates and slight rotation to the SW on the Aegean segment.
(3) The age of tear initiation, which is mainly Miocene, between 17 and 7 Ma in the east, but only as young as 5–0 Ma in the west.

Thus, the role of the Africa/Europe convergence direction along the microplates of the Alpine Tethyan system seems very critical, since: (i) the western Hellenic subduction zone is normal to the Aegean microplate boundary and it becomes highly oblique at its eastern segment until the Aegean/Anatolia transition zone, whereas (ii) the Anatolian/Cyprus segment moves to the west, sub-parallel to the plate boundary of the East Mediterranean basins and thus, the subduction process becomes inactive during the last few Myrs. In any case, the question about the causes of the slab tear along the Aegean/Anatolian transition zone in the Middle-Late Miocene remains to be determined. The existence of the Sinai microplate, incorporating the Eratosthenes Seamount and the Levantine Basin underlain by thinned continental crust south of Cyprus (Mascle et al. 2000; Netzeband et al. 2006; Granot 2016) may have affected the Aegean /Anatolian slab tear, which seems to occur approximately along the north prolongation of the boundary between the Herodotus Basin and the Sinai-Levantine block. In any case, recently presented reconstructions of the initiation of the modern Hellenic subduction since the Late Miocene have shown the almost stable position of the pre-Late Miocene East Mediterranean northern boundary east of Antalya. On the contrary, the Hellenic arc and trench system has migrated several hundred km to the WSW, up to its present position in the southern Ionian Sea with a simultaneous dextral rotation of ~50° (LePichon et al. 2019) (Fig. 11.9).

11.2 Neotectonics—Active Tectonics

11.2.1 Kinematics of the Hellenic Arc, Paleomagnetic Rotations and Tectonic Dipoles

The present geotectonic regime in the Hellenic arc and trench system is characterized by an asymmetry of motion, caused by the orthogonal geometry of the arc, with normal convergence at the Ionian Sea and sub-parallel strike slip motion at the Pliny and Strabo trenches (Le Pichon et al. 1981), (Fig. 11.10). Thus, compression prevails along the Hellenic Trench in the Ionian Sea, where NE–SW slip orientation, is observed, in the earthquake focal mechanisms along the subduction zone, contrary to a dominant left-lateral strike-slip motion with a minor compressional component in the NNE–SSW direction, observed in the area of Eastern Crete and the Dodekanese, north of the Pliny and Strabo trenches.

Another major tectonic feature of the Hellenic arc is its clockwise rotation of 40–60° since the Middle Miocene, determined from **paleomagmetic** measurements (Fig. 11.11) (Laj et al. 1982; Kissel and Laj 1988; Marton et al. 1990; Duermeijer et al. 1998, 1999, 2000; Kissel et al. 2003; VanHinsbergen et al. 2005, 2010; Broadley et al. 2006). These large rotations have been observed especially along its western zone within the external Hellenides from measurements in the Mio-Pliocene sediments. On the contrary, opposite rotations have been determined in other areas in the Southern Aegean Sea and Crete.

The clockwise rotation in western Greece may be due: (i) to rotations during Miocene thrust sheet emplacement within the thin-skinned tectonics, with a fore landward decrease and

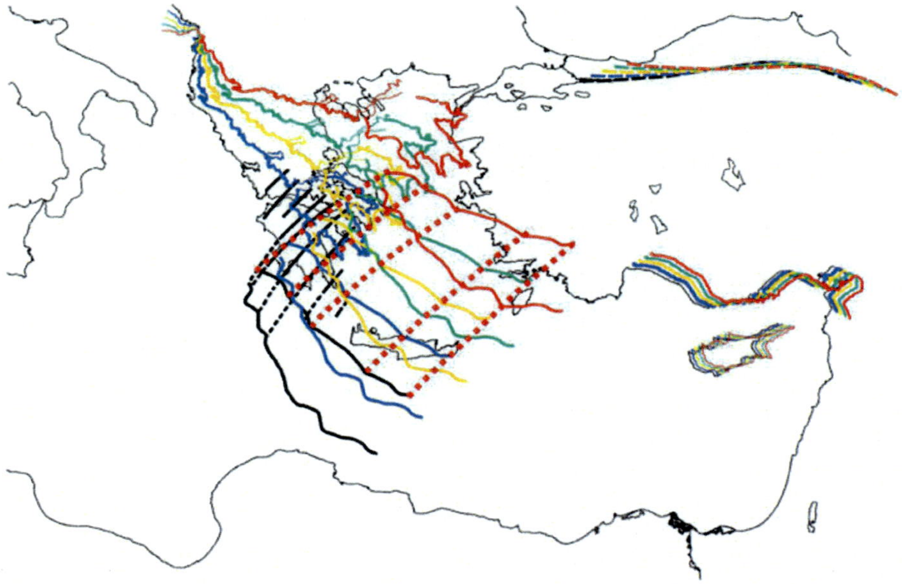

Fig. 11.9 Five successive reconstructions at equal angle intervals showing the flow lines from present to pre-subduction stage of the Hellenic arc (Late Miocene). Note the almost stable zone of Cyprus and opposite Taurus belt for the same period (after LePichon et al. 2019)

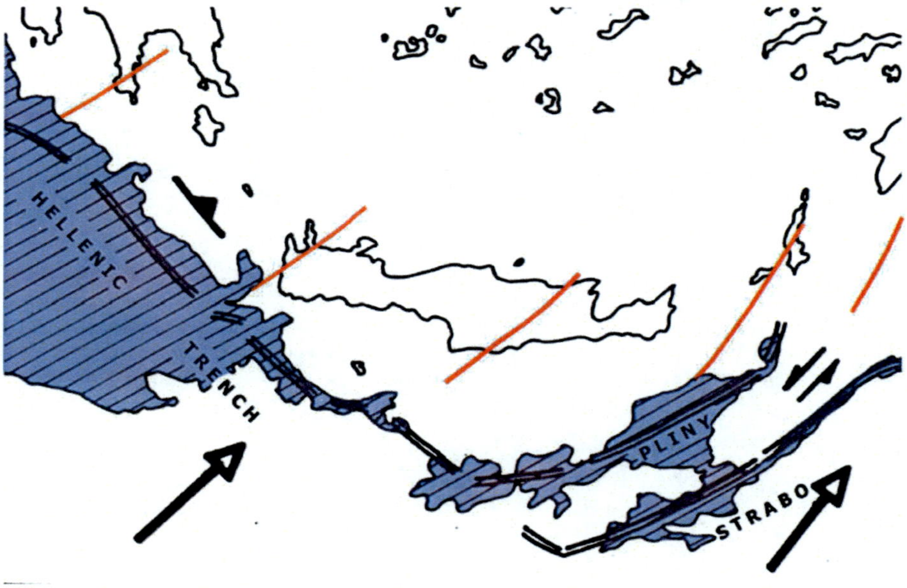

Fig. 11.10 Schematic representation of the Hellenic arc kinematics, where a normal subduction/overthrust is dominant in the western part of the tectonic contact, while a left-lateral strike-slip is dominant in the eastern part (according to the data provided by LePichon et al. (1979, 1981). The thin lines show the orientation of the main compressive stress, based on the fault plane solutions of the major intermediate depth earthquakes of the subduction zone

(ii) to thick-skinned post-Pliocene rotations, from complex interplay motions in the transition zone from the Anatolian westward extrusion and the Aegean extension (Broadley et al. 2006). The extreme value of 20° of clockwise rotation of the Zakynthos block since only 0.77 Ma (Duermeijer et al. 1999) should be regarded as a rather local effect, eventually related to salt tectonics. Overall, a 5°/Ma dextral rotation is estimated in the west and a similar sinistral rotation in the east (VanHinsbergen et al. 2005, 2010; Lepichon et al. 2016).

Extensional structures are dominant within the Aegean plate, which often accomodate also a significant horizontal slip component, producing block tilting around horizontal or plunging axes, as in the case of the Isthmus of Corinth (Freyberg 1973). At the same time, numerous tilt motions of fault blocks can be observed at a large scale, which systematically show a southward rotation, leading to the creation of **tectonic dipoles** (Mariolakos 1976). Thus, opposite vertical motion is observed at the ends/poles of large tectonic dipoles, such as: (i) the Oeti—Parnassos dipole in the Central Sterea Region, where the southward tilt has produced maximum elevation of Mount Oeti and maximum subsidence of the Sperchios valley. (ii) The Northern

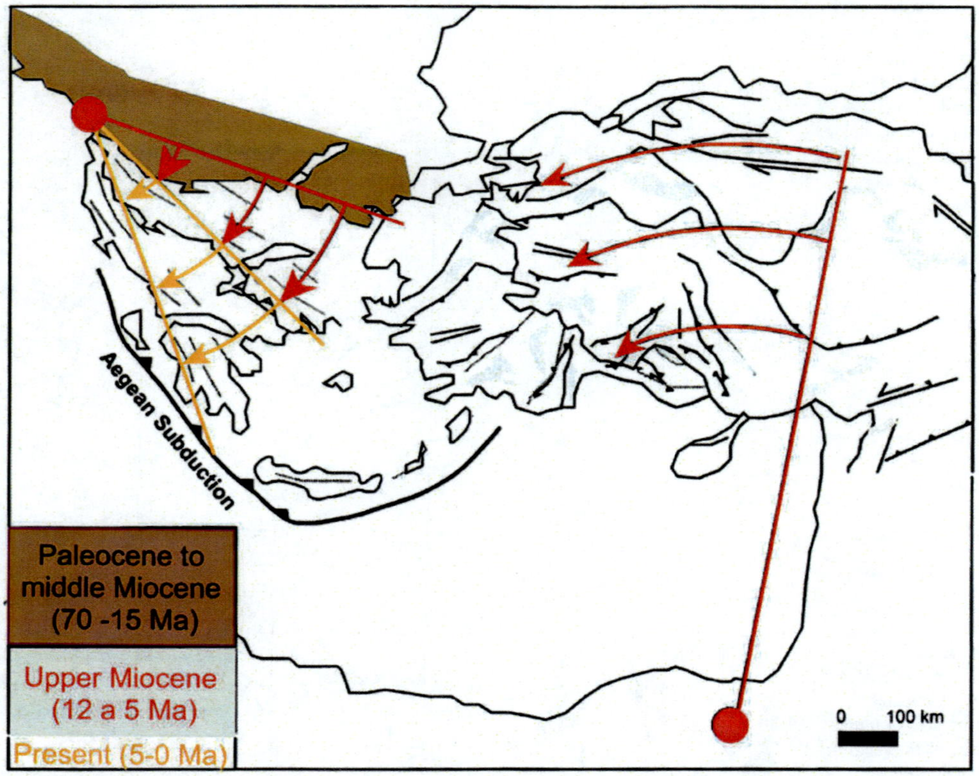

Fig. 11.11 Clock-wise paleomagnetic rotation of the western Aegean area around the Scutari pole in Northern Albania and opposite sense rotation of Anatolia around the Sinai pole modified from Kissel et al. (2003)

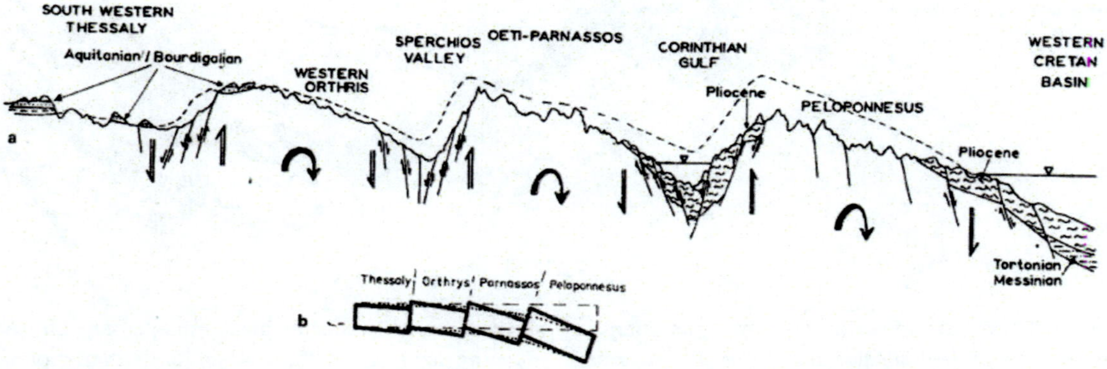

Fig. 11.12 The tectonic dipoles proposed by Mariolakos (1976) in the Central Sterea and Peloponnese (from Dermitzakis and Papanikolaou 1979). The gradually increasing southward tilt of the crustal blocks along the Hellenic chain is shown in **b**

Peloponnese dipole, where maximum elevation is observed in the Ziria mt and maximum subsidence in the adjacent Corinth Gulf (Fig. 11.12). These tectonic dipoles belong to the CHSZ representing transverse structures to the arc, created during the last few million years.

Several km of relative vertical motion is observed at the edges of the tectonic dipoles. Thus, 4.7 km of relative vertical motion occurs in the Corinth rift with 2.5 km elevation in Ziria mt, the northern edge of the Northern Peloponnese dipole and more than 2.3 km subsidence in the southern edge of the Oeti-Parnassos dipole (850 m depth plus more than 1.5 km sedimentary thickness in the Corinth Gulf). The relative motion is smaller, about 2.5 km in the Othris-Oeti tectonic boundary along the Sperchios River rift, where 2.2 km of elevation is observed at the northern edge of the Oeti dipole and a few hundred meters subsidence at the southern edge of the Othris dipole in the Sperchios Valley. Thus, an increase of the relative vertical displacement is observed along with the successive southward tilting of the tectonic dipoles from Thessaly to Peloponnese. This is followed by subsidence and submergence with deposition of marine Plio-Quaternary sediments in the south (Fig. 11.12). The overall resulting crustal deformation along the Hellenic chain indicates a brittle adjustment of a bending towards the south. This is also followed by the gradually reduced thickness of the crust, reported from the Central Pindos—Thessaly down to the central Peloponnese by 6–8 km (Makris 1973, 2010). This implies lower crustal flow and thinning towards the south along the Hellenic chain, as this has been modelled in the Corinth rift (Westaway 2002). It is remarkable that the recently described ENE–WSW along dip crustal fractures beneath the Hellenic subduction zone with gradual subsidence towards the SSE (Sachpazi et al. 2016) are compatible with the shallow crustal deformation of the tectonic dipoles (see also Fig. 4.17).

11.2.2 Active Tectonics and Crustal Structure

The differences of the crustal thickness within the Aegean plate and its boundaries with the adjacent plates signify the major tectonic processes that are taking place. A correlation of these crustal differences with the major neotectonic and active structures may clarify which processes are inactive and/or shallow, having no impact on the crustal structure. Thus, besides the crustal thinning observed along the Hellenic chain from central continental Greece to the Peloponnese, there are other more important crustal differences mainly on both sides of the Cretan basin (Makris 1977a, b, 2010; Makris et al. 2013) (Fig. 11.13).

Detailed analysis of the crustal structure above the subduction zone in the area of Crete and the Cretan Basin, using wide aperture seismic data both onshore and offshore, has shown that the continental crust beneath Crete gets its maximum thickness of 32 km immediately north of the northern coastal zone and then very rapidly is thinned out to its half thickness of about 16 km beneath the southern part of the Cretan basin, within a horizontal distance of less than 20 km (Bohnhoff et al. 2001) (Fig. 11.14). It is characteristic that the upper continental crust beneath Crete gets a

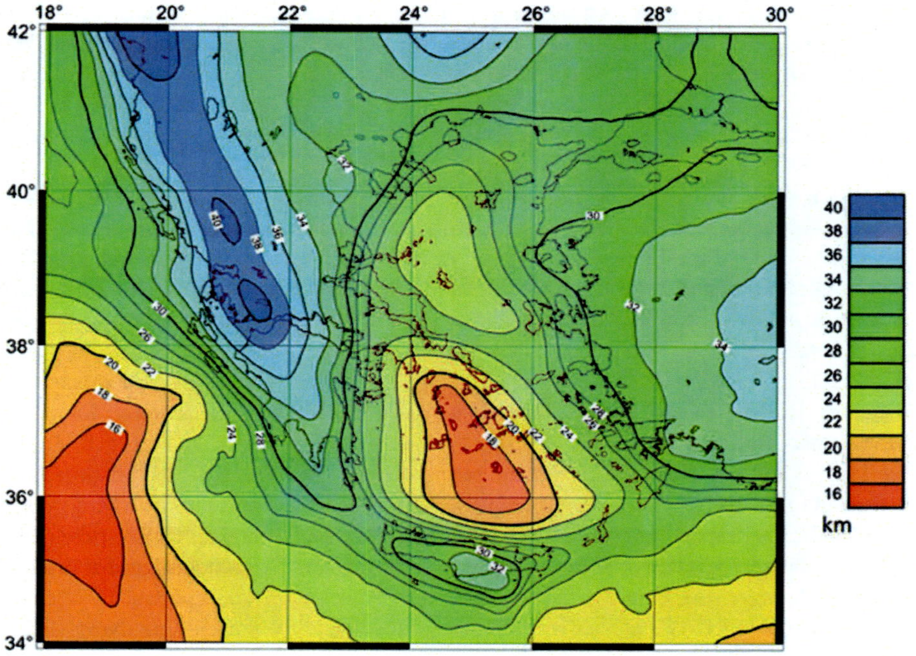

Fig. 11.13 Map showing the thickness of the Hellenic crust (from Makris et al. 2013). A significant reduction of 8–14 km crustal thickness is observed in the Southern Aegean on both sides of the Cretan basin

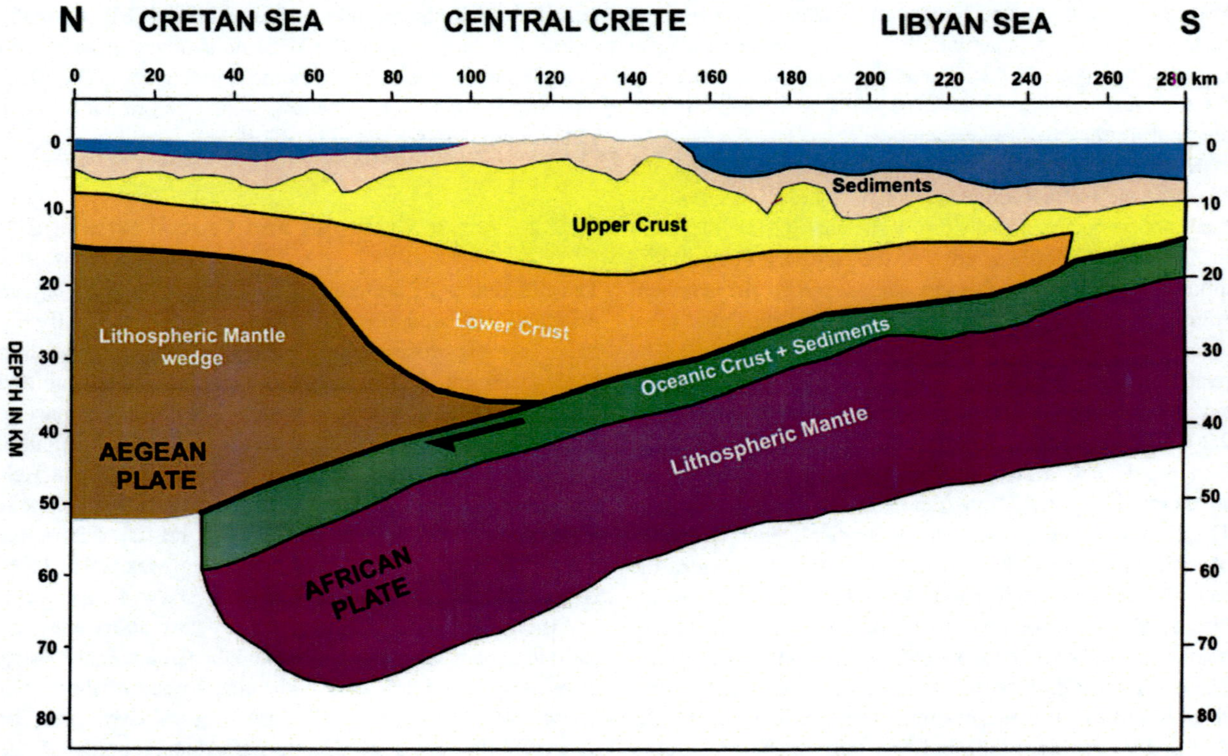

Fig. 11.14 N–S tectonic profile across Crete based on the seismic investigation of the Hellenic subduction zone using wide aperture seismic data (based on Bohnhoff et al. 2001)

maximum thickness of about 13 km, which becomes only 5 km beneath the Cretan basin (where the upper 2–3 km are soft sediments and sea water), whereas the lower continental crust is about 18 km thick beneath Crete, but only 8 km beneath the Cretan basin. The top of the subducting slab was detected at about 25 km depth beneath the Ptolemeus trough along the southern Cretan coastline and at about 45 km depth beneath the northern Cretan coastline. The upper mantle beneath the Cretan basin occurs between 15 km and 50–60 km, with an overall thickness of about 40 km in the mantle wedge between the subducting slab and the thin continental crust of the Cretan basin.

The northern boundary of the Aegean plate in the North Aegean is not evident in the crustal structure, since there is no crustal difference in the ENE–WSW direction (Fig. 11.13). In the eastern prolongation of the plate boundary in the Sea of Marmara there is a minor ENE–WSW trend with reduced crustal thickness of only 2 km (28 km vs. 30 km). This implies that the plate boundary is deformed exclusively by strike slip motion without causing any significant impact on the crustal thickness. On the contrary, the approximately 10 km difference of the crustal thickness between the Pindos chain and the west Aegean coastal zone (Fig. 11.13) reflect the ongoing plate convergence and subduction of the East Mediterranean crust beneath the Hellenic margin. Additionally, the crustal thickness variations along the western Hellenic margin in the Ionian Sea indicate that the backstop area, developed west of the Hellenic trench has very thin continental crust of a few km, since the total thickness together with the underlying Ionian oceanic crust is 22–26 km (Fig. 11.13). On the contrary, east of the Hellenic trench the crust rapidly thickens to 30–34 km.

The crustal extension of the Aegean microplate is conducted through a combination of the arc parallel structures and the transverse structures developed within the CHSZ. In a transverse section from the Southern Peloponnese to the Central Aegean, successive arc parallel neotectonic horsts and grabens can be observed (Papanikolaou et al. 1988), related to large extensional detachment faults, with a total slip of several kilometers up to a few tens of km (Papanikolaou and Royden 2007; Papanikolaou 2010). These faults often separate the underlying metamorphics from the overlying non-metamoprhics, representing the marginal faults of the neotectonic basins (see also Fig. 5.5). It is interesting that detailed GPS data have shown that there is an important E–W extension in southern Peloponnese of about 3–5 mm/year (Hollenstein et al. 2008; Chousianitis

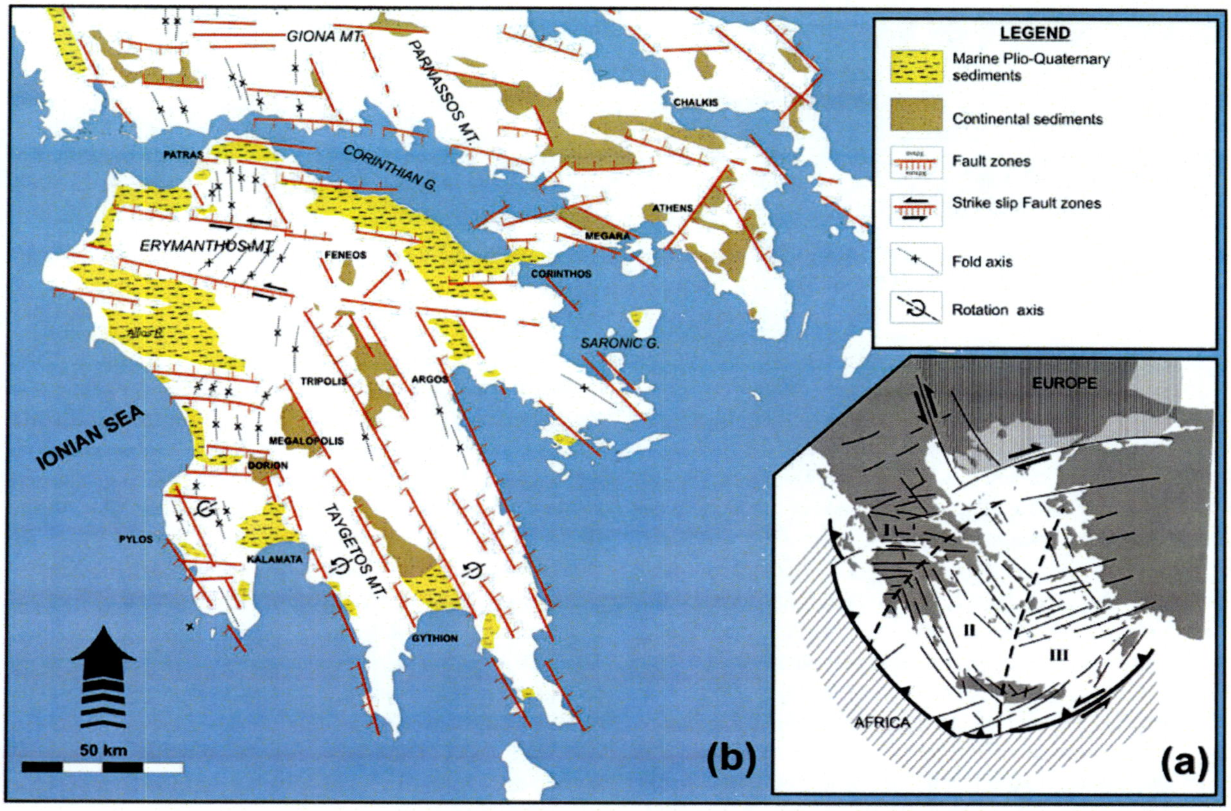

Fig. 11.15 Neotectonic map of the major marginal faults of the post-Alpine basins in the Southern continental Greece (**b**) and the neotectonic Aegean model (**a**). Three segments have been broadly distinguished, based on the different fault orientations, kinematics and with large variations also in their seismic potential (from Mariolakos et al. 1985)

et al. 2015). In the eastern Aegean Sea, Roche et al. (2019) have described E–W extensional detachments of Late Miocene age producing a N–S extension, which is present also in the current seismic activity, e.g. the 6.6 magnitude Kos 2017 earthquake, where a 35 km E–W oriented fault, dipping to the south was activated (Ocakoglu et al. 2018).

11.2.3 Neotectonic Deformation and Seismo-Tectonics

The major neotectonic faults, usually active since the Late Miocene, represent in their majority the **marginal faults** of the post-Alpine sedimentary basins. Simultaneously, they are among the most **seismically active faults** in the Hellenic arc. Systematic analysis of these faults showed that three major neotectonic segments can be distinguished in the region of the Hellenic arc, showing a different fault fabric (Mariolakos and Papanikolaou 1981, 1984; Mariolakos et al. 1985) (Fig. 11.15):

(1) The Central Greece area, with the exception of Attica, Southern Evia and the southeast half of the Peloponnese are characterized by transverse E–W trending normal or oblique faults, which are seismically very active.
(2) Attica, Southern Evia, the southeast half of the Peloponnese and the Cyclades, with the exception of the Amorgos Island, are characterized by arc parallel NW–SE trending normal faults with much lower seismic activity, in comparison to the previous ones.
(3) Amorgos, Ikaria, Samos and the Dodecanese islands are characterized by arc parallel ENE–WSW trending normal or strike slip faults, that accommodate high seismicity.

It is noteworthy that in Greece, primary surface ruptures associated with the activation of major active faults that lead to catastrophic earthquakes have been described already since the nineteenth century, such as the 1861 Aegion earthquake (M ~ 6.7) in Northern Peloponnese (Schmidt

Fig. 11.16 The open folding-bending of the Pleistocene beds in the Paliki Peninsula of Western Cephalonia, with a N–S orientation of the fold axis (from Papanikolaou and Triantaphyllou 2013)

1867) and the 1894 Atalanti earthquakes (M ~ 6.4, M ~ 6.8) (Locris) (Papavasiliou 1894; Skouphos 1894; Mitzopoulos 1895; Pantosti et al. 2001). These old seismic events activated E–W trending normal faults that belong to the CHSZ (Papanikolaou and Royden 2007), as this was also observed in the 1981 earthquake sequence of the Alkyonides in the Perachora peninsula, East Corinth Gulf (M = 6.7, M = 6.4, M = 6.2) (Mariolakos et al. 1981; Jackson et al. 1982) and the 1999 M = 5.9 Athens-Parnitha earthquake, that damaged the western and northwestern suburbs of Athens causing 140 casualties (Papanikolaou et al. 1999; Papadimitriou et al. 2002; Ganas et al. 2004).

Apart from the extensional structures, which are expressed through normal or oblique-slip faults, there are also some compressive zones with activation of thrust faults, mainly in the front of the Hellenic arc, such as for example in Cephalonia (Mercier et al. 1979). Here, folds "in-the-making" can be observed, from the folding of marine beds of Lower-Middle Pleistocene age (Papanikolaou and Triantaphyllou 2013) (Fig. 11.16). This surficial bending-folding is developed towards the deeper tectonic layers of the area into a more intense folding, representing an evolving growth folding. This compressive deformation is related with simultaneous brittle deformation, including reverse faults in combination with strike-slip faults, like the 2014 M = 6.0 and M = 5.9 Cephalonia earthquakes accompanied by significant uplift of the Paliki peninsula (Valkaniotis et al. 2014; Boncori et al. 2015).

The correlation of the active tectonic structures with the earthquake fault plane solutions provide a concise view of the **seismotectonics** in the Hellenic arc and the Aegean (Fig. 11.17) (Taymaz et al. 1991; Jackson 1994; Kiratzi and Louvari 2003; Shaw and Jackson 2010). The fault plane solutions show the general distribution of compression in the circumference of the arc and in the subduction zone, where, however, 80% of the plate convergence is aseismic (Shaw and Jackson 2010). Extensional zones with normal and oblique slip faulting dominate in the interior of the Aegean microplate with different extension orientations (e.g. Kiratzi and Louvari 2003). Thus, E–W trending normal faults are observed in central continental Greece and N–S trending normal faults in the northern part of the Pindos chain and in the southern Peloponnese. At the same time, strike slip fault plane solutions with right-lateral motion are observed along the northern boundary of the microplate in the Northern Aegean area both in the North Aegean and the Skyros Basins and at the western extremity of the CHSZ in the Cephalonia fault. Fault plane solutions with left-lateral strike slip motion are observed along the southeast part of the arc in Eastern Crete—Dodecanese (Kiratzi and Louvari 2003). The fault plane solutions along the subduction zone show the co-existence of thrusting and strike slip faulting in the along dip NE–SW transcurrent faults (Sachpazi et al. 2016) (see also Fig. 4.17). The neotectonic distribution of normal, reverse and strike slip faulting has shown the dominance of compressive structures at the front of the Hellenic arc and of the extensional structures at the back arc area, with minor distinct strike slip zones (Mercier 1979; Angelier 1979; Mariolakos and Papanikolaou 1984).

The **deformation rates** are impressive in some major active tectonic zones of Greece. The Corinth Rift is among the most rapidly extending regions of the Earth's continental crust, with 10–15 mm/year of north–south extension (1.1 m of extension over 90 years, Billiris et al. 1991; Briole et al. 2000). Strain is released through intense seismic activity where more than a dozen of strong earthquakes (M > 6.0) have occurred over the last 150 years. Thus, towards the central part of the Corinth rift, the uplift rate of the Pleistocene marine terraces in the Northern Peloponnese, is 1.5–2.0 mm/year (Armijo et al. 1996; DeGelder et al. 2019). This rate corresponds to a slip rate of 5–8 mm/year of the fastest active faults along the southern coasts of the Corinthian gulf. Towards the eastern and western part of the Gulf the uplift rate based on coral U/Th dating, diminishes to 0.5 mm/yr (Roberts et al. 2009) and 0.7 mm/year (Houghton et al. 2003) respectively, indicating also lower fault slip-rates. It has to be clarified that uplift rates and fault slip-rates vary along strike the active faults, so these rates should not be regarded as constant in the study sites (e.g. Morewood and Roberts 1999; Roberts et al. 2009). The fault

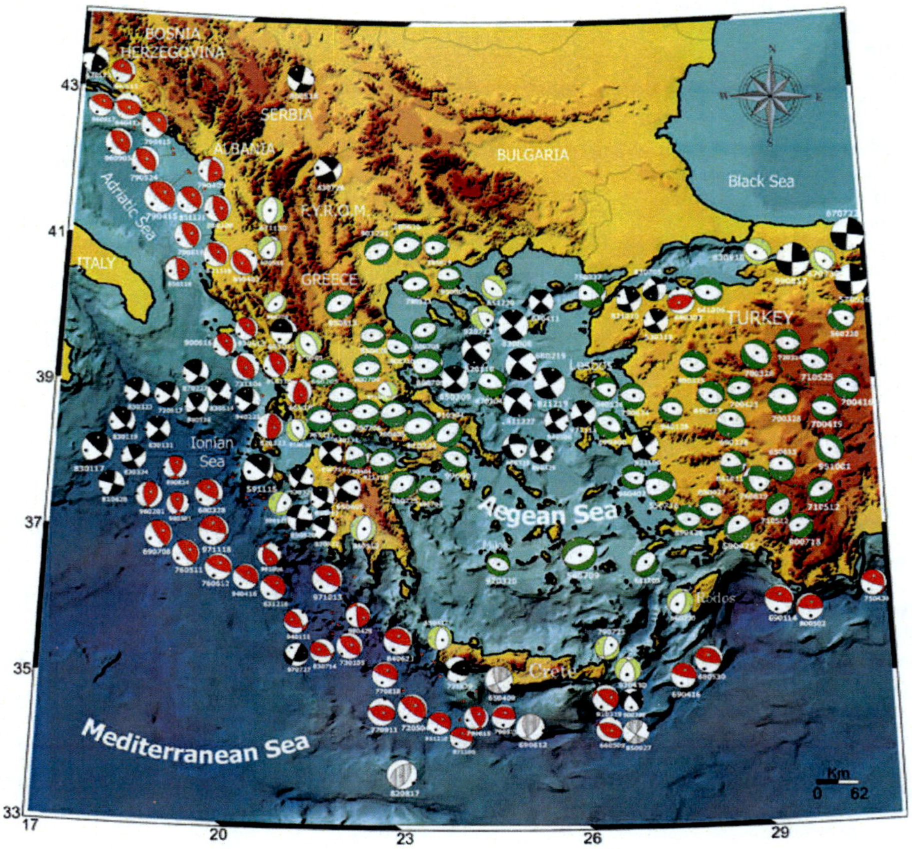

Fig. 11.17 Map of the distribution of the fault plane solutions of the earthquakes in the Hellenic region (from Kiratzi and Louvari 2003). The compressive mechanisms corresponding to thrust faults are shown in red. The extensional mechanisms corresponding to normal faults are shown in green, whereas the mechanisms corresponding to strike-slip faults are shown in black

segments in the Perachora peninsula have slip-rates ranging from 0.6 to 2.0 mm/year (Collier et al. 1998; Mechernich et al. 2018) and were activated in the 1981 earthquake sequence (Ms 6.7, 6.4, 6.3), through a stress transfer and triggering mechanism (Hubert et al. 1996). These events produced spectacular primary surface ruptures and dramatic coastal uplift and subsidence (Mariolakos et al. 1981; Jackson et al. 1982).

The deformation in the northern boundary of the Aegean microplate in the Northern Aegean Basin is also highly significant, as shown by its southern marginal fault, which has a length of 160 km and its vertical throw reaches 5–6 km during the Pliocene—Quaternary, although it is mainly a strike-slip fault (Papanikolaou et al. 2002, 2006). The submarine escarpment to the north of the Sporades and Limnos islands, is up to 1.5 km high, corresponding to the upper third of the fault surface. This indicates that the sedimentation rate of the North Aegean Basin is much lower than the throw rate of its southern marginal Fault. This fault, is the largest seismically active fault in the Aegean Sea (Papanikolaou and Papanikolaou 2007) (Fig. 11.18a) and has an average slip rate of its normal component about 1 mm/year, while the strike-slip rate is definitely quite higher (around 12–15 mm/year). The overall structure indicates a gradual shift from strike slip to normal faulting as approaching the west, since the width and the depth of the basin becomes significantly larger towards the west. As a result, its' westernmost segment is regarded as a potential tsunamogenic source for the Holocene tsunami sediments traced in the Thermaikos Gulf (Reicherter et al. 2010), including a tsunami candidate described by Herodotus during the Greek-Persian war in 479 BC in Potidea (Mathes-Schmidt et al. 2019).

Another 120 km long tectonic zone is mapped in the southeast Cyclades, extending from the Christiana volcanic islands to Santorini and Amorgos islands towards the ENE (Nomikou et al. 2018a). This fault zone hosts the larger intra-plate earthquake epicenter of magnitude 7.4 and the largest tsunami generated in the Mediterranean Basin during the twentieth century (Galanopoulos 1957; Ambraseys; 1960; Okal et al. 2009). At its southwestern part it comprises the volcanic structures of the Christiana–Santorini–Kolumbo volcanic center (Nomikou et al. 2013). The fault segment that was activated in 1956 is more than 40 km long, running

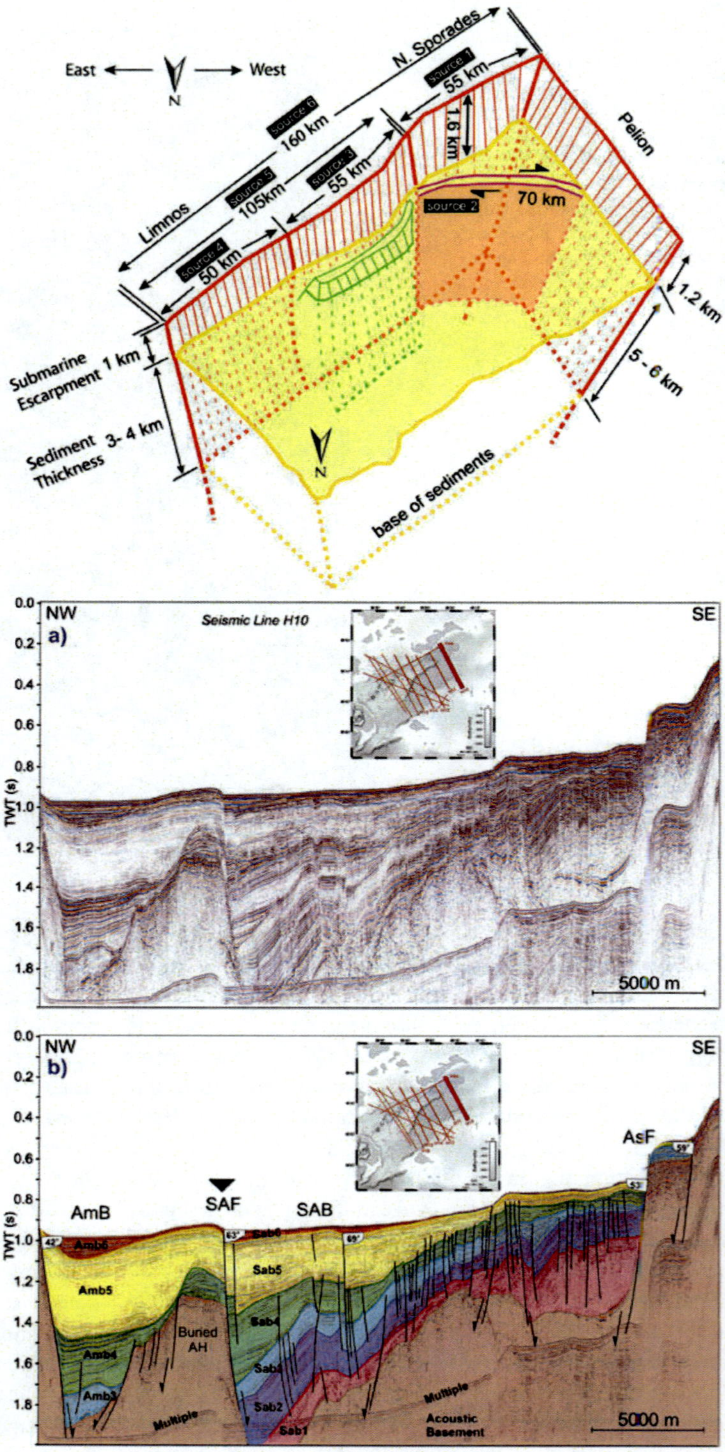

Fig. 11.18 a Stereographic 3D diagram of the largest marginal faults of the Northern Aegean basin. The southern boundary of the basin has a length of 160 km and a throw of about 5–6 km, which may generate a multi-segment earthquake rupture that can result into an earthquake of magnitude 7.6 (from Papanikolaou and Papanikolaou 2007). **b** NW–SE striking multi-channel reflection seismic line across Amorgos Basin (AmB) and Santorini–Anafi Basin (SAB). Upper part shows seismic data, lower part shows interpretation of seismic line HH10. The Amorgos Fault occurs at the NW edge of the profile with indication of its 42° dip towards the SE. The Santorini–Anafi Fault (SAF) dips with 63° also towards the SE, whereas the Astypalaea Fault (AsF) dips with 53° towards the NW. The overall structure is a NE–SW tectonic graben, filled with ~700 m of sediments (from Nomikou et al. 2018). Note the absence of the lower stratigraphic formations Sab 1 and Sab2 from the base of the Amorgos Basin. The Anhydros Horst (AH) is buried below the upper formations of Sab5 and Amb5

immediately south of the Amorgos steep coastline and has formed a submarine escarpment of 600 m (Fig. 11.18b). The fault has disrupted 600–650 m of sediments in the Amorgos–Anafi Basin occurring above the subsided Alpine basement. Taking into account also the 700–800 m onshore throw up to the top of the Amorgos Island mountain range then the total throw of the Amorgos Fault exceeds 1.8 km. Another sub-parallel fault (the Anafi fault) is observed within the basin with several hundred meters of throw. Detailed survey at the base of the Amorgos fault has indicated the existence of 9 m of recent movement on the fault plane, probably representing the 1956 event (Nomikou et al. 2018a). It is remarkable that the sedimentary sequence within the Amorgos-Anafi basin show internal onlaps and extreme variations of sediment thickness with maximum values along the subsiding edges of the neotectonic blocks due to growth faulting. Additionally, different age of unconformable deposition in the adjacent basins is deduced from the absence of the Sab1 and Sab2 lower formations from the Amorgos basin. Thus, the Amorgos marginal fault of the Amorgos Basin should be much younger than the Anafi fault.

The distinction of the seismically active faults from the inactive faults is of major importance both for scientific reasons but also for practical reasons, regarding the safety of the cities and infrastructures. Thus, a systematic neotectonic mapping was undertaken during the 1980s and 1990s from the Greek universities and research Institutes, co-ordinated and financed by the Earthquake Planning and Protection Organization of Greece. A similar project was undertaken by the National Centre for Marine Research regarding submarine neotectonic maps at scale 1/100,000 (see also Fig. 11.23). Thus, several neotectonic maps were elaborated and published during the 1990s from seismically active zones of Greece, such as the Corinth area (Papanikolaou et al. 1996) (Fig. 11.19). This neotectonic map shows the eastern part of Corinth, ruptured by several sub-parallel E–W normal faults, which are bordered to the west by a prominent NE–SW trending Nemea-Corinth fault zone. The western part of Corinth forms a non-disrupted neotectonic block where several Middle-Late Pleistocene marine terraces are observed, without presence of E–W active faults. It is remarkable that east of the Nemea-Corinth fault, the Aghios Vassilios southern marginal fault zone of the Corinth graben is active, contrary to its western prolongation in the Nemea–Stymfalia E–W trending fault zone, which is inactive.

11.2.4 Paleoseismology and Seismic Hazard

Recent developments in the field of earthquake geology and **paleoseismology** have provided the possibility of defining the position and the magnitude of the imminent major earthquakes, based on the length and the surface rupture parameters of the active faults, by applying empirical formulas (Wells and Coppersmith 1994; Pavlides and Caputo 2004; Wesnousky 2008), and decoding the seismic history of each fault, at least for the Middle-Upper Holocene (e.g. Pavlides 2003; Koukouvelas et al. 2010). As already mentioned, the integrity of the data of the historic seismicity for the Hellenic region is limited, even though we have the richest archive since antiquity (550 BC). Despite the long earthquake historical catalogue in Greece since 550 BC, it is considered complete for events $M \geq 7.3$ since 1500 and for $M \geq 6.5$ and $M \geq 5.5$ only since 1845 and 1911 respectively (Papazachos et al. 2000a). Thus, the completeness period is a small fraction of the period covered by the historical record, but is important since it is used as input data in the probabilistic seismic hazard assessment. Historic earthquake catalogues are generally too short compared to the recurrence intervals of particular faults both worldwide (e.g. Scholz 2002) and in Greece (e.g. Chatzipetros et al. 2005). Therefore, the sample from the earthquake historical record is clearly incomplete and a large number of faults would not have been ruptured during the completeness period of the historical record, implying that fault specific approaches are of decisive value for seismic hazard assessment (Papanikolaou et al. 2015a). The latter has been demonstrated in the region of Attica where 24 active faults have been mapped, but only 9 strong historical events have affected the Attica region (Deligiannakis et al. 2018). In particular, only four of these events can be clearly attributed to certain faults due to large epicentral and magnitude uncertainties of historical events. In the offshore settings the incompleteness is even higher. For example in the Skyros basin, 19 active faults have been mapped, based on high resolution bathymetric data and seismic reflection profiles, which can generate earthquakes stronger than $M = 6.0$ (8 of them $M > 7.0$) (Papanikolaou et al. 2019b). It is interesting to note that only 3 of these faults have been ruptured and recorded in the earthquake historical records, thus seismic hazard is severely underestimated since 16 of the 19 seismic sources were unknown.

The seismic history of active faults can be revealed through: (a) paleoseismic trenching supported by C^{14} radiocarbon dating of soil horizons (McCalpin 2009) such as the Skinos Fault in the Perachora Peninsula (Collier et al. 1998), (b) cosmogenic isotope dating which measure the cosmic rays over the fault surfaces (free face of postglacial scarps) as is the case of the Sparta fault in the eastern margin of the Taygetus mt (Benedetti et al. 2002). The terrestrial laser scanning technology (t-LiDAR) has provided new possibilities for extracting paleoevents on bedrock fault scarps by analyzing the surface roughness of the fault plane (e.g. Pisia fault, Wiatr et al. 2015) since the periodic exhumation of the fault plane by distinct surface faulting earthquakes results in differential grades of weathering (e.g. Carcaillet et al. 2008). T-Lidar can be combined with cosmogenic isotope dating offering important insights into the past earthquake history of a fault (e.g. defining the slip and

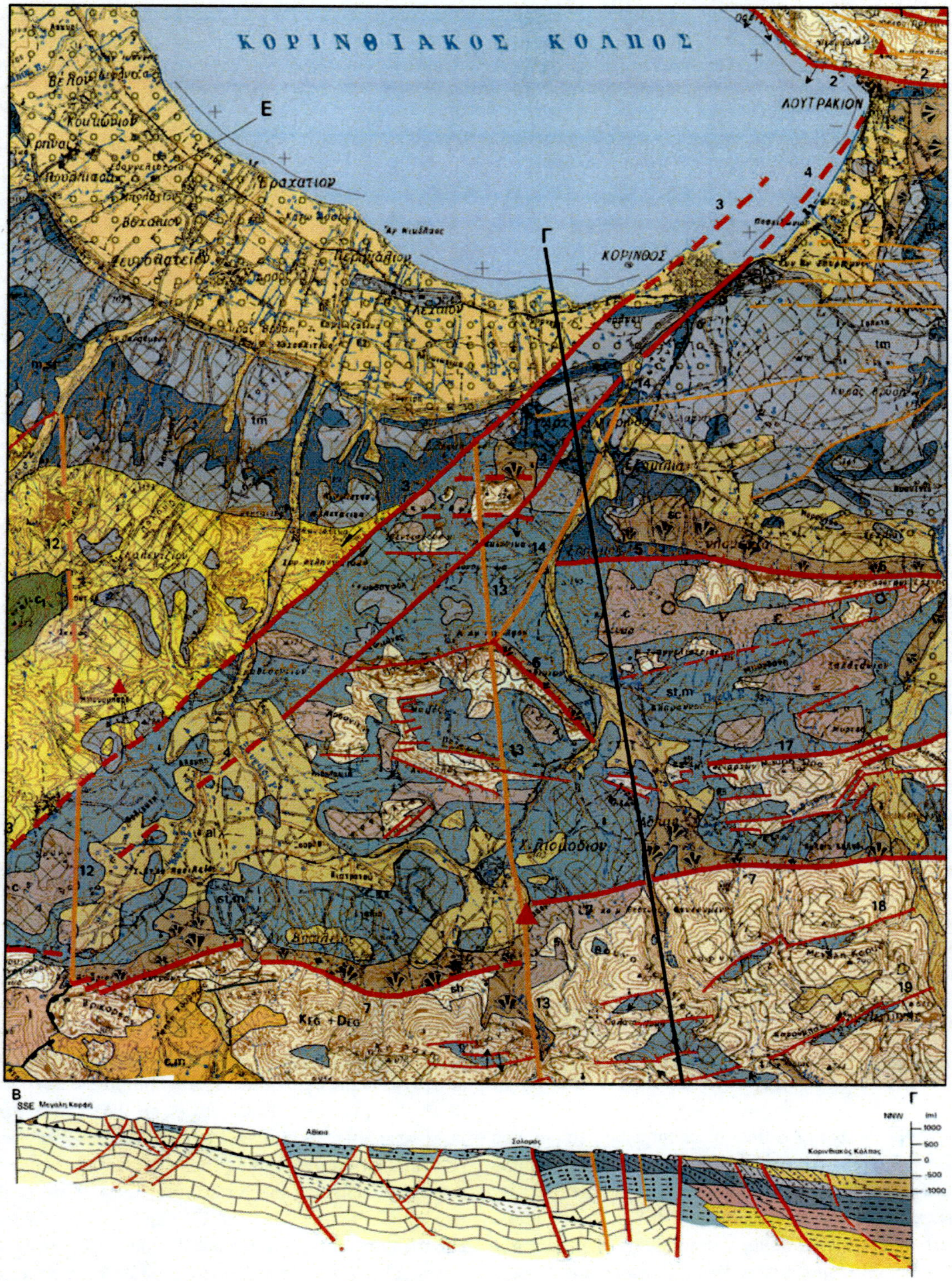

Fig. 11.19 Neotectonic map of the Corinth area (Northeastern part of the Korinthos sheet, at 1/100,000 scale, from Papanikolaou et al. 1996). Active faults are shown in red, probably active faults in orange and inactive faults in green. The faults are numbered and there is an information sheet for each one, including its length, throw, mechanical characteristics, seismic potential and seismic history

age of the last 7 events of the Pisia fault, Mechernich et al. 2018) or even fault interaction processes along and across strike active faults (Cowie et al. 2017).

Lastly, calculation of time dependent conditional probabilities of the activation of a fault is also possible, based on its characteristics during the Holocene seismic deformation, as well as the evaluation of the future intensities on maps, based on the analysis of the seismic response of the geological formations around the fault, e.g. in Sparta (Papanikolaou et al. 2013).

The paleo-seismological research has provided excellent results regarding the recurrence rate of the seismic activity of the active faults during the Holocene, with time periods ranging from a few hundred years to a few thousand years, thus greatly surpassing the uppermost limit of the historic seismicity recording in Greece. A characteristic recent case was the 6.6 magnitude earthquake of Kozani in 1995, which occurred in an area considered as non-seismic by the antiseismic regulation of the Earthquake Protection and Prevention Organisation of 1984. At the same time, the hazard identification of each fault is possible, based on the time elapsed since its last activation in relation to the recurrence rate. In this regard, for example the Sparta fault is considered as a mature fault, as it has exceeded its mean recurrence interval, since its last activation (M ~ 7.0) in 464 BC. On the other hand, the Atalanti fault, which was activated in 1894 (M ~ 6.4, M ~ 6.8), is considered as immature since paleoseismic trenching reveal that 1894-type earthquakes repeat every 660–1120 years (Pantosti et al. 2004).

In that case, the identification of the recurrence rate of a fault is necessary in relation to the slip rate, as expressed through the throw rate, as low throw rate faults tend to have longer recurrence intervals (of a few thousand years) while high throw rate faults have a short recurrence interval (of a few centuries) (Roberts and Michetti 2004; Roberts et al. 2004) (Fig. 11.20).

Therefore, fault slip-rates are of great importance since they govern earthquake recurrence and can be used as data input for fault specific seismic hazard maps (Roberts et al. 2004; Papanikolaou et al. 2013; Deligiannakis et al. 2018) with the additional possibility to classify the faults in categories based on terms of hazard.

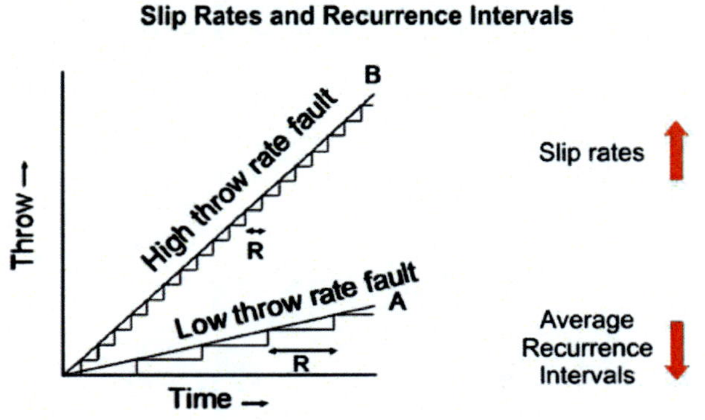

Fig. 11.20 The relation of the average recurrence interval of earthquakes of a fault with its slip-rate (from Roberts et al. 2004)

11.2.5 The Active Volcanoes

The outcrops of the modern Aegean volcanic arc are distinguished in four groups (Nomikou et al. 2013) (Fig. 11.21a):

(i) the northwestern group of **Methana,** comprises the outcrops of the Methana peninsula, but it extends also in the Sousaki area at the northwestern coastal area of the Saronikos Gulf as well as in the Aegina and Poros islands, where the volcanism was active throughout the Pliocene and ended in the Early Pleistocene (Pe-Piper and Piper 2002). Active fumaroles are still observed in the Sousaki volcanic area, where the volcanic rocks penetrated the Alpine basement (mainly Jurassic ophiolites and limestones) and the post-Alpine marine sediments of Pliocene-Lower Pleistocene age (Papastamatiou 1937). The Methana mountainous peninsula (750 m of altitude) is made of several volcanic domes, formed mainly during three major extrusive periods within the last 750 Ka (Hurni et al. 1995). It is noteworthy that pyroclastic rocks and pumices are found in very restricted outcrops, and the volcanism has been dominated by andesitic and dacitic lavas.

(ii) the **Milos** group, comprises also the outcrops of the surrounding islands of Kimolos, Antimilos and Polyaegos. The volcanic rocks form numerous outcrops of volcanic domes, craters and pyroclastic deposits within marine Plio-Quaternary sediments (Sonder 1925; Fytikas et al. 1986). The volcanic activity started during Pliocene in the Prophitis Ilias mt of western Milos, where the lavas are overlying the Upper Miocene–Lower Pliocene marine sediments and the underlying metamorphic basement, croping out along the southern coast of Milos Island. The recent volcanic eruptions of Trachilas (~400 Ka) and Fyriplaka (40–120 Ka) were sub-aerial, contrary to the previous submarine eruptions. Intense geothermal alterations occur with large deposits of industrial minerals under exploitation. Extended outcrops of opsidian occur in several locations and islets with pre-historic «industrial» settlements of neolithic tools, such as the Phylakopi site. High temperature vents and thermal springs (~100 °C) are traced in several

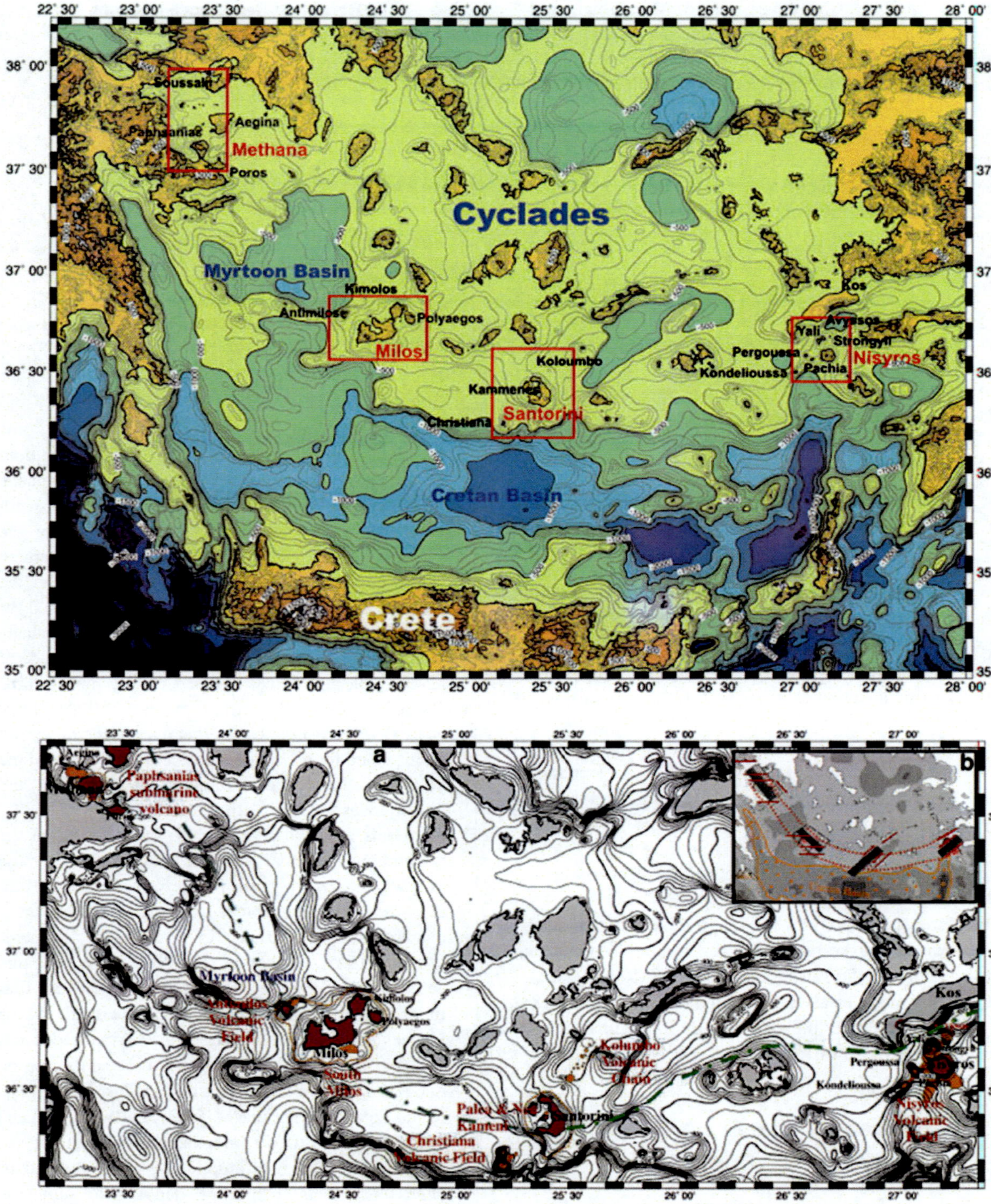

Fig. 11.21 a The four volcanic centres of the modern Aegean volcanic arc. **b** The submarine volcanic outcrops around the onshore outcrops of the volcanic islands (from Nomikou et al. 2013)

spots along the island with the Palaeochori coastal site being one of the most characteristic.

(iii) The **Santorini** group, comprises also the Christiana islands in the SW and the submarine Kolumbo volcanic crater in the NE. The volcanism on Santorini Island started in about 700–800 Ka and is characterized by major caldera forming Plinian volcanic eruptions, followed by intermediate periods of minor activity (Ktenas 1935; Druitt et al. 1989). The last «Minoan» eruption occurred 3600 years ago and destroyed the pre-existing Thera settlements in the archaeological site of Akrotiri (Friedrich et al. 2006). The volcanism started in the Christiana islands in Late Pliocene and ended in the Early Pleistocene. The post-eruptive volcanic activity formed the Kammeni islands in the centre of the caldera from 197 BC in the Palea Kammeni to the last eruption of the Liatsikas dome in 1950. The Kolumbo submarine volcano is younger, probably Late Pleistocene in age, with several eruptions building its 500 m height crater (see also Fig. 4.10). The last eruption occurred in 1650 AD causing 70 casualties.

(iv) The **Nisyros** group, comprises also the outcrops of Kos, Yali, Strongyli, Pachia and Pergoussa islands (Nomikou et al. 2018b). The older volcanic activity in the area is observed in several outcrops of Upper Miocene–Pliocene age in Kos Island. More recently, a major volcanic structure was formed at 500 Ka in the Kefalos peninsula of western Kos, followed by a huge eruption, known as the «Kos ignibrite eruption» in 161 Ka. This eruption covered the central Kos plateau of Antimachia by several tens of meters of pyroclastic rocks and pumice, spread all over the Dodekanese islands and the western coast of Minor Asia (e.g. Allen 2001). The volcanic activity continued in the Late Pleistocene with the formation of Nisyros Island, which is a strato-volcano, characterized by a caldera and post-caldera domes. Its last eruption occurred in 1873–1888 with phreato-magmatic explosions towards the bottom and the walls of the caldera (see also Fig. 4.8).

The volcanic outcrops of the modern Aegean volcanic arc are much larger, when the submarine volcanic formations are also taken into account. Since the first discovery of the Paphsanias submarine volcano, north of the Methana volcanic centre (Papanikolaou et al. 1988, 1989; Pavlakis et al. 1989) several other volcanic structures have been detected on the sea floor, usually around the volcanic islands (Nomikou et al. 2013) (Fig. 11.21b). Such submarine outcrops have been reported from Nisyros and surrounding islets (Papanikolaou et al. 1998; Papanikolaou and Nomikou 2001; Nomikou 2004; Nomikou and Papanikolaou 2010), Milos–Antimilos (Alexandri et al. 2001), Santorini–Kammenes (Nomikou et al. 2014) and Kolumbo (Nomikou et al. 2012).

n interesting feature of the submarine volcanoes is their **volcanic relief**, which exceeds 600–700 m height from the sea bottom (Nomikou and Papanikolaou 2010). The highest volcanic edifice is observed in the Antimilos Island, where besides the 750 m of the onshore outcrops (Marinos 1961) there are another 700–900 m submarine outcrops along its northern slopes (Alexandri et al. 2001). The Kolumbo Submarine relief is about 500 m (Nomikou et al. 2012) whereas, the Kammeni islands relief is more than 600 m (160 m onshore and 440 m offshore) (Nomikou et al. 2014). The Nisyros volcano and surrounding volcanic islets have an impressive volcanic relief up to 1400 m (700 m onshore the Prophitis Ilias dome and 600–700 m offshore) (Nomikou and Papanikolaou 2010). In the Methana peninsula the volcanic relief starts from about 400 m depth at the base of the Paphsanias submarine volcano (Papanikolaou et al. 1988) up to the 750 m elevation onshore, thus exceeding 1150 m.

Thus, the overall volcanic structures of the Aegean arc are even more impressive, reaching 1.5 km of volcanic relief, if they are observed from the sea bottom. This is observed mainly in the Methana and Antimilos **volcanic domes**, where no collapsing volcanic structures are observed. On the contrary, in **calderas**, like in the Santorini and Nisyros volcanoes or large **craters**, like in Milos and especially in the very geometric Kolumbo submarine volcano, the height of the volcanic relief is reduced, due to the loss of the pre-existing volcanic summit during the major eruptions. Additionally, extended thermal springs are observed offshore in almost all the volcanic islands and submarine vents (Varvavas and Cronan 2005; Kilias et al. 2013).

11.3 The Recent Paleogeographic Evolution and Its Impacts on Biodiversity

11.3.1 Late Pleistocene—Holocene Paleogeography

The change of the sea level due to the climate changes, which was about −125 m at the end of the last glacial period, had a major impact on the physico-geographical image of Greece. The 125 m isobath that was the former shoreline during the last glacial period only 18–20,000 years ago demonstrates some major changes in the recent palaogeography, particularly in the coastal zone. For example, during eustatic

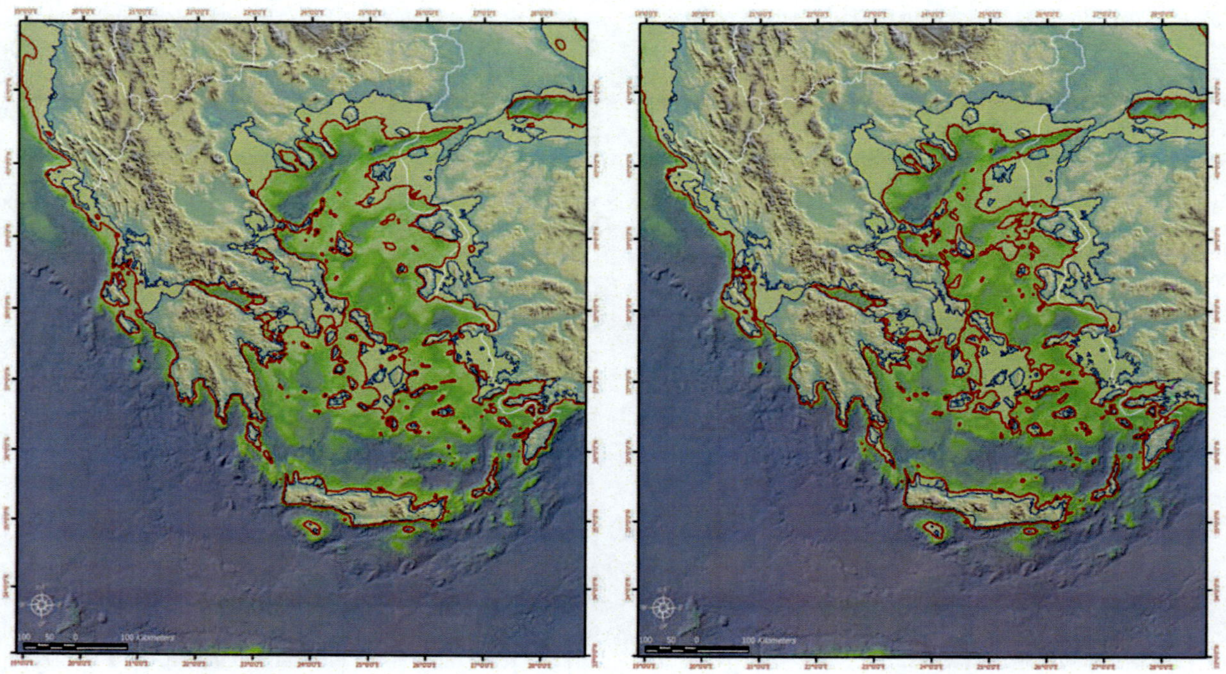

Fig. 11.22 Bathymetric maps of the Greek seas, outlining the 125 m and 200 m isobaths, where the low stand sea levels were during previous glacial periods

sea-level lowstands the Gulf of Corinth was a lake (Perissoratis et al. 2000; Roberts et al. 2009) similarly to the Evoikos Gulf (Perrissoratis and Van Andel 1991), whereas the region of the Cyclades was forming a continuous land (Fig. 11.22). Recent IODP drilling in the Corinth Gulf has revealed 11 climate driven environmental and sedimentary changes during the last 650 Ka, reflecting the periods of marine and lacustrine environment, due to the fluctuations of the sea level and the subsequent isolation of the Gulf from the adjacent Patraikos and Saronikos gulfs (McNeill et al. 2019).

In the Hellenic arc, due to the intense tectonic activity that cause significant uplift or subsidence, the active deformation is significant and the reconstruction of the paleogeography with the aforementioned simple outlining of the paleo-coastlines, based on the present isobaths is clearly false. The interplay between the eustatic sea-level fluctuations and the tectonic activity define both the geomorphic features such as notches (Flemming 1978; Pirazzoli et al. 1989) and marine terraces (Armijo et al. 1996) as well as the depositional history (e.g. Collier 1990). The interplay produces a complex paleoenvironment with major lateral and temporal variations even in neighboring localities, as it has been demonstrated from borehole data (Papanikolaou et al. 2015b). A characteristic example of such interplay driven by fault slip rates in the offshore setting concerns the Messiniakos Gulf (Fig. 11.23), where the paleo-coastline of the last glacial period was mapped at the 107 m isobath instead of the 125 m, due to its location in the uplifting frontal part of the Hellenic arc (Papanikolaou et al. 1988). At the same time, there are segments of the coastal zone with further post-glacial uplift of several meters, up to a few tens of meters, next to major marginal faults, such as the western coast of the Taygetus mt in the Kitries peninsula. Similar Holocene deformation structures detected from the mapping of the submarine paleocoastal zones have been described in the Saronic Gulf (Papanikolaou et al. 1989a), the Southern Evoikos Gulf (Papanikolaou et al. 1989b), the Argolikos Gulf (Papanikolaou et al. 1994), the Laconikos Gulf (Papanikolaou et al. 2001) and the Kyparissiakos Gulf (Papanikolaou et al. 2007).

The extensional collapse of the Aegean region during Quaternary destroyed the previously continuous continental area of Aegeis (Philippson 1901) and lateral migrations of mammals from Minor Asia to continental Greece gradually became impossible (Sakellariou and Galanidou 2016). Bathymetric analysis combined with slip rates of active faults in the Aegean area between Chios-Lesvos-Limnos and Evia-Pelion concluded that land bridges probably existed during the lowstands of MIS 4, 6 and 8 (Papanikolaou et al. 2019c). Thus, hominin migrations were possible through the North Aegean until the late most Middle Pleistocene around 140 Ka.

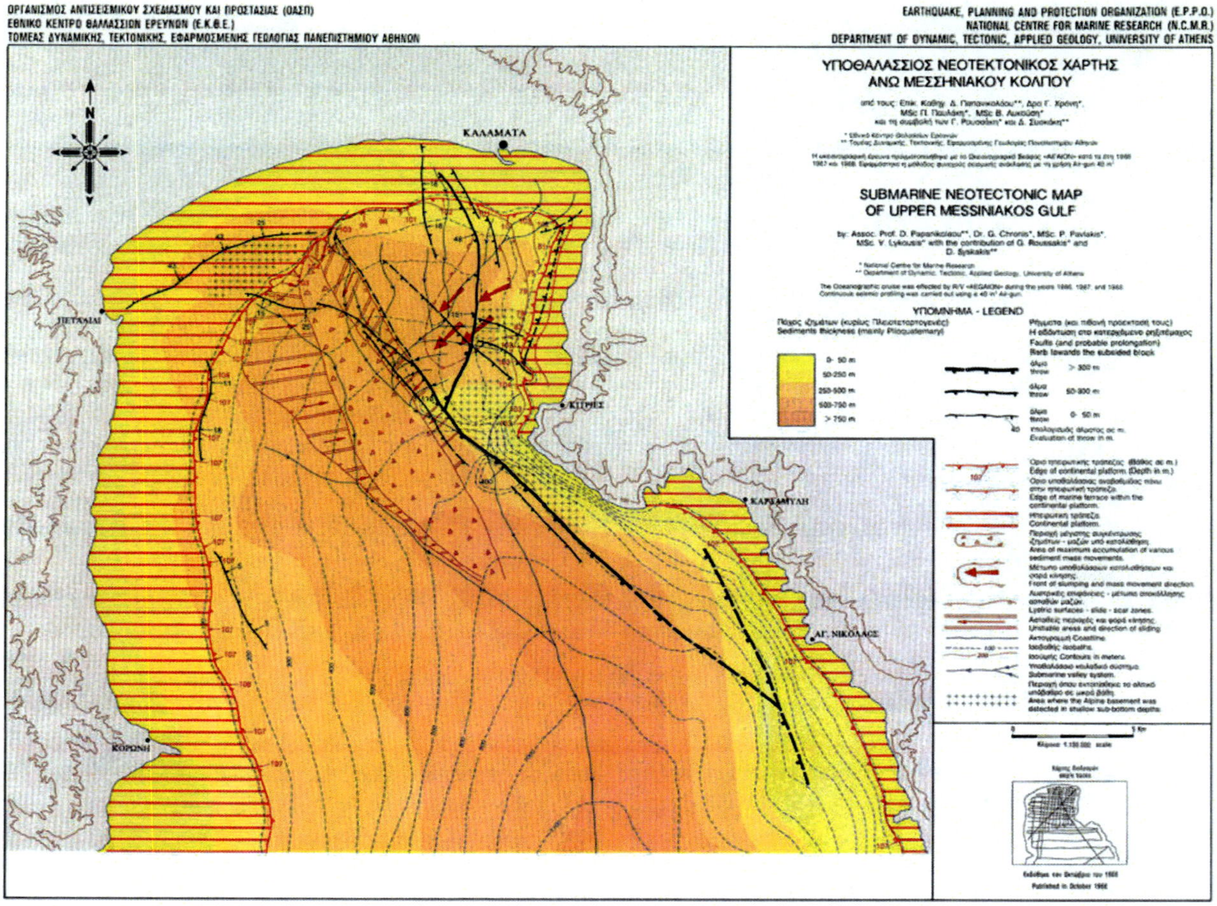

Fig. 11.23 Submarine Neotectonic Map of the Upper Messiniakos Gulf, showing the edge of the continental shelf, as defined through oceanographic bathymetric and litho-seismic research (from Papanikolaou et al. 1988). The depth of the paleocoast in the main southern tectonic block is 107 m, while at the smaller blocks to the north it is reduced to 103, 99, and 79 m, producing a relative Holocene uplift up to 28 m

11.3.2 Endemism and Biodiversity in the Hellenic Peninsula and the Aegean Archipelago

The aforementioned eustatic movement of the Holocene since the end of the last glacial period isolated many islands of the present geography of Greece and favored the development of specific adjustments to the plants and the animals of the region, resulting in the **endemic** phenomenon of the area. If this was only one isolated phenomenon, its impact would have been relatively limited. However, this phenomenon of the periodic fluctuation of the sea level due to eustatism has been repeated several times during the Quaternary (see also Fig. 5.11). This means that the organisms are adapted in a certain coastal or semi-mountainous or mountainous environment during one period, and then, they have to re-adapt into a new environment at higher altitudes of about 100–120 m. This process would have provoked a less dramatic change in a uniform environment, like a large continental area, but in a complex landscape like the Greek peninsula and the Aegean area it has resulted into significant impacts on **biodiversity.** The reason is that sea level change is accompanied by significant geographical changes, especially in the peninsulas and islands where some former continental areas become isolated, and prevent the migrations of the organisms. Especially in the islands the sea-level changes may result in dramatic reduction of their area during the high stands with immediate impact on their biodiversity (Weigelt et al. 2016; Fernandez-Palacios 2016). At the same time, the cold periods lower the levels of optimal growth of the organisms, which rise again during the warm periods. This leads to a huge impact on biodiversity, because both in

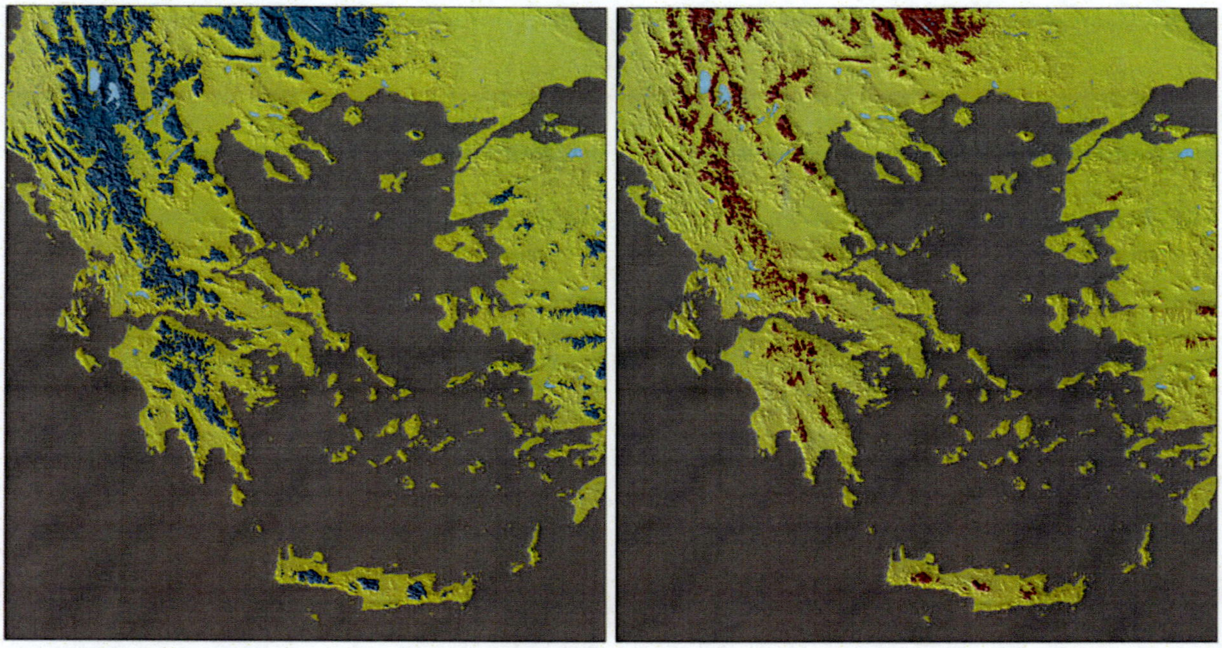

Fig. 11.24 Morphological maps of the Hellenic peninsula, by highlighting the altitudes exceeding 800 and 1,200 m, where endemic species may develop and survive during climatic changes

the island and the mountainous regions endemic species adjusted to delicate environments are created (Cowling et al. 1996, 2014).

Biodiversity depends on the diversity of the environments where organisms are growing and evolved (Arianoutsou-Faraggitaki et al. 2003). This means that in an area with environmental homogeneity, such as in a flat area, with similar geological bedrock and similar climate, a high biodiversity is not expected. Thus, it will be characterized either by rich vegetation, although comprising only a few species, or by poor vegetation of the steppe type. On the contrary, in areas with intense relief, both plain and mountainous species will be observed. Depending on the climate and the hydro-lithological regime, aquatic or hydrophobic species will be grown. Depending on the geological background, there will be plants that thrive in calcareous soils, or acidic, or basic, etc.

In Greece, there is a large morphological, geological, and climate diversity, due to the active tectonism and the ongoing orogenic process that was described in the previous chapters, but also because of the heavy impact of the recent climate changes. The allocation of the high altitudes in the Hellenic Peninsula creates isolated mountainous massifs, where many specific organisms may escape and survive during extreme climatic changes (Cowling et al. 2014) (Fig. 11.24).

This can be observed in thematic maps dealing with the distribution of endemic organisms and mainly of the plants, showing extreme concentrations on the island front of the arc and in the Aegean Sea, as well as in mountainous environments, isolated from other similar environments, occurring further north in the Balkans (Georghiou and Delipetrou 2010) (Fig. 11.25).

At the same time, the geological structure of the Hellenic mountains and islands, with outcrops of basic ophiolite rocks, such as in Northern Pindos mt, or metamorphic schists and acidic granitic rocks elsewhere, such as in the Cyclades, combined with the hydro-geological conditions of permeable or non-permeable formations and the different hydrochemistry of the water resources, supports a large diversity of geo-environments. In addition to this diversity, the climatic diversity should be combined, which is highly differentiated along the country. For example, rainfall is considerably higher in western Greece (almost by twofold) compared to eastern Greece. Moreover, average temperatures differ significantly as indicated by the numerous climatic zones that have been distinguished in Greece (Fig. 11.26).

The significant concentration of endemic species mentioned above, as shown in maps in the form of hot spots, concerns the already large biodiversity observed along the Alpine orogenic system in the Mediterranean, as compared mainly to Northern Europe in the interior of the European plate. The eastward continuation of the Alpine chain to the Iranides, the Afganides, and the Himalayas shows much smaller biodiversity probably because there is no eustatic

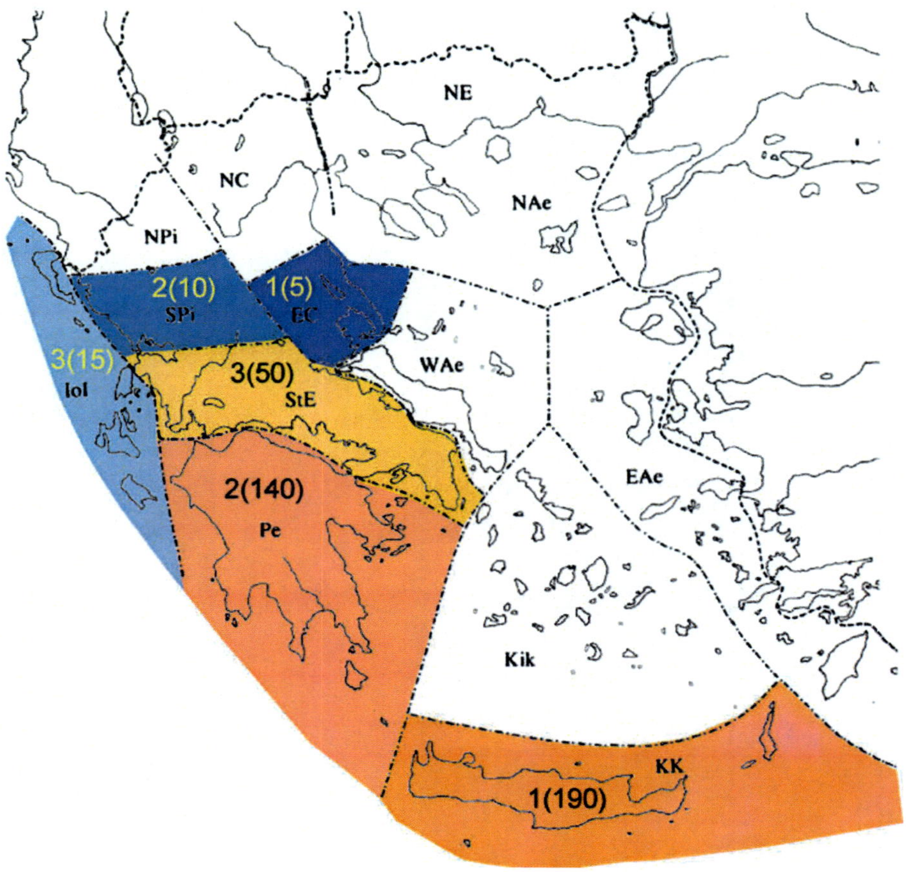

Fig. 11.25 Distribution of endemic species of the Hellenic flora in the plant counties of Greece. The maximum numbers can be observed in Crete, Peloponnese and Central Greece (from Georghiou and Delipetrou 2010)

impact, as is the case of the Mediterranean orogenic systems (Fig. 11.27).

It is remarkable that very recently it has been discovered that in the southernmost peninsula of Mani in continental Greece in the Apidima Cave there are fossils of Homo Sapiens of 210 Ka, the earliest found in Eurasia, much older than the 70–80 Ka, considered previously (Harvati et al. (2019). This new age of contemporary man in Europe changes several views about our provenance, evolution and co-existence with the previous hominin Homo Neadertalensis, whose extinction is dated around 30 Ka ago. Thus, instead of a short period of 30–40 thousands of years of parallel existence of the two recent hominins in the Aegean area during the last glacial period of Wurm, it is extended to more than 180 thousands of years earlier during the Riss glacial period.

In conclusion, the exceptional physico-geographic environment of the Hellenic-Aegean region, the active geodynamic processes, the past geodynamic processes that resulted in a variable puzzle of different rocks and the complex alternation of environments during the Quaternary have favored the growth of biodiversity in general, but also the evolution of man in particular, who has been distinguished due to the extraordinary cultural characteristics developed in this region especially since the Neolithic period. It is no coincidence that both **philosophical currents** and **science** itself were developed in the Aegean region, through the efforts of man to understand and pursue the harmony of nature, and finally, the reasoning of its mechanisms. It is in this charismatic area that **geomythology** has been created by the Ancient Greeks, with core concept the search for reasoning and explanation of the astonishing natural phenomena and especially of natural hazards within an environment under climatic change (Mariolakos 2019). The intervention of the «gods» where and when the logical understanding of the time could not explain the phenomena was a mental escape-way out, just like in the «deus ex machina» introduced also in the ancient tragedies, played at the theatres as the ultimate solution of the drama.

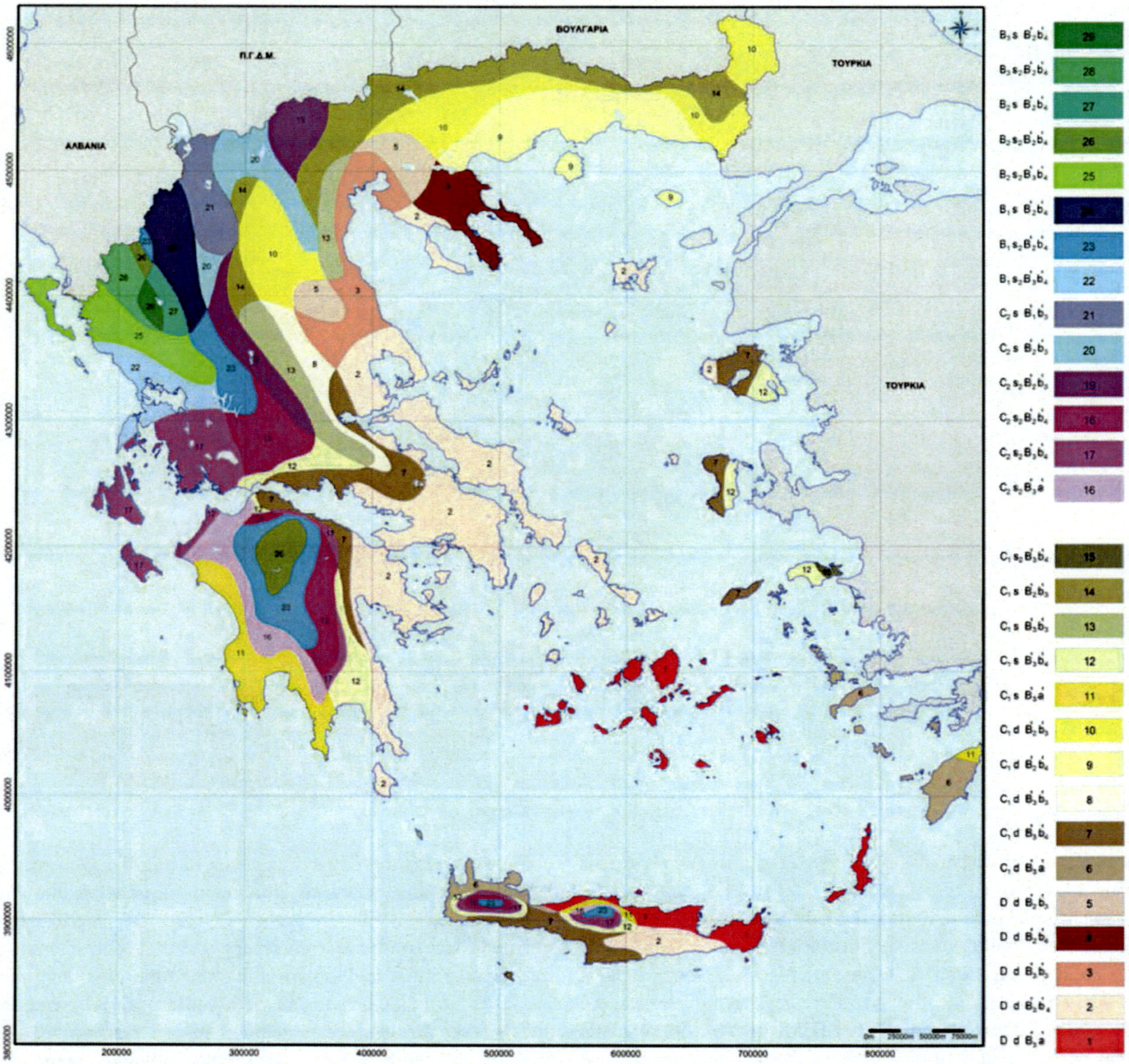

Fig. 11.26 Climatic classification of Greece according to Thornthwaite (from Karras 1973). *Arid climates*: 1. Arid thermal climate, but with intense effect of the sea on the configuration of its thermal character (South Cyclades and northern coast of the central and eastern Crete). 2. Very arid to arid thermal climate with effect of the sea (southern Thessaly, eastern Central Greece, Peloponnese, northern and central Aegean, western Chalkidiki, western Lesvos and southeastern Crete). 3. Very arid climate with small water excess during winter, with evapotranspiration 855–997 mm (Thessaly and western and northern coast of Thermaikos Gulf). 4. Very arid to arid climate, but with very intense effect of the sea on the configuration of its thermal character (Chalkidiki). 5. Arid to very arid climate, with moisture index −40 to −20, with small water excess during winter, with evapotranspiration 712–855 mm (northern and central Macedonia). 6. Arid to very arid climate, but with larger effect of the sea (Northwestern Crete and Dodecanese). 7. Arid climate with intense effect of the sea (central part of Central Greece, northern and eastern central Peloponnese, southern Crete, central Lesvos, western Chios and Ikaria). 8. Arid climate with evapotranspiration 855–997 mm (western Thessaly and central part of Central Greece). 9. Arid to semiarid climate with intense effect of the sea (eastern Macedonia and parts of Thrace). 10. Arid to semiarid climate with small water excess during winter and evapotranspiration 712–855 mm (northwestern Thessaly, western central and eastern Macedonia and Thrace). 11. Semiarid to arid climate with clear effects of the sea (south and western Peloponnese, Patras area, central Crete and northern Rhodes Island). 12. Semiarid to arid climate, with character depending on the sea (central part of the Central Greece, Panachaiko Mt, eastern Peloponnese, central and western Crete, Lesvos, Chios, western Samos). 13. Semiarid climate, with character not depending on the sea and potential evapotranspiration 855–997 mm

(eastern slopes of Pindos Mt and southern-central Macedonia). 14. Semiarid climate, with moisture index from −20 to 0, with moderate water excess during winter, potential evapotranspiration 712–885 mm (northwestern Thessaly, western and northern Macedonia and Thrace). 15. Semiarid to semihumid climate, with moisture index from −20 to 0, with large water excess during winter, potential evapotranspiration 855–997 mm (eastern Samos). *Humid climates*: 16. Semihumid to semiarid climate, with more effect from the sea (western Peloponnese, northern parts of the mountainous western and central Crete). 17. Semihumid climate, with evapotranspiration 855–997 mm (cnetral Greece, Cephalonia, Zakynthos, Northeastern part of the Central Peloponnese, mountainous parts of the central Crete). 18. Semihumid climate, with thermal character affected by the intense impact of the sea (Timphristos, Varsoussia, eastern part of the central Peloponnese, western part of the central Crete). 19. Subhumid to semihumid climate (eastern slopes of Pindos and Timphristos). 20. Subhumid to semihumid climate, with evapotranspiration 712–855 mm (southwestern part of Macedonia). 21. Humid to subhumid climate, with moisture index from 0 to 20, with moderate water shortage during summer and evapotranspiration 570–712 mm (northwestern part of Macedonia). 22. Humid climate, with evapotranspiration 855–997 mm (southwesten Epirus and Lefkada). 23. Humid climate, with larger water shortage during summer (Pindos, central parts of Peloponnese, mountainous areas of the western and central Crete). 24. Humid climate, with moisture index from 20 to 40, with moderate water shortage during summer and with evapotranspiration 712–855 mm. 25. Very humid to humid climate, with evapotranspiration 855–997 mm (Kerkyra, western and central Epirus). 26. Very humid to humid climate, with relatively larger water shortage during summer (internal part of the northern Epirus, central Peloponnese). 27. Very humid to humid climate, with moisture index from 40 to 60 (Central Epirus). 28. Very humid climate, with relatively large water shortage during summer. 29. Very humid climate, with moisture index from 60 to 80, with relatively moderate water shortage during summer and with evapotranspiration 712–855 mm.

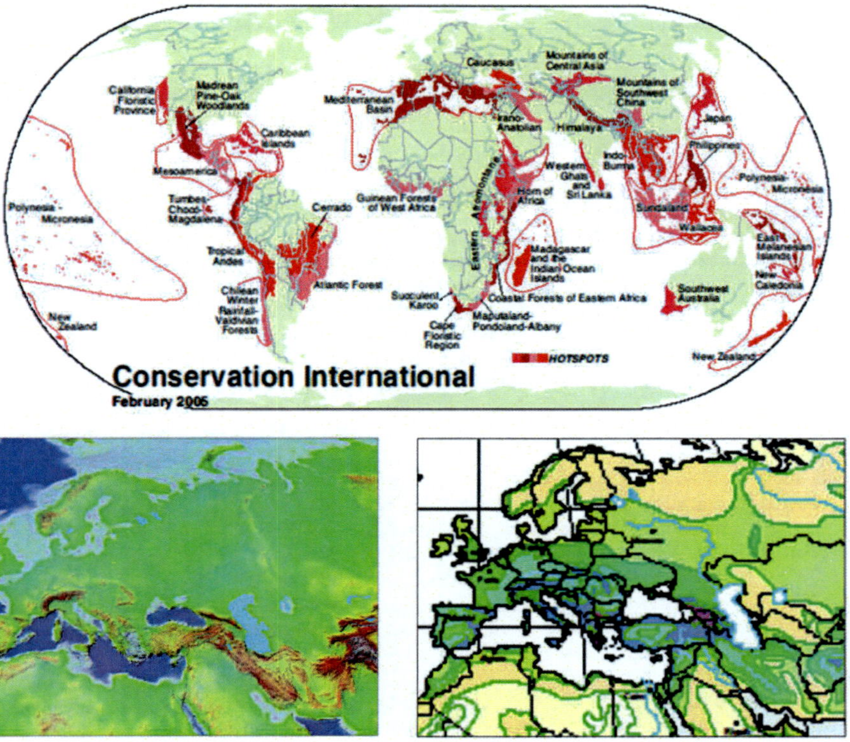

Fig. 11.27 Biodiversity distribution in the Earth (based on Myers et al. 2000). Eurasia shows high values in the circum-Mediterranean floral systems, along the mountain chains of the Tethyan Alpine orogenic system. The two maps focused on Europe and adjacent areas corresponding to the physical geography and the biodiversity respectively show their interrelation

References

Alexandri, M., Papanikolaou, D., Nomikou, P. & Ballas, D. 2001. Geological structure of the Myrtoon basin (Aegean Sea) based on swath bathymetry. Rapport Commission International pour l'exploration scientifique de la Mer Mediterranee (CIESM), 36, 3.

Allen, S. R. 2001. Reconstruction of a major caldera forming eruption from pyroclastic deposit characteristics: Kos Plateau Tuff, eastern Aegean Sea. J. Volcanol. Geothermal Res., 105, 141–162.

Ambraseys, N. 1960. The seismic sea wave of July 9, 1956, in the Greek Archipelago. J. Geophys. Res., 65, 1257–1265.

Angelier, J. 1979. Néotectonique de I' arc Egéen. Soc. Geol. Nord, Publ., 3, 1-417.

Arianoutsou-Faragitaki, M., Giannitsaros, A. & Koubli-Sovantzi, L. 2003. Terrestrial ecosystems of Greece. University of Athens, Dept. of Ecology ,316p (in greek).

Armijo, R., Meyer, B., King, G., Rigo, A. & Papanastasiou, D. 1996. Quaternary evolution of the Corinth Rift and its implications for the late Cenozoic evolution of the Aegean. Geoph. J. Intern., 12, 11–53.

Armijo, R., Meyer, B., Hubert, A. & Barka, A. 1999. Westward propagation of the North Anatolian fault into the northern Aegean: timing and kinematics. Geology, 27, 267–270.

Benedetti, L., Finkel, R., Papanastasiou, D., King, G., Armijo, R., Ryerson, F., Farber, D. & Flerit, F. 2002. Post-glacial slip history of the Sparta Fault (Greece) determined by 36 Cl cosmogenic dating: evidence for non-periodic earthquakes. Geoph. Res. Let., 29.

Beniest, A., Brun, J. P., Gorini, C., Crombez, V., Deschamps, R., Hamon, Y., Smit, J. 2016. Interaction between trench retreat and Anatolian escape as recorded by neogene basins in the northern Aegean Sea. Marine and Petroleum Geology, 77, 30–42.

Billiris, H., Paradissis, D., Veis, G., England, P., Featherstone, W., Parsons, B., Cross, P., Rands, P., Rayson, M., Sellers, P., Ashkenazi, V., Davison, M., Jackson, J. & Ambraseys, N. 1991. Geodetic determination of tectonic deformation in central Greece from 1900 to 1988. Nature, 350, 124–129.

Biryol, B.C., Beck, S.L., Zandt, G. & Ozacar, A.A. 2011. Segmented African lithosphere beneath the Anatolian region inferred from teleseismic P-wave tomography. Geophys. J. Int., 184(3), 1037–1057.

Bohnhoff, M., Makris, J., Papanikolaou, D. & Stavrakakis, G. 2001. Crustal investigation of the Hellenic subduction zone using wide aperture seismic data. Tectonophysics, 343, 239-262.

Boncori, J. P. M., Papoutsis, I., Pezzo, G., Tolanei, C., Atzori, S., Ganas, A., Karastathis, V., Salvi, S., Kontoes, C. & Antonioli, A. 2015. The February 2014 Cephalonia earhquake (Greece): 3D Deformation Field and Source Modelling from multiple SAR Techniques. Seimological Res. Let., 86/1, 124–137.

Briole, P., Rigo, A., Lyon-Caen, H., Ruegg, J. C., Papazissi, K., Mitsakaki, C., Baltodimou, A., Veis, G. & Deschamps, A. 2000. Active deformation of the Corith rift, Greece: Results from repeated Global Positioning System surveyd between 1990–995. J. Geophys. Res., 105, 25,605–25,625.

Broadley, L., Platzman, E., Platt, J., Papanikolaou, M. & Mathews, S. 2006. Paleomagnetism and the tectonic evolution of the Ionian zone, northwestern Greece. In Dilek & Pavlides (eds), Postcollisional tectonics and magmatism in the Mediterranean region and Asia, Geological Society of America, Special Paper 409, 137–155

Brunn, J.H. 1976. L' arc concave zagro-taurique et les arcs convexes taurique et égéen: collision et arcs induits. Bull. Soc. Géol. France, 18, 2, 481–497.

Brun, J.P., Faccenna, C., Gueydan, F., Sokoutis, D., Philippon, M., Kydonakis, K. & Gorini, C. 2016. The two - stage Aegean extension, from localised to distributed, a result of slab rollback acceleration. Canadian Journal of Earth Sciences, 53, 11, 1142–1157.

Brun, J.P. & Sokoutis, D. 2010. 45 m.y. of Aegean crust and mantle flow driven by trench retreat. Geology, 38, 815–818.

Carcaillet, J., Manighetti, I., Chauvvel, C., Schlagenhauf, A. & Nicole, J-M. 2008. Identifying past earthquakes on an active normal fault (Magnola, Italy) from the chemical analysis of its exhumed carbonate fault plane. Earth and Planetary Science Letters, 271, 145–158.

Channel J.E.T., D' Argenio, B. & Horvath, F. 1979. Adria, the AfricanPromontory in Mesozoic Mediterranean Paleogeography. Earth Science Reviews, 15, 213–292.

Chatzipetros, A., Kokkalas, S., Pavlides, S. & Koukouvelas, I. 2005. Palaeoseismic data and their implication for active deformation in Greece. J. Geodynamics, 40, 170–188.

Chousianitis, K., Ganas, A. & Evangelidis, C. P. 2015. Strain and rotation rate patterns of mainland Greece from continuous GPS data and comparison between seismic and geodetic moment release. J. Geophys. Res., 120, 3909–3931.

Collier, R. E. L. 1990. Eustatic and tectonic controls upon Quaternary coastal sedimentation in the Corinth basin, Greece. J. Geol. Soc. London, 147, 301–314.

Collier, R. E. L., Pantosti, D., D' Addezio, G., De Martini, P. M., Masana, E. & Sakellariou, D. 1998. Paleoseismicity of the 1981 Corinth earthquake fault. Seismic contribution to extensional strain in central Greece and implications for seismic hazard. J. Geoph. Res., 103, 30001–30019.

Cowie, P. A., Phillips, R. J. Roberts, G. P., McGaffrey, K., Zijerveld, L. J. J., Gregory, L. C., Faure Walker, J., Wedmore, L., Dunai, T., Binnie, S. A., Freeman, S. P., Wilcken, K., Shanks, R. P., Huismans, R. S., Papanikolaou, I., Michetti, A. M. & Wilkinson, M. 2017. Orogen-scale uplift in the central Italian Apennines drives episodic behaviour of earthquake faults. Scientific Reports, 7, 44858.

Cowling, R., Rundel, P., Lamont, B., Arroyo, M. & Arianoutsou, M. 1996. Plant diversity in mediterranean-climate regions. TREE, 11.

Cowling, R., Potts, A., Bradshaw, P., Colvilles, J., Arianoutsou, M., Ferrier, S., Forest, F., Fyllas, N., Hopper, S., Ojedas, F., Proches, S., Smith, J., RundellO, P., VassilakiS, E. & Zutta, B. 2014. Variation in plant diversity in mediterranean-climate ecosystems: the role of climatic and topographical stability. J. Biogeogr., https://doi.org/10.1111/bi.12429.

De Gelder, G. Fernandez-Blanco, D., Melnick, D., Duclaux, G., Bell, R., Munoz, J. J., Armijo, R. & Lacassin, R. 2019. Lithosphere flexure and rheology determined by climate cycle markers in the Corinth rift. Nature, Scientific Reports, 9: 4260.

Del Ben, A., Mocnik, A., Volpi, V. & Karvelis, P. 2015. Old domains in the south Adria plate and their relationship with the west Hellenic front. J. Geodynamics, 89, 15–28.

Deligiannakis, G., Papanikolaou, I. D. & Roberts, G. 2018. Fault specific GIS based Seismic Hazard Maps for the Attica Region, Greece. Geomophology, 306, 264–282.

Dermitzakis, M. & Papanikolaou, D. 1979. Paleogeography and Geodynamics of the Aegean Region during the Neogene. VII Int. Congress Medit. Neogene, Athens 1979, Ann. Géol. Pays Hellén., hors série IV, 245–289.

Dinter, D.A. & Royden, L.H., 1993. Late Cenozoic extension in NE Greece: Strymon Valley detachment and Rhodope Core Complex. Geology, 21, 45–48.

Drakatos, G., Karastathis, V., Makris, J., Papoulia, J. & Stavrakakis, G. 2005. 3D crustal structure in the neotectonic basin of the Gulf of Saronikos (Greece). Tectonophysics, 400, 55–65.

Druitt, T. H., Mellors, R. A., Pyle, D. M. & Sparks, R. S. J. 1989. Explosive volcanism on Santorini, Greece. Geological Magazine, 2, 95–126.

Duermeijer, C., Krijgsman, W., Langereis, C. & Ten-Veen, J. 1998. Post-early Messinian counterclockwise rotations on Crete:

Implications for the late Miocene to recent kinematics of the southern Hellenic arc. Tectonophysics, 298, 177–189.

Duermeijer, C., Krijgsman,, W., Langereis, C., Meulenkamp, J., Triantaphyllou, M. & Zachariasse, W. 1999. A late Pleistocene clockwise rotation phase of Zakynthos (Greece) and implications for the evolution of the western Aegean arc. Earth & Planetary Science Letters, 173, 315–331.

Duermeijer, C., Nyst, M., Meijer, C., Langereis, C. & Spakman, W. 2000. Neogene evolution of the Aegean arc: Paleomagnetic and geodetic evidence for a rapid and young rotation phase. Earth & Planetary Science Letters, 176, 509–525.

Evangelidis, C. P. 2017. Seismic anisotropy in the Hellenic subduction zone: Effects of slab segmentation and subslab mantle flow. Earth and Planetary Science Letters, 480, 97–106.

Faccenna, C., Jolivet, L., Piromallo, C. & Morelli, A., 2003. Subduction and the depth of convection in the Mediterranean mantle. J. Geoph. Res., 108.

Faccenna, C. et al 2014. Mantle dynamics in the Mediterranean. Rev. Geophys., 52, 283–332.

Fassoulas, C., Kilias, A. & Mountrakis, D., 1994. Postnappe stacking extension and exhumation of HP/LT rocks in the island of Crete, Greece. Tectonics, 13, 127–138.

Fernandez-Palacios, J. M. 2016. Island biodiversity shaped by sea-level shifts. Nature, 17880.

Finetti, I. & Del Ben, A. 2005. Crustal tectono-stratigraphic setting of the Adriatic Sea from new CROP seismic data. In CROP Project: Deep seismic exploration of the central Mediterranean and Italy. Atlases 1, 519–548, Elsevier, Amsterdam.

Finetti, I., Papanikolaou, D., Del Ben, A. & Karvelis, P. 1991. Preliminary geotectonic interpretation of the East Mediterranean Chain and the Hellenic Arc. Bull. Geol. Soc. Greece, 25/1, 509–526.

Flemming, N. C. 1978. Holocene eustatic changes & coastal tectonics in the northeast Mediterranean. Implications for models of crustal consumption. Phil. Trans. R. S. London, a, 289, 405–457.

Flerit, F., Armijo, R., King, G., & Meyer, B. 2004. The mechanical interaction between the propagating North Anatolian Fault and the back-arc extension in the Aegean. Earth and Planetary Science Letters, 224(3), 347–362.

Foumelis, M., Fountoulis, I., Papanikolaou, I & Papanikolaou, D. 2013. Geodetic evidence for passive control of a major Miocene tectonic boundary on the deformation field of Athens (Greece). Annals of Geophysics, 56, S0674, 1–9.

Freyberg, B. 1973. Geologie des Isthmus von Korinth. Erlanger Geol. Abh., 95, 1–183.

Friedrich, W. L. O., Kromer, B., Friedrich, M., Heinemeier, J., Pfeiffer, T. & Talamo, S. 2006. Santorini eruption radiocarbon dated to 1627-1600 B.C. Science, 312, 548.

Fytikas, M. Innocenti, F., Kolios, N., Manetti, P., Mazzuoli, R., Poli, G., Rita, F. & Villari, L. 1986. Volcanology and petrology of volcanic products from the island of Milos and neighboring islets. J. Volcan. Geothermal Res., 28, 297–317.

Galanopoulos, A. 1957. The seismic sea wave of 9th of July 1956. Proc. Acad. Athens, 32, 90–101 (in greek).

Galanopoulos, A. 1972. Plate tectonics in the area of Greece as reflected in the deep focus seismisity. Bull. Geol. Soc. Greece, IX, 2, 266–285.

Ganas, A., Pavlides, S., Sboras, S., Valkaniotis, S., Papaioannou, S., Alexandris, G. A., Plessa, A. & Papadopoulos, G. 2004. Active fault geometry and kinematics in Parnitha Mountain, Attica, Greece. J. Struct. Geol., 26, 2103–2118.

Georghiou, K. & Delipetrou, P. 2010. Patterns and traits of the endemic plants in Greece. Botan. J. Linnean Soc., 162, 130–422.

Govers, R. & Wortel, M.J.R. 2005. Lithosphere tearing at STEP faults: Response to edges of subduction zones. Earth & Planet. Sci. Lett., 236, 505–523.

Granot, R. 2016. Paleozoic oceanic crust preserved beneath the eastern Mediterranean. Nature Geoscience, 9, 701–705. https://doi.org/10.1038/NGEO2784.

Grasemann, B., Schneider, D., Stockli, D. & Iglseder, C. 2012. Miocene bivergent crustal extension in the Aegean: evidence from the western Cyclades (Greece). Lithosphere, 4, 1, 23–39.

Guillaume, B., Husson, L., Funiciello, F., Faccenna, C. 2013. The dynamics of laterally variable subductions: Laboratory models applied to the Hellenides. Solid Earth, 4 (2), pp. 17–200.

Harvati, K., Roding, C., Bosman, A., Karakostis, F., Grun R., Stringer, C., Karkanas, P., Thompson, N., Koutoulidis, V., Moulopoulos, L., Gorgoulis, G. & Kouloukoussa, M. 2019. Apidima cave fossils provide earliest evidence of Homo sapiens in Eurasia. Nature, 571, 500–504. https://doi.org/10.1038/s41586-019-1376-z

Hollenstein, CH., Muller, M.D., Geiger, A. & Kahle, H.G. 2008. Crustal motion and deformation in Greece from a decade of GPS measurements, 1993–2003. Tectonophysics, 449, 17–40.

Houghton, S.L., Roberts, G.P., Papanikolaou, I.D., Mcarthur, J.M. & Glimour, M. A. 2003. New 234U-230Th coral dates from thewestern Gulf of Corinth: Implication for extensional tectonics. Geophysical Research Letters, 29, https://doi.org/10.1029/2003GLO18112.

Hubert, A., King, G., Armijo, R., Meyer, B. & Papanastasiou, D. 1996. Fault-reactivation and rupture propagation of the 1981 Corinth earthquake sequence. Earth Plan. Sc. Let., 142, 573–585.

Hurni, L., Dietrich, V. J. & Gaitanakis, P. 1995. Geological maop of Greece, Methana (Saronic Gulf). 1/25,000 with explanatory text. Birkhauser Basel and IGME, Athens.

Jackson. J. 1994. Active tectonics of the Aegean region. Ann. Rev. Earth & Plan. Sci., 22, 239-271.

Jackson, J., Gagnepain, J., Houseman, G., King, G. C. P., Papadimitriou, P., Soufleris, C. & Virieux, J. 1982. Seismicity, normal faulting and the geomorphological development of the Gulf of Corinth (Greece): The Corinth earthquakes of February – March 1981. Earth Plan. Sc. Let., 57, 377–397.

Jolivet, L. & Faccenna, C. 2000. Mediterranean extension and the Africa - Eurasia collision. Tectonics, 19, 6, 1095–1106.

JoliveT, L., Faccenna, C., Huet, B., Labrousse, L., Le Pourhiet, L., Lacombe, O., Lecomte, E., Burov, E., Denele, Y., Brun, J.P., Philippon, M., Paul, A., Salaun, G., Karabulut, H., Piromallo, C., Monie, P., Gueydan, F., Okay, A.I., Oberhansli, R., Pourteau, A., Augier, R., Gadenne, L. & Driussi, O., 2013. Aegean tectonics: Strain localisation, slab tearing and trench retreat. Tectonophysics, 597–598, 1–33.

Jolivet, L., Faccenna, C. & Piromallo, C. 2009. From mantle to crust: Streching the Mediterranean. Earth & Planetary Science Letters, 285, (1-2) 198–209.

Jolivet, L., Goffe, B., Monie, P., Truffert, C., Patriat, M. & Bonneau, M.,1996. Miocene detachment in Crete and exhumation P–T–t paths of high pressure metamorphic rocks. Tectonics 15, 1129–1153.

Jolivet, L., Lecomte, E., Huet, B., Denele, Y., Lacompe, O., Labrousse, L., Le Pourhiet, L. & Mehl, C. 2010. The North Cycladic Detachment System. Earth & Plan. Sci. Let., 289, 87–104.

Jolivet, L., Menant, A., Sternai, P., Rabillard, A., Arbaret, L., Augier, R., Laurent, V., Beaudouin, A., Grasement, A., Huet, B., Labrousse, L. & Le Pourhiet, L. 2015. The geological signature of a slab tear below the Aegean. Tectonophysics, 659, 166–182.

Karras, G. 1973. Climatic classification of Greece according to Thornthwaite. PhD Thesis, University of Athens, 200 p (in greek).

Kilias, A. 1996. Tectonic evolution of the Olympus – Ossa Mountains: emplacement of the blueschists Unit in Eastern Thessaly and exhumation of Olympus-Ossa carbonate dome as a result of Tertiary extension (central Greece). Mineral Wealth, 96, 7–22.

Kilias, A.A., Tranos, M.D., Orozco, M., Alonso-Chaves, F.M. & Soto, J.I. 2002. Extensional collapse of the Hellenides: a review. Rev. Soc. Geol. España, 15, 129–139.

Kilias, A., Falalakis, G., Sfeikos, A., Papadimitriou, E., Vambaka, A. & Gkarlaouni, C. 2013. The Thrace basin in the Rhodope province of NE Greece – A tertiary supradetachment basin and its geodynamic implications. Tectonophysics, 595–596, 90–105.

Kiratzi, A. & Louvari, E. 2003. Focal mechanisms of shallow earthquakes in the Aegean Sea and the surrounding lands determined by waveform modelling: a new database. J. Geod., 36, 251–274.

Kissel, C. & Laj, C. 1988. The Tertiary geodynamic evolution of the Aegean arc: a paleomagnetic reconstruction. Tectonophysics, 146, 183–201.

Kissel, C., Laj, C., Poisson, A. & Gorur, N. 2003. Paleomagnetic reconstruction of the cenozoic evolution of the eastern Mediterranean. Tectonophysics, 362 (1–4), 199–217.

Koukouvelas, I., Kokkalas, A. & Zygouri, B. 2010. Geology and Earthquakes. Disigma Editions, Thessaloniki, 415 p (in greek).

Ktenas, C. 1935. Le groupe d' iles du Santorin. Contribution a l' etude des laves tertiaires et quaternaires de la mer Egee. Memoirs de l' Academie d' Athenes, I/4, 53p.

Laj, C., Jamet, M., Sorel, D. & Valente, J. 1982. First paleomagnetic results from Mio-Pliocene series of the Hellenic sedimentary arc. Tectonophysics, 86, 45-67.

Le Pichon, X. & Angelier, J. 1979.The Hellenic Arc and Trench system: a key to the neotectonic evolution of the Eastern Mediterranean area. Tectonophysics, 60, 1–42.

Le Pichon, X. & Angelier, J. 1981. The Aegean Sea. Phil. Trans. R. Soc. London, A300, 357–372.

Le Pichon, X., Angelier, J., Aubouin, J., Lyberis, N., MontI, S., Renard, V., Got, H., Hsü, K., Mart, Y., Mascle, J., Mathews, D., Mitropoulos, D., Tsoflias, P. & Chronis, G. 1979. From subduction to transform motion: A seabeam survey of the hellenic trench system. Earth & Plan. Sc. Let., 44, 441–450.

Le Pichon, X., Sengor, A. M. C. & Imren, C. 2019. A new approach to the opening of the Eastern Mediterranean Sea and the origin of the Hellenic subduction zone: Part 2: The Hellenic subduction zone. Canadian J. Earth Sciences, in print.

Le Pichon, X., Şengor, A.M.C., Kende, J., İmren, C., Henry, P., Grail, C., Karabulut, H. 2016. Propagation of a strike-slip plate boundary within an extensional environment: The westward propagation of the North Anatolian fault. Canadian Journal of Earth Sciences, 53 (11), 1416–1439.

Lister, G.S., Banga, G. & Feenstra, A., 1984. Metamorphic core complexes of Cordilleran type in the Cyclades, Aegean Sea, Greece. Geology 12, 221–225.

McCalpin, J. P. 2009. Paleoseismology. Academic Press, 2nd international Geophysics series, vol. 95, Elsevier Publishing, 647p.

Makris, J. 1973. Some geophysical aspects of the evolution of the Hellenides. Bull. Geol. Soc. Greece, X, 1, 206–13.

Makris, J. 1977a. Geophysical Investigations of the Hellenides. Hamburger Geophys. Einzelsch, 34.

Makris, J. 1977b. Seismic and Gravity studies in Greece and somegeodynamic implications. VI Coll. Geol. Aegean Region, Athens1977, Proc. I, 61–85.

Makris, J. 2010. Geophysical studies and tectonism of the Hellenides. Bull. Geol. Soc. Greece, 43, 32–45.

Makris, J., Papoulia, J. & Yegorova, T. 2013. A 3-D density model of Greece constrained by gravity and seismic data. Geophys. J. Int., 194, 1–17.

Marinos, G. 1961. The Antimilos volcano in Aegean Sea. Bull. Geol. Soc. Greece, 4/1, 38–50 (in greek).

Mariolakos, I., 1976. Thoughts and opinions about some problems of the geology and tectonics of Peloponnesus. Ann. Géol. Pays Héllen., 27, 215–313 (in greek).

Mariolakos, I. 2019. Geomythology. Livanis Editions, Athens (in greek).

Mariolakos, I. & Papanikolaou, D. 1981. The Neogene Basins of the Aegean Arc from the Paleogeographic and the Geodynamic point of view. Intern.Symp.(H.EA.T.) Hell. Arc and Trench, Athens 1981, Proceedings I, 383–399.

Mariolakos, I. & Papanikolaou, D., 1984. Deformation pattern and relation between deformation and seismicity in the Hellenic arc. Bull. Geol. Soc. Greece, 19, 59–76 (in greek).

Mariolakos, I., Papanikolaou, D. & Lagios, E., 1985.. A neotectonic geodynamic model of Peloponnesus based on morphotectonics, repeated gravity measurements and seismicity. Geol. Jb., B50, 3–17.

Mariolakos, I. Papanikolaou, D., Symeonidis, N., Karotsieris, Z. & Lekkas, S. 1981. The deformation of the area around the eastern Korinthian Gulf, affected by the earthquakes of February-March 1981. International Symposium H.E.A.T. (Hellenic Arc & Trench System), Athens 1981, Proceedings I, 400–420.

Marton, E., Papanikolaou, D. & Lekkas, E. 1990. Paleomagnetic results from the Pindos, Paxos and Ionian zones of Greece. Physics of the Earth & Planetary Interiors, 62, 60–69.

Mascle, J., Benkhelil, J., Bellaiche, G., Zitter, T., Woodside, J., Loncke, L. & Prismed II Scientific Party. 2000. Marine geologic evidence for a Levantine-Sinai plate, e new piece of the Mediterranean puzzle. Geology, 28, 779–782

Mathes-Schmidt, M., Papanikolaou, I., Reicherter, K. & Palikarakis, A. 2019. Event deposits in the Eastern Thermaikos Gulf and Kassandra (Northern Greece): evidence of the 479 BC «Herodotus tsunami». Zeitschrift fur Geomorphologie, 62, 101–125.

Mechernich, S., Schneiderwind, S., Mason, J., Papanikolaou, I. D., Deligiannakis, G., Palikarakis, A., Binnie, S. A., Dunal, T. J. & Reicherter, K. 2018. The seismic history of the Pisia Fault (Eastern Corinth rift, Greece) from fault plane weathering features and cosmogenic 36 Cl dating. J. Geophys. Res., Solid State, 123, 4266–4284.

McKenzie, D.P. 1970. Plate tectonics in the Mediterranean Region. Nature, 226, 239–243.

McKenzie, D.P. 1972. Active tectonics of the Mediterranean Region. Geoph. J.R. Astron.Soc., 30, 109–185.

McKenzie, D., 1978. Active tectonics of the Alpine - Himalayan belt: the Aegean sea and surrounding regions. Geophys.J.R. Astr. Soc., 55, (1), 217–254.

McKenzie, D., Davies, D. & Molnar, P. 1970. Plate tectonics of the Red Sea and East Africa. Nature, 226, 243–248.

McNeill, L. C., Shillington, D. J. et al 2019. High resolution record reveals climate-driven environmental and sedimentary changes in an active rift. Scientific Reports, 9:3116.

Mercier, J.L 1979. Signification néotectonique de l'Arc Egéen. Une revue des idées. Rev. Géol. Dyn. Géogr. Phys., 21, 1, 5–15.

Mercier, J.L, Delibassis, N., Gauthier, A., Jarrige, J.J., Lemeille, F., Philip, H., Sebrier, U. & Sorel, D., 1979. La néotectonique de l'Arc Egéen. Rev. Géol. Dyn. Géogr. Phys., 21, 1, 67–92.

Mitzopoulos, K. 1895. The Mega-Earthquake of Lokris in April 1894. Ethnikon Typografeion (National Imprimery), Athes, 40p (in greek).

Molnar, P. & Tapponier, P. 1977. Relation of the tectonics of easternChina to the India-Eurasia collision: Application of slip-line fieldtheory to large-scale continental tectonics. Geology, 5(4), 212–216.

Monopolis, D. & Bruneton, A. 1982. Ionian Sea (Western Greece): its structural outline deduced from drilling and geophysical data. Tectonophysics, 83, 227–242.

Morewood, N. C. & Roberts, G. P. 1999. Lateral propagation of the surface trace of the South Alkyonides fault, central Greece: Its impact on models of fault growth and displacement-length relationships. J. Struct. Geol. 21, 635–652.

Myers, N., Mittermeier, R. A., Mittermeier, C. G., Da Fonceka, G. A. B. & Kent, J. 2000. Biodiversity hotspots for conservation priorities. Nature, 403, 853–858.

Netzeband, G., Goel, K., Hubscher, C., Ben-Avraham, Z., Delghani, G., Gajewski, D. & Liersch, P. 2006. The Levantine basin – crustal structure and origin. Tectonophysics, 418, 167–188.

Nomikou, P. 2004. Contribution to the geodynamics of Dodekanese: the submarine area of Kos – Nisyros islands. PhD Thesis, University of Athens, 470 p (in greek).

Momikou, P. & Papanikolaou, D. 2010. The morphotectonic structure of the Kos-Nisyros-Tilos volcanic area based on onshore and offshore data. XIX Congress Carpatho-Balkan Geological Association, Thessaloniki, Proceedings, 99, 557–564.

Nomikou, P., Carey, S., Papanikolaou, D., Croff Bell, K., Sakellariou, D., Alexandri, M. & Bejelou, K. 2012. Submarine volcanoes of the Kolumbo volcanic zone, NE of Santorini Caldera, Greece. Global & Planetary Change, 90–91, 135–151.

Nomikou, P., Hubscher, C., Papanikolaou, D., Farangitakis, P., Runhau, M. & Lambridou, D. 2018. Expanding extension, subsidence and lateral segmentation within the Santorini-Amorgos basins during Quaternary: Implications for the 1956 Amorgos events, central-south Aegean Sea, Greece. Tectonophysics, 722, 138–153.

Nomikou, P., Papanikolaou, D., Alexandri, M., Sakellariou, D. & Rousakis, G. 2013. Submarine volcanoes along the Aegean Volcanic Arc. Tectonophysics, 597–598, 123–146.

Nomikou, P., Papanikolaou, D. & Dietrich, V. J. 2018. Geodynamics and Volcanism in the Kos-Yali-Nisyros Volcanic Field. In: Nisyros Volcano. The Kos-Yali-Nisyros Volcanic Field. Dietrich & Lagios (editors), 13–56, Springer.

Nomikou, P., Parks M., Papanikolaou D., Pyle D., Mather T., Carey S., Watts A., Paulatto M., Kalnins M., Livanos I., Bejelou K., Simou E., Perros I. 2014. The emergence and growth of a submarine volcano: The Kameni islands, Santorini (Greece). Geo. Res. J., Vol.1, 8–18.

Ocakoğlu N., Nomikou P., İşcan Y., Loreto M.F., Lampridou, D. 2018. Evidence of extensional and strike-slip deformation in the offshore Gökova-Kos area affected by the July 2017 Mw 6.6 Bodrum-Kos earthquake, eastern Aegean Sea. Geo-Marine Letters, https://doi.org/10.1007/s00367-017-0532-4, GMLE-D-17-00084.1.

Okal, E., Synolakis, C., Uslu, B., Kalligeris, N. & Voukouvalas, E. 2009. The 1956 earthquake and tsunami in Amorgos, Greece. Geophys. J. Int., 178, 1533–1554.

Pantosti, D., De Martini, P. M., Papanastassiou, D., Palyvos, N., Lemeille, F. & Stavrakakis, G. 2001. A reappraisal of the 1894 Atalanti earthquake surface ruptures, central Greece. Bull. Seismol. Soc. Am., 91, 760–780.

Pantosti, D., De Martini, P. M., Papanastassiou, D., Lemeille, F., Palyvos, N. & Stavrakakis, G. 2004. Paleoseismological trenching across the Atalanti Fault (central Greece): Evidence for the ancestors of the 1894 Earthquake during the Middle Ages and Roman Times. Bull. Seismol. Soc. Am., 94, 531–549.

Papadimitriou, P., Voulgaris, N., Kassaras, I., Kaviris, G., Delibassis, N. & Makropoulos, K. 2002. The Mw = 6.0, 7 September 1999 Athens earthquake. Natural Hazards, 27, 15–33.

Papanikolaou, D. 1980a. Contribution to the geology of the Aegean Sea. The island of Paros. Ann. Géol. Pays Hellén. 30/1, 65–96.

Papanikolaou, D. 1980b. Les écailles de Thymaena: témoins d' un mouvement tectonique miocène vers l'intérieur de l'arc égéen. C.R. Acad. Sc. Paris, 290, 307–310.

Papanikolaou, D. 1980c. The Metamorphic Hellenides. 26th Int. Geol. Congress., Paris, 1980, Abstracts, I, 371.

Papanikolaou, D. 1986b. The medial tectonometamorphic belt of the Hellenides. Bull. Geol. Soc. Greece, 20, 101–120 (in greek).

Papanikolaou, D. 1986c. Geology of Greece. Eptalofos Publications, 240 p. Athens (in greek).

Papanikolaou, D., 1987. Tectonic evolution of the Cycladic blueschist belt (Aegean Sea, Greece). In: Helgeson, H.C. (Ed.), Chemical Transport in Metasomatic Processes: NATO ASI Series, 429–450.

Papanikolaou, D. 1988. Introduction to the Geology of Crete and Field guide. IGCP No 276, October 1988 FieldMeeting in Crete, 3–36.

Papanikolaou, D. 1989. Occurrence of Arvi, Western Thessaly and Orliakas type formations in Argolis. Bull. Geol. Soc. Greece, 24, 71–84 (in greek).

Papanikolaou, D. 2010. Major paleogeographic, tectonic and geodynamic changes from the last stage of the Hellenides to the actual Hellenic arc and trench system. Bull. Geol. Soc. Greece, 43, 71–85.

Papanikolaou, D. 2015. Geology of Greece. Patakis Publ. Co., 443 p., Athens (in greek).

Papanikolaou, D., Lykoussis, V., Chronis, G. & Pavlakis, P. 1988a. A comparative study of neotectonic basins across the Hellenic arc: The Messiniakos, Argolikos, Saronikos and Southern Evoikos Gulfs. Basin Research, 1/3, 167–176.

Papanikolaou, D., Chronis, G., Pavlakis, P., Lykoussis, B., Roussakis, G. & Syskakis, D. 1988a. Submarine Neotectonic Map of Ano Messiniakos Gulf. Scale 1/100,000, Earthquake Planning & Protection Organization, National Centre Marine Research, University of Athens.

Papanikolaou, D., Alexandri, M. & Nomikou, P. 2006. Active Faulting in the North Aegean Basin. Geol. Soc. Amer. Sp. Paper, 409, 189–209.

Papanikolaou, D., Alexandri, M., Nomikou, P. & Ballas, D. 2002. Morphotectonic structure of the Western Part of the North Aegean Basin based on swath bathymetry. Marine Geology, 190, 465–492.

Papanikolaou, D., Chronis, G. & Metaxas, C. 1994. Neotectonic structure of the Argolikos Gulf. Bull. Geol. Soc. Greece, 30/2, 305–316.

Papanikolaou, D., Chronis, G., Pavlakis, P., Lykoussis, B., Roussakis, G. & Syskakis, D. 1989a. Submarine Neotectonic Map of Saronic Gulf. Scale 1/100,000, Earthquake Planning & Protection Organization, National Centre Marine Research, University of Athens.

Papanikolaou, D., Chronis, G., Pavlakis, P., Lykoussis, B., Roussakis, G. & Syskakis, D. 1989b. Submarine Neotectonic Map of Southern Evoikos Gulf. Scale 1/100,000, Earthquake Planning & Protection Organization, National Centre Marine Research, University of Athens

Papanikolaou, D. & Dermitzakis, M. 1981. Major changes from the last stage of the Hellenides to the actual Hellenic Arc and Trench system. Intern.Symp. on the Hellenic Arc and Trench (H.E.A.T.), Athens 1981, Proceedings, II, 57–73.

Papanikolaou, D., Fountoulis, I. & Metaxas, C. 2007. Active faults, deformation rates and Quaternary palaeogeography at Kyparissiakos Gulf (SW Greece) deduced from on-shore and off-shore data. Quaternary International, 142, 14–30.

Papanikolaou, D., Gouliotis, L. & Triantaphyllou, M. 2009. The Itea–Amfissa detachment: a pre-Corinth rift Miocene extensional structure in central Greece. Geol. Soc.,London, Sp. Publ., 311, 293–310.

Papanikolaou, D., Lekkas, E., Sideris, C., Fountoulis, I., Danamos, G., Kranis, C., Lozios, S., Antoniou, I., Vassilakis, E., Vasilopoulou, S., Nomikou, P., Papanikolaou, I., Skourtsos, E. & Soukis, K. 1999.

Geology and Tectonics of Western Attica in relation to the 7-9-1999 earthquake. European Centre for Prediction & Forecasting Earthquakes, Newsletter, 3, 30–34.

Papanikolaou, D., Logos, E., Lozios, S. & Sideris, C. 1996. Korinthos sheet. Neotectonic map of Greece, at scale 1/100,000. Earthquake Planning & Protection Organization, Geological Society of Greece/Tectonic Committee, University of Athens/Dynamic, Tectonic Applied Geology.

Papanikolaou, D., Metaxas, C. & Chronis, G. 2001. Neotectonic structure of Lakonikos Gulf. Bull. Geol. Soc. Greece, 34/1, 289–296.

Papanikolaou, D. & Nomikou, P. 2001. Tectonic structure and volcanic centres at the eastern edge of the Aegean volcanic arc around Nisyros Island. Bull. Geol. Soc. Greece, 34/1, 289–296.

Papanikolaou, D., Nomikou, P. & Lykousis, V. 1998. Monitoring of seismic activity along the Aegean volcanic arc: with focus on the eastern part of the arc in Kos and mainly Nisyros islands: 6. Submarine reconnaisance. European Centre for Prevention and Forecasting of Earthquakes, Newsletter, 2, 23–26.

Papanikolaou, D., Nomikou, E. & Papanikolaou, I. 2019a. The western transformation of the North Anatolian Fault in the North Aegean Basins. 23 ATAG, 20th Anniversary of 1999 Earthquakes, Istanbul.

Papanikolaou, D., Nomikou, P., Papanikolaou, I., Lampridou, D., Rousakis, G. & Alexandri, S. 2019b. Active tectonics and sesmic hazard in Skyros Basin, North Aegean Sea, Greece. Marine Geology, 407, 94–110.

Papanikolaou, D., Nomikou, P., Papanikolaou, I., Triantaphyllou, M., Galanidou, N., Dimiza, M. & Tzedakis, C. 2019c. Pleistocene palaeogeography and Palaeolithic archaeology. The tectonically active North Aegean Sea case study. 15th Congress Geol. Soc. Greece, Special Publications, 7, 82–83.

Papanikolaou, D. & Papanikolaou, I. 2007a. Geological, geomorphological and tectonic structure of NE Attica and seismic hazard implications for the northern edge of the Athens Plain. Bull. Geol. Soc. Greece, 40, 425–438.

Papanikolaou, D. & Royden, L. 2007. Disruption of the Hellenic arc: Late Miocene extensional detachment faults and steep Pliocene–Quaternary normal faults — or what happened at Corinth? Tectonics, 26.

Papanikolaou, D. & Triantaphyllou, M. 2013. Growth folding and uplift of Lower and Middle Pleistocene marine terraces in Kephalonia: implications to active tectonics. 4th Intern. INQUA meeting on Paleoseismology, Active Tectonics and Archaeoseismology, Aachen, Proc., 185–188.

Papanikolaou, D. & Vassilakis, E. 2010. Thrust faults and extensional detachment faults in Cretan tectono-stratigraphy: implications for Middle Miocene extension. Tectonophysics, 488, 233–247.

Papanikolaou, I. & Papanikolaou, D. 2007b. Seismic hazard senarios from the longest geologically constrained active fault of the Aegean.978 Quaternary Intern., 171–172, 31-44.

Papanikolaou, I. D., Van Balen, R., Silva, P. G. & Reicherter, K. 2015a. Geomorphology of Active Faulting and seismic hazardassessment: New tools and future challenges. Geomorphology, 237,1–13.

Papanikolaou, I. D., Triantaphyllou, M., Palikarakis, A. & Migiros, G. 2015b. Active faulting at the Corinth Canal based on surfaceobservations, borehole data and paleoenvironmental interpretations. Passive rupture during the 1981 earthquake sequence? Geomo-prphology, 237, 65–78.

Papanikolaou, I. D., Roberts, G. P., Deligiannakis, G., Sakellariou, A. & Vassilakis, E. 2013. The Sparta Fault, Southern Greece: From segmentation and tectonic geomorphology to seismic hazard and time dependent probabilities. Tectonophysics, 597–598, 85–105.

Papanikolaou, M., Papanikolaou, D. & Triantaphyllou, M. 2010. Post-Alpine Late Pliocene – Middle Pleistocene uplifted marine sequences in Zakynthos Island. Bull. Geol. Soc. Greece, 43/1, 475–485.

Papastamatiou, J. 1937. The volcanic rocks of eastern Crommyonia. Proceedings Acad. Athens, 12, 22–30, 104–111 (in greek).

Papavasiliou, S. 1894. Sur la nature de la grande crevasse produite a la suite du derniere tremblement de terre de Locride. C. R. Acad. Sci., Paris, 119, 112–114.

Papazachos, B., Comninakis, P.E., Karakaisis, G.F., Karakostas, B.G., Papaioannou, C.A., Papazachos, C.B. & Scordilis, E.M. 2000. A cataloque of earthquakes in Greece and surrounding area for the period 550 BC – 1999. Publ. Lab. Univ. Thessaloniki.

Papazachos, B.C., Karakostas, B., Papazachos, C.B. & Skordilis, E. 2000. The geometry of the Wadatti-Benioff zone and lithosperic kinematics in the Hellenic arc. Tectonophysics, 319, 275–300.

Pavlakis, P., Lykoussis, B., Papanikolaou, D. & Chronis, G. 1989. Discovery of a submarine volcanic centre in the West Saronic Gulf: The Paphsanias Volcano. Bull. Geol. Soc. Greece, 24, 59–70 (in Greek).

Pavlides, S. 2003. Earthquake Geology. Introduction to Neotectonics, Morphotectonics and Palaeoseismology. University Studio Press, Therssaloniki, 378 p (in Greek).

Pavlides, S. & Caputo, R. 2004. Magnitude versus faults' surface parameters: quantitative relationships from the Aegean region. Tectonophysics, 380, 159–188.

Pearce, D., Rondenay, S., Sachpazi, M., Charalampakis, M., Royden, L.H. 2012. Seismic investigation of the transition from continental to oceanic subduction along the western Hellenic subduction Zone. Journal of Geophysical Research: Solid Earth, 117 (7), art. no. B07306.

Pe-Piper, G. & Piper, D. 2002. The Igneous Rocks of Greece.The Anatomy of an Orogen. Gebrueder Borntraeger, Berlin/Stuttgart.

Perissoratis, C. & Van Andel, T. 1991. Sea-level changes and tectonics in the Quaternary extensional basin of the South Evoikos Gulf, Greece. Terra Nova, 3, 294–302.

Perissoratis, C., Piper, D. J. W. & Lykousis, V. 2000. Alternating marine and lacustrine sedimentation during the Late Quaternary in the Gulf of Corinth rift basin, central Greece. Marine Geol., 167, 391–411.

Philippson, A. 1901. Beiträge zur Kenntnis der griechischen Inselwelt. Peterm. Milt. Erganzunheft,134, 1–172, Gotha.

Philippon, M., Brun, J.P., Gueydan, F. & Sokoutis, D. 2014. The interaction between back-arc extension in the Aegean and Anatolia escape since Middle Miocene.Tectonophysics, 631, 176–188.

Pirazzoli, P. A., Montaggioni, L. F., Saliege, J. F., Segonzac, G., Thommeret, Y. & Vergnaud-Grazzini, C. 1989. Crustal block movements from Holocene shorelines: Rhodes Island (Greece). Tectonophysics, 170, 89–114.

Piromallo, C. & Morelli, A. 2003. P wave tomography of the mantle under the Alpine-Mediterranean area. Journal of Geophysical Research: Solid Earth, 108 (B2).

Reicherter, K., Papanikolaou, I., Roger, J., Mathes-Schmidt, M. Papanikolaou, D., Roessler, S. Gruetzner, C. & Stamatis, G. 2010. Holocene tsunamigenic sediments and tsunami modelling in the Thermaikos Gulf area (northern Greece). Zeitschrift fur Geomorphologie, 54, 99–126.

Roberts, G. P. & Michetti, A. M. 2004. Spatial and temporal variations in growth rates along active normal fault systems: an example from the Lazio-Abbruzo Apennines, central Italy. J. Struct. Geol., 26, 339–376.

Roberts, G.P., Cowie, P., Papanikolaou, I. & Michetti, A. 2004. Fault scaling relationships, deformation rates and seismic hazards: an example from the Lazio-Abruzzo Apennines, central Italy. J. Struct. Geology, 216, 377–398.

Roberts, G. P., S. L. Houghton, C. Underwood, I. Papanikolaou, P. A. Cowie, P. Van Calsteren, T. Wigley, F. J. Cooper, and

J. M. McArthur 2009. Localization of Quaternary slip rates in an active rift in 105 years: An example from central Greece constrained by 234U-230Th coral dates from uplifted paleoshorelines. J. Geophys. Res., 114, B10406, https://doi.org/10.1029/2008JB005818.
Roche, V., Jolivet, L., Papanikolaou, D., Bozkurt, E., Menant, A. & Rimmele, G. 2019. Slab fragmentation beneath the Aegean/Anatolia transition zone: insights from the tectonic and metamorphic evolution of the Eastern Aegean region. Tectonophysics, 754, 101–129.
Royden, L.H., 1993. Evolution of retreating subduction boundaries formed during continental collision. Tectonics, 12, 629–638.
Royden, L.H. & Papanikolaou, D.J., 2011. Slab segmentation and late Cenozoic disruption of the Hellenic arc. Geochemistry, Geophysics, Geosystems, 12, Q03010.
Sachpazi, M., Laigle, M., Charalampakis, M., Diaz, J., Kissling, E., Gesret, A., Becel, A., Flueh, E., Miles, P., Hirn, A. 2016. Segmented Hellenic slab rollback driving Aegean deformation and seismicity. Geophysical Research Letters, 43(2), 651–658.
Sakellariou, D. & Galanidou, N. 2016. Pleistocene submerged landscapes and Palaeolithic archaeology in the tectonically active Aegean Region. Sp. Publ. Geol. Soc., London, 411, 145–178.
Salaün, G., Pedersen, H. A., Paul, A., Farra, V., Karabulut, H., Hatzfeld, D., Papazachos, C.,Childs, D.M., Pequegnat, C., & Simbaad Team. 2012. High-resolution surface wave tomography beneath the Aegean-Anatolia region: constraints on upper-mantle structure. Geophysical Journal International, 190(1), 406–20.
Schmidt, J. F.J. 1867. Treatise on the 1867 Aegion Earthquake. Athens (National Observatory) (in Greek).
Scholz, C. H. 2002. The mechanics of earthquakes and faulting. Cambridge University Press, Cambridge, 471 p.
Sengor, A.M.C., Tuysuz, O., Imren, C., Sakinc, M., Eyidogan, H., Gorur, N., Le Pichon, X. & Rangin, C. 2005. The North Anatolian Fault: A new look. Ann. Rev. Earth Planet. Sci., 33, 37–112.
Shaw, B. & Jackson, J. 2010. Earthquake mechanisms and active tectonics of the Hellenic subduction zone. Geophys. J. Intern., 181, 966–984.
Skouphos, T. 1894. Die zwei grossen Erdbenen in Lokris am 8/20 und 15/27 April 1894. Z. Ges. Erdkd., 24, 409–474.
Sonder, R. 1925. Zur Geologie und Petrographie der Inselgruppe von Milos. Zeitschrift fur Volcanologie, 8, 181–229.
Tapponnier, P., 1977. Evolution tectonique du système alpin en Méditerranée: poinçonnement et écrasement rigide - plastique. Bull. Soc. Géol. France, 19, 3, 437–460.
Taymaz, T., J. A. Jackson, and McKenzie, D., 1991. Active tectonics of the north and central Aegean Sea. Geophysical Journal International, 106, 433–490.
Valkaniotis, S., Ganas, A., Papathanassiou, G. & Papanikolaou, M. 2014. Field observations of geological effects triggered by the January-February 2014 Cephalonia earthquakes (Ionian Sea, Greece). Tectonophysics.
Van Hinsbergen, D. J., Dekkers, M. J., Bozkurt, E., & Koopman, M. 2010. Exhumation with a twist: paleomagnetic constraints on the evolution of the Menderes metamorphic core complex, western Turkey. Tectonics, 29(3).
Van Hinsbergen, D.J.J., Langereis, C.G. & Meulenkamp, J. 2005. Revision of time, magnitude and distribution of Neogene rotations in the western Aegean region. Tectonophysics, 396, 1–34.
Van Hinsbergen, D.J.J. & Meulenkamp, J. 2006. Neogene supradetachment basin development on Crete (Greece) during exhumation of the South Aegean core complex. Basin Research, 18, 103–124.
Varnavas, S. P. & Cronan, D. S. 2005. Submarine hydrothermal activity off Santorini and Milos in the central Hellenic volcanic arc: a synthesis. Chemical Geology, 224/1–3, 40–54.
Weigelt, P., Steinbauer, M. J., Cabral, J. S. & Kreft, H. 2016. Late Quaternary climate change shapes island biodiversity. Nature, 17443.
Wells, D. L. & Coppersmith, K.J. 1994. New empirical relationships among magnitude, rupture length, rupture width and surface displacement. Bull. Seism. Soc. Am., 84, 974–1002.
Wernicke, B. 1981. Low-angle normal faults in the Basin and Range Province: Nappe tectonics in an extending orogen. Nature, 291, 645–648.
Wernicke, B. & Burchfiel, B.C. 1982. Modes of extensional tectonics. J. Struc. Geol., 4, 104–115.
Wesnousky, S.G., 2008. Displacement and Geometrical Characteristics of Earthquake Surface Ruptures: Issues and Implications for Seismic-Hazard Analysis and the Process of Earthquake Rupture. Bulletin of the Seismological Society of America 98, 1609–1632.
Westaway, R. 2002. The Quaternary evolution of the Gulf of Corinth, central Greece: coupling between surface processes and flow in the lower continental crust. Tectonophysics, 348, 269–318.
Wiatr, T., Papanikolaou, I., Fernández-Steeger, T. & Reicherter K. 2015. Bedrock fault scarp history: Insight from t-LiDAR backscatter behavior and analysis of structure changes. Geomorphology, 228, 421–431.

Bibliography

General Bibliography

Greek scientific journals that published and/or are publishing geological papers either in Greek or in other foreign languages have been the following:

- *Annales Geologiques des Pays Helleniques*, which is edited since 1942 by the Geological Department of the National & Kapodistrian University of Athens, in exchange with more than 400 scientific journals worldwide. In 2006 it was updated, publishing only in English language and it was renamed *«Hellenic Journal of Geosciences»* but a few years ago it was terminated, due to the economic crisis in Greece. It has published 45 volumes and has been widely known mainly from the PhD Theses (These d'Etat) of French colleagues, who worked in Greece in the 1950s, 1960s and 1970s. It also published proceedings of International Congresses like the RCMNS 1979, on the Mediterranean Neogene and IGCP results like the IGCP no 276 «Paleozoic geodynamic domains and their alpidic evolution in the Tethys», including also the Terrane maps of the Mediterranean (1996) and IGCP No 329 «Paleogeographic and paleoecologic evolution of Paratethyan basins during Neogene and their correlation to the global scales».
- *Bulletin of the Geological Society of Greece*, which is edited by the Geological Society of Greece since 1954, in exchange with more than 120 scientific journals worldwide. Since the 1980s it publishes the proceedings of the international congresses of the Society, which are usually organised every 2–3 years. More than 50 volumes have been published.
- *Praktika (Proccedings) of the Academy of Athens*, which is edited by the Academy of Athens since 1924. It contains a few geological papers together with other papers concerning different topics of Sciences (Mathematics, Physics, Chemistry, etc). During the last years very few papers are published mainly because there are no Academicians specialised in geology.
- *Geological Geophysical Research*, which was edited from the Institute of Geology and Mining Research (IGME), with publication of 23 volumes (1951–1981). This scientific journal of IGME was publishing papers referring to the results mainly from the scientists working in the Institute. It was accompannied by the Edition of the *Geological Maps of Greece, at scale 1/50.000.* Unfortunately, the journal stopped in 1981. Currently, the IGME only edits the geological maps, several of which, need to be updated.

Besides the above journals, which published more than 85% of the geological papers edited in Greece, a few more topical journals have emerged and published occasionally, usually not exclusively, geological papers but papers of wider interest, like the *«Mineral Wealth»*, edited by the Chamber of Mining Engineers of Greece, which contain also non reviewed papers, articles etc.

In conclusion, the publication of geological papers in Greek journals has been deteriorated and Greek geologists are mainly publishing in international peer reviewed scientific journals, following the international trend. The result is that papers that have been published in Greek journals, particularly in the past, are considered by some foreign geologists as «grey literature» and this has resulted in several ethic issues. It is quite common to find new papers by especially young researchers who totally ignore the previous literature on the Geology of Greece, offering a misleading viewpoint that geological knowledge started to build up in the 1990s, when the modern methods of bibliographic research were established.

Due to the large number of foreign geologists working in Greece a plethora of papers concerning the Geology of Greece have been published in foreign journals such as:

D. I. Papanikolaou, *The Geology of Greece*, Regional Geology Reviews,
https://doi.org/10.1007/978-3-030-60731-9

Bulletin Societe Geoloqigue de France, Comptes Rendus Academie des Sciences de Paris, Revue de Geographie Physique – Geologie Dynamique, Bulletin Societe Geologique du Nord (France), Zeitschrift Deutsche Geologische Geselltschaft, Geologisches Jahrbuch, Neus Jahrbuch der Geologischen – Paaleontologischen Abhandlungen/Monatshefte, Geologischen Rundschaw, Zeitschrift fur Geomorphologie (Germany), Journal (and Special Publications) Geological Society of London, Geological Magazine, Journal of Structural Geology, Terra Nova (United Kingdom), Eclogae Geologiae Helvetiae, Schweizerische Mineralogische und Petrographische Mitteilungen (Switzerland), Bolletino Societa Geologica Italiana, Rivista Italiana Paleontologia, Ofioliti (Italy), Bulletin (and Special Papers) of the Geological Society of America, Geology, Geosphere, American Journal of Science (USA), Canadian Journal of Earth Sciences and several other international journals like: Nature, Science, Tectonophysics, Earth & Planetary Science Letters, Journal of Geophysical Research, Tectonics, Geochemistry-Geophysics-Geosystems, Lithos, Marine Geology, Journal of Geodynamics, Geo-Marine Letters, Quaternary International, Basin Research, Geophysical Research Letters etc.

The above list of international Journals where papers related to the Geology of Greece have been published is also modified since the 1980s, because just like in Greece, several journals edited from Academic bodies, Geological Societies, Research Institutions etc have become inactive or changed their name and characteristics and, on the contrary, new journals edited by International Publishing Companies (Elsevier, Wiley, Springer, Blackwell etc) have emerged or became dominant.

Apart from the regular volumes published by the various scientific journals a lot of papers have been published in special issues, usually following some international congresses. Some of these volumes edited from various Academic Institutions have become rare and difficult to access electronically outside the libraries. A selection of such special issues is displayed below:

(1) Bulletin de la Societe Geologique de France, 18, 2, 1976. Comptes Rendus du Colloque sur la Geologie des regions Egeennes a Orsay.

(2) Reunions extraordinaires des Societes geologiques de France et de Grece en Grece (9–25 Septembre, 1976). Bulletin de la Societe Geologique de France, tome XIX, no 1, 1977, p. 1–116.

(3) VI Colloquium on the Geology of the Aegean Region (Athens, September 1977). Proceedings, vol. I, II, III, IGME, 1979, 1,405 p.

(4) Alps, Apennines, Hellenides, Inter-Union Comission on Geodynamics, Scientific Report, No 38. H. Closs, D. Roeder, K. Schmidt (editors). E. Schweizerbart'sche Verlags-buchandlung, Stuggart, 1978, p. 389–620.

(5) VII International Congress on Mediterranean Neogene, Athens, September 27–October 2, 1979. Annales Geologiques des Pays Helleniques, Hors Serie I, II, III, 1,461 p. and Proceedings IV, 1981, 368 p.

(6) International Symposium on the Hellenic Arc and Trench System (HEAT), Athens, April 1981, Proceedings vol. I, 439 p. and II, 445 p., 1982, National and Technical University of Athens.

(7) The Geological Evolution of the Eastern Mediterranean, Edinburgh, September 1982, Abstracts, 112 p. and Proceedings in: Geological Society of London, Special Publication, 17, 1984, 824 p.

(8) Special Volume dedicated to Prof. J. Papastamatiou, Geological – Geophysical Research, IGME, 1986, 478 p.

(9) Geologie von Griechenland, V. Jacobshagen et al., 1986, 363 p. Beitrage zur regionalen geologie der Erde, Band 19, Gebruder Borntraeger, Berlin.

(10) Anniversary volume for the 40 years of the Geological Society of Greece, «Development and Contribution of Geology», Special Publications of the Geological Society of Greece, no 2, 300 p.

(11) IGCP Project 276, «Terrane Maps and Terrane Descriptions», D. Papanikolaou & F.P. Sassi (editors), Annales Geologiques des Pays Helleniques, 37, 193–599, 1997, including three map sheets and one Table of tectono-stratigraphic diagrams.

(12) The Igneous Rocks of Greece, G. Pe-Piper & D. Piper, Beitrage zur Geologie der Erde, Band 30, Gedruder Borntraeger, Berlin, 2002, 573 p.

(13) The TRANSMED Atlas, Cavazza et al (editors), 2004, 141 p. + CD-ROM, Springer.

(14) Post-collisional Tectonics and Magmatism n the Mediterranean Region and Asia, Dilek & Pavlides (editors), Special Paper of the Geological Society of America, No 409, 2006, 644 p.

(15) Tectonic Development of the Eastern Mediterranean Region, Robertson & Mountrakis (editors), Special Publications Geological Society of London, 260, 2006.

(16) Collision and collapse at the Africa-Arabia-Eurasia subduction zone, VanHinsbergen, Edwards & Govers

(editors) Special Publications Geological Society of London, 311, 2009.

(17) Ophiolites and related Geology of the Balkan Region, Robertson, Karamata & Saric, (editors) Lithos, 108, 1–4, 280 p., 2009.

(18) The Aegean: A natural laboratory for Tectonics – Tectonometamorphics, Papanikolaou, Roberts & Royden (editors), Tectonophysics, 595–596, 2013, 262 p.

(19) The Aegean: A natural laboratory for Tectonics – Neotectonics, Papanikolaou, Roberts, Royden, (Editors), Tectonophysics, 597–598, 2013, 160 p.

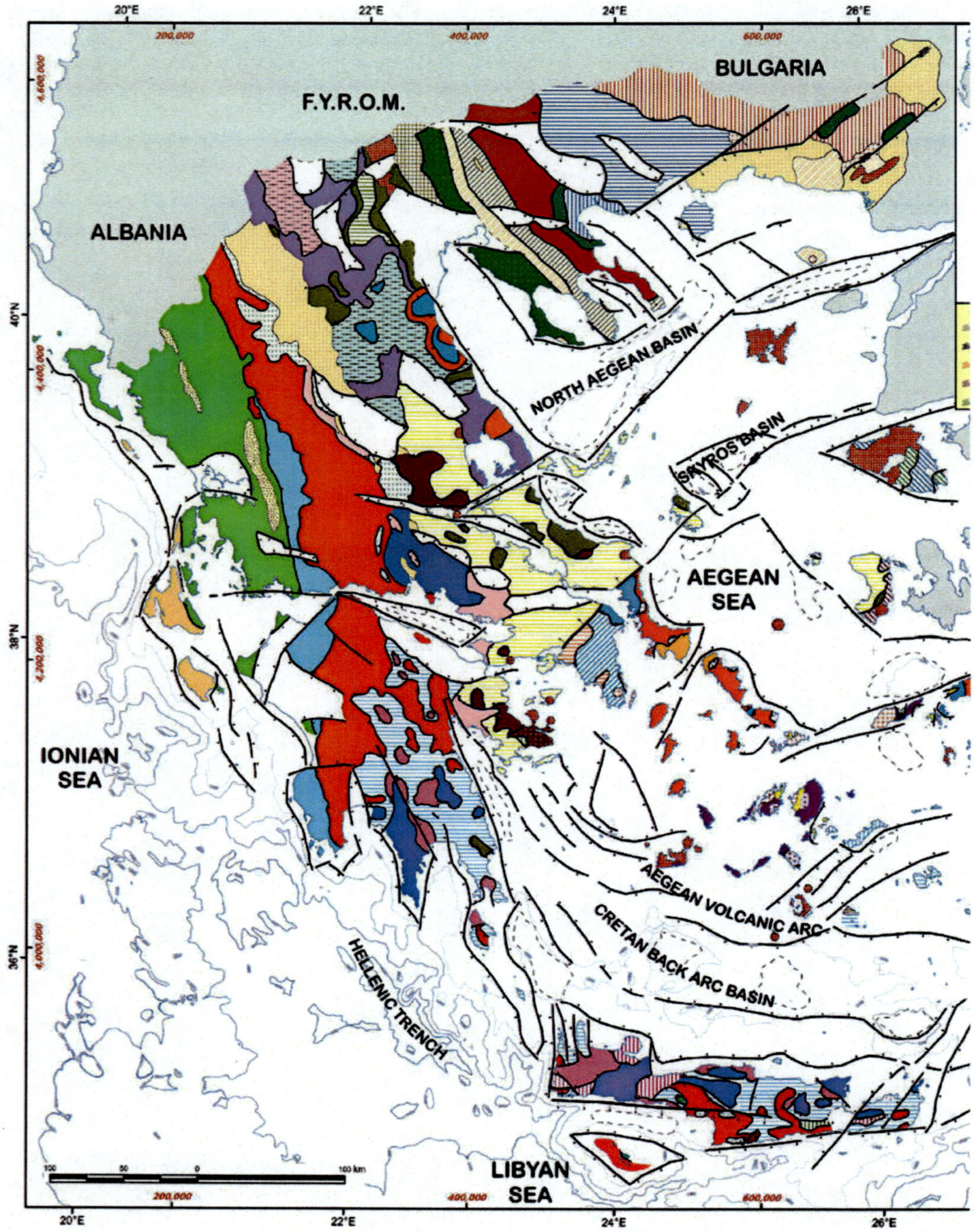
20°E
22°E
24°E
26°E
40°N
38°N
36°N
200,000
400,000
600,000
BULGARIA
F.Y.R.O.M.
ALBANIA
NORTH AEGEAN BASIN
SKYROS BASIN
AEGEAN SEA
IONIAN SEA
AEGEAN VOLCANIC ARC
CRETAN BACK ARC BASIN
HELLENIC TRENCH
LIBYAN SEA
100 km

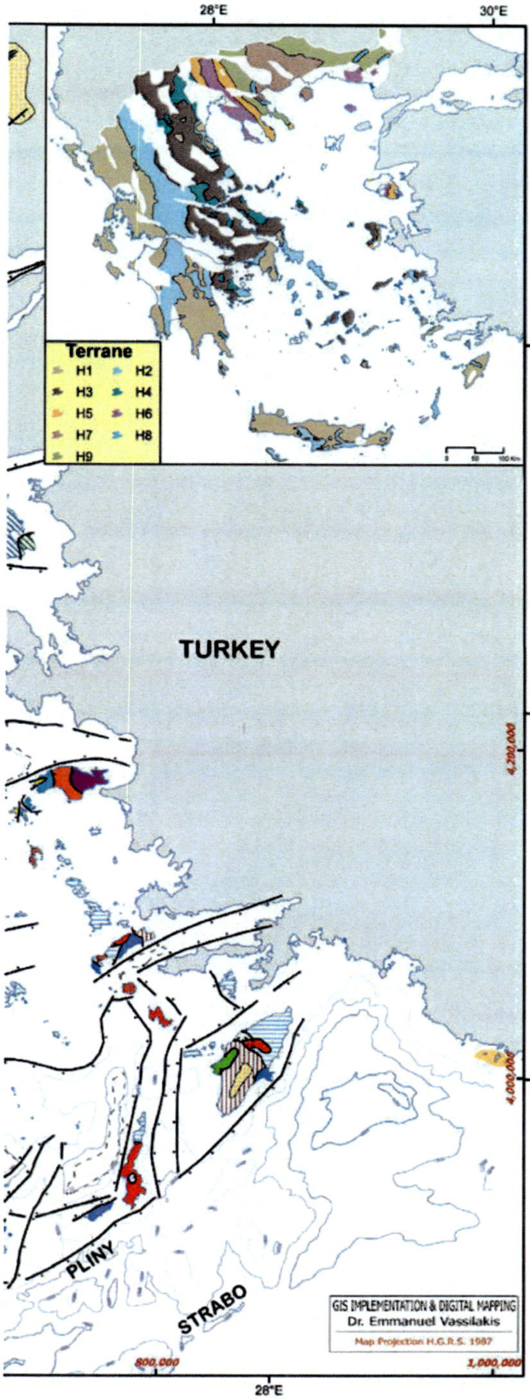
28°E
30°E
Terrane
H1
H2
H3
H4
H5
H6
H7
H8
H9
40°N
38°N
36°N
TURKEY
PLINY
STRABO
4,200,000
4,000,000
800,000
1,000,000
GIS IMPLEMENTATION & DIGITAL MAPPING
Dr. Emmanuel Vassilakis
Map Projection H.G.R.S. 1987

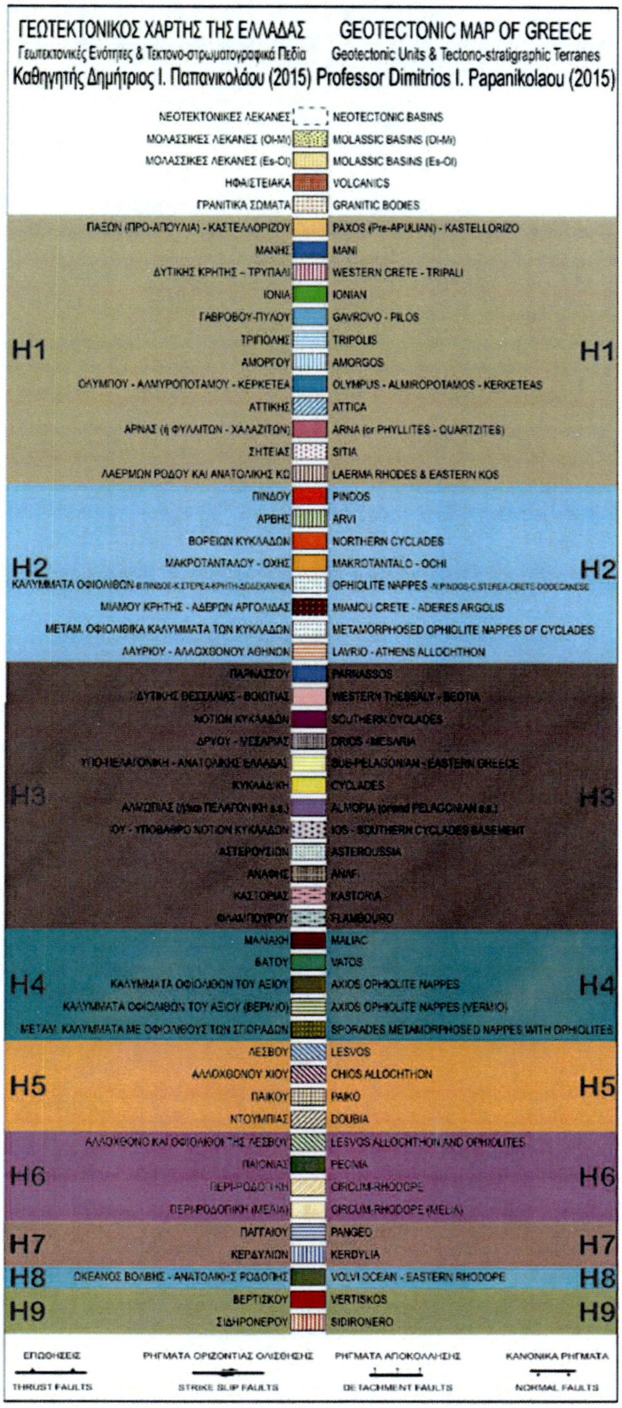
ΓΕΩΤΕΚΤΟΝΙΚΟΣ ΧΑΡΤΗΣ ΤΗΣ ΕΛΛΑΔΑΣ
GEOTECTONIC MAP OF GREECE
Γεωτεκτονικές Ενότητες & Τεκτονο-στρωματογραφικά Πεδία
Geotectonic Units & Tectono-stratigraphic Terranes
Καθηγητής Δημήτριος Ι. Παπανικολάου (2015)
Professor Dimitrios I. Papanikolaou (2015)
NEOTECTONIC BASINS
MOLASSIC BASINS (Ol-Mi)
MOLASSIC BASINS (Eo-Ol)
VOLCANICS
GRANITIC BODIES
H1
PAXOS (Pre-APULIAN) - KASTELLORIZO
MANI
WESTERN CRETE - TRIPALI
IONIAN
GAVROVO - PILOS
TRIPOLIS
AMORGOS
OLYMPUS - ALMIROPOTAMOS - KERKETEAS
ATTICA
ARNA (or PHYLLITES - QUARTZITES)
SITIA
LAERMA RHODES & EASTERN KOS
H2
PINDOS
ARVI
NORTHERN CYCLADES
MAKROTANTALO - OCHI
OPHIOLITE NAPPES
MIAMOU CRETE - ADERES ARGOLIS
METAMORPHOSED OPHIOLITE NAPPES OF CYCLADES
LAVRIO - ATHENS ALLOCHTHON
H3
PARNASSOS
WESTERN THESSALY - BEOTIA
SOUTHERN CYCLADES
DRIOS - MESARIA
SUB-PELAGONIAN - EASTERN GREECE
CYCLADES
ALMOPIA
IOS - SOUTHERN CYCLADES BASEMENT
ASTEROUSSIA
ANAFI
KASTORIA
FLAMBOURO
H4
MALIAC
VATOS
AXIOS OPHIOLITE NAPPES
AXIOS OPHIOLITE NAPPES (VERMIO)
SPORADES METAMORPHOSED NAPPES WITH OPHIOLITES
H5
LESVOS
CHIOS ALLOCHTHON
PAIKO
DOUBIA
H6
LESVOS ALLOCHTHON AND OPHIOLITES
PEONIA
CIRCUM-RHODOPE
CIRCUM-RHODOPE (MELIA)
H7
PANGEO
KERDYLIA
H8
VOLVI OCEAN - EASTERN RHODOPE
H9
VERTISKOS
SIDIRONERO
THRUST FAULTS
STRIKE SLIP FAULTS
DETACHMENT FAULTS
NORMAL FAULTS

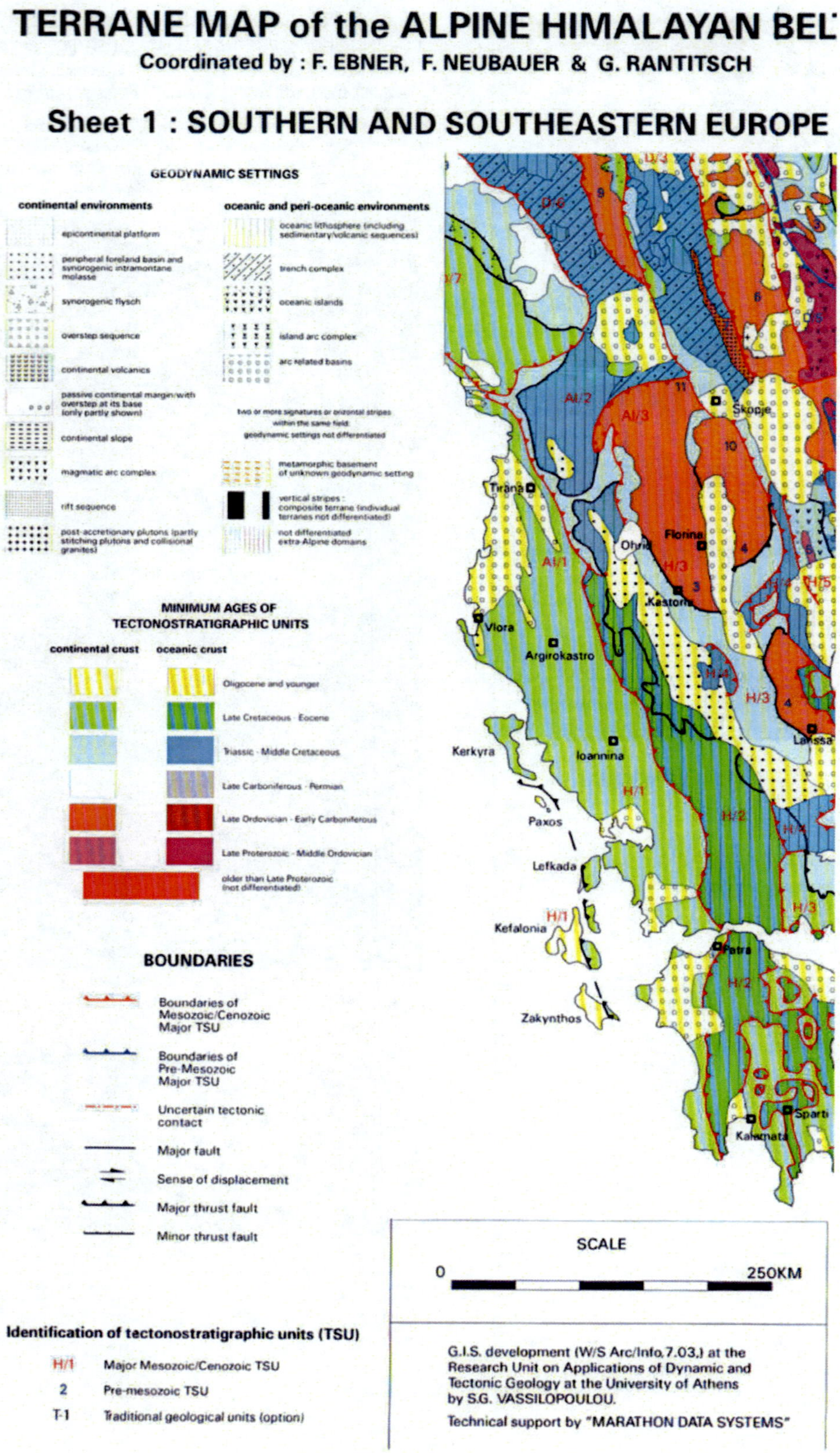
TERRANE MAP of the ALPINE HIMALAYAN BEL
Coordinated by : F. EBNER, F. NEUBAUER & G. RANTITSCH
Sheet 1 : SOUTHERN AND SOUTHEASTERN EUROPE
GEODYNAMIC SETTINGS
continental environments
epicontinental platform
peripheral foreland basin and synorogenic intramontane molasse
synorogenic flysch
overstep sequence
continental volcanics
passive continental margin with overstep at its base (only partly shown)
continental slope
magmatic arc complex
rift sequence
post-accretionary plutons (partly stitching plutons and collisional granites)
oceanic and peri-oceanic environments
oceanic lithosphere (including sedimentary/volcanic sequences)
trench complex
oceanic islands
island arc complex
arc related basins
two or more signatures or horizontal stripes within the same field: geodynamic settings not differentiated
metamorphic basement of unknown geodynamic setting
vertical stripes: composite terrane (individual terranes not differentiated)
not differentiated extra-Alpine domains
MINIMUM AGES OF TECTONOSTRATIGRAPHIC UNITS
continental crust
oceanic crust
Oligocene and younger
Late Cretaceous - Eocene
Triassic - Middle Cretaceous
Late Carboniferous - Permian
Late Ordovician - Early Carboniferous
Late Proterozoic - Middle Ordovician
older than Late Proterozoic (not differentiated)
BOUNDARIES
Boundaries of Mesozoic/Cenozoic Major TSU
Boundaries of Pre-Mesozoic Major TSU
Uncertain tectonic contact
Major fault
Sense of displacement
Major thrust fault
Minor thrust fault
Identification of tectonostratigraphic units (TSU)
H/1 Major Mesozoic/Cenozoic TSU
2 Pre-mesozoic TSU
T-1 Traditional geological units (option)
SCALE
0
250KM
G.I.S. development (W/S Arc/Info.7.03.) at the Research Unit on Applications of Dynamic and Tectonic Geology at the University of Athens by S.G. VASSILOPOULOU.
Technical support by "MARATHON DATA SYSTEMS"
Skopje
Tirana
Ohrid
Florina
Kastoria
Vlora
Argirokastro
Kerkyra
Ioannina
Larissa
Paxos
Lefkada
Kefalonia
Patra
Zakynthos
Kalamata
Sparti
Al/1
Al/2
Al/3
H/1
H/2
H/3
H/4
H/5

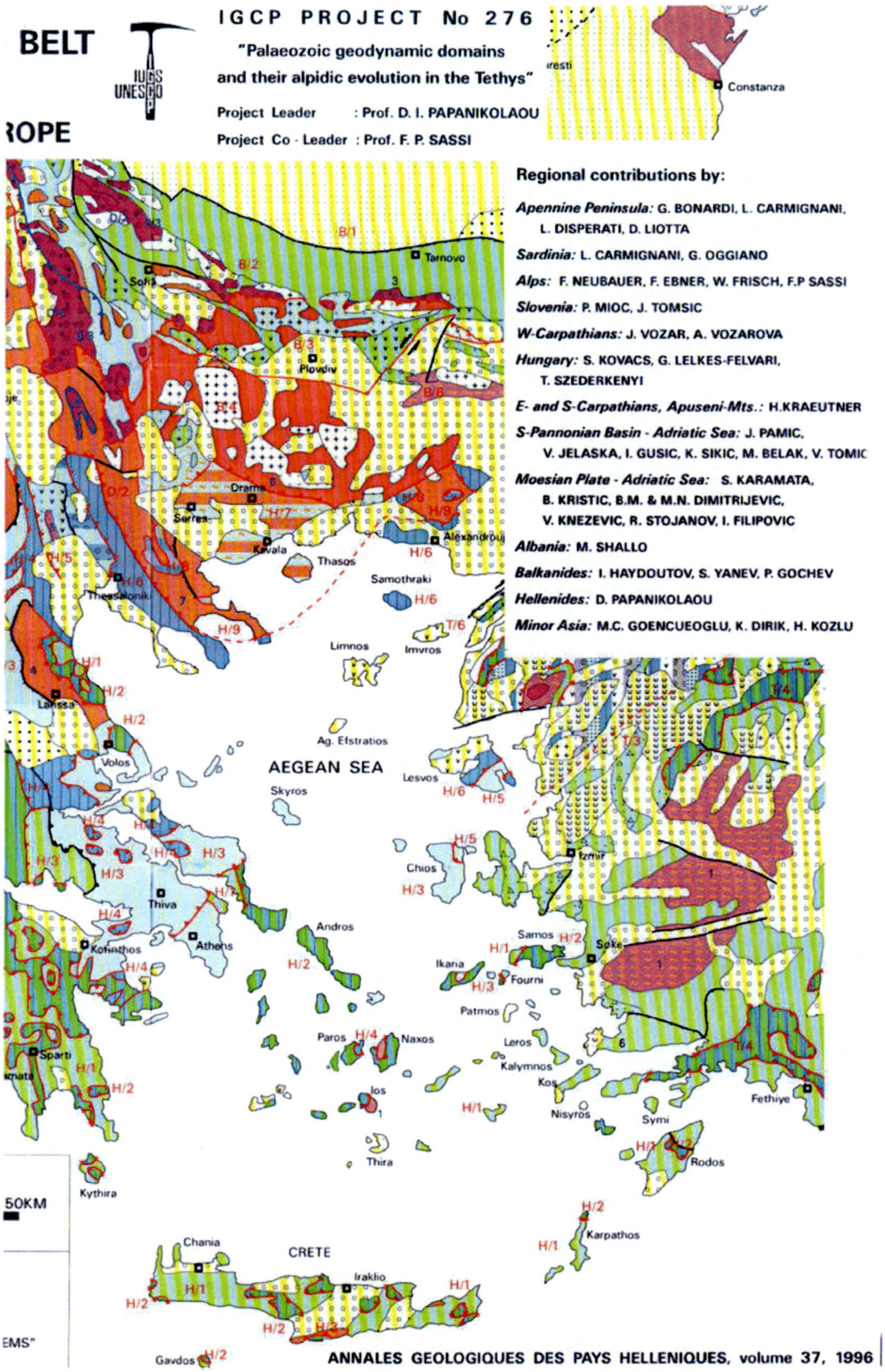

BELT
IUGS
UNESCO
ROPE
IGCP PROJECT No 276
"Palaeozoic geodynamic domains
and their alpidic evolution in the Tethys"
Project Leader : Prof. D. I. PAPANIKOLAOU
Project Co - Leader : Prof. F. P. SASSI
Constanza
Regional contributions by:
Apennine Peninsula: G. BONARDI, L. CARMIGNANI, L. DISPERATI, D. LIOTTA
Sardinia: L. CARMIGNANI, G. OGGIANO
Alps: F. NEUBAUER, F. EBNER, W. FRISCH, F.P SASSI
Slovenia: P. MIOC, J. TOMSIC
W-Carpathians: J. VOZAR, A. VOZAROVA
Hungary: S. KOVACS, G. LELKES-FELVARI, T. SZEDERKENYI
E- and S-Carpathians, Apuseni-Mts.: H.KRAEUTNER
S-Pannonian Basin - Adriatic Sea: J. PAMIC, V. JELASKA, I. GUSIC, K. SIKIC, M. BELAK, V. TOMIC
Moesian Plate - Adriatic Sea: S. KARAMATA, B. KRISTIC, B.M. & M.N. DIMITRIJEVIC, V. KNEZEVIC, R. STOJANOV, I. FILIPOVIC
Albania: M. SHALLO
Balkanides: I. HAYDOUTOV, S. YANEV, P. GOCHEV
Hellenides: D. PAPANIKOLAOU
Minor Asia: M.C. GOENCUEOGLU, K. DIRIK, H. KOZLU
Tarnovo
Sofia
Plovdiv
Drama
Serres
Kavala
Thasos
Samothraki
Thessaloniki
Limnos
Imvros
Larissa
Volos
Ag. Efstratios
AEGEAN SEA
Skyros
Lesvos
Chios
Izmir
Thiva
Athens
Korinthos
Andros
Samos
Soke
Ikaria
Fourni
Patmos
Leros
Kalymnos
Kos
Paros
Naxos
Ios
Thira
Nisyros
Symi
Rodos
Fethiye
Sparti
Kythira
50KM
Karpathos
Chania
CRETE
Iraklio
Gavdos
EMS"
ANNALES GEOLOGIQUES DES PAYS HELLENIQUES, volume 37, 1996